Modeling and Optimization Technologies

Volume 18

Series Editors

Srikanta Patnaik, SOA University, Bhubaneswar, India

Ishwar K. Sethi, Oakland University, Rochester, USA

Xiaolong Li, Indiana State University, Terre Haute, USA

Editorial Board

Li Chen, The Hong Kong Polytechnic University, Hung Hom, Hong Kong

Jeng-Haur Horng, National Formosa University, Yulin, Taiwan

Pedro U. Lima, Institute for Systems and Robotics, Lisbon, Portugal

Mun-Kew Leong, Institute of Systems Science, National University of Singapore, Singapore, Singapore

Muhammad Nur, Diponegoro University, Semarang, Indonesia

Luca Oneto, University of Genoa, Genoa, Italy

Kay Chen Tan, National University of Singapore, Singapore, Singapore

Sarma Yadavalli, University of Pretoria, Pretoria, South Africa

Yeon-Mo Yang, Kumoh National Institute of Technology, Gumi, Korea (Republic of)

Liangchi Zhang, The University of New South Wales, Kensington, Australia

Baojiang Zhong, Soochow University, Suzhou, China

Ahmed Zobaa, Brunel University London, Uxbridge, Middlesex, UK

The book series Modeling and Optimization in Science and Technologies (MOST) publishes basic principles as well as novel theories and methods in the fast-evolving field of modeling and optimization. Topics of interest include, but are not limited to: methods for analysis, design and control of complex systems, networks and machines; methods for analysis, visualization and management of large data sets; use of supercomputers for modeling complex systems; digital signal processing; molecular modeling; and tools and software solutions for different scientific and technological purposes. Special emphasis is given to publications discussing novel theories and practical solutions that, by overcoming the limitations of traditional methods, may successfully address modern scientific challenges, thus promoting scientific and technological progress. The series publishes monographs, contributed volumes and conference proceedings, as well as advanced textbooks. The main targets of the series are graduate students, researchers and professionals working at the forefront of their fields.

Indexed by SCOPUS. All books published in the series are submitted for consideration in Web of Science.

More information about this series at http://www.springer.com/series/10577

Srikanta Patnaik · Kayhan Tajeddini · Vipul Jain
Editors

Computational Management

Applications of Computational Intelligence in Business Management

Editors
Srikanta Patnaik
Department of Computer Science
and Engineering
Faculty of Engineering
SOA University
Bhubaneswar, Odisha, India

Vipul Jain
School of Management
Victoria University of Wellington
Wellington, New Zealand

Kayhan Tajeddini
Institute for International Strategy
Tokyo International University
Tokyo, Japan

Sheffield Business School
Sheffield Hallam University
Sheffield, UK

ISSN 2196-7326 ISSN 2196-7334 (electronic)
Modeling and Optimization in Science and Technologies
ISBN 978-3-030-72931-8   ISBN 978-3-030-72929-5 (eBook)
https://doi.org/10.1007/978-3-030-72929-5

© The Editor(s) (if applicable) and The Author(s), under exclusive license to Springer Nature Switzerland AG 2021, corrected publication 2021

This work is subject to copyright. All rights are solely and exclusively licensed by the Publisher, whether the whole or part of the material is concerned, specifically the rights of translation, reprinting, reuse of illustrations, recitation, broadcasting, reproduction on microfilms or in any other physical way, and transmission or information storage and retrieval, electronic adaptation, computer software, or by similar or dissimilar methodology now known or hereafter developed.

The use of general descriptive names, registered names, trademarks, service marks, etc. in this publication does not imply, even in the absence of a specific statement, that such names are exempt from the relevant protective laws and regulations and therefore free for general use.

The publisher, the authors and the editors are safe to assume that the advice and information in this book are believed to be true and accurate at the date of publication. Neither the publisher nor the authors or the editors give a warranty, expressed or implied, with respect to the material contained herein or for any errors or omissions that may have been made. The publisher remains neutral with regard to jurisdictional claims in published maps and institutional affiliations.

This Springer imprint is published by the registered company Springer Nature Switzerland AG
The registered company address is: Gewerbestrasse 11, 6330 Cham, Switzerland

Preface

With the continuous development in technology and strategies, dynamically changing business environment demands for automation of solutions. Over the last decade, there have been efforts by the researchers for automation of the systems to self-acquire information by investigating the problem environment and utilize this information to interpret valuable insights for taking decision for the real-world problems without human interventions. Computational intelligence (C.I.) is one such paradigm that stems from artificial intelligence, motivated by biological as well as linguistics computational models that are further developed to solve real-world problems in a human-like approach. Computational intelligence consists of a set of computational techniques that imitates human brain along with other aspects of nature to adopt the decision-making approach. Moreover, the three major pillars of computational intelligence are based upon fuzzy systems, neural networks and evolutionary computation. Another inclusion to the collection has been the swarm intelligence that takes inspiration from several survival strategies existing in nature and develops computational models for solving complex real-world problems that are hard or almost impossible to solve using traditional approaches. Some other approaches include support vector machines and expert systems. However, most of the computational intelligence techniques are claimed to be capable of addressing stochastic and uncertain environments where reasoning is a significant attribute to derive potential solutions. Moreover, the existence of randomness and uncertainty in various processes makes certain circumstances unpredictable, thus making the problem solving more complex to reason out. Also, its high accuracy and efficiency in performance of the computational intelligence techniques in such complex turn of events have helped it to gain fast attention of researchers and industry practitioners in accomplishing tremendous success. Some of the potential application areas are currently gaining huge attention for the successful applicability of computational intelligence range over pattern recognition, forecasting, system modeling, intelligent process control, customer relationship management, health care and biomedical applications.

With the development of many intelligent and smart applications and products, computational intelligence has become a significant attribute responsible for learning, reasoning and decision making. Further, with vast commercialization and hype in globalization, recently, focus of applicability of computational intelligence techniques has been exponentially shifted from traditional application domains to business application domain. Computational intelligence has now been successfully adopted in business management domain to address diverse range of business application problems such as production and operational problems, financial decisions, logistics, supply chain management, pricing, marketing and many more. Hence, computational management can be defined as the integration of computational intelligence techniques in business application domain providing unique stance to various decision-making approaches to solve real-time business problems. Computational management basically focuses on the computational aspects of business domain. C.I. techniques are further being utilized to design data collection-oriented business applications to overcome the challenges being faced in retrieving, extracting and integrating data where heterogeneous sources are involved. In addition, C.I. techniques are being used to develop shopping advisors, designing personalized product bundles, predicting market share on the basis of customer behavior, regulating market fluctuations, etc. They are also being used to design automated decision-making systems in stock trading and e-business sectors for providing profitable strategies. Thus, inclusion of C.I. techniques in business management domain will enhance not only productivity and performance but also profit along with market expansion.

This book attempts to cover all potential aspects of business management comprising theoretical and empirical models as well as simulations of applications. Out of around 80 different proposals received, we have selected only 31 most relevant chapters to publish in this volume on the basis of several metrics such as quality of content, recentness and coverage. This book is expected to provide valuable insights to academics, researchers and industry practitioners working in the area of computational management. The entire collection of chapters have been broadly divided into four major categories to be mentioned as (i) Computational Modeling, (ii) Management Optimization (iii) Computational Intelligence (iv) Computational Management Applications.

Part I: Computational Modelling

This part presents a collection of the chapters presenting various computational models that can help us observe and understand the behavior and complexity of the business domain to make significant decisions while analyzing potential consequences. Out of the complete collection of chapters, four chapters are best suited for this part as follows:

In Chapter 1 entitled "Computational Management—An Overview," Pragyan Nanda et al. are presenting an overview of several computational aspects of diverse business management applications where incorporating computational intelligence will provide better predictive accuracy and increased revenue and profit with smooth operations.

In Chapter 2 entitled "Mathematical and Computational Approaches for Stochastic Control of River Environment and Ecology: From Fisheries Viewpoint,"

Hidekazu Yoshioka presents a computational stochastic control framework for dynamic optimization of the river environment and its ecology focusing on fisheries problem in Japan. He provides several interesting illustrations in support of his work.

Chapter 3 in the volume has been compiled by Rojers P. Joseph and Arun T. M. entitled as "Models and Tools of Knowledge Acquisition." This chapter discusses the effectiveness of the emerging as well as novel techniques like Competitive Intelligence (CI), Artificial Intelligence (AI) and Application Programming Interfaces (APIs), being adopted widely for knowledge acquisition. Cost-saving, convenience in usage of the techniques and efficiency in acquisition of knowledge are being some of the evaluation metrics.

The next chapter (Chap. 4) in the collection implements an approximate entropy and sample entropy indicators extracted from the India VIX for studying the feasibility of portfolio rotation strategies based on style, size and time horizons. The authors further propose that their approach provides computationally supportive arguments in favor of a potentially beneficial alternative for portfolio managers and can be used to enhance portfolio returns and to mitigate risk in the context on Indian market. This chapter has been titled as "Profits Are in the Eyes of the Beholder: Entropy-Based Volatility Indicators and Portfolio Rotation Strategies" by the authors Abhijeet Chandra and Gaurav Jadhao.

The yet next chapter (Chap. 5) bearing the title "Asymmetric Spillovers Between the Stock Risk Series: Case of CESEE Stock Markets" by Tihana Škrinjarić aims to present a Vector Auto-Regression (VAR) model with multidirectional spill-over of shocks in return and risk series among the selected stock markets. It quantifies spill-over indices using various approaches discussed in the literature.

Chapter 6 in the collection which has been entitled as "Millennial Customers and Hangout Joints: An Empirical Study Using the Kano Quantitative Model" by the authors Siddharth Tiwari et al. presents an empirical study on the customers service experience on the basis of seven dimensions by collecting survey data from 224 respondents. They further explore the significance of the satisfaction model, the refined Kano in an attempt to associate it with customer value followed by demonstrating a new framework to illustrate service experiences along with the influential dimensions and suggestive actions.

Another chapter (Chap. 7) that has been selected for this volume deals with the effects of macroeconomic uncertainty by the authors Rangan Gupta and Xin Sheng. The authors have title this chapter as "The Effects of Oil Shocks on Macroeconomic Uncertainty: Evidence from a Large Panel Dataset of US States." It investigates the effects of oil price shocks on macroeconomic uncertainty using a large panel dataset of 50 US states. They examine both linear and nonlinear impulse response functions of uncertainty to oil shocks by using the local projection method.

Chapter 8 by Adarsh Anand et al. is entitled as "Understanding and Predicting View Counts of YouTube Videos Using Epidemic Modelling Framework." This chapter proposes a two-stage based epidemic modeling framework to understand the viral content of the YouTube platform having huge amount of views within short period of time. They have modeled awareness and viewing as the two different stages for the view counts and validated the model on several YouTube datasets.

Sarada Ghosh and Guru Prasad Samanta in Chap. 9 with the title "Gross Domestic Product Modeling Using "Panel-Data" Concept" attempt to predict the gross domestic product (GDP) using "panel-data" models. They have considered the gross domestic product for fifteen individual countries of Europe in different four years. They developed the panel data models using the investment, labor force growth and budget surplus as inputs for predicting the GDP of the countries to study the economic condition of Europe from the statistical point of view.

Chapter 10 in this part entitled as "Supply Chain Scheduling Using an EOQ Model for a Two-Stage Trade Credit Financing with Dynamic Demand" Alok Kumar discusses the optimum scheduling for a part of supply chain system using an EOQ model where the demand is dynamic, varies with time and one of the promotional efforts in the form of a two-stage trade credit is considered. The authors present sensitivity analysis of the parameters and its managerial implications in support of the applicability of the model.

The next chapter (Chap. 11) entitled "Software Engineering Analytics—The Need of Post COVID-19 Business: An Academic Review" by the authors Somayya Madakam and Rajeev K. Revulagadda explores the state of the art of COVID-19 across the globe with respect to business. They conduct data analysis on secondary data and reveal that software analytics technology is the only industry that is lightly affected. Thus, it is expected for rapid growth and can be seen as a solution provider for sustainability of software automation in terms of reduced cost, reduced human intervention and time management.

The last chapter (Chap. 12) in this part belongs to the authors Nteboheng Pamella Phadi and Sonali Das which has been entitled as "The Rise and Fall of the SCOR Model: What After the Pandemic?". In this chapter, the authors have explored the vulnerability of supply chain networks in the presence of the COVID-19. They present the SCOR model as an opportunity to attain agility and resilience through strategic and intelligent re-design while focusing on various performance attributes to recover from the pandemic.

Part II: Management Optimization

The second part deals with the optimization problems existing in the business management domain and different perspectives used to solve it.

Chapter 13 categorized to this part entitled "A Comparative Study on Multi-objective Evolutionary Algorithms for Tri-objective Mean-Risk-Cardinality Portfolio Optimization Problems" by Georgios Mamanis investigates several state of the arts for the evolutionary multi-objective optimization methods. It presents a comparative study of their efficiency in the context of multi-objective portfolio optimization problems. They focus on solving the mean-risk-cardinality portfolio optimization problem with six different risk measures.

Yet another chapter in this part (Chap. 14) is entitled as "Portfolio Insurance and Intelligent Algorithms" by Vasilios N. Katsikis and Spyridon D. Mourtas. The authors present intelligent heuristic algorithms for portfolio insurance optimization with the objective of minimizing the insurance cost. They incorporate modern heuristic-based optimization techniques for minimization. The proposed algorithm is claimed to be well-tuned for solving time-varying instance. It has been validated with several numerical illustrations.

Chapter 15 categorized to this part entitled as "On Interval-Valued Multiobjective Programming Problems and Vector Variational-Like Inequalities Using Limiting Subdifferential" by B. B. Upadhyay and Priyanka Mishra deals with the generalized vector variational-like inequalities, namely Minty and Stampacchia vector variational-like inequalities for interval-valued functions and interval-valued multi-objective programming problem. The authors establish equivalence among the solutions of considered vector variational-like inequalities and LU-efficient minimizers of orders of interval-valued multi-objective programming problems.

Again, the next chapter (Chap. 16) by Namita Srivastava entitled "Portfolio Optimization Using Multi Criteria Decision Making" deals with the financial decision making presenting an extended review of multi-criteria decision making (MCDM). It further explores uncertain factors influencing the decision-making process in portfolio optimization to bridge the gap between qualitative and quantitative methods.

One more chapter (Chap. 17) on multi-objective optimization has been shortlisted in this part by Vandana Y. Kakran and Jayesh M. Dhodiya which is entitled as "Uncertain Multi-objective Transportation Problems and Their Solution." This chapter focuses on studying the multi-objective transportation problem (MOTP) in the uncertain domain area. It further develops the expected value and dependent chance constraint models for uncertainty and converts them into deterministic form for ease of computation which is again converted into linear model using Charnes and Cooper's transformations. And finally, the chapter solves it using the fuzzy programming technique and weighted sum methodology. The authors have backed this with the help of numerical Illustrations.

Part III: Computational Intelligence

This part comprises the chapters that employs the core computational intelligence techniques for problem solving.

Chapter 18 of the collection entitled as "Agile Computational Intelligence for Supporting Hospital Logistics During the COVID-19 Crisis" by Rafael Tordecilla et al. presents a case study regarding the use of "agile" computational intelligence for supporting logistics in Barcelona's hospitals during the COVID-19 crisis in 2020. They have presented the problem of distributing protection items to health workers daily as multiple rich variants of vehicle routing and team orienteering problems. The chapter discusses some of the computational aspects of the employed flexible heuristic-based algorithms along with several computational experiments to generate high-quality solutions.

The next chapter (Chap. 19) is another multi-objective optimization-based work entitled "Multi-objective Assignment Problems and Their Solutions by Genetic Algorithm" by Anita Ravi Tailor and Jayesh M. Dhodiya which employs genetic algorithm to solve multi-objective assignment problems (MOAP) on the basis of aspiration levels and shape parameters of exponential membership function. The proposed approach has been supported by numerical illustrations.

Chapter 20 again dealing with multi-objective optimization problem has been entitled as "Role of Evolutionary Approaches to Solving Multi-objective Optimization Problems" by Surbhi Tilva and Jayesh M. Dhodiya. This chapter explores the foundation of multi-objective optimization problems using intelligent meta-heuristic-based evolutionary algorithms.

The next chapter (Chap. 21) entitled "Improving Financial Bankruptcy Prediction Using Oversampling Followed by Fuzzy Rough Feature Selection via Evolutionary Search" by Pankhuri Jain et al. presents a new methodology for improving the bankruptcy prediction performance of various machine learning algorithms. The authors suggest the adoption of oversampling technique for initial conversion of the imbalanced dataset of bankrupt and non-bankrupt into balanced dataset. They next use fuzzy rough feature selection technique via evolutionary search for extracting relevant and non-redundant features and finally evaluate the performance of various machine learning algorithms supported by experimental results.

Chapter 22 of this volume having the title "An Integrated Fuzzy MCDM Approach to Supplier Selection—Indian Automotive Industry Case" by Vijaya Kumar Manupati et al. proposes a robust supplier selection framework by the integration of fuzzy TOPSIS, fuzzy AHP and fuzzy VIKOR to identify the potential supplier for a bus body building unit in the Indian context. Later on, it uses the spearman rank correlation coefficient to test the reliability in the ranks generated by each technique.

Part IV: Computational Management Applications

The last part deals with the application areas of computational management.

Chapter 23 in this volume is entitled as "Decision Support System to Assign Price Rebates of Fresh Horticultural Products Based on Quality Decay" by Cláudia Matos et al. This chapter proposes a decision support system that assigns pricing for horticultural products with an objective of maximizing profit and minimizing waste on the basis of quality decay along time. Various parameters such as acquisition cost, stock maintenance, cycle time, and order quantity, deterioration coefficients based on quality decay and product sensitivity in price variation are taken into consideration to attain sustainability.

The next chapter (Chap. 24) in the last part by S. Sundarakamatchi and M. S. Gajanand entitled "Recommendation Engine for Stock Market Trading" presents a recommendation engine that employs technical analysis strategies to remove inherent biases. The resulting baseline list of stocks thus generated serves as evaluation basis to make the actual trades.

Chapter 25 in this volume has been entitled as "Cross-Listing Effect and Domestic Stock Returns: Some Empirical Evidence" by Naliniprava Tripathy et al. which explores the influence of ADRs listing on the stock returns using event study methodology. It further employs variance ratio and GARCH model to assess the influence of cross-listing of ADRs on the volatility of underlying domestic stocks. The findings may provide strong support to regulatory decision-makers of the country.

Nikhil Gupta et al. have entitled the next chapter (Chap. 26) as "Investigation on Supply Chain Vulnerabilities and Risk Management Practices in Indian Manufacturing Industries" that investigates the risk and vulnerabilities in supply chain management in Indian Manufacturing Industry scenario. They study the six driving factors: information, inventory, sourcing, pricing, facility and transportation. They further classify the risks influencing the factors: under inventory, the fluctuation of demand, capacity limitations, stock out of products, etc. The authors use failure mode and effect analysis (FMEA) to study the risk factors and DEMATEL analysis to study causal relations among risk clusters, and finally, fishbone diagram has been used to study the cause and effect relationship between the risk factors.

Chapter 27 in this volume entitled "Forecasting Long-term Electricity Demand: Evolution from Experience-Based Techniques to Sophisticated Artificial Intelligence (AI) Models" by Abhishek Das and Somen Dey addresses the electricity demand forecasting problem. This chapter presents a systematic evolution of the computational intelligence techniques with their merits, demerits and comparative performance along with case studies from the Indian power sector.

Multi-criteria decision-making approach is one of the most widely adopted techniques in management. Chapter 28 by Vinay Yadav and Subhankar Karmakar discusses its applicability in the sustainable municipal solid waste management systems which has been entitled as "Multi Criteria Decision Making for Sustainable Municipal Solid Waste Management Systems". The authors have proposed a two-stage model that selects appropriate combination of locations for MSW facilities with simultaneous consideration of all relevant economic, social, technical and environmental attributes.

The next chapter (Chap. 29) in this collection entitled "Inventive Investment Using Bigdata: Tools, Applicability and Challenges Associated" by Janibulul Bashir and Tahir Ahmad Wani investigates various big data and machine learning tools in terms of efficacy from different business perspectives, their applicability in diverse business scenarios and challenges encountered.

Chapter 30 of this volume by Sahana Prasad, entitled as "Computational Aspects of Business Management with Special Reference to Monte Carlo Simulation," explores the concepts of Monte Carlo simulation as applied to business management scenario. A few specific case studies have also been demonstrated to present its application and interpretation.

The last chapter (Chap. 31) of this volume is "Sustainability in Energy Economy and Environment: Role of AI Based Techniques" by Trina Som which explores the role of AI in energy economy sustainability. Conventional methods being lengthy and extensive search processes, supplemented with trial and error process, modern optimization techniques attempt to provide promising results on the basis of random search although still difficult to achieve best solutions. Thus, AI-based techniques embracing attributes such as partial truth, uncertainties, imprecision and approximation have been explored in this chapter in the context of energy economics along with relative strengths and weaknesses.

We hope that the hard works clubbed with the extended experiences and findings of the authors from diverse field working in a unified direction compiled together in this edited volume will provide a strong base for the naïve researchers and experts in the respective domains as well.

Bhubaneswar, India
Tokyo, Japan/Sheffield, UK
Wellington, New Zealand

Prof. (Dr.) Srikanta Patnaik
Prof. (Dr.) Kayhan Tajeddini
Dr. Vipul Jain

The original version of the book was revised: The author affiliation of the editor "Kayhan Tajeddini" has been corrected. The correction to the book is available at https://doi.org/10.1007/978-3-030-72929-5_32

About This Book

The rapid technological advancement has raised the need of adopting computational intelligence techniques to solve diverse range of complex business management problems such as production and operational issues, financial decisions, logistics and transportation issues, supply chain management and dynamic pricing and marketing campaigns. Computational management focuses on understanding the computational aspect of business domain problems and employs computational intelligence tools and techniques to face emerging challenges. This book provides useful insights to scholars, researchers and practitioners about many ongoing as well as many yet to be initiated developments and works in this field. The purpose of this book is to provide attention to diverse business problems by selecting chapters from almost all domains. This book can serve as a prescribed material to readers having interest in this field.

Contents

Part I Computational Modelling

1. Computational Management—An Overview 3
   Pragyan Nanda, Deepti Patnaik, and Srikanta Patnaik

2. Mathematical and Computational Approaches for Stochastic Control of River Environment and Ecology: From Fisheries Viewpoint 23
   Hidekazu Yoshioka

3. Models and Tools of Knowledge Acquisition 53
   Rojers P. Joseph and T. M. Arun

4. Profits Are in the Eyes of the Beholder: Entropy-Based Volatility Indicators and Portfolio Rotation Strategies 69
   Abhijeet Chandra and Gaurav Jadhao

5. Asymmetric Spillovers Between the Stock Risk Series: Case of CESEE Stock Markets 97
   Tihana Škrinjarić

6. Millennial Customers and Hangout Joints: An Empirical Study Using the Kano Quantitative Model 137
   Tiwari Siddharth, Yash Daultani, and R. Rajesh

7. The Effects of Oil Shocks on Macroeconomic Uncertainty: Evidence from a Large Panel Dataset of US States 159
   Rangan Gupta and Xin Sheng

8. Understanding and Predicting View Counts of YouTube Videos Using Epidemic Modelling Framework 177
   Adarsh Anand, Mohammed Shahid Irshad, and Deepti Aggrawal

9. Gross Domestic Product Modeling Using "Panel-Data" Concept 195
   Sarada Ghosh and G. P. Samanta

10 **Supply Chain Scheduling Using an EOQ Model for a Two-Stage Trade Credit Financing with Dynamic Demand** 219
Alok Kumar

11 **Software Engineering Analytics—The Need of Post COVID-19 Business: An Academic Review** 231
Somayya Madakam and Rajeev K. Revulagadda

12 **The Rise and Fall of the SCOR Model: What After the Pandemic?** 253
Nteboheng Pamella Phadi and Sonali Das

Part II Management Optimization

13 **A Comparative Study on Multi-objective Evolutionary Algorithms for Tri-objective Mean-Risk-Cardinality Portfolio Optimization Problems** 277
Georgios Mamanis

14 **Portfolio Insurance and Intelligent Algorithms** 305
Vasilios N. Katsikis and Spyridon D. Mourtas

15 **On Interval-Valued Multiobjective Programming Problems and Vector Variational-Like Inequalities Using Limiting Subdifferential** 325
B. B. Upadhyay and Priyanka Mishra

16 **Portfolio Optimization Using Multi Criteria Decision Making** 345
Namita Srivastava

17 **Uncertain Multi-objective Transportation Problems and Their Solution** 359
Vandana Y. Kakran and Jayesh M. Dhodiya

Part III Computational Intelligence

18 **Agile Computational Intelligence for Supporting Hospital Logistics During the COVID-19 Crisis** 383
Rafael D. Tordecilla, Leandro do C. Martins, Miguel Saiz, Pedro J. Copado-Mendez, Javier Panadero, and Angel A. Juan

19 **Multi-objective Assignment Problems and Their Solutions by Genetic Algorithm** 409
Anita R. Tailor and Jayesh M. Dhodiya

20 **Role of Evolutionary Approaches to Solving Multi-objective Optimization Problems** 429
Surbhi Tilva and Jayesh M. Dhodiya

21	**Improving Financial Bankruptcy Prediction Using Oversampling Followed by Fuzzy Rough Feature Selection via Evolutionary Search** .... Pankhuri Jain, Anoop Kumar Tiwari, and Tanmoy Som	455
22	**An Integrated Fuzzy MCDM Approach to Supplier Selection—Indian Automotive Industry Case** .... Vijaya Kumar Manupati, G. Rajya Lakshmi, M. Ramkumar, and M. L. R. Varela	473

Part IV Computational Management Applications

23	**Decision Support System to Assign Price Rebates of Fresh Horticultural Products Based on Quality Decay** .... Cláudia Matos, Vinicius Maciel, Carlos M. Fernandez, Tânia M. Lima, and Pedro D. Gaspar	487
24	**Recommendation Engine for Stock Market Trading** .... S. Sundarakamatchi and M. S. Gajanand	499
25	**Cross-Listing Effect and Domestic Stock Returns: Some Empirical Evidence** .... Naliniprava Tripathy, Amit Tripathy, and Deepak Tandon	517
26	**Investigation on Supply Chain Vulnerabilities and Risk Management Practices in Indian Manufacturing Industries** .... Nikhil Gupta, R. Rajesh, and Yash Daultani	535
27	**Forecasting Long-term Electricity Demand: Evolution from Experience-Based Techniques to Sophisticated Artificial Intelligence (AI) Models** .... Abhishek Das and Somen Dey	553
28	**Multi Criteria Decision Making for Sustainable Municipal Solid Waste Management Systems** .... Vinay Yadav and Subhankar Karmakar	587
29	**Inventive Investment Using Bigdata: Tools, Applicability and Challenges Associated** .... Janibul Bashir and Tahir Ahmad Wani	599
30	**Computational Aspects of Business Management with Special Reference to Monte Carlo Simulation** .... Sahana Prasad	629
31	**Sustainability in Energy Economy and Environment: Role of AI Based Techniques** .... Trina Som	647

Correction to: Computational Management C1
Srikanta Patnaik, Kayhan Tajeddini, and Vipul Jain

Part I
Computational Modelling

Chapter 1
Computational Management—An Overview

Pragyan Nanda, Deepti Patnaik, and Srikanta Patnaik

Abstract Nowadays, almost all businesses whether small, medium or large are under the tremendous pressure of global competition. With the emergence of internet and e-commerce, geographical barriers have been almost removed, thus, making customers more demanding. With global competitions, customers are showered with alternative options for even the tiniest product. They can no more be controlled by the take it or leave it approach. They demand highest quality product at lowest possible rate and even within the minimum possible time. Agility and flexibility have become the only survival strategies. Business managers in such scenarios are putting all their effort in resolving market-driven issues by identifying them and responding in a timely manner to convert them into opportunities. Amidst these competitions and demand uncertainties, depending on intuition based decision making approach to face new challenges may lead to destructive consequences. Therefore, in the data and technology driven era, business managers must shift their focus to facts and evidences for making faster and rewarding decisions. In such a scenario, deploying Computational Intelligence techniques is providing promising insights to support business managers in better decision making thus gaining competitive advantages over others. Successful organizations are now making planned efforts in adopting computational intelligence techniques to solve complex business problems in practice. Some of the complex application areas where employing computational intelligence has already been successful include customer relationship management, identifying up-selling and cross-selling opportunities, increasing conversion rate by targeting potential customers, social media analytics, risk mitigation, fraud detection, sentiment analysis, targeting marketing campaigns, product pricing, personnel recruiting, employee retention and financial planning etc. In today's fast changing uncertain business environment hardly any facet of business is remaining untouched by computational intelligence. Hence, this chapter makes an attempt to discuss the application of computational intelligence techniques in business management. Various underlying principles

P. Nanda (✉) · D. Patnaik
Interscience Institute of Management and Technology, Bhubaneswar, Odisha, India

S. Patnaik
Siksha O Anusandhan Deemed to be University, Bhubaneswar, Odisha, India

of computational intelligence are discussed followed by the computational aspects and selective applications.

Keywords Uncertainty · Computational intelligence · Deductive reasoning · Decision making

1.1 Introduction

With the wide-spread adoption of new-age technologies like Cloud Computing, Internet of Technology (IoT), Artificial Intelligence (A.I.), Big Data, Data Science, Computational Intelligence, Machine Learning etc., technological advancement is no more confined to engineering domain. Artificial Intelligence is no more confined to the building of cyborgs or robots thus breaking the oldest notion of limitation. It is now making way for its optimal applications endorsing the efficiency of computational intelligence and machine learning in businesses to raise it to the next higher level of success. In other words, it can be mentioned here that A.I. and computational intelligence are playing the role of game-changers in the new age technology embracing business world. In the current tech-driven world, almost all departments of any sector are being handled through software enabled functions to support crucial decision making processes. Most of the organizations are either now successfully adopting artificial intelligence, computational intelligence and machine learning in various forms to support several strategies and add value to diverse applications or in the planning phase to deploy it [1–3]. With the exponential growth of the global A.I. Market, business intelligence is one of the most widely deployed interactive tools that provide solutions along with actionable insights and visualizations to create value. Some of the common areas of business management broadly employing these techniques include identification of potential consumers using insightful visualizations while developing targeted marketing campaigns, building sentiment analysis based models to understand the behavior of customers for launching new products in e-commerce market, ideating new concepts and solutions to incorporate A.I., computational intelligence and machine learning techniques to support organizational strategies in real-time decision makings, validating new metrics and so on. With the growth and expansion of business, almost all organizations face operational issues related to different aspects ranging from procuring raw materials, supplier selection to manpower allotment, scheduling of relevant processes and managing finances and final finished product distribution. Each operational process involves certain level of computational aspects which forms the foundation of making crucial decisions with respect to the prevailing conditions for carrying out smooth execution. Additionally, value chain analysis, market research and assessment of competitors are some other issues requiring attention. Since most of the business processes have computational aspects hence, they can be represented mathematically so as to make it easy to incorporate intelligence based approaches to the systems. Thus, this management of the computational aspects of various management problems that leads to

Figure 1.1 presents a word cloud showcasing Computational Management and its related peripherals consisting of the tools, techniques and applications areas.

Fig. 1.1 Computational management and its peripherals

the conceiving of the term Computational Management. Figure 1.1 presents a word cloud showcasing Computational Management and its related peripherals consisting of the tools, techniques and applications areas.

Business Management as it is widely known has been quite challenging since beginning due to the existence of uncertainty and ambiguity while making most of the crucial decisions. Uncertainty and ambiguity exists in most of the business processes irrespective of its association with financial sector. In business world, uncertainty defined as the lack of certainty in any consequent event may prevail due to many technical as well as natural reasons [4, 5]. Some of which include (i) market dynamics: the pace and abruptness at which the market changes, (ii) unpredictable nature of fast occurring changes across different sectors (iii) Fast declining of highly volatile markets such as automotive and real estate, (iv) High interconnectedness of certain markets over geographical regions leading to increased level of dependency which may get affected in case of international conflicts (v) lack of finance that leads to unknown and unexpected outcomes (vi) cyber fraud and attacks. Other natural causes may include occurrence of any natural climatic disaster affecting means of communications and transportation, outbreak of pandemics such as COVID-19 bringing everything to stand-still, poor performance of market economy following disruption due to any social tragic event or crisis, industrial mishaps etc. Also the traditional approach of "One size fits all" approach of employing generalized strategy for all scenarios has been proved to be inadequate to support decision making. In order to deal with the uncertainty, ambiguity and volatility existing in business environments on regular basis, business owners, managers and leaders rely on understanding the computational aspects of the systems to sort out the complexity overhead involved in it. Initially, managers have to narrow down to potential strategies by analyzing several influential factors. Then, they further consider different potential scenarios and prepare probable survival plans to be executed in case of occurrence of such an event thus converting the crisis into opportunities. Against the traditional approaches that suggests playing safe, current intelligent approaches are usually being built around embracing uncertainty, focusing on reducing the severity and minimizing overall loss.

Thus, the computational aspect of various systems in the dynamic business environment plays a significant role in formulating proactive solution strategies. In the age of digital transformation, where data and information of the complex environments is at the ease of access to almost every stake holder, business organizations prefer to invest in adoption of artificial intelligence and computational intelligence in a broader sense in the form of analysis tools. This adoption of next-generation intelligent tools and techniques enables the system to automate processes to provide clarity by formulating new strategies. This is further backed by extracting relevant information and demographics from uncertain events, thus, making it agile and risk-tolerant to reap rewarding opportunities from crisis. With the extracted information and demographics current trends and potential future demands can be identified. Further analysis of influential factors such as demand elasticity, inventory capacity, competitor's capacity, its expansion plan can showcase the impact on overall performance [6–11].

Thus, adoption of computational intelligence techniques, contribute to stabilize the business environments amid uncertainty and ambiguity. This chapter discusses various computational aspects of business management from several application points of view along with their limitations and risk. Section 1.2 enumerates base principles of computational intelligence, Sect. 1.3 discusses the computational aspects of business management, Sect. 1.4 identifies the benefits and limitations of computational management, Sect. 1.5 elaborates selective application areas of computational management and Sect. 1.6 summarizes the concepts and techniques along with application areas.

1.2 Computational Intelligence Techniques

Computational Intelligence (C.I.), a sub-branch of A.I. has emerged as a result of advancements in the field of algorithmic development with the intent of solving highly complex problems with the help of explorative logics and deductive reasoning like human-brain. It involves the study of various mechanisms that facilitates adaption to dynamically changing and complex problem environments by projecting intelligent behavior. These mechanisms exhibit various relevant abilities such as to learn from experience, generalizing learning and applying it to certain set of circumstances, discover significant factors influencing new circumstances, abstract needs, associate the factors and requirements and adapt to new challenges [12]. Combining all these together, intelligence can be stated as an ability to observe, understand, think, reason, comprehend and learn from experience. Modeling of these abilities computationally into algorithms, programs codes, and systems lead to the evolution of computational intelligence. Systems designed to incorporate these abilities are popularly known as intelligent systems, thus, intelligent systems are a result of successful studying and

1 Computational Management—An Overview

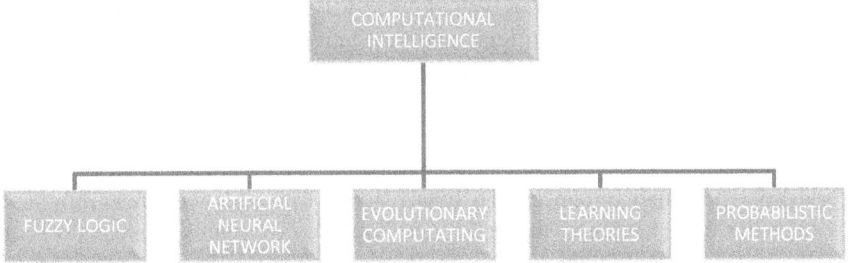

Fig. 1.2 Underlying principles of computational intelligence

modeling of various mechanisms involved in accomplishing several processes going-on in human and nature into natural and biological intelligence. Although there is no specific definition of computational intelligence, in simple words computational intelligence can be stated as the ability of modern day computers trained to learn with the help of experimental observations and data. This learning can be further improvised with the integration of feedbacks and error checks to get approximately accurate solutions.

Computational Intelligence techniques are empowered to utilize incomplete and inexact information to suggest set of actionable possibilities to adapt to the complex problem environments. Most of the C.I. techniques built are mostly inspired by nature itself or various processes existing in the nature driving the life-cycle of all organisms living in nature including animals, birds, insects and even human-beings. The computational aspects of these processes have been studied and researched to find resemblance with complex real-world problem scenarios for modeling it mathematically. Next, the influential factors are then identified and mapped to the variables of the stochastic problem environment to drive optimal solutions. Real-world problems are stochastic in nature due to the existence of uncertainty and ambiguity thus, making it complex to address by traditional approaches. Since, C.I. techniques have been designed to incorporate this inherent uncertainty and ambiguity; they are powered to provide potential solutions to real-world problems. However, computational intelligence techniques are built around five major complimentary techniques including the (i) Fuzzy Logic (ii) Artificial Neural Networks (iii) Evolutionary Computing (iv) Learning Theories (v) Probabilistic Methods [13–15]. These principles are discussed and shown in Fig. 1.2.

1.2.1 Fuzzy Logic

Business world now consists of massive amount of data and information about products, services, demands, market fluctuations, real-time traffic, customer preferences etc. This data and information is available in heterogeneous and qualitative form and consists of uncertainty in verbal as well as numerical form. Moreover, since conventional methods rely on Boolean logic and qualitative data is hard to be processed using Boolean logic, hence they fall short to deal with real-time decision making. Fuzzy logic is best suited for modeling the complex processes of real-life problems. It is capable of handling incompleteness, in contrast to other conventional methodologies that require exact information for computations. Moreover, most of the real world problems cannot be converted to binary 0 and 1 s it enables computer programs to deal with natural languages through codes. Fuzzy logic resembles human decision making and deals with approximate reasoning which is not crisp. It considers all the intermediate possibilities between binary 0 and 1 [16–18]. Fuzzy logic works on assumptions that supports the blurry values between the binary values of true and false such as may be true, partially true, may be false, partially false or true to a certain degree of percentage. Thus fuzzy logic maintains elasticity in decision making considering the imprecision, incompleteness and vagueness existing in verbal uncertainties. Basic steps in fuzzy logic incorporation include defining linguistic variables from subjective information, fuzzification, and construction of inference rule base followed by de-fuzzification and final decision is inferred from it. Business management experts are adopting fuzzy logic to deal with qualitative matters that presents subjective information through linguistic variables providing effective models.

1.2.2 Artificial Neural Networks

Artificial Neural Network is modeled around human-neurons consisting of three major components, namely, the axon responsible for conducting signal and communicating information, the cell body responsible for information processing and the synapse responsible for controlling signals. Due to its dynamic nature, most of the business processes that involve crucial decision making usually reflect non-linear behavior. Since decision making forms an integral part of business management, thorough analysis of all available information is a must. Several fields of business management such as enterprise management, market research, manufacturing and procurement, production operation management, supply chain management, logistics and transportation, warehouse management and fulfillment, retail, trading and forecasting etc. exhibits non-linearity in nature due to inherent uncertainty. In such a complex environment, Artificial Neural Networks (ANN) proves to be a powerful tool with its capability to learn from past mistakes to minimize errors and improve prediction accuracy [19–24]. ANN with its layered intelligent computational features provides strong support to decision making process by analyzing complex environments and their influential factors for potential solutions. Conventional logical methods are not capable of managing such scenarios. The steps of ANN involves defining number of layers to be deployed first, followed by initialization of randomized weights, then identification of activation rate and error rate along with the bias computes output and compares with the expected output to improve accuracy and goes on repetitively until convergence is attained. The final decision will be generated once validation phase is over. Thus, adoption of ANN in business management opens new opportunities and scopes for improvisation of accuracy in making impactful decisions such as amount of products to be stocked up, managing finance, pricing and discounts to be bundled together for larger purchases, analyzing previous sales to forecast demand, and selecting suppliers on the basis of their loyalty and price offerings incorporating the feedback based learning system. Some of the common application areas include identifying targeted customers, stock purchase, forecasting demand and sales, fraud prevention, loan approval, hiring talents, market analysis etc.

1.2.3 Evolutionary Computing

Evolutionary Computation is a powerful tool comprising of different intelligent tools such as Evolutionary Strategies, Genetic Algorithms, Genetic Programming, and Evolutionary Programming. They exhibit a search based optimization approach that imitates the natural evolution mechanism by Charles Darwin [25–32]. The basic steps involve population initialization followed by fitness based selection process for individuals and further crossover the inherited information and then mutate to induce variations and then again select next generation repetitively to follow the steps until convergence. Although evolutionary computation techniques are highly popular in technical sector due to their potentiality, they have found their recognition in business management domain too. Multi objective optimization problems such as production planning and scheduling problems, optimal transportation planning, improved logistic system, cost optimization etc. where imprecision, uncertainty, arbitrariness dwells, evolutionary computation techniques can be adopted.

1.2.4 Learning Theories

In human beings and other organisms, learning is a psychological process that brings together various past experiences and draws cognitive inferences on the basis of deductive reasoning of cause and effects to enhance the knowledge and develops skills to predict further outcomes in similar scenario/environment. However, computational learning theory is completely based upon assumptions that support the system to draw inferences following generalized principles from available data. Learning theory plays a major role in building the computational intelligence tools and techniques. Some of the widely adapted learning theories include Exact Learning approach, Bayesian Inference, Algorithmic Learning and Approximate Learning theory.

1.2.5 Probabilistic Methods

Probabilistic methods also play a significant role in modeling computational intelligence for business management applications by expressing information. It is best suited for treating numerical uncertainty existing in business management processes. The main goal of probabilistic method is to evaluate several possible outcomes of an intelligent system on the basis of randomness and generate the best possible solutions [33–35].

1.3 Computational Aspects of Business Management

Although in this era of data, predominant adoption of artificial intelligence and computational intelligence by business managers and leaders is still in its infant stage, but most of the eminent firms and organizations are quite excited about deploying it into diverse business functions and reap phenomenal rewards. With the exponentially increasing barrage of data and constantly fluctuating preferences of consumers has raised the complexity of businesses making it quite difficult for conventional methods to drive the growth of the organizations. Radical changes in technology have unbolted new realms of opportunities by incorporating intelligent techniques into conventional methods to generate actionable insights [36]. The availability of valuable data in vast amount can be converted into new business values and unfold new expansion directions. Extensive incorporation of computational intelligence in business can be done by employing intelligent computer software with underlying intelligent algorithms and techniques that can analyze and manipulate the computational aspects according to requirement to increase efficiency and productivity along with improving customer experience, thus, boosting revenue and driving the growth of the business. However, to incorporate computational intelligence, business managers and leaders need to understand the computational aspects of the various business functions and processes thoroughly, dependability on various factors and components, relationships between the functions and processes and their influences etc. with a view of reducing operational costs while increasing productivity. Again, all the computational aspects of various business functions and their feasibility to adopt computational intelligence are yet to be explored. Still, some of the business functions which have already explored the computational aspects and deployed computational intelligence are discussed below.

1.3.1 Marketing

Marketing is one of the most relevant business functions to increase sales and revenue. It refers to actions or functions related to advertising/promoting products and services on the basis of market research to sell it. Understanding the computational aspects such as demographics representing customer interest and preferences, their browsing history, analyzing consumer recent purchase behavior will help in customizing promotional strategies for the targeted customers [37]. Integrating intelligent tools and techniques helps in building powerful recommendation engines and chat-bots to suggest customized product and service ranges according to the customer's need. Further, computational intelligence techniques can be adopted to design effective marketing campaigns for future products on the basis of response data from customer reviews.

1.3.2 Sales

Sales plays an important role in any business organization irrespective of the product or service relevance and contributes extensively towards revenue generation. Understanding the computational aspects of several factors that directly or indirectly influence the sale can help in adopting computational intelligence techniques into sales functions. C.I. techniques helps in improving the accuracy in forecasting future demands from various valuable insights generated from massive sales data. It can also provide new leads to potential future customers, thus helping professionals to follow-up them and increase the conversion rate.

1.3.3 Customer Relationship Management (CRM)

Computational intelligence techniques can also be employed to automate the process of customer relationship management (CRM) by using chat-bots. C.I. techniques use Big Data tools to collect data about customers' shopping behavior through various loyalty schemes and reward plans. Chat-bots also interact with customers' shopping online and tailors strategies to enhance customer experience.

1.3.4 Finance

In the finance department, adoption of C.I. will help in automating most crucial yet relevant functions, enhancing productivity and accuracy. Since book-keeping, keeping data about loans, interest rates, past credit data, evaluating credit scores, record maintaining, investment and stock trading advisory are some of the most important tasks of the financial department involving high-level computational aspects where small errors may cost huge such that making errors cannot be afforded, hence, automating the computational aspects of various processes will reduce the chances of loss [38]. Again, process automation system reduces cost, time and errors in performing repetitive tasks allowing professionals focus on other significant activities. Computational intelligence powered financial-advisor bots are capable of monitoring real-time financial status of the organizations thus helping professionals in making better financial decisions. Financial sectors are also incorporating computational intelligence to identify frauds and aim to reduce risks by securing transactions.

1.3.5 Human Resources

Computational Intelligence techniques can be employed to automate many human resource activities with several computational aspects such as evaluating qualitative as well as quantitative talents on the basis of various metrics, for internships and recruitment. Again repetitive tasks involved in this processes can be automated. These intelligent tools and techniques can be incorporated to screening bio-data, call candidate and schedule or even interview the candidates for open positions. It can also help in providing personalized experiences as per requirement.

1.3.6 Manufacturing

In manufacturing sector, with the availability of real-time data, computational aspects of several processes can be explored for employing intelligent techniques. Supply chain management, procuring raw materials, supplier selection, inventory tracking and warehouse management etc. are some of the processes where computational intelligence techniques are being successfully incorporated for automation. Using intelligent tools and techniques production process can be speed up or slow down depending on real-time demand data prediction. Further, addition of specific sensors along with computational intelligence techniques can support in predicting maintenance requirements and quality control.

1.3.7 Maintenance and Service

Buildings are also now a day empowered by A.I. Computational aspects of buildings such as number of occupant population, capacity, usage etc. can be analyzed to optimize the energy consumption in buildings. Customization of lighting, temperature cooling and heating systems can enhance the stay experience of the occupants. Based on the collected data intelligent building monitoring systems can adjust several functions as needed.

1.3.8 Research and Development

Finally, research and development sector can extend hands to innovate by building domain specific tools for separate sectors as required such as automotive, health care, pharmaceuticals, fashion and clothing sector and lastly the financial sector. It keeps improving product and service experience on the basis of the data

1.4 Benefits and Limitations of Computational Intelligence in Business

Like all other tools and techniques, computational intelligence techniques also have their set of benefits and limitations enumerated as follows.

Benefits of Computational Intelligence in Business

1. Process automation is one of the most significant benefits of incorporating computational intelligence in business domain.
2. Increased productivity and efficiency due to automation of repetitive tasks.
3. Improvised and targeted marketing campaigns due to analysis of market research data.
4. Enhanced customer experience about products/services due to personalization.
5. Increased revenue and profit generation by reducing costs and increasing sales.
6. Enhanced Fraud detection and security monitoring services can be empowered by computational intelligence techniques.
7. Higher Customer conversion rate.

Limitations of Computational Intelligence in Business

1. Although data is available in abundance but still there is lack of proper data for specific new domain to the problem is not available some time.
2. Algorithmic bias existing in software may cause severe irreversible issues
3. Limited computational capacity posing as constraint, while dealing with large datasets
4. Security and privacy issues will always prevail.

1.5 Applications of Computational Intelligence in Business

The applicability of the computational intelligence techniques are rapidly growing with the enormous efforts of the researchers in this field. This soft computing technique is used to address complex real-time problems, where traditional modeling or mathematical approaches fail due to its complexity and stochasticity. The computational intelligence has a wide range of applications in various fields of study namely;

computer science, healthcare sector, social science, power systems, signal and image processing and business process etc. With the use of the techniques of computational intelligence there is a notable change in addressing the needs of the business. The present day business is witnessing an effective support tool in the form of CI techniques. Almost all the techniques of CI are used for the business, of which artificial intelligence (AI) is the popular technique. The AI has surpassed the human intelligence process by predicting future decisions and extracting human insights and recognizing the patterns. The approach towards the business is completely changed with the emergence of AI in the business operations. For any business there exists three needs that should be addressed and AI supports in this. The needs which can be supported by AI are business process automation, data analysis and customer-employee engagements.

1.5.1 Applications of Computational Intelligence in the Field of Marketing

With the use of the computational intelligence techniques the business process needs and requisites are addressed. The most common task in the business process is back office administration and financial activities, with robotic process automation (RPA) techniques the digital and physical tasks can be automated. The robotic process automation uses software bots to automate. Those tasks which are routine and repetitive in nature and which are usually performed by the knowledge workers can be automated. RPA is the least expensive and easily implemented techniques yielding quick and high rate of return and suits to multi-back end system. With the increase in the volume of data the analysis and interpretation demanded a change in the technique. Hence machine learning (ML) algorithms were used for the pattern recognition of the voluminous data. The ML applications facilitates in predicting the customer choice of buying, fraud detection, etc. The ML models are trained with some set of data and further the model gets better with the use of new data for prediction.

The CI optimizes the customer relationship management process analyses the customer sentiments and answer accordingly with the help of relevant key word searches. By deploying the Natural Language Processing (NLP) techniques the AI interprets the words and analyzes the data. Also it helps the analysts to generate useful insights by application of contextual and reasoning algorithms while focusing on growing needs of the data. The customer experience improvement was also been able to be focused by the advent of CI. With the emergence of CI the dynamics of the business world has underwent a tremendous change. The CI based application helps in improvising the performance of the marketing field. With the technological advancements, today's business demands digital marketing upon traditional method of marketing. Since the data is on a higher side the usage of CI helps the marketing managers in many possible ways. The marketing managers get guidance in the form of [1]. The implementation of CI in marketing is not done completely till date and many software in this regard has come into existence. Since marketing is such a field which involves both qualitative as well as quantitative aspects, thus, enabling the scope for growth of CI and where only econometrics is not sufficient [2]. The major applications of CI in the field of marketing has enhanced and upgraded the outdated methods of marketing [39]. The creation of customized marketing campaigns by analyzing the data was the advent of CI for making the companies a successful one. In addition to this it also improves the yield management by dynamic pricing and excellent customer service. In order to solve the issues and lead the market, marketing management need decision makers related to market. Precisely CI is a knowledge-driven tool which assists in decision making through examining the information. The effectiveness of the market enhanced with CI application in the business by developing the sales and marketing strategies that drive considerable improvement in business performance. The adaption of CI enables the marketing managers to understand the marketing, sales and operation trends. With CI in the field of marketing has led to achieving of better competitive advantage over the other companies also it found to be very effective in handling the marketing processes and improvised data analysis [40].

1.5.2 Computational Intelligence and Human Resource

The key to success of any organization is the human resource since they are the source of knowledge which every organization should and must draw on. Hence acquiring and retention of employees play a vital role in the present day scenario. And moreover the recruitment process is a very time consuming process therefore the human resource planning has tremendously changed with the application of CI. The traditional method of conducting interviews included the recruiters initial contacting like scrutinizing the CV's of the candidates, give feedback to the rejected employees and conduct the interview for the selected ones [41]. The piling up of irrelevant resumes and time spent in the interview process for retention and appraisal was done easy with intelligent recruitment. It made the assessment of employees an effortless task while analyzing which set of employees needs to be trained, which employee needs to be motivated and the identifying the deserving employee who must be rewarded [42].

The human resource management is and has been a strategic asset where employees are the key assets and acquiring them plays an important role [43]. The CI driven recruitments assistants have been introduced in the form of chat bots to enable connection and up-to-date connection possibilities with candidates through mails or text messages [44]. The recruitment system demands modernization in order to ease the organizations and save the expenditures involved in this process [45]. The recruitment systems include these following steps for a quick recruitment process. This recruitment system facilitates the managers to find the eligible job applicants quickly with efficient and effective manner and moreover with the application of CI the paperwork is reduced and the pooling of the candidates is found to be on a larger side giving scope for better selection [45]. In addition to this the candidate's personality traits, values as well as attitude can be accessed with the skimming of the data posted by the candidates in their respective social media. Hence it can be said that CI is helping the recruiter's to meet the traditional aspect of recruitment process. Furthermore the scope for future improvements of the rejected candidates can be possible with the feedback provided by the system. Apart from the other benefits that the CI driven organizations enjoy—the decreased manual work in the process yields more time to focus on the potential candidates.

The recruitment system framework, shown in Fig. 1.3 facilitates the managers to find the eligible job applicants quickly with efficient and effective manner and moreover with the application of CI the paperwork is reduced and the pooling of the candidates is found to be on a larger side giving scope for better selection. In addition to this the candidate's personality traits, values as well as attitude can be accessed with the skimming of the data posted by the candidates in their respective social media. Hence it can be said that CI is helping the recruiter's to meet the traditional aspect of recruitment process. Furthermore the scope for future improvements of the rejected candidates can be possible with the feedback provided by the system. Apart from the other benefits that the CI driven organizations enjoy—the decreased manual work in the process yields more time to focus on the potential candidates. As for any adaption in the technology there exist pros as well as cons. Along with the above benefits this application may result in losing of jobs to automation since technology is replacing human involvement. Furthermore the authors of a journal whose work is published in *Strategic HR Review* argue that programs have the ability to bypass the candidate's name, gender and age which are the primary source of biasness.

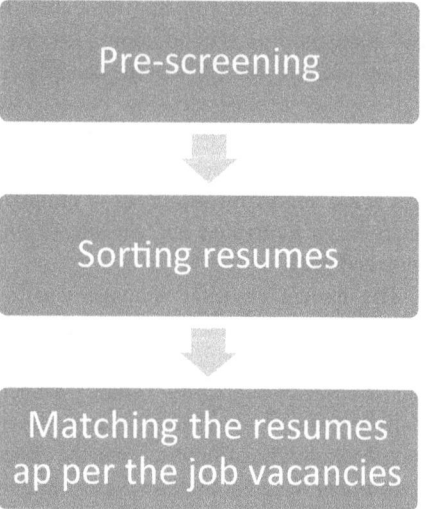

Fig. 1.3 The recruitment system framework of an organization incorporating C.I

1.5.3 Computational Intelligence and Finance

The versatility of CI is also witnessed in the financial sector. Finance is the backbone of the country's economy. The financial markets play a vital role in the present day modern society with regard to economical and social organization. The conservative thought of saving only in banks has completely replaced with investing in financial markets. This market is completely noisiest and volatile in nature. The success or failure of the investment made by the investor is dependent on the quality of information he is using to support his decision and how rapidly the decision is being made. Owing to its practicality the financial market movement analysis is widely studied. With the technological support the financial sector is completely changed. The financial sector with implementation of CI is benefitting greatly. CI is used in different sectors that are from finance sector to agriculture to healthcare sector. It can be an alternative for the finance problems in price forecasting and market efficiency. The primary motto behind the application of CI is to simplify the problem of predicting future market movements to a pattern recognition problem whereby inputs can be derived from historical data and the output from the past data [46]. From the past literature we can derive a lot many approaches to address the problem, may it be a statistical approach or a computational intelligence algorithm, the financial data is defined as a time series. Where the time series is set of organized numerical data observations recorded in sequence with time. This time series allows modeling the series behavior as well as prediction of the future behavior of the time series. This is only possible with the help of several time series tools. Those tools facilitate simple and easy ways to perform data mining tasks and the extracted features of time series results in maximization of investor's profit.

Despite of the fact that the share price cannot be predicted accurately due to the stochastic environment prevailing in the stock market there is no well established or tested methodology for prediction. CI enumerates for guiding the investors for building an intelligent trading system. Its two main features for its use are price prediction and risk mitigation. The broader application of CI in the financial sector is:

- Improving the credit decision
- Identification of threats to financial institutions
- Addressing the financial gaps.

The credit worthiness of the borrowers is the primary step in this process. The credit rating agencies and lenders analyze the data to know the credit worthiness of the borrower. The traditional way of generation of credit scores were formal identification along with bank transactions, credit history and income statement. Now with the application of CI the detailed history of the borrowers can be traced from alternate sources such as public data, satellite images, company registries and social media [47]. The biggest threat that is identified with the financial institutions is cyber fraud, money laundering, etc. that needs to be checked. To combat these threats the institutions undertakes the know-your-customer (KYC) to verify the identity of the customers. The use of CI certainly helped in fraud detection and the only reason stated for the use of technology.

There exists a global trade finance gap driven by relatively high cost of firm creditworthiness assessment. The application of ML algorithms to the alternate sources creates financial solutions and as well provides a guide of the financial products to the customers. The trust issues can be addressed with the use of chat-bots and language processing. Still some issues remains underexplored in the present emerging markets such as the financial literacy barriers in assessing the financial services for customers [47]. Even though there are many advantages some loopholes of the technology exists such as cost involved in automation and customer's illiteracy in technology adaption, unreliable internet connectivity etc.

1.5.4 Computational Intelligence in Operations Management

In this present trend of digital transformation and the nature of CI techniques yet another prominent sector in the business process is witnessing a change. It is the production and operations management sector, which emphasizes on design, planning, manufacturing, quality control, energy system and scheduling. Due to the increase in complexity and need of higher efficiency with greater flexibility and better product quality and lower cost of production gave rise to the change in industrial manufacturing. The implementation of this technique significantly contributed in improving the control and manufacturing systems. The need of the hour is smart production systems with the state-of-the-art solution to increase the production as well as quality and sustainability. From the sustainability viewpoint the CI techniques application has the potential for addressing the issue of scarce resource i.e., optimum utilization and increase in productivity. Moreover, the CI techniques present a wide range of applications providing opportunities for sustainable development in sectors, including inventory and supply chain management and predictive maintenance. For this purpose ML is the tool that has been successfully utilized in many processes for optimization, manufacturing applications and predictive maintenance in various industries [48]. Lieber et al. have proposed an approach to improve the quality pf process and products in steel industry. The ML techniques were found beneficial for improved quality control optimization in this system [49, 50]. Furthermore they have also described the main areas in sustainable manufacturing and objective that can be derived with the application:

- The key objective of the supply chain management is the availability of the product in the appropriate place at specified time and with the use of the ML application there will be improved transparency, accelerates decision-making and accurate demand forecasting.
- In case of quality control it improves response time and allows eliminating possible failures.

- Since predictive maintenance detects possible production malfunctions causing product quality issues ML creates an accurate forecasting as when the machinery needs a repair.
- Yet another key area in sustainable management is energy consumption and the objective of this area is to strike a balance energy usage. This tool improves excessive use of certain materials, unnecessary scrap waste, inefficient supply chain management, logistics, and unequal distribution of energy resources.

1.6 Conclusion and Future Scope

With the changing trends in the society the technology need to be advanced to reap the benefits. The application of CI has given enormous results but in some sectors it needs to be implemented. The main reason behind not choosing the automation of the sectors is lack of technical knowledge, internet uncertainty, cost aspect and habituated to traditional way. In order to boost the applicants the programming of the tools need to be made even simpler. Successful adoption of computational Intelligence is just a beginning to embrace the exponentially increasing opportunities in intelligent business process management. Organizations that do not want to held back and miss out on this revolution must adopt computational intelligence techniques in various applications. Computational intelligence can help business managers in identifying hidden patterns in market evolutions in order to deploy the learned strategies of the winners to quantify uncertainty and mitigate risks. Moreover, computational intelligence itself, encapsulates some powerful principles such as fuzzy logic, artificial neural network, evolutionary computation etc., to help managers process data faster and generate actionable insights for better decision making. Although C.I. has been around the corner for a while, still there are many computational aspects of business management which are yet to be explored and employ the intelligent technique for problem solving. These potential areas can be explored and impact of computational intelligence along with the influential factors in future.

References

1. Sterne J (2017) Artificial intelligence for marketing: practical applications. Wiley
2. Wierenga B (2010) Marketing and artificial intelligence: great opportunities, reluctant partners. In: Marketing intelligent systems using soft computing. Springer, Berlin, pp 1–8
3. Russell SJ, Norvig P (2016) Artificial intelligence: a modern approach. Malaysia
4. Courtney H, Kirkland J, Viguerie P (1997) Strategy under uncertainty. Harv Bus Rev 75(6):67–79
5. Fosbrook B (2017) How scenarios became corporate strategies: alternative futures and uncertainty in strategic management
6. Kopa M, Wiesemann W (2017) Special issue on the 12th international conference on computational management science
7. Aggarwal R (2019) Analysis on practical and computational aspects of optimization theory. Homepage 9(4):2. https://www.ijmra.us

8. Vuong QH (2019) Computational entrepreneurship: from economic complexities to interdisciplinary research. Problems Perspect Manage 17(1):117–129
9. de Araújo SA, de Barros DF, da Silva EM, Cardoso MV (2019) Applying computational intelligence techniques to improve the decision making of business game players. Soft Comput 23(18):8753–8763
10. Chung W, Mustaine E, Zeng D (2020) A computational framework for social-media-based business analytics and knowledge creation: empirical studies of CyTraSS. Enterp Inf Syst 1–23
11. Hauke J, Lorscheid I, Meyer M (2018) Individuals and their interactions in demand planning processes: an agent-based, computational testbed. Int J Prod Res 56(13):4644–4658
12. Engelbrecht AP (2007) Computational intelligence: an introduction. Wiley
13. Rutkowski L (2008) Computational intelligence: methods and techniques. Springer Science & Business Media
14. Siddique N, Adeli H (2013) Computational intelligence: synergies of fuzzy logic, neural networks and evolutionary computing. Wiley
15. Abbas A, Zhang L, Khan SU (2015) A survey on context-aware recommender systems based on computational intelligence techniques. Computing 97(7):667–690
16. Bezděk V (2014) Using fuzzy logic in business. Procedia Soc Behav Sci
17. Valášková K, Kliještik T, Mišánková M (2014) The role of fuzzy logic in decision making process. In: 2014 2nd international conference on management innovation and business innovation, vol 44, pp 143–148
18. Şen Z (2017) Intelligent business decision-making research with innovative fuzzy logic system. Int J Res Innov Commer 1(1):93–111
19. Sharma V, Rai S, Dev A (2012) A comprehensive study of artificial neural networks. Int J Adv Res Comput Sci Softw Eng 2(10)
20. Tkáč M, Verner R (2016) Artificial neural networks in business: two decades of research. Appl Soft Comput 38:788–804
21. Levine DS, Chen KY, AlQaudi B (2017) Neural network modeling of business decision making. In: 2017 International joint conference on neural networks (IJCNN), May 2017. IEEE, pp 206–213
22. Trad A, Kalpić D (2017) A neural networks portable and agnostic implementation environment for business transformation projects the basic structure. In: 2017 IEEE international conference on computational intelligence and virtual environments for measurement systems and applications (CIVEMSA), June 2017. IEEE, pp 153–158
23. Mirzaey M, Jamshidi MB, Hojatpour Y (2017) Applications of artificial neural networks in information system of management accounting. Int J Mechatron Electr Comput Technol 7(25):3523–3530
24. Li Y, Jiang W, Yang L, Wu T (2018) On neural networks and learning systems for business computing. Neurocomputing 275:1150–1159
25. Biethahn J, Nissen V (eds) (2012) Evolutionary algorithms in management applications. Springer Science & Business Media
26. Burduk A, Musiał K (2016) Optimization of chosen transport task by using generic algorithms. In: IFIP international conference on computer information systems and industrial management. Springer, Cham, pp 197–205
27. Mahammed N, Benslimane SM, Ouldkradda A, Fahsi M (2018, July) Evolutionary business process optimization using a multiple-criteria decision analysis method. In: 2018 International conference on computer, information and telecommunication systems (CITS), July 2018. IEEE, pp 1–5
28. Rabe M, Goldsman D (2019) Decision making using simulation methods in sustainable transportation. In: Sustainable transportation and smart logistics. Elsevier, pp 305–333
29. Ojstersek R, Brezocnik M, Buchmeister B (2020) Multi-objective optimization of production scheduling with evolutionary computation: a review. Int J Ind Eng Comput 11(3):359–376
30. Tan W, Huang L, Kataev MY, Sun Y, Zhao L, Zhu H, Guo K, Xie N (2020) Method towards reconstructing collaborative business processes with Cloud services using Evolutionary Deep Q-Learning. J Ind Inf Integr 100189

31. Jose S, Vijayalakshmi C (2020) Design and analysis of multi-objective optimization problem using evolutionary algorithms. Procedia Comput Sci 172:896–899
32. Fernandez E, Navarro J, Solares E, Coello CC (2020) Using evolutionary computation to infer the decision maker's preference model in presence of imperfect knowledge: a case study in portfolio optimization. Swarm Evol Comput 54:100648
33. Sinulingga U, Napitupulu N, Manurung A (2019) A usage of probabilistic methods in decision making for wholesalers' problem. J Phys Conf Ser 1235(1):012121 (IOP Publishing)
34. Xu Z, He Y, Wang X (2019) An overview of probabilistic-based expressions for qualitative decision-making: techniques, comparisons and developments. Int J Mach Learn Cybern 10(6):1513–1528
35. Guo X (2020) Probabilistic forecasting in decision-making: new methods and applications. Doctoral dissertation, UCL (University College London)
36. Groesser SN, Jovy N (2016) Business model analysis using computational modeling: a strategy tool for exploration and decision-making. J Manag Control 27(1):61–88
37. Yang Y, Yang YC, Jansen BJ, Lalmas M (2017) Computational advertising: a paradigm shift for advertising and marketing? IEEE Intell Syst 32(3):3–6
38. Alrabiah A (2018) Optimal regulation of banking system's advanced credit risk management by unified computational representation of business processes across the entire banking system. Cogent Econ Financ 6(1):1486685
39. Forrest E, Hoanca B (2015) Artificial intelligence: marketing's game changer. In: Trends and innovations in marketing information systems. IGI Global, pp 45–64
40. Shahid MZ, Li G (2019) Impact of artificial intelligence in marketing: a perspective of marketing professionals of Pakistan. Glob J Manage Bus Res
41. O'Donovan D (2019) HRM in the organization: an overview. In: Management science. Springer, Cham, pp 75–110
42. Baron IS, Agustina H (2018) The challenges of recruitment and selection systems in Indonesia. J Mgt Mkt Rev 3(4):185–192
43. Bas A (2012) Strategic HR management: strategy facilitation process by HR. Procedia Soc Behav Sci 58:313–321
44. Upadhyay AK, Khandelwal K (2018) Applying artificial intelligence: implications for recruitment. Strat HR Rev
45. Dickson DR, Nusair K (2010) An HR perspective: the global hunt for talent in the digital age. Worldwide Hosp Tourism Themes 2(1):86–93
46. Teixeira LA, De Oliveira ALI (2010) A method for automatic stock trading combining technical analysis and nearest neighbor classification. Expert Syst Appl 37(10):6885–6890
47. Biallas M, O'Neill F (2020) Artificial intelligence innovation in financial services
48. Pérez-Ortiz M, Jiménez-Fernández S, Gutiérrez PA, Alexandre E, Hervás-Martínez C, Salcedo-Sanz S (2016) A review of classification problems and algorithms in renewable energy applications. Energies 9(8):607
49. Lieber D, Stolpe M, Konrad B, Deuse J, Morik K (2013) Quality prediction in interlinked manufacturing processes based on supervised & unsupervised machine learning. Procedia CIrp 7:193–198
50. Hernández-Julio YF, Paba MÁJ, Narváez NEL, Hernández HM, Bernal WN (2017) Framework for the development of business intelligence using computational intelligence and service-oriented architecture. In: 2017 12th Iberian conference on information systems and technologies (CISTI), June 2017. IEEE, pp 1–7

Chapter 2
Mathematical and Computational Approaches for Stochastic Control of River Environment and Ecology: From Fisheries Viewpoint

Hidekazu Yoshioka

Abstract We present a modern stochastic control framework for dynamic optimization of river environment and ecology. We focus on fisheries in Japan and show several examples of simplified optimal control problems of stochastic differential equations modeling fishery resource dynamics, reservoir water balance dynamics, benthic algae dynamics, and sediment storage dynamics. These problems concern different phenomena with each other, but they all reduce to solving degenerate parabolic or elliptic equations. Optimal controls and value functions of these problems are computed using finite difference schemes. Finally, we present a higher-dimensional problem of controlling a dam-reservoir system using a semi-Lagrangian discretization on sparse grids. Our contribution shows the state-of-art of modeling, analysis, and computation of stochastic control in environmental engineering and science, and related research areas.

Keywords Inland fisheries · Stochastic control · Dynamic programming · Numerical schemes

2.1 Introduction

Rivers are a part of hydrological cycles as well as a part of human lives. Flowing waters in rivers are stored by dam-reservoir systems and utilized as primary water resources for drinking, irrigation, hydropower generation, and so on [58]. Operation of dam-reservoir systems should harmonize the water use with river water environment and ecosystems. Regulated flows released from a dam alter its downstream flow regimes and often threaten riparian habitats and aquatic species [27, 70]. Therefore, exploring a unified framework to balance ecological and human dimensions is of high importance in river environmental management.

H. Yoshioka (✉)
Graduate School of Natural Science and Technology, Shimane University and Fisheries Ecosystem Project Center, Shimane University, Nishikawatsu-cho 1060, Matsue 690-8504, Japan
e-mail: yoshih@life.shimane-u.ac.jp

Stochastic optimal control as a branch of modern mathematical sciences plays a central role in analysis and management of noise-driven dynamical systems [42]. Noises in the context of environmental and ecological management arise from stochastic river water flows [56] and highly nonlinear and possibly unresolved biological phenomena such as biological growth phenomena [80]. Stochastic differential equations (SDEs) [42], which are formally seen as ordinary differential equations (ODEs) driven by noises, serve as an efficient mathematical tool for modeling and controlling noisy system dynamics.

Problems related to river environmental and ecological dynamics are not the exception that the stochastic control applies. However, such an outlook has not been paid much attention before the author and his co-workers started modeling, analysis, and computation of inland fishery resource management in rivers in Japan. Fishery resources management problems in seas have conventionally been studied as both stochastic and deterministic optimal control problems [20, 32, 33, 50], possibly because of their huge impacts on food and economy worldwide. By contrast, the problems in inland waters usually have smaller impacts; nevertheless, they have been serving as unique elements to sustain local ecosystems, societies, and sometimes ecological education [76]. In addition, as we will demonstrate in this chapter, there exist many interesting management problems specific to inland fishery resources.

The objective of this chapter is to present a unified mathematical framework and specific applications of the stochastic control to river environmental and ecological problems with a focus on inland fisheries management. Problems we focus on concern management of the diadromous fish *Plecoglossus altivelis altivelis* (*P. altivelis*, Ayu) in Japan as one of the most common inland fishery resources in the country [1, 37]. The unique life history of *P. altivelis* is explained later, but what is important here is that managing the fish requires considering not only its life history, but also surrounding environmental conditions from multiple sides. This motivates us to separately study sub-control problems, such as fish growth [77], benthic algae management [73], sediment storage management [74], and dam-reservoir system management [86], which have different characteristics with each other but can be handled by a unified framework based on dynamic programming and viscosity solutions: appropriate weak solutions to degenerate elliptic equations [3, 16].

We show that each control problem reduces to solving a corresponding degenerate elliptic (or parabolic, hyperbolic) equation often called Hamilton–Jacobi-Bellman (HJB) equation. This is conducted either analytically or numerically, but usually the latter is employed in applications because of the nonlinearity and nonlocality of the HJB equations [7, 22, 24, 34, 51]. We approach the specific problems numerically using finite difference and semi-Lagrangian schemes.

We also consider a coupled higher-dimensional problem where resource and environmental dynamics should be managed concurrently. Conventional numerical schemes encounter a huge computational cost when there are more than three to four state variables. This issue is called the curse of dimensionality. Computational costs of the conventional numerical methods for stochastic control problems increase exponentially when the total number of state variables increases linearly. We use a semi-Lagrangian scheme [8] on sparse grids [10] to alleviate this issue.

2.2 Stochastic Control

2.2.1 Stochastic Differential Equation

We briefly and formally present a basic framework of the stochastic control. Interested readers should read the textbooks and reviews of stochastic control and its applications [12, 42, 71, 72]. Throughout this chapter, each problem is considered based on a standard complete probability space [42]. The explanation in this section is formal, and coefficients and parameters appearing in the third section will be specified in each control problem.

We consider continuous-time dynamics. We denoted time as $t \geq 0$. The total number of state variables is denoted as $M \in \mathbb{N}$. The state variables are assumed to be càdlàg and are represented as a vector $\mathbf{X}_t = [X_{i,t}]_{1 \leq i \leq M}$. Its range is Ω, which is bounded or unbounded, and is problem-dependent. We assume that the process $\mathbf{X} = (\mathbf{X}_t)_{t \geq 0}$ is a jump-diffusion process governed by a system of SDEs driven by compound Poisson jumps. From a mathematical side, it is more convenient to consider a generic Lévy process as a driving noise process [42]. However, problems considered in this chapter require only compound Poisson processes (and Brownian motions if necessary).

We only use jump processes in the mathematical modeling discussed later but incorporating diffusive dynamics does not encounter difficulties in most cases. Both Brownian and jump noises are considered in this sub-section for the sake of explanation. Recall that the continuous-time Markov chains are represented using appropriate Poisson processes [71].

The N-dimensional standard Brownian motion is denoted as $\mathbf{B}_t = [B_{i,t}]_{1 \leq i \leq N}$ at time t. In addition, the K-dimensional compound Poisson process is denoted as $\mathbf{P}_t = [P_{i,t}]_{1 \leq i \leq K}$ at time t. Each $P_{i,t}$ is mutually independent with each other. For each $1 \leq i \leq K$, the jump intensity of $P_{i,t}$ is denoted as $\lambda_i > 0$, and the probability density of the jump size of $P_{i,t}$ as p_i. We assume $\mathbf{X}_t \in \Omega$ a.s. for $t \geq 0$. The control process $\mathbf{u} = (\mathbf{u}_t)_{t \geq 0}$ is assumed to have a compact range U and progressively measurable with respect to a natural filtration generated by the processes $(\mathbf{B}_t)_{t \geq 0}$ and $(\mathbf{P}_t)_{t \geq 0}$. These assumptions are rather standard. The admissible set of $\mathbf{u}$ is denoted as U. The system of Itô's SDEs governing $\mathbf{X}$ is set as

$$\mathrm{d}\mathbf{X}_t = b(t, \mathbf{X}_t, \mathbf{u}_t)\mathrm{d}t + \sigma(t, \mathbf{X}_t)\mathrm{d}\mathbf{B}_t + l(t, \mathbf{X}_{t-}, \mathbf{u}_{t-})\mathrm{d}\mathbf{P}_t, \quad t > 0 \quad (2.1)$$

with an initial condition $\mathbf{X}_0 \in \Omega$. Here, $\mathbf{X}_{t-}$ is the left limit of $\mathbf{X}_t$ at time t, and each coefficients b, σ, l are assumed to have suitable dimensions and chosen so that, for each $\mathbf{u}$, the system (2.1) has a path-wise unique solution. Sufficient conditions for the unique existence are found in textbooks such as [12, Chapter 4]. We assume that only b and l are modulated by $\mathbf{u}$ to simplify the explanation.

2.2.2 Performance Index and Value Function

The performance index is a functional of t, $\mathbf{X}$, and $\mathbf{u} \in U$, to be minimized with respect to $\mathbf{u}$ by a decision-maker: the controller of the dynamics (2.1). The expectation conditioned on $\mathbf{X}_t = \mathbf{x}$ is denoted as $\mathbb{E}^{t,x}$. The terminal time is denoted as $T > 0$, which is possibly unbounded. The performance index $\phi = \phi(t, \mathbf{X}, \mathbf{u})$ conditioned on $\mathbf{X}_t = \mathbf{x}$ is set as

$$\phi(t, \mathbf{X}, \mathbf{u}) = \mathbb{E}^{t,x}\left[\int_t^T f(s, \mathbf{X}_s, \mathbf{u}_s)e^{-\delta(s-t)}ds + g(T, \mathbf{X}_T)e^{-\delta(T-t)}\right]. \quad (2.2)$$

Here, $\delta \geq 0$ is the discount rate to represent myopia of the decision-maker; a larger δ means that he/she is more myopic (e.g., Bian et al. [5]). The coefficients f, g are sufficiently regular and bounded in $[0, T] \times \Omega \times U$ and $\Omega \times U$, respectively. The first and second terms of (2.2) mean the cumulative utility/disutility and the penalty incurred at $T > 0$. For a problem with an unbounded T where b, σ, l, f are time-independent and $g \equiv 0$, ϕ is alternatively set as

$$\phi(\mathbf{X}, \mathbf{u}) = \mathbb{E}^x\left[\int_0^\infty f(\mathbf{X}_s, \mathbf{u}_s)e^{-\delta s}ds\right], \quad (2.3)$$

provided that the right-hand side exists, where $\mathbb{E}^x$ represents $\mathbb{E}^{0,x}$. Notice that in both the finite-horizon and infinite-horizon cases, we assume $\mathbf{X}_t \in \Omega$ a.s. for $t \geq 0$ so that the problem is not terminated in $[0, T]$.

The value function is the minimized ϕ with respect to $\mathbf{u} \in U$:

$$\Phi(t, \mathbf{X}) = \inf_{\mathbf{u} \in U} \phi(t, \mathbf{X}, \mathbf{u}) \text{ for } T < +\infty \left(\Phi(\mathbf{X}) = \inf_{\mathbf{u} \in U} \phi(\mathbf{X}, \mathbf{u}) \text{ for } T = +\infty\right). \quad (2.4)$$

A minimizer in (2.4), if it exists, is called an optimal control $\mathbf{u} = \mathbf{u}^*$. Without loss of generality, we consider Markov controls formally represented in a feedback form $\mathbf{u}_t^* = \mathbf{u}^*(t, \mathbf{X}_t)$ in the finite-horizon case and $\mathbf{u}_t^* = \mathbf{u}^*(\mathbf{X}_t)$ in the infinite-horizon case. The Markov control assumption is not restrictive in applications [42].

2.2.3 HJB Equation

The HJB equation is a nonlinear and nonlocal degenerate parabolic differential equation governing the value function Φ. Based on a dynamic programming principle [42, 45, 62], the HJB equation in the finite-horizon case is

$$-\frac{\partial \Phi}{\partial t} + \delta\Phi - \frac{1}{2}\sigma_{ik}\sigma_{kj}\frac{\partial \Phi}{\partial x_i \partial x_j}$$
$$- \inf_{\mathbf{u} \in U}\left\{b_i \frac{\partial \Phi}{\partial x_i} + f - \lambda_i \int \{\Phi - \Phi(t, \mathbf{x}(1 + l(t, \mathbf{u}, \mathbf{x})))\} p_i(z_i) dz_i\right\} = 0 \quad \text{in } [0, T) \times \Omega \quad (2.5)$$

subject to the terminal condition $\Phi = g$ in Ω at $t = T$, where the Einstein's convention has been used in (2.5). Some boundary conditions may be prescribed along the boundary of Ω. This is not explicitly considered here but will be prescribed in each problem below when necessary. The coefficients of the dynamics (2.1) and the performance index (2.2) are inherited in this HJB equation. The equation in the infinite-horizon case is derived by simply omitting the first term of (2.5) and the terminal condition, and setting the domain of the equation as Ω.

2.2.4 Remarks

We close this section with remarks on HJB equations. A candidate of the optimal control $\mathbf{u}_t^*$ is formally derived as

$$\mathbf{u}^*(t, \mathbf{x}) = \arg\min_{\mathbf{u} \in U}\left\{b_i(t, \mathbf{x}, \mathbf{u})\frac{\partial \Phi}{\partial x_i}(t, \mathbf{x}, \mathbf{u}) + f(t, \mathbf{x}, \mathbf{u})\right\}, \quad (2.6)$$

implying that solving the stochastic control problem ultimately reduces to finding Φ by solving (2.5). This point is an advantage of the stochastic control model because we can obtain the optimal control $\mathbf{u}^*$ for all the possible states if we can solve the HJB equation once, depending on the performance index ϕ that can be determined flexibly. A disadvantage is that solving an HJB equation is not always easy, but the equation can be resolved numerically for moderately small systems having two to three state variables.

Key mathematical problems in the stochastic control modeling are unique existence and regularity of the HJB Eq. (2.5). If the coefficients of the problem satisfy certain boundedness and regularity assumptions, then an HJB equation would admit a unique classical solution satisfying the equation pointwise (e.g., Bian et al. [6], Pham [44]). However, this is not the case in general. We often must seek for solutions in a weaker sense because of the loss of regularity of solutions where the regularity of some of the coefficients in the problem drop.

The most plausible candidate of weak solutions can be viscosity solutions [16], where one-sided HJB equations (Formally, "=" in (2.5) is replaced by "$\leq$" or "$\geq$") are satisfied by appropriate sufficiently smooth test functions. The most useful point in relying on the concept of viscosity solutions is that solutions to an HJB equation need not be continuously differentiable, and even not required to be continuous in some cases (e.g., Touzi [62]). Notice that a classical solution is a viscosity solution (the converse statement is false). In this view, it is natural to analyze HJB equations from a viscosity viewpoint. The definition of viscosity solutions is not presented here but will be found in the references of the separate problems below. The existence and uniqueness of solutions to HJB equations are in general proven with the help of a comparison argument stating that a viscosity sub-solution is always not greater than a viscosity super-solution [16].

The control variable employed in the explanation above was assumed to be continuous-time; however, there exist other classes of control variables of importance in both theory and applications. Such examples include impulse controls and singular controls [42, Chaps. 8 and 9] and partial observation controls [46]. In these problems, the basic strategy is the same: set dynamics, set a performance index, derive an optimality equation, and finally solve the equation. Their optimality equation is more complicated and more difficult to manage. We discuss one related simple but a delicate example at the end of Sect. 2.3.

In most cases, HJB equations are not solvable analytically, and must be solved numerically. Convergent numerical schemes can be developed based on the monotonicity, stability, and consistency requirements [4]. Unfortunately, these schemes do not always exhibit satisfactory accuracy in applications. There have been recent progresses on high-resolution schemes for computing viscosity solutions [23, 47, 49]. Although not discussed here, variational characterizations of HJB equations are also possible [31], with which weak differentiability of solutions can be analyzed.

2.3 Specific Problems

2.3.1 "Non-renewable" Fishery Resource Management

In this section, we present four problems related to inland fisheries and river environmental management. Each problem is formulated under a simplified setting to clearly present the structure of the problem, value function, and optimal control. Different numerical schemes are used in different problems, so that readers can understand that many schemes are available for computing HJB equations.

The first problem is harvesting a "non-renewable" fishery resource in a finite horizon. The fish considered here is *P. altivelis*: a major inland fishery resources in Japan as a core of the aquatic ecosystem as well as an indispensable element shaping regional culture [77]. Harvesting the fish is mostly for recreational purposes in the country. A recent survey revealed that the fish is the third most popular inland fishery resource in Japan [39, 40]. The fish is diadromous (especially, born in the sea and grow up in a river) having a unique one-year life history [38, 83]. Adult fishes grow up in a mid-stream river from spring to the coming summer. During this period, they eat rock-attached algae like diatoms. They migrate together to the downstream river in the coming autumn for spawning. After that, the adult fishes die and the hatched larvae drift toward the sea, and grow up by feeding on plankton until coming spring. In the spring they migrate together toward the mid-stream river.

Conventionally, the harvesting season of the fish in a river in Japan is the early summer (June to July) to the late coming autumn (October to November) with slight seasonal differences among different rivers in the country. The fish as a fishery resource is renewable unless its life cycles are not terminated due to an extinction. However, it can be a non-renewable resource from the spring to the coming autumn during which the fish does not spawn and their population monotonically decreases by harvesting, natural deaths, and predation by the waterfowls such as the great cormorant *Phalacrocorax carbo* [69].

To achieve a sustainable fisheries management of the fish, one has not to harvest the fish too much, while the predation from the other species like the waterfowls should be mitigated if possible, in order to increase the population that can be potentially harvested. The decision-maker, which is a local fishery cooperative and its union members, should therefore concurrently consider both harvesting and protection of the fish.

The formulation below follows Yoshioka and Yaegashi [77, 79] but is modified so that the model becomes simpler but still non-trivial. The problem here considers harvesting a non-renewable (lumped) population in a finite interval $[0, T]$ with $T > 0$. The population at time t is denoted as X_t. Physically, the population is the total number of individual fishes, and should be an integer variable. However, we regard it as a real variable assuming that the population is sufficiently large. This theoretical assumption allows us to formulate the mathematical modeling here in a more tractable manner. We assume that the population does not increase during $[0, T]$. We also assume that harvesting a larger individual fish is more profitable and contributes to a larger utility of the decision-maker. In this view, the body weight of the fish should also be considered. Biologically, it is quite natural to consider that individual fishes have different growth curves. Such phenomena can be described with the individual-based models [14] but can be more complicated than the model presented below. Therefore, we assume that the decision-maker knows the mean growth curve of the individual fishes, and set it as a Lipschitz continuous function U_t at time t. This is assumed to be a given positive and increasing variable. Logistic-like, sigmoid-shape growth curves [81] can be used to represent U_t. The (mean) total biomass at t is $X_t U_t$.

We assume that the population decreases due to natural death, predation, or harvesting. The natural deaths are assumed to be due to both time-continuous and catastrophic events, where the latter is described by a compound Poisson process [52] $P = (P_t)_{t \geq 0}$. The natural (and gradual) mortality rate is denoted as $R > 0$, the predation pressure as $p > 0$, the harvesting pressure as $h = (h_t)_{t \geq 0}$, and the jump intensity of the catastrophic deaths as $\kappa > 0$. The jump size of P is in $\gamma \in [\underline{\gamma}, \overline{\gamma}]$ with constants $0 < \underline{\gamma} < \overline{\gamma} < 1$ and is generated by a probability density $g = g(\gamma)$. At each jump time t, the population decreases as $X_t = (1 - \gamma)X_{t-}$. The harvesting rate is assumed to be a continuous-time variable having the compact range $[0, \overline{h}]$ with $\overline{h} > 0$. The decision-maker is able to reduce the predation pressure from p to $p(1 - u_t)$ with another control $u = (u_t)_{t \geq 0}$, the resource protection, valued in $[0, 1]$. For example, fishery cooperatives and/or residents can control the waterbird population by executing countermeasures such as the shooting and fireworks [60, 69]. The present model therefore has the two control variables: h and u.

The governing SDE of the population dynamics of the fish is set as

$$dX_t = -X_{t-}((R + p(1 - u_t) + h_t)dt + dP_t) \text{ for } t > 0 \text{ with } X_0 = x \geq 0. \quad (2.7)$$

The population dynamics are assumed to be density-independent for simplicity. The SDE (2.7) represents the population decrease by each of the factors explained above.

The decision-maker decides the controls (h, u) to dynamically minimize

$$\phi(t, X; h, u) = \mathbb{E}^{t,x}\left[\int_t^T (-w_1 h_s U_s X_s + w_2 p u_s U_s X_s) ds - w_3 U_T X_T\right], t > 0 \quad (2.8)$$

where $w_1, w_2, w_3 > 0$ are weighting factors. The first to the third terms of (2.8) represent the cumulative utility by harvesting, the cumulative cost by taking the protection measure, and the terminal utility gained if the population remains in the river at the terminal time $t = T$. We are implicitly assuming that the harvesting cost is much smaller than the other terms. This assumption is reasonable because harvesting the fish *P. altivelis*, being different from the fisheries carried out in seas, in general uses only fishing rods [1] and/or casting nets [59] but not ships.

The last term is motivated by the fact that a larger fish is more successful in the spawning [83] and by the assumption that a more successful spawning enhances its life cycle sustainability, and thus sustainable fisheries of the fish. The liner dependence of each term on the biomass is an assumption to simplify the problem. It is more realistic to consider that the utility by the harvesting is concave with respect to $h_s U_s X_s$. Under certain conditions, each term can be replaced by concave alternatives [79] at the expense of increasing the model complexity.

The value function $\Phi = \Phi(t, x)$ as the minimized ϕ with respect to the controls (h, u) is governed by the HJB equation

$$-\frac{\partial \Phi}{\partial t} - \inf_{h,u}\left\{\begin{array}{l}-x((R + p(1 - u) + h))\frac{\partial \Phi}{\partial x} - w_1 h U_t x + w_2 p u U_t x \\ -\kappa \int \{\Phi - \Phi(t, x(1 - \gamma))\} g(\gamma) d\gamma\end{array}\right\} = 0 \quad (2.9)$$

in $[0, T) \times (0, +\infty)$, subject to $\Phi_{t=T} = -w_3 U_T x$ in $(0, +\infty)$ and $\Phi_{x=0} = 0$ in $[0, T)$ with a polynomial growth in the far field $x \to +\infty$. The boundary and terminal conditions coincide at $(t, x) = (T, 0)$.

The HJB Eq. (2.9) has a smooth solution of the form $\Phi(t, x) = \Psi(t)x$ with $\Psi : [0, T] \to \mathbb{R}$ if it solves the first-order ODE

$$-\frac{d\Psi}{dt} - \inf_{h,u}\left\{\begin{array}{l}-((R + \kappa \int \gamma g(\gamma)d\gamma + p(1 - u) + h))\Psi \\ -w_1 h U_t + w_2 p u U_t\end{array}\right\} = 0 \text{ in } [0, T) \quad (2.10)$$

subject to the terminal condition $\Psi(T) = -w_3 U_T$. It is elementary to check that, with any smooth solution Ψ to (2.10), the function $\Psi(t)x$ solves (2.9). By the compactness of the ranges of the controls and the Lipschitz continuity of U, the ODE (2.10) admits a unique smooth solution. Then, $\Psi(t)x$ is continuously-differentiable with respect in t and x, which is therefore a smooth (and thus viscosity) solution to our HJB Eq. (2.9). As a byproduct, it is enough to estimate the net mortality rate $R + \kappa \int \gamma g(\gamma) d\gamma$ but not the gradual and catastrophic ones separately. Based on this finding, in what follows, we replace $R + \kappa \int \gamma g(\gamma) d\gamma$ by R as a net mortality.

In summary, we can reduce an HJB equation to an ODE. We find optimal controls (h^*, u^*) as functions of t as follows:

$$h_t^* = \arg\min_{h \in [0,\bar{h}]}\{h(-w_1 U_t - \Psi)\} \text{ and } u_t^* = \arg\min_{u \in [0,1]}\{pu(w_2 U_t + \Psi)\}. \quad (2.11)$$

This control policy is advantageous from a practical viewpoint since it is free from the population, which is difficult to accurately estimate in real-world problems.

Now, the control problem reduced to solving an ODE having a smooth solution. It can be discretized with any ODE solvers such as the classical explicit and implicit Euler methods, Runge-Kutta methods, and Multi-stage methods. Here, we use the four-stage Runge-Kutta scheme. The computation here proceeds in a time-backward manner since we are considering a terminal value problem.

Parameter values are chosen $T = 150$ (day), $R = 0.01$ (1/day), $p = 0.01$ (1/day), $\bar{h} = 0.02$ (1/day), $w_1 = 3$, $w_2 = 2$, and different values of w_3. The length of the time interval is determined based on the assumption that the problem is considered from spring to autumn. For example, it corresponds to the beginning of May to the end of coming September. As a growth curve U, we use the classical logistic model with the growth rate 0.045 (1/day), the maximum body weight 90 (g), and the initial weight 6 (g), considering the previous study results [77, 79]. Using slightly different parameter values do not critically affect the results presented below. The time increment for numerical discretization of (2.10) is 0.01 (day), which is sufficiently smaller than T.

The main interest of the computation here is to see whether the real-world harvesting policy of the fish [77, 79] emerges in the presented model:

$$h_t^* = \begin{cases} 0 & (0 \leq t \leq T_0) \\ > 0 & (T_0 < t \leq T) \end{cases} \quad (2.12)$$

with exactly one $T_0 \in (0, T)$. Figures 2.1 and 2.2 show the optimal controls and the resulting population N, the body weight U, and the function Ψ for $w_3 = 1$ and 3, respectively. The ranges of these variables have been normalized to the unit interval [0, 1] in the figure panels for the convenience of presentation. As implied in Fig. 2.1, there is a parameter range that the control policy of the form (2.12), which is employed in the real world, is optimal. In this case, the countermeasure to reduce the predation pressure should be taken during the early stage. This strategy is in accordance with the fact that the mass release of the fish into a river is carried out in spring (at $t = 0$) and/or summer during which waterfowls frequently predate the fish [60].

Choosing a larger value of the parameter $w_3 = 3$ induces the optimal control to take the countermeasure both the early and late stages because of the preference

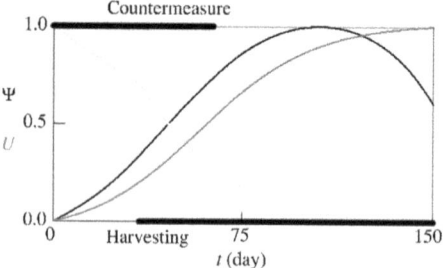

Fig. 2.1 The optimal controls controlled population X, the body weight U, and the function Ψ ($w_3 = 1$). The bold lines in the upper and lower edges of the figure panel represent the time intervals during which taking countermeasure and harvesting are possible, respectively. Notice that the ranges of these variables have been normalized to the unit interval [0, 1]

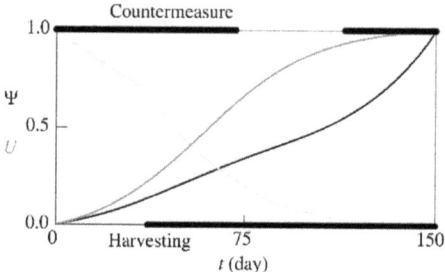

Fig. 2.2 The optimal controls controlled population X, the body weight U, and the function Ψ ($w_3 = 3$). The same figure legends are the same with Fig. 2.1

of the decision-maker to make the terminal population larger. Finally, although not presented in the figure, choosing a larger value of w_3 such as $w_3 = 10$ leads to the optimal policy $u_t^* \equiv 1$ with $h_t^* \equiv 0$ that can maximally reduce the population decrease, but without harvesting. This is an unrealistic situation.

2.3.2 Dam-Reservoir System Management

The second problem concerns an environmentally-friendly dam-reservoir system management receiving stochastic inflows. Operating a dam-reservoir system often include multiple purposes like water resources supply, hydropower generation, and downstream environmental management, which is therefore a complex optimization problem [43]. The stochastic control approach can work as an efficient mathematical tool for dam-reservoir systems management [65, 75, 86]. We consider a dam-reservoir system receiving stochastic inflows, as a model case to find its environmentally-friendly operation policy.

We consider the water balance dynamics of a reservoir given by

$$dY_t = (Q_t - q_t)dt \text{ for } t > 0 \text{ with } Y_0 = y \geq 0, \qquad (2.13)$$

where the water volume in the reservoir at time t is denoted as Y_t and the outflow discharge at t as q_t (the control variable). The range of Y_t is $\Omega = [0, \overline{Y}]$ with the prescribed volume $\overline{Y} > 0$. The inflow discharge Q follows a continuous-time Markov chain ϑ having the transition matrix $\mathbf{S} = [s_{i,j}]_{1 \leq i,j \leq I}$ with some $I \in \mathbb{N}$ [71]. The inflow discharge of the ith regime is denoted as $Q(i) > 0$. Seasonality of river flows is not considered here for convenience.

The admissible range of the control q must be carefully determined for rigorous mathematical modeling. Naively, the admissible range of q is $A = [\underline{q}, \overline{q}]$ with $0 \leq \underline{q} < \overline{q}$ imposed by technical and/or operational restrictions. We set $\underline{q} < \min_i Q(i)$ and $\overline{q} > \max_i Q(i)$ assuming a satisfactory ability of the system to handle the stochastic inflows. The range of q does not have to be modulated if $Y_t \in (0, \overline{Y})$, while it has to be if $Y_t = 0, \overline{Y}$. For example, if $q_t > Q_t$ and $Y_t = 0$, then we possibly encounter the unphysical state $Y_{t+\varepsilon} < 0$ for some $\varepsilon > 0$. We set A as $A = [\underline{q}, Q_t]$ ($Y_t = 0$) and $A = [Q_t, \overline{q}]$ ($Y_t = \overline{Y}$). This modification is physically appropriate, while it incurs a discontinuity of the range of controls. This difficulty can be overcome by accordingly modifying the corresponding HJB equation at the boundaries. For related problems, see Picarelli and Vargiolu [48].

The decision-maker decides the control q to minimize the performance index

$$\phi(\vartheta, Y; q) = \mathbb{E}^{i,y}\left[\int_0^\infty e^{-\delta s}\left(\frac{1}{2}(q_s - Q_s)^2 + \frac{a}{2}(\hat{q} - q_s)_+^2 + f(V_s)\right)ds\right], \qquad (2.14)$$

where $\delta > 0$ is the discount rate, $\hat{q} > 0$ is the environmental flow, and $a > 0$ is the weighting factor. The first through third terms represent penalization of the water balance condition, drawdowns from the environmental requirement with the threshold discharge $\hat{q}$ below which the downstream aquatic environment is severely affected, and a penalization of large or small water volumes, respectively. The coefficient $f \geq 0$ is Lipschitz continuous. The second term is relevant especially if the decision-maker concerns thick growth of the nuisance green filamentous algae in dam-downstream rivers due to too small river discharge [17, 35, 73].

The value function $\Phi_i = \Phi(i, y)$ as the minimized ϕ with respect to the outflow discharge q is governed by the HJB equation

$$\delta\Phi_i + \sum_{1 \leq j \leq I, j \neq i} s_{ij}(\Phi_i - \Phi_j)$$
$$- \inf_{q \in A}\left\{(Q(i) - q)\frac{d\Phi_i}{dy} + \frac{(q - Q(i))^2}{2} + \frac{a(\hat{q} - q)_+^2}{2} + f(v)\right\} = 0(\geq 0) \quad \text{in } D \qquad (2.15)$$

with $D = \{1, 2, 3, \ldots, I\} \times \Omega$ and the equality "=" is necessary only for $0 < y < \overline{Y}$. The controlled discharge at the boundaries $y = 0, \overline{Y}$ must satisfy the modified range as discussed above; instead, (2.15) is relaxed to use "$\geq$" at $y = 0, \overline{Y}$. This formulation, which is physically relevant as explained above, harmonizes with the constrained viscosity solution approach [29]. By the degenerate ellipticity of the HJB Eq. (2.15), the compactness of A, the signs $s_{ij} \geq 0 (i \neq j)$, and the continuity of f, the comparison argument (e.g., [29, Theorem 2.1]) proves the existence of at most one constrained (continuous) viscosity solution.

The local Lax-Friedrichs scheme enhanced by the fifth-order Weighted Essentially Non-Oscillatory (i.e., WENO) reconstruction [26] combined with the fast sweeping [87] is applied to (2.15). This is a high-resolution scheme that has successfully been applied to degenerate elliptic equations like HJB equations. Due to the regularity deficit, meaning that solutions to the HJB equations are not smooth, the scheme seems to be less than fifth-order accurate.

We demonstrative a computational example assuming an existing dam-reservoir system in Japan ($\overline{Y} = 6 \times 10^7$ (m^3)). The domain Ω is normalized to [0, 1], and Y, q in the computation are normalized using $\overline{Y}$ (m^3) and 86,400 (s). The parameters are the same with Yoshioka [75] where we do not consider the hydropower production here: $\underline{q} = 1.0$ (m^3/s), $\overline{q} = 200$ (m^3/s), $\delta = 0.01$ (1/day), $\hat{q} = 10$ (m^3/s), $f = C(0.2\overline{Y} - v)_+^2 + C(v - 0.8\overline{Y})_+^2$ with $C = 5 \times 86,400/\overline{Y}$, and $a = 0.20$. The weighting factors a, C have been determined to balance each term. The Markov chain ϑ employed here is the non-parametric one [75] with $I = 61$ and the discharge of ith regime as $Q(i) = 1.25 + 2.5i$ (m^3/s). The domain Ω is uniformly discretized with 401 vertices. The error tolerance of the fast marching is 10^{-15} in the l^∞ norm. The initial guess is $\Phi \equiv 0$.

The numerical solution has been successfully computed using the scheme (Fig. 2.3) with 8,000 iterations despite that the error tolerance of the convergence is very small. The results also present the computed optimal control $q = q^*$ normalized with respect to the inflow discharge Q. The decision-maker can dynamically decide the outflow discharge by referring this computational result.

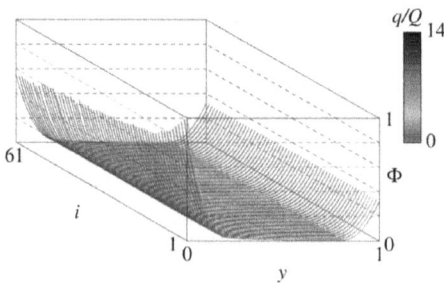

Fig. 2.3 The computed normalized value function (solid curves) and the optimal normalized discharge (color contour)

2.3.3 Algae Growth Management

The third problem is cost-efficient management of algae population in a river downstream of a dam. As explained above, thick growth of the nuisance green filamentous algae is one of the most serious environmental problems in modern river management [17, 35]. Here, we solely focus on a hydraulic control of nuisance benthic algae population dynamics following the previous study [73, 78] under a simpler setting. The model below is formulated as a deterministic control problem, but the stochastic algae population dynamics will be considered in a coupled problem later.

We consider population dynamics of benthic algae in a lumped manner, and the population, such as the biomass per unit area of the riverbed, at time t is denoted as Z_t. The river discharge at time t as the control variable to be optimized by operating the dam is denoted as q_t having the range $[\underline{q}, \overline{q}]$ with $0 < \underline{q} < \overline{q}$. We assume that the population dynamics follow a control-dependent logistic model [73]:

$$dZ_t = \left[rZ_t\left(1 - \frac{Z_t}{K(q_t)}\right) - \alpha q_t Z_t \right] dt, \; t > 0, \; Z_0 = z > 0, \tag{2.16}$$

where $r > 0$ is the intrinsic growth rate, $\alpha > 0$ is the proportional coefficient of detachment, and $K : [\underline{q}, \overline{q}] \to \mathbb{R}_+$ is the environmental capacity as a positive, Lipschitz continuous, and increasing. The first and second terms in the right-hand side of (2.16) represents the growth and the detachment by hydraulic disturbance of the population, respectively.

The unique characteristic of the model (2.16) is that the environmental capacity is population-dependent. As discussed in Yoshioka [73] based on the survey results [25, 64], population abundance of benthic algae and submerged vegetation can be a unimodal convex function of the flow velocity. This unimodal nature is considered because of the balance between the physical disturbance and nutrient flux transport by the river flow: too small flow discharge triggers poor algae growth while too large flow discharge triggers the algae detachment. We set the linear function $K(q_t) = K_0 + K_1 q_t$ with $K_0, K_1 > 0$ as the simplest control-dependent model. For more details, the readers should see Yoshioka [73]. With this specification, it is natural to set the range of the population as $\Omega = [0, K_0 + K_1 \overline{q}]$.

The performance index contains the disutility triggered by the algae population (first term) and the penalization of the deviation between the target and realized discharge of the dam placed upstream (second term):

$$\phi(Z; q) = \mathbb{E}^z \left[\int_0^\infty \left(Z_s^m + \frac{a}{2}(q_s - \hat{q})^2 \right) e^{-\delta s} ds \right], \tag{2.17}$$

where $\delta > 0$ is the discount rate, $a > 0$ is the weighting factor, $\hat{q} \in (\underline{q}, \overline{q})$ is the target discharge, and $m > 0$ is the shape parameter. The corresponding HJB equation governing the value function $\Phi = \Phi(z)$, the minimized ϕ by choosing q, is

$$\delta\Phi - \inf_{q\in[\underline{q},\overline{q}]}\left\{\left(rz\left(1-\frac{z}{K(q)}\right) - \alpha qz\right)\frac{d\Phi}{dz} + \frac{a}{2}(q-\hat{q})^2 + z^m\right\} = 0 \text{ in } \Omega, \tag{2.18}$$

which is satisfied up to the boundary points. Notice that we can alternatively set the formal boundary condition $\Phi(0) = 0$ by taking the limit $z \to +0$ in (2.18). The HJB Eq. (2.18) has at most one viscosity solution that is continuous, by the comparison argument [73], owing to the regularity of the coefficients. We see that the uniqueness in the viscosity sense still holds true if the monomial term Z_s^m in (2.17) is replaced by a discontinuous one such as $\chi_{\{Z_T \geq \bar{z}\}}$ with some $\bar{z} > 0$. The comparison argument in the discontinuous case can follow with a special auxiliary function in the contradiction argument (e.g., [11, Theorem 11.4]).

We proceed to numerical computation of the HJB Eq. (2.18). We use the monotone finite difference scheme combining the one-sided upwind discretization for the controlled part and the exponential discretization for the non-controlled part [73]. This scheme is monotone, stable, and consistent; it is therefore convergent in a viscosity sense [4]. A drawback of the scheme is that it is at most first-order accurate due to the monotonicity; however, it can be combined with the policy iteration algorithm: a fast Newton-like method for solving discretized HJB equations [2]. The tri-diagonal nature of the coefficient matrix of the discretized HJB equation enhances computational efficiency by using the Thomas method [61].

The domain Ω is normalized to the unit interval $[0, 1]$. The computational condition is specified below: $r = 1$, $K_0 = 0.4$, $K_1 = 0.3$, $\alpha = 0.5$, $\delta = 2.0$, $\underline{q} = 0.1$, $\overline{q} = 2.0$, $\hat{q} = 1.0$, $a = 0.1$, and $m = 0.5$. We get the maximum value of the environmental capacity $K_0 + K_1\overline{q} = 1$ being consistent with the length of the normalized domain defined above. The previous study assumed the convex case $m = 2.0$, while the computation here assumes the concave and less regular case; the latter is expected to be concave at least near $z = 0$ by invoking the asymptotic analysis result [73]. The domain Ω is divided with 501 equidistant vertices and the policy iteration is terminated if the error tolerance 10^{-14} in the l^∞ norm is satisfied.

We briefly analyze the value function Φ (Fig. 2.4) and the optimal control q^* (Fig. 2.5) for different values of the weighting factor a. Each computation terminated with less than or equal to five iterations starting from the initial guess $\Phi = 0$ in Ω. Figure 2.4 shows that the profiles of the computed value functions Φ are concave as

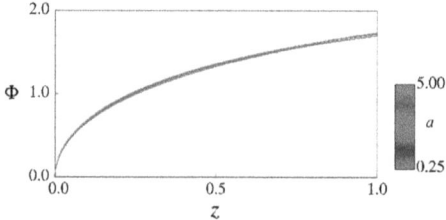

Fig. 2.4 The computed value functions Φ for the weighting factors $a = 0.25i$ ($i = 1, 2, 3, \ldots, 20$)

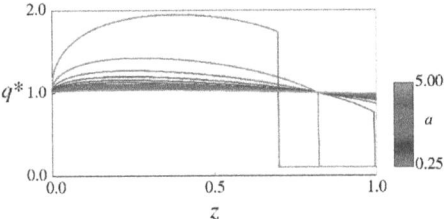

Fig. 2.5 The computed optimal control q^* for the weighting factors $a = 0.25i$ ($i = 1, 2, 3, \ldots, 20$)

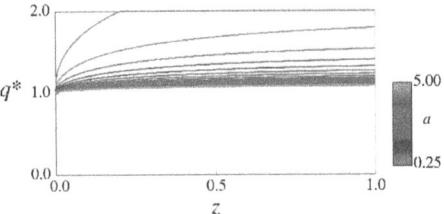

Fig. 2.6 The computed optimal control q^* for the weighting factors $a = 0.25i$ ($i = 1, 2, 3, \ldots, 20$) with the control-independent environmental capacity with $K_0 = 1$ and $K_1 = 0$

predicted by the asymptotic analysis result, and are difficult to distinguish with each other under the specified computational condition. Figure 2.5 implies a transition of the optimal control q^* between $a = 0.75$ and $a = 1.00$ such that q^* seems to be continuous and close to the target value $\hat{q} = 1$ for relatively large $a \geq 1$, while it is discontinuous and clearly different from $\hat{q} = 1$ for smaller a. This kind of sudden transition is due to that the minimizer of the second term of (2.18), assuming that it is an interior solution $\left(q^* \in (\underline{q}, \overline{q})\right)$, is a solution to a non-monotone third-order polynomial. This kind of phenomenon does not occur if $K_0 = 1$ and $K_1 = 0$, as demonstrated in Fig. 2.6. The non-trivial profile of the optimal control q^* in the present control-dependent model is due to considering the balance between the physical disturbance and nutrient transport, which is not considered in the standard control-independent model.

2.3.4 Sediment Storage Management

The last example is a sediment replenishment problem in a dam-downstream river. This model problem is slightly different from the previous ones because the decision-maker can intervene only discretely and randomly, whereas he/she can control the target dynamics continuously in the previous problems. It is still not always possible to collect information of environmental and ecological dynamics under natural environment and are often provided only discretely like weekly or monthly [55, 68].

The sediment replenishment problem in a dam-downstream river considered here has a simple environmental background; sediment supply, which usually occurs along a natural river, is stopped at a dam. A critical issue is that the benthic community in the dam-downstream river is critically affected by the absence of the sediment supply [19, 41, 54]. Some environmental managers encountering this issue have been trying to replenish earth and soils from outside the river [57]. However, cost-effectiveness of the sediment replenishment has only recently been considered from a mathematical side by the authors [74, 81, 82, 84].

Assume that we can place a sediment lump in a dam-downstream reach. The storable amount of the sediment is $\overline{W} > 0$. The amount of stored sentiment at time t is denoted as W_t with the range $\Omega = [0, \overline{W}]$. Assume a constant river flow having a sufficiently large discharge such that the sediment is transported toward downstream. The transport rate as a function of the hydraulic variables is then considered as a constant $S > 0$ [67, 74]. The corresponding sediment storage dynamics are

$$dW_t = -S\chi_{\{W_t>0\}}dt \text{ for } t > 0 \text{ with } W_0 = w \in \Omega. \tag{2.19}$$

This is a discontinuous dynamical system [15] where the drift coefficient is discontinuous when $W_t = 0$. It has a unique continuous solution in the Filippov sense: $W_t = \max\{0, W_0 - St\}$ ($t \geq 0$) owing to the one-sided Lipschitz property of (2.19).

The decision-maker can replenish sediment storage to the maximum value impulsively. This assumption is valid if the sediment replenishment can be conducted with a much shorter time than the time-scale of the dynamics. We further assume that the intervention can be carried out only discretely and randomly, and the chances of the interventions are identified as jump times of a Poisson process $N = (N_t)_{t \geq 0}$ with the intensity $\lambda > 0$. The inverse λ^{-1} then gives the mean as well as standard deviation of the intervals between successive interventions. The quantity λ^{-1} is therefore the characteristic time scale of interventions.

The amount of sediment replenished at time t is $\eta = (\eta_t)_{t \geq 0}$ (For its deeper mathematical details, see [63]). Clearly, the state that the decision-maker should avoid is the sediment depletion $W_t = 0$. In addition, sediment replenishment can be costly. We assume that the decision-maker must pay both fixed cost $d > 0$ (e.g., labor cost) and the proportional cost $c\eta_t$ when he/she replenishes sediment.

The performance for this problem is set as

$$\phi(W, \eta) = \mathbb{E}^w \left[\int_0^\infty \chi_{\{W_s=0\}} e^{-\delta s} ds + \int_0^\infty (c\eta_s + d)\chi_{\{\eta_s>0\}} e^{-\delta s} dN_s \right] \tag{2.20}$$

with $\delta > 0$. The first term penalizes the sediment depletion, while the second is the incurred costs. By a dynamic programming argument [63], the value function $\Phi = \Phi(w)$ as the minimized ϕ with respect to η is governed by

$$\delta \Phi + S\chi_{\{w>0\}} \frac{d\Phi}{dw} + \lambda \left(\Phi - \min_{\eta \in \{0, 1-w\}} \{\chi_{\{\eta>0\}}(c\eta + d) + \Phi(w + \eta)\} \right) = \chi_{\{w=0\}} \text{ in } \Omega. \quad (2.21)$$

No boundary condition is necessary since the HJB equation is assumed to be satisfied up to the boundary points. The optimal replenishment policy is

$$\eta_t^* = \eta^*(W_t) = \arg\min_{\eta \in \{0, 1-w\}} \{\chi_{\{\eta>0\}}(c\eta + d) + \Phi(W_t + \eta)\}. \quad (2.22)$$

This applies only at the randomly arriving intervention chances.

Despite that the coefficients in (2.21) are non-smooth, Yoshioka [74] proved that the HJB equation has a smooth exact solution, which is actually the value function under certain conditions. The optimal replenishment policy is

$$\eta^*(W_t) = (1 - W_t)\chi_{\{W_t \leq \overline{w}\}}, \quad (2.23)$$

representing a strategy that the sediment should be replenished if its storage is below the threshold $\overline{w}$. Finding the optimal control thus reduces to finding $\overline{w}$ if it exists.

We present numerical examples with a regularized counterpart where the characteristic function $\chi_{\{w=0\}}$ is replaced by $\chi_\varepsilon(w) = \max\{0, 1 - w/\varepsilon\}$ with a regularization parameter $0 < \varepsilon < 1$. Similarly, we replace $\chi_{\{w>0\}}$ by the regularized one $1 - \chi_\varepsilon(w)$. This regularization, which has not been used so far, is introduced for the two reasons. The first reason is to guarantee the unique existence of continuous viscosity solutions. The verification argument [74] implies that a smooth solution is a value function; however, the unique existence of (viscosity) solutions was not discussed. The regularization method does not critically affect the dynamics and optimal control, as demonstrated below, while it guarantees that the Eq. (2.21) has Lipschitz continuous coefficients if $\varepsilon > 0$, and we see that the comparison theorem applies [11, Chap. 3] with a slight adaptation to the presented model. The second reason is to improve convergence of numerical schemes expecting that viscosity solutions have higher regularity with smoother coefficients.

We present a computational example of the HJB Eq. (2.21) with the regularization method to numerically show that the optimal policy is the threshold type and the impacts of the regularization. We choose the following parameter values: $S = 0.1$, $\lambda = 0.1$, $c = 0.5$, $d = 0.4$, and $\delta = 0.1$. The HJB Eq. (2.21) is numerically discretized using the semi-Lagrangian scheme with the third-order WENO interpolation [13]. Its advantage is high stability owing to the semi-Lagrangian nature combined with the high-order and sharp interpolation by the WENO reconstruction. Its disadvantage is that a careful scaling required between the temporal increment Δt and spatial increment Δx for convergence [13]. In our case, the scheme is 1.5th-order accurate if Δt is proportional to $(\Delta x)^{1.5}$. We use $\Delta x = 1/300$ and $\Delta t = 30(\Delta x)^{1.5}$. The computation starts from the initial guess $\Phi \equiv 0$ and employs a value function iteration with the error tolerance of 10^{-10} in the l^∞ norm.

We examine the regularization parameters $\varepsilon = 0.1, 0.05, 0.01, 0.005$, and 0.001. Figures 2.7 and 2.8 show the computed value functions Φ and the auxiliary variables

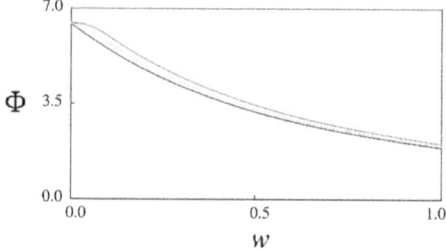

Fig. 2.7 The computed value functions Φ for $\varepsilon = 0.1$ (Red), 0.05 (Green), and 0.01 (Blue)

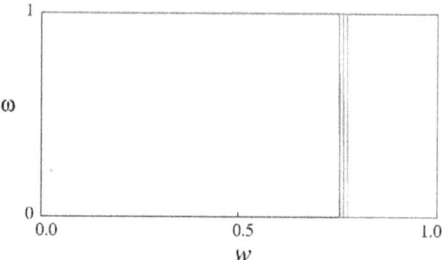

Fig. 2.8 The computed auxiliary functions ω for $\varepsilon = 0.1$ (Red), 0.05 (Green), and 0.01 (Blue)

ω: $\omega = 1$ if $\eta^* = 1 - w$ and $\omega = 0$ otherwise. The results with $\varepsilon = 0.005$ and 0.001 are very close to that with $\varepsilon = 0.01$, and are therefore not plotted in the figures. Regularizing the non-smooth coefficients locally influence the solution shape near $x = 0$ and globally affects the magnitude in the domain. The impacts of regularization get smaller as ε decreases. It should also be noted that Fig. 2.8 implies that the optimal control is indeed the threshold type under the regularization. The computational results suggest choosing $\varepsilon = O(\Delta x)$ to compute reasonably accurate numerical solutions. We expect the convergence of numerical solutions under this limit; this is an open issue.

2.4 Coupled Problem

2.4.1 Overview

We discussed only 1-D examples above. In this section, we present a demonstrative example of formulation and computation of a coupled fisheries and environmental problem. This is a control problem of a dam-reservoir system and its downstream environment where the water balance dynamics receiving regime-switching stochastic inflows, nuisance benthic algae population dynamics, and lumped sediment storage dynamics are concurrently considered. The fishery resource dynamics are evaluated in a performance index.

Due to high-dimensional nature of the problem having the four state variables in addition to the time variable, numerical computation of its associated HJB equation is very costly if we rely on a standard numerical method like a finite difference scheme and a semi-Lagrangian scheme on a conventional structured grid. More specifically, such conventional numerical methods encounter exponential increase of the computational cost with respect to the increase of the increase of the degree of freedom in the target system: the curse-of-dimensionality. We employ the sparse grid technique to alleviate this computational issue [10].

2.4.2 Control Problem

We present a management problem concerning a dam-reservoir system. This is a control problem based on the above-discussed separate problems. As explained above, regulated river flows downstream of a dam encounter a critical reduction of the sediment transport and associated environmental problems like the thick growth of nuisance benthic algae. The latter potentially leads to the reduction of the growth rate of fishery resources like *P. altivelis*. From a fisheries viewpoint, mitigating the environmental impacts on the growth of the fish can be addressed at least by controlling the outflow discharge of the dam and/or replenishing the sediment.

The system dynamics to be controlled have the four state variables: the inflow of the reservoir as a continuous-time Markov chain ϑ and the three continuous-time variables, which are the water volume of the reservoir X_1, the sediment storage X_2, and the nuisance benthic algae population X_3. The number of regimes of α is represented by $I \in \mathbb{N}$ and the range of the variable X_j as $\Omega_j = [0, \overline{X}_j]$ ($j = 1, 2, 3$). The governing system of X_j ($j = 1, 2, 3$) as follows:

$$\begin{aligned} \mathrm{d}X_{1,t} &= (Q_t - q_t)\mathrm{d}t \\ \mathrm{d}X_{2,t} &= -S(q_t)\chi_{\{X_{2,t}>0\}}\mathrm{d}t + \eta_t \mathrm{d}N_t \\ \mathrm{d}X_{3,t} &= \left[rX_{3,t}\left(1 - \frac{X_{3,t}}{K(q_t)}\right) - \alpha(X_{2,t})q_t X_{3,t}\right]\mathrm{d}t \end{aligned} \quad \text{for } t > 0 \qquad (2.24)$$

with initial conditions $X_{j,0} \in \Omega_j$ ($j = 1, 2, 3$) and $\alpha_0 \in i$, where the same notations of the parameters with the previous section are utilized except for the transport rate S and detachment coefficient α. The transport rate S was assumed to be a constant in the previous section but is now a non-negative coefficient depending on the outflow discharge. The detachment coefficient α was also set as a constant, but now as a function $\alpha(X_{2,t})$ depending on the sediment storage. We assume $\alpha(0) = 0$ because the algae detachment is not significant if the river flow does not contain soil particles (bedload) [84, Chap. 3]. Set the sequence representing the regimes as $\Theta = \{i\}_{1 \leq i \leq I}$.

The control variables of the present problem are the outflow discharge $q = (q_t)_{t \geq 0}$ as in the second model problem and the sediment replenishment $\eta = (\eta_t)_{t \geq 0}$ as in the fourth problem. We assume that, at each time t, the outflow discharge is valued in the discrete set $A_t = \{a_j Q_t\}_{0 \leq j \leq n}$ with some $n \in \mathbb{N}$ and some real a_j. The range A_t is modified appropriately if $X_{t,1} = 0$ or $X_{t,1} = \overline{X}_1$ as discussed in Sect. 3.2. The reservoir volume X_1 is then a.s. confined in Ω_1. As in Sect. 3.4, the replenishment is specified so that it equals either 0 (Do nothing) or $\overline{X}_2 - X_{2,t}$ (Fully replenish, only when $\overline{X}_2 < X_{2,t}$) at each jump time t of the Poisson process N representing the a sequence of chances to replenish the sediment.

We consider a performance index containing a penalization of the state variables deviating from a compact safe region $\Omega_S \subset \Omega_1 \times \Omega_2 \times \Omega_3$ (first term) at a terminal time $T > 0$, a penalization of the deviation of the outflow discharge from the inflow (second term), and a cumulative sediment replenishment cost (third term):

$$\phi(t, \vartheta, X_1, X_2, X_3; \eta, q)$$
$$= \mathbb{E}^{t,i,x_1,x_2,x_3} \left[\begin{array}{l} \int_t^T e^{-\delta(s-t)} f(X_{1,s}, X_{2,s}, X_{3,s}) ds \\ \int_t^T e^{-\delta(s-t)} \frac{1}{2} \left(1 - \frac{q_s}{Q_s}\right)^2 ds + \int_t^T (c\eta_t + d) \chi_{\{\eta_t > 0\}} e^{-\delta(s-t)} dN_s \end{array} \right], \quad (2.25)$$

with the discount rate $\delta \geq 0$, coefficients of proportional and fixed costs $c, d > 0$, and a continuous function f such that it equals 0 if its arguments fall in the safe region Ω_S and positive otherwise. This performance index means that the river environmental condition should be in the safe region by controlling the system dynamics. The interval $[0, T]$ can be chosen as the time around which the fish P. altivelis in the dam-downstream river stars to significantly growing up. At that time, the algae population, sediment storage, and reservoir water volume, should be in some safe region; otherwise, the growth of the fish can be critically affected. However, adjusting the river environmental condition to the required state by operating the dam-reservoir system and/or replenishing the sediment is costly. The problem considered here is only one example of the coupled modeling, but many other situations can be covered by modifying the performance index.

The value function is the minimized ϕ with respect to q and η:

$$\Phi(t, i, x_1, x_2, x_3) = \Phi_i(t, x_1, x_2, x_3) = \inf_{q, \eta} \phi(t, \vartheta, X_1, X_2, X_3; \eta, q). \quad (2.26)$$

The associated HJB equation that governs Φ is

$$-\frac{\partial \Phi_i}{\partial t} + \delta \Phi_i + \sum_{1 \leq j \leq I, j \neq i} s_{ij} (\Phi_i - \Phi_j) - f(x_1, x_2, x_3)$$

$$- \inf_q \left\{ \begin{array}{l} (Q(i) - q) \frac{\partial \Phi}{\partial x_1} - S(q) \chi_{\{x_2 > 0\}} \frac{\partial \Phi}{\partial x_2} + \left(r x_3 \left(1 - \frac{x_3}{K(q)}\right) - \alpha(x_2) q x_3 \right) \frac{\partial \Phi}{\partial x_3} \\ + \frac{1}{2} \left(1 - \frac{q}{Q(i)}\right)^2 \end{array} \right\}$$

$$+ \lambda \left(\Phi - \min_{\eta \in \{0, \overline{X}_2 - x_2\}} \left\{ \chi_{\{\eta > 0\}} (c\eta + d) + \Phi(t, i, x_1, x_2 + \eta, x_3) \right\} \right) = 0 \quad (2.27)$$

in $(0, T) \times \Theta \times \Omega_1 \times \Omega_2 \times \Omega_3$, subject to

$$\Phi(T, i, x_1, x_2, x_3) = 0 \text{ in } \Theta \times \Omega_1 \times \Omega_2 \times \Omega_3. \quad (2.28)$$

It seems to be hopeless to analytically handle the minimization term (2.27) because of its complexity, implying that we have to tackle this issue numerically.

2.4.3 Numerical Scheme

We use the semi-Lagrangian scheme on sparse grids [8]. The sparse grid technique does not utilize the conventional fully tensorized computational grids but only grids containing selected basis functions to balance computational costs and accuracy. Several sparse grids are available depending on the problem to be considered [10, 28, 53]. Computational complexity of spars grids is quasi-optimal in terms of theoretical computational efficiency with respect to L^2- and L^∞-norms, and are more suited to approximating higher-dimensional functions [10, Sect. 2.3].

We employ the modified sparse grids [28] that are comparably accurate with the standard ones [10] but have a smaller number of the grid points on boundary. This means that the former exhibit higher accuracy in the interior of the domain if the function to be approximated are sufficiently smooth near and along the boundary. The implementation method and a series of theoretical analysis results on this sparse grid technique are available in Kang and Wilcox [28]. Later, Fig. 2.9 shows that the sparse grid we use is indeed different from the standard grids.

A crucial point of the sparse grid technique is its flexibility to approximate value functions arising in high-dimensional problems; our HJB equation is not an exception. Another important point is that its approximation error in the L^2-sense is the order of $N^{-2}(\log N)^{d-1}$ that is slightly worse than the order N^{-2} of fully-tensorized

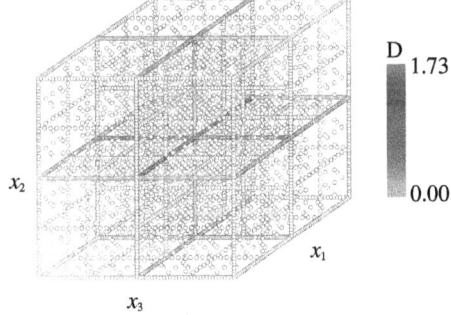

Fig. 2.9 The sparse grid. The color represents the distance ("D" in the figure) from the origin

(conventional) grids, where $N = 2^{l-d}$ is the degree-of-freedom per dimension, $l \in \mathbb{N}$ ($l \geq d$) is discretization level of the sparse grids, and d is the total number of dimensions. The grid becomes finer as the discretization level l increases. The implementation cost of the sparse grids is scaled as $N(\log N)^d$, while that of the standard grids as N^d, implying that the total cost of the former is far smaller than that of the latter in return for the slight accuracy deterioration.

Functions approximated on sparse grids should be at least twice partially differentiable in the weak sense [10, Theorem 3.8]. Too much irregular functions, like that having a discontinuity, should not be handled directly on sparse grids. In addition, the sparse grid technique does not guarantee monotonicity of the interpolations on them, meaning that some limitation method may be necessary in applications especially if the function to be approximated are not smooth [66]. We do not use limitation methods here but will be investigated in future.

2.4.4 Computational Conditions

The coefficient and parameter values are specified considering the approximation ability of the sparse grids. There are the four coefficients to be specified: K, S, α, and f. For simplicity, we set the constant environmental capacity case $K = 1$ with which the fastest logistic growth of the nuisance algae occurs. In addition, we normalize each Ω_j to the unit interval [0, 1] by appropriately normalizing each state variable. The Markov chain ϑ utilized here is based on the same data and identification method but has the smaller total number of regimes with that in Sect. 3.2. Here, we set $I = 21$ and $Q(i) = 2.5 + 5(i - 1)$ (m^3/s).

The transport rate S is based on the semi-empirical physical formulae based on the [67, 74] with the following specified hydraulic condition: river width 30 (m), longitudinal riverbed slope 0.001, soil (sediment) particles diameter 0.006 (m), Manning's roughness coefficient 0.03 (m$^{-1/3}$/s), the density of water 1,000 (kg/m^3), the soil density 2,600 (kg/m^3), and the gravitational acceleration coefficient 9.81 (m/s^2). Using these values, the transport rate S (m^3/day) is calculated as a function of the outflow discharge q (m^3/s) under the normalization of the state variables as

$$S(q) = \overline{X}_3^{-1} \times A \times \max\{Bq^{0.6} - C, 0\}^{1.5} \, (1/\text{day}) \tag{2.29}$$

with the constants $A = 3.82 \times 10^4$ (m^3/s), $B = 1.31 \times 10^{-2}$ (s/m^3), and $C = 4.7 \times 10^{-2}$. The maximum volume of the storable sediment in the river reach downstream of the dam, which is $\overline{X}_3$, is assumed to be 200 (m^3).

The coefficient α of the algae detachment is a function of the sediment storage x_2 such that the detachment occur if the sediment is not depleted ($x_2 > 0$). In addition, several field survey results suggest that the channel bed disturbance by the bedload transport is a driver of the algae detachment [30, 36]. As a simple model, we propose to set $\alpha(x_2) = \alpha_0 x_2^m$ (s/m^3/day) with positive constants α_0 and m. We set $\alpha_0 = 0.1$

(s/m³/day) and $m = 0.5$. The growth rate r is 0.5 (1/day) based on the previous studies [73, 78].

The coefficient f of the terminal condition must be carefully specified because of the approximation ability of the sparse grids. We consider f of the form

$$f = \begin{bmatrix} \max\{x_1 - \overline{x}_1, 0\}^p/(1-\overline{x}_1)^p + \max\{\underline{x}_1 - x_1, 0\}^p/\underline{x}_1^p \\ + \max\{\underline{x}_2 - x_2, 0\}^p/\underline{x}_2^p + \max\{x_3 - \overline{x}_3, 0\}^p/(1-\overline{x}_3)^p \end{bmatrix} (1/\text{day}) \quad (2.30)$$

with positive constants $\overline{x}_1, \underline{x}_1, \underline{x}_2, \overline{x}_3$, and p. This f means that the reservoir volume should be neither too small nor too large, the algae population should not be large, and the sediment storage should not be small. We choose $p = 3$ with which the smoothness $f \in C^2(\Omega_1 \times \Omega_2 \times \Omega_3)$ holds true and therefore this f can be approximated on the sparse grids [10, Theorem 3.8]. We set $\overline{x}_1 = 0.8$, $\underline{x}_1 = 0.2$, $\underline{x}_2 = 0.2$, $\overline{x}_3 = 0.8$. The coefficients $\{a_j\}_{0 \le j \le n}$ of the control set A_t is set as $a_0 = 1$, $a_1 = 1.0/2.0$, $a_2 = 2.0$, $a_3 = 1.0/3.0$, and $a_4 = 3.0$. The coefficients on the replenishment are set as $c = 0.15$ and $d = 0.05$.

The time horizon is set as $T = 60$ (day) assuming a management problem in a growth period (late spring to the coming summer in a year) of the fish *P. altivelis*. The time increment for the temporal integration of the semi-Lagrangian scheme is set as 0.005 (day), and we used the modified sparse grid of the level 11 in the sense of Kang and Wilcox [28] for discretization of the 3-D space $\Omega_1 \times \Omega_2 \times \Omega_3$. The total number of grid points is 6,017, and the minimum distance among the grid points is 1/256 (Fig. 2.9). The corresponding full-tensor grid requires $O(10^7)$ grid points, which is greater than that utilized here. Furthermore, we have 21 regimes, essentially implying that the total numbers of grid points are $O(10^5)$ and $O(10^8)$; the former is about $O(10^3)$ times smaller than the latter, demonstrating high efficiency of the sparse grid technique.

2.4.5 Numerical Computation

The obtained numerical solution at several time steps for small and large inflow discharges are presented to analyze the optimal controls as functions of the state variables. Figures 2.10, 2.11 and 2.12 show the computed Φ, η^*, and q^* of the relatively low and high flow regimes at several time instances. The computational results suggest that replenishing the sediment in this case is in general optimal if the sediment storage is not the full. A large water storage in the reservoir or a small alga population under a relatively high flow regime is found to be another possible condition where replenishing the sediment is not optimal.

The optimal level of the outflow discharge highly depends on the state variables, and it seems to be not easy to find a simple law governing it. Nevertheless, at least theoretically, the decision-maker can decide the control variables at each time instance by referring to the computational results over the phase space. A technical

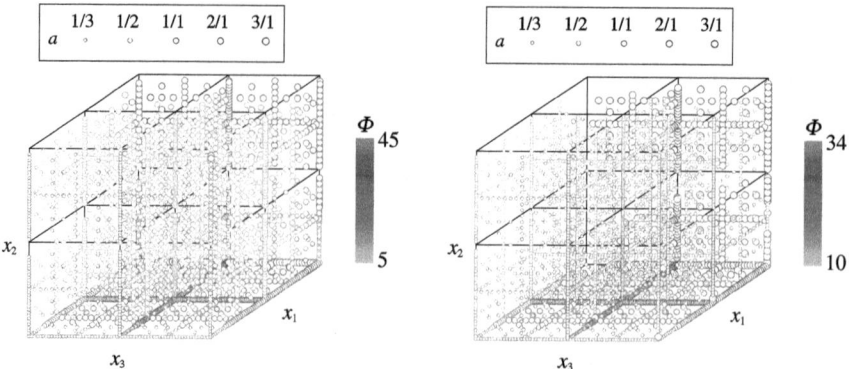

Fig. 2.10 The computed optimal controls at the time $t = 0$ (day) for the relatively low (Left: $i = 3$) and high (Right: $i = 13$) inflow regimes. Only the points where the sediment should be replenished ($\eta^* > 0$) are plotted

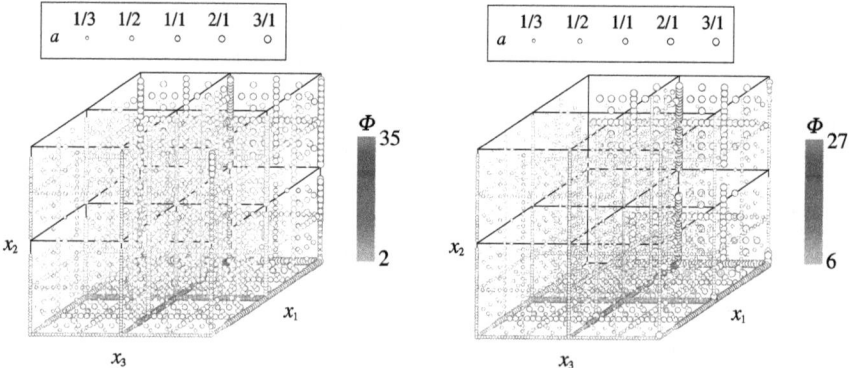

Fig. 2.11 The computed optimal controls at the time $t = 30$ (day) for the relatively low (Left: $i = 3$) and high (Right: $i = 13$) inflow regimes. The same legends with Fig. 2.10

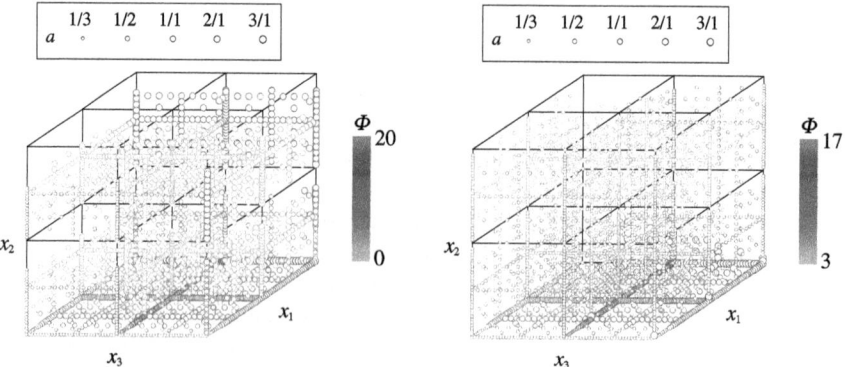

Fig. 2.12 The computed optimal controls at the time (day) for the relatively low (Left: $i = 3$) and high (Right: $i = 13$) inflow regimes

issue is that visualizing a high-dimensional data is usually a difficult task. In our example, each figure panel of Figs. 2.10, 2.11 and 2.12 is only the plot at some instance and some regime. Establishment of an effective visualization technique for high-dimensional data would be an interesting research topic related to optimization and control of many engineering problems. Utilizing some explainable artificial intelligence technique may be beneficial for better understanding the high-dimensional data like the output of this stochastic control model.

2.5 Conclusions

We analyzed independent stochastic control problems on fishery and environmental problems from a unified viewpoint. All the problems were based on a dynamic programming argument, which reduces solving a stochastic control problem to solving some HJB equation. We finally considered a coupled problem utilizing a semi-Lagrangian scheme with the sparse grid technique.

The problems we handled are only the tip of the iceberg: many other interesting topics remain to be addressed. For example, a stochastic control problem with more than one decision-makers would arise when considering integrated management of a watershed where many industries, such as agriculture, fisheries, urban planning, coexist. A massive numerical computation architecture as well as a smart numerical technique like the sparse grid technique should be combined to tackle this advanced issue. Assessing viability of system control strategies is also an important topic in applications [21]. Viability analysis of high-dimensional deterministic dynamical systems have been analyzed [9], but the analysis of stochastic dynamical systems seems to be far less analyzed. We are currently tackling this issue by using the proposed framework based on the dynamic programming methodology and numerical schemes. Employing a multi-objective formulation [18] would facilitate modeling and analysis of the stochastic viability.

Evaluating economic impacts of inland fisheries through river environmental management would be essential for achieving Sustainable Development Goals (SDGs) and symbiosis of humans and environment. A key would be setting a utility function to be optimized from both mathematical and engineering viewpoints. A collaboration of science and engineering will be necessary to resolve these complex but emergent issues. We hope that the presented mathematical framework would become a cornerstone to approach these issues.

Acknowledgements JSPS KAKENHI 19H03073, Kurita Water and Environment Foundation Grant 19B018 and 20K004, and grants from MLIT Japan for surveys of the landlocked *P. altivelis* and management of seaweeds in Lake Shinji support this research. The author gratefully thanks to Dr. Srikanta Patnaik for his invitation to this volume.

References

1. Aino S, Yodo T, Yoshioka M (2015) Changes in the composition of stock origin and standard length of ayu *Plecoglossus altivelis altivelis* during the Tomozuri angling season in the Nagara River, central Japan. Fish Sci 81:37–42
2. Alla A, Falcone M, Kalise D (2015) An efficient policy iteration algorithm for dynamic programming equations. SIAM J. Sci. Comput. **37**, A181–A200; Allawi MF, Jaafar O, Hamzah FM, El-Shafie A (2019) Novel reservoir system simulation procedure for gap minimization between water supply and demand. J Clean Prod 206:928–943
3. Azimzadeh P, Bayraktar E, Labahn G (2018) Convergence of implicit schemes for Hamilton-Jacobi-Bellman quasi-variational inequalities. SIAM J Control Optim 56:3994–4016
4. Barles G, Souganidis PE (1991) Convergence of approximation schemes for fully nonlinear second order equations. Asymptot Anal 4:271–283
5. Bian B, Dai M, Jiang L, Zhang Q, Zhong Y (2011) Optimal decision for selling an illiquid stock. J Optim Theory Appl 151:402–417
6. Bian B, Miao S, Zheng H (2011) Smooth value functions for a class of nonsmooth utility maximization problems. SIAM J Fin Math 2:727–747
7. Biswas IH, Chowdhury I, Jakobsen ER (2019) On the rate of convergence for monotone numerical schemes for nonlocal Isaacs equations. SIAM J Numer Anal 57:799–827
8. Bokanowski O, Garcke J, Griebel M, Klompmaker I (2013) An adaptive sparse grid semi-Lagrangian scheme for first order Hamilton-Jacobi Bellman equations. J Sci Comput 55:575–605
9. Botkin N, Turova V, Diepolder J, Bittner M, Holzapfel F (2017) Aircraft control during cruise flight in windshear conditions: viability approach. Dyn Games Appl 7:594–608
10. Bungartz HJ, Griebel M (2004) Sparse grids. Acta Numer 13:147–269
11. Calder J (2018) Lecture notes on viscosity solutions. Lecture notes at University of Minnesota. https://www-users.math.umn.edu/~jwcalder/viscosity_solutions.pdf
12. Capasso V, Bakstein D (2005) An introduction to continuous-time stochastic processes. Birkhäuser, Boston
13. Carlini E, Ferretti R, Russo G (2005) A weighted essentially nonoscillatory, large time-step scheme for Hamilton-Jacobi equations. SIAM J Sci Comput 27:1071–1091
14. Champagnat N, Ferrière R, Méléard S (2008) From individual stochastic processes to macroscopic models in adaptive evolution. Stoch Model 24(sup1):2–44
15. Cortes J (2008) Discontinuous dynamical systems. IEEE Control Syst Mag 28:36–73
16. Crandall MG, Ishii H, Lions PL (1992) User's guide to viscosity solutions of second order partial differential equations. Bull Am Math Soc 27:1–67
17. Cullis JDS, McKnight DM, Spaulding SA (2015) Hydrodynamic control of benthic mats of *Didymosphenia geminata* at the reach scale. Can J Fish Aquat Sci 72:902–914
18. Désilles A, Zidani H (2019) Pareto front characterization for multiobjective optimal control problems using Hamilton-Jacobi Approach. SIAM J Control Optim 57:3884–3910
19. Doretto A, Bo T, Bona F, Apostolo M, Bonetto D, Fenoglio S (2019) Effectiveness of artificial floods for benthic community recovery after sediment flushing from a dam. Environ Monit Assess 191, Article No. 88
20. do Val JBR, Guillotreau P, Vallée T (2019) Fishery management under poorly known dynamics. Eur J Oper Res 279:242–257
21. Doyen L, Thébaud O, Béné C, Martinet V, Gourguet S, Bertignac M, Fifas S, Blanchard F (2012) A stochastic viability approach to ecosystem-based fisheries management. Ecol Econ 75:32–42
22. Dumitrescu R, Reisinger C, Zhang Y (2019) Approximation schemes for mixed optimal stopping and control problems with nonlinear expectations and jumps. Appl Math Optim (in press)
23. Falcone M, Paolucci G, Tozza S (2020) Multidimensional smoothness indicators for first-order Hamilton-Jacobi equations. J Comput Phys 409:109360

24. Ferretti R, Sassi A (2018) A semi-Lagrangian algorithm in policy space for hybrid optimal control problems. ESAIM: Control Optim Calculus Variations 24:965–983
25. Inui R, Akamatsu Y, Kakenami Y (2016) Quantification of cover degree of alien aquatic weeds and environmental conditions affecting the growth of *Egeria densa* in Saba River, Japan. J JSCE 72:I_1123–I_1128 (in Japanese with English Abstract)
26. Jiang GS, Peng D (2000) Weighted ENO schemes for Hamilton-Jacobi equations. SIAM J Sci Comput 21:2126–2143
27. Jiang H, Simonovic SP, Yu Z, Wang W (2020) A system dynamics simulation approach for environmentally friendly operation of a reservoir system. J Hydrol 587:124971
28. Kang W, Wilcox LC (2017) Mitigating the curse of dimensionality: sparse grid characteristics method for optimal feedback control and HJB equations. Comput Optim Appl 68:289–315
29. Katsoulakis MA (1994) Viscosity solutions of second order fully nonlinear elliptic equations with state constraints. Indiana Univ Math J 43:493–519
30. Katz SB, Segura C, Warren DR (2018) The influence of channel bed disturbance on benthic Chlorophyll a: a high resolution perspective. Geomorphology 305:141–153
31. Krylov NV (2018) Sobolev and viscosity solutions for fully nonlinear elliptic and parabolic equations. American Mathematical Society, Rhode Island
32. Kvamsdal SF, Poudel D, Sandal LK (2016) Harvesting in a fishery with stochastic growth and a mean-reverting price. Environ Resource Econ 63:643–663
33. Kvamsdal SF, Sandal LK, Poudel D (2020) Ecosystem wealth in the Barents Sea. Ecol Econ 171:106602
34. Lai J, Wan JW, Zhang S (2019) Numerical methods for two person games arising from transboundary pollution with emission permit trading. Appl Math Comput 350:11–31
35. Lessard J, Hicks DM, Snelder TH, Arscott DB, Larned ST, Booker D, Suren AM (2013) Dam design can impede adaptive management of environmental flows: a case study from the Opuha Dam, New Zealand. Environ Manage 51:459–473
36. Luce JJ, Steele R, Lapointe MF (2010) A physically based statistical model of sand abrasion effects on periphyton biomass. Ecol Model 221:353–361
37. Miyadi D (1960) The Ayu. Iwanami Shoten, Tokyo (in Japanese)
38. Murase A, Ishimaru T, Ogata Y, Yamasaki Y, Kawano H, Nakanishi K, Inoue K (2020) Where is the nursery for amphidromous nekton? Abundance and size comparisons of juvenile ayu among habitats and contexts. Estuar Coast Shelf Sci 241:106831
39. Nakamura T (2019) Numbers of anglers for the seas, inland waters, and inland fish species of Japan. Nippon Suisan Gakkaishi 85:398–405 (in Japanese with English Abstract)
40. Nakamura T (2020) Numbers of potential recreational anglers for the seas, inland waters, and inland fish species of Japan. Nippon Suisan Gakkaish, 19-00034 (in Japanese with English Abstract)
41. Nukazawa K, Kajiwara S, Saito T, Suzuki Y (2020) Preliminary assessment of the impacts of sediment sluicing events on stream insects in the Mimi River, Japan. Ecol Eng 145:105726
42. Øksendal B, Sulem A (2019) Applied stochastic control of jump diffusions. Springer, Cham
43. Olden JD, Naiman RJ (2010) Incorporating thermal regimes into environmental flows assessments: modifying dam operations to restore freshwater ecosystem integrity. Freshw Biol 55:86–107
44. Pham H (2002) Smooth solutions to optimal investment models with stochastic volatilities and portfolio constraints. Appl Math Optim 46:55–78
45. Pham H (2009) Continuous-time stochastic control and optimization with financial applications. Springer, Berlin, Heidelberg
46. Pham H, Tankov P (2009) A coupled system of integrodifferential equations arising in liquidity risk model. Appl Math Optim 59:147
47. Picarelli A, Reisinger C (2020) Probabilistic error analysis for some approximation schemes to optimal control problems. Syst Control Lett 137:104619
48. Picarelli A, Vargiolu T (2020) Optimal management of pumped hydroelectric production with state constrained optimal control. J Econ Dyn Control 103940 (in press)

49. Rathan, S, L 1-type smoothness indicators based WENO scheme for Hamilton-Jacobi equations. Int J Numer Methods Fluids 92:1927–1947 (In press)
50. Reed WJ (1988) Optimal harvesting of a fishery subject to random catastrophic collapse. Math Med Biol 5:215–235
51. Salgado AJ, Zhang W (2019) Finite element approximation of the Isaacs equation. ESAIM: Math Model Numer Anal 53:351–374
52. Schlomann BH (2018) Stationary moments, diffusion limits, and extinction times for logistic growth with random catastrophes. J Theor Biol 454:154–163
53. Shen J, Yu H (2010) Efficient spectral sparse grid methods and applications to high-dimensional elliptic problems. SIAM J Sci Comput 32:3228–3250
54. Smolar-Žvanut N, Mikoš M (2014) The impact of flow regulation by hydropower dams on the periphyton community in the Soča River, Slovenia. Hydrol Sci J 59(5):1032–1045. https://doi.org/10.1080/02626667.2013.834339
55. Sohoulande CD, Stone K, Singh VP (2019) Quantifying the probabilistic divergences related to time-space scales for inferences in water resource management. Agric Water Manag 217:282–291
56. Song X, Zhong D, Wang G, Li X (2020) Stochastic evolution of hydraulic geometry relations in the lower Yellow River under environmental uncertainties. Int J Sedim Res 35:328–346
57. Stähly S, Franca MJ, Robinson CT, Schleiss AJ (2019) Sediment replenishment combined with an artificial flood improves river habitats downstream of a dam. Sci Rep 9:1–8
58. Steinfeld CMM, Sharma A, Mehrotra R, Kingsford RT (2020) The human dimension of water availability: influence of management rules on water supply for irrigated agriculture and the environment. J Hydrol 588:125009
59. Tago Y (2003) Number of Ayu caught by cast and tenkara nets at several depths in a concrete pond. Suisanzoshoku 51:225–226 (in Japanese with English Abstract)
60. Takai N, Kawabe K, Togura K, Kawasaki K, Kuwae T (2018) The seasonal trophic link between Great Cormorant *Phalacrocorax carbo* and ayu *Plecoglossus altivelis altivelis* reared for mass release. Ecol Res 33(5):935–948
61. Thomas LH (1949) Elliptic problems in linear difference equations over a network. Watson Sci. Comp. Lab. Rep, Columbia University, New York
62. Touzi N (2012) Optimal stochastic control, stochastic target problems, and backward SDE. Springer, New York
63. Wang H (2001) Some control problems with random intervention times. Adv Appl Probab 33:404–422
64. Wang H, Li Y, Li J, An R, Zhang L, Chen M (2018) Influences of hydrodynamic conditions on the biomass of benthic diatoms in a natural stream. Ecol Ind 92:51–60
65. Ware A (2018) Reliability-constrained hydropower valuation. Energy Policy 118:633–641
66. Warin X (2014) Adaptive sparse grids for time dependent Hamilton-Jacobi-Bellman equations in stochastic control. arXiv preprint arXiv:1408.4267
67. Wong M, Parker G (2006) Reanalysis and correction of bed-load relation of Meyer-Peter and Müller using their own database. J Hydraul Eng 132:1159–1168
68. Wu Y, Chen J (2013) Investigating the effects of point source and nonpoint source pollution on the water quality of the East River (Dongjiang) in South China. Ecol Ind 32:294–304
69. Yaegashi Y, Yoshioka H, Unami K, Fujihara M (2018) A singular stochastic control model for sustainable population management of the fish-eating waterfowl Phalacrocorax carbo. J Environ Manage 219:18–27
70. Yang B, Dou M, Xia R, Kuo YM, Li G, Shen L (2020) Effects of hydrological alteration on fish population structure and habitat in river system: a case study in the mid-downstream of the Hanjiang River in China. Glob Ecol Conserv 23:e01090
71. Yin GG, Zhu C (2009) Hybrid switching diffusions: properties and applications. Springer, New York, Dordrecht, Heidelberg, London
72. Yong J (2020) Stochastic optimal control-A concise introduction. Math Control Related Fields (in press). https://doi.org/10.3934/mcrf.2020027

73. Yoshioka H (2019) A simplified stochastic optimization model for logistic dynamics with control-dependent carrying capacity. J Biol Dyn 13:148–176
74. Yoshioka H (2020a) River environmental restoration based on random observations of a non-smooth stochastic dynamical system. arXiv preprint arXiv:2005.04817
75. Yoshioka H (2020b) Mathematical modeling and computation of a dam-reservoir system balancing environmental management and hydropower generation. In: 2020 7th International Conference on Power and Energy Systems Engineering (CPESE 2020), September 26–29, 2020, Fukuoka Institute of Technology, Fukuoka, Japan (Accepted on May 15, 2020). Full paper, 4 p. (In press)
76. Yoshioka H, Yaegashi Y (2017) Stochastic optimization model of aquacultured fish for sale and ecological education. J Math Ind 7:1–23
77. Yoshioka H, Yaegashi Y (2018) Mathematical analysis for management of released fish. Optimal Control Appl Methods 39:1141–1146
78. Yoshioka H, Yaegashi Y (2018) Robust stochastic control modeling of dam discharge to suppress overgrowth of downstream harmful algae. Appl Stoch Model Bus Ind 34:338–354
79. Yoshioka H, Yaegashi Y (2018) Stochastic differential game for management of non-renewable fishery resource under model ambiguity. J Biol Dyn 12:817–845
80. Yoshioka H, Yaegashi Y, Yoshioka Y, Hamagami K, Fujihara M (2019) A primitive model for stochastic regular-impulse population control and its application to ecological problems. Adv Control Appl Eng Ind Syst 1:e16
81. Yoshioka H, Yaegashi Y, Yoshioka Y, Tsugihashi K (2019) A short note on analysis and application of a stochastic open-ended logistic growth model. Lett Biomath 6:67–77
82. Yoshioka H, Yaegashi Y, Yoshioka Y, Hamagami K (2019) Hamilton–Jacobi–Bellman quasi-variational inequality arising in an environmental problem and its numerical discretization. Comput Math Appl 77:2182–2206
83. Yoshioka H, Tanaka T, Aranishi F, Izumi T, Fujihara M (2019) Stochastic optimal switching model for migrating population dynamics. J Biol Dyn 13:706–732
84. Yoshioka H, Hamagami K, Tujimura Y, Yoshioka Y, Yaegashi Y (2020) Chapter 3. Hydraulic control of the benthic alga Cladophora glomerata. In: Ayu and river environment in Hii River, Japan—research results from 2015 to 2020. Published by Laboratory of Mathematical Sciences for Environment and Ecology, Shimane University (in Japanese). Available at https://www.hiikawafish.jp/date/200300ayutokankyo.pdf
85. Yoshioka H, Yaegashi Y (2020) Optimally controlling a non-smooth environmental system with random observation and execution delay. In: 18th International Conference of Numerical Analysis and Applied Mathematics (ICNAAM2020), September 17–23, Sheraton Hotel, Rhodes, Greece (in press)
86. Yoshioka H, Yoshioka Y (2019) Modeling stochastic operation of reservoir under ambiguity with an emphasis on river management. Optim Control Appl Methods 40(4):764–790
87. Zhang YT, Zhao HK, Qian J (2006) High order fast sweeping methods for static Hamilton-Jacobi equations. J Sci Comput 29:25–56

Chapter 3
Models and Tools of Knowledge Acquisition

Rojers P. Joseph and **T. M. Arun**

Abstract The growth of communication channels, personal computers and the internet has radically altered the importance and use of knowledge within an economy, leading to the emergence of a knowledge economy. Digital technologies have transformed the way firms acquire and process external knowledge for effective strategy-making. In a knowledge/digital economy, firms use novel techniques such as Competitive Intelligence (CI) as well as emerging technologies such as Application Programming Interfaces (APIs) and Artificial Intelligence (AI) for knowledge acquisition. In this paper we discuss how CI, AI, and API enhance the effectiveness of the knowledge acquisition process by firms. Overall, the tools, models, and emerging technologies used in knowledge acquisition and knowledge management lead to cost savings at various levels, greater convenience for users, and more effective acquisition and use of knowledge by firms.

Keywords Knowledge/digital economy · Competitive intelligence · Artificial intelligence · Application programming interface · External knowledge sourcing

3.1 Introduction

Knowledge is identified as a vital force for enhancing the competitiveness of firms in the new millennium. Knowledge acquisition has emerged as a prominent theme in organizations ever since digitalization began transforming the global business landscape through e-commerce and other digital models. This transformation has led to the emergence of a digital economy, amply supported by the accompanying information explosion and the fast-paced growth of information and communication infrastructure [49]. Digital economy is an economy of abundance, as it is not constrained by physical resources, unlike the traditional industrial age economy. Enabled by content digitization, the digital economy, in turn, has led to the formation of a knowledge economy in which the production and consumption of information

R. P. Joseph (✉) · T. M. Arun
Indian Institute of Management Rohtak, Rohtak, Haryana 124010, India
e-mail: rojers.joseph@iimrohtak.ac.in

is fundamental to all economic activities. In a knowledge economy, information is treated not merely as a tool, but as a product in itself. More than ever before, firms are becoming increasingly dependent on the knowledge component in their value chain for growth.

3.1.1 Knowledge Economy

The knowledge-based economy is an economy where more effective acquisition, creation, dissemination, and use of knowledge by people and firms lead to greater economic and social development [45]. The growth of communication channels, personal computers and the internet has radically altered the importance of knowledge within an economy, with the knowledge component within every unit of the production cost increasing every year. Newer business models that challenge the industrial approach to producing knowledge goods are emerging. Firms in a knowledge-based economy best utilize the knowledge inputs to create new products and services through innovation. Apple, 3 M and General Electric are examples of firms that make the best use of intangible knowledge assets to greatly enhance shareholder value.

As the firms in a knowledge economy are becoming increasingly dependent on the intangible knowledge component to build competitive advantage, newer ways to acquire, process and utilize knowledge have emerged. Knowledge acquisition has become a critical activity by firms to compete effectively in the marketplace. As a result, new disciplines such as competitive intelligence (CI) that are based on knowledge acquisition, processing, and utilization, came into existence in business management in the twenty-first century. The use of artificial intelligence (AI) in business decision making is gaining increased acceptance among business firms-big as well as small. Knowledge acquisition, processing, and utilization has become an important activity in business firms.

The rest of this paper is structured as follows. Section two discusses the relevant literature on knowledge acquisition, followed by section three that details the importance of knowledge component in a firm's strategic decisions. Section four provides an overview of the sources and techniques of knowledge acquisition. The role of technology in knowledge acquisition and management is discussed in section five followed by the final section that concludes the paper.

3.2 Literature Review

From a business perspective, knowledge acquisition refers to a firm's efforts in obtaining knowledge from external sources [13]. However, from a computing perspective, the knowledge acquisition process comprises the extraction, structuring, and organization of domain knowledge [41]. Research has shown that knowledge and intellectual capital make a significant contribution towards value creation by firms [4,

21]. Knowledge acquisition is a critical component of the knowledge management process that is aimed at generating, appropriating and using knowledge to develop competitive advantage by firms [40].

3.2.1 The Role and Importance of Knowledge Acquisition

To acquire knowledge and intellectual capital, companies expend a great deal of resources and efforts for building systems. However, such systems are often inefficient and do not create the intended value [38], even though such investments are found to positively impact the economy at a macro level [6]. Moreover, firms need to address several significant questions that arise while planning for knowledge acquisition. For instance, in a strategic partnership that involves knowledge-sharing, shielding the partners of their knowledge base is generally considered important [18]. However, Tapscott and Ticoll [43] suggest that companies need not always view transparency as a threat, but as an opportunity to build trustful partnerships with external entities such as suppliers for acquiring critical knowledge. For instance, in complex technological domains such as telecommunications, patent pools are created by two or more firms to license out their patents among themselves or to third parties.

Another critical question in knowledge acquisition is whether the company should develop the knowledge base internally or acquire the required knowledge from outside. For instance, the technology needed for a new product can be developed internally or sourced from outside [2, 33], a decision largely governed by the firm's strategic intent in which a greater emphasis is placed on value than cost savings [35]. Moreover, the advent of digital technologies, particularly the accelerated growth of AI and Machine Learning (ML), has significantly transformed the knowledge acquisition techniques and processes in use.

3.2.2 Digitizing Knowledge Acquisition

A new paradigm in knowledge acquisition is the use of digital methods to acquire knowledge quickly and efficiently, primarily the role of AI in the knowledge acquisition process. AI is a branch of computer science that aims to create independent thinking systems designed to mimic human thinking and problem solving processes [17, 39]. The advent of ML and natural language processing (NLP) have enabled the processing of both structured data like numbers and unstructured data like images and speech by unsupervised machines. AI is related to knowledge acquisition in two key ways. First, it reduces the efforts involved in the knowledge acquisition process. Second, knowledge acquisition is a key step in the development of 'expert systems' in AI [25]. Businesses have since adopted AI assisted systems [36] and recent research has called for more AI adoption in industries including marketing [44], hospitality [26], and telecommunications [47] among others.

Technology has significantly reduced the costs associated with knowledge acquisition, especially from customers. For instance, smartphone penetration currently stands at 3.18 Billion users globally, which enables easy collection of information from users [29]. However, this also leads to significant ethical and privacy concerns. Several organizations were caught engaging in unethical CI by spying on their consumers. For instance, Forbes recently reported that the popular Chinese micro-video sharing social media application, Tik-Tok collects data from users' mobile clipboards without their consent [12]. Incidents like these have a massive reputation deteriorating effect on an organization, which can even result in it being accused of spying for a government, leading to its ban [7]. However, despite the downsides, mobile phones continue to be an extremely cost effective mode of information extraction. AI/ML coupled with NLP and image processing technologies provides access to new forms of information like text and speech, which was hitherto difficult to extract without an actual human being present for analysing information [28].

With the recent advances in AI, businesses have also started exploring how AI can be integrated with human systems to advance knowledge management and decision making [27]. However, there are also contradicting voices that say that AI cannot replace complex knowledge management processes and can only aid human work [32]. Also, the ethical issues in knowledge acquisition can be a significant barrier towards acceptance of these practices by stakeholders. However, either way, AI can and does help in the knowledge acquisition as well as analysis. Section 3.5 deals with how AI and information APIs are used by firms to manage knowledge and also deliver unique values to customers.

The following section discusses the meaning, sources and techniques of knowledge acquisition in detail.

3.3 Knowledge Management and Firm Strategy

Knowledge management may be defined as the strategy of ensuring appropriate knowledge to relevant stakeholders inside the organization to help improve firm performance [30]. For business firms, knowledge acquisition as well as knowledge management (KM) are closely linked to corporate strategy [3]. For instance, competitive intelligence (CI), which refers to a company' efforts to gather and analyze publicly unavailable information about its industry and competitors, is increasingly used by established firms. CI can help a company develop or refine its strategy. The intangible nature of the knowledge component in a firm's strategy makes it difficult for competitors to imitate and thus, knowledge is widely recognized as a critical source of competitive advantage. The significant role that knowledge resources play in a firm's strategic decisions is discussed here.

3.3.1 Strategic Knowledge—An Internal and External Perspective

"If you know the enemy and know yourself, you need not fear the result of a hundred battles. If you know yourself but not the enemy, for every victory gained you will also suffer a defeat. If you know neither the enemy nor yourself, you will succumb in every battle."[16], p. 11].

The above statement was made by the great Chinese General Sun Tzu, in his magnum opus, the Art of War. Reliable information or intelligence is paramount in making any business decision and constitutes an important strategic issue [34]. Information is vital in any game of strategy. This understanding has led to the creation and growth of the field of CI.

3.3.2 Competitive Intelligence

CI may be defined as the ethical sourcing, processing, and storing of information about the strategic moves and activities of your competitors and markets for easy and efficient dissemination at various levels of the organization to safeguard the firm against competitive threats [37]. Since then, CI has evolved into an industry unto itself with the creation of multiple specialist businesses that help organizations find, collect and analyse the collected information. CI collects information on three key factors, customer, competitors and markets. It is estimated that just the market research sector of CI alone was worth more than $20 Billion in 2015 [15].

However, collecting a large amount of data does not imply that it is useful for the firm. Knowledge management, which refers to proper management of the collected information to analyse and provide strategic insights, is vital for firms in formulating an effective strategy. Thus, by their very nature, both CI and KM are symbiotic in their relationship and are strategic to a firm's performance. While, CI is more outside looking and helps bring knowledge in to the organization, KM helps manage all the knowledge inside the organization for its use in decision making at various levels. The summary of the relationship between these two factors are provided in Fig. 3.1. The figure highlights that a firm requires both internal and external knowledge to make business decisions. While CI gets external information on competitor actions, market demands, and new technologies in the industry, KM helps to assimilate this knowledge with internal KM systems of the firm.

Acquiring relevant knowledge is arguably one of the hardest tasks in the KM-CI process. Some pertinent questions guide this process: How to know what knowledge to acquire, Where to find the knowledge? How to extract this knowledge? How to analyse the acquired knowledge? and How to know if the knowledge is valid? Thus, knowledge acquisition primarily consists of four broad steps: (1) planning knowledge acquisition, (2) extracting knowledge, (3) analysing knowledge, and (4) verifying knowledge [22, 23].

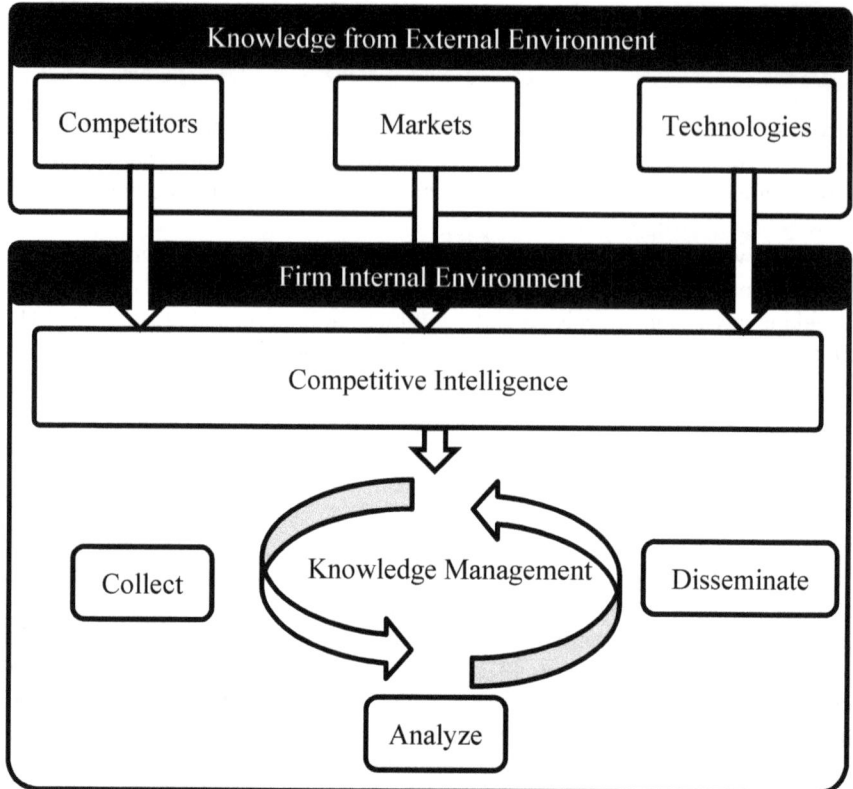

Fig. 3.1 Knowledge acquisition through competitive intelligence and knowledge management. *Source* Created by authors

Another critical development in the domain of knowledge acquisition and management revolves around AI. Current research in AI around building models for time, space, substance, and processes may in the near future become valuable for analysing domains and constructing robust domain knowledge bases [5]. Knowledge acquisition as an area finds its evolution in information systems literature. As we will see in the following sections, this is even truer today with technologies like AI and Machine Learning (ML) revolutionizing the acquisition process. In other words, the process of knowledge acquisition involves elicitation, collection, analysis, modelling and validation of knowledge. A detailed description of each of the steps can be found in Liou et al. [23] which is still relevant today. The following section highlight the importance of considering the current business landscape as a part of a knowledge economy where knowledge is the prime commodity.

3.4 Knowledge Acquisition: Sources and Techniques

A vast body of literature suggests that firms acquire new knowledge from external sources, mainly from smaller, entrepreneurial firms [38]. With the advent of digitization and the emergence of a knowledge economy, the way firms acquire knowledge has undergone significant transformation over the past few decades. This section focuses on the major external sources of knowledge and the different techniques available for firms to acquire knowledge externally.

3.4.1 External Sources of Knowledge

The external sources of knowledge include customers, suppliers, competitors, creditors, partners, and external experts, among others [8] (refer to Fig. 3.2). Customer knowledge is part of the relational intellectual capital of a firm. The external knowledge regarding customers can be divided into knowledge about customers and knowledge from customers [14]. Customer knowledge is a critical enabler of the customer relationship management (CRM) process in firms, aimed at supporting business activities to build competitive advantage.

The knowledge about suppliers is used to understand how a supplier can match the requirement of an organization. Reliable supply partners can provide important insights about quality, delivery, financial risks involved and so on. Competitors too can be important business partners for a firm. Techniques such as competitive intelligence (CI) are used for collecting, organizing, analysing and presenting data about competitors. IT systems such as data mining and analysis, document management systems, and expert systems are commonly used for this purpose [41]. The confidence vested by creditors and rating agencies in a firm is an indicator of its financial success and creditworthiness. Often, firms engage in partnerships to acquire patented technologies or technical know-how. However, it is important to collect information to understand the ability of partners to perform their roles before entering into partnerships.

Fig. 3.2 Major external sources of knowledge. *Source* Created by authors

3.4.2 Techniques of Knowledge Acquisition

The techniques of knowledge acquisition can be broadly classified into three: basic techniques, group techniques and supplementary techniques [24]. A description of these techniques is provided in the following subsections (refer to Fig. 3.3 for a diagrammatic representation).

3.4.2.1 Basic Techniques

The basic techniques include unstructured interviewing, structured interviewing, and observations. Interviews are unstructured when predetermined questions are not used. However, the interviewers may have certain ideas about the topic on which they plan to collect data through interviews. In a structured interview, all interviewees are asked exactly the same questions in a predetermined order. Observation technique involves gathering knowledge by observing the phenomenon under investigation. The technique focuses on observing the human behaviour, the phenomenon, and human interactions related to the phenomenon.

3.4.2.2 Group Techniques

The group techniques of knowledge acquisition consist of brainstorming, nominal group technique (NGT), Delphi technique, consensus, and computer-aided group sessions. Brainstorming is a technique used to generate ideas through freewheeling discussions among the members of a group or designated individuals. NGT is a technique used to reach decisions quickly through discussions in small groups. It involves identifying problems, generating solutions and arriving at decisions without delay. In Delphi technique, consensus is sought from a panel of experts over a series of questions and discussions. Selecting a suitable facilitator, identifying experts with knowledge in the relevant domain, and defining the problem correctly are key to success while using Delphi. In consensus decision making, people who are experts

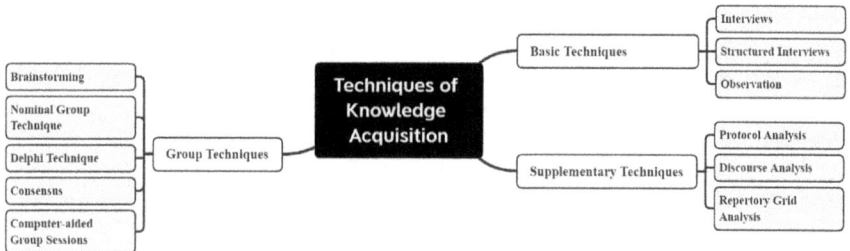

Fig. 3.3 Techniques of knowledge acquisition. *Source* Adapted from Liou [24]

in a narrow and well defined domain reach a consensus about the accuracy of solutions available to resolve a problem. This requires communicating the details of the problem and its possible solutions among the experts where solutions within the domain are not the subject of common sense. When decision making involves groups of people aided by systems that support decisions rather than individuals supported by systems, it is referred to as computer-aided group sessions [31].

3.4.2.3 Supplementary Techniques

The supplementary techniques of knowledge acquisition are used for data collection in specific contexts. They include protocol analysis, discourse analysis, and repertory grid analysis. In protocol analysis, researchers obtain verbal reports from participants as a source of data. The subjects are trained to think aloud as they solve a given problem and their verbal data is collected for analysis. Discourse analysis is a method used to analyse written, spoken or sign language in relation to its social context. It is often referred to as analysis of language beyond sentence and can provide important clues about various aspects related to society and culture. As technique for knowledge acquisition, repertory grid analysis is gaining popularity among information system researchers. Devised by George Kelly in 1955, it is an interviewing technique used for eliciting personal constructs about what people think regarding a given topic. It is generally used in preliminary studies that will be followed by a further qualitative or quantitative inquiry.

3.5 Role of Technology in Knowledge Acquisition and Management

Technological advancements have significantly reduced the cost of knowledge acquisitions with more and more business models focusing on information acquisition and consolidation through Big Data [19]. For instance, it is estimated that, every day, Facebook alone processes content amounting to 2.5 billion pieces and more than 500 Terabytes of data [10]. Further, this data is repurposed and reused for other uses like serving advertisements. The huge amount of useful data that companies sit on has enabled firms to package these data consumption by other organizations and users through Application Program Interfaces (API) and consequently led to the creation of an API economy [42, 48]. Further, newer and better Artificial Intelligence and Machine Learning (AI/ML) techniques of analysing this data are now available that not only help in the analysis phase, but also contributes to other stages in the knowledge acquisition process.

3.5.1 The Emergence of API Economy

An application programming interface (API) can be thought of as a set of data and related functions of a data that are available for outside users. An excellent example of an API is the Google sign-in feature in other apps. A user can sign-in and use an application of a vendor without going through the cumbersome registration process, provided she allows the application to acquire her data from Google. When this is done, the user gains convenience, Google gains information about the user's app usage and the application gets the user information in a standardized format which is easy for processing. This is further supported by technologies like AI [20].

Further, APIs can also be used to avail information analysis services that might otherwise be unavailable. For instance, Amazon Web Services offers machine learning APIs under the name Amazon SageMaker that may be used by any of their subscribers to analyse and draw inferences from their data. This, reduces the financial investments required by their subscribers to acquire machine learning skills to analyse their data. Overall, APIs result in more convenience for users, cost savings, and easy acquisition of knowledge by firms and help promote the growth of an API economy.

3.5.2 Artificial Intelligence

Perhaps, the greatest contribution of AI to knowledge acquisition is in the analysis and validation part of the acquisition. AI/ML based algorithms provide better pattern detection and validation of data. Recently AI has also been used to detect manipulation in acquired data particularly in industries like financial services [9] and insurance [11] among others. The studies report that in regular scenarios, validating data usually takes significant manual effort, time and cost. Further, outputs and inferences from all these processes can enhance the algorithm that plans the data collection, as AI by definition is a learning system that improves itself from previous processes.

The second contribution that AI adoption makes to a firm is the development of expert systems from the outputs of knowledge acquisition. Knowledge acquisition through AI/ML processes enable firms to develop an expert system, which is an information system that consolidates expert knowledge acquired through knowledge acquisition to create a 'virtual' expert based on the knowledge to interface with a user. It consists of primarily three components: (1) Knowledge Base, (2) Inference Engine and (3) User Interface. These systems interface with customers or other stakeholders without human intervention. An example of such a system may be chatbots used by companies in customer relationship management [1]. These systems interact with users and learn from them to further strengthen the expert system. Further, as discussed in the previous section several business models are possible through the API access of such systems.

3 Models and Tools of Knowledge Acquisition

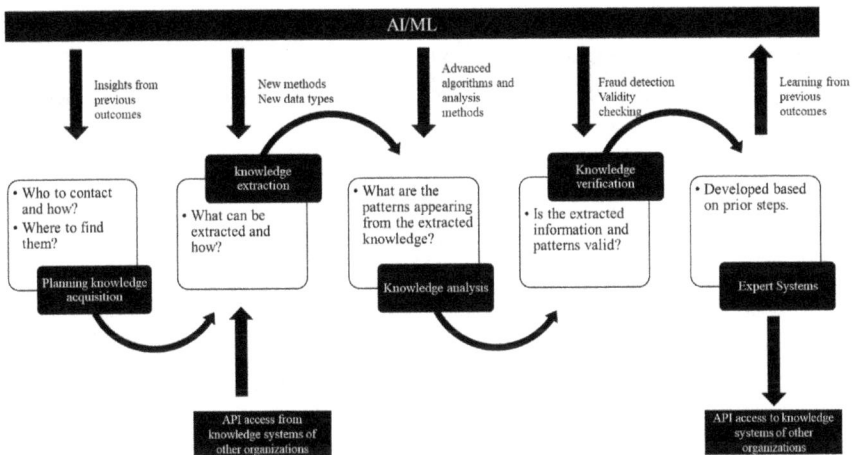

Fig. 3.4 AI/ML and API helping the knowledge acquisition steps. *Source* Created by authors

The summary of the interaction between AI/ML, API, and knowledge acquisition steps is presented in Fig. 3.4. The representation can be considered as a simplified model of implementing an AI/ML powered information and knowledge management system. AI/ML system requires a large amount of current as well as previous data to function. It also requires complex algorithms to analyse the captured data. Both these factors can quickly add up to prohibitively large costs. Using AI/ML and APIs can help reduce these costs.

The data for processing usually comes from various internal and external APIs. For instance, a firm may have a chatbot powered by a service like chatbot.com.[1] The APIs from the service handle the data collection and interfacing with the customer. This takes care of planning as well as extraction of data. The collected data can then be processed by an in-house team, or can be outsourced to be processed by well-trained ML algorithms belonging to Google, Amazon or Microsoft. These services may also provide the necessary verification required to ensure the genuineness of data. The outcomes of the analysis can be used to further strengthen the chatbot system to handle future queries better. Further, if the chatbot is sufficiently trained for a particular context, say airline booking, the service can be processed and sold to other airline companies by building APIs to access the service from the outside. This, essentially allows a firm to tap into the capabilities of another firm in real time as well as contributing capabilities simultaneously. Variety of businesses like price comparison engines, and booking aggregators rely heavily on the above processes to acquire knowledge and analyse the model. A real life example of how knowledge acquisition and management is carried out with the help of AI by a business is illustrated in the box titled: Knowledge Management Strategy at Spartan Race.

---

[1] https://www.chatbot.com/

Knowledge Management Strategy at Spartan Race

Spartan Race, headquartered in Boston, US, organizes series of obstacle races for customers across 30 countries. The series comprises extreme wellness events such as obstacle course races of varying distances and difficulty levels organized in the US, Canada, Australia, South Korea, and several countries in Europe. Spartan follows a customer-centric self-service strategy that is based on a knowledge management approach. The company recognizes that presenting knowledge that matches the customer's unique needs is vital to the success of its strategy. To implement its strategy, Spartan established around forty help centres across countries. In each country, the help centre is customized to meet the specific queries of customers in that country. The gathered knowledge is disseminated to each user group within a centre, namely volunteers, racers and so on using some of the best practices in organizing content, including the use of labels. This ensures that each user can find the right content that meets his or her specific needs.

In addition, Spartan uses AI for effective knowledge management. To tackle the large volume of FAQs that Spartan receives during chat, it incorporated Answer Bot in the Web Widget to offer self-service along with live support. AI allows Spartan to provide knowledge that is customized to each customer's requirement as well as the country's context. The company also avails data on how users are interacting with Answer Bot to continuously upgrade the help centre database. As a small company, Spartan found Answer Bot highly useful during the time of races when there is a sudden spike in customer queries. This helps users get answers quickly and in real time, as manpower is harder to find during weekends when most of the races happen. Agents can also devote more time to productive activities by deflecting repetitive questions, which leads to greater engagement by them. Significantly, Spartan registered a 9.5% reduction in chat volume after introducing Answer Bot. This also led to saving about three hours per day by its support team engaged in live chat, increasing productivity.

Source Adapted from Wren [46]

3.6 Conclusion

This paper discusses the meaning and importance of knowledge acquisition for business firms. It explains how knowledge acquisition is critical for firms and how it is linked to strategy making in business firms. The paper also discusses the various sources of external knowledge as well as the tools and techniques commonly used for it, including competitive intelligence (CI). The chapter also discusses in detail the use of technological tools such artificial intelligence (AI) and application programming interface (API) in knowledge acquisition. The developed framework that summarizes the use of this technology also highlights how boundaries between firms are becoming thin and how collaboration and pooling of capabilities are used by both new and old firms for creating values for themselves, their customers and also other firms. This line of argument opens up interesting future research opportunities to understand these inter-organizational relationships. Further, intelligent use of various APIs and AI can also help new firms to create new value and value delivery mechanisms. This also warrants future research attention. The chapter concludes with the case of a firm that has used AI and API in its knowledge acquisition and management process for making more effective strategic decisions. Future researchers can also examine the impact of such technologies on various key success factors like customer engagement, satisfaction, and profits among others.

References

1. Adam M, Wessel M, Benlian A (2020) AI-based chatbots in customer service and their effects on user compliance. Electron Markets, pp 1–19
2. Appleyard MM (1998) Cooperative knowledge creation: the case of buyer-supplier co-development in the semiconductor industry
3. Bierly P, Chakrabarti A (1996) Generic knowledge strategies in the US pharmaceutical industry. Strateg Manag J 17(S2):123–135
4. Blair MM, Wallman SM (eds) (2000) Unseen wealth: report of the brookings task force on intangibles. Brookings Institution Press
5. Breuker J, Wielinga B (1989) Models of expertise in knowledge acquisition. In: Studies in computer science and artificial intelligence, vol 5. North-Holland, pp 265–295
6. Brynjolfsson E, Hitt LM (2000) Beyond computation: information technology, organizational transformation and business performance. J Econ Perspect 14(4):23–48
7. Bureau ET (2020) Chinese apps banned in India: India bans 59 Chinese apps including TikTok, WeChat, Helo—The Economic Times. Economic Times. https://economictimes.indiatimes.com/tech/software/india-bans-59-chinese-apps-including-tiktok-helo-wechat/articleshow/76694814.cms
8. Chan JO (2009) Integrating knowledge management and relationship management in an enterprise environment. Commun IIMA 9(4):4
9. Cirqueira D, Nedbal D, Helfert M, Bezbradica M (2020) Scenario-based requirements elicitation for user-centric explainable AI. In: International cross-domain conference for machine learning and knowledge extraction. Springer, Cham, pp 321–341
10. Constine J (2012) How big is Facebook's data? 2.5 billion pieces of content and 500+ terabytes ingested every day. https://techcrunch.com/2012/08/22/how-big-is-facebooks-data-2-5-billion-pieces-of-content-and-500-terabytes-ingested-every-day
11. Dhieb N, Ghazzai H, Besbes H, Massoud Y (2020) A secure AI-driven architecture for automated insurance systems: fraud detection and risk measurement. IEEE Access 8:58546–58558
12. Doffman Z (2020) Warning—apple suddenly catches tiktok secretly spying on millions of iPhone users. Forbes. https://www.forbes.com/sites/zakdoffman/2020/06/26/warning-apple-suddenly-catches-tiktok-secretly-spying-on-millions-of-iphone-users/#173162d634ef
13. Gamble PR, Blackwell J (2001) Knowledge management: a state of the art guide. Kogan Page Publishers
14. Gebert H, Geib M, Kolbe L, Riempp G (2002) Towards customer knowledge management: integrating customer relationship management and knowledge management concepts. In: The second international conference on electronic business (ICEB 2002), pp 296–298

15. Gilad B (2015) Companies collect competitive intelligence but don't use it. Harv Bus Rev
16. Giles L (2001) Sun Tzu on the art of war. IndyPublish.com
17. Haenlein M, Kaplan A (2019) A brief history of artificial intelligence: on the past, present, and future of artificial intelligence. Calif Manage Rev. https://doi.org/10.1177/0008125619864925
18. Inkpen AC (1998) Learning and knowledge acquisition through international strategic alliances. Acad Manag Perspect 12(4):69–80
19. Jiang J, Ji S, Long G (2020) Decentralized knowledge acquisition for mobile internet applications. World Wide Web, pp 1–17
20. Lee K, Ha N (2018) AI platform to accelerate API economy and ecosystem. In: 2018 International conference on information networking (ICOIN). IEEE, pp 848–852
21. Lev B (2000) Intangibles: management, measurement, and reporting. Brookings Institution Press
22. Liou YI, Nunamaker JF Jr (1993) An investigation into knowledge acquisition using a group decision support system. Inf Manage 24(3):121–132
23. Liou YI, Weber ES, Nunamaker JF Jr (1990) A methodology for knowledge acquisition in a group decision support system environment. Knowl Acquis 2(2):129–144
24. Liou YI (1990) Knowledge acquisition: issues, techniques, and methodology. In: Proceedings of the 1990 ACM SIGBDP conference on trends and directions in expert systems, pp 212–236
25. Liou YI (1992) Knowledge acquisition. ACM SIGMIS Database: The DATABASE for Advances in Information Systems 23(1):59–64. https://doi.org/10.1145/134347.134364
26. Lu L, Cai R, Gursoy D (2019) Developing and validating a service robot integration willingness scale. Int J Hosp Manag. https://doi.org/10.1016/j.ijhm.2019.01.005
27. Metcalf L, Askay DA, Rosenberg LB (2019) Keeping humans in the loop: pooling knowledge through artificial swarm intelligence to improve business decision making. Calif Manage Rev. https://doi.org/10.1177/0008125619862256
28. Nadkarni PM, Ohno-Machado L, Chapman WW (2011) Natural language processing: an introduction. J Am Med Inform Assoc 18(5):544–551
29. Newzoo (2019) Global Mobile Market Report. https://resources.newzoo.com/hubfs/Reports/2019_Free_Global_Mobile_Market_Report.pdf?utm_campaign=Mobile_Report_Launch_2019&utm_medium=email&_hsmi=76926953&_hsenc=p2ANqtz--L1_j59zvsoS1Hh7Y7baruFKUpud3YkRouhY7C3dQp3IELIWTio6PlrPMC6HCgTNfngzvwLFwFjl2ef23swz
30. O'dell C, Grayson CJ (1998) If only we knew what we know: identification and transfer of internal best practices. Calif Manage Rev 40(3):154–174
31. O'Donnell S (1996) An introduction to group decision making and group decision-support systems. In: Computer aided decision support in telecommunications. Springer, Dordrecht, pp 183–207
32. Pettersen L (2019) Why artificial intelligence will not outsmart complex knowledge work. Work, employment and society. https://doi.org/10.1177/0950017018817489
33. Pitt M, Clarke K (1999) Competing on competence: a knowledge perspective on the management of strategic innovation. Technol Anal Strateg. Manage. 11(3):301–316
34. Prescott JE (1999) The evolution of competitive intelligence: designing a process for action. Propos Manage 37–52
35. Quinn JB (1999) Strategic outsourcing: leveraging knowledge capabilities. MIT Sloan Manag Rev 40(4):9
36. Ransbotham S, Kiron D, Gerbert P, Reeves M (2017) Reshaping business with artificial intelligence. MIT Sloan Mangement Review and The Boston Consulting Group
37. Rouach D, Santi P (2001) Competitive intelligence adds value: five intelligence attitudes. Eur Manag J 19(5):552–559
38. Russ M, Jones JG, Jones JK (2008) Knowledge-based strategies and systems: a systematic review. In: Knowledge management strategies: a handbook of applied technologies. IGI Global, pp 1–62
39. Russell S, Norvig P (2002) Artificial intelligence: a modern approach
40. Schauer H, Schauer C (2008) Modeling techniques for knowledge management. In: Knowledge management strategies: a handbook of applied technologies. IGI Global, pp 91–115

41. Shang Y (2005) Expert systems. In: The electrical engineering handbook, pp 367–377. https://doi.org/10.1016/B978-012170960-0/50031-1
42. Tan W, Fan Y, Ghoneim A, Hossain MA, Dustdar S (2016) From the service-oriented architecture to the web API economy. IEEE Internet Comput 20(4):64–68
43. Tapscott D, Ticoll D (2003) The naked corporation: how the age of transparency will revolutionize business. Simon and Schuster
44. Wirth N (2018) Hello marketing, what can artificial intelligence help you with? Int J Mark Res 60(5):435–438
45. World Bank (2007) Building knowledge economies: advanced strategies for development. WBI Development Studies. Washington DC. https://openknowledge.worldbank.org/handle/10986/6853
46. Wren H (2020) 5 of the best knowledge management examples. Zendesk, The Library, Blog. https://www.zendesk.com/blog/3-best-knowledge-management-examples/
47. Yigzaw M, Hill S, Banser A, Lessa L (2010) Using data mining to combat infrastructure inefficiencies: the case of predicting non-payment for Ethiopian telecom. AAAI spring symposium—technical report
48. Zachariadis M, Ozcan P (2017) The API economy and digital transformation in financial services: the case of open banking. SSRN Electron J. https://doi.org/10.2139/ssrn.2975199
49. Zimmermann HD (2000) Understanding the digital economy: challenges for new business models. AMCIS 2000 Proceedings. Paper, 402

Chapter 4
Profits Are in the Eyes of the Beholder: Entropy-Based Volatility Indicators and Portfolio Rotation Strategies

Abhijeet Chandra and Gaurav Jadhao

Abstract Literature suggests that traders and investors in financial markets perceive changes in expected market volatility represented by the implied volatility index such as the VIX for timing their strategies of portfolio rotation. Researchers have successfully employed the entropy-based measures to study financial time-series to address the issue of nonlinearity and the restrictions associated with theoretical probability distributions. In this study, we implement the approximate entropy (ApEn) and the sample entropy (SaEn) indicators—computed from the India Volatility Index (India VIX)—to study the feasibility of portfolio rotation strategies based on style, size and time horizons. We compute the approximate and the sample entropies, and the India VIX. We find that ApEn and SaEn capture the higher order movements better than the change in India VIX, implying a better indicator of volatile market. Between ApEn and SaEn, the later reflects the fluctuations better. Our findings provide computationally supportive arguments in favour of a potentially beneficial alternative for portfolio managers. Practitioners can use this approach to enhance portfolio returns and to mitigate risk in the context on Indian market.

Keywords India volatility index · Approximate entropy · Sample entropy · Portfolio rotation · Trading strategy

JEL Code C63 · G11 · G17

A. Chandra (✉)
Vinod Gupta School of Management, Indian Institute of Technology Kharagpur, Kharagpur 721302, West Bengal, India
e-mail: abhijeet@vgsom.iitkgp.ac.in

G. Jadhao
Financial Engineering, Indian Institute of Technology Kharagpur, Kharagpur 721302, India

BlackRock, Inc., Mumbai 400063, India

4.1 Introduction

Active trading in financial market using portfolio rotation strategies is a seasoned phenomenon. In the beginning of the twentieth century, Keynes studied various assets and their returns to find variations across business cycles. The recommendations, based on his research, included a trading strategy popularly known as active investment philosophy. The trading strategy proposed by this investment approach was based on a constant switching across assets with varying maturity periods. Their forecast estimates captured the changes in the prevailing interest rates [1, 2]. Such a strategy is closely related to many market timing strategies of recent times. Although the empirical literature suggests that asset rotation does work in favour active traders, this phenomenon in financial markets contradicts several hypotheses proposed by the theorists and the empiricists. Among major ones are the efficient market hypothesis, the random walk theory, and the no arbitrage theory. The empirical research shows conflicting results. However, active market traders employ asset rotation strategies in their day-to-day operations in order to generate abnormal returns. Particularly, the literature on financial economics dealing with market timing argue that the volatility, as well as its sign, of a risky asset are predictable. In such cases, traders can formulate a strategy that bases on the forecasts of the asset volatility and its signs, and then rotates across assets exploiting relative asset returns or relative volatility. It is, therefore, of both theoretical and practical interests for researchers and practitioners alike to examine the viability of an asset rotation strategy that aims to use statistical stylized facts for trading with the objective to generate superior economic gains. We present our empirical results of these simulated trading strategies based on asset rotation to support these arguments.

Volatility Index (VIX henceforth) is often indicated as investors fear gauge, primarily due to its ability to measure the expected volatility in the stock market, both up-side and down-side. It is observed that at a low VIX level, investors in the market tend to be optimistic and complacent, rather than fearful in the market. This connotes that investors perceive no or low potential risks. On the contrary, a high level of VIX value signifies investors' perception about a significant risk. In this case, we can expect that the market under this circumstance would move sharply, either upward or downward.

Previous studies show that a higher percentage change in the India VIX has the potential to be used as an indicator to switch between large-cap and mid-cap portfolios in order to generate positive portfolio returns [3, 4]. In our paper, we study portfolio rotation strategies with respect to style, timing, and size employing two improvised indicators. The indicators we employ here are a sample entropy (SaEn) and an approximate entropy (ApEn), derived from the observed time-series values of the India Volatility Index (India VIX). The India VIX level and its percentage changes tend to suggest the expected quantum of change in the underlying volatility, but these observed changed do not show us about the level of stochastic characteristics found within the same series. Entropy indicators show this level of randomness and uncertainty, independently from the predictable components of the volatility changes.

This uncertain component could affect the market risk premium and the discount rates used to find the worth of value and growth stocks and thus lead to a state in which value outperforms growth stock portfolios (see, for example [3, 5]).

The economic justification of the effectiveness of style-, timing-, and size-based portfolio rotations has attracted considerable interest. Empirical literature shows that market timing and style-based rotation is appropriate, particularly in time-series studies in the financial economics. Rebalancing of portfolio using investing style tends to outperform significantly typical buy-and-hold approaches (see [6–9]).

Copeland and Copeland [10] suggest that the portfolios consisting of value stocks outperform (underperform) the portfolios consisting of growth stocks when there is positive (negative) changes in the volatility index. The volatility index is a better leading indicator for market timing than macro factors that are released to the public with substantial time delay. They particularly reported trading strategies using shorter time periods were economically significant. Boscaljon et al. [11] adapted a similar approach and show that the results are statistically significant for holding periods longer than 30 days. In our study, we revisit this strategy by introducing two new indicators of uncertainty: the sample entropy (SaEn) and the approximate entropy (ApEn). We apply these improvised indicators to examine the trading strategies in the context of the Indian stock market.

Pincus [12] states that entropy can detect the extent of irregularity, instead of simply explaining the deviations from a mean, unlike what most of the volatility measures intend to explain. We show that the entropy-based indicators are significantly and strongly related to the portfolio rotation strategies based on style and size. The relationship is stronger for entropy-based strategies than the strategies based on the VIX percentage changes. Also, we have extended the scope of our study to the returns on portfolios with size-, timing-, and style-based strategies across various asset classes. We have investigated the empirical relationships and then the performance evaluation of three unique portfolio rotation strategies based on the percentage change in the VIX, the sample entropy (SaEn), and the approximate entropy (ApEn) by conducting linear regression analyses and simulations.

4.2 Entropy and Its Applications to the VIX

There have been several studies that have found compelling evidence that the implied volatility measurements such as the volatility index (VIX) can effectively predict expected volatility in the market, more so in the short run [13]. However, Becker et al. suggest that forecasting can improve with the use of the VIX. In these studies, the VIX appears to be the most significant approach or even a reasonably good approach to perform successful market timing strategies. Some studies suggest that the old VIX was unable to estimate the expected changes in volatility while the new VIX helps in enhancing the predictive performance based on unexpected changes in risk aversion [14, 15].

The VIX is used to represent the expected volatility in the market. Investors tend to respond to this measure as an indicator of investment risk. Consequently, an increase in volatility will influence investors' choices with respect to asset allocation by altering equity investments or changing negatively and substantially their risk exposure. While the VIX is often considered the fear index, it effectively acts as an indicator of the underlying sentiment of the market, relating to the level of uncertainty in the concerned equity market. The VIX can be used for multiple applications in securities markets. Few such applications include forming portfolios in terms of style-, size-, timing-, horizon-based rotations.

In recent times, entropy has been applied to the study of financial markets (see, for example [4, 12, 16–18], among others). Entropy concepts helps us to achieve a measure of the nonlinear dynamics in the financial time-series. This function is attributed to the fact that entropy is not restricted by the underlying assumptions regarding any theoretical probability distributions [19]. The fundamental idea behind this argument is that securities that are more volatile exhibit a greater entropy state than more stable securities. Two fundamentally different phenomena are likely to exist and show that time-series data tend to deviate from a constant state. These two phenomena are that series tend to, first, exhibit larger standard deviations (can be expressed by the VIX as an implied volatility measure), and, second, appear highly irregular trends (estimated by an entropy).

The two phenomena are originally not mutually exclusive and as such researchers have been using them to characterize two very different characteristics of randomness that are related to the fluctuations in a time-series, such as security prices. The standard deviation denotes the extent of variance from the centrality while entropy provides a useful metric for categorizing the extent of irregularity or complexity of the data set. Evaluating the subtle but complex shifts in the data series is a primary goal for exploring the potential information contained therein.

Approximate entropy is found robust to the outliers as well as it can be modelled with a time-series with 50 or more observations [12]. Another alternative measure of system complexity that is often used in this regard is the sample entropy. The literature is replete with detailed discussion of these alternative measures of entropy [20, 21]. Additionally other entropy measures, termed superinformation [22], as well as Thannon, Renyi, and Tsallis (TRT) entropy [19], have been applied to analysing time series in the financial markets and their underlying nonlinear dynamics. Within this study, we employed ApEn and SaEn models, which consist of three inputs for (i) time series (TS), (ii) matching template length (M), and (iii) matching tolerance level (R).

In statistics, approximate entropy (ApEn) is a computation-based approach that is used to quantify the extent of regularity and unpredictability of fluctuations in a time series over time periods. Usually, the presence of repetitive patterns of fluctuations in a time series data set allows it to be forecasted with more accuracy than in case of a time series in which such patterns are not observed. The approximate entropy connotes the possibilities of two or more consecutive similar patterns of observations. A time series that contains several repetitive patterns is less likely to carry the characteristics of the approximate entropy; on the contrary, a less predictable process has a higher approximate entropy.

The advantages of the approximate entropy include:

(a) *Lower computational requirements*: the approximate entropy can easily be designed to model for data samples that are smaller (for example, a data set with n < 50). It can also be deployed in real time.
(b) *Less affected by noise in the data*: Even if the data set is noisy, we can compare the approximate entropy measure with the noise in the data to establish the prevalence of true information.

Similar to approximate entropy, the sample entropy measures complexity. However, it differs from approximate entropy such that it excludes self-similar patterns. The sample entropy is an improvisation over the approximate entropy and used primarily for examining the complexity level of a physiological time-series signal, and in the process, is able to diagnose diseased state of the sample data set. This entropy measure performs better irrespective of data length and is suitable for a trouble-free implementation. The sample entropy is also computationally different from the approximate entropy. Unlike the sample entropy, the approximate entropy compares the template vector with other vectors, including a comparison with itself.

We compute the approximate entropy (*ApEn*), the sample entropy (*SaEn*), and the India VIX change for the sample period (see Fig. 4.1. We observe that the *ApEn* and the *SaEn* capture the higher order movements better than the change in the India VIX, implying a better indicator of volatile market. Between the *ApEn* and the *SaEn*, the later reflects the fluctuations better.

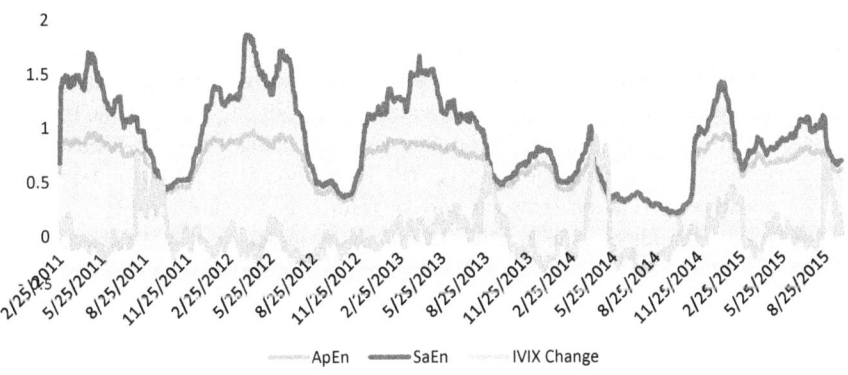

Fig. 4.1 Time-series trends of the ApEn, the SaEn, and the India VIX Change

4.3 Literature Review

The recent literature on computational finance has paid considerable attention to study the style-/size-based portfolio rotation feasibility utilizing the VIX changes as an indicator. However, there has been a considerably small amount of work in the context of the Indian financial markets. The most recent studies (see, for example [3, 4, 23]) show empirical evidences that higher changes in the volatility index is more likely to act as a signal that can be used to switch across different portfolios, say, comprising of mid-cap and large-cap stocks. Such a strategy successfully yields positive portfolio returns.

Historically, the volatility-induced switching of portfolios based on style and size characteristics has been popular since early 1990s. Researchers identified the structural and return-related factors and used it to derive successful trading strategies based on size and value premium for small companies [24]. Fama and French [25] used these factors on size and value premium to integrate into asset pricing framework. The research on incorporating factor premium has shaped our understanding of portfolio construction using portfolio-specific characteristics. However, there are contrary evidences to the above proposition as well. The critiques argue that the effects of size, value or other factors are attributed to the inability of the market to price an asset efficiently, and not due to those factors [26, 27]. It is also suggested that, in the times of capital market equilibrium when majority of investors are assumed to have constant risk aversion, the market risk premium is found to be linearly related to the market portfolio [28]. In a related research, Copeland and Copeland [10] combined these ideas together and proposed a trading strategy using volatility indices as a signal to change between portfolios.

In another line of research, financial economists have studied volatility-based investment and trading decisions where portfolio optimization and strategies are formulated using volatility as one of the parameters. To incorporate volatility into portfolio strategy decisions, several estimates of volatility are found in the relevant literature. Nelson [29] suggested to estimate volatility using the generalized autoregressive conditional heteroscedasticity (GARCH) method as an estimator. It is based on the negative correlation between current returns and future returns volatility. According to Fleming et al. [30], the VIX is a better indicator of implied volatility. These volatility estimates are used frequently for formulating trading strategies. Levis and Liodakis [31] researched portfolio rotation and showed the profitability of size and value/growth rotation strategies in the United Kingdom. Their findings provide strong empirical evidence for the trading strategy based on size but not on style. Bauer et al. [32] examined a similar framework in order to explore whether or not size and value premium are predictable through a timing strategy in the Japanese market. The results of their study suggest that there is sufficient predictability under low transaction cost levels while is it difficult when transaction costs are high.

Sectoral rotation strategies for holding optimal portfolios has been well studied in recent years. Orozco [33] evaluated sector rotation strategies to study whether these strategies outperform a traditional approach such as index investing. Their results supported the efficacy of rotation strategies for both portfolio managers and marginal investors in terms of earning higher returns with lower exposures. Kinlaw et al. [34] propose two measures for managing trading strategies based on market bubbles. These measures taken together have the potential to locate market bubbles across sectors and factors as and when they emerge, and help in formulating trading strategies using these signals.

Studies show that, in case of an increase (decrease) in the volatility index (VIX), the value portfolios tend to out-perform (under-perform) the growth portfolios. This finding connotes the VIX as a better leading indicator for timing stock market and useful for profitable trading strategies [11]. Some recent works have incorporated the measures of sample entropy and approximate entropy to study the feasibility of the style-based rotation strategies [35]. In their analysis, they have used the daily returns on iShare Large-cap Value ETF (IWD) and Growth ETF (IWF) to measure the difference between value and growth portfolios. They have employed the ApEn and the SaEn models of entropy in order to test the predictability of large shifts in the VIX series which then can be used for style rotation timing and growth portfolios. We have extended this analysis in the Indian context by implementing these trading strategies for size-based portfolios across various asset classes.

Other related research that addresses the portfolio timing strategies using alternative methods including the market phases such as bullish and bearish markets [36], broad market conditions such as crisis, growth, and stability [37], and computational tools such as exponential Lévy processes [38]. However, our study focuses on implementing the stock-specific characteristics for portfolio rotation strategies and entropy-based measures to assess the signalling and market timing.

4.4 Research Objectives

With the background, the primary objectives of our research study are as following:

1. To find a better measure to signal the expected changes in the financial market index in the context of the Indian stock market;
2. To evaluate the entropy-based measures to determine the return generation process in different asset classes; and
3. To examine if the entropy-based measures provides more robust rotation signals for switching between portfolios.

The portfolios proposed for implementing the switching and rotation strategies are based on:

1. Style: the portfolios with value stocks versus the portfolios with growth stocks,
2. Size: the portfolios comprising large-cap stocks versus those comprising small-cap stocks, and
3. Time/investment horizon-based rotation of portfolios.

Table 4.1 Propositions considered in the study

Proposition	Sources
Market risk premium is directly related to the market variance	Merton [28], Chandra and Thenmozhi [3], Shaikh [23]
Market timing provides the highest value-addition in a highly volatile market environment	Liu et al. [37]
Market timing and/or style rotation is appropriate in various time periods and outperform naïve buy-and-hold approaches	Arshanapalli et al. [39], Holmes and Faff [7], Puttonen and Seppa [9], Maio [8], Kanojia and Arora [36]
Changes in expected volatility has been used to time the portfolio rotation strategies, and a successful market timing strategy requires prediction accuracy	Copeland and Copeland [10], Boscaljon et al. [11], Jadhao and Chandra [4]
A higher shift in the value of India VIX can be used as a signal to switch between portfolios of varying sizes, such as large and mid-cap, to achieve positive portfolio returns	Bagchi [40], Chandra and Thenmozhi [3], Chakrabarti and Kumar [41]
Entropy-based measures can successfully be applied to examine financial time-series to address issues of non-linearity and the restrictions associated with their theoretical probability distributions	Pincus and Kalman [16], Aksaraylı and Pala [42]

4.5 Theoretical Framework

Our research study employs the following theoretical framework for pinning down different propositions that we consider in this study (Table 4.1).

The structure of the research is presented in Fig. 4.2. Here, we show the research process used to specify stylized facts of the assets under consideration, construct portfolios, formulate trading strategies, carry out the simulations, and exhibit the computational results with relevant outcome and interpretations.

India VIX was used to formulate a market timing strategy for portfolio rotation, as proposed by Copeland and Copeland [10], for BAARA value and growth indices implemented in US market [11]. Approximate and sample entropies was developed by Steve M. Pincus and initially used to analyse medical data such as heart rate. Efremidze et al. [35] used the sample entropy and the approximate entropy for style rotation in the context of ETFs (iShare large-cap value and growth ETFs).

4.6 Data Variables, Measurements, and Sources

For the purpose of analysis and simulation, we used the following data sets:

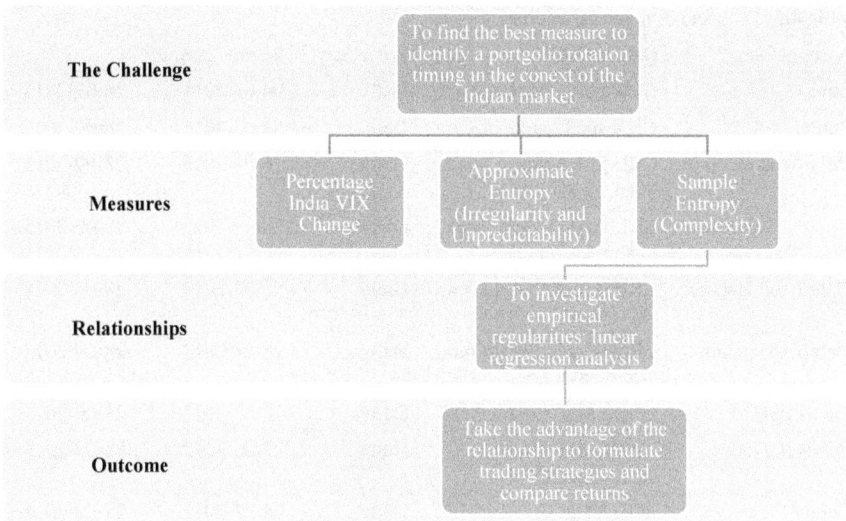

Fig. 4.2 Workflow and theoretical framework of the research

The data on the above-mentioned indices were obtained from the Bloomberg database. We considered the differences in sample periods as part of the limitation owing to data availability. Further analyses of the data was carried out using R programming language. The data series were processed for approximate entropy and sample entropy. We withheld the preliminary results in this regard, citing space constraint.

4.7 Methods and Measures

To investigate empirical regularities and then the performance of three strategies of trading, using either of the three indicators, namely, the percentage change in the IVIX, the sample entropy (SaEn), and the approximate entropy (ApEn), we conduced linear regression analysis. The sample period used in the regression estimates for various asset classes is mentioned in Table 4.2.

We started with the following baseline specification of the regression estimate:

$$R_{1t} - R_{2t} = \alpha + \beta * X_{t-1} + \varepsilon \tag{4.1}$$

where, X_{t-1} is obtained by converting India VIX into a time-series of the percentage change in the India VIX or its approximate entropy (ApEn) or its sample entropy (SaEn). The baseline model is modified for several asset markets as discussed above.

Table 4.2 Data and sample characteristics

Asset market	Data variable	Frequency	Sample period	
Equity	Nifty (large-cap) index	Daily	04-03-2011	28-09-2015
Equity	CNX small-cap index	Daily	04-03-2011	28-09-2015
Debt/fixed income	CRISIL long-term bond index (AAA rated)	Daily	26-08-2013	28-09-2015
Debt/fixed income	CRISIL short-term bond index (AAA rated)	Daily	26-08-2013	28-09-2015
Debt/fixed income	CRISIL debt hybrid fund (60 + 40)	Daily	26-08-2013	28-09-2015
Debt/fixed income	CRISIL debt hybrid fund (75 + 25)	Daily	26-08-2013	28-09-2015
Funds (ETFs)	iShare MSCI India ETF	Daily	22-08-2012	28-09-2015
Funds (ETFs)	iShare MSCI India small-cap ETF	Daily	22-08-2012	28-09-2015
Equity	India VIX	Daily	04-03-2011	28-09-2015

The dependent variable (indicated by $R_{1t} - R_{2t}$) is measured as the return on the portfolio proxied by a representative index (see Table 4.3).

We have obtained the series for ApEn and SaEn by coding their respective algorithms in the R, using the inputs from Table 4.4.

We calculate the percentage change in the India VIX as following. The percentage changes in the volatility index (i.e., India VIX or IVIX) is expressed as the difference between the IVIX at time t and the 75-day (about three months) historical moving

Table 4.3 Model variables for baseline equation

Model specification	Asset market	Proxy index	Model variable
Model 1	Equity	Nifty large-cap index	R_{1t}
	Equity	CNX small-cap index	R_{2t}
Model 2	Debt/fixed income	CRISIL long-term bond index (AAA-rated)	R_{1t}
	Debt/fixed income	CRISIL short-term bond index (AAA-rated)	R_{2t}
Model 3	Debt/fixed income	CRISIL debt hybrid fund index (60 + 40)	R_{1t}
	Debt/fixed income	CRISIL debt hybrid fund index (75 + 25)	R_{2t}
Model 4	Fund (ETFs)	iShare MSCI India ETF	R_{1t}
	Funds (ETFs)	iShare MSCI India Small-cap ETF	R_{2t}
	Equity	India volatility index (IVIX)	Underlying for deriving three indicators for X_{t-1}

Table 4.4 Entropy model inputs

Time-series (TS)	Matching template length (M)	Matching tolerance level (R)
Running 120-days of daily India VIX series	2	20% of the standard deviation of the time-series (TS) over the matching template length (M)

average of the volatility index, divided by the 3-month historical moving average of the underlying volatility index (i.e., IVIX). The mathematical expression of the above calculation is as follows:

$$\Delta IVIX_t = \frac{\left(IVIX_t - \frac{1}{n}\sum_{i=1}^{n=75} IVIX_{t-i}\right)}{\frac{1}{n}\sum_{i=1}^{n=75} IVIX_{t-i}} \quad (4.2)$$

The above-calculated value is used in the analyses that follows.

4.8 Empirical Results

We start with the regression estimates. The return premium on the pair of proxy portfolios is regressed upon the proposed indicators, namely, the India VIX Change, the Approximate Entropy, and the Sample Entropy. The results are presented as following:

Model 1:

In the Sub-model, *d1vc* connotes that the trading strategy consists of 1-day holding period and X(t − 1) series has IVIX Change series (indicated by *vc*). Subsequently, numeric terms 1, 2, 5, 10, ... Stand for holding period for respective trading strategies, and *vc* stands for percentage IVIX Change, *a* stands for approximate entropy and *s* stands for sample entropy.

The regression estimates presented in Table 4.5 suggest that the slope is consistently positively across all the holding periods. We observe a strong linear relationship between large cap minus small cap portfolio returns with entropy indicators for holding periods of more than 40.

Model 2:

In the Sub-model, *d1vc* connotes that the trading strategy consists of 1-day holding period and X(t − 1) series has IVIX Change series (indicated by *vc*). Subsequently, numeric terms 1, 2, 5, 10, ... Stand for holding period for respective trading strategies, and *vc* stands for percentage IVIX Change, *a* stands for approximate entropy and *s* stands for sample entropy.

From the regression estimates presented in Table 4.6, we observe that the slope (coefficient of X(t − 1)) is not consistently positively across all the holding periods

Table 4.5 Regression estimates for nifty large-cap Index *minus* CNX small-cap index returns regressed upon change in India VIX, approximate entropy, and sample entropy

		Coefficients	Standard error	t Stat	P-value	Significance
D1vc	Intercept	−4.99179E−06	0.000244591	−0.020408711	0.983720888	0
d1vc	IVIX Change	0.00152988	0.001197362	1.277708434	0.201611012	0
d2vc	Intercept	1.38823E−05	0.000374252	0.037093596	0.970416844	0
d2vc	IVIX Change	0.002500948	0.00183188	1.365235626	0.172447172	0
d5vc	Intercept	2.43746E−05	0.000624759	0.039014318	0.968885768	0
d5vc	IVIX Change	0.00392494	0.003058422	1.283321775	0.199638813	0
d10vc	Intercept	9.79395E−05	0.000861634	0.113667109	0.909521595	0
d10vc	IVIX Change	0.002755346	0.004218011	0.653233399	0.513736852	0
d20vc	Intercept	0.00061189	0.001231789	0.496749478	0.619461	0
d20vc	IVIX Change	0.003540341	0.006030052	0.587116146	0.55724135	0
d30vc	Intercept	0.001515222	0.001579081	0.959558853	0.337479685	0
d30vc	IVIX Change	−0.001586803	0.007730175	−0.205273938	0.83739456	0
d40vc	Intercept	0.499246662	0.003045013	163.9554871	0	111
d40vc	IVIX Change	0.081712043	0.014755208	5.537844266	3.83034E−08	111
d50vc	Intercept	−0.000656233	0.002228415	−0.29448412	0.76844366	0
d50vc	IVIX Change	0.039595621	0.010798218	3.666866174	0.0002573	111
d60vc	Intercept	−0.00065208	0.002428217	−0.268542643	0.788332101	0
d60vc	IVIX Change	0.072610208	0.0117664	6.170978921	9.53136E−10	111
d90vc	Intercept	0.002132833	0.00294996	0.72300426	0.469831244	0
d90vc	IVIX Change	0.126394636	0.014294605	8.842121398	3.67101E−18	111
d1a	Intercept	−0.000211213	0.000825597	−0.255830592	0.798129569	0
d1a	ApEn	0.000236957	0.00120771	0.196203448	0.844487236	0
d2a	Intercept	−0.00036691	0.001265346	−0.289967947	0.771895509	0
d2a	ApEn	0.000411827	0.001850988	0.222490488	0.823973456	0
d5a	Intercept	−0.001115243	0.002130682	−0.523420832	0.600787047	0
d5a	ApEn	0.001415279	0.00311683	0.454076505	0.649863504	0
d10a	Intercept	−0.002302429	0.002933021	−0.785002689	0.432621331	0
d10a	ApEn	0.003093567	0.004290517	0.721024382	0.471048	0

(continued)

4 Profits Are in the Eyes of the Beholder: Entropy-Based …

Table 4.5 (continued)

		Coefficients	Standard error	t Stat	P-value	Significance
d20a	Intercept	−0.005261469	0.004182953	−1.257836003	0.208718507	0
d20a	ApEn	0.007477942	0.006118958	1.22209403	0.221934357	0
d30a	Intercept	−0.013262629	0.005306462	−2.499335281	0.012587782	11
d30a	ApEn	0.019779528	0.007762463	2.548099659	0.010966459	11
d40a	Intercept	0.515314892	0.010175384	50.64328608	1.6586E−289	111
d40a	ApEn	−0.024184833	0.014884877	−1.624792277	0.104493818	0
d50a	Intercept	−0.037330474	0.007305955	−5.109595204	3.80522E−07	111
d50a	ApEn	0.056526815	0.010687385	5.289115876	1.48274E−07	111
d60a	Intercept	−0.046626829	0.008017966	−5.815293745	7.93006E−09	111
d60a	ApEn	0.070996745	0.011728937	6.053127062	1.94712E−09	111
d90a	Intercept	−0.09474313	0.00959188	−9.877430652	4.21329E−22	111
d90a	ApEn	0.149445032	0.014031308	10.65082656	2.81827E−25	111
d1s	Intercept	−0.00027	0.000611	−0.44066	0.659549	0
d1s	SaEn	0.000234	0.000615	0.381018	0.703264	0
d2s	Intercept	−0.00045	0.000937	−0.48561	0.627342	0
d2s	SaEn	0.000393	0.000942	0.416833	0.676882	0
d5s	Intercept	−0.00113	0.001577	−0.71956	0.471948	0
d5s	SaEn	0.001038	0.001587	0.654362	0.513016	0
d10s	Intercept	−0.00243	0.00217	−1.12173	0.262222	0
d10s	SaEn	0.002368	0.002184	1.084486	0.278387	0
d20s	Intercept	−0.00478	0.003095	−1.54302	0.123113	0
d20s	SaEn	0.004837	0.003114	1.553158	0.120674	0
d30s	Intercept	−0.01067	0.003925	−2.71802	0.006671	111
d30s	SaEn	0.011349	0.003949	2.873803	0.004134	111
d40s	Intercept	0.515061	0.007523	68.46159	0	111
d40s	SaEn	−0.0171	0.00757	−2.25832	0.024121	11
d50s	Intercept	−0.02917	0.005394	−5.40848	7.8E−08	111
d50s	SaEn	0.031614	0.005427	5.82527	7.48E−09	111
d60s	Intercept	−0.03555	0.00592	−6.0056	2.59E−09	111
d60s	SaEn	0.038794	0.005956	6.512999	1.12E−10	111
d90s	Intercept	−0.06381	0.007127	−8.95341	1.44E−18	111
d90s	SaEn	0.073253	0.007171	10.21555	1.82E−23	111

Statistical Significance: 0 is for $p > 0.1$, 1 for $p < 0.10$, 11 for $p < 0.05$, 111 for $p < 0.01$

Table 4.6 Regression estimates for CRISIL long-term bond index *minus* CRISIL short-term bond index returns regressed upon change in India VIX, approximate and sample entropy

Sub model	Variable	Coefficients	Standard error	t Stat	P-value	Significance
d1vc	Intercept	4.57E−05	6.82E−05	0.671012	0.502529	0
d1vc	IVIX Change	−0.00016	0.000266	−0.60931	0.542599	0
d2vc	Intercept	8.75E−05	0.000106	0.829303	0.407336	0
d2vc	IVIX Change	−0.00042	0.000411	−1.01625	0.31001	0
d5vc	Intercept	0.00035	0.000187	1.874381	0.061471	1
d5vc	IVIX Change	0.000554	0.000727	0.761368	0.446803	0
d10vc	Intercept	0.000663	0.00022	3.020128	0.002658	111
d10vc	IVIX Change	0.001422	0.000856	1.661741	0.097203	1
d20vc	Intercept	0.001073	0.000295	3.634429	0.000308	111
d20vc	IVIX Change	0.001835	0.001151	1.594634	0.111438	0
d30vc	Intercept	0.001438	0.000368	3.906337	0.000107	111
d30vc	IVIX Change	0.0007	0.001435	0.487846	0.625877	0
d40vc	Intercept	0.046966	0.000749	62.74568	1.3E−236	111
d40vc	IVIX Change	0.007206	0.002918	2.469718	0.013861	11
d50vc	Intercept	0.001006	0.000624	1.610774	0.107872	0
d50vc	IVIX Change	−0.00974	0.002433	−4.00197	7.25E−05	111
d60vc	Intercept	0.000378	0.000759	0.498106	0.618632	0
d60vc	IVIX Change	−0.01409	0.00296	−4.76094	2.54E−06	111
d90vc	Intercept	−0.00104	0.001047	−0.99293	0.321233	0
d90vc	IVIX Change	−0.00604	0.004081	−1.4806	0.139354	0
d1a	Intercept	0.000229	0.000204	1.120745	0.262944	0
d1a	ApEn	−0.00034	0.000352	−0.95968	0.337691	0
d2a	Intercept	0.00043	0.000316	1.360689	0.174236	0
d2a	ApEn	−0.00063	0.000545	−1.16387	0.24504	0
d5a	Intercept	0.000834	0.000559	1.490881	0.136635	0
d5a	ApEn	−0.00087	0.000964	−0.90737	0.364659	0
d10a	Intercept	0.00167	0.000658	2.535781	0.011529	11

(continued)

Table 4.6 (continued)

Sub model	Variable	Coefficients	Standard error	t Stat	P-value	Significance
d10a	ApEn	−0.00181	0.001135	−1.59812	0.110659	0
d20a	Intercept	0.002931	0.000883	3.319649	0.000968	111
d20a	ApEn	−0.00336	0.001522	−2.2098	0.02758	11
d30a	Intercept	0.003909	0.001098	3.561219	0.000405	111
d30a	ApEn	−0.0045	0.001892	−2.38068	0.017661	11
d40a	Intercept	0.024644	0.001987	12.40199	6.45E−31	111
d40a	ApEn	0.040937	0.003425	11.95327	4.41E−29	111
d50a	Intercept	0.001826	0.001901	0.96039	0.337332	0
d50a	ApEn	−0.00168	0.003277	−0.51255	0.608498	0
d60a	Intercept	−0.00064	0.002328	−0.27647	0.782303	0
d60a	ApEn	0.001608	0.004012	0.400704	0.688812	0
d90a	Intercept	−0.0026	0.003145	−0.82806	0.408037	0
d90a	ApEn	0.002748	0.00542	0.507084	0.612323	0
d1s	Intercept	0.000126	0.000174	0.722825	0.470131	0
d1s	SaEn	−0.00012	0.000241	−0.51012	0.610195	0
d2s	Intercept	0.000238	0.00027	0.881917	0.378253	0
d2s	SaEn	−0.00023	0.000373	−0.62264	0.533809	0
d5s	Intercept	0.000456	0.000477	0.957807	0.338631	1
d5s	SaEn	−0.00015	0.000659	−0.23071	0.817635	0
d10s	Intercept	0.001063	0.000562	1.892559	0.059004	11
d10s	SaEn	−0.00058	0.000778	−0.74623	0.455883	1
d20s	Intercept	0.002246	0.000753	2.980634	0.003019	111
d20s	SaEn	−0.00174	0.001043	−1.66499	0.096553	0
d30s	Intercept	0.003247	0.000936	3.468377	0.00057	111
d30s	SaEn	−0.00271	0.001295	−2.09118	0.037026	0
d40s	Intercept	0.028263	0.001688	16.74284	4.15E−50	111
d40s	SaEn	0.028233	0.002336	12.08657	1.27E−29	111
d50s	Intercept	0.001072	0.00162	0.661958	0.508309	0
d50s	SaEn	−0.00025	0.002241	−0.11066	0.91193	11
d60s	Intercept	−0.00143	0.001981	−0.72393	0.469453	0
d60s	SaEn	0.002511	0.002742	0.916053	0.360089	11
d90s	Intercept	−0.00404	0.002675	−1.51133	0.131349	0
d90s	SaEn	0.004424	0.003701	1.195239	0.23257	0

Statistical Significance: 0 is for $p > 0.1$, 1 for $p < 0.10$, 11 for $p < 0.05$, 111 for $p < 0.01$

Implying that the difference in returns of Fixed income portfolios are not positively related to the % change in VIX, ApEn and SaEn. We also don't observe a significant linear relationship between long term bond and short term bond portfolio returns with entropy indicators for various holding periods.

Model 3:

Statistical Significance: 0 is for $p > 0.1$, 1 for $p < 0.10$, 11 for $p < 0.05$, 111 for $p < 0.01$

In the Sub-model, $d1vc$ connotes that the trading strategy consists of 1-day holding period and $X(t-1)$ series has IVIX Change series (indicated by vc). Subsequently, numeric terms 1, 2, 5, 10, ... Stand for holding period for respective trading strategies, and vc stands for percentage IVIX Change, a stands for approximate entropy and s stands for sample entropy.

From the regression estimates presented in Table 4.7, we observe that the slope is consistently positively across all the holding periods. We also see a strong linear relationship between debt hybrid 60 + 40 fund minus 75 + 25 fund returns with entropy indicators for holding periods of more than 10. We can observe this consistence for all three indicators (IVIX Change, ApEn and SaEn).

Model 4:

In the Sub-model, $d1vc$ connotes that the trading strategy consists of 1-day holding period and $X(t-1)$ series has IVIX Change series (indicated by vc). Subsequently, numeric terms 1, 2, 5, 10, ... Stand for holding period for respective trading strategies, and vc stands for percentage IVIX Change, a stands for approximate entropy and s stands for sample entropy.

From the regression estimates presented in Table 4.8, we observe that the slope is consistently positively across all the holding periods. We see a strong linear relationship between iShares MSCI India ETF minus iShares MSCI India Small-cap ETF returns with entropy indicators for holding periods of more than 10 days. In all numerical results, we observe that we will not be able to incorporate fixed income based portfolios to device trading strategies based on percentage change in VIX entropy based indicators on it.

4.9 Formulation and Testing of Trading Strategies

Based on the relationship between the portfolio return differentials and our proposed market timing indicators, we exploit the above-mentioned results to derive trading strategies based on various threshold[1] changes in our proposed indicators, namely the

---

[1]The thresholds used for signalling using the India VIX change are: -30, -20, -10, 10, 20, 30, 40, 50, 60, 70, 80 and 90 percentages. Also, multiples of SD in the ApEn and the SpEn considered are -1.75, -1.5, -1.25, -1, -0.75, -0.5, -0.25, 0.25, 0.5, 0.75, 1, 1.25, 1.5.

Table 4.7 Regression estimates for CRISIL debt hybrid fund (60 + 40) *minus* CRISIL debt hybrid fund (75 + 25) returns regressed upon change in India VIX, approximate entropy, and sample entropy

Sub model	Variable	Coefficients	Standard error	t Stat	P-value	Significance
d1vc	Intercept	5.16E−05	6.99E−05	0.739112	0.460192	0
d1vc	IVIX Change	−0.00032	0.000272	−1.18168	0.237906	0
d2vc	Intercept	0.000108	0.000102	1.059807	0.289753	0
d2vc	IVIX Change	−0.00052	0.000399	−1.3109	0.190505	0
d5vc	Intercept	0.000278	0.000152	1.82532	0.06856	1
d5vc	IVIX Change	−0.00058	0.000593	−0.97691	0.329094	0
d10vc	Intercept	0.000509	0.000209	2.434578	0.015264	11
d10vc	IVIX Change	−0.00216	0.000815	−2.65136	0.008276	111
d20vc	Intercept	0.000884	0.000278	3.177511	0.001579	111
d20vc	IVIX Change	−0.00394	0.001084	−3.63533	0.000307	111
d30vc	Intercept	0.0013	0.000332	3.91607	0.000103	111
d30vc	IVIX Change	−0.0055	0.001294	−4.24984	2.56E−05	111
d40vc	Intercept	0.036802	0.000613	60.07724	2E−228	111
d40vc	IVIX Change	−0.00775	0.002388	−3.24674	0.001247	111
d50vc	Intercept	0.002615	0.000367	7.117407	3.92E−12	111
d50vc	IVIX Change	−0.00668	0.001432	−4.6621	4.04E−06	111
d60vc	Intercept	0.003194	0.000392	8.151545	3.01E−15	111
d60vc	IVIX Change	−0.00889	0.001527	−5.82179	1.05E−08	111
d90vc	Intercept	0.004649	0.000462	10.07234	8.12E−22	111
d90vc	IVIX Change	−0.01865	0.001799	−10.3654	6.71E−23	111
d1a	Intercept	0.000288	0.000209	1.376126	0.16941	11
d1a	ApEn	−0.00044	0.000361	−1.21453	0.225127	0
d2a	Intercept	0.000584	0.000307	1.905761	0.057265	0
d2a	ApEn	−0.00088	0.000528	−1.66427	0.096697	0
d5a	Intercept	0.001358	0.000454	2.993175	0.0029	111
d5a	ApEn	−0.00199	0.000782	−2.539	0.011424	0

(continued)

Table 4.7 (continued)

Sub model	Variable	Coefficients	Standard error	t Stat	P-value	Significance
d10a	Intercept	0.002534	0.000624	4.064375	5.61E−05	111
d10a	ApEn	−0.00374	0.001075	−3.48226	0.000542	111
d20a	Intercept	0.004151	0.00083	5.003258	7.87E−07	111
d20a	ApEn	−0.00605	0.00143	−4.2282	2.81E−05	111
d30a	Intercept	0.006074	0.000986	6.157906	1.53E−09	111
d30a	ApEn	−0.00883	0.0017	−5.19353	3.03E−07	111
d40a	Intercept	0.031901	0.001842	17.32257	8.2E−53	111
d40a	ApEn	0.008815	0.003174	2.777308	0.005691	111
d50a	Intercept	0.010945	0.001051	10.41181	4.5E−23	111
d50a	ApEn	−0.01535	0.001812	−8.47313	2.8E−16	111
d60a	Intercept	0.013695	0.001104	12.4073	6.14E−31	111
d60a	ApEn	−0.01936	0.001902	−10.1767	3.36E−22	111
d90a	Intercept	0.021394	0.001295	16.52125	4.43E−49	111
d90a	ApEn	−0.03095	0.002232	−13.8695	3.83E−37	111
d1s	Intercept	0.000224	0.000178	1.257024	0.209342	0
d1s	SaEn	−0.00026	0.000247	−1.07128	0.28457	0
d2s	Intercept	0.000433	0.000261	1.658729	0.097809	0
d2s	SaEn	−0.0005	0.000362	−1.37333	0.170278	0
d5s	Intercept	0.000994	0.000387	2.566689	0.010563	1
d5s	SaEn	−0.00109	0.000536	−2.02621	0.043284	0
d10s	Intercept	0.001857	0.000533	3.482618	0.000541	111
d10s	SaEn	−0.00206	0.000738	−2.79195	0.005443	111
d20s	Intercept	0.003301	0.000709	4.653898	4.2E−06	111
d20s	SaEn	−0.00369	0.000981	−3.76458	0.000187	111
d30s	Intercept	0.005112	0.000842	6.074006	2.5E−09	111
d30s	SaEn	−0.00582	0.001165	−4.99421	8.22E−07	111
d40s	Intercept	0.031517	0.00156	20.2042	1.64E−66	111
d40s	SaEn	0.007829	0.002159	3.627154	0.000317	111
d50s	Intercept	0.009285	0.0009	10.31948	9.95E−23	111
d50s	SaEn	−0.01013	0.001245	−8.1374	3.34E−15	111
d60s	Intercept	0.011647	0.000946	12.31426	1.48E−30	111
d60s	SaEn	−0.01285	0.001309	−9.81482	6.98E−21	111
d90s	Intercept	0.01812	0.001115	16.24568	8.29E−48	111
d90s	SaEn	−0.02054	0.001543	−13.3078	1.01E−34	111

Statistical Significance: 0 is for $p > 0.1$, 1 for $p < 0.10$, 11 for $p < 0.05$, 111 for $p < 0.01$

Table 4.8 Regression estimates for iShare MSCI India ETF *minus* iShare MSCI India small-cap ETF returns regressed upon the change in India VIX, the approximate entropy, and the sample entropy

Sub model	Variable	Coefficients	Standard error	t Stat	P-value	Significance
d1vc	Intercept	−0.00072	0.000623	−1.16394	0.245012	0
d1vc	IVIX Change	0.001925	0.002407	0.799663	0.424293	1
d2vc	Intercept	−0.00148	0.000801	−1.84422	0.065753	0
d2vc	IVIX Change	0.00364	0.003096	1.175745	0.240267	11
d5vc	Intercept	−0.00353	0.001084	−3.25628	0.001207	111
d5vc	IVIX Change	0.006379	0.004189	1.522761	0.128462	11
d10vc	Intercept	−0.00677	0.0014	−4.83755	1.76E−06	111
d10vc	IVIX Change	0.009786	0.00541	1.808864	0.071084	11
d20vc	Intercept	−0.01277	0.001911	−6.68516	6.28E−11	111
d20vc	IVIX Change	0.023328	0.007385	3.159012	0.001681	111
d30vc	Intercept	−0.0186	0.002405	−7.73171	6.05E−14	111
d30vc	IVIX Change	0.031716	0.009295	3.412141	0.000698	111
d40vc	Intercept	0.00797	0.006057	1.316016	0.188783	11
d40vc	IVIX Change	0.050369	0.023407	2.151851	0.031896	11
d50vc	Intercept	−0.03045	0.003068	−9.92504	2.79E−21	111
d50vc	IVIX Change	0.07035	0.011858	5.932768	5.63E−09	111
d60vc	Intercept	−0.03606	0.003191	−11.3001	1.77E−26	111
d60vc	IVIX Change	0.105459	0.012333	8.55117	1.56E−16	111
d90vc	Intercept	−0.04888	0.003817	−12.8047	1.36E−32	111
d90vc	IVIX Change	0.139161	0.014754	9.432279	1.6E−19	111
d1a	Intercept	−0.00142	0.001882	−0.756	0.450009	11
d1a	ApEn	0.001322	0.00321	0.411982	0.680532	0
d2a	Intercept	−0.00288	0.002422	−1.19115	0.234169	0
d2a	ApEn	0.002659	0.00413	0.643799	0.520006	0
d5a	Intercept	−0.00676	0.003278	−2.06302	0.039636	1
d5a	ApEn	0.006044	0.00559	1.081191	0.280143	0
d10a	Intercept	−0.01209	0.004234	−2.85534	0.004481	111
d10a	ApEn	0.00992	0.007221	1.37384	0.170118	0
d20a	Intercept	−0.02232	0.00581	−3.84195	0.000138	111
d20a	ApEn	0.017995	0.009908	1.816175	0.069953	1
d30a	Intercept	−0.0339	0.007311	−4.63738	4.53E−06	111
d30a	ApEn	0.028667	0.012468	2.299359	0.021903	11
d40a	Intercept	0.076293	0.018095	4.216279	2.96E−05	111
d40a	ApEn	−0.12184	0.030858	−3.94835	9.02E−05	111

(continued)

Table 4.8 (continued)

Sub model	Variable	Coefficients	Standard error	t Stat	P-value	Significance
d50a	Intercept	−0.06103	0.009472	−6.44327	2.79E−10	111
d50a	ApEn	0.057493	0.016154	3.5591	0.000408	111
d60a	Intercept	−0.07531	0.010143	−7.42537	5.01E−13	111
d60a	ApEn	0.074278	0.017297	4.294406	2.11E−05	111
d90a	Intercept	−0.13814	0.01174	−11.7673	2.48E−28	111
d90a	ApEn	0.165701	0.02002	8.276854	1.2E−15	111
d1s	Intercept	−0.00167	0.001598	−1.0452	0.296447	0
d1s	SaEn	0.001445	0.002172	0.665247	0.506205	1
d2s	Intercept	−0.00323	0.002057	−1.57247	0.116486	0
d2s	SaEn	0.002685	0.002794	0.960886	0.337082	11
d5s	Intercept	−0.00744	0.002781	−2.67532	0.007715	111
d5s	SaEn	0.005933	0.003779	1.570082	0.11704	11
d10s	Intercept	−0.01279	0.003591	−3.56146	0.000405	111
d10s	SaEn	0.00913	0.00488	1.871028	0.061936	11
d20s	Intercept	−0.02293	0.004925	−4.65616	4.15E−06	111
d20s	SaEn	0.01559	0.006692	2.329503	0.020237	11
d30s	Intercept	−0.03341	0.006199	−5.39004	1.1E−07	111
d30s	SaEn	0.022674	0.008422	2.692109	0.007343	111
d40s	Intercept	0.066365	0.015358	4.32118	1.88E−05	111
d40s	SaEn	−0.08481	0.020868	−4.06419	5.61E−05	111
d50s	Intercept	−0.05873	0.008021	−7.32149	1.01E−12	111
d50s	SaEn	0.043527	0.010899	3.993794	7.5E−05	111
d60s	Intercept	−0.07136	0.008586	−8.31162	9.31E−16	111
d60s	SaEn	0.054806	0.011666	4.697939	3.42E−06	111
d90s	Intercept	−0.11967	0.010024	−11.9393	5.02E−29	111
d90s	SaEn	0.108017	0.013619	7.931233	1.48E−14	111

Statistical Significance: 0 is for $p > 0.1$, 1 for $p < 0.10$, 11 for $p < 0.05$, 111 for $p < 0.01$

India VIX change (IVIX), the approximate entropy (ApEn), and the sample entropy (SaEn), for different holding periods.[2]

We then implemented size/time rotation trading strategies and evaluated their performances. After generating the time series of both SaEn and ApEn based on the inputs given in Table 4.4, we tested the time series properties of each entropy indicator against value minus growth returns. For the two entropy-based trading strategies, we produced several signal thresholds based on a mean plus multiple standard deviations of prior 140-days of each entropy series. We chose 140-days, i.e., about seven months,

---

[2] Holding periods of 2, 5, 10, 20, 30, 40, 50, 60 and 90 days are taken for the analyses.

in order to capture sufficient seasonality of the data. The signal was then lagged by 60 days (considering highest cross-correlations between the entropy and the value minus growth return series). This means that there is a substantial time lag between entropy signal (it is a strong leading indicator) and size/time performance. We tested several other lags between 1 and 50 days, but overall their performance was inferior to a 60-day lag.

We also created trading signal thresholds using the different levels of the percentage change in the VIX. This strategy was used in previous literature [10, 11], and we used it as a benchmark for the performance of our entropy-based trading strategies. In addition to the previously used criteria for the evaluation of trading rules, we also calculated average transaction returns for each trading rule by dividing the cumulative return on number of round-trip transactions.

We provide the details of the trading strategies that are tested using the three market timing indicators (see Table 4.9). A detailed set of all the trading strategies tested as part of the research is provided in the Appendix 1. Further we filter the top 10 strategies based on threshold change is the three indicators, i.e., the India VIX

Table 4.9 Comparative analysis of different trading strategies based on the proposed indicators

Sr No.	Signal (Threshold change in)	Portfolios	No of strategies which gave average transaction return greater than 1%*	No of total strategies tested
1	VIX change	CNX Nifty -CNX Small-cap last	47	127
2	VIX change	CRISIL Debt Hybrid 60+40 Fund Index - CRISIL Debt Hybrid 75+25 Fund Index	23	127
3	VIX change	iShares MSCI India ETF - iShares MSCI India Small-Cap ETF	56	127
4	ApEn	CNX Nifty -CNX Small-cap last	62	130
5	ApEn	CRISIL Debt Hybrid 60+40 Fund Index - CRISIL Debt Hybrid 75+25 Fund Index	12	130
6	ApEn	iShares MSCI India ETF - iShares MSCI India Small-Cap ETF	52	130
7	SaEn	CNX Nifty -CNX Small-cap last	61	130
8	SaEn	CRISIL Debt Hybrid 60+40 Fund Index - CRISIL Debt Hybrid 75+25 Fund Index	20	130
9	SaEn	iShares MSCI India ETF - iShares MSCI India Small-Cap ETF	76	130

Note Grey shades indicate an abnormally positive/negative results

Table 4.10 Performance of the thirty best trading strategies tested with three indicators and their comparative performance with respect to the benchmark index

Sr. No	Portfolio_Indicator	Holding period indicator-threshold	No. of days (invested)	No of transactions	Cumulative return (in %)	SD	Sharpe ratio	Coefficient of variation	Max draw down
1	iShare Msci_apen	40d.75sd	66	21	6.8908	13%	0.5197	1.2699	-148%
2	iShare Msci_apen	40d-2sd	17	2	1.9509	1%	1.9583	0.0868	-29%
3	iShare Msci_apen	90d-2sd	17	2	1.7362	2%	0.7156	0.2376	-57%
4	iShare Msci_apen	40d-1.75sd	20	3	1.8459	1%	2.1380	0.0935	-31%
5	iShare MSCI_VIX	90d-30p	27	4	4.3600	3%	1.4244	0.1896	-56%
6	iShare MSCI_VIX	60d-30p	27	4	2.9878	4%	0.7818	0.3454	-97%
7	Nifty VIX	40d-0p	526	87	257.4820	9%	29.5217	0.1782	-53%
8	Nifty VIX	90d-30p	27	4	5.1619	6%	0.9307	0.2901	-60%
9	Nifty VIX	40d10p	116	50	64.1476	12%	5.4682	0.2121	-51%
10	Nifty VIX	40d60p	14	8	7.6473	8%	0.9435	0.1484	-42%
11	Nifty VIX	40d30p	49	26	24.0062	11%	2.0973	0.2336	-49%
12	Nifty VIX	40d20p	43	27	22.2770	13%	1.7184	0.2502	-52%
13	Nifty VIX	40d40p	19	13	10.1779	16%	0.6541	0.2905	-50%
14	Nifty VIX	40d70p	7	5	3.6207	5%	0.7962	0.0879	-25%
15	Nifty_apen	40d.5sd	113	24	60.8888	13%	4.7846	0.2362	-52%
16	Nifty_apen	40d.75sd	138	33	75.8128	14%	5.5310	0.2495	-53%
17	Nifty_apen	40d1sd	139	29	64.0776	7%	8.9866	0.1547	-47%
18	Nifty_apen	40d1.25sd	51	10	21.9138	3%	6.3436	0.0804	-26%
19	Nifty_apen	40d.25sd	74	21	41.2925	14%	2.9512	0.2507	-55%
20	Nifty_apen	90d-1.5sd	66	4	2.2360	9%	0.2367	2.7879	-169%
21	Nifty_apen	40d1.5sd	1	1	0.4365	0%	0.0000	0.0000	0%
22	Nifty_apen	50d-.75sd	25	2	4.0442	10%	0.3989	1.6045	-151%
23	Nifty_apen	10d-2.25sd	11	1	0.3357	2%	0.1719	0.6399	-121%
24	Nifty_spem	40d.5sd	93	15	49.4292	14%	3.4724	0.2678	-52%
25	Nifty_spem	90d-1.25sd	87	3	9.1733	10%	0.9360	0.9294	-129%
26	Nifty_spem	40d.75sd	93	18	46.6585	11%	4.1541	0.2239	-51%
27	Nifty_spem	40d.5sd	78	18	43.5275	13%	3.4226	0.2279	-52%
28	Nifty_spem	40d1.25sd	73	16	36.0337	11%	3.3595	0.2173	-44%
29	Nifty_spem	40d1sd	61	18	28.3159	8%	3.5540	0.1716	-45%
30	Nifty_spem	40d1.5sd	27	12	14.1210	10%	1.3641	0.1979	-41%
31	Nifty(d)	Benchmark			0.244333161	1%	-0.404961872	53.55443786	-263%
32	iShare MSCI India ETF (d) (large and mid size comp)	Benchmark			0.02681725	2%	-0.402173174	456.8742144	-212%
33	iShare MSCI India Small cap ETF (d)	Benchmark			-0.193896738	1%	-0.687276418	-49.28200844	-197%

change, the ApEn and the SpEn, respectively, and calculate the following performance measures and compare them with benchmark (the CNX Nifty) performance over the same period (refer to Table 4.10):

1. Cumulative return
2. Standard deviation
3. Sharpe ratio[3]
4. Coefficient of variation
5. Maximum drawdown.

The entropy-based trading strategies provide better performance than the India VIX percentage change–based trading strategies according to the criteria of average transaction returns. This leads us to argue that the entropy measures of the VIX provide more accurate rotation signals than the VIX percentage change thresholds in the context of Indian markets.

Moreover the VIX series based on sample entropy as an indicator used on the portfolio iShares MSCI India ETF—iShares MSCI India Small-Cap ETF provided an exceptionally large number of strategies giving higher per transaction returns, which leads us to argue that sample entropy is a more robust measure to decide portfolio rotation signal. The hybrid fund portfolio tested has given abnormally lower number of portfolios with average transaction return greater than 1%. Hence we exclude this type of hybrid portfolios from our further comparative analysis study.

---

[3] For the purpose of computing the Sharpe ratio, we use the risk-free rate (Rf) = 2% for benchmarking or buy-and-hold strategies, and Rf = 0% for zero investment strategies.

4.10 Summary and Concluding Remarks

Our results provide comparative performance of the chosen-best thirty strategies on the basis of the average transaction return with the benchmarks (i.e., Nifty Index, iShare MSCI India ETF, and iShare MSCI India Small-cap ETF). We observe that strategies based on entropy indicators have shown significant risk-adjusted returns as seen by performance and risk measures of cumulative return (particularly, natural logarithmic cumulative returns), standard deviation, Sharpe ratio, coefficient of variation, and maximum drawdown.

Amongst the entropy-based measures, the sample entropy measure has yielded better portfolio rotation timing signals to obtain superior portfolio strategies, when compared to the approximate entropy. Hence, we conclude that the sample entropy computed over the India VIX series is a better measure to signal the expected changes in a financial market index in the context of the Indian stock market. These measures determine asset returns in asset classes, such as equity and exchange-traded funds (ETFs). However, our results cannot support using these strategies for hybrid portfolios and portfolios consisting debt securities as significantly.

This approach provides practitioners including portfolio managers and investors a potential feasible alternative to traditional portfolio rotation strategies. Our proposed approach helps in enhancing portfolio returns and achieving outstanding results along with mitigating risk in the context of the Indian stock market. Our results need to be taken with care, however, as these strategies may be sample-specific and, thus, could benefit from more back testing on other sample periods and sub-periods. As a caution, the robustness of our methodology would benefit from testing it on other markets and asset classes.

Appendix 1: Trading Strategies Using India VIX Change and the Entropies (Both Approximate and Sample Entropies) Tested Through Simulations

See Tables 4.11 and 4.12.

Table 4.11 Trading strategies based on sample entropy and approximate entropy

10d.25sd	1d1.25sd	20d-1sd	30d.75sd	40d-1.75sd	5d.5sd	60d1.5sd
10d-0.25sd	1d-1.25sd	20d-2sd	30d-0.75sd	40d1sd	5d-0.5sd	60d-1.5sd
10d.5sd	1d1.5sd	2d.25sd	30d1.25sd	40d-1sd	5d.75sd	60d-1.75sd
10d-0.5sd	1d-1.5sd	2d-0.25sd	30d-1.25sd	40d-2sd	5d-0.75sd	60d1sd
10d.75sd	1d-1.75sd	2d.5sd	30d1.5sd	50d.25sd	5d1.25sd	60d-1sd
10d-0.75sd	1d1sd	2d-0.5sd	30d-1.5sd	50d-0.25sd	5d-1.25sd	60d-2sd
10d1.25sd	1d-1sd	2d.75sd	30d-1.75sd	50d.5sd	5d1.5sd	90d.25sd
10d-1.25sd	1d-2sd	2d-0.75sd	30d1sd	50d-0.5sd	5d-1.5sd	90d-0.25sd
10d1.5sd	20d.25sd	2d1.25sd	30d-1sd	50d.75sd	5d-1.75sd	90d.5sd
10d-1.5sd	20d-0.25sd	2d-1.25sd	30d-2sd	50d-0.75sd	5d1sd	90d-0.5sd
10d-1.75sd	20d.5sd	2d1.5sd	40d.25sd	50d1.25sd	5d-1sd	90d.75sd
10d1sd	20d-0.5sd	2d-1.5sd	40d-0.25sd	50d-1.25sd	5d-2sd	90d-0.75sd
10d-1sd	20d.75sd	2d-1.75sd	40d.5sd	50d1.5sd	60d.25sd	90d1.25sd
10d-2sd	20d-0.75sd	2d1sd	40d-0.5sd	50d-1.5sd	60d-0.25sd	90d-1.25sd
1d.25sd	20d1.25sd	2d-1sd	40d.75sd	50d-1.75sd	60d.5sd	90d1.5sd
1d-0.25sd	20d-1.25sd	2d-2sd	40d-0.75sd	50d1sd	60d-0.5sd	90d-1.5sd
1d.5sd	20d1.5sd	30d.25sd	40d1.25sd	50d-1sd	60d.75sd	90d1.75sd
1d-0.5sd	20d-1.5sd	30d-0.25sd	40d-1.25sd	50d-2sd	60d-0.75sd	90d1sd
1d.75sd	20d-1.75sd	30d.5sd	40d1.5sd	5d.25sd	60d1.25sd	90d-1sd
1d-0.75sd	20d1sd	30d-0.5sd	40d-1.5sd	5d-0.25sd	60d-1.25sd	90d-2sd

Note: In the above table, d stands for number of days of holding period, sd denotes the standard deviation, and p stands for a percentage change in the India VIX levels

Table 4.12 Trading strategies based on India VIX changes

10d-0p	1d30p	20d70p	30d30p	40d70p	5d-10p	60d40p
10d10p	1d-30p	20d80p	30d-30p	40d80p	5d20p	60d50p
10d-10p	1d40p	20d90p	30d40p	40d90p	5d-20p	60d60p
10d20p	1d50p	2d-0p	30d50p	50d-0p	5d30p	60d70p
10d-20p	1d60p	2d10p	30d60p	50d10p	5d-30p	60d80p
10d30p	1d70p	2d-10p	30d70p	50d-10p	5d40p	60d90p
10d-30p	1d80p	2d20p	30d80p	50d20p	5d50p	90d-0p
10d40p	1d90p	2d-20p	30d90p	50d-20p	5d60p	90d10p
10d50p	20d-0p	2d30p	40d-0p	50d30p	5d70p	90d-10p
10d60p	20d10p	2d-30p	40d10p	50d-30p	5d80p	90d20p
10d70p	20d-10p	2d40p	40d-10p	50d40p	5d90p	90d-20p
10d80p	20d20p	2d50p	40d20p	50d50p	60d-0p	90d30p
10d90p	20d-20p	2d60p	40d-20p	50d60p	60d10p	90d-30p
1d-0p	20d30p	30d-0p	40d30p	50d70p	60d-10p	90d40p
1d10p	20d-30p	30d10p	40d-30p	50d80p	60d20p	90d50p
1d-10p	20d40p	30d-10p	40d40p	50d90p	60d-20p	90d60p
1d20p	20d50p	30d20p	40d50p	5d-0p	60d30p	90d70p
1d-20p	20d60p	30d-20p	40d60p	5d10p	60d-30p	90d80p
						90d90p

Note: In the above table, d stands for number of days of holding period, sd denotes the standard deviation, and p stands for a percentage change in the India VIX levels

References

1. Mini PV (1995) Keynes; investments: their relation to the general theory. Am J Econ Sociol 54(1):47–56
2. Skidelsky R (1992) John Maynard Keynes: the economist as saviour, 1920–1937. Allen Lane, UK
3. Chandra A, Thenmozhi M (2015) On asymmetric relationship of India volatility index (India VIX) with stock market return and risk management. Decision 42(1):33–55
4. Jadhao G, Chandra A (2017) Application of VIX and entropy indicators for portfolio rotation strategies. Res Int Bus Financ 42(3):1367–1371
5. Efremidze L, DiLellio JA, Stanley D (2013) Using VIX entropy indicators for style rotation timing. J Invest 23(3):130–143
6. Arshanapalli BG, Switzer LN, Panju K (2004) Equity-style timing: a multi-style rotation model for the Russel large-cap and small-cap growth and value style indexes. J Portfolio Manage 8(2):9–23
7. Holmes K, Faff R (2008) Style drift, fund flow and fund performance: new cross-sectional evidence. Financ Serv Rev 16:55–71
8. Maio P (2013) The "Fed model" and the predictability of stock returns. Rev Financ 17(4):1489–1533
9. Puttonen V, Seppä T (2007) Do style benchmarks differ? J Asset Manage 7:425–428
10. Copeland MM, Copeland TE (1999) Market timing: style and size rotation using VIX. Financ Anal J 55:73–81
11. Boscaljon B, Filbeck G, Zhao X (2011) Market timing using the VIX for style rotation. Financ Serv Rev 20:35–44
12. Pincus S (2008) Approximate entropy as an irregularity measure for financial data. Economet Rev 27:329–362
13. Chang P-C, Fan C-Y, Lin J-L (2011) Trend discovery in financial time series data using a case based fuzzy decision tree. Expert Syst Appl 38(5):6070–6080
14. Arak M, Mijid N (2006) The VIX and VXN volatility measures: fear gauges or forecasts? Deriv Use Trading Regul 12:14–27
15. Goldwhite P (2009) Diversification and risk management: what volatility tells us. J Investing 18(3):40–48
16. Pincus S, Kalman RE (2004) Irregularity, volatility, risk, and financial market time series. Proc Natl Acad Sci USA 101:13709–13714
17. Maasoumi E, Racine J (2002) Entropy and predictability of stock market returns. J Econometrics 107(1–2):291–312
18. Molgedey L, Ebeling W (2000) Local order, entropy and predictability of financial time series. Eur Phys J B Condens Matter Complex Syst 15:733–737
19. Bentes SR, Menezes R, Mendes DA (2008) Long memory and volatility clustering: is the empirical evidence consistent across stock markets? Phys A Stat Mech Appl 387(15):3826–3830
20. Richman JS, Moorman JR (2000) Physiological time-series analysis using approximate entropy and sample entropy. Am J Physiol Heart Circ Physiol 278(6):H2039–H2049
21. Thuraisingham RA, Gottawald GA (2006) On multiscale entropy analysis for physiological data. Phys A Statis Mech Appl 366(1):323–332
22. Bose R, Hamacher K (2012) Alternate entropy measure for assessing volatility in financial markets. Phys Rev E 86:056112
23. Shaikh I (2018) Investors' fear and stock returns: evidence from national stock exchange of India. Eng Econ 29(1)
24. Chan KC, Chen N-F (1991) Structural and return characteristics of small and large firms. J Financ 46(4):1467–1484
25. Fama EF, French KR (1993) Common risk factors in the returns on stocks and bonds. J Financ Econ 33(1):3–56

26. Haugen RA, Baker L (1996) Commonality in the determinants of expected stock returns. J Financ Econ 41(3):401–439
27. Lakonishok J, Shleifer A, Vishnu RW (1994) Contrarian investment, extrapolation, and risk. J Financ 49(5):1541–1578
28. Merton R (1980) On estimating the expected return on the market: an exploratory investigation. J Financ Econ 9:323–361
29. Nelson DB (1991) Conditional heteroskedasticity in asset returns: a new approach. Econometrica 59(2):347–370
30. Fleming J, Ostdiek B, Whaley RE (1995) Predicting stock market volatility: a new measure. J Futures Markets 15(3):265–302
31. Levis M, Liodakis M (1999) The profitability of style rotation strategies in the United Kingdom. J Portfolio Manage 26(1):73–86
32. Bauer R, Guenster N, Otten R (2004) Empirical evidence on corporate governance in Europe: the effect on stock returns, firm value and performance. J Asset Manage 5:91–104
33. Orozco JL (2016) Portfolio asset allocation on a sector rotation strategy triggered by Fed's discount rate. Appl Econ Theses 16
34. Kinlaw W, Kritzman M, Turkington D (2019) Crowded trades: implications for sector rotation and factor timing. J Portf Manage 45(5):46–57
35. Efremidze L, DiLellio JA, Stanley DJ (2014) Using VIX entropy indicators for style rotation timing. J Investing 23(3):130–143
36. Kanojia S, Arora N (2018) Investments, market timing, and portfolio performance across Indian bull and bear markets. Asia-Pacific J Manage Res Innov 13(3–4):98–109
37. Liu F, Tang X, Zhou G (2019) Volatility-managed portfolio: does it really work? J Portf Manage 46(1):38–51
38. Lozza SO, Angelelli E, Ndoci A (2019) Timing portfolio strategies with exponential Lévy processes. CMS 16:97–127
39. Arshanapalli B, Switzer LN, Hung LTS (2004) Active versus passive strategies for EAFE and the S&P 500 indices. J Portf Manage 28:17–29
40. Bagchi D (2012) Cross-sectional analysis of emerging market volatility index (India VIX) with portfolio returns. Int J Emerg Mark 7(4):383–396
41. Chakrabarti P, Kumar KK (2020) High-frequency return-implied volatility relationship: empirical evidence from Nifty and India VIX. J Dev Areas 54(3)
42. Aksaraylı M, Pala O (2018) A polynomial goal programming model for portfolio optimization based on entropy and higher moments. Expert Syst Appl 94:185–192

Chapter 5
Asymmetric Spillovers Between the Stock Risk Series: Case of CESEE Stock Markets

Tihana Škrinjarić

Abstract Asymmetric behaviour of stock market variables is nothing new in theory and practice. However, the methodology which captures this behaviour should be developed and used properly, so that potential investors can benefit from it in the best possible way. This chapter deals with asymmetric spillovers between risk series of selected Central Eastern and South-Eastern European markets. There are several contributions to the literature. Firstly, the chapter gives an extensive and exhausting literature overview so that the advantages and shortfalls previous studies can be observed. This is helpful for future research and empirical applications. Secondly, this chapter deals with 15 CESEE at once, so that the full potential of the mentioned markets can be exploited. Thirdly, the empirical part of the analysis extensively deals with the simulation of investment strategies which are based on the results of the estimation, so that (potential) investors could obtain insights into the usefulness of the applied methodology. The methodology consists of VAR (vector autoregression) modelling with the inclusion of the spillover index methodology Diebold Yilmaz (Econ J 119(534):158–171 2009 [36]], (Int J Forecast 28(1):57–66 2012 [38]) and the extension to asymmetric spillovers in Baruník et al. (J Financial Markets 27:55–78 2016 [9]). The results of the analysis indicate that using the output from the asymmetric spillover behaviour, the investor could obtain better portfolio risk and return compared to the strategies when such information is not included.

Keywords Spillovers · Realized volatility · Emerging markets · Vector autoregression

---

T. Škrinjarić (✉)
Department of Mathematics, Faculty of Economics and Business, University of Zagreb, Zagreb, Croatia
e-mail: tskrinjar@net.efzg.hr

5.1 Introduction

Today's portfolio management is a very difficult task to conduct. The investor needs to have different knowledge. Mostly in the area of finance and quantitative tools, as well as in tracking major economic, political, and other movements in many different countries, etc. The stylized facts of financial, and especially stock markets include the asymmetric behaviour of the return series, their inter-relationships, and especially the risk series asymmetries as well [73]. Thus, many models have been developed to capture such behaviour, as previous models which were based on the assumption on the symmetric behaviour is very restrictive [58]. It is not rational to assume that investors react equally to positive versus negative news. Quite the opposite is true, the stock market reacts more to bad news [81]. Furthermore [117] found that if the return and volatilities series on stock markets exhibit negative correlation, the market reacts stronger to negative news, which indicates asymmetric behaviour.

This chapter is focusing on the asymmetries in such behaviour. There are several reasons on why. Such behaviour is widely documented in the literature. Early work includes research that found asymmetric behaviour of volatility of stocks and portfolios [20, 40, 44, 49, 87, 98]. Newer findings include findings such as: higher volatility during market upturns and downswings [16, 67, 117], return series and innovations in the expected volatility being negatively correlated [34], asymmetric attention on stock markets causes asymmetric volatility behaviour [4], asymmetric correlations between return series decrease portfolio diversification [2], and the literature is still growing. Since the spillovers of volatilities between asset returns are important within the risk valuation and diversification purposes [46], the asymmetries should be modelled and recognized properly. Although many (M)GARCH (multivariate generalized autoregressive conditional heteroskedasticity model) have been developed and empirically tested over the years,[1] they still have some disadvantages in applications. Some of them include the problem of limited number of variables that could be included in the analysis, not being able to capture asymmetric spillover in a way the methodology in this chapter will be able to capture, etc. Thus, the main methodology which will be utilized in this chapter is the VAR (vector autoregression) model with the extension of the [35, 36] spillover index, which is able to capture the spillovers of shocks between variables of interest. However, since the original spillover methodology in the papers of [35, 36] cannot capture the asymmetric spillovers, we follow the extension of those indices to the asymmetric spillovers as in [10]. In this paper, the authors distinguish good and bad volatility series, which are the main definitions of asymmetric behaviour of the volatility series and spillovers between them. Thus, the main goal of this chapter is to explore asymmetric spillovers between volatility series of selected stock markets, to achieve better portfolio diversification over time. The main markets which will be in focus are Central and Eastern European (CEE) and South and Eastern European (SEE) markets, alongside several developed ones (German, Russian and the US one), which were found in previous literature to be mostly related to the CEE/SEE ones. Since previous literature mostly stops after the

---

[1] As a starting point, please see [13, 101].

estimation of the econometric model, this chapter fills the gap in the literature by utilizing the results of the estimation in the second step of the empirical analysis. Namely, the information about the spillover behaviour is then used to form trading strategies so that the benefits in terms of portfolio risk and return can be obtained.

There are several reasons why the existing research should be extended with this one which focuses on CEE and SEE markets. Trenca et al. [111] state that the mentioned markets are less researched compared to the mature ones, and the investors on such markets are often characterized by less-informed investors. By extending the existing research with this one, such investors can better utilize the insights into the characteristics of risk and returns, as well as shock spillovers from one market to another. This helps the diversification purposes for international investors. The country-specific factors which dominate on CEE and SEE markets still exist [8], and as [50] state: attractive risk-adjusted return. Some recent research even found that the more developed and mature markets are not strongly integrated with CEE and SEE one: [26, 32, 51, 68]. Özer et al. [123] even found that the intra-regional diversification benefits still exist between the mentioned markets. Not only from the investor's point of view, but there are also macroeconomic reasons why these markets should be investigated. Markets of CEE and SEE countries still have untapped economic potentials, which could be exploited [28, 92]. If we get a better understanding of such markets, better coordination between them can be made, which can enhance the further development of such markets [43, 94]. Finally, previous research found that there exists asymmetric behaviour in the case of CEE markets [25]. Thus, further and deeper analysis is needed.

The contributions of this research are as follows. Firstly, the literature overview section is exhaustive in terms of including all of the papers which were found during the research and writing process, with a critical overview. This is usually not the case in existing research. Secondly, the empirical part of the research tried to include the longest time possible, with the inclusion of as many CEE and SEE countries possible. In that way, different market dynamics and political and economic changes over time were included in the analysis. This gives better insights into the usefulness of such an approach in dynamic portfolio management. Thirdly, the results obtained from the empirical estimation are utilized in simulations of possible investment strategies, so that comparisons can be made in terms of portfolio characteristics which are important for investors. Using the results from estimations to explore investment possibilities is also rarely found in the literature.

The rest of this chapter is structured as the following description goes. The second section focuses on a critical overview of previous related research in terms of methodology, similar stock markets, and similar research questions. The third section describes the main methodology used in the study. The empirical analysis is given in the fourth section, which is divided into several subsections: main estimation results, robustness checking and implementation in simulating the investment strategies. The final, fifth section concludes the chapter.

5.2 Related Literature Review

The interconnectedness, spillovers, and contagion between financial markets have been researched in the last two decades fairly extensively. The most common methodological approaches include the multivariate GARCH, vector autoregression (VAR), cointegration approach (Johansen approach or VEC, vector error correction) and related models and methods. This chapter belongs to the group of research that focus on spillover methodology. Thus, the majority of the overview will be focused on such related work. Minor comments will be given on the research which deals with markets that are in the focus of this research but do not answer specific questions that are in focus here. Based on this section, the analysis in the third section could be done with better understanding on the relationship between the markets of interest in specific periods.

5.2.1 (M)GARCH Approaches to Modelling

The first group of papers is probably the most common one in literature. Here, the MGARCH approach is utilized to obtain information about the correlations between markets and shock spillovers between risk series. Some authors observe the interactions between more developed markets and the CESEE ones; others focus solely on the CESEE markets. The findings have changed over the years, due to some markets and economies being more connected to the more developed ones more (due to entering the European Union, European monetary union, etc.). Some earlier work includes the following. We do not include earlier research, as not much data was available for many of the CEE and SEE markets, due to them not being established. Other earlier work which includes some of the CEE and/or SEE markets is as follows, where interested readers can obtain information about earlier developments regarding the topics investigated here and in related literature: [12, 15, 23, 27, 42, 48, 49, 56, 62, 63, 70, 75, 84–86, 96, 97, 106, 110, 115, 116, 122].

Bein and Tuna [14] examined the Hungarian, Polish and Czech markets to empirically study the impact of the sovereign debt crisis of GIPSI countries. Furthermore, the UK, German and French markets were included as well, with the DCC multivariate model as the main methodological approach. The included period which was divided before and after the crisis ranges from 2009 to 2013. Here, the main findings indicated that GIPSI countries except for Greece have significant spillovers to CEE countries. Similar is true for the three developed markets and their correlation with the CEE ones. The correlations were found to be great even before the sovereign debt crisis. Thus, the authors concluded that the three observed CEE countries provide fewer diversification benefits for investors who also observe other mentioned markets in this analysis. Amonlirdviman and Carvalho [119] observed the period from 2010 until 2015 and the returns and risks of Austrian, German,

Russian, Polish, and Turkish markets via MARMA-GARCH (multivariate autoregressive moving average) models. The main idea here vas to investigate the linkages among returns and transmission of volatilities between the observed markets. The results indicated that there exists a significant co-movement of return series, with the Turkish and Russian markets being more volatile compared to the rest of them. The Polish market was the only one to be found which had the characteristic of a positive correlation between the expected risk and expected return. Dedi and Škorjanec [32] conducted a similar analysis, with the same methodology, but for the following markets: Croatian, US, UK, German, Austrian, Czech, Hungarian and Polish (period: 2011−2017). The authors did find some significant co-movements between the observed markets, but still, some diversification possibilities were found as well. Univariate and multivariate GARCH models were used in the study of [124] over the US, UK, Hungarian, German, Austrian, Polish, and Czech stock market data. The authors considered a long period of analysis (1997–2016). In that way, different events could have been taken into consideration. This research was more focused on the Croatian data and how different movements affected the integration of the Croatian market to the rest of the mentioned ones. Some of the findings include the following. The Russian crisis caused the Croatian market to disintegrate from other markets; with an increase of the correlations of the Croatian market to others in the post-crisis period. A BEKK model was utilized in [72] on the data for the Croatian, Great Britain, Hungarian, American, German, Serbian and Montenegrin stock markets to observe the spillovers from more developed to the Western Balkan markets. In the observed period from 2005–2015, the authors obtained the following results. There exist regional spillovers between the Serbian, Croatian and Montenegrin stock markets, as well as those markets and the more developed ones. However, the responses of the three Balkan markets were in some cases late, which indicates possible exploitation purposes by the investors. A newer study of [64] applied the BEKK, CCC, and DCC models over data for Hungarian, Romanian, Croatian, Czech, and Polish markets (from 2008 to 2017). The correlations between the observed markets were found to be significant, with cross-volatility spillovers being greater than own-volatility ones. This indicated lower diversification possibilities on the observed markets.

As can be seen in this first group of papers, authors often observe only several CEE and/or SEE markets at once. Comparisons cannot always be made, as different periods are used in the empirical analyses. Furthermore, the main approach is to observe how much do movements in more developed markets spillover and shape return and risk series of the CEE and SEE markets. These movements are measured in the GARCH specification of the model, where risk series from more developed markets are exogenous variables. Less work is done which focuses on the intraregional spillovers. Furthermore, the research which focuses solely on the CEE and/or SEE markets also does not include the majority of the countries which belong to this region. The reasoning could lie in data unavailability. Only several authors utilize the asymmetric approach when looking at the research from the methodological point of view. The few which have utilized asymmetric or nonlinear models find that the behaviour of the observed variables and their interactions is not symmetrical in good and bad times. Finally, the analysis usually stops when the main econometric model is estimated, with some recommendations for international investors, without trying to implement the results in a way in which investors would be interested.

5.2.2 Causality Testing and Cointegration

Another popular approach in empirical modelling is the VAR model, or sometimes VEC, where authors test for cointegration between stock markets; look at how the stock markets interact with macroeconomic variables, etc. The authors usually use monthly data, as the VAR models ask for different data characteristics compared to the (M)GARCH models.

Early work includes: [30, 31, 33, 47, 52, 61, 69, 71, 79, 80, 83, 86, 88, 89, 95, 104, 105, 118–109, 114, 115, 121]. Stoica et al. [102] focused on the CEE and SEE markets (Czech Republic, Bulgaria, Slovakia, Poland, Romania), and the German and French as more developed ones. The observed period was from 2000 to 2010, and GARCH models with the VAR ones were combined. The results, again, are not surprising: the CEE and SEE markets were receivers of shocks from the German and French markets, except for the Slovakian market, which was already found in other related papers. Moreover, the intra-region spillovers of shocks were greater than the innovation shocks received from the German and French markets. A combination of the Markov regime switching (MRS) methodology with the VEC model is found in [93]. Here, authors observe the interaction between the Balkan markets (Croatian, Bulgarian, Bosnian and Serbian) with the US, UK and Japanese market. For the observed period (from 2009 to 2016), the results have shown that there exist a short- and long-term relationship between the observed markets. However, due to the changes of the regimes, small diversification possibilities exist sometimes. [26] explained why the research should focus on asymmetric behaviour on financial markets, with an empirical study of the asymmetric causality test for the Hungarian, Slovakian, Czech, Turkish, German and Polish stock markets (period: 1995–2014). Main findings of this research have shown that there exists asymmetric causality from the Czech to the Hungarian and Polish markets; as well as from the German to the rest of the markets. Some benefits of diversification still exists which could be interesting to the international investors. Finally, a newer study of [5] uses the Zivot-Andrews unit root test (for structural breaks in the series) so that the pre- during and post-crisis period can be detected in Balkan stock indices (Serbian, Macedonian and Croatian). The whole time span was from 2006 until 2017. Simple correlation and Granger tests were employed in each subsample. The results of this analysis indicate that the Croatian market causes the other two ones, with bi-directional causation found for all of the pairs of return series. However, the analysis was static and symmetric. The most recent study is of [22], in which authors combined the PCA (principal component analysis), VAR and asymmetric DCC GARCH models. The sample included the CEE markets of Lithuania, Bulgaria, Latvia, Croatia, Estonia, Romania, Hungary, Poland and Czech Republic, combined with the German, American and the UK markets (period: 2000–2016). The dynamic part of the analysis indicates short-run cointegration relationships between the CEE and more developed markets, with correlation analysis showing that contagion effects were present in data as well.

This group of papers mostly utilizes monthly data, as VAR and VEC models assumes some data characteristics which should hold which are different compared to daily data and GARCH models. Some authors provide cointegration tests which are focused on long-run relationships between markets, but use short time spans for the analysis (e.g. several years). This approach is either misunderstood in the research or some other problems exist which forced the authors to utilize a methodology which should be used on a longer time span. Furthermore, all of the research within this group does not simulate any trading strategy at all, which could have been based on the results of the estimated models. The results here are not surprising, that the more developed and mature markets affect the CEE and/or SEE ones, with the cointegration between them being significant, which diminishes diversification possibilities. Although earlier work found that the correlations are lower and cointegration was lower as well, several major political and economic events have contributed to the greater integration between the CEE/SEE markets and the more developed ones, but between the CEE/SEE markets as well. Majority of research has a static approach of the analysis, which means that the entire sample is used for the purpose of model estimation. Changes in the economies and on financial markets for surely affect the coefficients of econometric models. Thus, some dynamics should be included as well. Although the diversification benefits have decreased over the decades, there still exist some, which should be properly investigated, which is enabled via the next step in the VAR modelling, the spillover index estimation. The next subsection gives an overview of the related work in this area.

5.2.3 Spillover Methodology Approach

The most interesting group of papers is the following one, which utilizes a similar methodology, with a focus on the asymmetries in the data as well. The VAR models are used to estimate [35, 36] spillover indices. Furthermore, the majority of research observes spillovers between countries (i.e. country indices) for international diversification purposes. Fewer papers exist which observe less developed markets and spillovers between the different types of financial assets. Different types of financial markets have been explored, such as the stock market, sovereign debt market, financial institutions (mostly banks). Furthermore, initial papers focused on more developed countries and regions. Newer ones include developing markets as well. Newer studies started to include markets such as the CEE and SEE ones. But still, there is more work to be done, especially regarding the developing markets.

The first paper on this topic is the seminal paper of [35], where the spillover index was developed. The research was divided into two parts, the first one theoretically develops the index with interpretations and the second part is the empirical analysis. The authors have observed 19 different stock markets (1992–1997 was the observed period), where separate VAR models and indices were estimated for the return and risk series. The used stock markets were more developed ones and none of the CEE and SEE markets were included in the study. The main results indicated that the shocks in return series have greater spillovers between the countries when compared to shocks in risk series. Yilmaz [120] has focused on ten East Asian stock markets and used the same methodology as the previously mentioned paper. For the time from 1992 until 2009, the author obtained the following results. Spillovers between the return and risk series change over time, with an increase of the spillovers themselves during market crashes and crises. Diebold and Yilmaz [38] applied the same methodology for selected markets (Argentinean, Chilean, US, Mexican, and Brazilian markets), where the same conclusions were found as in [120]. These three papers estimated the original VAR model. Then, the Choleski decomposition was made in the second step. Since these papers represent the beginning of the spillover methodology, they are relatively simple compared to the later ones. However, they represent the foundation for later work, which has extended the analysis to tailor specific questions. Due to the drawbacks of the Choleski decomposition (variable ordering in the procedure) [36] have developed a G-VAR (generalized) framework with nonlinear impulse response functions (IRFs) and spillover indices. The focus was on the US bond, foreign exchange, stock, and commodities market, for the period from 1999 to 2010. Here, the main findings indicated that the volatility spillovers between these markets are limited until the global financial crisis. This research was useful for investors who aim to form portfolios with different financial assets, compared to the previous three, in which spillovers were observed between countries. Diebold and Yilmaz [37] was another paper that contributed to the methodological part of the spillover indices, in which the authors propose connectedness measures based on network topology. In the second part of the paper, an empirical analysis is provided, where the US financial institutions' stock return and volatility series (for the period 1999–2010) is used. Both static and dynamic approaches were made in the research.

A work that is close to this research is [66], in which the author utilized the spillover indices methodology for the return series and BEKK GARCH model for volatility series for the CEE, SEE, and developed markets. The sample included the Macedonian, Polish, Hungarian, Croatian, Serbian and Czech markets, alongside the German and UK one (from 2005 to 2014). Main findings of this research are: correlations between the markets are high and stable; with an increase during the financial turmoil; the Macedonian and Serbian markets were mostly isolated from more developed markets; the spillover index was greatest during financial disruptions on the markets; and the CEE and SEE markets were net receivers of shock spillovers compared to developed markets, with the Serbian and Macedonian markets being greatest net receivers from others. Alter and Beyer [1] utilized the spillover index methodology to estimate spillovers between sovereign credit markets and banks in the Euro area. This research is important for policymakers and the creditworthiness of countries and their bank systems. The period of the analysis was from 2009 to 2012, with the extension of the [36] methodology to the estimation of contagion indices. The estimated contagion index has fluctuated within a stable interval over the observed period, with high values around important policy events (specific dates in 2010, 2011, and 2012). CDS (credit default swap) spreads were in the centre of attention in [60]. CEE and SEE countries were included in the analysis, with the rest of the European countries (period: 2007–2012, in total 24 countries). Panel GLS (generalized least squares) error correction was applied, with the spillover index methodology. The authors observed the dynamics of the fundamentals which have affected the CDS spreads, with the spillover indices being estimated between the countries so that the net emitters and receivers of shocks in Europe can be detected. Advanced Europe has provided shocks to the Euro area periphery countries rather than vice versa. A new study of CDSs is found in [21], for more interested readers. The African stock markets were in the focus of [103]. The authors analysed how did the regional, global, commodity, and exchange rate shock spillover to the selected African stock markets. Thus, the analysis can be extended not solely between stock markets, but between financial markets and macroeconomic variables as well. Research of [39] included all 28 EU members in the VAR model, by observing return and volatility spillovers among the countries for the period from 2005 to 2015. The analysis was done from a macroeconomic perspective, with comments on the effects of the Eurozone debt crisis. The results were not surprising; the CEE and SEE stock markets were net receivers of shocks from other EU members. However, there are no simulated trading strategies found in this research.

Regional analysis can be found as well. Belke and Dubova [17] observed the US, Japanese, the UK markets alongside the Euro area stock and bond markets. The US market was found to be dominant as a shock emitter, the international spillovers were found to be very strong across asset classes, and the integration of the markets has increased over time (observed period was: 1995–2016). Volatility spillovers were in the focus of [45]. In this research, the authors extended the spillover index by constructing spillover indices based on the DCC GARCH model. The stock indices of Latin America were used, alongside the US market (from 2003 until 2016) as a control variable for the global factors. Not surprisingly, the spillover indices change substantially over time, especially during the financial crisis and the European sovereign bond crisis. Another paper that focused on the volatility spillovers was [82]. Greek, Irish, Portuguese, Spanish and Italian markets (GIPSI) were in the focus of this research, with the rolling-window spillover index and the network connectedness estimation. Global and regional stock market data was included as well. Authors found strong spillovers in the observed period (2002–2016), with increases during the crises periods, which is in favour of conclusions of financial contagion evidence. Sehgal et al. [99] focused on 13 global financial markets (more developed ones) in the period from 1999 to 2017. Both risk and return series are examined via the spillover methodology, with the block-aggregation technique included. Greater spillovers were found for risk series, with regional integration being stronger compared to global integration. There is less work done regarding the CEE and SEE markets. One recent work includes [127], in which authors utilized the Diebold and Yilmaz spillover index for selected CESEE countries (Bulgaria, Slovakia, Czech Republic, Croatia, Poland, Slovenia, Ukraine, and Hungary), in the period 2012–2019. Some countries were found to be a net emitter of shocks of volatilities across countries, such as Slovenia and the Czech Republic, and some were net receivers (Ukraine and Croatia). Some guidelines are given for international investors based on the results. However, no trading strategies were found in the paper. Some previously mentioned papers have utilized the asymmetric approaches in the analysis, due to financial market characteristics which are often specific, with asymmetric behavior being one of them. Investors do not observe the right and left tail of return distributions in the same way. Thus, using standard deviation as risk measures has been often criticized in the previous literature [29, 41]. That is why research has been emerging which observes asymmetric spillovers, in terms of calculating the realized volatility measure and dividing it into positive and negative volatility (more details will be given in the methodology section).

Baruník et al. [10] suggested using the realized variance as the measure of volatility, with an application on 21 US stocks (observed period: 2004–2011). The research showed how good and bad volatility is transmitted at different magnitudes, which change over time. The intramarket connectedness of the observed stocks has increased during the financial crisis. Baruník et al. [11] observed realized volatility as a risk measure and divided it into bad and good volatility. The intra-day data was used for most traded currencies (Canadian dollar, Australian dollar, Euro, British pound, Swiss franc, and Japanese yen against the American dollar) in the period 2007–2015. The authors found that negative spillovers are related to the sovereign debt crisis of the Eurozone, whereas positive spillovers were correlated to the subprime crisis. Bevilacqua et al. [19] has observed asymmetry in volatility spillovers between the stock, credit, and exchange markets between developed EU countries (Germany, France, Italy, and the UK) and selected CEE ones (Slovakia, Poland, Hungary, and Czech Republic). For the period from 2008 to 2017, the author wanted to explore spillovers in volatility series concerning good or bad news and concerning leverage of the asymmetry. The greatest spillovers were found on the stock markets, with net emitters of volatility being the developed markets. The currency markets were mostly under the effects of the Brexit vote. Finally, asymmetries were found in the study, which indicates that analysis and policymakers alongside international investors should consider such results. BenSaïda e al. [18] has focused on major stock markets (US, Canadian, UK, French, German, Italian, Japanese) for the period 2000–2017. Not surprisingly, the results of this research are similar compared to previous ones, there exist asymmetric spillovers on the observed markets, which are time-varying, with bad events having greater effects on the bad spillovers. Furthermore, the US market was the net emitter of bad shocks in the model, whereas the French and Canadian being the ones that emitted good volatilities to others in the study. Ma et al. [118] have extended the research on the relationship between oil and stock markets by focusing on asymmetric spillovers between the volatility series of oil and stock prices. Again, the good and bad volatilities were defined and observed for the WTI futures prices, S&P 500, and Shanghai stock market indices. This paper included both the VAR model with spillover indices and the asymmetric generalized DCC GARCH model (with robustness checking, with Markov switching VAR). The bad volatility series had greater spillover effects over the observed period (2007–2016), with all of the spillovers being time-varying.

Some recent research is the following. Forbes et al. [59] combined the EPU (economic policy uncertainty) dynamics with the stock market one (the US, Australian market, Canadian, Chinese, Japanese, and the UK one). The idea was to observe if there exist asymmetric spillovers of shocks between the EPU index and stock market volatilities. Again, the bad volatility series were more connected to the EPU indices, especially during the debt crisis and trade negotiations (problematic periods). Finally, a sector analysis was done in [100], which focused on 28 Chinese sectors, for the period from 2000 to 2019. Similar conclusions are obtained here as in previous research. However, no trading strategies were found to be simulated as help for potential investors.

The application of this methodology has extended to other areas in economics, some of which can be found in [7], where authors briefly mentioned different applications [1], where sovereign credit markets and banks of Eurozone were used [24], and Škrinjarić, Lovretin Golubić and Orlović [126], where exchange rate volatilities were observed [74], in which author observed the Eurozone stock, bond, exchange rate, and money markets [112], where authors focus on spillovers between the financial and the real economy, or [126], where authors observe EPU and stock market interactions. It is easily seen that the areas of applications of the spillover methodology are rapidly growing, due to the straightforward interpretations and relatively easy estimations of needed variables. The results can be used in a variety of areas, especially regarding (international) portfolio management and for policymakers as well.

5.3 Methodology Description

5.3.1 VAR Models and Spillover Index

The main methodology applied in this part of the chapter is the VAR (vector autoregression) model, with the extension to the spillover indices. The details are given in [35, 36, 79–78, 113], which this study follows. A VAR model of order p, VAR(p) of N variables can be written in the following form:

$$\boldsymbol{y}_t = \boldsymbol{v} + \boldsymbol{A}_1 \boldsymbol{y}_{t-1} + \boldsymbol{A}_2 \boldsymbol{y}_{t-2} + \cdots + \boldsymbol{A}_p \boldsymbol{y}_{t-p} + \boldsymbol{\varepsilon}_t. \tag{5.1}$$

where $\boldsymbol{y}_t$ is the vector of N variables, $\boldsymbol{A}_i$ are the coefficient matrices, of order N, with $i \in \{1, 2, ..., p\}$, $\boldsymbol{v}$ is the vector of intercepts in the model, with $\boldsymbol{\varepsilon}_t$ being the innovation processes vector in the model. It holds that $E(\boldsymbol{\varepsilon}_t) = \boldsymbol{0}$, $E(\boldsymbol{\varepsilon}_t \boldsymbol{\varepsilon}_t') = \Sigma_\varepsilon < \infty$ and for $t \neq s$ it holds $E(\boldsymbol{\varepsilon}_t \boldsymbol{\varepsilon}_s') = 0$. A compact form of the VAR(p) model can be denoted with a VAR(1) as follows:

$$Y_t = v + AY_{t-1} + \varepsilon_t, \tag{5.2}$$

in which $Y_t = \begin{bmatrix} y_t \ y_{t-1} \ \cdots \ y_{t-p} \end{bmatrix}'$, $v = \begin{bmatrix} v \ 0 \ \cdots \ 0 \end{bmatrix}'$, $A = \begin{bmatrix} A_1 & A_2 & \cdots & A_{p-1} & A_p \\ I_N & 0 & \cdots & 0 & 0 \\ 0 & I_N & & \vdots & \vdots \\ \vdots & & \ddots & \vdots & \vdots \\ 0 & 0 & \cdots & I_N & 0 \end{bmatrix}$ and $\varepsilon_t = \begin{bmatrix} \varepsilon_t \ 0 \ \cdots \ 0 \end{bmatrix}'$. As spillover indices are estimated from IRFs (impulse response function) and the forecast error variance decomposition (FEVD), the MA(∞) representation of model (5.2) is observed as following:

$$Y_t = \mu + \sum_{i=1}^{\infty} A^i \varepsilon_{t-i}, \tag{5.3}$$

where it is assumed that the model is stable ($\det(I_{Np} - Az) \neq 0$ for $|z| \leq 1$), and $\mu \equiv (I_{Kp} - A)^{-1} v$, i.e.:

$$Y_t = \Phi(L)\varepsilon_t, \tag{5.4}$$

where $\Phi(L)$ is the polynomial of the lag operator L, in which the values $\phi_{jk,i}$ are the impulse responses of variables in the model. The variance-covariance matrix Σ_ε can be orthogonalized via the Choleski decomposition procedure (such that $E(P^{-1}\varepsilon_t P^{-1}\varepsilon_s') = 0$, where P^{-1} is a lower triangular matrix). Another approach is to utilize the VAR framework in which decompositions of forecast-error variance are made based on generalized IRFs and GFEVDs (generalized forecast error variance decomposition[2]). As the generalized approach does not depend on variable ordering, majority of existing research uses this approach, as well as this chapter. In the first step, the h-step forecast of every variable in the model is done, by calculating the difference:

$$Y_{t+h} - E(Y_{t+h}) = \sum_{i=0}^{h-1} \Phi_i \varepsilon_{t+h-i}, \tag{5.5}$$

and in the next step, the mean squared error of (5.5) is estimated:

$$E[Y_{t+h} - E(Y_{t+h})]^2 = \sum_{k=1}^{N} (\phi_{jk,0}^2 + \cdots + \phi_{jk,h-1}^2). \tag{5.6}$$

---

[2] For details, please see [91].

Now, the variance decomposition of every element i in vector Y_t, $\omega_{jk,h}$, is estimated as:

$$\omega_{jk,h} = \sigma_j^{-1} \sum_{i=0}^{h-1} \left(e'_j \Phi_i \sum_\varepsilon e_k\right)^2 / e'_j \Phi_i \sum_\varepsilon \Phi'_i e_j, \tag{5.7}$$

where the numerator is the shocks-contribution which happen in variable k to the j-th variable forecast error, the denominator is the MSE (mean squared error) forecast of the j-th variable, e_j and e_k are the j-th and k-th column in matrix I_{Np}.

Based on all decompositions which are estimated via (5.7), the [35, 36] spillover index, SI, is estimated as the fraction in the total forecast error variance of the h-step ahead forecast error variance of the j-th variable which is a result of shocks in the rest of the variables in the model:

$$S = \sum_{\substack{j,k=1 \\ j \neq k}}^{N} \omega_{jk,h} \bigg/ \sum_{i=0}^{h-1} \sum_{j,k=1}^{N} \omega_{jk,h} 100\%. \tag{5.8}$$

Besides the total spillover index in (5.8), the directional "to" and "from" indices can be estimated, as follows:

$$S_{j\cdot,h} = \frac{1}{N} \sum_{\substack{k=1 \\ j \neq k}}^{N} \omega_{jk,h} 100\%, \tag{5.9}$$

and

$$S_{\cdot j,h} = \frac{1}{N} \sum_{\substack{k=1 \\ j \neq k}}^{N} \omega_{k j,h} 100\% \tag{5.10}$$

Furthermore, the net spillover index for every variable can be estimated as the difference between the "to" and "from" indices; as well as the pairwise indices can be estimated between every pair of two variables in the model. All of the indices can be put in a table, called the spillover table, so that the interpretations can be easier. Finally, the dynamic analysis is often conducted, so that the changes over time can be observed. The procedure is to estimate a rolling window VAR model, with overlapping windows and estimate the spillover indices for every window.

5.3.2 Asymmetric Spillovers

Since this chapter allows for an asymmetric behaviour of volatility spillovers, we follow [10, 11] and [19], where the following measures are used as developed in [3] and Barndorff-Nielsen [9]. Based on the intraday data on returns r_i, the realized variance on day t can be calculated as:

$$RV_t = \sum_{i=1}^{n} r_i^2. \tag{5.11}$$

This realized variance can be decomposed into realized semivariances due to positive or negative returns:

$$RV_t^+ = \sum_{i=1}^{n} r_i^2 I_{r_i > 0} \tag{5.12}$$

and

$$RV_t^- = \sum_{i=1}^{n} r_i^2 I_{r_i \leq 0}, \tag{5.13}$$

where $I_{r_i > 0}$ and $I_{r_i \leq 0}$ are indicator functions which refer to returns which are positive or non-positive respectively. As we do not have intraday data, the [90] measure of range volatility will be used instead of formula (5.11):

$$RV_t = \frac{(\log H_t - \log L_t)^2}{4 \log 2}, \tag{5.14}$$

where H and L denote highest and lowest price respectively. This measure is an unbiased estimator of daily volatility and efficient one.

Now, the data is separated based on positive and negative semivariances, and separate VAR models are estimated for both, with the spillover indices as described in the previous section. If we denote with S^- and S^+ the total spillover indices for the negative and positive semivariance models respectively, then the spillover asymmetry measure SAM can be calculated as:

$$SAM = S^+ - S^-, \tag{5.15}$$

which can, of course be estimated on a rolling window basis. If the value of SAM is zero, then there are no asymmetries in spillovers. Positive values mean that the spillovers from RV^+ are larger than RV^- (and vice versa). Similarly, the directional spillovers can be observed as well, to see the degree of the asymmetric behaviour. Denote with $S_{j\cdot,h}^+$ and $S_{j\cdot,h}^-$ the "to" positive and negative spillovers. Then, the

$SAM_{j\cdot,h}$ is the asymmetry measure for directional spillovers received by variable j from all other variables, calculated as:

$$SAM_{j\cdot,h} = S^+_{j\cdot,h} - S^-_{j\cdot,h} \tag{5.16}$$

The "from" measure $SAM_{\cdot j,h}$ is calculated in a similar fashion.

5.4 Empirical Results

5.4.1 Variable Description

For the empirical analysis, daily data on the following indices were collected from [65]: BIRS (Bosnia and Herzegovina), BSE SOFIX (Bulgaria), CROBEX (Croatia), PX (Czech Republic), DJ Estonia Total Market (Estonia), DAX (Germany), BUMIX (Hungary), DJ Latvia Total Market (Latvia), DJ Lithuania Total Market (Lithuania), WIG (Poland), BET (Romania), Belex 15 (Serbia), SBITOP (Slovenia), MOEX (Russia), Cyprus Main Market (Cyprus), PFTS (Ukraine) and S&P500 (USA). The main idea was to collect as much data possible, concerning the time-series aspect, but countries as well. Thus, the mentioned countries had the most data available, with a starting point of 1 December 2020. The last data point is for the date 16 July 2020. The developed markets of Russia, the USA, and Germany were the ones found in the literature review which affect the CEE and SEE markets the most. All daily indices data is used to calculate daily return series, which were used to calculate the average monthly return series, as well as monthly RV series (realized volatility, positive and negative realized volatility). Every series has 94 data points in total. Descriptive statistics for all series are given in Table 5.1.[3] For easier readings throughout the results, all of the indices are named after the county which they represent: BIH (Bosnia and Herzegovina), BULG (Bulgaria), CRO (Croatia), CYP (Cyprus), CZE (Czech Republic), EST (Estonia), GER (Germany), HUNG (Hungary), LAT (Lativa), LITH (Lithuania), POL (Poland), ROM (Romania), RUS (Russia), SERB (Serbia), SVK (Slovakia), SLO (Slovenia), UKR (Ukraine) and USA (United States of America).

---

[3] The correlations between the return series are shown in Table 5.7 in the Appendix.

Table 5.1 Descriptive statistics for monthly return series, RV, RV$^+$ and RV$^-$

Return	BIH	BULG	CRO	CYP	CZE	EST	GER	HUNG	LAT
Min	−0.0060	−0.0129	−0.0105	−0.0166	−0.0097	−0.0102	−0.0088	−0.0064	0.0000
Median	−0.0002	0.0000	0.0001	−0.0006	0.0002	0.0003	0.0001	0.0003	0.0000
Mean	−0.0002	0.0001	−0.0001	−0.0008	0.0000	0.0003	0.0003	0.0005	0.0000
Max	0.0051	0.0066	0.0037	0.0153	0.0050	0.0053	0.0069	0.0121	0.0013
SD	0.00181	0.00,231	0.00,189	0.00381	0.00213	0.00191	0.00233	0.00218	0.00014
RV	BIH	BULG	CRO	CYP	CZE	EST	GER	HUNG	LAT
Min	0.001	0.006	0.003	0.002	0.006	0.003	0.011	0.005	−6.16
Median	0.009	0.017	0.011	0.073	0.023	0.009	0.033	0.022	0.29
Mean	0.013	0.027	0.025	0.137	0.034	0.013	0.049	0.042	0.382
Max	0.096	0.366	0.817	1.493	0.529	0.166	0.784	0.518	6.974
SD	0.014	0.043	0.086	0.212	0.058	0.018	0.081	0.079	1.561
RV+	BIH	BULG	CRO	CYP	CZE	EST	GER	HUNG	LAT
Min	0.0000	0.0040	0.0030	0.0020	0.0060	0.0030	0.0120	0.0050	0.0020
Median	0.0080	0.0170	0.0090	0.0790	0.0180	0.0090	0.0320	0.0200	0.0060
Mean	0.0150	0.0270	0.0150	0.1380	0.0290	0.0110	0.0430	0.0430	0.0090
Max	0.1120	0.2800	0.3080	1.2850	0.5330	0.1410	0.4740	0.9010	0.0610
SD	0.0180	0.0360	0.0330	0.2010	0.0560	0.0150	0.0510	0.1040	0.0100
RV−	BIH	BULG	CRO	CYP	CZE	EST	GER	HUNG	LAT
Min	0.0002	0.0044	0.0041	0.0019	0.0047	0.0022	0.0067	0.0042	0.0018
Median	0.0077	0.0135	0.0091	0.0623	0.0220	0.0081	0.0307	0.0183	0.0051
Mean	0.0126	0.0257	0.0199	0.1190	0.0398	0.0142	0.0547	0.0385	0.0315
Max	0.1543	0.3966	0.5586	1.9770	0.5269	0.1904	1.0940	0.8141	2.1060
SD	0.0182	0.0469	0.0582	0.2372	0.0757	0.0228	0.1152	0.0987	0.2167
Return	LITH	POL	ROM	RUSS	SERB	SVK	SLO	UKR	USA
Min	−0.0077	−0.0084	−0.0103	−0.0044	−0.0096	−0.0067	−0.0082	−0.0061	−0.0052
Median	0.0000	0.0005	0.0003	0.0004	0.0002	0.0001	0.0003	0.0007	0.0005
Mean	0.0001	0.0002	0.0001	0.0003	0.0002	0.0003	0.0003	0.0004	0.0004
Max	0.0051	0.0050	0.0050	0.0063	0.0047	0.0153	0.0053	0.0057	0.0087
SD	0.00216	0.00241	0.00193	0.00182	0.00202	0.00318	0.00235	0.00190	0.00236
RV	LITH	POL	ROM	RUSS	SERB	SVK	SLO	UKR	USA
Min	0.007	0.006	0.004	0.001	0.006	0.001	0.008	0.003	0.010
Median	0.020	0.020	0.021	0.010	0.021	0.014	0.030	0.015	0.036
Mean	0.028	0.034	0.024	0.014	0.03	0.041	0.043	0.029	0.055
Max	0.392	0.524	0.178	0.043	0.351	0.650	0.519	0.519	0.472
SD	0.041	0.063	0.021	0.011	0.039	0.105	0.056	0.056	0.071
RV+	LITH	POL	ROM	RUSS	SERB	SVK	SLO	UKR	USA
Min	0.0050	0.0060	0.0030	0.0000	0.0040	0.0000	0.0080	0.0020	0.0080
Median	0.0170	0.0170	0.0210	0.0140	0.0220	0.0120	0.0250	0.0140	0.0330
Mean	0.0230	0.0290	0.0270	0.0190	0.0300	0.0370	0.0380	0.0260	0.0510
Max	0.3410	0.3040	0.1430	0.1420	0.3000	0.6970	0.5630	0.5710	0.4720
SD	0.0360	0.0400	0.0200	0.0190	0.0350	0.1050	0.0590	0.0610	0.0700
RV−	LITH	POL	ROM	RUSS	SERB	SVK	SLO	UKR	USA
Min	0.0081	0.0037	0.0028	0.0000	0.0051	0.0000	0.0055	0.0021	0.0097

(continued)

Table 5.1 (continued)

Return	LITH	POL	ROM	RUSS	SERB	SVK	SLO	UKR	USA
Median	0.0222	0.0196	0.0149	0.0065	0.0186	0.0073	0.0310	0.0133	0.0335
Mean	0.0330	0.0368	0.0199	0.0109	0.0276	0.0391	0.0436	0.0300	0.0511
Max	0.4276	0.6765	0.1917	0.0559	0.3869	1.5910	0.4742	0.4760	0.4491
SD	0.0463	0.0773	0.0218	0.0115	0.0434	0.1735	0.0527	0.0534	0.0676

Note All values for RV, RV+ and RV− series are multiplied with 1000. SD denotes standard deviation, Min and Max denote minimal and maximal value

5.4.2 Static Results

All of the variables were tested to be stationary via ADF (augmented Dickey-Fuller) test and are used in the VAR models in levels.[4] Thus, a VAR model was estimated for the entire sample for return series, then two other models for RV+ and series, i.e. 3 VAR models in total. The lag length was chosen so that the diagnostics of the model are fulfilled, with the stability of the model in mind. It was sufficient to estimate VAR(1) for all 3 models. Based on the estimated VAR model, the spillovers were estimated and are shown in Tables 5.2, 5.3 and 5.4, with $h = 12$. Table 5.2 depicts the spillover table for the return series, whereas Tables 5.3 and 5.4 show the spillover tables for the RV+ and RV− series respectively. Bolded numbers in each table are the total spillover indices. It can be seen that the RV series have greater values of the total spillovers (and individual values of spillovers), which is in line with previous literature that greater spillovers are found for volatility series compared to the return series. Furthermore, the results of the tables are interpreted as follows. If we focus on the first row of Table 5.2, it shows that 59.94% of the variance of the return series for BIH is due to shocks in the same return series, 2.04% of the BIH return variance is explained via shocks in the BULG return series, etc. The columns titled FROM_av and FROM_sum show how much on average and in total does a variance of a return series receive shocks from other countries. The rows titled TO_av and TO_sum give similar information, but how much on average and in total do shocks in return series of one country spill over to other countries' variances of the return series. In that way, information can be obtained on: how much a country return receives shocks from others, how much it gives shocks to others, is the country net emitter or receiver of shocks, how much is the country return series connected to other countries, is a country return series more "closed" in terms of the values of how much it receives shocks from others, etc. All this information is relevant to international investors. This is true for both return and RV tables. There are several important insights an investor can obtain from the spillover tables in Tables 5.2, 5.3, and 5.4, regarding building the portfolio and for diversification purposes. By focusing on the return spillover table, the greatest receivers of shocks on average were CZE, UKR, and SERB. CZE receives shocks mostly from SLO, UKR, and SERB. Then, UKR receives shocks from SLO, CZE, and SERB, whereas SERB receives shocks from CZE, SLO, and CRO in the greatest manner. This gives information about the connectedness of several markets

---

[4] Due to many variables used in the study, results are omitted but are available upon request.

Table 5.2 Spillover table, return series, full sample

Return	BIH	BULG	CRO	CYP	CZE	EST	GER	HUNG	LAT	LITH	POL	ROM	RUSS	SERB	SVK	SLO	UKR	USA	FROM_av	FROM_sum
BIH	59.94	2.04	6.97	0.24	2.48	0.84	3.02	0.46	4.93	1.16	1.44	3.40	1.99	5.02	0.36	1.11	2.68	1.92	2.36	40.06
BULG	0.68	30.40	8.96	3.38	5.66	7.06	5.17	3.63	0.28	0.57	2.62	8.55	0.54	8.62	0.94	4.73	7.36	0.85	4.09	69.60
CRO	2.68	7.13	25.47	1.76	4.63	8.19	3.84	3.18	0.42	2.01	8.31	6.91	0.70	7.18	0.21	4.74	7.62	5.01	4.38	74.53
CYP	0.29	6.02	4.00	44.23	5.28	2.12	4.81	1.99	0.14	1.17	5.18	3.70	1.04	8.69	4.87	1.89	3.91	0.66	3.28	55.77
CZE	0.27	3.99	4.36	2.36	22.34	7.34	7.80	1.95	0.50	2.62	4.46	5.76	0.69	8.85	0.36	11.00	9.86	5.48	4.57	77.66
EST	0.98	5.69	8.36	1.60	7.82	23.83	5.28	5.46	0.33	0.86	3.34	6.37	1.58	7.00	0.24	6.04	6.64	8.59	4.48	76.17
GER	1.61	5.08	3.98	1.26	10.34	6.66	29.67	2.08	0.24	2.69	4.46	7.28	1.51	8.24	0.54	5.74	5.14	3.48	4.14	70.33
HUNG	0.71	5.36	6.49	0.83	4.04	10.17	3.07	42.63	1.03	0.46	1.98	3.26	1.65	5.58	1.50	4.23	5.67	1.32	3.37	57.37
LAT	4.80	0.32	0.34	0.14	0.12	0.34	0.45	1.86	85.67	0.66	0.81	0.21	0.19	0.52	1.37	1.27	0.72	0.22	0.84	14.33
LITH	1.34	1.04	1.26	1.41	2.55	0.49	0.50	0.42	2.86	76.67	3.50	0.97	0.92	0.38	0.32	1.60	1.90	1.87	1.37	23.33
POL	1.73	2.58	8.83	2.58	7.24	4.32	4.10	1.40	1.03	3.43	29.85	6.48	0.13	5.95	0.17	6.10	11.05	3.03	4.13	70.15
ROM	1.92	7.81	9.28	2.26	7.14	6.73	6.75	1.65	0.13	0.92	8.62	27.56	1.14	7.80	0.26	3.51	5.18	1.37	4.26	72.44
RUSS	2.10	1.90	1.49	1.21	2.64	5.42	2.75	8.18	2.18	0.19	0.39	1.07	56.15	4.33	1.85	2.54	2.88	2.71	2.58	43.85
SERB	1.46	6.37	7.34	3.10	9.53	6.98	6.50	2.90	0.44	1.47	5.10	6.96	1.03	23.62	0.09	7.18	7.93	1.99	4.49	76.38
SVK	0.08	3.27	0.40	7.89	0.88	0.85	0.22	1.37	1.93	0.33	0.17	0.24	3.45	3.72	73.55	0.12	1.28	0.24	1.56	26.45
SLO	0.26	3.61	4.72	1.10	12.27	6.33	4.29	2.04	0.30	2.99	4.83	3.31	0.64	7.57	0.14	25.57	14.09	5.94	4.38	74.43
UKR	1.01	4.97	6.74	2.01	9.87	6.39	3.73	2.25	0.84	4.24	7.66	3.86	1.09	7.35	0.12	12.59	23.39	1.90	4.51	76.61
USA	1.44	0.89	4.67	0.08	7.89	13.20	4.36	1.27	1.45	2.68	1.89	1.43	1.58	2.64	0.13	8.30	3.24	42.85	3.36	57.15
TO_av	1.37	4.00	5.19	1.95	5.90	5.50	3.92	2.48	1.12	1.67	3.81	4.10	1.17	5.85	0.79	4.86	5.71	2.74	–	–
TO_sum	23.36	68.06	88.20	33.22	100.38	93.44	66.65	42.08	19.04	28.45	64.78	69.78	19.88	99.44	13.47	82.69	97.13	46.56	–	**58.70**
NET	16.70	1.54	−13.67	22.55	−24.20	−23.12	−9.29	−27.75	4.29	41.70	7.67	−25.93	56.50	−73.00	60.96	−6.08	−39.98	10.59	–	–

Note FROM_av denotes spillover from other countries on average, FROM_sum denotes total spillover from other countries, TO_av denotes spillover to other countries on average, TO_sum denotes total spillover to other countries

5 Asymmetric Spillovers Between the Stock … 115

Table 5.3 Spillover table, RV+ series, full sample

RV+	BIH	BULG	CRO	CYP	CZE	EST	GER	HUNG	LAT	LITH	POL	ROM	RUSS	SERB	SVK	SLO	UKR	USA	FROM_av	FROM_sum
BIH	58.13	0.87	2.85	0.32	2.65	2.33	3.14	1.10	0.70	0.79	7.64	4.46	2.11	1.39	2.25	2.58	4.99	1.71	2.46	41.87
BULG	0.39	12.71	9.01	1.18	7.42	8.51	9.27	3.70	3.12	0.82	6.46	5.78	0.06	8.92	0.11	8.46	8.30	5.78	5.13	87.29
CRO	0.20	7.08	10.47	0.70	9.01	8.94	9.72	3.56	3.57	1.32	6.90	6.49	0.09	8.35	0.07	9.17	8.94	5.42	5.27	89.53
CYP	0.11	4.35	3.29	56.10	2.73	2.75	4.18	0.89	1.17	0.14	2.86	1.30	4.44	3.69	1.63	2.54	2.86	4.99	2.58	43.90
CZE	0.26	6.69	9.99	0.57	11.31	8.55	9.78	3.14	3.42	0.80	7.48	6.79	0.01	8.19	0.02	9.36	8.73	4.91	5.22	88.69
EST	0.14	6.80	9.32	0.63	7.86	11.13	9.27	4.31	4.32	1.48	7.18	5.97	0.04	8.50	0.02	9.16	9.06	4.80	5.23	88.87
GER	0.25	7.23	9.67	0.79	8.57	8.85	10.33	3.43	3.44	1.12	7.08	6.63	0.04	8.27	0.11	9.36	8.88	5.97	5.27	89.67
HUNG	0.24	6.64	7.69	0.27	6.54	9.23	7.82	21.27	3.27	0.34	5.41	4.12	0.12	7.07	0.33	7.99	7.69	3.95	4.63	78.73
LAT	0.28	5.83	7.59	0.88	8.00	7.73	7.55	3.05	13.82	3.13	8.35	4.80	0.30	6.64	0.02	8.26	8.64	5.16	5.07	86.18
LITH	0.53	0.42	1.52	0.74	0.28	1.88	0.94	0.12	5.01	82.52	0.31	2.92	0.39	0.75	0.22	0.33	0.69	0.42	1.03	17.48
POL	0.36	6.16	8.27	0.65	8.06	8.27	8.73	2.77	4.34	0.89	12.39	7.03	0.11	7.84	0.01	8.94	10.40	4.77	5.15	87.61
ROM	0.25	6.01	8.87	0.53	8.04	7.71	9.09	2.47	2.63	1.28	8.01	14.70	0.27	7.94	0.18	8.91	8.31	4.81	5.02	85.30
RUSS	1.11	1.21	0.86	2.14	0.49	0.86	1.03	0.12	1.57	0.31	1.74	1.60	78.22	1.25	3.00	0.60	0.58	3.31	1.28	21.78
SERB	0.11	7.66	9.23	1.01	7.99	9.14	9.24	3.48	3.54	1.07	7.11	6.51	0.04	11.46	0.06	8.59	8.58	5.19	5.21	88.54
SVK	0.28	0.92	0.89	17.43	0.63	1.05	1.20	0.58	0.73	0.65	0.56	1.00	5.77	1.21	57.96	1.27	1.32	6.55	2.47	42.04
SLO	0.12	6.70	9.18	0.42	8.41	8.99	9.67	3.59	3.84	0.73	7.45	6.85	0.13	7.91	0.06	10.63	9.57	5.76	5.26	89.37
UKR	0.15	6.66	9.09	0.55	7.95	8.93	9.20	3.52	4.20	1.10	8.71	6.30	0.02	7.98	0.02	9.65	10.70	5.28	5.25	89.30
USA	0.01	6.86	8.13	4.06	6.42	6.91	9.06	2.57	3.94	0.34	5.85	5.13	0.44	6.80	1.72	8.35	7.69	15.73	4.96	84.27
TO_av	0.28	5.18	6.79	1.93	5.95	6.51	6.99	2.49	3.10	0.96	5.83	4.92	0.84	6.04	0.58	6.68	6.78	4.63	–	–
TO_sum	4.80	88.08	115.42	32.87	101.07	110.63	118.90	42.41	52.78	16.29	99.07	83.68	14.36	102.71	9.82	113.51	115.23	78.78	–	72.25
NET	37.07	−0.79	−25.89	11.03	−12.20	−20.96	−40.17	43.77	−35.30	71.31	−13.77	−61.90	−74.18	−60.67	79.55	−24.21	−30.96	5.49	–	–

Note FROM_av denotes spillover from other countries on average, FROM_sum denotes total spillover from other countries, TO_av denotes total spillover to other countries on average, TO_sum denotes total spillover to other countries

Table 5.4 Spillover table, RV− series, full sample

RV−	BIH	BULG	CRO	CYP	CZE	EST	GER	HUNG	LAT	LITH	POL	ROM	RUSS	SERB	SVK	SLO	UKR	USA	FROM_av	FROM_sum
BIH	59.12	2.64	1.06	0.98	1.09	1.40	1.09	0.38	10.68	1.56	2.74	2.56	8.36	0.89	2.17	1.19	1.88	0.22	2.40	40.88
BULG	0.64	27.54	7.71	1.08	6.95	6.73	7.10	0.95	4.31	6.08	5.32	2.82	0.13	7.60	0.05	5.55	5.92	3.54	4.26	72.46
CRO	0.02	3.70	10.64	0.20	9.50	9.24	9.33	0.68	4.34	9.43	6.33	4.62	0.07	8.87	0.03	9.10	9.25	4.64	5.26	89.36
CYP	0.63	1.96	1.27	62.07	1.84	1.95	1.61	1.31	0.93	1.12	2.12	0.70	4.01	3.27	4.08	1.41	1.16	8.56	2.23	37.93
CZE	0.02	3.07	9.44	0.29	10.24	9.34	9.33	0.55	3.78	9.60	7.00	4.45	0.09	8.39	0.04	9.83	9.60	4.93	5.28	89.76
EST	0.02	2.89	9.59	0.33	9.77	10.63	9.16	0.55	4.14	9.37	6.72	4.24	0.06	8.72	0.11	9.48	9.43	4.78	5.26	89.37
GER	0.15	3.03	9.46	0.41	9.53	8.96	10.87	0.57	4.16	9.52	6.40	4.71	0.07	8.38	0.08	9.26	9.25	5.20	5.24	89.13
HUNG	0.40	1.96	4.04	0.72	3.28	3.15	3.39	58.47	2.34	3.18	1.72	0.94	1.15	3.70	1.35	4.38	4.16	1.66	2.44	41.53
LAT	0.18	2.40	8.66	0.55	8.35	8.72	8.35	0.63	17.17	7.95	5.65	3.57	0.68	6.90	0.04	7.82	8.01	4.39	4.87	82.83
LITH	0.01	2.78	9.60	0.17	9.76	9.16	9.46	0.54	3.96	10.50	6.83	4.72	0.09	8.26	0.02	9.62	9.70	4.80	5.26	89.50
POL	0.05	2.83	8.16	0.36	8.87	8.29	8.04	0.35	3.36	8.77	12.76	5.24	0.19	8.28	0.05	8.71	10.24	5.42	5.13	87.24
ROM	0.34	2.13	7.89	0.14	7.62	7.12	8.30	0.44	2.90	8.05	7.30	18.59	0.15	7.12	0.56	8.11	8.07	5.18	4.79	81.41
RUSS	1.70	0.90	0.47	2.80	0.54	0.83	1.05	0.33	2.78	0.34	0.19	0.53	81.95	0.57	3.09	0.37	0.33	1.22	1.06	18.05
SERB	0.04	4.29	9.42	0.84	9.02	9.02	8.80	0.64	3.49	8.59	6.89	4.29	0.04	11.14	0.02	8.46	8.72	6.30	5.23	88.86
SVK	1.93	0.19	0.35	16.64	0.36	1.13	0.22	0.47	1.35	0.23	1.02	1.08	7.81	0.39	64.38	0.27	0.39	1.80	2.10	35.62
SLO	0.02	2.42	9.20	0.25	9.98	9.22	9.23	0.75	3.77	9.61	7.07	4.78	0.10	8.12	0.05	10.36	9.72	5.34	5.27	89.64
UKR	0.02	2.44	9.25	0.17	9.68	9.12	9.10	0.72	4.10	9.62	8.05	4.63	0.09	8.25	0.03	9.62	10.27	4.83	5.28	89.73
USA	0.16	2.32	7.18	3.74	7.64	7.13	7.52	0.60	3.64	7.37	6.89	4.50	0.03	9.28	0.42	8.01	7.51	16.07	4.94	83.93
TO_av	0.37	2.47	6.63	1.75	6.69	6.50	6.53	0.61	3.77	6.49	5.19	3.43	1.36	6.29	0.72	6.54	6.67	4.28	–	–
TO_sum	6.31	41.94	112.75	29.69	113.77	110.51	111.09	10.43	64.02	110.40	88.24	58.38	23.12	107.00	12.21	111.20	113.35	72.80	–	**69.04**
NET	34.57	30.53	−23.39	8.23	−24.40	−21.38	−69.56	72.40	25.48	−23.16	−6.83	−40.34	65.74	−71.38	77.42	−21.47	−29.42	11.13	–	–

Note FROM_av denotes spillover from other countries on average, FROM_sum denotes total spillover from other countries, TO_av denotes spillover to other countries on average, TO_sum denotes total spillover to other countries and investors should be careful when putting the stocks from the mentioned markets in the same portfolio. Since the greatest spillovers are found between the mentioned country returns, this could be helpful when the investor is forecasting a bull market, to include those countries in the portfolio, as they will increase the portfolio value. The opposite is true if the investor is forecasting a bearish market. A similar analysis is done by observing the greatest net emitters of shocks to others: CZE, SERB, and UKR. The CZE series gives shocks mostly to SLO, GERM, and UKR, the SERB series gives shocks to CZE, BULG, and CYP, and finally, the UKR series gives shocks mostly to POL, SLO, and CZE. This means that the mentioned subsets of markets are connected more between them, rather than to others.

Next, by focusing on the RV tables, the greatest receivers of spillovers on average were GER, CRO, and UKR. The GER series received shocks mostly from UKR, EST, and SERB, the CRO series received shocks from GERM, SLO, and CZE, whereas UKR received shocks from SLO, GERM, and CRO in the greatest manner. On the other side, the greatest emitters of shocks on average were CRO, GERM, and SLO. The CRO series gave shocks mostly to CZE, GERM, and EST, GERM gave shocks to CRO, CZE, and SLO, and finally, the SLO shocks spilled over mostly to UKR, CZE, and GERM. By observing the percentages of variances of return series which are the least explained by shocks in the same country, the top 3 countries are UKR, SERB, and CZE, and the opposite is true for LAT, LITH, and SVK. For the RV series, the smallest variance shares of the same country are found for CRO, SLO, and UKR, whereas the opposite is true for LITH, RUS, and BIH. This means that the stock markets of those countries which have the greatest percentages of their variance series explained by shocks in the same market are the most closed ones, and could be potentially beneficial for diversification possibilities. These countries are Latvia, Lithuania, Slovakia, Russia and Bosnia, and Herzegovina. The results are in line with [127], where authors found that the SLO market was the net emitter of volatility shocks, and the CRO and UKR markets being found the net receivers of shocks. Furthermore, the results regarding BIH being a closed market is in line with [125], where this market was found to be more affected by the shocks on the same market compared to shocks from its surroundings. The great link between the SLO and GERM in both directions is found in [57] as well, with the GER affecting the CZE being confirmed here as in [56]. The results that the SVK is another closed market is in line with [102].

If we focus on the asymmetries between the RV series, the SAM measures were calculated for every country to see which country had greater positive or negative spillovers. The results are shown in Table 5.5. Positive values of the SAM measures indicate that a country has emitted more shocks to others or has received from others during positive volatility and vice versa. In that way, a country that receives and gives shocks more when the volatility is a positive one, this country could be useful for those investors interested in obtaining better portfolio value. The most vulnerable countries, which received mostly negative volatility shocks are HUNG, BULG, and ROM. This should warn investors that these countries are not safe to have in their portfolios when the market is going down, as they will absorb negative volatility the most. Countries that give the greatest positive shocks are HUNG, GER, and CRO.

Table 5.5 SAM measures for every country, full sample

SAM:	BIH	BULG	CRO	CYP	CZE	EST	GER	HUNG	LAT
To	0.098	−2.367	−0.028	0.567	3.263	0.951	−0.336	−3.355	3.404
From	−2.019	0.671	1.328	−1.024	−1.929	0.223	1.556	1.627	−4.449
SAM:	LITH	POL	ROM	RUS	SERB	SVK	SLO	UKR	USA
To	0.221	−0.572	−1.906	0.798	0.888	0.447	0.415	0.373	0.167
From	0.940	0.628	0.627	−0.516	0.121	−1.747	0.783	0.939	−0.785

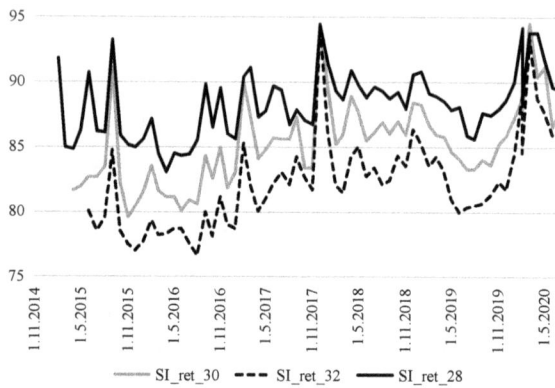

Fig. 5.1 Total spillover index, rolling windows, return series

However, all of this analysis so far was static, as it included the entire sample. Since investors have to restructure their portfolios on a dynamic basis, the next subsection deals with dynamic results.

5.4.3 Dynamic Analysis

All 3 VAR models were re-estimated on a rolling window basis, with the length of the window equal to 30 months.[5] Additionally, for the robustness checking (next subsection), the window lengths of 32 and 28 months were chosen as well. Figure 5.1 depicts the total spillover index between the return series in the observed period. The investor can observe increases or decreases in the total index to tailor his portfolio accordingly. This is maybe more important for the RV series, due to them indicating risk spillovers over time. However, the values of all indices in Fig. 5.1 indicate an increase of the spillover in the observed period, which is in line with previous literature that the integration among stock markets is rising. A great increase is visible in, e.g., the beginning of 2020 due to the COVID crisis. Figures 5.2 and 5.3 depict the total spillover indices for RV+ and RV− series respectively. Here it can be seen

---

[5] Due to quick changes in stock markets, the chosen lengths are appropriate, as this is in accordance with Škrinjarić [128, 129] and [6].

Fig. 5.2 Total spillover, rolling windows, RV+ series

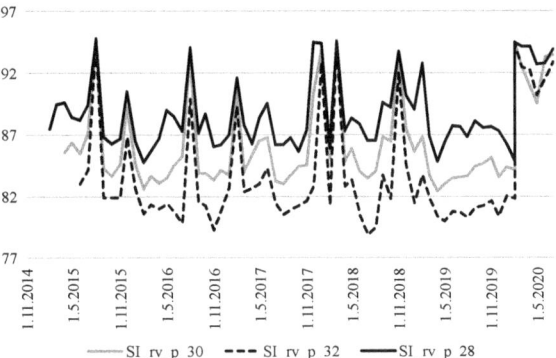

Fig. 5.3 Total spillover, rolling windows, RV− series

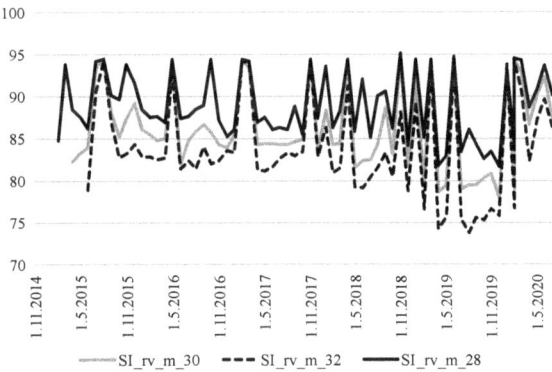

that the volatility of the spillover indices is greater compared to those in Fig. 5.1. Furthermore, the total index has greater values for both RV series compared to the return series, as this is in line with previous literature as well, that the risk series between stock markets are more relaxed compared to the return series. That is why based on these results investors should pay more attention to the spillovers of the RV series compared to the return ones.

To observe whether there exist asymmetries in the spillovers over time, the rolling SAM measure was calculated based on the formula (5.14) and is shown in Fig. 5.4. This dynamic is giving the investor insights into when good and bad volatility spillovers take place. An investor could base his strategy solely based on results in Fig. 5.4, e.g. he could opt to invest in the observed markets when the SAM measure is positive. If the measure is negative, he could either sell the portfolio or hold it until the measure becomes positive again. Finally, as the volatility of all SAM measures is great, we have calculated the cumulative value for each SAM over time, to observe whether positive or negative SAM is stronger in the observed period. This is depicted in Fig. 5.5. It is obvious that the majority of the time the SAM measure is of negative value. This indicates that the negative volatility was present more in the observed period, compared to the positive one. This is, again, in line with previous literature:

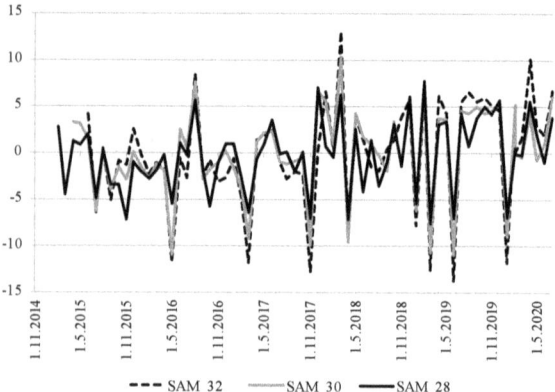

Fig. 5.4 Rolling SAM measure, total spillover index for RV+ and RV−

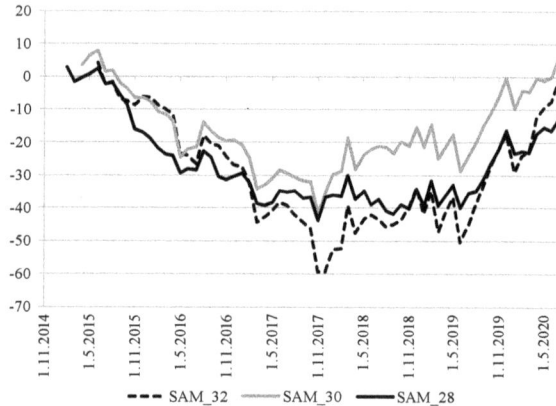

Fig. 5.5 Cumulative values of rolling SAM measure from Fig. 5.4

the integration among the observed markets is growing, and the market is bearish a lot of the time.

5.4.4 Robustness Checking

Before moving to the simulation of the strategies, the robustness of the results was checked so that we can be confident in using such results. Diebold and Yilmaz [35, 36] suggest to change the length of the rolling windows for the estimation of the dynamic indices. The idea is that although the values of the indices will be on different levels, their dynamics should be the same over time. That is why all of the previously observed indices in the previous subsection were estimated with the window length of 28, 30 and 32. As it is seen throughout Figs. 5.1, 5.2, 5.3, 5.4 and 5.5, the dynamics of all series is almost the same. Thus, this gives us confidence that such results can be used in portfolio strategies. The simulation results are given in the next subsection.

5.4.5 Simulation of Investment Strategies

For the purpose of simulating trading strategies based on previous results, two benchmark strategies were simulated in the first step. One refers to the minimal portfolio variance from the Markowitz portfolio selection model, in which the portfolio variance is minimized every month, in the following manner:

$$\min_{w_t} w_t' \sum\nolimits_t w_t$$
$$s.t.\ w_t' e = 1$$
$$w_t' \geq 0, \qquad (5.17)$$

where w_t is the vector of weights in every month t, $\sum_t$ is the variance-covariance matrix in month t, e is the vector of unit values. The second benchmark portfolio consists of equal weights of every county. Thus, the first portfolio will be called "Min var", due to the nature of the objective function in every month, and the second one will be called "Eq_weights".

Then, several strategies were simulated based on the rolling window results from all three VAR models. Their description is as follows. For every VAR model (the return based one, and the RV+ and RV− ones), the investor collects the net spillover indices of every country. The main model is the one in which the rolling window length is 30 months.

- "return_based_3" – Net indices from the return series VAR model are extracted. The investor compares the net index of a country i in month t to the zero value. If the net index is positive, this means that the shocks in return series of this country have spilled over to other return series in a greater manner compared to the shocks this return series i has received from others in that month t. Then, if the net index i is positive in month t, the investor looks if the return series of the same country is positive or negative in the same month. If it is positive, this means that shocks in that return series will have positive spillovers to others in month t. The investor buys those indices which have positive net indices and positive return series in month t. If this criterion is not satisfied in the next month, the investor sells those indices and buys those indices which satisfy the criterion. However, a ranking of the net spillover indices is done, in a way that the greatest positive net spillover index is the first one, then the second greatest is the second one, etc. The investor limits his buying and selling to the top three indices in every month.

- "return_based_5" – the same as the previous strategy, but the investor limits his buying and selling to the top five indices in every month.
- "return_based_10" – the same as the previous strategy, but the investor limits his buying and selling to the top ten indices in every month.
- "rv_p_3" – Net indices from the RV+ series VAR model are extracted. The investor compares the net index of a country i in month t to the zero value. If the net index is positive, this means that the shocks in good volatility series of this country have spilled over to other return series in a greater manner compared to the shocks this RV+ series has received from others in the same month t. Then, if the net index i is positive in month t, the investor looks if the return series of the same country is positive or negative in the same month. If it is positive, this means that shocks in the RV+ series will have positive spillovers to other RV+ series in month t. The investor buys those indices which have positive net indices and positive return series in month t. If this criterion is not satisfied in the next month, the investor sells those indices and buys those indices which satisfy the criterion. However, a ranking of the net spillover indices is done, in a way that the greatest positive net spillover index is the first one, then the second greatest is the second one, etc. The investor limits his buying and selling to the top three indices in every month.
- "rv_p_5" – the same as the previous strategy, but the investor limits his buying and selling to the top five indices in every month.
- "rv_p_10" – the same as the previous strategy, but the investor limits his buying and selling to the top five indices in every month.
- "return_and_rv_p_10" – a combination of using the net indices from the return and RV+ models simultaneously. The investor observes if the net index of country i of the return model is positive and if the net index of country i of the RV+ model is positive. If both indices are positive, then the investor buys the country index. The opposite is true if any of the net indices are of negative value. The investor limits his buying and selling to the top ten indices every month, so that better diversification possibilities are in place, compared to limiting himself to only three or five indices.
- "rv_m_3" – Net indices from the RV− series VAR model are extracted. The investor compares the net index of a country i in month t to the zero value. If the net index is positive, this means that the shocks in bad volatility series of this country have spilled over to other return series in a greater manner compared to the shocks this RV− series has received from others in the same month t. The idea is that it is better to use net emitters of shock, rather than net receivers, as this would reduce the diversification possibilities. Then, if the net index i is positive in month t, the investor looks if the return series of the same country is positive or negative in the same month. The investor buys those indices which have positive net indices and positive return series in month t. If this criterion is not satisfied in the next month, the investor sells those indices and buys those indices which satisfy the criterion. However, a ranking of the net spillover indices is done, in a way that the greatest positive net spillover index is the first one, then the second greatest is the second one, etc. The investor limits his buying and selling to the top three indices in every month.

- "rv_m_5" – the same as the previous strategy, but the investor limits his buying and selling to the top five indices in every month.
- "rv_m_10" – the same as the previous strategy, but the investor limits his buying and selling to the top five indices in every month.
- "return_and_rv_m_10" – a combination of using the net indices from the return and RV– models simultaneously. The investor observes if the net index of country i of the return model is positive and if the net index of country i of the RV– model is positive. If both indices are positive, then the investor buys the country index. The opposite is true if any of the net indices are of negative value. The investor limits his buying and selling to the top ten indices every month, so that better diversification possibilities are in place, compared to limiting himself to only three or five indices.

All strategies start with a unit value invested in them. The portfolio values of the return and RV+ models are shown in Fig. 5.6, whereas RV– (with the return model again) are compared in Fig. 5.7. Both benchmark strategies are included in Figs. 5.6 and 5.7 for easier comparison. In both figures, it is obvious that strategies that combined the return and RV results simultaneously outperform all other strategies. The values of both portfolios increase over time much better compared to all other strategies. Although the majority of strategies did end up with an increase in the portfolio value at the end of the observed period, the mentioned two obtained substantial results. This could indicate that using both return and risk series in evaluating the portfolio dynamics over time is important to include in the analysis. When looking at the rest of the strategies, of course, the return based ones did provide better portfolio values the majority of the time. This is not surprising, as the return based strategies did look at best return performance. However, investors are not interested solely in the portfolio return series, but other portfolio performance measures are

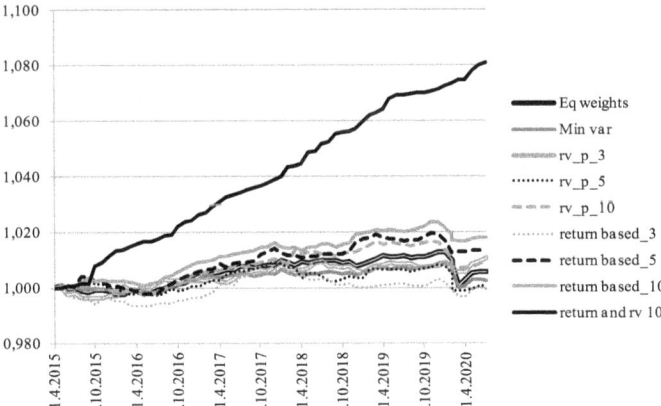

Fig. 5.6 Simulated portfolio values, benchmark portfolios, return based portfolios and RV+ based portfolios

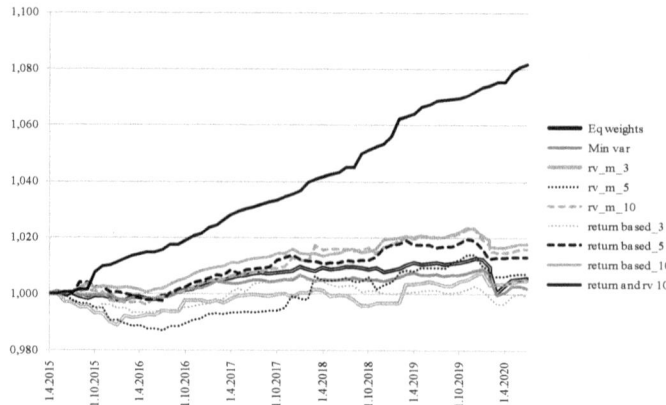

Fig. 5.7 Simulated portfolio values, benchmark portfolios, return based portfolios and RV− based portfolios

looked at. That is why based on all strategies, several performance measures have been calculated. They are shown in Table 5.6.

Table 5.6 includes several return and risk measures, as well as the certainty equivalent ("utility" row), which is estimated as $CE \approx \mu - 0.5\gamma\sigma^2$, where μ denotes the portfolio return, γ the coefficient of absolute risk aversion, σ^2 portfolio risk [53, 54]. For an average investor, the value of γ was chosen to be 2. It is obvious that the best performing portfolio is the combination of observing the return and RV− estimation results. Thus, the investor could benefit in terms of portfolio return, risk, and the utility he receives by implementing such strategies. Thus, it is advisable that future research, as well as practical implementations, utilize the results from the return and RV− models simultaneously. In that way, the best potential could be obtained. Furthermore, the structure of this portfolio was observed, it is given in Fig. 5.8. In the majority of cases, there are only several markets in the portfolio which satisfy the aforementioned criteria to enter the portfolio. The country index which enters this portfolio in the most cases is LITH, LAT, POL, UKR, and SLO, whereas the least amount of entering to the portfolio is found for SVK, BULG, ROM, RUS, and BIH.

Since the previously observed strategies were based on all of the net indices, alongside the return dynamics over time, they could be rather complicated if the investor is looking at stocks as investment instruments. The analysis can become very cumbersome. That is why another approach in the simulation was made, with a simple basis: the SAM measure which was calculated via formula (5.14) was observed in every month t. Positive value of this measure indicates that the spillovers of RV+ series are greater than spillovers of the RV− in month t, and this means that the good volatility spillovers were greater than the bad ones. In that case, the investor is looking at the positive return series of country i, and invests into those country indices which have positive returns in month t alongside positive SAM value. Although the investor does not receive full information about the spillovers among the country indices themselves, at least he is following the positive trends

5 Asymmetric Spillovers Between the Stock …

Table 5.6 Portfolio characteristics, return and risk measures

Portfolio:	Eq weights	Min var	rv_p_3	rv_p_5	rv_p_10	return_3	return_5	return_10	return and rv_p 10	rv_m_3	rv_m_5	rv_m_10	return and rv_m 10
Return													
Mean	0.00008	0.00003	0.00122	0.00025	0.00000	0.00015	0.00012	0.00016	−0.00001	0.00008	0.00009	0.00022	0.00123
Median	0.00027	0.00016	0.00089	0.00036	0.00001	0.00004	0.00013	0.00000	0.00000	0.00000	0.00003	0.00023	0.00094
Max	0.00246	0.00230	0.00625	0.00316	0.00275	0.00382	0.00510	0.00362	0.00290	0.00662	0.00801	0.00535	0.00625
Min	−0.00716	−0.00614	0.00000	−0.00217	−0.00286	−0.00380	−0.00310	−0.00330	−0.00581	−0.00474	−0.00460	−0.00395	0.00000
SD	0.00127	0.00109	0.00113	0.00097	0.00116	0.00130	0.00119	0.00118	0.00132	0.00157	0.00178	0.00135	0.00120
Skew	−2.902	−2.750	1.875	−0.336	−0.197	−0.087	0.376	−0.217	−1.626	0.768	0.940	0.307	2.419
Kurt	17.478	17.360	7.889	3.988	3.566	4.612	7.536	4.986	8.230	7.996	8.995	6.475	9.770
Total return	0.0049	0.0024	0.0101	−0.0027	0.0065	−0.0001	0.0068	0.0146	0.0772	0.0052	0.0042	0.0132	0.0828
Portfolio risk													
Mean	0.00001	0.00001	0.00003	0.00002	0.00006	0.00003	0.00002	0.00005	0.00003	0.00007	0.00005	0.00002	0.00003
Median	0.00001	0.00001	0.00002	0.00001	0.00002	0.00001	0.00001	0.00002	0.00002	0.00002	0.00001	0.00001	0.00001
Maximum	0.00001	0.00001	0.00030	0.00011	0.00127	0.00038	0.00008	0.00034	0.00015	0.00133	0.00127	0.00015	0.00018
Minimum	0.00001	0.00001	0.00001	0.00001	0.00001	0.00001	0.00001	0.00001	0.00001	0.00001	0.00001	0.00001	0.00001
Utility	0.00008	0.00003	0.00016	−0.00001	0.00012	0.00000	0.00015	0.00025	0.00122	0.00008	0.00009	0.00022	0.00123
VaR	−0.00006	−0.00002	−0.00008	0.00007	−0.00009	0.00011	−0.00010	−0.00022	−0.00117	0.00003	0.00000	−0.00019	−0.00118
RpR	8.00	3.00	43.67	12.50	0.00	5.00	6.00	3.20	−0.33	1.14	1.80	11.00	41.00

Note Utility is the certainty equivalent. VaR is value at risk estimated based on 95% confidence, RpR is the return per one unit of risk

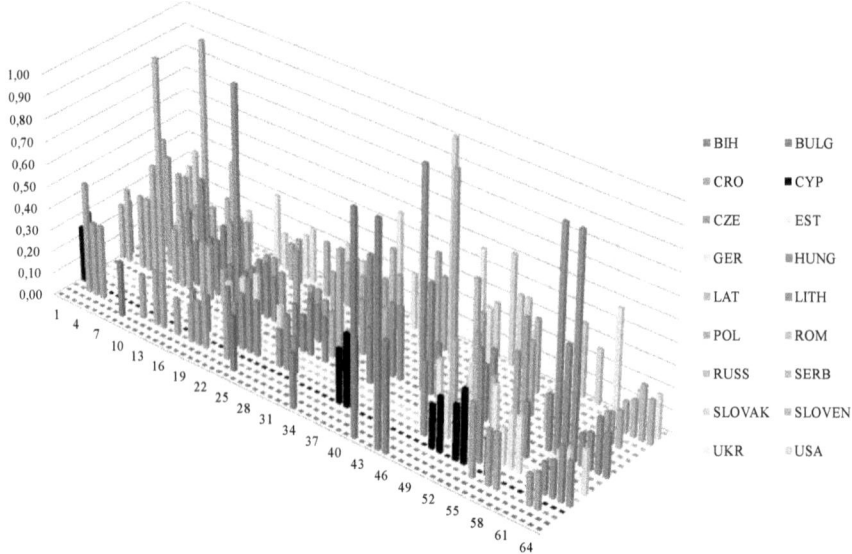

Fig. 5.8 Portfolio weights of the best performing strategy (return_and_rv_m_10)

and values. The simulated portfolio called SAM is depicted in Fig. 5.9, in which it can be seen that it also obtains good portfolio value over time, although not as good as the two best performing strategies (return series combined with the RV+ or RV− ones). However, the portfolio value of the SAM strategy is better than other strategies which were examined in Figs. 5.6 and 5.7. This means that simpler strategies based on the asymmetric results could be a starting point for future analysis as well.

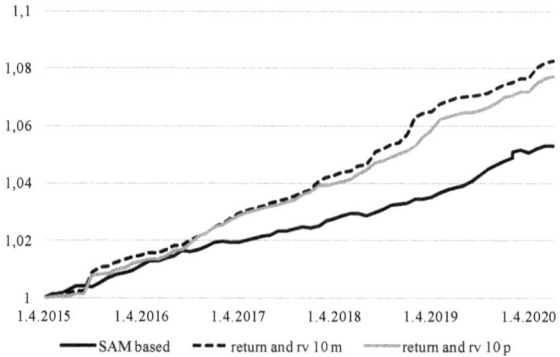

Fig. 5.9 Simulated SAM based portfolio value compared to best portfolios

Fig. 5.10 Simulated SAM based portfolio risk compared to best portfolios

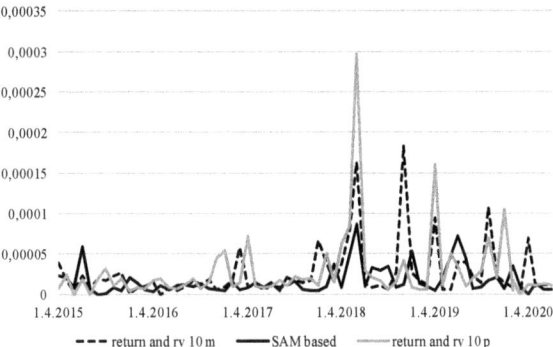

Furthermore, as again, investors do not observe only the portfolio value over time but are interested in the risk mitigation, we have calculated the portfolio risk series for all three portfolios from Fig. 5.9. The risk series in terms of the portfolio standard deviation measure is shown in Fig. 5.10. The SAM based portfolio obtains smaller risk over the observed period. Thus, although this portfolio did not obtain better portfolio value over time, it provides better diversification possibilities compared to the other two portfolios in Figs. 5.9 and 5.10. This gives faith that simple strategies based on the estimation results from the previous subsection could provide good portfolio value over time, with good portfolio diversification.

5.5 Discussion and Conclusion

This chapter dealt with an important topic of the asymmetric behaviour of spillovers between return and risk series of stock markets. Two contributions were made: the first one which refers to a systematic and critical overview of the related literature, in which great insights into the observed stock markets were made. The second one refers to the detailed empirical analysis in which dynamic estimation was made and the results were used in simulating the investment strategies which could have been taken into consideration. The findings of this chapter have significant implications in practice for international investors, portfolio diversification, as well as for macroeconomic policy decisions and guidelines. International investors could utilize the used methodology in their portfolio selection and rebalancing over time. The approach used in this study does not need to be the only one used, but it can be a part of the total portfolio selection procedure. It could be used as a starting point to get insights into which countries are net emitters or receivers of bad and good volatilities respectively. Then, the selection process of the investment instruments could be based on different criteria compared to those utilized in this study. On the other side, the investor could use other methodologies as a first step to select which financial instruments are interesting to him based on his criteria, and the second step could include the VAR estimation results as obtained here. There are many possibilities, and they depend on the investor's goals, funds, and preferences. The implications for portfolio diversification possibilities are also important, as the results here show that portfolios can be constructed in a way that a reasonable return is obtained, without bearing a great risk. The asymmetric behaviour can be extracted and studied more carefully so that the diversification of the portfolio could be achieved in a better way when compared to using standard symmetric models. Finally, policymakers could also obtain important information from utilizing such models, as better knowledge about their stock markets could provide them opportunities to tailor policies and measures to develop the markets more and to make them more attractive for potential international investors.

Some of the shortfalls include the following. A relatively short period was used in the analysis, due to data unavailability. Future research should try to find longer periods so that more insights can be obtained from greater crises such as the Eurozone debt crisis, the financial crisis of 2008, etc. Furthermore, a monthly analysis was made. Some investors are interested in the dynamics every week. Thus, this could be another way to go in future work. Then, the simulated portfolios are rather simple, as they were based on the net spillover indices for every country. Future work could consider how to observe pairwise spillover indices simultaneously. However, this presents a difficult task, as a strategy should take into consideration many indices at once. Presenting such an analysis for an investor could present a problem. Although, it could provide better results in terms of achieving better portfolio risk or return characteristics. Then, other popular approaches of trading can be included as well, to find which combination of strategies can yield the best results.

Since some potential in utilizing the results of dynamic spillovers between the return and asymmetric volatility series was found in this research, the results in this chapter can be a stepping stone for future related analysis. In that way, the investor goals could be achieved even better.

Appendix

See Table 5.7.

Table 5.7 Correlation of return series, total sample

	BIH	BULG	CRO	CYP	CZE	EST	GER	HUNG	LAT	LITH	POL	ROM	RUSS	SERB	SVK	SLO	UKR	USA
BIH	1	−0.019	0.131	0.131	0.048	0.013	0.114	0.039	−0.187	−0.131	0.095	0.112	0.195	0.053	0.081	0.001	0.126	−0.139
BULG	−0.019	1	0.467	0.356	0.375	0.461	0.410	0.276	0.009	0.056	0.347	0.494	0.159	0.507	0.193	0.351	0.410	0.133
CRO	0.131	0.467	1	0.314	0.461	0.579	0.382	0.271	−0.081	−0.061	0.529	0.545	0.148	0.557	0.001	0.447	0.533	0.358
CYP	0.131	0.356	0.314	1	0.273	0.203	0.202	0.161	−0.044	0.058	0.312	0.257	0.196	0.33	0.377	0.181	0.261	0.025
CZE	0.048	0.375	0.461	0.273	1	0.626	0.580	0.208	0.022	0.116	0.513	0.492	0.169	0.679	0.101	0.728	0.669	0.455
EST	0.013	0.461	0.579	0.203	0.626	1	0.476	0.396	−0.038	0.008	0.453	0.482	0.252	0.607	−0.030	0.559	0.556	0.567
GER	0.114	0.41	0.382	0.202	0.58	0.476	1	0.211	0.006	0.014	0.405	0.405	0.214	0.518	−0.028	0.475	0.464	0.368
HUNG	0.039	0.276	0.271	0.161	0.208	0.396	0.211	1	−0.165	0.033	0.133	0.139	0.172	0.257	−0.076	0.190	0.233	0.172
LAT	−0.187	0.009	−0.081	−0.044	0.022	−0.038	0.006	−0.165	1	0.071	−0.04	0.014	0.004	0.045	0.047	−0.059	−0.013	−0.020
LITH	−0.131	0.056	−0.061	0.058	0.116	0.008	0.014	0.033	0.071	1	0.025	0.01	−0.028	−0.016	0.067	0.088	−0.026	0.057
POL	0.095	0.347	0.529	0.312	0.513	0.453	0.405	0.133	−0.040	0.025	1	0.479	0.065	0.494	0.078	0.479	0.568	0.234
ROM	0.112	0.494	0.545	0.257	0.492	0.482	0.405	0.139	0.014	0.01	0.479	1	0.092	0.521	0.025	0.333	0.327	0.186
RUSS	0.195	0.159	0.148	0.196	0.169	0.252	0.214	0.172	0.004	−0.028	0.065	0.092	1	0.192	0.144	0.178	0.208	0.197
SERB	0.053	0.507	0.557	0.33	0.679	0.607	0.518	0.257	0.045	−0.016	0.494	0.521	0.192	1	0.096	0.595	0.591	0.298
SVK	0.081	0.193	0.001	0.377	0.101	−0.03	−0.028	−0.076	0.047	0.067	0.078	0.025	0.144	0.096	1	0.013	0.004	0.022
SLO	0.001	0.351	0.447	0.181	0.728	0.559	0.475	0.19	−0.059	0.088	0.479	0.333	0.178	0.595	0.013	1	0.757	0.485
UKR	0.126	0.410	0.533	0.261	0.669	0.556	0.464	0.233	−0.013	−0.026	0.568	0.327	0.208	0.591	0.004	0.757	1	0.340
USA	−0.139	0.133	0.358	0.025	0.455	0.567	0.368	0.172	−0.020	0.057	0.234	0.186	0.197	0.298	0.022	0.485	0.340	1

References

1. Alter A, Beyer A (2014) The dynamics of spillover effects during the European sovereign debt turmoil. J Bank Finance 42:134–153
2. Amonlirdviman K, Carvalho C (2010) Loss aversion, asymmetric market comovements, and the home bias. J Int Money Finance 29:1303–1320
3. Andersen T, Bollerslev T, Diebold F, Labys P (2001) The distribution of realized exchange rate volatility. J Am Stat Assoc 96(453):42–55
4. Andrei D, Hasler M (2015) Investor attention and stock market volatility. Rev Financ Stud 28(1):33–72
5. Angelovska J (2017) The impact of financial crises on the short-term interaction between balkan stock markets. UTMS J Econom 8(2):53–66
6. Antonakakis N, André C, Gupta R (2016) Dynamic spillovers in the United States: stock market, housing, uncertainty, and the macroeconomy. South Econ J 83(2):609–624
7. Arčabić V, Škrinjarić T (2019) Synchronization and spillovers of business cycles in the European Union. In: Šimurina J, Načinović Braje I, Pavić I (eds) Proceedings of FEB zagreb international odyssey conference on economics and business. Faculty of Economics & Business, University of Zagreb, Opatija, pp 113–127
8. Baele L, Bekaert G, Schäfer L (2015) An anatomy of central and eastern european equity markets. In: Columbia Business School Working Paper, No. 15–71. Columbia Business School, Singapore
9. Barndor-Nielsen OE (2002) Econometric analysis of realized volatility and its use in estimating stochastic volatility models. J Roy Stat Soc Ser B (Stat Methodol) 64(2):253–280
10. Baruník J, Kočenda E, Vácha L (2016) Asymmetric connectedness on the US stock market: bad and good volatility spillovers. J Financial Markets 27:55–78
11. Baruník J, Kočenda E, Vácha L (2017) Asymmetric volatility connectedness on the forex market. J Int Money Finance 77:39–56
12. Baumöhl E (2013) Stock market integration between the CEE-4 and the G7 markets: asymmetric DCC and smooth transition approach. MPRA Paper No. 43834
13. Bauwens L, Laurent S, Rombouts JVK (2006) Multivariate GARCH models: a survey. J Appl Econometrics 21(1):79–109
14. Bein MA, Tuna G (2015) Volatility transmission and dynamic correlation analysis between developed and emerging european stock markets during sovereign debt crisis. J Econom Forecast 2:61–80
15. Beirne J, Caporale GM, Schulze-Ghattas M, Spagnolo N (2010) Global and regional spillovers in emerging stock markets: a multivariate GARCH-in-mean analysis. Emerg Markets Rev 11(3):250–260
16. Bekaert G, Wu G (2000) Asymmetric volatility and risk in equity markets. Rev Financ Stud 13:1–42
17. Belke A, Dubova I (2017) International spillovers in global asset markets, Ruhr Economic Papers, No. 696, RWI—Leibniz-Institut für Wirtschaftsforschung, Essen
18. BenSaïda A (2018) Good and bad volatility spillovers: An asymmetric connectedness. J Financ Markets 43:78–95
19. Bevilacqua M (2018) Asymmetric volatility spillovers between developed and developing European countries. MNB Working Papers, No 2018/2
20. Black F (1976) Studies of stock price volatility changes. In: Proceedings of the 1976 meetings of the american statistical association, business and economical statistics section, pp 177–181
21. Bostanci G, Yilmaz K (2020) How connected is the global sovereign credit risk network? J Bank Financ 113(105761) forthcoming
22. Boţoc C, Anton SG (2020) New empirical evidence on CEE's stock markets integration. World Economy 2020(00):1–18
23. Brzeszczyński J, Welfe A (2007) Are there benefits from trading strategy based on the returns spillovers to the emerging stock markets? Emerg Markets Financ Trade 43(4):74–92

24. Bubák V, Kočenda E, Žikeš F (2011) Volatility transmission in emerging European foreign exchange markets. J Bank Finance 35(11):2829–2841
25. Caraiani P (2012) Nonlinear dynamics in CEE stock markets indices. Economics Letters 114:329–331
26. Cevik EI, Korkmaz T, Cevik E (2017) Testing causal relation among central and eastern European equity markets: evidence from asymmetric causality test. Econom Rese-Ekonomska Istraživanja 30(1):381–393
27. Chelley-Steeley PL (2005) Modeling equity market integration using smooth transition analysis: a study of eastern european stock markets. J Int Money Finan 24(5):818–831
28. Chen M-P, Chen P-F, Lee C-C (2014) Frontier stock market integration and the global financial crisis. North Am J Econom Finan 29:84–103
29. Cox LAT Jr (2008) Why risk is not variance: an expository note. Risk Anal 28(4):925–928
30. Czerny A, Koblas M (2008) Stock markets integration and the speed of information transmission. Czech J Econom Finance 58:2–20
31. Dajčman S, Festić M (2012) The interdependence of the stock markets of slovenia, the Czech Republic and hungary with some developed european stock markets—The effects of joining the European Union and the Global Financial Crisis. Romanian J Econom Forecasting 15(4):163–180
32. Dedi L, Škorjanec D (2017) Volatilities and equity market returns in selected central and Southeast European countries. Ekonomski pregled 68(4):384–398
33. Demian C-V (2011) Cointegration in Central and East European markets in light of EU accession. J Int Financ Markets, Inst Money 21(1):144–155
34. Dennis P, Mayhew S, Stivers C (2006) Stock returns, implied volatility innovations, and the asymmetric volatility phenomenon. J Financ Quant Analys 31(2):381–406
35. Diebold FX, Yilmaz K (2009) Measuring financial asset return and volatility spillovers with application to global equity markets. Econ J 119(534):158–171
36. Diebold FX, Yilmaz K (2012) Better to give than to receive, predictive directional measurement of volatility spillovers. Int J Forecast 28(1):57–66
37. Diebold FX, Yilmaz K (2014) On the network topology of variance decompositions, measuring the connectedness of financial firms. J Econometr 182:119–134
38. Diebold FX, Yilmaz K (2011) Equity market spillovers in the Americas. In: Alfaro R (ed) Financial stability, monetary policy, and central banking. Santiago, Bank of Chile Central Banking Series, vol 15, pp 199–214
39. Dumitrescu S (2015) European equity market return. Volatility Liquidity Spillover Dyna During Eurozone Debt Crisis, Financ Stud 19(2):30–50
40. Engle RF, Ng VK (1993) Measuring and testing the impact of news on volatility. J Finan 48:1749–1778
41. Fabozzi FJ, Markowitz HM (eds) (2002) The theory and practice of investment management. Wiley, Hoboken, New Jersey
42. Federova E (2011) Transfer of financial risk in emerging Eastern European stock markets, a sectoral perspective. BOFIT Discussion Papers 24. Bank of Finland
43. Ferreira P (2018) What guides central and eastern european stock markets? A view from detrended methodologies. Post-Communist Economies 30(6): 805–819
44. French KR, Schwert GW, Stambaugh R (1987) Expected stock returns and volatility. J Financ Econ 19:3–29
45. Gamba-Santamaria S, Gomez-Gonzalez JE, Hurtado-Guarin JL, Melo-Velandia LF (2017) Stock market volatility spillovers, evidence for Latin America. Finan Res Lett 20:207–216
46. Garcia R, Tsafack G (2011) Dependence structure and extreme comovements in international equity and bond markets. J. Bank Finan. 35:1954–1970
47. Gilmore CG, McManus MG (2002) International portfolio diversification, US and Central European equity markets. Emerg Markets Rev 3:69–83
48. Gjika D, Horváth R (2013) Stock market comovements in Central Europe, evidence from the asymmetric DCC model. Econ Model 33:55–64

49. Glosten LR, Jagannathan R, Runkle DE (1993) On the relation between the expected value and the volatility of the nominal excess return on stocks. J Finan 48:1779–1801
50. Golab A, Allen DE, Powell R (2015) Aspects of volatility and correlations in european emerging economies. In: Finch N (ed) Emerging markets and sovereign risk. Palgrave Macmillan, UK
51. Guidi F, Ugur M (2014) An analysis of South-Eastern European stock markets, evidence on cointegration and portfolio diversification benefits. J Int Financ Markets, Inst Money 30:119–136
52. Guidi F, Gupta R (2010) Cointegration and conditional correlations among German and Eastern Europe equity markets. Munich Personal RePEc Archive. MPRA Paper No. 21732, Munchen, Germany
53. Guidolin M, Timmermann A (2007) Asset allocation under multivariate regime switching. J Econom Dynam Control 31:3503–3544
54. Guidolin M, Timmermann A (2008) International asset allocation under regime switching, skew, and kurtosis preferences. Rev Financ Stud 21:889–935
55. Gębka B, Serwa D (2007) Intra- and inter-regional spillovers between emerging capital markets around the world. Res Int Bus Finan 21(2):203–221
56. Harrison B, Moore W (2009) Spillover effects from London and Frankfurt to central and eastern European stock markets. Appl Financ Econom 19(18):1509–1521
57. Harrison B, Moore W (2010) Non-linear stock market co-movement in central and East European countries. In: Matousek R (ed) Money, banking and financial markets in central and Eastern Europe. Palgrave macmillan studies in banking and financial institutions. Palgrave Macmillan, London
58. Hatemi-J A (2012) Asymmetric causality tests with an application. Empirical Econom 43:447–456
59. He F, Wang Z, Yin L (2020) Asymmetric volatility spillovers between international economic policy uncertainty and the U.S. stock market. North Am J Econom Finan 51(101084):1–29
60. Heinz FF, Sun Y (2014) Sovereign CDS Spreads in Europe, the role of global risk aversion, economic fundamentals, liquidity, and spillovers. IMF Working Paper, WP/14/17
61. Herrmann, S, Jochem A (2003) The international integration of money markets in the Central and East European accession countries, deviations from covered interest parity, capital controls and inefficiencies in the financial sector. Discussion paper 08/03, Economic Research Centre, Deutsche Bundesbank, Frankfurt, Germany
62. Horváth R, Petrovski D (2013) International stock market integration, Central and South Eastern Europe compared. Econom Syst 37(1):81–91
63. Horváth R, Petrovski D (2012) International stock market integration, Central and South Eastern Europe compared, William Davidson Institute Working Paper Number 1028
64. Hung NT (2020) An analysis of CEE equity market integration and their volatility spillover effects. Eur J Manage Bus Econom 29(1):23–40
65. Investing (2020). https://www.investing.com
66. Ivanov M (2014) Volatility spillovers and stock market co-movements among Western, Central and Southeast European stock markets. Eur J Econom Manage 1(1):47–72
67. Jackwerth JC, Vilkov G (2014) Asymmetric volatility risk, evidence from option markets. Available at SSRN, http://ssrn.com/abstract=2325380
68. Kamisli M, Kamisli S, Ozer M (2015) Are volatility transmissions between stock market returns of Central and Eastern European countries constant or dynamic? Evidence from MGARCH models. The Conference Proceedings of 10th MIBES Conference, Larisa, 15–17 October 2015, 190-203
69. Karagoz K, Ergun S (2009) Stock market integration among Balkan countries, Proceedings of 4th Annual MIBES international conference, 276–286
70. Kasch-Haroutounian M, Price S (2001) Volatility in the transition markets of central Europe. Appl Financ Econom 11(1):93–105
71. Kenourgios D, Samitas A (2011) Equity market integration in emerging balkan markets. Res Int Bus Finan 25(3):296–307

72. Latinović M, Bogojević Arsić V, Bulajićc M (2018) Volatility spillover effect in Western Balkans. Acta Oeconomica 68(1):79–100
73. Longin FM, Solnik B (2001) Extreme correlation of international equity markets. J Finan 56(2):649–676
74. Louzis DP (2013) Measuring return and volatility spillovers in Euro area financial markets. Bank of Greece Working Paper, No. 154
75. Lucey BM, Voronkova S (2006) The relations between emerging European and developed stock markets before and after the Russian crisis of 1997–1998. In: Batten JA, Kearney C (eds) Emerging European financial markets, independence and integration post-enlargement, Vol. 6, International Finance Review Book Series, Elsevier, The Netherlands, pp 383–413
76. Lütkepohl H (2006) New introduction to multiple time series analysis. Springer, Berlin
77. Lütkepohl H (1993) Introduction to multiple time series analysis. Springer-Verlag
78. Lütkepohl H (2010) Vector autoregressive models. Economics Working Paper ECO 2011/30, European University Institute
79. MacDonald R (2001) Transformation of external shocks and capital market integration. In: Schöder M (ed) The new capital markets in central and eastern Europe. Springer-Verlag, Berlin
80. Mandaci PE, Torun E (2007) Testing integration between the major emerging markets. Central Bank Rev 1:1–12
81. Medovikov I (2016) When does the stock market listen to economic news? New evidence from copulas and news wires. J Bank Finan 65:27–40
82. Mensi W, Boubaker FZ, Al-Yahyaee KH, Kang SH (2018) Dynamic volatility spillovers and connectedness between global, regional, and GIPSI stock markets. Finan Res Lett 25:230–238
83. Moore T (2007) Has entry to the European Union altered the dynamic links of stock returns for the emerging markets? Appl Financ Econom 17(17):1431–1446
84. Morales L, Andreosso-O'Callaghan B (2014) The global financial crisis, world market or regional contagion effects? Int Rev Econom Fin 29:108–131
85. Morales L, Andreosso-O'Callaghan, B (2010) The global financial crisis, world market or regional contagion effects?. MFA (Midwest Finance Association) Annual Meeting, February 24th-27th February (2010) Las Vegas. Nevada, USA
86. Munteanu A, Filip A, Pece A (2014) Stock market globalization, the case of emerging european countries and the US. Procedia Econom Fin 15:91–99
87. Nelson DB (1991) Conditional Heteroskedasticity in asset returns, a new approach. Econometrica 59:347–370
88. Olbrys J, Majewska E (2013) Granger causality analysis of the CEE stock markets including nonsynchronous trading effects. Argumenta Oeconomica 2(31):151–172
89. Onay C (2006) A co-integration analysis approach to European Union integration, the case of acceding and candidate countries. Eur Integr Online Papers 10:2–11
90. Parkinson M (1980) The extreme value method for estimating the variance of the rate of return. J Bus 53:61–65
91. Pesaran HM, Shin Y (1998) Generalized impulse response analysis in linear multivariate models. Econom Lett 58:17–29
92. Pop C, Bozdog D, Calugaru A (2013) The Bucharest Stock Exchange case, is BET-FI an index leader for the oldest indices BET and BET-C? Int Bus Res Teach Pract 7(1):35–56
93. Radovanov B, Marcikić A (2016) Integration and regional position of stock markets, a case of South East European countries. In: Symposium proceedings—XV international symposium Symorg 2016, Reshaping the future through sustainable business development and entrepreneurship, pp 598–603
94. Rozlucki W (2010) The creation of exchanges in countries with communist histories. In: Exchanges Regulated (ed) Harris, L. Oxford University Press, Dynamic Agents of Economic Growth, Oxford
95. Samitas A, Kenourgios D, Paltalidis N (2006) Short and long run parametric dynamics in the Balkans stock markets. Int J Bus Manage Econom 2(8):5–20

96. Scheicher M (2001) The comovements of stock markets in Hungary, Poland and the Czech Republic. Int J Finan Econom 6(1):27–39
97. Schotman PC, Zalewska A (2006) Non-synchronous trading and testing for market integration in central european emerging markets. J Empir Finan 13(4):462–494
98. Schwert GW (1990) Stock volatility and the crash of '87. Rev Financ Stud 3:77–102
99. Sehgal S, Saini S, Deisting F (2019) Examining dynamic interdependencies among major global financial markets. Multinat Finan J Multinat Finan J 23(1–2):103–139
100. Shen Y-Y, Jiang Z-Q, Ma J-C, Wang G-J, Zhou W-X (2020) Sector connectedness in the Chinese stock markets. Cornell University paper, arXiv.2002.09097
101. Silvennoinen A, Teräsvirta T (2009) Multivariate GARCH models. In: Andersen TG, Davis RA, Kreiss J-P, Mikosch ThV (eds) Handbook on financial time series, pp 201–229
102. Stoica O, Perry MJ, Mehdian S (2015) An empirical analysis of the diffusion of information across stock markets of Central and Eastern Europe. Prague Econom Papers 24(2):192–210
103. Sugimoto K, Matsuki T, Yoshida Y (2014) The global financial crisis, an analysis of the spillover effects on African stock markets. Emerg Markets Rev 21:201–233
104. Svilokos T (2012) Capital market cointegration of old and new EU Member States. Econom Res -Ekonomska Istraživanja 25(1):313–336
105. Syllignakis MN, Kouretas GP (2010) German, US and central and eastern european stock market integration. Open Econom Rev 21:607–628
106. Syllignakis MN, Kouretas GP (2011) Dynamic correlation analysis of financial contagion, Evidence from the Central and Eastern European markets. Int Rev Econom Finan 20(4):717–732
107. Syriopoulos T (2004) International portfolio diversification to Central European stock markets. Appl Financ Econom 14(17):1253–1268
108. Syriopoulos T (2007) Dynamic linkages between emerging European and developed stock markets, Has the EMU any impact? Int Rev Financ Analys 16:41–60
109. Syriopoulos T, Roumpis E (2009) Dynamic correlations and volatility effects in the Balkan equity markets. J Int Financ Markets, Inst Money 19(4):565–587
110. Trenca I, Petria N, Dezsi E (2014) Linkages among the stock markets of Eastern Europe. Revista Econom 66(1):91–104
111. Trenca I, Petria N, Pece AM (2015) Empirical inquiry of gregarious behavior, evidence from European emerging markets. Revista Econom 67(2):143–160
112. Uluceviz E, Yilmaz K (2018) Measuring real-financial connectedness in the U.S. Economy. Koc University-TUSIAD Economic Research Forum, Working Paper No. 1812
113. Urbina J (2013) Financial spillovers across countries, Measuring shock transmissions. MPRA Working paper
114. Vizek M, Dadić T (2006) Integration of Croatian. CEE and EU Equity Markets, Cointegration Approach, Ekonomski Pregled 57(9–10):631–646
115. Voronkova S (2004) Equity market integration in central european emerging markets, a cointegration analysis with shifting regimes. Int Rev Financ Analys 13(5):633–647
116. Wang P, Moore T (2008) Stock market integration for the transition economies, time-varying conditional correlation approach. The Manchester School 76:116–133
117. Wu G (2001) The determinants of asymmetric volatility. Rev Financ Stud 14(3):837–859
118. Xu W, Ma F, Chen W, Zhang B (2019) Asymmetric volatility spillovers between oil and stock markets, Evidence from China and the United States. Energy Econom 80:310–320
119. Yavas BF, Dedi L (2016) An investigation of return and volatility linkages among equity markets, a study of selected European and emerging countries. Res Int Bus Finan 37:583–596
120. Yilmaz K (2010) Return and volatility spillovers among the East Asian equity markets. J Asian Econom 21(3):304–313
121. Yuce A, Simga-Mugan C (2000) Linkages among Eastern European stock markets and the major stock exchanges. Russian and East European Finan Trade 36(6):54–69
122. Égert B, Kočenda E (2007) Interdependence between Eastern and Western European stock markets, evidence from intraday data. Econom Syst 31:184–204

123. Özer M, Kamenković S, Grubišić Z (2020) Frequency domain causality analysis of intra- and inter-regional return and volatility spillovers of South-East European (SEE) stock markets. Econom Res-Ekonomska Istraživanja 33(1):1–25
124. Šikić L, Šagovac M (2017) An international integration history of the Zagreb Stock Exchange. Publ Sect Econom 41(2):227–257
125. Škrinjarić T (2019) Stock market stability on selected CEE and SEE markets, a quantile regression approach. Post-Commun Econom 32(3):352–375
126. Škrinjarić T, Orlović Z (2020) Economic policy uncertainty and stock market spillovers, case of selected CEE markets. Mathematics 8(7):1–33
127. Škrinjarić T, Šego B (2019) Risk connectedness of selected CESEE stock markets, spillover index approach. China Finance Review International 10(4):447–472
128. Škrinjarić, T (2020) Return, risk and market index online volume search interdependence, shock spillover approach on Zagreb Stock Exchange. Economic Review. Forthcoming
129. Škrinjarić, T, Lovretin Golubić Z, Orlović Z (2020) Shock spillovers between exchange rate return, volatility and investor sentiment. In: Šimurina J, Načinović Braje I, Pavić I (eds) Proceedings of FEB Zagreb 11th international odyssey conference on economics and Business, Zagreb, Faculty of Economics and Business, 2020, pp 358–372

Chapter 6
Millennial Customers and Hangout Joints: An Empirical Study Using the Kano Quantitative Model

Tiwari Siddharth, Yash Daultani, and R. Rajesh

Abstract The ascent of the service sector in almost every economies of the world has created among practitioners a major share of interest in service operations. In practice, the impact of the service expertise on consumers' satisfaction is vital for service marketers. Many service providers wanted the utilization of varied improvement actions to boost their performance. To be really successful, a service provider must create value for its customers. The needs of an individual have changed drastically and as an effort to achieve this requirement, fulfillers are often traced to the traditional times of human origin. Customers continuously foresee to having a brand new and completely different experience after they walk into a store and once, they fail to get so, disappointment takes over. Today, what folks very want doesn't seem to be merchandise, however satisfying experiences. Experiences are attained through series of activities. The need of human interface is apparent in any service settings. We study empirically using a survey of about 224 respondents, based on the seven dimensions the service experience of customers from the view of customer journey. Additionally, in this chapter we explore the concept of the importance–satisfaction model, the refined Kano, and attempt to associate it with customer value. The results of the study demonstrate a new framework, illustrating the service experience, the dimensions influencing the service experience, a way to choose applicable practical actions, and the way it is linked to customer value. We use the students' profile and age group of the students as the control variables for this study.

Keywords Consumer satisfaction · Customer value · Kano model · I-S (importance of satisfaction) · Empirical study

---

T. Siddharth · R. Rajesh (✉)
ABV-Indian Institute of Information Technology and Management, Gwalior, India
e-mail: rajesh@iiitm.ac.in

Y. Daultani
Indian Institute of Management Lucknow, Lucknow, India

6.1 Introduction

It is quite evident that, the café industry stands a challenging area that strains the establishment of high level customer service and uninterrupted quality improvement. As lifestyles change, the frequency of a persons' dining out becomes more. The customers' desire new flavour, cosy ambience satisfying the need of making pleasant memories. To summarise, an excellent on the whole dining experience. The café dine-out experience is encompassed of together tangible and intangible elements. While, tangible essentials can effortlessly be enhanced, the intangible portion of the eating place facility requires extensive thought. Researchers have explained in their studies the capability to bring high quality service will deliver long term financial viability and sustainable business triumph [1]. Henceforth, the eateries that offer customers with superior services can improve a stronger spirited position in today's vibrant market-place.

This chapter is based on the idea of keeping the customer at the centre. It involves research using the analytical tool to deep dive into the study by understanding the customer's preference and needs when it comes to choosing a weekend joint. This study would be of a great help to the owner of the hangout joints in India. They can anticipate the demand and tweak on what needs to be different from others. As for the customer needs, this study will enable the users to gain knowledge about what kind of hangout joint is available, the trending places around and what is popular among the crowd and choose according to their preferences. The café has experienced greater than before rivalry and rising opportunities of customers which concerns their general service quality. There has remained a necessity to inspire local utilization, attracting the advent of guests, and identify the customers' wants to meet their desires.

6.1.1 Motivations of the Study

Over the period, the consumers have developed shopping as a habit and a mode of their leisure time activity. With increased comforts and changes in lifestyle, reflected by mode of travel, outfits, living environments, food habits, and shopping environments, new outlook has propelled at the namesake of outing. There is a rising demand for lifestyle products, eating out, and entertainment; thus, these expectations have generated exciting doing out, shopping in various categories and formats of retail outlets. The term "client experience" has received an increasing attention from marketing choice makers and scholastic researchers. According to Schmitt [2], 'Experiences' are personal events that occur in response to several stimulations. These experiences involve the whole living beings and often result from direct observation and/or contribution in the events, whether they are real, illusory or virtual.

6.1.2 Background of the Study

Choi and Chu [3] suggests that, to be victorious in the hospitality industry, the service providers must present better-quality customer worth. As per a managerial standpoint, client value has to be designed, resources should be organized to accomplish the required level and human resources set in place to instrument the strategy. In a café business, value is carried through several key qualities which include menu surroundings physical environment and services. Thus, value is a package of tangibles and intangibles contradictory in arrangement transverse to folks. Essentially, for executives, client value is a business creation to be carried to clienteles, whilst for the patron this is a service to be practiced. Cafés may intend to put across a precise figure and elevate a firm level of anticipation. The clienteles may recognize and infer this in a different way. Service delivery can be observed from two varied standpoints: that of the service providers or else that of the customers' and the standpoints would be anticipated to vary. Exclusively, in the hospitality zone (e.g.: Hangout Joint business), there are approximately struggle tangled in the assessment of service delivery. Most customers' may observe a service as being reasonably insignificant, whereas for others it is crucial.

This research explores the above matters with a vision to expand the thoughtfulness of the groups of customers' value based on customer service occurrence. In this regard, the key determination of the research is to investigate and incorporate SERVQUAL, Importance–Satisfaction (I–S) model, the Kano model [4], in order to conclude suitable realistic activities and its connected client value. The numerous categories of client value delimited in these methods influence client attainment, client continuance, and client brim indirectly and directly. As per a managerial standpoint, client value has to be designed, resources should be organized to accomplish the required level and human resources set in place to instrument the strategy. For the client value is a survived experience and is usually a trade-off among profit and expenses. In a café business, value is carried through several key qualities which includes menu surroundings physical environment and services. Thus, value is a package of tangibles and intangibles contradictory in arrangement transverse to folks. The combined model of generating client value offered in this research can be a valuable orientation for service providers that demand to produce superior value for clienteles and henceforth for the firms themselves. The remnants of the research are planned as follows. Section 6.2 presents a detailed literature review, which is followed by the model development in Sect. 6.3. Section 6.4 details the results and related discussions, followed by the conclusions and scope of future works in Sect. 6.5.

6.2 Literature Review

We present the literature review of the relevant articles in Table 6.1.

Table 6.1 Literature on customer satisfaction analysis techniques

S. No.	Author and year	Contribution	Analysis techniques	Findings
1	Sukwadi et al. [5]	Identified and analysed the concept of value to summarize and link with the multiple tools and techniques	SERVQUAL, Kano Model, blue ocean strategy, and I-S model	Argued that when customers' experiences are at centre, there are considerable differences amid cafés and consumers' opinions on what constitutes enjoyable experiences
2	Sathish and Venkatesakumar [6]	Analysed the levels of satisfaction among the youth and developed the model of customer experience management	Customer experience management	Demonstrated the growing demand for life style products, consumption out, and activities
3	Marković et al. [7]	Observed the customer expectations analysis on restaurants as per their performance on all the parameters	Dineserv approach	Observed the gap between customers' outlook and their discernment of delivered service and commented this as the chief indicator for assessing service quality of service providers
4	Mey et al. [8]	Observed on the major youth of Asia in comparison to non-Asian youth considering customer satisfaction point	SERVQUAL	Perceived that the smallest expectations and lowest perceptions of their hotel stay in Malaysia were from Malaysian hotel guests
5	Large and König [9]	Developed an instrument for the measurement of the internal service quality of purchasing departments	Enhanced SERVQUAL	Offered a procedure to improve the internal customer orientation of purchasing departments
6	Patrício et al. [10]	Discussed the applicability of SERVQUAL to restaurant services	SERVQUAL	Noticed that the emergence of a factor that reflects individual attention given to customers; is aligned with current marketing trends

6.3 Model Development

6.3.1 Research Gap

Researchers have focused majorly on the top resorts, hotels and restaurants; whereas this research will shift the locus of area to the ground level covering all types of hangout joints i.e. cafés, pubs, clubs, sport bars, food courts and restaurants. Most of the research has been based on the demographics of non-Asian geographical areas, while this study will have its roots in the Indian context. As trends in millennial are changing on a fast pace, earlier researches have become redundant day by day.

6.3.2 Research Objectives

(i) To analyse the millennial customer's mind-set and demands for choosing a hangout joint for meeting each other, (ii) To understand how to create a loyal customer and analyse the ways to provide customer satisfaction, (iii) To derive managerial insights for owners and managers of hangout joints for designing and improving their services.

6.3.3 The Kano Model

6.3.3.1 Kano Model

In the quality management extent, many methods are accessible for examining and measuring the service quality routine of the service providers. The Kano model has been typically applied within the quality management area more specifically. Many researchers [4, 11] would agree that Kano model has the reward by categorizing customer desires. Kano model was inspire from Herzberg's motivator–hygiene model as level (or "two-way") model [12]. This Kano model is shown in Fig. 6.1 and the quality attributes are sorted into five categories:

1. Attractive quality attributes: attributes that give satisfaction if present, but that result in no dissatisfaction if absent;
2. One-dimensional quality attributes: attributes characterized by a linear relationship between the customers' perception of satisfaction and the degree of fulfilment of the attributes;
3. Must-be quality attributes: attributes whose absence will result in customer dissatisfaction, but whose presence does not significantly contribute to the customer satisfaction;
4. Indifferent quality attributes: attributes that result in neither satisfaction nor dissatisfaction, regardless of being fulfilled or not; and

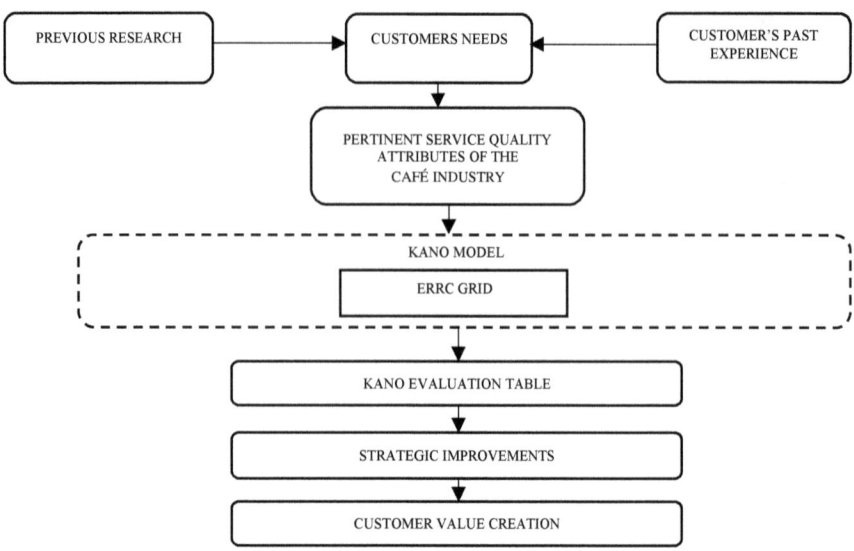

Fig. 6.1 Kano framework. Adapted from [5]

5. Reverse quality attributes: attributes that result in dissatisfaction when fulfilled and in satisfaction when not fulfilled.

In classifying customer needs, trade-offs are sometimes necessary. If some service attributes cannot be met instantaneously for technical or fiscal reasons, service providers must deliberate other standards that have the utmost effect on customer satisfaction. To resolve this problematic, Kano model which is considering the degree of position of attributes is proposed. Consuming obtained the consequences for degree of importance, the qualities are then classified into two broad categories: (i) "high" if the degree of importance is greater than the mean and, (ii) 'low' if the degree of importance is fewer than the mean. Tables 6.2, 6.3 and Fig. 6.2 present the redefined categories of quality attributes obtained by refining the Kano model.

The categories shown in Table 6.2 is helpful in evaluating the categories.

Table 6.2 Categories in refined Kano model

Kanos' categories	High important	Low important
Attractive	High attractive	Less attractive
One dimensional	High value added	Necessary
Must-be	Critical	Necessary
Indifferent	Potential	Care free

Adopted from [12]

Table 6.3 Categories in refined Kano model

Eliminate	Reduce
Factors or elements that no-longer have value or may even detract from value for customers	Attributes that have been over designed in the race of competition or those have less attraction of customers
Raise	Create
Attributes that can result in significant value for customers or those that have high attraction to customers	Factors that can discover new sources of value for customers or those that can create new demand and attract noncustomers

Adopted from [12]

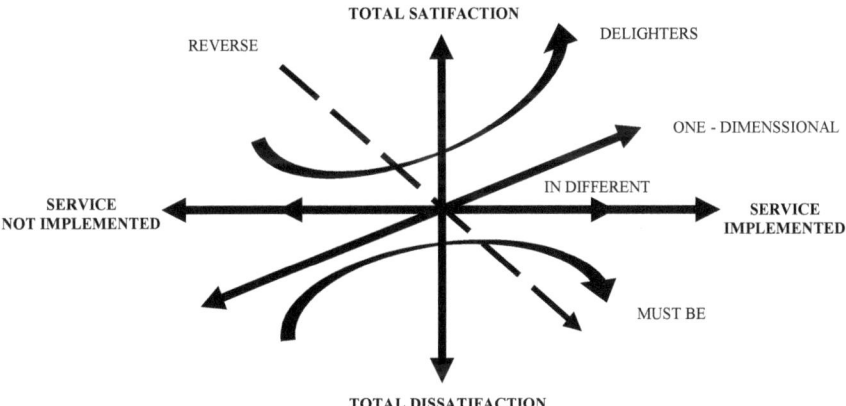

Fig. 6.2 Kano model. Adopted from [12]

6.3.3.2 ERRC Model

Kim et al. [13] utilized a four-action framework referred to as the 'eliminate reduce raise create grid' (ERRC). The ERRC applied actions reconstruct client value perception by responding four features of the grid in Table 6.3.

6.3.4 Research Design

6.3.4.1 Factor Analysis

Factor analysis is used in the fields of psychology, education and is the method of understanding sympathetic self-reporting questionnaire. Initially, factor analysis reduces a huge number of variables into a slighter set of variables. Furthermore, it establishes fundamental dimensions among measured variable and latent construct, thus allowing the creation and modification of theory. Then, it provides construct rationality indication of self-reporting scales. Factor analysis is normally used in the fields of psychology, education and is measured the method of outstanding for sympathetic self-reporting questionnaire. Factor analysis is statistical techniques that has many uses, three of which determination be fleetingly. Initially, factor analysis reduces a huge number of variables into a slighter set of variables.

6.3.4.2 Quantitative Analysis of Kano Model

Kano model is used as a start line of the projected measuring, which comprises conducting pilot research, emerging and administrating the Kano questionnaire, yet analysing questionnaire results to spot Kano categories for every CR. Afterward gaining the classification results, the projected four-step approach is applied to quantify the Kano model. The model for the study is shown in Fig. 6.3. Figure 6.3 explains about the model of the study where in we have decided the factors on the left and the variables to make the decision on the right-hand side. To calculate the customer satisfaction index.

- Calculating the CS and DS Values

The projected quantitative analysis of Kano model flinches with devious two significant values i.e. the amount of customer satisfaction (CS) and the degree of customer dissatisfaction (DS). Meanwhile dissimilar customers usually have dissimilar needs and expectations, calculating CS and DS value can reproduce the average impact of a

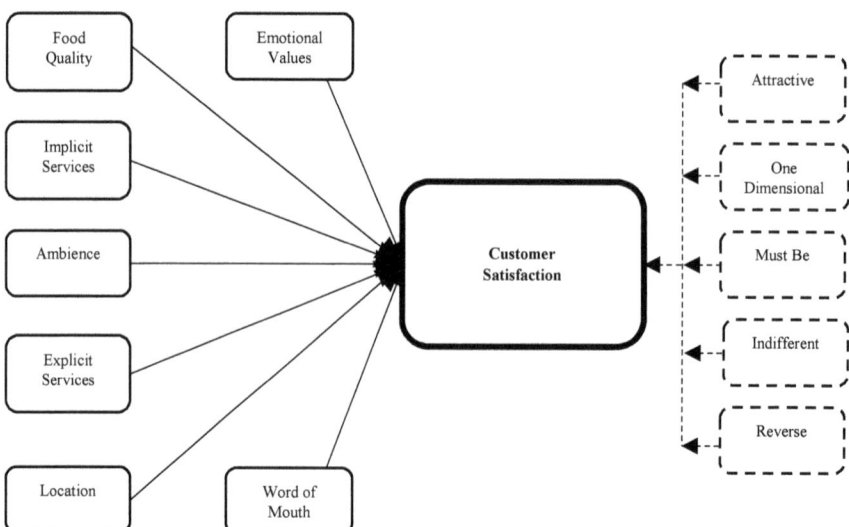

Fig. 6.3 Model for the study

CR on the satisfaction of all customers. To compute the CS value for CRi (CS_i), it is essential to add all the responses with satisfaction fundamentals attractive and one-dimensional and division it by the total number of attractive (f_A), one-dimensional (f_O), must-be (f_M) and indifferent responses (f_I):

$$CS_i = (f_A + f_O)/(f_A + f_O + f_M + f_I) \quad (6.1)$$

Likewise, the DS value for CRi (DS_i) is considered by adding all responses with dissatisfaction elements one-dimensional and must-be and dividing it by the entire number of attractive, one-dimensional, must-be and indifferent responses:

$$DS_i = -(f_O + f_M)/(f_A + f_O + f_M + f_I) \quad (6.2)$$

- Determining the CS and DS Points

In the preceding step, two standing values (CS and DS) were calculated for 3 Kano categories of CRs. When significant CS and DS values are obtained, the terms of "the existence of a certain CR or its sufficiency and non-fulfilment of a CR" are used to define the level of fulfilment of a certain CR.

- Plotting the Relationship Curves

After observing the CS and DS points, the assembly curves among customer satisfaction and CR execution are plotted, which is shown in Fig. 6.4. The *x axis* represents CR fulfilment level starting from 0 to 1. The *y axis* represents the degree of customer satisfaction or dissatisfaction starting from −1 to 1.

- Identifying the S-CR Relationships Function

As per the above figure, it can be understood that the relationships among customer satisfaction and CR fulfilment (S-CR) can be roughly quantified by a suitable function. Normally speaking, the S-CR relationship function can be articulated as $S = f(x, a, b)$, where S signifies the degree of customer satisfaction, x signifies the fulfilment level of CRs extending from 0 to −1, a and b are modification parameters for diverse Kano quantitative categories of CRs.

- One Dimensional Attribute

About the one dimensional CRs, the relationship curve can be exclusively identified, since for any two separate points, there is only one line through them. The relationship function can be expressed as $S = a_1 x + b_1$, where a_1 is the slope of the line and b_1 is the DS value when the CR fulfilment level (x) is equal to 0. Relieving the CS and DS points, that is, (1, CS_i) and (0, DS_i), into the equation, it gives that $a_1 = CS_i + DS_i$ and $b_1 = DS_i$. Therefore, the S-CR function for one-dimensional attributes is:

$$S_i = CS_i - DS_i x_i + DS_i \quad (6.3)$$

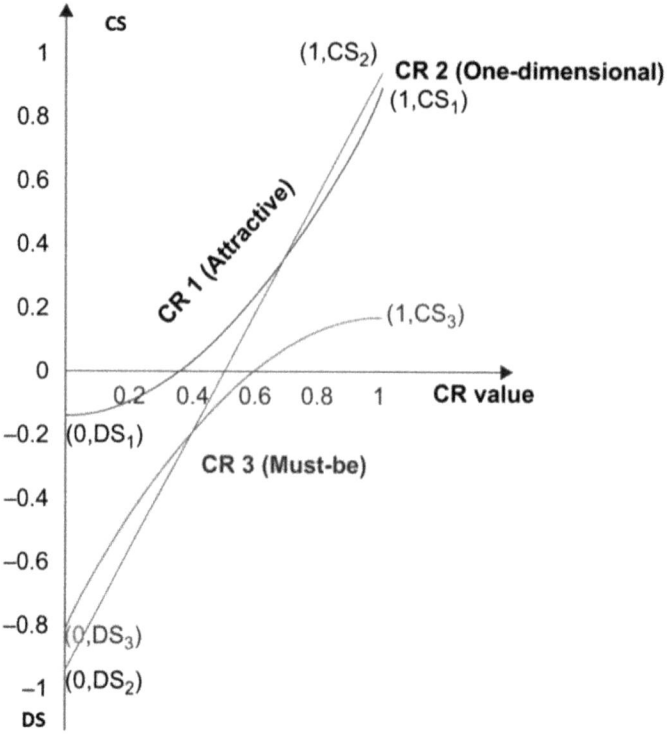

Fig. 6.4 Relations among customer satisfaction and CR fulfilment

- Attractive Attribute

The S-CR function of attractive attribute can be projected by an exponential function $S = a_2 e^x + b_2$. Analogous to the instance of one dimensional attributes, a_2 is a constraint for regulating the slope of the curve and b_2 is for regulating the vertical of the relationships curve in the Kano diagram. Relieving the CS and DS point into the equation bounces that $a_2 = (CS_i - DS_i)/(e - 1)$ and $b_2 = -(CS_i - eDS_i)/(e - 1)$. Consequently, the S-CR function for attractive attributes is:

$$S_i = [(CS_i - DS_i)/(e - 1)]e_i^x - (CS_i - eDS_i)/(e - 1) \qquad (6.4)$$

- Must Be Attribute

Likewise, the S-CR function of must be attribute can also be estimated by an exponential function, $S = a_3(-e^{-x}) + b_3$. Relieving the CS and DS points, $(1, CS_i)$ and $(0, DS_i)$, into the equation gives that:

$$a_3 = [e(CS_i + DS_i)/(e - 1)] \qquad (6.5)$$

$$\text{and,} \quad b_3 = (eCS_i + DS_i)/(e-1) \quad (6.6)$$

Thus, the S-CR function for must-be attributes is:

$$S_i = [e(CS_i + DS_i)/(e-1)]e - x + (eCS_i + DS_i)/(e-1) \quad (6.7)$$

6.3.5 Survey Instrument

Since the review of literature, 21 self-reported items were adapted for this research from multiple literatures which are referred in reference. The questionnaire utilized in the current research consisted of three items for the dimension of *Ambience*, three items for the measurement of *Food Quality*, five items for the measurement of *Implicit Services*, three items for the measurement of the *Word Of Mouth*, two items for the measurement for the *Explicit Services*, two items for the measurement of the *Location* and three items for the measurement of *Emotional Values*. A five point Likert scale (1 = Strongly Disagree, 2 = Disagree, 3 = Neutral, 4 = Agree, 5 = Strongly Agree) was used for each item.

6.3.5.1 Questionnaire Analysis

This study will be directed by using a three stage research methodology. Initially, a group list of all the applicable service quality attributes connected to café industry was derived from SERVQUAL dimensions and identified based on some previous literature review. Afterward gathering professional advices and inputs from academic supervise. This research is explanatory exploration with a quantitative approach. The population in this study is a consumer of multiple cafes located in Central India. In this study there were 21 question items and 224 samples were taken for the study. The questionnaire was given to consumers who were in multiple cafes also it was promoted through multiple social media channel (Facebook, Instagram, Twitter and TikTok) to get the reliable responses of the youth and genX. Data will be taken by giving questionnaires to respondents using a Likert scale of 1 to 5. Number 1 states strongly disagree with number 5 which states strongly agree. In this study, the research concept that was built was the effect of Ambiance, Food Quality, Location, Word of Mouth, Emotional Values, Implicit Services and Explicit Services as per the model used in Fig. 6.3 on customer satisfaction.

- Ambience of a café in measuring Customer Satisfaction

Ambience is very important in your customer's experience, the purchasers will form an impact of your business supported the atmosphere of your offices and outlets, this impression can either be positive or negative and can ultimately have a bearing on their overall experience along with your brand. The three-item self-reported measurement of Ambience scale from Parasuraman et al. [14] was accustomed measure customer's satisfaction level in this chapter.

- Food Quality of a café in measuring Customer Satisfaction

The quality of service or quality of food have a significantly influence on customer satisfaction. In accumulation, service quality dimensions also customer satisfaction have a significantly influence on customer retention. The three-item self-reported measurement of Food Quality scale from Khan et al. [15] was used to measure customer's satisfaction level in this chapter. These items were unswervingly related to ambience to improve the customer experience had a positive relationship with Customer's Satisfaction.

- Location of a café in measuring Customer Satisfaction

It is clear that a café's location needs to be convenient for your customers, to build a good reputation for the café's name, location is extremely important. The two-item self-reported dimension of Location scale from Schoefer [16] was used to degree customer's satisfaction level in this chapter. However, three items of Location for interacting with customers is directly related to measuring customer satisfaction in Indian Café environment were selected for this chapter. These items were directly related to ambience to enhance the customer experience had an optimistic relationship with Customers' Satisfaction.

- Word of Mouth of a café in measuring Customer Satisfaction

Word of mouth also known as viva-voce, is the short-lived of evidence from person to person using oral communiqué, which might be as modest as telling somebody the time of diurnal. The three item self-reported dimension of Word of Mouth scale from Parasuraman et al. [14] was used to measure customer's satisfaction level in this chapter. These items were related to ambience to enhance the customer experience had an optimistic relationship with Customers' Satisfaction.

- Emotional Value of a café in measuring Customer Satisfaction

Contains in the worth gain from clients feelings or emotional situations when espousing and using a service. The three-item self-reported dimension of Emotional Value scale from Yuan and Wu [17] was used to degree customer's satisfaction level in this chapter. Originally, their customer satisfaction scale consisted of twelve items with three different categories. However, three items of Emotional Value for interacting with customers is directly related to measuring customer satisfaction in Indian Café setting were designated for this chapter.

- Implicit Services of a café in measuring Customer Satisfaction

Psychological benefits or extrinsic features which the consumer may sense only vaguely The five item self-reported measurement of Implicit Services scale from

Parasuraman et al. [14] was used to measure customer's satisfaction level in this chapter. These items were related to ambience to augment the customer experience had a positive relationship with Customer's Satisfaction.

- Explicit Services of a café in measuring Customer Satisfaction

The two item self-reported dimension of Explicit Services scale from Parasuraman et al. [14] was used to degree customer's satisfaction level in this chapter. Originally, their customer satisfaction scale consisted of multiple items with five servqual categories. These items were straight related to ambience to enhance the customer experience had a positive relationship with Customer's Satisfaction.

6.3.5.2 Data Analysis

The main persistence of this research chapter was to inspect the suitability of the items and the internal consistency of the constructs that the instruments measure. For these motives, an exploratory factor analysis was initially conducted to assess the factor structure of the scale. Next, a reliability analysis on items was performed to test the reliability of the questionnaire set, and then Kano Quantitative Analysis using Kano Evaluation has been performed on the data set.

6.3.5.3 Exploratory Factor Analysis

EFA is generally performed within the primary phases of emerging a replacement or revised instrument [18], the exploratory correlational analysis was conducted through use of the Statistical Package for the Social Sciences (SPSS, version 23). We calculate the Kaiser-Meyer-Olkin (KMO) Measure of Sampling Adequacy (KMO) and Bartlett's Test of Sphericity to verify that the data collected for an exploratory correlational analysis were appropriate. The KMO test was wont to verify the sampling adequacy for the analysis, and Bartlett's test of sphericity was wont to determine if correlations between items were sufficiently large for EFA. If the results of the initial EFA show items which are loading on the incorrect factors or cross-loading on multiple factors, those items are to be deleted.

6.3.5.4 Reliability Test

In this research, the model is measured and was tested using SPSS Version 23.0. We check the validity and reliability of the collected data, where validity test is employed to check the validity of the questionnaire statement and the questionnaire reliability test is conducted to check the consistency of the questionnaire given to respondents. The internal consistency of data was tested by means of Cronbach's alpha for each competency in SPSS. Good internal consistency implies that if, a contributor who responses a survey item positively is supplementary likely to answer other items inside the survey definitely.

6.3.5.5 Kano Quantitative Analysis

The key target of the study is to inspect what are the key customer wants or requirements (CRs) for their customer loyalty level. Targeted customers are the campus students, who are a fundamental client segment in the café service sector. The study is additionally appeared by four phases including leading preliminary research, processing survey results and applying Kano quantitative analysis. Kanos' evaluation table is according to Table 6.2.

Conducting the Preliminary Study

The Questionnaire were since refined and assembled into seven categories; i.e., Ambiance, Food Quality, Location, Word of Mouth, Emotional Values, Implicit Services and Explicit Services. Hence, the Kano questionnaire is examining these CRs and their impact on customer satisfaction.

Processing Survey Results

The survey received 224 answers, on behalf of the youth of the central Indian public characterized by numerous disciplines and different ways of life style. For implementing the Kano model and classification depictions discussed before, the demographics of the survey are abbreviated in Table 6.4. Rendering to survey results, only one attribute, of having loyalty benefits is named as in-different traits. In this way, it would not be contained in the additional assessment of Kano model in the resulting segment because of their unimposing impact on customer satisfaction.

Table 6.4 Demographic summary

	Group	Frequency (N-224)	%
Gender	Male	118	52.679
	Female	106	47.321
Age group	15–21 years	45	20.089
	22–30 years	168	75.000
	Others	11	4.911
Education level	Junior high school	2	0.893
	Senior high school	17	7.589
	College and university	147	65.625
	Post graduate	57	25.446
	Others	1	0.446

Implementing Kano Quantitative Analysis Model

Based on the survey results from above stage, the anticipated quantitative analysis of Kano model is then applied to perceive S-CR relationship function for various CRs, which can obviously mirror the effect of disparate CRs on customer satisfaction with any café. In view of the last Kano association, proper equations are then set to process the estimations a, b and to direct the fundamental function for each CR. All the S-CR functions are observed. The research study of café design proves that the planned methodology can be emphatically acknowledged to recognize the assorted connections among customer satisfaction and CR fulfilment level.

6.4 Results and Discussion

6.4.1 Exploratory Factor Analysis Results

We conduct an EFA on 21 items and used a quartimax rotation using SPSS version 23.0. EFA is a statistical method working to surge the reliability of the scale by classifying unsuitable items that can be detached. In this research, the seven factors (i.e., Ambience, Food Quality, Location, Word of Mouth, Emotional Values, Implicit Services and Explicit Services) are used to control the pattern of the construction in the 21-item of empirical study of café customer's satisfaction instrument. And we construct a scree plot, and conduct KMO and Bartlett's test, and observe the total variance explained and the rotation matrix. All the Demographic details of the responses are mentioned in Table 6.4. The total variance explained is shown in Table 6.5 and the factor rotation matrix is shown in Table 6.6.

The Kaiser-Meyer Olkin Measure confirmed the sampling adequacy for the analysis, KMO $= 0.764$ which is overhead Kaiser's suggested threshold of 0.6. Bartlett's test of sphericity is equal to 1283.87, $p < 0.000$ at degree of freedom of 210, designated that correlations among items were sufficiently large for EFA. The initial 21-item structure explained 66.91% of the variance in the pattern of relationships among the items, which is also satisfactory in EFA.

6.4.2 Reliability Analysis Results

An item analysis was showed to test the reliability of each factor of the desires of the millennial for the hang out joints instrument. And the internal consistency is found to be satisfactory and it ranges from 0.6 to 0.9. All seven factors on this scale had high rating for reliability. The Cronbach's alpha for Ambience, Food Quality, Location, Word of Mouth, Emotional Values, Implicit Services and Explicit Services were also tested separately and was observed to be satisfactory.

Table 6.5 Total variance explained as in EFA

Total variance explained

component	initial eigenvalues			Extraction sums of squared loadings			Rotation sums of squared loadings		
	Total	% of variance	Cumulative %	Total	% of variance	Cumulative %	Total	% of variance	Cumulative %
1	4.908	21.337	21.337	4.908	21.337	21.337	3.839	16.690	16.690
2	2.988	12.991	34.328	2.988	12.991	34.328	1.901	8.267	24.957
3	2.169	9.430	43.758	2.169	9.430	43.758	1.893	8.230	33.188
4	1.374	5.972	49.730	1.374	5.972	49.730	1.878	8.164	41.352
5	1.107	4.815	54.545	1.107	4.815	54.545	1.568	6.818	48.170
6	1.005	4.371	58.916	1.005	4.371	58.916	1.490	6.480	54.650
7	0.923	4.014	62.930	0.923	4.014	62.930	1.449	6.300	60.950
8	0.917	3.986	66.915	0.917	3.986	66.915	1.372	5.965	66.915
9	0.826	3.591	70.507						
10	0.768	3.341	73.848						
11	0.715	3.108	76.956						
12	0.663	2.884	79.840						
13	0.640	2.784	82.624						
14	0.589	2.560	85.184						
15	0.540	2.346	87.530						
16	0.472	2.051	89.580						
17	0.451	1.960	91.540						
18	0.414	1.802	93.341						
19	0.363	1.579	96.920						
20	0.341	1.484	99.404						

(continued)

Table 6.5 (continued)

Total variance explained

component	initial eigenvalues			Extraction sums of squared loadings			Rotation sums of squared loadings		
	Total	% of variance	Cumulative %	Total	% of variance	Cumulative %	Total	% of variance	Cumulative %
21	0.323	1.405	100.000						

Extraction Method: Principal Component Analysis.

Table 6.6 Factor rotation matrix

	Factor						
	1	2	3	4	5	6	7
A1			0.529				
A2			0.581				
A3			0.705				
FQ1	0.593						
FQ2	0.386						
FQ3	0.470						
IS1				0.791			−0.132
IS2				0.499			
IS3				0.315			
IS4				0.613			
IS5				0.670			
ES1							0.671
ES2							0.565
L1	−0.023				0.588		
L2					0.421		
WOM1						0.669	
WOM2						0.506	
WOM3						0.644	
EV1		0.434					
EV2		0.616					
EV3		0.592					

Extraction Method: Principal Axis Factoring.
Rotation Method: Quartimax with Kaiser Normalization.

6.4.3 Kano Quantitative Analysis Results

Constructed on the survey outcomes, the anticipated quantitative analysis of Kano model is formerly implemented to detect S-CR relationship functions for diverse CRs, which can evidently imitate the impact of diverse CRs on customer satisfaction by café evidence. By the Kano model and category definitions conversed former, the survey result are analysed and summarized in Table 6.7.

Kano model delivers an effective method in categorizing diverse CRs into diverse categories founded on their influence on customer satisfaction. So, the results clearly show that the output framework can be used and utilized in any sort of café designing and making a customer experience excellent and to achieve the desired customer satisfaction level. This study provides the one dimensional, attractive and must be attributes of the café, which can be implemented in any Indian conditions and scenarios. So, this research and results proves that the attributes chosen in the research

Table 6.7 Kano quantitative analysis results for calculating CS and DS

Customer requirements	A	O	M	I	R	Total (N)	Category	CS	DS
A1	103	94	20	5	2	224	A	0.887387387	0.513513514
A2	117	67	31	5	4	224	A	0.836363636	0.445454545
A3	141	69	10	1	3	224	A	0.950226244	0.357466063
FQ1	118	92	9	3	2	224	A	0.945945946	0.454954955
FQ2	81	109	30	3	1	224	O	0.852017937	0.623318386
FQ3	86	115	20	1	2	224	O	0.905405405	0.608108108
IS1	113	100	9	1	1	224	A	0.955156951	0.488789238
IS2	79	123	17	4	1	224	O	0.905829596	0.627802691
IS3	62	118	34	6	4	224	O	0.818181818	0.690909091
IS4	71	130	19	2	2	224	O	0.905405405	0.671171171
IS5	25	101	64	27	7	224	O	0.580645161	0.760368664
ES1	39	95	37	5	48	224	O	0.761363636	0.75
ES2	16	35	31	65	77	224	R	0.346938776	0.448979592
L1	20	42	89	56	17	224	M	0.299516908	0.632850242
L2	30	66	113	13	2	224	M	0.432432432	0.806306306
WOM1	50	140	29	3	2	224	O	0.855855856	0.761261261
WOM2	46	132	38	3	5	224	O	0.812785388	0.776255708
WOM3	68	125	24	5	2	224	O	0.869369369	0.671171171
EV1	41	67	101	12	3	224	M	0.488687783	0.760180995
EV2	27	68	104	22	3	224	M	0.429864253	0.778280543
EV3	50	100	68	4	2	224	O	0.675675676	0.756756757

is reliable and can positively influence the customer experience and satisfaction levels. Also, these results can be implemented in I-S Model to achieve the conclusion or for an ERRC Grid analysis. Hence, we can say that the One-Dimensional attributes, Attractive attributes can be enhancing and retained.

6.5 Conclusions and Future Scope

6.5.1 Conclusions

By studying consumer view on in service experience of nowadays, this research enlarges the existing literature, in which a supposed growth in customer value is in attention. This research explores what is essentially behind the notion of such a surge, from service providers as well as from customer's point of view. This research has shown that when customers experience is in emphasis, there are considerable variances among cafés and consumers' opinions on what constitutes pleasurable experience and how these might have been brought in café surroundings. However, cafes' are strongly absorbed on finding suitable action of attractive consumer experience, customers description of outstanding experiences is still to a huge extent, established by values. This research has industrialized a valuable methodology for analysing service attributes on a strategic basis. The customer value model projected here is based on, the refined Kanos' model and mutual with the quantitative evaluation method. By utilizing this analytic methodology, service providers can make strategic improvement results with respect to the service offers.

The customer value model projected here is based on, the refined Kano model and by utilizing this analytic methodology, service providers can make strategic improvement results with respect to the service offers. As an outcome, the service providers can significantly grow the customers' satisfaction and reduce the operating costs, henceforth there is an increase in the overall customers' value. This eventually ends up with several competitive advantages for service providers. This research responds to the categorical call to discover value creation in a café sector by giving a framework for understanding how a café may create value for customers. Per se, this research contributes to the service quality discipline in three ways. Initially, this research explicitly associates service experience and customer value creation with in the café industry. Secondly, the framework established during this study clearly defines the broad range of critical service attributes, which service providers can manage in their pursuit and is an important creation for his or her customers. Thirdly, the most important part is based on the model the owner can understand the related customer value that could be obtained from the customer margin, which is littered with both customer retention and customer acquisition. The S-CR relationship function supports in understanding the customer needs in an exceedingly best precise way which will help us in understand the satisfaction level on the later process.

6.5.2 Future Scope

Future work might be coordinated to the current angle by incorporating Kano model into some product configuration tools, similar to Quality Function Deployment (QFD). Future work could represent considerable authority in improving several suspicions by inferring progressively strong information in deciding the type of relationship curves of attractive and must-be attributes objectively. Also, further study in this area is required on inspecting the effect of this framework and the manner in which it can raise the image/brand and profit through value creation. In contrary, for future research, the customer experience management metrics can be additionally incorporated to deliver the feedback information and to deal with the customer experience. This input data can be frequently incorporated with the organization's metrics systems (i.e. its business balanced scorecard and individual service standards).

References

1. Keiser TC (1988) Strategies for enhancing service quality. J Serv Mark 2:65–70
2. Schmitt B (1999) Experiential marketing. J Mark Manag 15(1–3):53–67
3. Choi TY, Chu R (2001) Determinants of hotel guests' satisfaction and repeat patronage in the Hong Kong hotel industry. Int J Hosp Manag 20(3):277–297
4. Yang CC (2005) The refined Kano's model and its application. Total Qual Manag Bus Excell 16(10):1127–1137
5. Sukwadi R, Yang CC, Fan L (2012) Capturing customer value creation based on service experience–a case study on news café. J Chin Inst Ind Eng 29(6):383–399
6. Sathish AS, Venkatesakumar R (2011) Coffee experience and drivers of satisfaction, loyalty in a coffee outlet-with special reference to "café coffee day" . J Contemp Manag Res 5(2):1
7. Marković S, Raspor S, Šegarić K (2010) Does restaurant performance meet customers' expectations? An assessment of restaurant service quality using a modified DINESERV approach. Tourism Hosp Manage 16(2):181–195
8. Mey LP, Akbar AK, Yong DGF (2006) Measuring service quality and customer satisfaction of the hotels in Malaysia: Malaysian, Asian and non-Asian hotel guests. J Hosp Tour Manag 13(2):144
9. Large RO, König T (2009) A gap model of purchasing's internal service quality: concept, case study and internal survey. J Purch Supply Manag 15(1):24–32
10. Patrício V, Leal RP, Pereira ZL (2006) Applicability of SERVQUAL in restaurants: an exploratory study in a Portuguese resort. Enterp Work Innov Stud 2:127–136
11. Baek SI, Seung KP, Weon SY (2009) Understanding key attributes in mobile service: Kano model approach. In: Smith MJ, Salvendy G (eds) Human interface, Part II. HCII, pp 355–364
12. Sauerwein E, Bailom F, Matzler K, Hinterhuber HH (1996) The Kano model: how to delight your customers. Int Work Semin Prod Econ 1(4):313–327
13. Kim WGK, Ng CYN, Kim Y (2009) Influence of institutional DINESERV on customer satisfaction, return intention and word-of-mouth. Int J Hosp Manag 28:10–17
14. Parasuraman A, Zeithaml VA, Berry LL (1985) A conceptual model of service quality and its implications for future research. J Mark 49(4):41–50
15. Khan SA, Liang Y, Shahzad S (2015) An empirical study of perceived factors affecting customer satisfaction to re-purchase intention in online stores in China. J Serv Sci Manag 8(03):291
16. Schoefer K (2008) The role of cognition and affect in the formation of customer satisfaction judgements concerning service recovery encounters. J Consum Behav Int Res Rev 7(3):210–221
17. Yuan YHE, Wu CK (2008) Relationships among experiential marketing, experiential value, and customer satisfaction. J Hosp Tourism Res 32(3):387–410
18. Wetzel A (2011) Factor analysis methods and validity evidence: a systematic review of instrument development across the continuum of medical education. Ph.D. thesis, Virginia Commonwealth University. https://scholarscompass.vcu.edu/etd/2385

Chapter 7
The Effects of Oil Shocks on Macroeconomic Uncertainty: Evidence from a Large Panel Dataset of US States

Rangan Gupta and Xin Sheng

Abstract This study investigates the effects of oil price shocks on macroeconomic uncertainty using a large panel dataset of 50 US states. We examine both linear and nonlinear impulse response functions of uncertainty to oil shocks by using the local projection method. We disaggregate oil shocks according to their origin into the oil supply (production), economic activity (aggregate demand), oil inventory (speculative demand), and oil market-specific (consumption demand) shocks. We also consider the spillover effects across the measures of uncertainty in the US when estimating the impulse responses of uncertainty to various types of oil shocks. Our results show that uncertainty is affected by both supply and demand-side oil shocks and the effects of oil shocks on uncertainty are contingent on the states of oil dependence.

Keywords Oil shocks · Macroeconomic uncertainty · Local projections · Impulse response functions · Oil dependence

JEL Codes C32 · Q41

R. Gupta
Department of Economics, University of Pretoria, Pretoria 0002, South Africa
e-mail: rangan.gupta@up.ac.za

X. Sheng (✉)
Lord Ashcroft International Business School, Anglia Ruskin University, Chelmsford CM1 1SQ, UK
e-mail: xin.sheng@aru.ac.uk

© The Author(s), under exclusive license to Springer Nature Switzerland AG 2021
S. Patnaik et al. (eds.), *Computational Management*, Modeling and Optimization in Science and Technologies 18, https://doi.org/10.1007/978-3-030-72929-5_7

7.1 Introduction

Over the last decade, there has been a dramatic increase in macroeconomic volatility and uncertainty in the global economy. As a consequence, the important role of volatility and uncertainty in the economy has gained growing attention from researchers and policymakers in recent years.[1] Existing studies in the field generally find that uncertainty is a big driver of macroeconomic fluctuations at the country-level. Realizing a tremendous amount of existing heterogeneities that exist within regions of a national economy, for instance, within the states of the United States (US) for which regional-level data is consistently available for long spans, recent studies (see, e.g., [19, 20]) have also highlighted the heterogenous but negative impact of uncertainty on regional macroeconomic variables.

In the meanwhile, there has been a rising consensus that uncertainty represents an endogenous response to unexpected changes in other macroeconomic variables (e.g., aggregate demand or supply shocks), contributing to amplifying the effects of these macroeconomic shocks [18, 20]. As a result, an important question for policymakers, both at national and regional levels, is to decide the possible driving forces of uncertainty, which in turn will provide an indication on the direction to which the economy is headed, and thus help to determine the appropriate economic policy responses. In this regard, several studies (see, e.g., [2, 5, 10, 13–16]) at the country-level show that oil price shocks, particularly the aggregate demand shocks, are major driving forces of macroeconomic uncertainty, with both direct and indirect transmission channels related to consumption, investment, production, inflation and the size of the public sector.

Under this background, this study contributes to the literature in the field by analyzing the predictive role of disaggregated oil price shocks (i.e., oil supply shocks, oil-specific consumption demand shocks, oil inventory demand shocks, and global economic activity shocks) as calculated by Baumeister and Hamilton [3], for the uncertainty levels of the 50 states of the US and District of Columbia, as developed by Mumtaz [19], and Mumtaz et al. [21]. We use a panel data set-up which allows for dependence across the US states, and predict the future path of the regional uncertainties based on impulse response functions (IRFs) derived by feeding the oil shocks into the local projections (LP) method-based linear model of Jordà [11]. Given the extent of regional heterogeneities in terms of the underlying state of each of the cross-sectional unit, national-level results are not guaranteed to be translated to regional levels, and hence makes our analysis of tremendous value to regional policymakers. Moreover in this regard, realizing that the states differ in terms of their oil dependence (as measured by the difference between oil consumption and production over oil consumption in each US state), we also analyze the impact of oil shocks in a nonlinear version of the LP model, by making the IRFs contingent on whether the state is a net producer or net consumer.

---

[1] See e.g., Gupta et al. [7–9] and Chuliá et al. [4] for reviews of the relevant literature.

To the best of our knowledge, this is the first attempt to predict regional uncertainties within the US, based on oil shocks, and also deducing the predictability by making the analysis conditional on the heterogeneity associated in terms of the oil dependency of the states. The remainder of the paper is organized as follows: Sect. 7.2 outlines the data and methodologies, with Sect. 7.3 discussing the results, and Sect. 7.4 concluding the paper.

7.2 Data and Methodology

The uncertainty data used in this study are obtained from the h-step-ahead (h = 1, 2, 3, 4) forecasts of a factor-augmented vector autoregression (FAVAR) model utilized by Mumtaz [19] and Mumtaz et al. [21].[2] The uncertainty measures are computed using the seasonally adjusted employment growth rates, unemployment rates, house price growth rates, and the real per-capita growth rates of personal incomes, benefit incomes, dividend incomes, social insurance contributions, and other incomes. The dataset includes quarterly macro uncertainty measures at four forecast horizons (i.e. h = 1, 2, 3, 4 are corresponding to forecasts in 3, 6, 9, and 12 months) for the 50 US states and D.C. from 1977:Q2 to 2015:Q3. The measures developed by Mumtaz [19] and Mumtaz et al. [21] are basically following Jurado et al. [12], who provide estimates of macroeconomic uncertainty as average time-varying variances in the unpredictable components of the real and financial variables. Mumtaz [19] and Mumtaz et al. [21] improve the estimates by filtering out the effects of idiosyncratic uncertainty and measurement error, which we use for robustness analysis. Also, note that the state-level uncertainty measures starting in 1977 use 8 macroeconomic and financial data series, while there is another version starting in 1990 which incorporates 21 series, and provides a broader measure of uncertainty, which we also use for robustness check of our main analysis.

Using the method proposed by Baumeister and Hamilton [3], the real price of oil, the production of global crude oil, real economic activity, and crude oil inventories over the same sample period are included into a structural vector autoregression (SVAR) model to generate the oil supply, oil consumption demand, economic activity, and oil inventory shocks, respectively.[3] The data are available from the Federal

---

[2] See online technical appendix of Mumtaz [19] and Mumtaz et al. [21] for details of the uncertainty construction procedures. The MATLAB codes to estimate the FAVAR model and the underlying data to generate the oil shocks is available from the website of Professor Haroon Mumtaz at: https://sites.google.com/site/hmumtaz77/research-papers?authuser=0.

[3] The real price of oil is obtained by deflating the nominal spot prices of West Texas Intermediate (WTI) with the U.S. consumer price index (CPI). The real economic activity is proxied by the industrial production index for the OECD and BRICS countries, and Indonesia. The crude oil inventories are calculated by producing the US crude oil stocks by the ratio of OECD to US petroleum inventories. The MATLAB codes to estimate the SVAR and the underlying data to generate the oil shocks is available for download from the research segment of the website of Professor Christiane Baumeister at: https://sites.google.com/site/cjsbaumeister/research.

Reserve Bank of St Louis, the US Energy Information Administration (EIA), the Federal Reserve Economic Data (FRED) and the OECD Main Economic Indicators (MEI) databases. The oil shocks estimates are available at a monthly frequency, which we then convert to quarterly data to match the uncertainty measures by using three-month averages comprising a specific quarter.

To test the linear response of economic uncertainty to oil price shocks, this study uses the LP method of Jordà [11]. The baseline model of calculating LP IRFs in panel dataset can be specified as follows:

$$U_{i,t+s} = \alpha_{i,s} + Oil\ Shock_{i,t} \beta_s + \epsilon_{i,t+s}, \text{ for } s = 0, 1, 2, \ldots, H \quad (7.1)$$

where $U_{i,t}$ is the uncertainty of state i in time t, s is the length of forecast horizons, H represents the maximum length of forecast horizons,[4] $\alpha_{i,s}$ measures the fixed effect for panel data and β_s captures the responses of economic uncertainty in time $t + s$ to an identified oil price shock in time t. The LP IRFs are computed as a series of β_s which are estimated separately at each horizon (s).[5]

We also examine whether the impacts of oil price shocks on economic uncertainty are regime-dependent and contingent on the oil dependence of US states. Oil dependency, measured by a ratio of the difference between the oil consumed and oil produced over oil consumed, can be calculated based on the raw oil consumption and production data from the EIA. Following the method proposed by Auerbach and Gorodnichenko [1], a smooth transition function is employed to switch the oil dependence of US states between high and low-regimes. The nonlinear model is specified as follows:

$$U_{i,t+s} = (1 - F(z_i))\left[\alpha_{i,s}^{R_H} + Oil\ Shock_{i,t} \beta_s^{R_H}\right]$$
$$+ F(z_i)\left[\alpha_{i,s}^{R_L} + Oil\ Shock_{i,t} \beta_s^{R_L}\right]$$
$$+ \epsilon_{i,t+s}, \text{ for } s = 0, 1, 2, \ldots, H \quad (7.2)$$

where $z_{i,t} \sim N(0, 1)$ is the switching variable measuring oil dependence in US states, with positive values of $z_{i,t}$ indicating high oil dependence in State i.[6] The smooth transition function $F(z_{i,t}) = \exp(-\gamma z_{i,t})/1 + \exp(-\gamma z_{i,t})$, $\gamma > 0$, where $F(z_{i,t})$ is bounded between 0 and 1. The values of $F(z_{i,t})$ are close to 1 if State i is in low oil dependence regime (denoted by R_L), and 0 otherwise (denoted by R_H).

---

[4]The maximum forecast horizons under invesitagion are 20 quarters in this research, which are corresponding to 5-year forecast horizons.

[5]Further details about the LP method have been discussed by Jordà [11], but due to considerations of the space limit, we are not presented in this study.

[6]According to Ravn and Uhlig [23], we de-trend the variable using the Hodrick-Prescott filter with a smoothing parameter lambda equals 1600 for quarterly data.

To test the robustness of our results, we also consider the spillover effects of US aggregate macroeconomic and financial uncertainties as obtained by Jurado et al. [12] and Ludvigson et al. [18] using 134 macro variables and 148 financial variables respectively,[7] and real estate uncertainty, as derived by Nguyen Thanh et al. [22] using 40 real estate based variables,[8] on the responses of state uncertainty to oil shocks. The model specified in Eq. (7.2) can be extended as follows:

$$U_{i,t+s} = (1 - F(z_i))\left[\alpha_{i,s}^{R_H} + Oil\ Shock_{i,t}\beta_s^{R_H}\right]$$
$$+ F(z_i)\left[\alpha_{i,s}^{R_L} + Oil\ Shock_{i,t}\beta_s^{R_L}\right]$$
$$+ X_{i,t}\gamma_s + \epsilon_{i,t+s}, \text{ for } s = 0, 1, 2, \ldots, H \quad (7.3)$$

where $X_{i,t}$ is a vector of control variables, including macroeconomic uncertainty (MU), financial uncertainty (FU), and real estate uncertainty (REU) in the US at the aggregated level.

7.3 Results and Analysis

7.3.1 Linear Impulse Responses

Figure 7.1 reports the linear impulse response functions (IRFs) of economic uncertainty (h = 1) to the disaggregated oil shocks over 20 quarters (i.e., 5 years) for our model defined in Eq. (7.2). The grey areas represent the two-standard error bands and are constructed using panel corrected standard errors.

Our results show that macroeconomic uncertainty decreases immediately after a positive economic activity shock for the first 6 quarters. This finding is in line with the existing literature that reports the negative responses of uncertainty to economic activity shocks (e.g., [2, 13–16]). This negative relationship between macroeconomic uncertainty and economic activity shock can be caused by the increase in global aggregate demand that produces positive economic outlooks and reduces uncertainty [10].

---

[7] These indices are available from the website of Professor Sydney C. Ludvigson: www.sydneyludvigson.com/data-and-appendixes.
[8] The REU index data are available from the website of Professor Johannes Strobel: https://sites.google.com/site/johannespstrobel/.

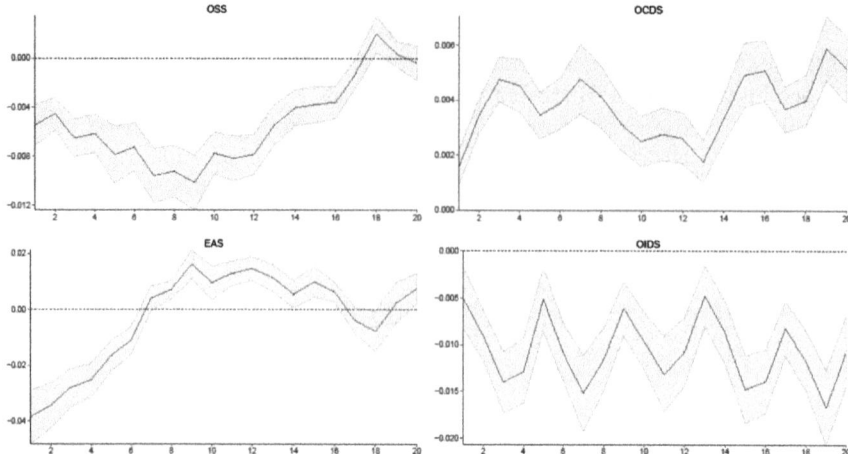

Fig. 7.1 Responses of macroeconomic uncertainty to the structural oil price shocks. *Note* OSS represents the oil supply shocks; EAS represents the global economic activity shocks; OCDS represents the oil-specific consumption demand shocks; OIDS represents the oil inventory demand shocks

Our results also show that an oil consumption demand shock has positive and statistically significant impacts on uncertainty. This finding aligns with Kang and Ratti [13–15] and Kang et al. [16], who show that consumption demand shocks trigger significant increases of uncertainty.

We find negative and statistically significant responses of macroeconomic uncertainty to a positive oil inventory demand shock. This result partially aligns with the finding from Sheng et al. [24] who find that oil inventory demand shocks can exert the negative but short-lived influences over economic uncertainty at the aggregate country-level based on evidence of a panel of 45 economies. However, the results seem to contradict to the finding from Baumeister and Hamilton [3], who indicate that positive inventory demand shocks can exert negative impacts on economic activities given their speculative nature, which results in increased macroeconomic uncertainty.

Furthermore, our results show that the oil supply shock can exert a lasting and negative influence on uncertainty and a positive oil supply shock tends to reduce uncertainty for up to 18 quarters. This negative and statistically significant relationship between oil supply shocks and uncertainty has not yet been documented in the existing literature. There are only a few previous studies about the relationships between economic uncertainty and oil supply shocks and the results reported are mixed. Our results are in contrast with Antonakakis et al. [2] and Kang and Ratti [13–15] who find little evidence that an oil supply shock influences economic uncertainty. The results seem to contradict to Su et al. [25] and Kang et al. [16] who report empirical evidence for the positive impact of oil supply shocks on macroeconomic uncertainty. Thus, such peculiar effects of oil supply shocks on uncertainty reported in our study require attention for further investigations and highlight the need of employing the extended models in the next subsections for the sensitivity and robustness tests.

7.3.2 Responses of Uncertainty Contingent on High- Versus Low-Oil Dependence States

Figure 7.2 reports estimated results for the IRFs of economic uncertainty (h = 1) to the disaggregated oil price shocks in 20 quarters for both high-oil dependence states (as shown in the 1st column) and the low-oil dependence states (as shown in the 2nd column) using the nonlinear regime-switching model defined in Eq. (7.2).

We find evidence of a persistent and profound positive effect of the oil consumption demand shock on economic uncertainty in high oil dependence states, whilst this effect is less significant in low oil dependence states. This pattern can be explained by the work of Kilian and Murphy [17], who suggest that the real price of oil reacts positively to oil-specific demand shocks. As a result, positive oil consumption shocks leading to high oil prices can be regarded as bad news for high oil dependence states since they reduce investors' expectations on economic activity and corporate revenues and tend to rise economic uncertainty.

Comparing to low-oil dependence states, we find macroeconomic uncertainty is more sensitive to economic activity shocks in high-oil dependence states. This finding is in line with the work of Sheng et al. [24] who report that macroeconomic uncertainty of oil importers is more likely to be affected by oil shocks than that of

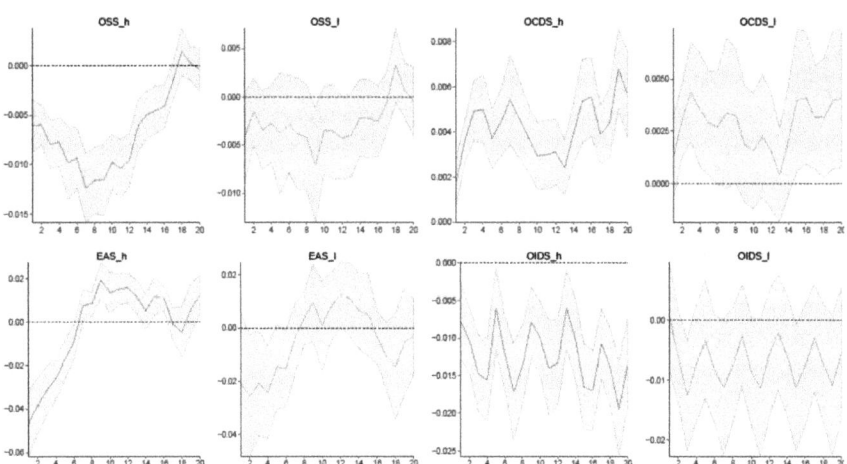

Fig. 7.2 Responses of macroeconomic uncertainty to the structural oil price shocks in the high- (*h*) oil-dependence States and low- (*l*) oil-dependence states. *Note* See Notes to Fig. 7.1 *h* and *l* indicate the high-oil dependence states and low-oil dependence states, respectively

oil exporters. In addition, our results indicate that the negative response of macroeconomic uncertainty to an oil inventory demand shock is more likely to be observed in high oil dependence states.

Our nonlinear impulse response results indicate that the negative effect of an oil supply shock on macroeconomic uncertainty can become less persistent for the low-oil dependence states comparing to the high-oil dependence states. As reported in Fig. 7.2, an oil supply shock exerts negative and statistically significant influences on economic uncertainty up to the 18th quarter for high-oil dependence states, whereas these impacts are very short-lived small in size and are only statistically significant in the 9th quarter for the low oil dependence states. Given recent evidence for the significant spillover effect across different measures of economic uncertainty in the US [6], it is important to estimate the IRFs of uncertainty to the various oil shocks after controlling for these spillover effects, which we turn to next.

7.3.3 Responses of Uncertainty at Different Horizons Conditioning on Aggregate US Uncertainty Spillovers

To examine if our results reported in Sect. 3.2 are robust to the influence of US uncertainty spillovers, we include a set of control variables, including US macroeconomic uncertainty, financial uncertainty, and real estate uncertainty at the aggregate country level, into the regime-switching model specified in Eq. (7.2). We also examine the impacts of oil shocks on measures of macroeconomic uncertainty at different forecasting horizons, i.e., over 1, 2, 3, and 4 quarters ahead (as shown in Figure 7.3a–d). The Figures report the estimated IRFs of uncertainty (for $h = 1, 2, 3, 4$) to oil shocks over 20 quarters in high- (in the 1st column) and the low- (in the 2nd column) oil dependence states using the model specified in Eq. (7.3).

The various macroeconomic uncertainty measures are available for four forecasting horizons of 1-, 2-, 3-, and 4-quarter-ahead, which enables us to test the robustness of our results in terms of responses of uncertainty at different horizons (e.g., short, medium- and long-term) to oil shocks.[9] Our results show that economic uncertainty is more responsive to oil price shocks in high-oil dependence states comparing to low-oil dependence states across all the four horizons and the IRFs at different horizons conditioning on aggregate US uncertainty spillovers are consistent with results reported in the last sub-section.

In high oil dependence states, our results indicate that the oil supply shock exerts statistically significant and negative impacts on economic uncertainty across all the

---

[9]We also used the alternative datasets of Mumtaz [19, 21] and Mumtaz et al. [18, 20] based on the metric of uncertainty which filters out the effects of idiosyncratic uncertainty and measurement errors over the same time period, and a broader measure of uncertainty starting in 1990, which uses 21 rather than 8 state-level variables. We found that the results are qualitatively similar and are included in Figs. 7.4 and 7.5 in the Appendix of this study.

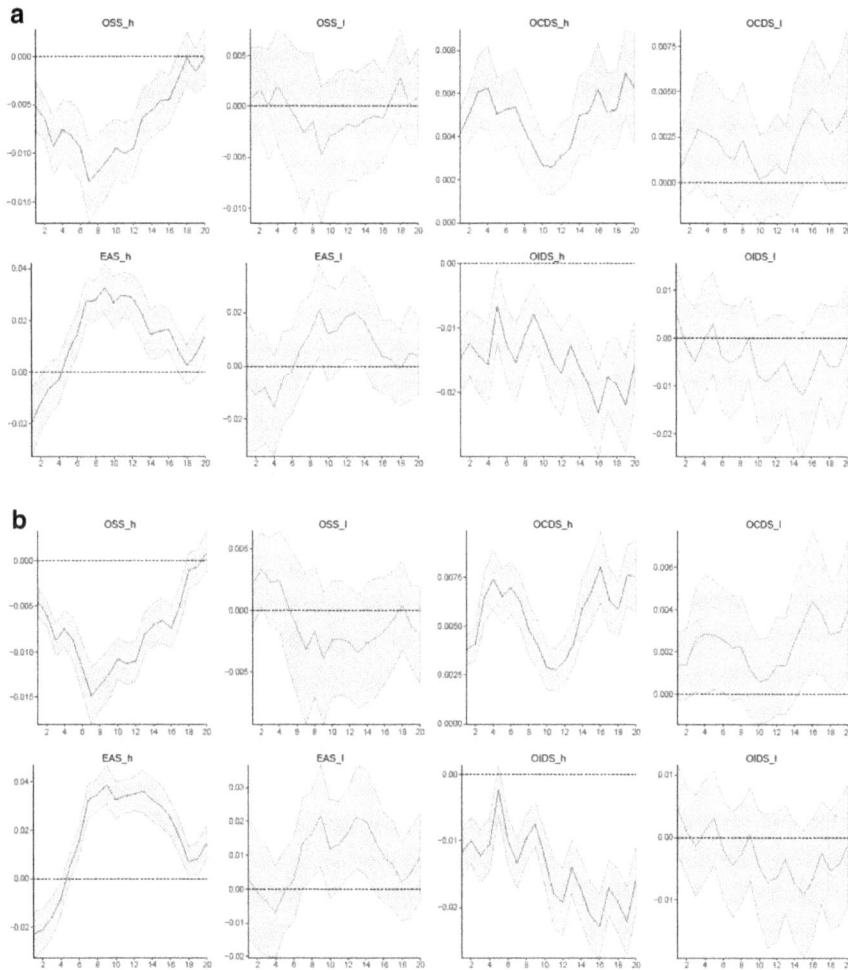

Fig. 7.3 Responses of macroeconomic uncertainty at different horizons to the four oil structural shocks after controlling for US uncertainty spillovers, **a** responses of uncertainty at horizon 1-quarter-ahead, **b** responses of uncertainty at horizon 2-quarter-ahead, **c** responses of uncertainty at horizon 3-quarter-ahead, **d** responses of uncertainty at horizon 4-quarter-ahead

four horizons, showing that an upward oil supply shock can cause a decline in uncertainty. Thus, oil supply shocks can be interpreted as good news for those states heavily relying on oil-importing and triggers a decline in uncertainty at different horizons. However, we find this effect is not significant for the low oil dependence states.

We find that uncertainty responses to oil consumption demand shocks are robust to the influence of US uncertainty spillovers in both high- and low-oil states. The patterns of IRFs are quantitively the same as the ones reported in the last sub-section. We find that economic uncertainty reacts positively to oil consumption demand

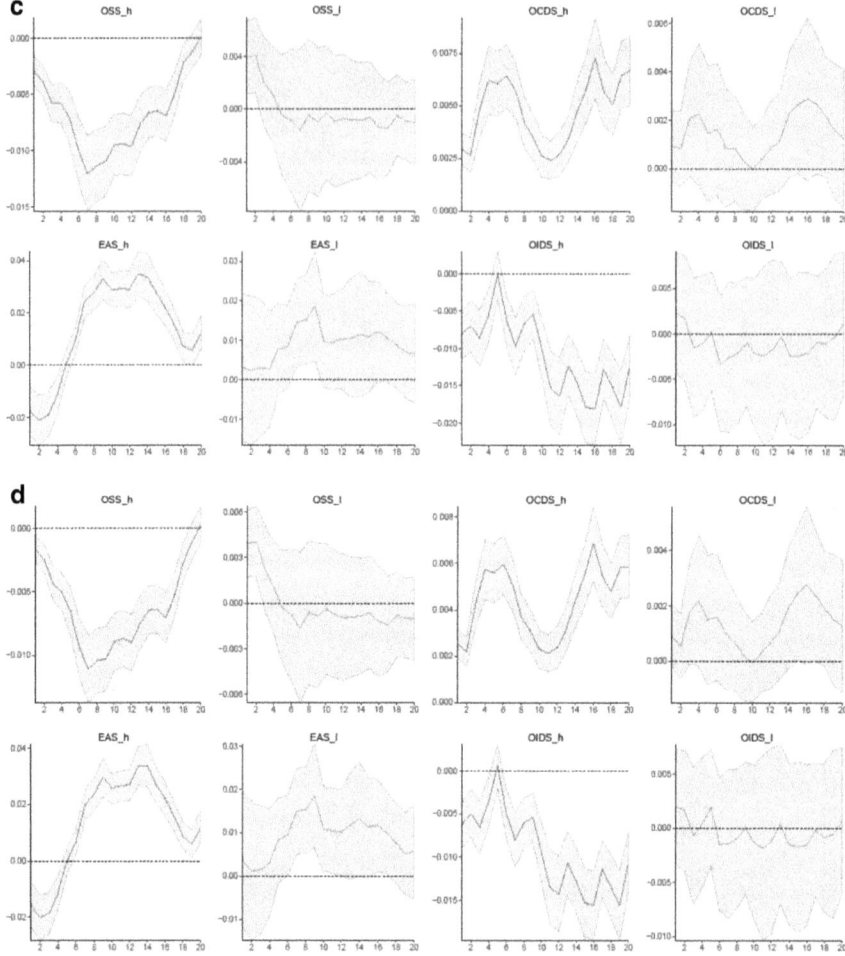

Fig. 7.3 (continued)

shocks in high oil dependence states, whilst the responses are less significant in low oil dependence states.

We also find that economic uncertainty is more responsive to the economic activity shock in the high-oil dependence states comparing to the high-oil dependence states. However, it is noted that the negative response of uncertainty to economic activity shock becomes less in size in the high-oil dependence states, while a small positive but significant response appears in the low-oil dependence states.

In addition, our robustness test results indicate that the negative impact of oil inventory demand shocks on macroeconomic uncertainty becomes not significant for low oil dependence states. This finding is partially in line with Baumeister and Hamilton [3], who find that the oil inventory demand shock seems to be less an important factor in affecting oil price movements and economic activity.

7.4 Conclusion

In this study, we investiage the impacts of four structural oil price shocks (i.e., the economic activity shocks, oil supply shocks, oil-specific consumption demand shocks and oil inventory demand shocks) on macroeconomic uncertainty in the US. Using the local projections (LP) methodology, the research contributes to the literature by examining both linear and nonlinear impulse responses of economic uncertainty to oil price shocks by utilizing a large panel dataset of 50 US states at the quarterly frequency. Our results indicate that economic uncertainty is affected by both the supply-side and demand-side oil shocks and the impacts of oil price shocks on economic uncertainty are state-dependent on oil dependence. Our results show that macroeconomic uncertainty is more responsive to oil price shocks in high-oil dependence states comparing to low-oil dependence states. Our results are robust to the influence of US uncertainty spillovers (e.g., macroeconomic uncertainty, real estate uncertainty and financial uncertainty) and various uncertainty measures at different forecasting horizons (e.g., in 3, 6, 9, and 12 quarters). The findings of this research have great implications for policymakers. For the future direction of studies, it would be of great value to extend our research for the out-of-sample forecast analysis.

Appendix

See Figs. 7.4 and 7.5.

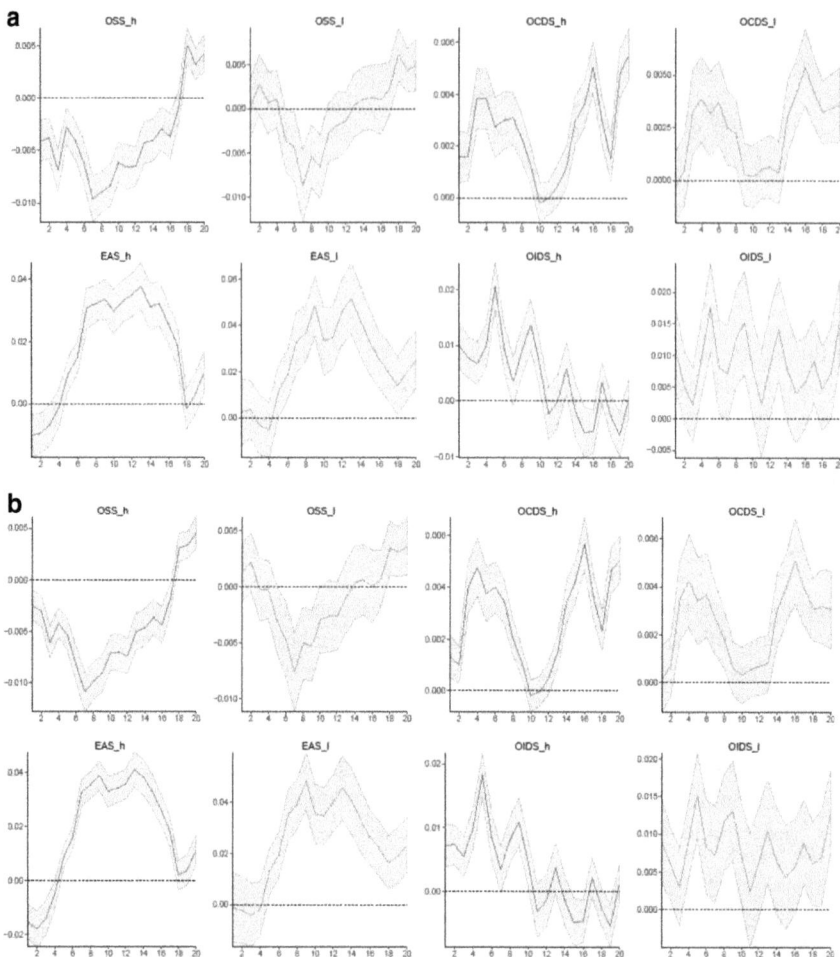

Fig. 7.4 Responses of macroeconomic uncertainty at different horizons to the four oil structural shocks after controlling for US uncertainty spillovers using the alternative measure of uncertainty as developed by Mumtaz [19] and Mumtaz (2018a): **a** responses of uncertainty at horizon 1-quarter-ahead, **b** responses of uncertainty at horizon 2-quarter-ahead, **c** responses of uncertainty at horizon 3-quarter-ahead, **d** Responses of uncertainty at horizon 4-quarter-ahead

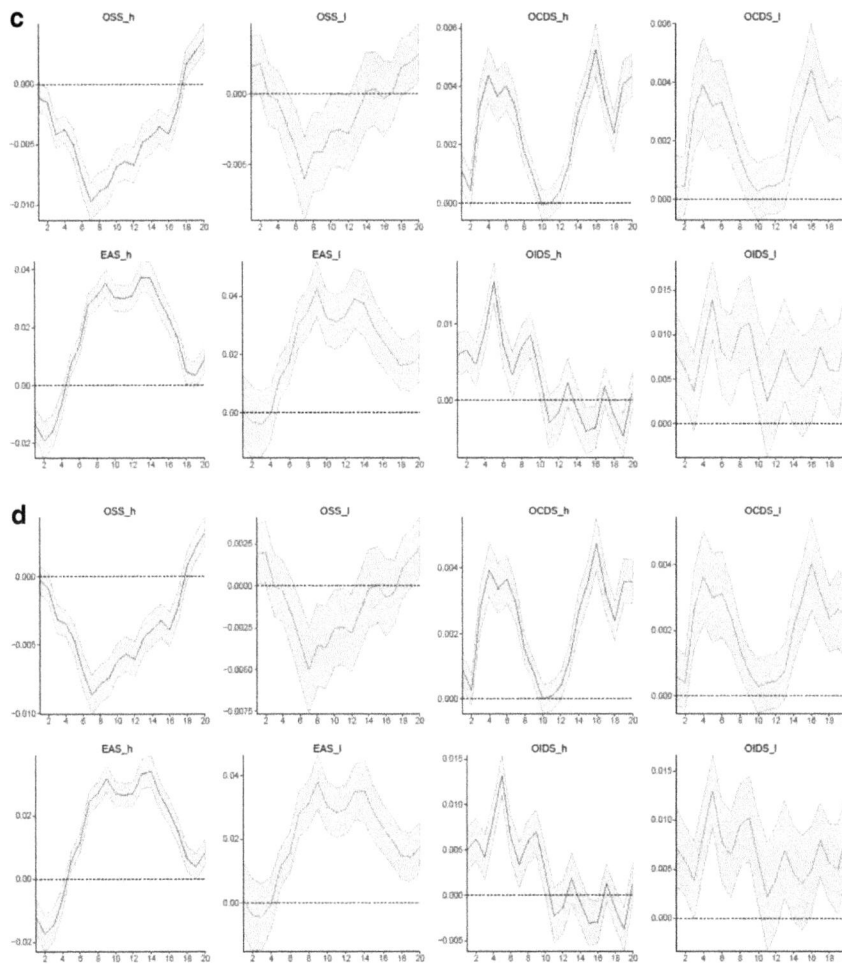

Fig. 7.4 (continued)

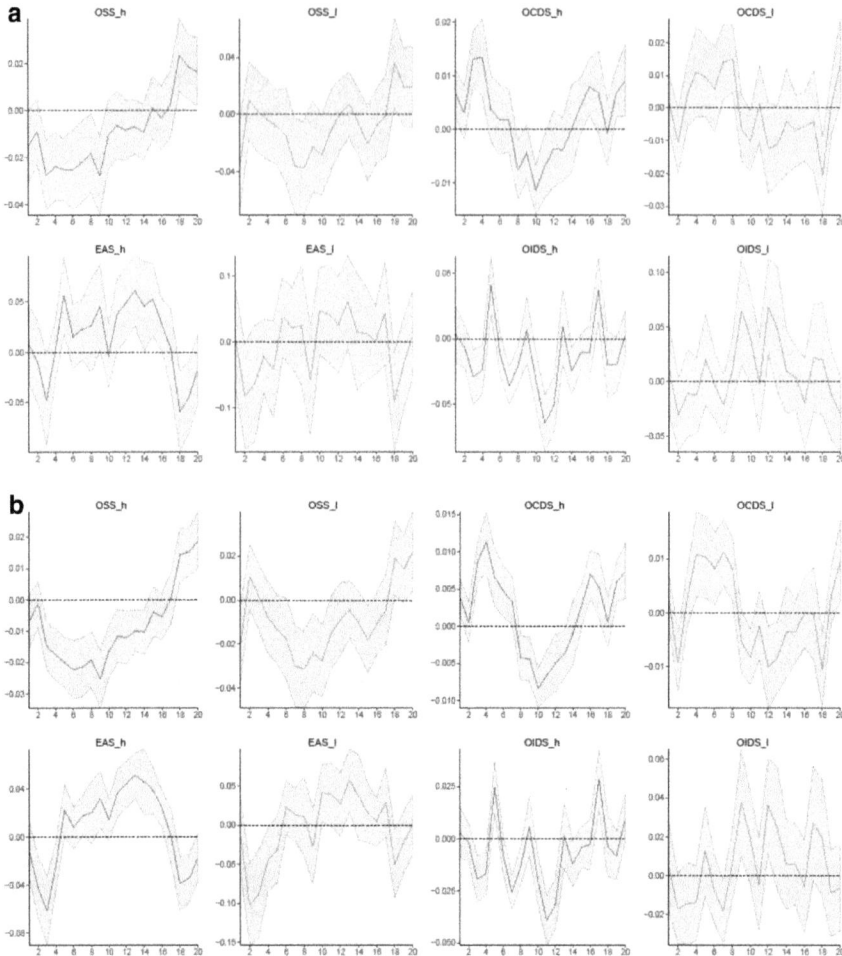

Fig. 7.5 Responses of macroeconomic uncertainty at different horizons to the four oil structural shocks after controlling for US uncertainty spillovers using the alternative broad measure of uncertainty starting in 1990 as developed by Mumtaz [19] and Mumtaz (2018a): **a** responses of uncertainty at horizon 1-quarter-ahead, **b** responses of uncertainty at horizon 2-quarter-ahead, **c** responses of uncertainty at horizon 3-quarter-ahead, **d** responses of uncertainty at horizon 4-quarter-ahead

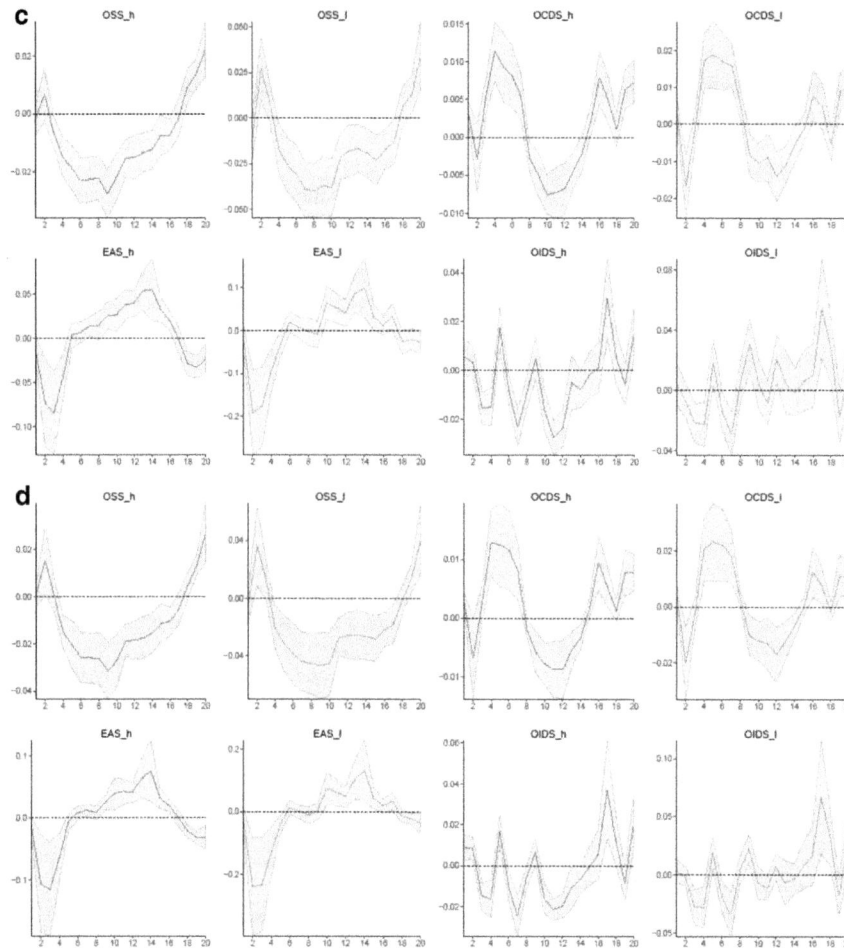

Fig. 7.5 (continued)

References

1. Auerbach AJ, Gorodnichenko Y (2013) Fiscal multipliers in recession and expansion. In: Fiscal policy after the financial crisis, p 63
2. Antonakakis N, Chatziantoniou I, Filis G (2014) Dynamic spillovers of oil price shocks and economic policy uncertainty. Energy Econ 44:433–447
3. Baumeister C, Hamilton JD (2019) Structural interpretation of vector autoregressions with incomplete identification: revisiting the role of oil supply and demand shocks. Am Econ Rev 109(5):1873–1910
4. Chuliá H, Gupta R, Uribe JM, Wohar ME (2017) Impact of US uncertainties on emerging and mature markets: evidence from a quantile-vector autoregressive approach. J Int Financ Markets Inst Money 48(C):178–191
5. Degiannakis S, Filis G, Panagiotakopoulou S (2018) Oil price shocks and uncertainty: how stable is their relationship over time? Econ Model 72(C):42–53
6. Gabauer D, Gupta R (2020) Spillovers across macroeconomic, financial and real estate uncertainties: a time-varying approach. Struct Change Econ Dyn 52:167–173
7. Gupta R, Lau C-K-M, Wohar ME (2019) The impact of US uncertainty on the Euro area in good and bad times: evidence from a quantile structural vector autoregressive model. Empirica 46:353–368
8. Gupta R, Ma J, Risse M, Wohar ME (2018) Common business cycles and volatilities in US states and MSAs: the role of economic uncertainty. J Macroecon 57:317–337
9. Gupta R, Olasehinde-Williams G, Wohar ME (2020) The impact of US uncertainty shocks on a panel of advanced and emerging market economies. J Int Trade Econ Dev. https://doi.org/10.1080/09638199.2020.1720785
10. Hailemariam A, Smyth R, Zhang X (2019) Oil prices and economic policy uncertainty: evidence from a nonparametric panel data model. Energy Econ 83:40–51
11. Jordà Ò (2005) Estimation and inference of impulse responses by local projections. Am Econ Rev 95(1):161–182
12. Jurado K, Ludvigson SC, Ng S (2015) Measuring uncertainty. Am Econ Rev 105(3):1177–1216
13. Kang W, Ratti RA (2013) Structural oil price shocks and policy uncertainty. Econ Model 35:314–319
14. Kang W, Ratti RA (2013) Oil shocks, policy uncertainty and stock market return. J Int Finan Markets Inst Money 26:305–318
15. Kang W, Ratti RA (2015) Oil shocks, policy uncertainty and stock returns in China. Econ Transit 23(4):657–676
16. Kang W, Ratti RA, Vespignani JL (2017) Oil price shocks and policy uncertainty: new evidence on the effects of US and non-US oil production. Energy Econ 66:536–546
17. Kilian L, Murphy DP (2012) Why agnostic sign restrictions are not enough: understanding the dynamics of oil market VAR models. J Eur Econ Assoc 10(5):1166–1188
18. Ludvigson SC, Ma S, Ng S (forthcoming) Uncertainty and business cycles: exogenous impulse or endogenous response? Am Econ J Macroecon
19. Mumtaz H (2018) Does uncertainty affect real activity? Evidence from state-level data. Econ Lett 167:127–130
20. Mumtaz H, Musso A (2019) The evolving impact of global, region-specific, and country-specific uncertainty. J Bus Econ Stat. https://doi.org/10.1080/07350015.2019.1668798
21. Mumtaz H, Sunder-Plassmann L, Theophilopoulou A (2018) The state-level impact of uncertainty shocks. J Money Credit Bank 50(8):1879–1899
22. Nguyen Thanh B, Strobel J, Lee G (2018) A new measure of real estate uncertainty shocks. Real Estate Econ. https://doi.org/10.1111/1540-6229.12270
23. Ravn MO, Uhlig H (2002) On adjusting the Hodrick-Prescott filter for the frequency of observations. Rev Econ Stat 84(2):371–376
24. Sheng X, Gupta R, Ji Q (2020) The impacts of structural oil shocks on macroeconomic uncertainty: evidence from a large panel of 45 countries. Energy Econ 91(C)

25. Su Z, Lu M, Yin L (2018) Oil prices and news-based uncertainty: novel evidence. Energy Econ 72:331–340

Chapter 8
Understanding and Predicting View Counts of YouTube Videos Using Epidemic Modelling Framework

Adarsh Anand, Mohammed Shahid Irshad, and Deepti Aggrawal

Abstract YouTube is one of the giants in social media arena. The Google owned website provides its users the facility to upload and view videos on its platform. YouTube's ever increasing popularity and influence on the society makes it an active area of research. The spread of information through certain videos is so rapid that it can be easily compared to an outbreak of an infection and hence the term 'viral videos' aptly describes them. The spread of infection in the masses has been mathematically explained by scientists through epidemic modelling. Drawing parallels with an epidemic breakout, in the current proposal we have tried to capture view-count growth patterns. To better understand the viral content, a two-stage modelling framework inculcating Susceptibility (awareness) and Infection (viewing) has been proposed. The models have been validated on various YouTube datasets with good results.

Keywords Epidemic modelling · Time-Lag · Viewership rate · View-count · YouTube

8.1 Introduction

Human kind has always been interested in understanding the natural phenomenon since the pre-historic time. Earlier we tried to explain these phenomena as heavenly events. The position of stars and season change was one of the first observational predictions made by human kind. Around 3500–3000 BCE, ancient Egyptian and Mesopotamian explained their observations using Earth as the center of the universe. One correct prediction encourages the scholars to find other patterns and observations which can benefit the humans. Mathematics, astronomy and medicines blossomed

---

A. Anand (✉) · M. S. Irshad
Department of Operational Research, Faculty of Mathematical Sciences, University of Delhi, Delhi 110007, India

D. Aggrawal
USME, DTU, East Delhi Campus, Delhi 110095, India

© The Author(s), under exclusive license to Springer Nature Switzerland AG 2021
S. Patnaik et al. (eds.), *Computational Management*, Modeling and Optimization in Science and Technologies 18, https://doi.org/10.1007/978-3-030-72929-5_8

during this period. Greeks were first to develop theories and laws behind the observations. They used natural philosophy to explain the physical changes of substance. Mathematical physics was born in this period which clarifies the physical experience and numerical relations. After Romans' conquest over Greeks, they used their scholars to further pursue the scientific advancement to remain in power. After the fall of Western Romans, knowledge of Greek theories deteriorated in Western Europe but was preserved by Islamic world in the early centuries of Middle Ages i.e. 400 to 1000 CE [34]. Greek and Islamic inquires revived "natural philosophy" in Western Europe from 10th to thirteenth century. From 17th to nineteenth century the modern science made its way. The world started using equipment like telescope, microscope to closely examine the phenomena. Observation through equipment prompted new branches of science known as chemistry and biology. With industrialization in nineteenth century applied sciences came into the picture. With time new streams came into the picture but modern science is majorly divided into three branches:

1. Natural Science (Physics, Biology, Chemistry etc.): This deals with the study of nature, natural phenomena and living organisms.
2. Formal Science (Mathematics, Logic, Statistics etc.): This deals with theoretical systems and logics.
3. Social Science (Psychology, Sociology, Law etc.): This deals with the study of an individual as well as society behavior.

Social science is newest among all the sciences. Initially the study of social science was mainly categorized as a part of humanities, but in the last few decades the scenario changed rapidly. In 1990 Wynne published a paper where he studied people's attitude towards learning and utilization of science and quoted that "those who promote scientific literacy in public ignore the social dimension of the science" [46]. Anthropology itself from natural science uses various social aspects of humans. In late nineteenth century and early twentieth century social anthropology established itself as social-science discipline. The ignorance of human behavior while studying marketing would have cost the researchers dearly. In 1990, Lauterborn proposed 4-C's (Convenience, Consumer, Cost and Communication) as an alternative for 4 P's (Place, Product, Price and Promotion) to target the mass instead of niches [32]. Similarly, in social science to quantify the observations and better representation various mathematical and statistical tools have been utilized by the researchers. Technology has now become an integral part of human life. New innovation and inventions changed the human behavior entirely. The origin of Internet has revolutionized the media industry. In the era of televisions and newspaper it was very easy for government to censor and have control on what information or knowledge is received by the population of the country. It was also easy to tackle the misleading and fake news from getting broadcasted. But with the launch of social network and media sites, everyone has got a platform where they can express their views and experiences. With evolution of these platforms, the users have more and more ways to express their views. Initially the ways were limited to texts only, then one could share images, and now videos can be uploaded.

With the launch of YouTube in 2005, the media industry got revolutionized. It had altogether opened a new market and platform for the companies, government and other organization to communicate with their potential consumers or voters. The reach of internet has been increasing with time and so are the number of netizens. YouTube became instant hit among the consumers because they don't have to adopt the premium service policy. YouTube is free to watch and developed its business model across advertisement. YouTube is one of the highest generators of internet traffic [2, 18, 30, 35]. Massive user base has attracted the advertisers on this platform. Now the YouTube has such a huge database that some of its popular channel videos garner more than a million views in less than 24 h. These types of videos are usually referred as viral videos. Word "viral" is from biology (Nature Science), which implies something which spreads effectively and rapidly. This viral behavior of videos is exactly same as the behavior of pathogens (infectious agents) in an epidemic. In the recent past, various social science researchers have used epidemic modelling analogy to depict and predict the diffusion on information through social media platforms. Also, there has been plenty of work done as far as view-count prediction is considered. Some important works are discussed in the next section.

8.1.1 Literature Review

There are various reasons to determine the view-count pattern for a YouTube video but most important is to optimally charge for the advertisement by YouTube from uploader's view-point and to find the optimal time frame in which the advertisement is viewed by maximum number of the viewers from advertiser's view-point. In recent times, various efforts have been made by researchers to precisely model the view-count growth of a video and its active life-cycle. Cheng et al. [14] provided mathematical equations to calculate the active life cycle of video based on certain parameters. Richier et al. [39] proposed six models to study the view-count growth patterns for the videos, three each for viral and non-viral videos. Khan and Vong [28] measured what proportion of traffic on YouTube is coming from other social media platforms using Webomatrics. Even the impact of YouTube recommendation system has been analyzed by Zhou et al. [48] as well as by Portilla et al. [38]. Three video-count prediction models were proposed by Aggrawal et al. [2] on the basis of increasing potential viewers and repeat viewing. Interpretive Structural modelling has been used by Bisht et al. [10] to find the most influencing and influenced attributes which leads user to view the video. Irshad et al. [26] have showed that total view-count of the videos is not always because of virality but also because of the existing potential viewers (Subscribers) of the channel. In their research, it has also been depicted that the view-count rate changes after a particular time point in video's life cycle. Yu et al. [47] have also showed that a video can have multiple phases in its life cycle. Deep learning models have been used by Cheng and Tsai [13] to perform

Table 8.1 Comparison between proposed and existing work

Criteria → Models↓	Social media	View-count prediction	Epidemic framework	Time lag
Richier et al. [39]	✓	✓	✗	✗
Bisht et al. [10]	✓	✗	✗	✗
Aggrawal et al. [2]	✓	✓	✗	✗
Irshad et al. [26]	✓	✓	✗	✗
Yu et al. [47]	✓	✓	✗	✗
Goldenberg et al. [22]	✓	✓	✗	✗
Bauckhage et al. [8]	✓	✗	✗	✗
Proposed work	✓	✗	✓	✗

sentiment analysis. To detect fake news, Cui et al. [16] also used sentiment analysis. Much research has been done using epidemic modelling too. Goffman and Newill [21] had used SIR and SIS techniques to discuss the analogy of information and diseases diffusion. Goldenberg et al. [22] had used SIR model to depict the network effect of word-of-mouth (WOM) for e-business and net related activities. Leskovec et al. [33] have used network SIS model to maximize the sales of a particular products in certain commodities through e-marketing, whereas [42] have used regression models to detect the initial outbreak of epidemics. Bauckhage et al. [8] have implemented SIR models to predict the view-count growth for YouTube videos. In this proposal, we have used time lagged SI analogy to predict the view-count of YouTube videos. Table 8.1 depicts how the proposed work is different from existing work pertaining to social media.

The organization of the chapter is as follows: Mathematical modelling is dealt in Sect. 8.2. This section is further divided into five subsections: Sect. 2.1 discusses the building blocks of epidemic modelling. Section 2.2 discusses the model assumptions. Section 2.3 is the one in which mathematical formulation for susceptible population has been generated, whereas, in Sect. 2.4 we inculcate the impact of time lag in conversion of susceptible population into infected population and mathematical formulation has been generated accordingly. The last Sect. 2.5 deals with the special cases of proposed model. Parameter estimation and goodness of fit is presented in Sect. 8.3. Section 8.4 discusses the results and their interpretation.

8.2 Building Block for Proposed Epidemic Modelling

8.2.1 Epidemic Modelling

Most of the chronic diseases spread instantaneously in a particular region. Usually this type of sudden outbreak of single disease in a particular region is termed as epidemic. These epidemics are a major cause of mortality and morbidity in developing nations. New epidemics like COVID-19, Ebola, Zika, MERS and SARS emerged whereas old epidemics like cholera, yellow fever and plague have returned [45].

Various researches have been carried out in the field of bio-technology and pharmaceuticals to find the cure of these epidemic diseases. At the same time researchers from the formal science and nature science worked together putting their respective knowledge together to come up with various mathematical models known as epidemiology models [20, 25, 36, 41]. In early 1760, Daniel Bernoulli created a model to evaluate the impact of smallpox vaccination on healthy people [9]. In 1906, Hamer developed a discrete time model to estimate the recurrence of measles [23]. For malaria, Ross developed a differential equation model to capture the incidence and its control [40]. These mathematical models help organization like W.H.O. and government agencies in designing the action plan for controlling epidemics. They provide an estimate of infected people, healthy people and susceptible people from the population. These mathematical models have been termed as epidemic models. In twentieth century, the mathematical epidemiology has grown exponentially and there are several epidemic models for a single epidemic having different infection spreading rate due to different geographical regions, ethnicity, immunity etc. Out of all available literature, the first edition of Bailey's book [6] is surely an important landmark. There are specific models for various well known epidemics such as rubella, HIV/AIDS, diphtheria, herpes, malaria, syphilis etc. There are various books on epidemic modelling [4, 7, 11, 12, 15, 37] which tells us the extent to which research is done in this field. In the most recent COVID-19 pandemic too, epidemiology has proven helpful in understanding the spread of the infection and thus undertaking necessary measures to avoid it [3, 5, 29, 44].

The epidemiological classes are often labeled as Healthy/ Passive Immune (M), Susceptible (S), Exposed (E), Infected (I) and Recovered (R). Each class represents different subsets of population. M is the class of infants who have antibodies from maternal bodies which protects them from diseases in the initial life span. This protection from diseases is known as passive immunity. Susceptible (S) is the class of infants who don't have antibodies or they disappeared with the growth, and so these adults may get exposed to the infectious agents. These are the people who can get infected and further act as infectious agents. From S a proportion of population enters the exposed (E) class. E is a class who had already been exposed to the infectious agents and infected but are not infectious. Infected (I) is a class of population who are infected with the epidemic and also acts as an infectious agent. If a person from infected (I) class meets any healthy person from population then this healthy person will now belong to susceptible (S) class. Last class is recovered (R), which has the proportion of infected population who has permanent infection-acquired immunity from the epidemic. The flow of the population is not always the same as shown in Fig. 8.1. There are various flow patterns of population depending upon the type of epidemic. The recurrence of epidemic is a big issue for these classical models. Depending upon the type of epidemic the researchers had designed various epidemic models such as SI, SIS, SEIS, SIR, SEI, SIRS, SEIR, MSEIR and MSEIRS

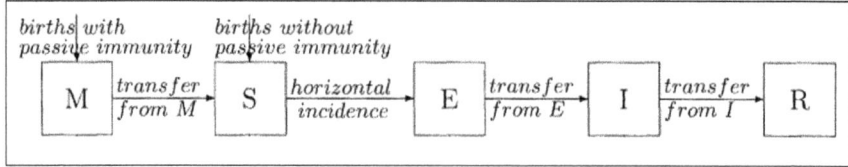

Fig. 8.1 General transfer diagram of population in epidemic modelling

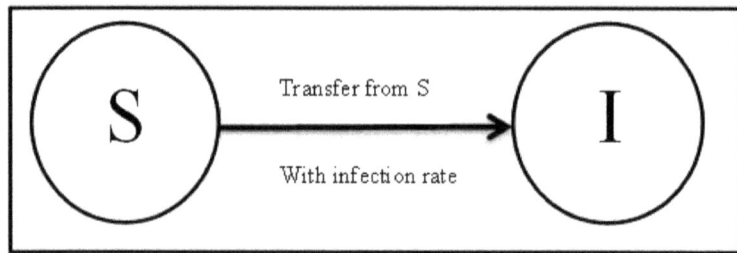

Fig. 8.2 Flow diagram of SI analogy

[24]. Figure 8.1 depicts the MSEIR model, but we have found various cases where following recovery the patient again gets infected after some time, for example in the cases of tuberculosis. Despite of different variants of epidemic models the SI, SIR and SIS are most common used models to describe an epidemic.

As discussed earlier a good number of proposals have been made in context of epidemic modelling as far as social media is concerned [8, 21, 22, 33, 42]. In the present work, using similar analogy of S-I, a view count prediction model is presented. Figure 8.2 shows the transition diagram of general population from social media point of view.

Here the susceptible are those viewers who are aware about the video, either by notification feature provided by the YouTube because they are the subscribers of that channel or by word-of-mouth, but they have not yet watched the video. These aware viewers are considered as susceptible (S) for the video who may view the video. Infected (I) are those viewers who have watched the videos and contribute to the view-count of the videos. By estimating the number of infected viewers we can predict the expected view-count on the video. There is existing literature in marketing which deals with the time lag in diffusion of product i.e. there is a time lag between consumer awareness and adoption process [1, 31, 43]. We have proposed mathematical formulation which have analogy of SI flow as well as discuss the time lag phenomena in diffusion of information.

8.2.2 Assumptions

The general sets of assumption that have been considered are:
1. The expected number of total viewers is fixed.
2. The number of people who get infected at any time t is directly dependent on remaining number of expected viewers who are susceptible.
3. Susceptible and Infection process are connected to each other. No viewer directly watches the video without getting into the susceptible population.
4. There is a time lag between becoming aware (Susceptible) and actually viewing (Infected) the video.
5. Infection takes place only in the case of positive word-of-mouth (WOM).

8.2.3 Modelling Information Diffusion as Susceptibility Process

During this stage the viewer get informed about the videos. The viewer can know about the video either by word-of-mouth or because he is the subscriber of that channel. This leads to a dilemma of instantly watching that video or to watch it after some time. During this time he can also access the comment section and gather more information about the video which helps in making the final decision.

As per the flow shown in Fig. 8.2, we assume that whole population is either divided into susceptible or infected but in case of YouTube there may be a population which is not aware yet. So as per assumption "the number of viewers to become aware (susceptible) at any time t is directly proportional to the remaining number of expected number of viewers". Using this assumption, the awareness process can be mathematically expressed as:

$$\frac{dS(t)}{dt} = \alpha[v - S(t)] \quad (8.1)$$

where v is the expected number of total potential viewers; $S(t)$ is the aware (Susceptible) viewers by time t and α is the rate of awareness (susceptibility). On solving Eq. 8.1 using initial condition at $t = 0$, $S(t) = 0$, we get:

$$S(t) = v\left(1 - e^{-\alpha t}\right) \quad (8.2)$$

Equation 8.2 represents the total aware viewers by time t who will further decide whether to watch the video or not. It is assumed that whole population will be aware as $t \to \infty$ and the information regarding the new video is received in the same manner (either by word-of-mouth or through subscription). The susceptible (awareness) model is obtained by using non-decreasing mean value function which demonstrates exponential growth process of susceptibility (awareness).

8.2.4 Modeling Viewing as Infection Process

Viewing is the process which matters the most because it increases the actual viewcount of the videos. The time lag between the viewers' awareness and viewing is taken care of while forming the mathematical equation for viewing process. Distributed time lag approach has been used in model for the assertion of time delay. Definite interval of past is measured through appropriate memory kernel. As per our knowledge this is the first attempt where we are taking finite time interval between susceptibility and infection (viewing) process. Even in real life we normally watch videos after some time due to personal or professional reasons. So the functional relationship between the aware viewers and viewing can be described as follows.

The potential viewers who become aware in susceptibility phase go on to become the actual viewers in the infected phase i.e. viewed the video. As per SI analogy only those who are susceptible gets infected. Therefore, only aware viewers can view the video. Hence, Eq. 8.2 can be considered as an upper limit for maximum number of viewers who can view the video. Thus the equation for infection (viewing) process can be presented as:

$$\frac{dI(t)}{dt} = b[S(t) - I(t)] \tag{8.3}$$

where $I(t)$ denotes the count of actual viewers who have viewed the video by time t and b is the rate of viewing (infection).

As discussed above, distributed time lag approach has been especially used in information diffusion method to model the time lag between viewer awareness and actual viewing of the video. Using the work presented by Diamond Jr. [19], it has been assumed that the potential viewer who wished to view the video in the past but viewed the video in the present is relevant to present viewing (infection) process. Cushing [17] advocated for continuously distributed time lag rather than discrete time lags. To model this phenomenon, the role of time lag has been determined in Eq. 8.3 as a weighted response measured over a finite interval of past time through an appropriate memory kernel. Hence, considering time lag, Eq. 8.3 is remodeled as an integro-differential equation as:

$$\frac{dI(t)}{dt} = b \int_0^t K(t-\tau)[S(t) - I(t)]d\tau \tag{8.4}$$

In this chapter, we have restricted ourselves to using a particular weak memory kernel, i.e.

$$K(t) = we^{-wt} \tag{8.5}$$

where, w denotes the rate the past has bearing on the present. In kernel function, the system's time scale can be defined as w^{-1}. The weighted response of the kernel

(Eq. 8.5) gets biased with passage of time and decreases exponentially. This means that more remote the past less reliable the kernel (Eq. 8.5). This is referred to as the weak generic delay kernel [17]. This particular form of kernel function has been cause it is more suited for information diffusion situation, as the positive awareness about the video has better chance of getting viewed in the near future than after a long delay.

In the current proposal, the kernel function denotes the homomorphism among the aware viewer at the time t and the aware viewer at the time $t - \tau$ who have not yet watched the video. Substituting the value of $S(t)$ from Eq. 8.2 and kernel function from Eq. 8.5 into Eq. 8.4, it gets transformed into:

$$\frac{dI(t)}{dt} = b \int_0^t w e^{-w(t-\tau)} \left[v\left(1 - e^{-\alpha t}\right) - I(\tau) \right] d\tau \qquad (8.6)$$

After solving Eq. 8.6 at initial condition $I(t = 0) = 0$ using Laplace transformation, we found the actual number of viewers who had viewed the video at time t is:

$$I(t) = v\left(1 - \left(1 - \frac{w}{2A}\right) e^{\frac{-w}{2}t}\right.$$
$$\left. - \left(e^{-\alpha t} - e^{\frac{-w}{2}t}\left(\cos At + \frac{1}{A}\left(\frac{w}{2} - \alpha\right) \sin At\right)\right) \frac{bw}{\alpha(\alpha - w) + bw}\right) \qquad (8.7)$$

where $A = \sqrt{bw - \frac{w^2}{4}}$.

Hence, Eq. 8.7 represents S-I based diffusion models by unifying the concept of time lag and different growth rate in information diffusion process, i.e. rate of susceptibility (α) and infection (b) are different.

8.2.5 Particular Case

8.2.5.1 Same Rate of Viewing Without Time Lag

In the above proposed two stage information diffusion model, there are fair chances that the awareness about the video and viewing of the video might happen at the same rate. Then, we can find the actual number of viewers from Eq. 8.3, by putting the value of $S(t)$ in it. Thus, the actual number of viewers who had viewed the video at time t is:

$$I(t) = v\left(1 - (1 + bt)e^{-bt}\right) \qquad (8.8)$$

Special model can be derived from the proposed generic model for two stage information diffusion model. The proposed information diffusion model (Eq. 8.7) can be formulated as new information diffusion model (Eq. 8.8), if we drop the time lag parameter along with letting the values of susceptibility and infection rate to be equal i.e. $\alpha = b$.

8.2.5.2 Different Rate of Viewing Without Time Lag

From assumptions, if we drop assumption (4) the S-I model for information diffusion in Eq. 8.7 can be obtained by taking limit $w \to \infty$, then kernel $K(t)$ will be replaced by delta function [27]. So we can consider the memory less kernel function which is a limiting case of Eq. 8.5. This means viewing is free of past impact i.e. instantaneous viewing happens. In this particular case we have considered different rate of susceptibility and infection i.e. rate of susceptibility and infection will not be equal throughout the video life cycle. Initially the rate of susceptibility will be higher than rate of infection where as in the later stage of video life cycle the rate of infection will be higher than rate of susceptibility. Thus, Eq. 8.9 denotes the total number of viewers by time t obtained from Eq. 8.3 by using initial condition $I_2(t) = 0$ at $t = 0$, when $b \neq \alpha$:

$$I(t) = v\left(1 - \frac{1}{b-\alpha}\left(be^{-\alpha t} - \alpha e^{-bt}\right)\right) \tag{8.9}$$

8.3 Numerical Illustration

In this section, we have validated our proposed framework on five different view-count datasets. The view-count for the videos has been collected manually for more than a month at the interval of every 12 h. The details of the video URL's have been provided in Table 8.2.

We have estimated parameters of all the models i.e. Model I, II and III which are generic two stage model (Eq. 8.7) and particular cases (Eqs. 8.8, 8.9) respectively

Table 8.2 URL's of the YouTube Video

Video I.D	URL
1	https://www.youtube.com/watch?v=r6FxROAHJH4
2	https://www.youtube.com/watch?v=9rOlvj6f6Yk
3	https://www.youtube.com/watch?v=2J2yXSLgKko&t=19s
4	https://www.youtube.com/watch?v=CY7K2VFyUeo
5	https://www.youtube.com/watch?v=oB94lvJbETE

(Table 8.3) on the five datasets. The values have been estimated using the Non Linear Regression module on IBM SPSS package and the results have been shown in Tables 8.4, 8.5 and 8.6.

The performance of the proposed models has been judged using various comparison criteria like R-Squared, Variance, Bias, M.S.E and R.M.P.S.E. Tables 8.7, 8.8 and 8.9 represent the comparison parameters of all datasets for all the models. As we can see the values of the comparison criteria are quite good for model I on all the five datasets under consideration. The performance of model II and III varies from dataset to dataset but they are never as good as model I.

For a better visual understanding, Figs. 8.3, 8.4, 8.5, 8.6 and 8.7 have been used to show a comparison between the actual view-count data and view-count values obtained through the 3 proposed models. As can be seen in Fig. 8.3, all the 3 proposed models are able to predict the view-count values quite close to the original data. In Fig. 8.4, it can be seen only model I is a close fit to the original data throughout, while models II and III could make better predictions in the later time period only. For dataset 3, as depicted in Fig. 8.5, the view-counts are growing at a fluctuating rate, with a steep rise in some time period and a slow riser in others. In this case, both model I and II perform well but model III is able to project the correct picture only in the initial time points. In Fig. 8.6 we can see that model I is the best throughout the video cycle. Model II and III except for the initial few time points also predict quite close to the original data. Figure 8.7 shows overlapping of all the 4 timelines depicting the original data and the three models on almost all the time points. This goodness-of-fit shows that the models predictive capability is quite high for this particular dataset.

8.4 Results and Interpretation

The work presented in this chapter uses an analogy from epidemic modeling to mathematically model the view-count growth patterns of viral YouTube videos. Three models have been proposed- one generic and 2 particular cases of it. The proposed models consider the spread of information by viewing a video and the movement of susceptible population to an infected group. The proposed models were validated on 5 manually collected view-counts datasets. We can see from the tables that the generic model is performing better than the particular cases i.e. Model I is performing better for all datasets on approximately all comparison parameters. This implies that there is actually a time lag between the awareness and the viewing of the video. The presence of definite time delay has its own benefits, this gives the government and other organizations to tackle fake news and remove any such content from the platform before its diffusion. As we can see in special case model where there is no time lag and the rate of susceptibility and infection is same, the results are not as good in comparison to other models. Even so, these models are helpful in predicting view-count of the videos which has two fold benefits, both for YouTube and the advertiser. YouTube can charge for advertisement as per the video growth rate. Also, advertisers

Table 8.3 Proposed models

	Description	Equation
Model I	With time-lag and different rates	$I(t) = v\left(1 - (1 - \frac{w}{2A})e^{\frac{-w}{2}t} - \left(e^{-\alpha t} - e^{\frac{-w}{2}t}(\cos At + \frac{1}{A}(\frac{w}{2} - \alpha)\sin At)\frac{bw}{\alpha(\alpha - w) + bw}\right)\right)$
Model II	With same rates and without time lag	$I(t) = v(1 - (1 + bt)e^{-bt})$
Model III	With different rates and without time lag	$I(t) = v\left(1 - \frac{1}{b-\alpha}(be^{-\alpha t} - \alpha e^{-bt})\right)$

8 Understanding and Predicting View Counts of YouTube ...

Table 8.4 Parameter estimation for dataset 1 and 2

DS 1	Model I	Model II	Model III	DS 2	Model I	Model II	Model III
v	5440.802	5271.342	5385.395	v	1959.523	1800.154	1841.764
b	0.495	0.24	0.117	b	0.141	0.295	0.99
α	0.99	–	0.99	α	0.99	–	0.138
w	0.161	–	–	w	0.103	–	–

Table 8.5 Parameter estimation for dataset 3 and 4

DS 3	Model I	Model II	Model III	DS 4	Model I	Model II	Model III
v	4795.198	3099.488	2376.269	v	1745.978	1396.012	1578.024
b	0.393	0.06	0.049	b	0.275	0.109	0.99
α	0.99	–	0.467	α	0.99	–	0.043
w	0.021	–	–	w	0.05	–	–

Table 8.6 Parameter estimation for dataset 5

DS 5	Model I	Model II	Model III
v	20,572.92	19,961.78	20,259.17
b	0.345	0.31	0.159
α	0.99	–	0.99
w	0.185	–	–

Table 8.7 Comparison criteria for dataset 1 and 2

DS 1	Model I	Model II	Model III	DS 2	Model I	Model II	Model III
R-Square	0.94	0.91	0.96	R-Square	0.96	0.62	0.73
Variance	284.32	344.53	243.28	Variance	63.09	196.06	167.60
Bias	−3.65	−39.60	−28.56	Bias	0.42	−25.59	−25.17
M.S.E	79,660.64	112,326.19	55,904.69	M.S.E	3924.16	35,942.98	25,798.53
R.M.P.S.E	284.35	346.80	244.95	R.M.P.S.E	63.09	197.73	169.48

Table 8.8 Comparison criteria for dataset 3 and 4

DS 3	Model I	Model II	Model III	DS 4	Model I	Model II	Model III
R-Square	0.96	0.96	0.85	R-Square	0.98	0.89	0.95
Variance	32,276.88	42,089.78	134,426.78	Variance	3073.34	18,887.02	8489.78
Bias	16.48	−43.52	−76.55	Bias	2.62	−34.21	−19.16
M.S.E	28,527.88	32,582.09	104,626.67	M.S.E	2999.80	15,050.05	7241.80
R.M.P.S.E	32,276.88	42,089.80	134,426.80	R.M.P.S.E	3073.34	18,887.05	8489.80

Table 8.9 Comparison criteria for dataset 5

DS 5	Model I	Model II	Model III
R-Square	0.93	0.85	0.91
Variance	952.05	1423.21	1117.25
Bias	−6.35	−173.16	−149.96
M.S.E	893,148.83	1,906,222.10	1,162,688.29
R.M.P.S.E	952.07	1433.71	1127.27

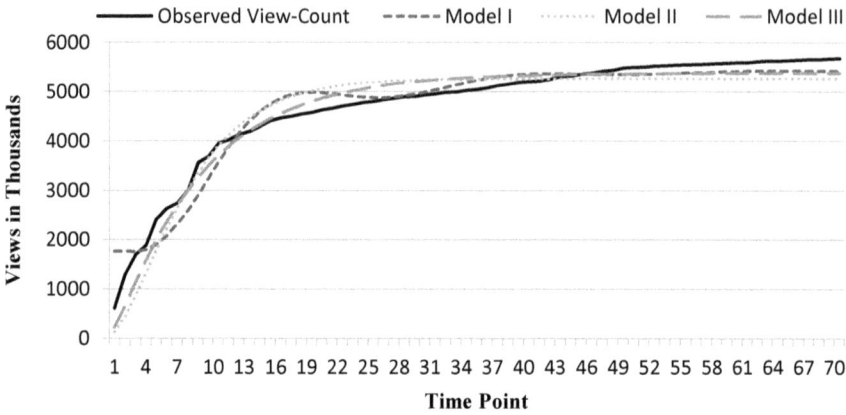

Fig. 8.3 Graphical representation for Dataset 1

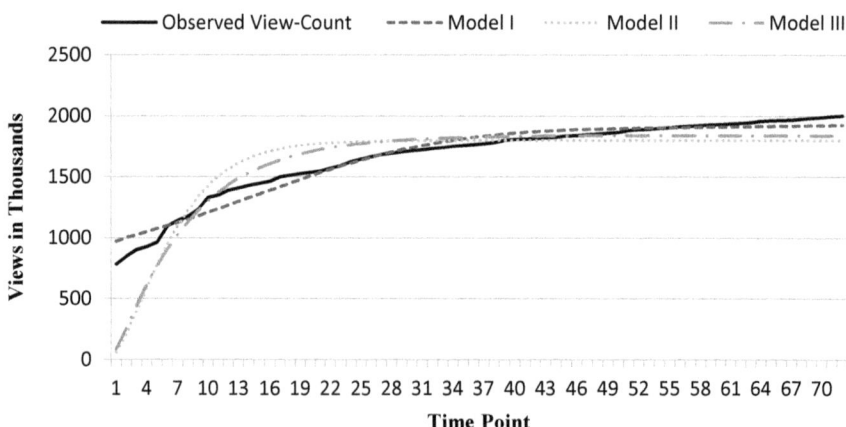

Fig. 8.4 Graphical representation for Dataset 2

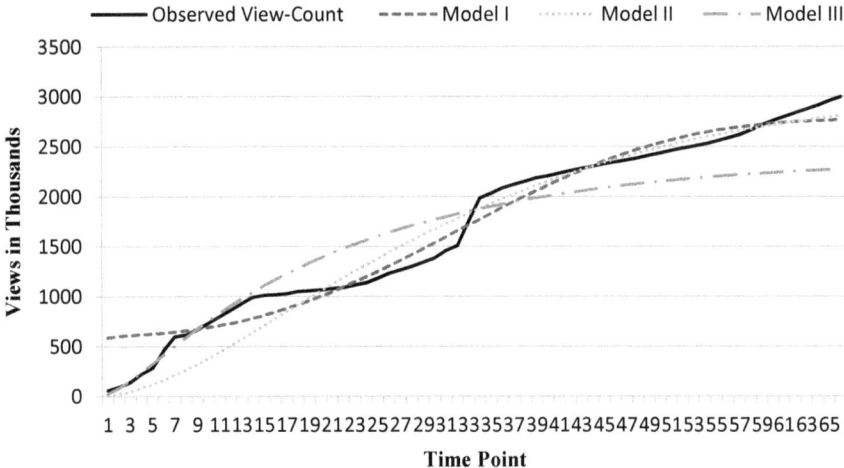

Fig. 8.5 Graphical representation for Dataset 3

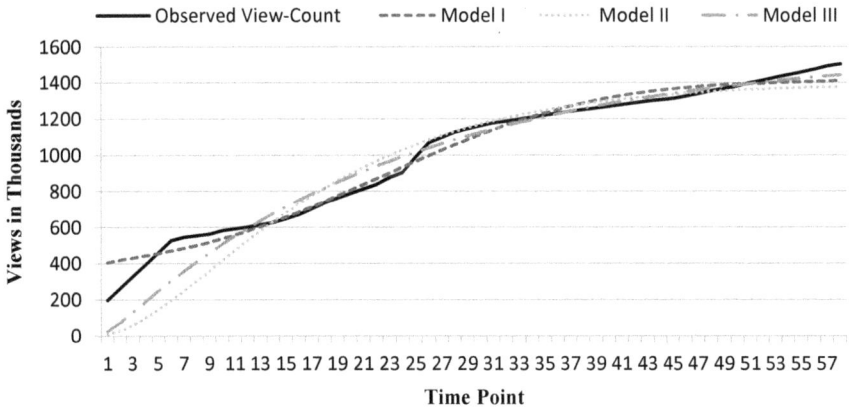

Fig. 8.6 Graphical representation for Dataset 4

can decide when to advertise on the video so as to garner maximum impact from it. The time lag between the awareness and the viewing also tells us about the viewer behavior. These observations tells us that even after initial delay, the video can still become viral as viewers may start viewing it after some time because of its socio-economic factor or other external factors. Thus, this work provides a mathematical explanation for the viewer's behavior which can be used to garner more views on a video, effective information sharing and prevent the spread of fake news.

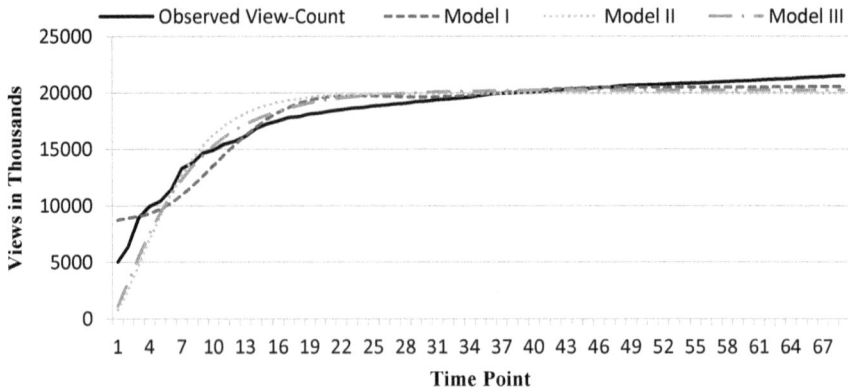

Fig. 8.7 Graphical representation for Dataset 5

References

1. Aggarwal R, Singh O, Anand A, Kapur PK (2019) Modeling innovation adoption incorporating time lag between awareness and adoption process. Int J Syst Assu Eng Manage 10(1):83–90
2. Aggrawal N, Arora A, Anand A, Irshad MS (2018) View-count based modeling for YouTube videos and weighted criteria–based ranking. Adv Math Tech Eng Sci CRC Press, pp 149–160
3. Ajbar A, Alqahtani RT (2020) Bifurcation analysis of a SEIR epidemic system with governmental action and individual reaction. Adv Differ Equ 2020(1):1–14
4. Anderson RM, May RM (1985) Helminth infections of humans: mathematical models, population dynamics, and control. In: Advances in parasitology. Academic Press, vol 24, pp 1–101
5. Bagal DK, Rath A, Barua A, Patnaik D (2020) Estimating the parameters of susceptible-infected-recovered model of COVID-19 cases in India during lockdown periods. Chaos, Solitons Fractals 140:110154
6. Bailey NTJ (1975) The mathematical theory of infectious diseases and its applications. Charles Griffin and company, Great Britain
7. Bartlett MS (1960) Stochastic population models. In: ecology and epidemiology (No. 04, QA27 18, B37).
8. Bauckhage C, Hadiji F, Kersting K (2015) How viral are viral videos? In: Ninth international AAAI conference on web and social media, pp 22–30
9. Bernoulli D (1760) Trial of a new analysis of mortality caused by smallpox, and the advantages of inoculation to prevent it. History Acad Roy Sci (Paris) Mem 1–45
10. Bisht M, Irshad MS, Aggarwal N, Anand A (2019) Understanding popularity dynamics for YouTube videos: an interpretive structural modelling based approach. In: 2019 amity international conference on artificial intelligence (AICAI) IEEE, pp 588–592
11. Busenberg S, Cooke K (1993) Vertically transmitted diseases. Biomathematics 23
12. Capasso V (1993) Mathematical structures of epidemic systems, vol 97. Lecture Notes in Biomathematics. Springer, Berlin
13. Cheng LC, Tsai SL (2019) Deep learning for automated sentiment analysis of social media. In: Proceedings of the 2019 IEEE/ACM international conference on advances in social networks analysis and mining, pp 1001–1004
14. Cheng X, Dale C, Liu J (2008) Statistics and social network of youtube videos. In: 2008 16th international workshop on quality of service. IEEE, pp 229–238
15. Cliff AD, Haggett P, Smallman-Raynor M (2004) World atlas of epidemic diseases. Arnold, London, pp 56–60

16. Cui L, Wang S, Lee D (2019) SAME: sentiment-aware multi-modal embedding for detecting fake news. In: Proceedings of the 2019 IEEE/ACM international conference on advances in social networks analysis and mining, pp 41–48
17. Cushing JM (1975) An operator equation and bounded solutions of integro-differential systems. SIAM J Math Anal 6(3):433–445
18. de Bérail P, Guillon M, Bungener C (2019) The relations between YouTube addiction, social anxiety and parasocial relationships with YouTubers: a moderated-mediation model based on a cognitive-behavioral framework. Comput Hum Behav 99:190–204
19. Diamond AM Jr (2005) Measurement, incentives and constraints in Stigler's economics of science. Eur J History of Econom Thought 12(4):635–661
20. Garrett L (1995) Swine flu and legionnaires' disease. The coming plague, The American Bicentennial
21. Goffman W, Newill VA (1964) Generalization of epidemic theory: an application to the transmission of ideas. Nature 204(4955):225–228
22. Goldenberg J, Libai B, Muller E (2001) Talk of the network: a complex systems look at the underlying process of word-of-mouth. Mark Lett 12(3):211–223
23. Hamer WH (1906) Epidemic disease in England: the evidence of variability and of persistency of type. Bedford Press
24. Hethcote HW (2000) The mathematics of infectious diseases. SIAM Rev 42(4):599–653
25. Huxsoll DL (1996) The hot zone: Richard Preston. Random House, New York 1994:300
26. Irshad MS, Anand A, Bisht M (2019) Modelling popularity dynamics based on youtube viewers and subscribers. Int J Math Eng Manage Sci 4(6):1508–1521
27. Karmeshu (1982) Time-lag in a diffusion-model of information. Math Modell 3(2):137–141
28. Khan GF, Vong S (2014) Virality over YouTube: an empirical analysis. Internet Res 24(5):629–647
29. Khyar O, Allali K (2020) Global dynamics of a multi-strain SEIR epidemic model with general incidence rates: application to COVID-19 pandemic. Nonlinear Dynam pp 1–21
30. Klobas JE, McGill TJ, Moghavvemi S, Paramanathan T (2018) Compulsive YouTube usage: a comparison of use motivation and personality effects. Comput Hum Behav 87:129–139
31. Lal VB, Kaicker S (1988) Modeling innovation diffusion with distributed time lag. Technol Forecast Soc Chang 34(2):103–113
32. Lauterborn B 1990 New marketing litany: four Ps passed: C-words take over. Advertising Age, October, 1
33. Leskovec J, Adamic LA, Huberman BA (2007) The dynamics of viral marketing. ACM Trans Web (TWEB) 1(1):5
34. Lindberg DC (2010) The beginnings of Western science: the European scientific tradition in philosophical, religious, and institutional context, prehistory to AD 1450. University of Chicago Press.
35. Malik H, Tian Z (2017) A framework for collecting youtube meta-data. Procedia Comput Sci 113:194–201
36. Oldstone MB (2009) Viruses, plagues, and history: past, present, and future. Oxford University Press
37. Perelson AS (1989) Mathematical and statistical approaches to AIDS epidemiology. In: Castillo-Chavez C (eds) Lecture Notes in Biomath. Springer-Verlag, NY, pp 350–370
38. Portilla Y, Reiffers A, Altman E, El-Azouzi R (2015) December. a study of YouTube recommendation graph based on measurements and stochastic tools. In: 2015 IEEE/ACM 8th international conference on utility and cloud computing (UCC). IEEE, pp 430–435
39. Richier C, Altman E, Elazouzi R, Altman T, Linares G, Portilla Y (2014) Modelling view-count dynamics in youtube. arXiv preprint arXiv: 1404.2570
40. Ross R (1911) The prevention of malaria, 2nd edn. Murray, London
41. Shilts R (1987) And the band played on. St. Martin's, New York
42. Shtatland ES, Shtatland T (2008) Early detection of epidemic outbreaks and financial bubbles using autoregressive models with structural changes. Proceedings of the NESUG 21

43. Singh O, Kapur PK, Sachdeva N, Bibhu V (2014). Innovation diffusion models incorporating time lag between innovators and imitators adoption. In: Proceedings of 3rd international conference on reliability, infocom technologies and optimization. IEEE, pp 1–6
44. Srivastava V, Srivastava S, Chaudhary G, Al-Turjman F (2020) A systematic approach for COVID-19 predictions and parameter estimation. Personal Ubiquitous Comput 1–13
45. World Health Organization (2018) Managing epidemics: key facts about major deadly diseases. World Health Organization
46. Wynne B (1990) The blind and the blissful. The Guardian, 11 April p 28
47. Yu H, Xie L, Sanner S (2015) The lifecycle of a YouTube video: Phases, content and popularity. In: Ninth international AAAI conference on web and social media
48. Zhou R, Khemmarat S, Gao L (2010) The impact of YouTube recommendation system on video views. In: Proceedings of the 10th ACM SIGCOMM conference on Internet measurement. ACM, pp 404–410

Chapter 9
Gross Domestic Product Modeling Using "Panel-Data" Concept

Sarada Ghosh and G. P. Samanta

Abstract In this work, we want to predict gross domestic product (GDP) using 'panel-data' models. Gross domestic product for fifteen individual countries of Europe in different four years (1995, 2000, 2005 and 2010) under various components of GDP are used here. The panel-data models have been developed using the investment, labour force growth and budget surplus as inputs in order to predict the GDP of the countries which help us to perceive the economic condition of Europe from the statistical point of view. The results indicate that the random effects model is more appropriate than fixed effects model. In addition, this chapter also contains a panel data regression model with cross-sectional dependence, heteroscedasticity and also considers serially correlated disturbances for random effects.

Keywords Gross domestic product · Panel data · Fixed effect · Random effect · Cross-sectional dependence · Serial-correlation · Heteroscedasticity

9.1 Introduction

The gross domestic product (GDP) is one of the most important measurement in the economics of a country. GDP is a primary indicator that decides the market value of all the goods and services produced in a country over a specific time period. GDP growth rate is the rate at which we can show the changes of GDP from one year to another year. The GDP represents the health of economics of a country. It helps to create a clear view about the economic status of a country which can help to develop a country in many ways. So, if the GDP growth is strong, then organizations or firms can able to hire more workers and can afford to pay higher salaries and wages, which leads to more spending by consumers on goods and services. There are many countries in Europe but hereby we consider the GDP of the fifteen countries in different time periods and identifying components affecting GDP of the countries whose GDP

S. Ghosh · G. P. Samanta (✉)
Department of Mathematics, Indian Institute of Engineering Science and Technology, Shibpur, Howrah 711103, India
e-mail: gpsamanta@math.iiests.ac.in

make a deep impression in the economics of the Europe. It was only in the 1980s that the European panels began setting up. In 1989, a special section of the European Economic Review published papers using the German Socio-Economic [33] Panel.

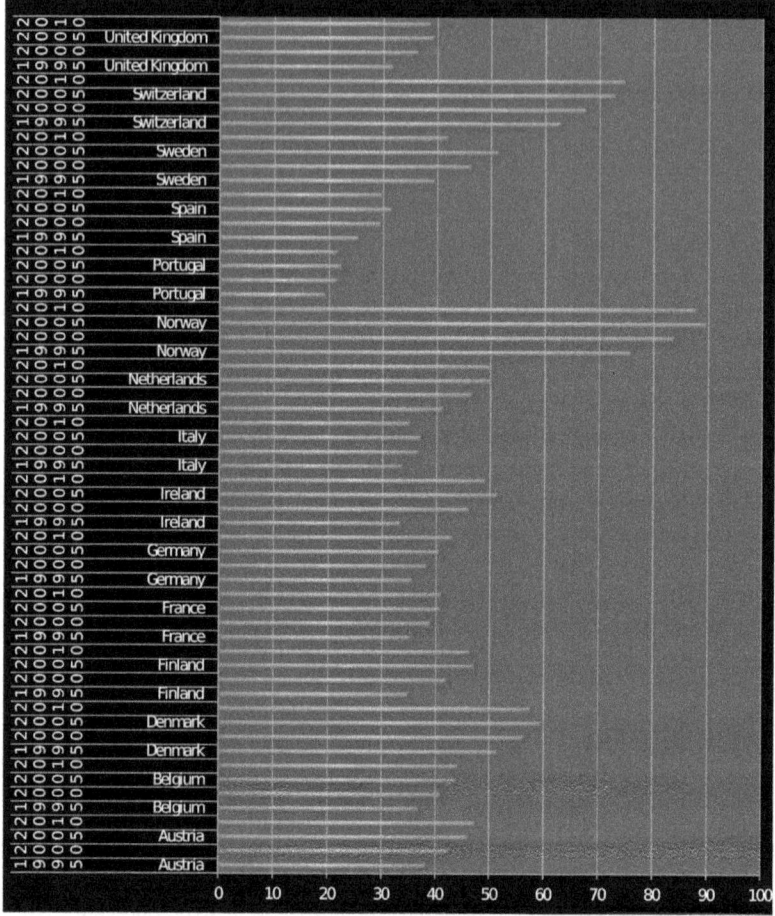

Series 1:: Initial GDP of different countries in Europe for several years

There are different factors which affect enormously to the GDP of any country. Moreover, the development of specific GDP components and related indicators, such as those for economic output, consumption, as well as data on the distribution of income and savings, can give valuable insights into the main drivers of economic activity [15]. Hereby, we highlight three main factors affecting GDP i.e., investment, labour force growth and budget surplus. The sources of investment are coming from various purposes in a country. Gross domestic investment is the purchase of apparatus by firms, the purchase of all newly produced structures, and changes in business inventories. It also consists of the consumption of fixed capital which is relatively stable component of GDP. Labour is the amount of physical, mental and social effort used to produce goods and services in an economy. Generally, growth and engagement in the labour force depends on manpowers, different skills etc., maintaining a balance in economics of a country. The government's budget can be either in surplus or in deficit: a surplus occurs when a government taxes more than it spends and a deficit occurs when the government spends more than it taxes. A budget surplus often refers to the financial state of government of a country. Generally, it is preferable to use the term 'savings' instead of 'budget surplus'. A budget surplus is an indicator of a healthy economy, though a country not having a budget surplus does not mean the economy of that country is not being run efficiently. So, it is not necessary for a government of a country to keep maintenance a surplus. A surplus indicates that the government has extra funds that can be allocated to pay debts (that is, reducing the interest payable) and thereby helping the economy in the future. As for example, a budget surplus can reduce taxes, start new programs and to augment existing public programs such as social security or medicare and other necessary purposes for development of the country. Apart from this, running a budget surplus carries a number of advantages: reducing the public debt, funding in military, paying for infrastructure, energy, and public works, wages, implementation of policies, or be saved to spend in the future once a deficit occurs. A budget surplus occurs due to a reduction in costs and spending or both. An increase in taxes can also lead to a surplus. A surplus decreases consumer demand, lowers consumer prices, thereby slows down the economy.

In this chapter, we have attempted to focus on predicting GDP profile. There are several ways to predict GDP growth. Here we have used the panel data concept [31, 32, 40, 44]. The term 'panel data' refers to the pooling of observations or features of both time-series and cross-sectional data [39]. This can be achieved by surveying a number of observations and following them over time. Moreover, an important advantage of the panel-data model is to get valuable information about relationships between different observations that can be extracted [31]. Gross domestic product (GDP) varies year to year because of different effects related to GDP. Here we have considered periodic observations of a set of variables characterizing the cross-sectional units over a particular time-span. Our purpose is to predict gross domestic product using panel-data concept under the various components of GDP which helps to make out the economic status across 15 countries in Europe.

9.2 Materials

This work is performed using the data of fifteen different countries of Europe. It is observed that the GDP of those countries make a highly influence in the economics status of Europe. The data for different years for Austria, Belgium, Denmark, Finland, France, Germany, Ireland, Italy, Netherlands, Norway, Portugal, Spain, Sweden, Switzerland and United Kingdom (in Fig. 9.1) are obtained through the website http://forums.eviews.com/viewtopic.php?t=17667#p56348 It is noted that United States is not included in this study and we only consider three main factors (investment,

Fig. 9.1 Location of study area in European countries

labour force growth and budget surplus) which affect deeply to GDP. The GDP of these countries is the central measure of national accounts, which summarises the economic position of Europe. We surveyed the GDP of those countries and the effecting factors of GDP from 1995 to 2010 years (in Series 1).

Let us consider the panel data model [6]:

$$Y_{it} = \beta_0 + X_{it}^{Tr}\beta + \varepsilon_{it}, \ i = 1, 2, \ldots, N; \ t = 1, 2, \ldots, T, \ [X_{it}^{Tr} = \text{Transpose of } X_{it}], \tag{9.1}$$

where i indicates country and t denotes time. The i subscript, therefore, represents the cross-section dimension whereas t subscript denotes the time-series dimension. β_0 is the intercept which is independent of i and t, β is of order $k \times 1$ and X_{it} is the ith observation on k explanatory variables and lastly ε_{it} denotes the error components of the model.

9.3 Statistical Preliminaries and Methods

In statistics and econometrics [19] cross-sectional data is a kind of data which is collected by observing many subjects (such as individuals, countries) at the same point of time, or without regard to differences in time. A time series is a series where data points are listed in time order. Panel data is a combination of these two types and so it is usually called as cross-sectional time-series data. Generally, panel data regression models can be categorised into two ways: (a) static and (b) dynamic models. Each of these can be identified in two sub-categorises, i.e., (i) balanced (or, complete), (ii) unbalanced (or, incomplete). Same temporal length for all individuals is taken in (i), whereas in (ii), different temporal lengths are utilized. The application of panel-data analysis in the field of GDP is very important in statistical point of view and it is also helpful for economic research [16, 44]. In this work we have considered the fixed and random effects in the panel data model. First, we have considered fixed effects which is used for analyzing the influence of variables that differ in time period. These effects explore the relationship between predictor and outcome variables within an entity (such as country, person, company) which has its own individual characteristics separately. When a model is good fitted, it evaluates the effects and includes relevant interactions [23, 25]. But it is not necessary that the entity always impact on the predictor variable. Here we have considered the following fixed-effects model [6, 9, 27]:

$$Y_{it} = \alpha_i + X_{it}^{Tr}\beta + e_{it}, \ i = 1, 2, \ldots, N; \ t = 1, 2, \ldots, T, \ [X_{it}^{Tr} = \text{Transpose of } X_{it}], \tag{9.2}$$

where α_i is the unknown intercept for each entity, i.e., individual intercept (fixed for given N). Y_{it} is the dependent variable, where $i=$ entity and $t=$ time. X_{it} represents independent variable which is $k \times 1$ regressor matrix, β is the coefficient for that independent variable which is a matrix of order $k \times 1$, e_{it} is the error term, where $E(X_{it}e_{it}) = O$ must hold. The vital presumption of the fixed-effects model is that the time-invariant characteristics are unique to the individual and should not be correlated with other individual characteristics. It is noted that the fixed-effects model has the power to control all time-invariant differences between the individuals. The estimated coefficients of the fixed-effects models cannot be biased due to omitting time invariant characteristic. Next, we consider the random effects which is the contrast across entities assumed to be random and uncorrelated with predictor or explanatory (independent) variables included in the model. The model be considered here such as [6, 9, 27]:

$$Y_{it} = \beta_0 + X_{it}^t\beta + \mu_i + e_{it}, \text{ where } i = 1, 2, \ldots, N; \ t = 1, 2, \ldots, T \tag{9.3}$$

where μ_i is specific for individual i and $\mu_i \sim \text{iid}(0, \sigma_\mu^2)$, i.e., the μ's of different individuals are independent, have a mean zero and their distribution is assumed to be not too far away from normality. The overall mean is captured in β_0. Y_{it} is the dependent variable where $i=$ entity and $t=$ time. X_{it} represents independent variable which is $k \times 1$ regressor matrix, β is the coefficient for that independent variable which is a matrix of order $k \times 1$ and e_{it} is the error term with $e_{it} \sim$ iid $(0, \sigma_u^2)$. From model (3) the value of θ can be determined as follows [34]:

$$\theta = \frac{var(e)}{var(e) + T var(\alpha)}, \qquad (9.4)$$

where T does not have to be the same for each individual. If $\theta = 1$, the fixed-effects and the random-effects estimator are identical. The term θ provides a measure of relative sizes of the within and between component variances. Normally, θ lies between 0 and 1. It is assumed that each entity error term has no correlation with the predictors which allows for time invariant variables to play an important role as independent(explanatory) variables. In random effects the individual characteristics that may or may not effect the predictor variables. So, there are some trouble if some variables may not be available, then leading to omitted variable bias in the model.

In this work the panel data models are elaborated using investment, labour force growth and budget surplus for different years predicting the gross domestic product (GDP). The relation can be formulated in a panel data model as follows:

$$Y_{GDP} = \beta_0 + \beta_1 X_{inv} + \beta_2 X_{lf} + \beta_3 X_{bs} + \mu_i + \lambda_t \qquad (9.5)$$

where Y_{GDP} is the gross domestic product; β_0 is the general intercept; X_{inv} is investment, X_{lf} is labour force growth and X_{bs} is budget surplus of the different countries in several time periods; μ_i and λ_t represent unobservable individual and time effects respectively and β_1, β_2 and β_3 are coefficients of independent variables. In a fixed-effects model, the μ_i's are assumed to be fixed. However, the main problem with the fixed-effects model is its specification with too many parameters, resulting in heavy loss of degrees of freedom. But in the random-effects model μ_i are assumed to be random. It is perceived that panel data assigns more informative data, more variability, less collinearity among the variables, more degrees of freedom and more efficiency. Here we have discussed about fixed effects and random effects models and their underlying assumptions such that residual sum of squares, R-squared etc. In statistics, residual sum of squares (RSS) is the sum of squares of the differences between the observed y-value and the predicted y-value. It is also known as the sum of squared residuals (SSR) or the sum of squared errors of prediction (SSE). It is a measure of inconsistency between the data and an estimation model. If the value of RSS is small, then it indicates that a tight fit of the model to the data. It is also used as an optimality basis in model selection and given by [6]:

$$RSS = Y^T Y - Y^T X (X^T X)^{-1} X^T Y = Y^T [I - X(X^T X)^{-1} X^T] Y = Y^T [I - H] Y \qquad (9.6)$$

where I be the $n \times n$ identity matrix and H is called hat matrix, or the projection matrix in linear regression: $H = X(X^T X)^{-1} X^T$. Here Y is an $n \times 1$ vector of dependent variable observations and each column of the $n \times k$ matrix X is a vector of observations on one of the k explanators.

Definition 1 In a standard regression model: the sum of the squares of the deviations of the predicted values from the mean value of a response variable is called the explained sum of squares (*ESS*).

$$\therefore ESS = \sum_{i=1}^{n}(\widehat{y}_i - \bar{y})^2.$$

Definition 2 Total sum of squares (*TSS*)=*ESS* + *RSS*.

It is noted that the *ESS* measures how much variation there is in the modelled values and this is compared to the *TSS*, which measures how much variation there is in the observed data, and to the *RSS*, which measures the variation in the modelling errors.

The *R*-squared is commonly used to examine how differences in one variable can be explained by a difference in a second variable. Moreover, *R*-squared gives the percentage from zero to hundred of the variation in *y* explained by *x* variables. The range is 0 to 1 (i.e. 0% to 100% of the variation in *y* can be explained by the *x* variables). *R*-squared shows how the dataset of points is well fitted to a curve or a line. Similarly, adjusted *R*-squared also indicates the goodness of fitting of a curve or line, but adjusts for the number of terms in a model. If we add more and more predictors to a model, then *R*-squared will be increased. Generally, *R*-squared always increases. So, it can appear to be a better fit with the more terms by adding to the model but this can be completely misleading. If the model is over fitted then a misleadingly high *R*-squared value can lead to misleading projections. The value of adjusted *R*-squared will be gradually decreasing by adding more useless variables to a model and it will be increased by adding more useful variables. *R*-squared will always be greater than or equal to Adjusted *R*-squared. The formula is [6]:

$$R_{adj}^2 = 1 - \frac{(1-R^2)(n-1)}{n-k-1} \tag{9.7}$$

where *R*-squared is sample *R*-squared, *k* be the number of predictors and *n* be the total sample size.

But now we have to choose one of them between fixed and random effects. In panel data analysis the Hausman test [26] can help us to choose between fixed-effects model or a random-effects model. The null hypothesis is taken as the preferred model is random effects versus the alternative (the fixed effects). The fixed-effect assumption is that the individual-specific effects are correlated with the independent variables where as the random-effect is the contrast across entities assumed to be random and uncorrelated with the independent variables included in the model. The appropriate statistic for this hypothesis is the χ^2-statistic. If χ^2-statistic greater than the critical value, then the null hypothesis is rejected, i.e., if the *p*-value is less than 0.05, we have to reject the null hypothesis otherwise it is accepted.

There are few tests for testing cross-sectional dependence in panel data. We use Breusch-Pagan LM test applied when T > N and Pesaran CD test applied when N

> T. Since from our data we get N > T, we use Pesaran CD test to test the cross-section dependence. The use of the CD test is explained by applying it to study the degree of dependence in per capital output innovations across countries within a given region and across countries in different regions. Cross-sectional dependence can lead to bias in tests results known as "contemporaneous correlation". In this work, the null hypothesis: there is no cross-section dependence, and the alternative hypothesis: presence of cross-section dependence. If the p-value is less than 0.05, we have to reject the null hypothesis otherwise it is accepted.

If random effects are present, it may affect tests for residual serial-correlation. For solving such kind of problem, we have to use a joint test, that has power against both alternatives. Bera et al. [10] propose locally robust tests both (i) first-order serial-correlation in residuals and (ii) individual random effects. For testing purpose of residual serial-correlation (or individual random effects) locally robust versus individual random effects (serial-correlation) for panel models and joint test of serial-correlation and the random effect specification is used in this work [3]. These tests are robust versus local misspecification of the alternative hypothesis. The null hypothesis is no random effects allowing for local departures from the assumption of no serial-correlation in residuals. We use serial-correlation (also called autocorrelation) tests for macro panels with long time series (over 20–30 years). In this context it is mentioned that serial-correlation is a relationship between a given variable and itself over various time intervals, it is often found in repeating patterns when the level of a variable effects its future level. The error terms are called serially correlated if error terms from different (usually adjacent) time periods (or cross-section observations) are correlated. In this work—the null hypothesis: there is no serial-correlation, and the alternative hypothesis such as: serial-correlation in idiosyncratic errors. It is mentioned that the 'idiosyncratic error' term is used to describe error from panel data that both changes over time and across units (individuals, firms, cities, etc.). We cannot accept null hypothesis (if p value <0.05).

In statistics the most important part is the variability quantified by the variance or any other measure of statistical dispersion (like moments etc.). If the variance is constant then it is referred as homoscedasticity (in Fig. 9.2) otherwise it is heteroscedasticity (in Fig. 9.3). We can deal simultaneously with heteroscedastic as well as serially correlated disturbances in the work in panel data analysis.

In a statistical analysis, heteroscedasticity occurs when the variance of the errors varies across observations [37, 38]. Breusch-Pagan test [12] helps the presence of heteroscedasticity in a model. In this work the null hypothesis is that the variance is constant, i.e., the presence of homoscedasticity and the alternative hypothesis is not the presence of homoscedasticity, i.e., the variance of error terms is not constant (presence of heteroscedasticity). If p-value is greater than 0.05, then we can failed to reject the null hypothesis or alternatively we can say that if p-value is less than 0.05, then heteroscedasticity is present in the model. In a panel data model for the standard error component: it is conjectured that the disturbances have homoscedastic variances and constant serial-correlation through the random individual effects [6, 31]. These assumptions may be restrictive in many applications of panel data. It can be shown that the model of standard error components may be extended to

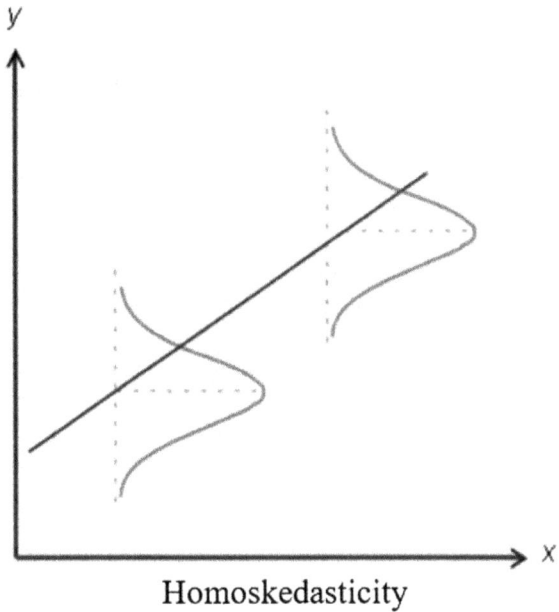

Fig. 9.2 Homoscedasticity

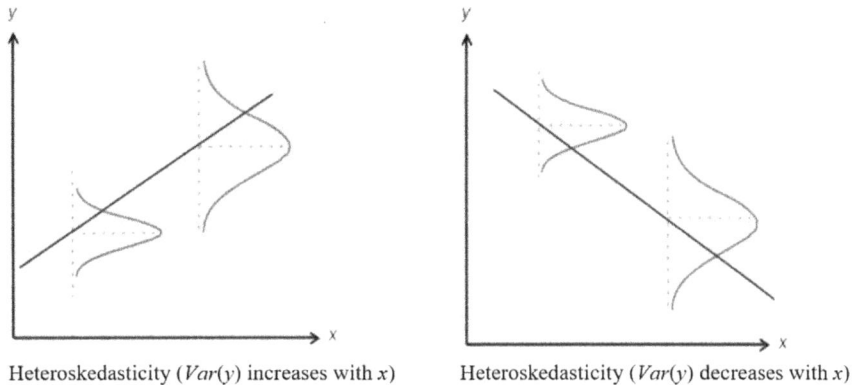

Fig. 9.3 Heteroscedasticity (var(y) increases with x [left]; var(y) decreases with x [right])

take into account serial-correlation [5] and may also be generalized to take into account heteroscedasticity [2, 7]. Generally, serial-correlation is disregarded when handling with heteroscedasticity and heteroscedasticity will be ignored when dealing with serial-correlation. The Lagrange Multiplier (LM) test [5, 13] which tests the presence of serial-correlation together with random individual effects jointly [7, 30] assuming homoscedasticity of the disturbances. After that LM statistic was derived which tests for homoscedasticity of the disturbances in the work of a one-way random effects for underlying model purpose. The latter LM test was derived by Baltagi et al. [8] which assumes no serial-correlation in the remainder disturbances and it allows for first order serial-correlation in the remainder disturbances [5]. The work of [8] simultaneously deals with heteroscedastic as well as serially correlated disturbances in the context of panel data model.

The choice between fixed and random effects models, the random effects model is preferable for the underlying dataset based on Hausman test. Therefore, we have to extend with random effects model. The models termed as panel by the econometricians have counterparts in the statistical field on mixed models or hierarchical models, or models for longitudinal data, although there are differences in jargon and more substantial distinctions [36]. In general, econometrics deal with non-experimental data and great emphasis is put on specification procedures and misspecification testings. Model specifications tend therefore to be very simple, while great attention is put on the issues of endogeneity of the regressors, dependence structures in the errors and robustness of the estimators under deviations from normality. The preferred approach is often semi-parametric or non-parametric. Heteroskedasticity-consistent techniques are becoming standard practice both in estimation and testing. Fitting correlated data is uttermost important for using in case of extensional random effects model. The model is defined as:

$$y_i = X_i \beta + Z_i v_i + e_i \tag{9.8}$$

where y_i is a $n_i \times 1$ vector of responses for subject i, X_i is known as $n_i \times p$ design matrix, β is a $p \times 1$ vector of population parameters, Z_i is known as $n_i \times r$ matrix, v_i is a $r \times 1$ vector which represents individual effects and lastly e_i denotes $n_i \times 1$ vector of random residuals. The errors follow a first order autoregressive (AR1) process:

$$e_k = \rho e_{k-1} + \varepsilon_k \tag{9.9}$$

where ε_k is assumed to be iid $N(0, \sigma_\varepsilon^2)$ and ρ is autocorrelation coefficient with $|\rho| < 1$. For comparison of models, we have used AIC (Akaike information criterion), BIC (Bayesian information criterion) and log-likelihood criteria.

Definition 3 Akaike information criterion (AIC): The Akaike information criterion is an estimator of out-of-sample prediction error that provides relative quality of statistical models for a given data-set. Given a collection of models for the data, AIC estimates the quality of each model, relative to each of the other models **(author?)** [24].

The AIC value of the model is such as:

$$AIC = 2k - 2\ln(\hat{L}) \tag{9.10}$$

where k denotes the number of estimated parameters in the model and $\hat{L}$ is the maximum likelihood estimator (MLE) for the model.

Definition 4 Bayesian information criterion (BIC): In statistics, the Bayesian information criterion (also known as Schwarz information criterion) is a criterion for selection of a model, among a finite set of models. The model that has lowest BIC is preferred. It is based on (in part) the MLE and it is closely related to AIC.
The BIC value of the model is formally defined as:

$$BIC = k \ln(n) - 2\ln(\hat{L}) \tag{9.11}$$

where $\hat{L}$ denotes the MLE of the model, n be the number of data points (or equivalently, the sample size) in observed data (x) and lastly k be the number of parameters estimated by the model.

Definition 5 Log-likelihood: The log-likelihood function is typically used to derive the maximum likelihood estimator of the parameter. The estimator is obtained by finding the parameter that maximizes the log-likelihood of the observed sample. In statistical field, it measures the goodness of fit of a statistical model to a sample of data for given values of the unknown parameters [21, 22]

9.4 Results

First we perform the plotting of the dataset, when plotting multiple datasets, the Wolfram language chooses a different colour for each dataset automatically which gives us the clear view. The countries are taken in the following order: 1. Austria, 2. Belgium, 3. Denmark, 4. Finland, 5. France, 6. Germany, 7. Ireland, 8. Italy, 9. Netherlands, 10. Norway, 11. Portugal, 12. Spain, 13. Sweden, 14. Switzerland and 15. United Kingdom.

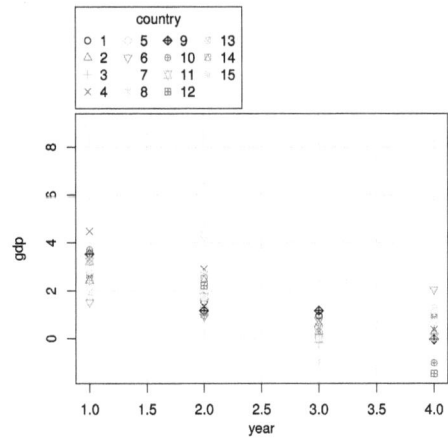

library(foreign)
Panel=read.csv("C:\\Users\\Acer-PC\\Desktop\\New folder (2)\\new.csv")
library(car)
scatterplot(gdp~year| country, boxplots=FALSE, smooth=TRUE, reg.line=TRUE, data=Panel)

Computation 1 : Scatter plot (time vs GDP)

As mentioned earlier here we wish to predict GDP under the influence of different factors such as investment, labour force growth and budget surplus using panel-data concept which helps to realize the economic status across 15 countries in Europe. For each model, investment, labour force growth and budget surplus are considered as independent variables whereas GDP as dependent variable. Firstly we have considered fixed-effects and then random-effects model. The basic use of *plm* package is to indicate the model formula, the data and the model to be estimated. For example, fixed-effects model and random-effects model are estimated using *plm* package.

```
library(plm)

fixed <- plm(gdp~inv+lf+bs, data = Panel,index=c("year","country"),model="within")

summary(fixed)

Oneway (individual) effect Within Model

Call:

plm(formula = gdp ~ inv + lf + bs, data = Panel, model = "within",
    index = c("year", "country"))

Balanced Panel: n=4, T=15, N=60

Residuals :

    Min.   1st Qu.   Median    3rd Qu.    Max.

-1.956606 -0.687389 -0.025913 0.535807 3.512373

Coefficients :

       Estimate Std. Error  t-value  Pr(>|t|)

inv -0.077912  0.059339 -1.3130  0.1948

lf   0.842681  0.178475  4.7216  1.755e-05

bs   0.032031  0.028195  1.1361  0.2610

Total Sum of Squares:   71.448

Residual Sum of Squares: 49.536

R-Squared:    0.30669

Adj. R-Squared: 0.2282

F-statistic: 7.81482 on 3 and 53 DF, p-value: 0.00020655
```

9 Gross Domestic Product Modeling Using "Panel-Data" Concept

Computation 2 : Fixed effects in panel data model

Here the panel data is balanced (a balanced data set is a set that consists of all elements observed in all time frame whereas if some entities do not possess an observation at certain time points while some other entities do, it is called unbalanced). The number of panels is denoted by n, the number of years is denoted by T and the number of total observations is represented by N. The summary shows the effects (in deviation from the overall intercept), their standard errors and the test of equality to the overall intercept and shows the residual sum of squares and also calculate the p-value.

```
random <- plm(gdp~inv+lf+bs, data =Panel,index=c("year","country"),model="random")
summary(random)
```

Oneway (individual) effect Random Effect Model

(Swamy-Arora's transformation)

Call:

plm(formula =gdp ~inv +lf +bs, data =Panel, model ="random", index=c("year", "country"))

Balanced Panel: n=4, T=15, N=60

Effects:

	var	std.dev	share
idiosyncratic	9.346e-01	9.668e-01	1
individual	8.058e-17	8.977e-09	0

theta: 6.661e-16

Residuals :

Min.	1st Qu.	Median	3rd Qu.	Max.
-2.97930	-1.08363	-0.14213	0.96214	4.80991

Coefficients :

	Estimate	Std. Error	t-value	Pr(>|t|)
(Intercept)	2.395966	1.882991	1.2724	0.2084799
inv	-0.082314	0.089762	-0.9170	0.3630633
lf	1.004524	0.268218	3.7452	0.0004275
bs	0.025377	0.042639	0.5952	0.5541407

Total Sum of Squares: 162.71

Residual Sum of Squares: 127.45

R-Squared: 0.21669

Adj. R-Squared: 0.17473

F-statistic: 5.16394 on 3 and 56 DF, p-value: 0.0031969

Computation 3 : Random effects in panel data model

For a random-effects model, the summary method provides information about the variance of the errors of the components. Here the panel data is balanced; the number of panels is denoted by n, the number of years is denoted by T and the number of total observations is represented by N. Interpretation of the coefficients is tricky because they include both within-entity and between-entity effects. The summary shows the effects of idiosyncratic (refers to the observation-specific zero-mean random-error term), individual and also calculate the value of the theta, their standard errors and the test of equality to the overall intercept and shows the residual sum of squares and also calculate the p-value. $Pr(> |t|)=$ Two-tail p-values test the hypothesis that each coefficient is different from zero. Now we have to find which is choosen between fixed and random-effects model. The Hausman test decides this decision in the model.

```
phtest(fixed,random)

Hausman Test

data: gdp ~inv + lf + bs

chisq = 0.82873, df = 3, p-value = 0.8426

alternative hypothesis: one model is inconsistent
```

Computation 4 : Fixed vs Random effects in panel data model

For Hausman test we have to first run a fixed-effects model and then save its estimates, then run a random-effects model and save its estimates, finally executes this test. We have to choose fixed-effects if the p-value is significant (for example < 0.05), if not then we have to use random-effects. Therefore, random-effects model is better than fixed-effects model (since $p = 0.8426$).

```
confint(fixed)
            2.5%        97.5%
inv  -0.19421469  0.03839076
lf    0.49287563  1.19248557
bs   -0.02322904  0.08729156

confint(random)
                  2.5%        97.5%
(Intercept) -1.29462849  6.08656105
inv         -0.25824316  0.09361616
lf           0.47882698  1.53022083
bs          -0.05819458  0.10894804
```

9 Gross Domestic Product Modeling Using "Panel-Data" Concept

Computation 5 : Confidence intervals for fixed and random-effects model

In this computation, we find 95% confidence level for both (fixed and random-effects) models.

fixed.time <- plm(gdp ~ inv + lf+ bs+factor(year), data=Panel, index=c("country", "year"), model="within")

plmtest(random, c("time"), type=("bp"))

Lagrange Multiplier Test - time effects (Breusch-Pagan) for balanced panels

data: gdp ~ inv + lf

chisq = 1.2461, df = 1, p-value = 0.2643

alternative hypothesis: significant effects

Computation 6 : Random vs Time-fixed effects in panel data model

Next, for choosing random effects or time-fixed effects we have to use Lagrange multiplier (LM) test. Here, null hypothesis be: no time-fixed effects needed and the alternative hypothesis be time-fixed effects is present. We have to choose time-fixed effects if the p-value is significant (for example < 0.05), if not then we have to use random-effects. So, in this work, no need to use time-fixed effects since p-value is greater than 0.05.

pcdtest(random, test =c("cd"))

Pesaran CD test for cross-sectional dependence in panels

data: gdp ~inv +lf +bs

z = 1.4953, p-value = 0.1348

alternative hypothesis: cross-sectional dependence

Computation 7 : Testing for cross-sectional dependence

Pasaran CD tests are useful to test whether the residuals are correlated across entities or not. Here, the null hypothesis is 'residuals are not correlated' and the alternative hypothesis is 'cross-sectional dependence'. If p-value is less than 0.05, then we cannot accept the null hypothesis. Otherwise if p-value is greater than 0.05, it confirms that 'there is no cross-section dependence'.

pbsytest(gdp~inv + lf+ bs, data=Panel,test="j")

Baltagi and Li AR-RE joint test - balanced panel

data: formula

chisq = 139.42, df = 2, p-value < 2.2e-16

alternative hypothesis: AR(1) errors or random effects

pbsytest(gdp~inv + lf+ bs, data=Panel)

Bera, Sosa-Escudero and Yoon locally robust test - balanced panel

data: formula

chisq = 0.035192, df = 1, p-value = 0.8512

alternative hypothesis: AR(1) errors sub random effects

pbsytest(gdp~inv + lf+ bs, data=Panel,test="re")

Bera, Sosa-Escudero and Yoon locally robust test (one-sided)

balanced panel

data: formula

z = 10.923, p-value < 2.2e-16

alternative hypothesis: random effects sub AR(1) errors

Computation 8 : Locally robust tests for serial-correlation or random effects

The choice between fixed and random effects specifications is based on Hausman test. After comparing the two estimators, the more efficient random effects estimator is chosen in this work. So, we have to extend with various tests of random effects. Locally robust tests are very useful for serial-correlation or random effects for residual serial-correlation (or individual random effects) locally robust versus individual random effects (serial-correlation) for panel models and joint test of serial-correlation and the random effects specification by Baltagi and Li [3]. If random effects are present, it may affect tests for residual serial-correlation. Here, the alternative hypothesis is the presence of AR(1) errors or random effects and the null hypothesis is there is no effect. If p-value is greater than 0.05, then we cannot reject the null hypothesis. Here, the p-value is less than 0.05 which confirms the presence of the underlying effects. Next, we have to extend our work to test for random effects which is executed in the one sided version taking attention that the variance of the random effects must be non-negative and for this purpose we accept the alternative hypothesis, i.e., random effects sub AR(1) errors since p-value is less than 0.05 (Computation 8). Bera et al. [10] derived locally robust tests both for first-order serial-correlation in residuals and for individual random effects. It is to compute a one-sided test which is expected to lead to a more powerful test (asymptotically $N(0,1)$ distributed) since

Computation 8 is true in this work otherwise we have to accept the two-sided test (asymptotically chi-squared distributed).

> pbgtest(random)
>
> Breusch-Godfrey/Wooldridge test for serial correlation in panel models
>
> data: gdp ~ inv + lf + bs
>
> chisq = 20.447, df = 4, p-value = 0.0004075
>
> alternative hypothesis: serial correlation in idiosyncratic errors

Computation 9 : General serial-correlation in panel data model

Generally, the test for serial-correlation is applicable to macro panels. Here, the null hypothesis is 'not presence of serial-correlation' and the alternative hypothesis is 'serial-correlation in idiosyncratic error'. If p-value is greater than 0.05, then we cannot reject the null hypothesis. Here, the p-value is less than 0.05 which confirms the presence of 'serial-correlation in idiosyncratic error'.

> u_root <- pdata.frame(Panel, index = c("country", "year"))
>
> library(tseries)
>
> adf.test(u_root$gdp, k=2)
>
> Augmented Dickey-Fuller Test
>
> data: u_root$gdp
>
> Dickey-Fuller = -5.2094, Lag order = 2, p-value = 0.01
>
> alternative hypothesis: stationary

Computation 10 : Unit roots in panel data model

In general, the Dickey-Fuller test is used to check for stochastic trends. The null hypothesis is 'the presence of unit root' (i.e., non-stationary) in the series and the alternative hypothesis is that 'unit root is absent', i.e., either stationarity, trend stationarity or explosive root depending on the test used in this work. If unit root is present, we have to take the first difference of the variable. We have to accept null hypothesis (since $p > 0.05$). In this work, the p-value is less than 0.05 which signifies that no unit roots are present here.

> library(lmtest)
>
> bptest(gdp~inv+lf+bs+factor(country), data = Panel, studentize=F)
>
> Breusch-Pagan test
>
> data: gdp ~ inv + lf + bs + factor(country)
>
> BP = 28.765, df = 17, p-value = 0.03676

Computation 11 : Heteroskedasticity in panel data model

Now, we test heteroskedasticity which generally occurs when the variance of the distribution is not constant. The Breusch-Pagan test [12] decides about the presence of heteroskedasticity or not with the help of p-value. Since p-value is less than 0.05, we cannot accept the null hypothesis, confirming 'presence of heteroskedasticity'. We can also deal simultaneously with heteroskedastic as well as serially correlated disturbances in the context of a panel data regression model.

```
reAR1ML <- lme(gdp~inv + lf+ bs, data=Panel,random=~1|country,
correlation=corAR1(0,form=~year|country))

reML=lme(gdp~inv + lf+ bs, data=Panel, random=~1|country)

anova(reML, reAR1ML)
```

	Model	df	AIC	BIC	logLik	Test	L.Ratio	p-value
reML	1	6	237.5947	249.7468	-112.7974			
reAR1ML	2	7	234.4555	248.6330	-110.2278	1 vs 2	5.139231	0.0234

Computation 12: An LR test for serial-correlation and one for random effects

The choice between fixed and random effects specifications is based on Hausman test, comparing the two estimators under the null hypothesis of no significant difference. If this is not rejected, the more efficient random effects estimator is chosen in this work. Since, random effect models are preferable for the underlying dataset, therefore, we have to extend with random effects model. We have computed the AR(1) test on extensions of the random effects model in this work. The LR (likelihood ratio) test for random effects compares between two underlying models, i.e., AR(1) test or auto regressive test on extensions of the random effects model and only extensions of the random effects model. Between them, AR(1) test on extensions of the random effects model is better than other one since it's AIC and BIC value is low (i.e., the AIC values are 234.45 and 237.59 and the BIC values are 248.63 and 249.75 with or without auto correlated error extensions of the random effects model respectively) and log-likelihood is high (-110.23 and -112.80 with or without auto correlated error extensions of the random effects model respectively).

9.5 Concluding Remarks

In this work we have analyzed the effects of explanatory variables on the responsible variable (GDP) based on the data set of fifteen European countries. We have explained that how the data is being modelled and the response allows a natural interpretation of model parameters in the panel data model. In the fixed-effects model it is evident that the model is balanced. The coefficient of investment, labour force growth and budget surplus indicates (by computation 2) how much Y changes overtime, on an average, when X increases by one unit. The residuals vary from -1.9567 to

3.5124 (by computation 2) for the fixed-effects model. It is also concluded (as per computation 3) that the random-effects model the interpretation of the coefficients is tricky since both the within-entity and between-entity effects are included [42]. Here, the coefficient of investment, labour force growth and budget surplus represents the average effect of X over Y when X changes across time and between countries by one unit. The residuals vary from -2.9793 to 4.8099 (as per computation 3) for the random-effects model.

There are many advantages of using panel data [31, 35]. Panel data suggests that individuals, states or countries are heterogeneous. Generally, time-series and cross-section separately does not have any controlling heterogeneity. But the econometric [19] interest on panel data is the outcome of different stimulations in the panel data. First of all, the desire of utilizing panel data for controlling unobserved time-invariant heterogeneity in cross-sectional models [4]. The time-series studies are afflicted with multicollinearity but panel data provide more variability, more informative data, less collinearity among the variables, more efficiency and more degrees of freedom. Panel data are also skillfully accepted to study the duration of economic states of the country and if these panels are long enough, they can shed light on the speed of adjustments to the economic status. So, we can say panel data are better able to study the dynamics of adjustment [10]. Economists studying workers' levels of satisfaction run into the problem of anchoring in a cross-section study [43] in Chap. 11). But Panel data are better able to recognize and measure effects that are simply not identified in pure cross-section or pure time-series data. The model also permit to establish and test more complicated behavioral models than purely cross-section or time-series data [2, 17]. But there are some limitations in panel data. First of all, one of the main problems is data collection, designing the panel surveys and data management. These problems arise when for incomplete account of the population of interest or does not response due to lack of cooperation of the respondent or because of interviewer error or sometimes respondent does not recall correctly or frequency of interviewing, interview spacing, reference period, the use of bounding and time-in-sample bias. Sometimes, reported behavior and actual behavior are not always the same or often the researchers are biased or ignoring some related factors. Measurement errors is another problem which may arise because of faulty responses due to deliberate distortion of responses, memory errors, unclear questions, inappropriate information, misrecording of responses or error in data preparation and interviewer effects [20, 29]. Apart from this, over generalization is another problem or sometimes many failures occur to recognize invalid data. Besides self selectivity [28] or non-response are the major problems in the studies. For any typical micro panels including annual data which cover a very short time span for each individual is one of the troubles in panel data analysis. Sometimes, misleading inference happens due to unaccounted cross country dependence in the macro panel data [6]. Panel data provides several advantages worth its cost. The use of panel data for that situation under consideration has to know its limitations. The fixed-effects model allows the individual or time or both specific effects to be correlated with independent variables X_{it}. So, it does not require any investigator to model their correlation patterns. But the number of unknown parameters increases with the number of sample observations and in that case when T or N for λ_t is finite, it introduces the classical incidental parameter problem which is the main disadvantage of the fixed-effects model in the panel data analysis. Besides, the fixed-effects estimator does not allow the estimation of the coefficients that are time-invariant. So, these are another drawback of fixed-effects model.

But here since the random-effects model is more appropriate, so we have to specify a conditional density of α_i given $X_i^T = (X_{it}, \ldots, X_{iT}), f(\alpha_i|X_i)$, while α_i are unobservable. We consider that the condition density $f(\alpha_i|X_i)$ is similar with the marginal density $f(\alpha_i)$.

In panel data analysis, the Hausman test helps us to choose between fixed-effects model or a random-effects model. After testing, we get the p-value as 0.8426 (by computation 4). Since it is greater than 0.05, so we accept the null hypothesis, i.e., we have to choose the random-effects model and this model is also better than time-fixed effects model (as per computation 6). We can also conclude that the number of parameters stay constant when sample size increases which allows the derivation of efficient estimators that make use of both within and between variation (as per computation 4). Apart from this, we can also draw the inference that it allows the estimation of the impact of time-invariant variables (as per computation 4). In panel data analysis, the Pasaran CD tests help us to test the cross-sectional dependence. After testing, we get the p-value as 0.1348 (as per computation 7). Since the value is greater than 0.05, so we accept the null hypothesis, i.e., there does not exist any cross-sectional dependence. It is also concluded from computation 8 that the presence of AR(1) errors or random effects model exists, so, furthermore addition of general serial-correlation depicts an important fact in this work. Now, let us consider Wooldridge or Breusch-Godfrey test for the serial-correlation purpose [14]. It can be shown (by computation 9) that the p-value is 0.0004075 which is less than 0.05. So, we have to reject the null hypothesis. Therefore, we also conclude that there exists serial-correlation in idiosyncratic errors. Next, Dickey-Fuller test is used to explore for stochastic trends and from p-value it is also concluded that the roots are stationarity in this work. We have considered the presence of heteroscedasticity in panel data model. In this work, the Breusch-Pagan test [12] helps a lot to understand the presence of heteroscedasticity. Heteroscedasticity is a systematic pattern in the errors where the variance of the error terms differ across observations heteroscedasticity. Here the p-value is 0.03676 (as per computation 11) which is less than 0.05. So we have to reject the null hypothesis and decide that the heteroscedasticity is present in the panel data model. So, we can also conclude that the variances of the errors are not constant (see computation 11). Therefore, heteroscedasticity and serial-correlation both (as per computations 9 and 11) are present in a random-effects panel data model and we can manage it simultaneously. Since random effects model is more efficient by Hausman test, we proceed with AR(1) test on extensions of the random effects model which is better than other underlying model since it's AIC and BIC value is low (by computation 12) and log-likelihood is high (as per computation 12) than other one.

Each of the regression coefficients is indicative of the impact of a change in the explanatory variable on the GDP, it would not be valid for a particular country in the sample, unless that country closely resembles the average country with regard to the economic structure summarized by the values of the explanatory variables. The labour force is strongly and positively correlated with GDP in the fixed and random-effects models. On the other hand, the investment appears to have negative correlation with the GDP. The measure of openness affects GDP through investment. The results of fixed and random-effects models show a one percent increase in GDP that leads to a rise in budget surplus by around 0.3% points. We can obtain the information about the structure of the individual and time dimensions of panel data (computation 2 and 3). We can also conclude the final market value of goods and services produced by labour of all but only of residents of a country over a given period of time. We can also measure the annual growth and expansion of the economy and can also scrutinize the relative economic strength of a country. The annual improvement in the standard of living of the average citizen of a country can also be measured. As per our computations, if the investment is lowered, budget surplus and labour force are continuously increased, then the economy of the country will be benefited. With respect to policy choices, it can be concluded that financial conditions of industrialized countries (covering a large part of the Europe) played an important role in assessing business cycles. Moreover, it can also be concluded that businesses invest less money into capital expansion. Likewise, the consumers buy fewer homes (as per computations 2 and 3). The main objective of this work is to introduce about European financial markets and the impact of financial openness on growth in business cycles. It can also be concluded that financial crisis of any country does not effect the other European countries (as per computation 7).

There are huge differences in the summary effects of two different models. Under the fixed-effects model, the assumption is that the size of true effect is same for all studies and the effect size varies only for due to sampling error (error in estimating the effect size). Therefore, whenever assigning weights to the various studies, it can be largely ignored the information in the smaller studies since we have been informed in better way about the same effect size in the larger studies. By contrast, in case of the random-effects model, the aim is not only estimate one true effect, but also estimate the mean of a distribution of effects. We want to be sure that all these effect sizes are represented in the summary estimate since each study provides information about various effect sizes. Apart from this we can say that in case of the fixed-effects model, the only source of uncertainty is within-study (estimation or sampling) error but for the random-effects model, similar source of uncertainty present together with an additional source (i.e., between studies variance). So, the standard error(s.e.), variance and confidence interval (C. I.) for the summary effect will always be larger (or wider) under the random-effects model than fixed-effects model (unless two models are same, as per computation 5). In this chapter, the standard error is 0.06, 0.18 and 0.03 for investment, labour force and budget surplus respectively for the fixed-effects model, and 0.09, 0.27 and 0.04 for the random-effects model. Random-effects model is more easily justified than the fixed-effects model because we should not assume a common effect size in the model. Under the fixed-effects model, it is also assumed that all dispersions in observed effects are due to sampling error, but in case of the random-effects model we allow some of that dispersions reflect real differences in effect size across studies. The impact of investment on long run economic growth, under both the random and fixed-effects model, is the same. The effect on GDP of a one percent fall in gross domestic investment is around 0.08 percent. Results under the fixed-effects model show that a one percent rise in the labour force towards GDP leads to 0.84 percent rise in growth of GDP. The impact is more stronger under the random-effects model. In this work, we have also determined some of the necessary determinants of sustainable economic growth in developing countries. The results provide useful insights into those three main factors (investment, labour force and budget surplus) which are related with GDP. Since we only analyze the relationship between GDP and its three main factors in this work, future research can extend the analytical framework in various ways particularly by considering the overall components of GDP by including financial systems and financial stability variables of the European countries. Clearly, future extended research work will be more captivating in econometrical studies.

References

1. Anderson TW, Hsiao C (1981) Estimation of dynamic models with error components. J Am Stat Ass 76:598–606
2. Baltagi BH, Griffin JM (1988) A generalized error component model with heteroscedastic disturbances. Int Econom Rev 29:745–753
3. Baltagi B, Li Q (1991) A joint test for serial correlation and random individual effects. Stat Probab Lett 11(3):277–280
4. Baltagi BH, Levin D (1992) Cigaretto taxation: raising revenues and reducing consumption. Struct Change Econom Dynam 3(2):321–335
5. Baltagi BH, Li Q (1995) Testing AR(1) against MA(1) disturbances in an error component model. J Econometrics 68(1):133–151
6. Baltagi BH (2005) Econometries analysis of panel data. Wiley, Chichester
7. Baltagi BH, Bresson G, Pirotte A (2006) Joint LM test for heteroskedasticity in a one-way error component model. J Econometrices 134:401–417
8. Baltagi BH, Jung BC, Song SH (2008) Testing for heteroscedasticity and serial correlation in a random effects Panel Data model. J Econometrics 154:122–4
9. Bell A, Fairbrother M, Jones K (2019) Fixed and random effects models: making an informed choice. Qual and Quant 53(2):1051–1074
10. Bera AK, Sosa-Escudero W, Yoon M (2001) Tests for the error component model in the presence of local misspecification. J Econometrics 101(1):1–23
11. Bertrand M, Duflo E, Mullainathan S (2003) How much should we trust differences-in-differences estimates? Quart J Econom 119(1):249–275
12. Breusch TS, Pagan AR (1979) A simple test for heteroskedasticity and random coefficient variation. Econometrica 47:1287–1294
13. Breusch TS, Pagan AR (1980) The lagrange multiplier test and its application to model specification in econometrics. Rev Econom Stud 47:239–254
14. Breusch TS, Godfrey LG (1981) A review of recent work on testing for autocorrelation in dynamic simultaneous models. In: Currie DA, Nobay R, Peel D (eds) Macroeconomic analysis: essays in macroeconomics and economics. Croom Helm, London
15. Campos NF, Coricelli F, Morettie L (2019) Institutional integration and economic growth in Europe. J Monetary Econom 103:88–104
16. Clarke P, Crawford C, Steele F, Vignoles A (2013) Revisiting fixed- and random-effects models: some considerations for policy-relevant education research. Edu Econom 23(3):1–19

17. Cornwell C, Schmidt P, Sickles R (1990) Production frontiers with cross-sectional and time-series variation in efficiency levels. J Econometrics 46(1–2):185–200
18. Croissant Y, Millo G (2008) Panel data econometrics in R: The plm package. J Statist Softw 27(2):
19. Davidson R, MacKinnon JG (1993) Estimation and inference in econometrics. Oxford University Press, New York, pp 320:323
20. Duncan G, Hill DH (1985) Panel data from a time series of cross-sections. J Econometrics 30:109–126
21. Fisher RA (1992) Statistical methods for research workers. In: Kotz S, Johnson NL (eds) Breakthroughs in Statistics. Springer Series in Statistics (Perspectives in Statistics). Springer, New York, NY. https://doi.org/10.1007/978-1-4612-4380-9_6
22. Ghosh S, Samanta GP (2019a) Fitting cumulative logit models for ordinal response variables in retail trends and predictions. Int J Stat Econom 20(1):32–49
23. Ghosh S, Samanta GP (2019b) Statistical modeling for cancer mortality. Lett Biomath 6(2). https://doi.org/10.1080/23737867.2019.1581104
24. Ghosh S, Samanta GP (2019c) Model justification and stratification for confounding of chlamydia Trachomatis disease. Lett Biomath 6(2). https://doi.org/10.1080/23737867.2019.1654418
25. Ghosh S, Samanta GP, Mubayi A (2020) Regression approaches of survival data in the presence of competing risks: an application to COVID-19; COVID-19 competing risks, COVID-19 ARCHIVES. Lett Biomath https://lettersinbiomath.journals.publicknowledgeproject.org/index.php/lib/article/view/307
26. Greene WH (2008) Econometric analysis, 6th edn. N.J, Upper saddle river, Prentice Hall, p 2008
27. Gurka MJ, Kelley GA, Edwards LJ (2012) Fixed and random effects models, wiley interdisciplinary reviews. Comput Stat 4(2):181–190
28. Hausman J, Wise D (1979) Attrition bias in experimental and panel data: the gary income maintenance experiment. Econometrica 47(2):455–473
29. Herriot RA, Spiers EF (1975) Measuring the impact of income statistics of reporting differences between the current population survey and administrative sources. In: Proceedings of the social statistics section, American Statistical Association, pp 147–158
30. Hoboken NJ, Baltagi BH, Li Q (1991) A joint test for serial correlation and random individual effects. Stat Probab Lett 11:277–280
31. Hsiao C (2003) Analysis of panel data, 2nd edn. Cambridge University Press, London
32. Hsiao C (2005) Why panel data? Singapore Econom Rev 50(2):1–12
33. Hujer R, Schneider H (1989) The analysis of labor market mobility using panel data. Eur Econom Rev 33(2–3):530–536
34. Johnston L (1984) Marxism and capitalist possession. The Sociolog Rev 32(1):18–37
35. Klevmarken NA (1989) Panel studies: what can we learn from them? Introd Eur Econom Rev 33:523–529
36. Laird NM, Ware JH (1982) Random-effects models for longitudinal data. Biometrics 38:963–74
37. Lejeune B (2006) A full heteroscedastic one-way error components model for incomplete panel: maximum likelihood estimation and lagrange multiplier testing, CORE discussion paper 9606, Universite Catholique de Louvain, pp 1–28
38. Li Q, Stengos T (1994) Adaptive estimation in the panel data error component model with heteroscedasticity of unknown form. Int Econom Rev 35:981–1000
39. Lloyd T, Morrisey O, Osei R (2001) Problems with pooling in panel data analysis for developing countries: the case of aid and trade relationships. Centre for Resin Economic Development (CREDIT), University of Nottingham Research Paper No. 01/14
40. Mahabbati R, Izady A, Mousavi Baygi M, Davary K, Hasheminia SM (2017) Daily soil temperature modeling using 'panel-data' concept. J Appl Stat 44(8):1385–1401. https://doi.org/10.1080/02664763.2016.1214240
41. Maddala GS (1971) The use of variance components models in pooling cross-section and time series data. Econometrica 39:341–358

42. Wang N, Zhang J, Xu L, Qi J, Liu B, Tang Y, Jiang Y, Cheng L, Jiang Q, Yin X, Jin S (2020) A novel estimator of between-study variance in random-effects models BMC Genomics, vol. 21, Article number: 149
43. Winkelmann L, Winkelmann R (1998) Why are the Unemployed so Unhappy? Evidence from Panel Data. Economica 65:1–15
44. Wooldridge J (2002) Econometric analysis of cross section and panel data. MIT Press

Chapter 10
Supply Chain Scheduling Using an EOQ Model for a Two-Stage Trade Credit Financing with Dynamic Demand

Alok Kumar

Abstract The effective supply chain scheduling is a crucial task in business management which can be determined by developing the optimum schedules. Here, this paper develops the optimum schedules using an EOQ model with dynamic demand pattern because in this era of globalization and dynamic environment the Economic Order Quantity (EOQ) model loses its importance when it is based upon the constant demand pattern. Therefore, it becomes indispensable to develop the EOQ model under an environment of dynamic demand pattern. Here, the dynamic demand pattern includes the relevant parameters which varies with time. The effects of such parameters are necessary to incorporate in determining the optimum schedules and hence the optimum inventory levels. Also, to establish a product in the market and to increase its customer base one can take the help of promotional efforts in the form of trade credit financing. This paper discusses the optimum scheduling for a part of supply chain system using an EOQ model where the demand is dynamic varies with time and one of the promotional effort in the form of a two-stage trade credit is considered. The applicability of the model can be well understood through the sensitivity analysis of the parameters and its managerial implications.

Keywords EOQ · Two-stage trade credit · Dynamic demand · Supply chain scheduling

10.1 Introduction

The retailer's procurement policy of inventory greatly affects the supply chain system. Therefore, to maintain the efficient supply chain system it becomes important to maintain the optimal inventory policy. Here, the optimal inventory policy is dependent upon the demand function which is based upon innovation diffusion criterion. There are several authors who have discussed the demand behaviour (adoption behaviour) of new products which are based upon innovation diffusion criterion as described

---

A. Kumar (✉)
FORE School of Management, New Delhi 110016, India
e-mail: alok@fsm.ac.in

below. The pure innovation aspect of the product adoption was explained by Fourt and Woodlock [9]. The Bass [6] model explains the combination of both the innovation and the imitation aspects of product adoption. A dynamic potential adopter diffusion model for the diffusion of oral contraceptives was discussed by Sharif and Ramanathan [19]. The effect of advertising under hazard rate demand framework for technology diffusion was discussed by Horsky and Simon [11]. The impact of optimal pricing and quality policies on diffusion of technology products was discussed by Teng and Thompson [21]. The inventory models based on dynamic demand function pertaining to the innovation diffusion criterion are few in numbers which are discussed as follows. The EOQ model based on demand influenced by dynamic innovation effect has been discussed by Chanda and Kumar [7]. Chanda and Kumar [8] discussed the EOQ model where demand influenced by innovation diffusion criterion under inflationary condition. The innovation diffusion based demand with dynamic potential market size for an EOQ model where parameters are explained with fuzzy nature was discussed by Aggarwal et al. [2]. The inventory model where demand depends on dynamic advertising expenditure was explained by Aggarwal and Kumar [3]. Aggarwal and Kumar [4] discussed an economic order quantity model with innovation diffusion criterion under influence of price dependent potential market size. This paper develops an EOQ model with time dependent dynamic demand under a two-stage trade credit policy where supplier offers a credit period to the retailer and the retailer also offers a credit period to the customer. The paper is organized as follows. Literature review which mainly deals with the authors' work pertaining to the trade credit period is given in Sect. 10.2. Assumptions and Notations which are useful to develop the model under the restricted conditions are depicted in Sect. 10.3. In Sect. 10.4, a detailed formulation of the model development is given. Numerical illustrations with sensitivity analysis is given in Sect. 10.5. The graphical representations showing convexity of the cost function is given in Sect. 10.6. The managerial implications along with conclusion is explained in Sect. 10.7.

10.2 Literature Review

The model discussed in this paper incorporates the two-stage trade credit financing. There are several authors who have developed the inventory models by integrating the concept of either one-stage or two-stage trade credit financing. The EOQ model with permissible delay in payments was first studied by Goyal [10] and thereafter Goyal's work was extended by several authors. The Goyal [10] model was extended by Aggarwal and Jaggi [1], Hwang and Shinn [12] by developing deterministic inventory model with a constant deterioration rate. Teng et al. [20] develops a two levels of trade credit inventory model where demand is an increasing function of time where the supplier offers a permissible delay linked to order quantity and the retailer provides a downstream trade credit period to its customers. Kumar and Aggarwal [13] discusses an optimal replenishment policy for an inventory model with demand follows innovation diffusion process under permissible delay in payments. Wu and

Zhao [23] developed a Supplier–buyer deterministic inventory model with trade credit and shelf-life constraint. Yang et al. [25] developed an optimal replenishment model under two-level trade credit in which the credit periods offered are explained as upstream trade credit and downstream trade credit where the upstream trade credit is linked to order quantity. This paper also incorporates the concept of limited storage capacity. Quin and Liu [17] discusses a single supplier single retailer optimal replenishment policy where ramp type demand and production rate dependent upon demand have been considered. They have considered the production rate which will be high or low depending upon the market demand. Thangam [22] developed an inventory model with two-level trade credit financing where it takes into the effect of order cancelations during advance sales period. The model incorporates a price discount program announces by the retailer during advance sales period to promote the sales. Here, the author discusses that the customer who gets an item has allowed paying on or before the permissible delay period which is accounted from the buying time rather than the start period of inventory sales. Kumar et al. [15] developed a production inventory model under a two-level trade credit financing in which the supplier offers the retailer a full trade credit period and the retailer offers the customers a partial trade credit period. The authors have considered the perishable items and hence uses the preservation technology for the purpose of decay in potential worth of items. To reduce the ordering cost learning by doing phenomenon is used and the demand function has been taken as the function of selling price and trade credit.

Majumdar et al. [16] discussed an economic production quantity model with partial trade credit policy under fuzzy environment using Generalized Hukuhara derivative approach. Wu et al. [24] developed an inventory model under downstream partial trade credits to credit-risk customers for deteriorating items using discounted cash-flow approach. Kumar and Chanda [14] developed an EOQ model for dynamic pricing-advertising environment under permissible delay in payments. Banu and Mondal [5] developed an inventory model where the demand function is influenced by customers' credit under a two-level trade credit period environment. Due to the dynamic nature of the credit period offered by the supplier to the retailer (M), they have considered M as a q-fuzzy number wherein the model has been explained by taking two separate cases such as crisp and fuzzy cases. Banu and Mondal [5] develops a deterministic EOQ model in which the supplier offers credit period to the retailer and in turn the retailer also offers credit period to the customers where demand function is dependent on the length of the customer's credit period and the retailer's ordering cost per order depends on the number of replenishment cycles. Saxena et al. [18] developed a single retailer and a single supplier a finite planning horizon inventory model under inflation where it focuses on green inventory supply chain.

10.3 Assumptions and Notations

10.3.1 Assumptions:

i. Demand rate follows innovation diffusion process and is governed by the innovation effect only.
ii. The replenishment rate is instantaneous.
iii. Shortages are not allowed.
iv. Lead time is zero.
v. The coefficient of innovation is constant over the planning period.
vi. There is only one product bought per new adopter.
vii. The supplier offers credit period to the retailer and the retailer also offers credit period to the customer.
viii. The potential market size is dynamic and varies with time.
ix. $M \geq N$.

10.3.2 Notations

A	Ordering cost per order.
C	Cost per unit item.
I	Inventory carrying charge per unit time.
T	Cycle time.
Q	Order quantity.
$\lambda(t)$	Demand rate at time t.
I_c	Interest charged per \$ per unit time.
I_e	Interest earned per \$ per unit time.
M	Retailer's trade credit period offered by the supplier to the retailer.
P	Selling price per unit item.
p	Coefficient of innovation.
$\overline{N}(t)$	Potential market size at time t.
N_0	Initial potential market size.
N	Trade credit period for the customer offered by the retailer.
$N(t)$	Cumulative number of adopters at time t.
$K_1(T)$	Total cost of the system per unit time for $M \leq T$.
$K_2(T)$	Total cost of the system per unit time for $N \leq T \leq M$.
$K_3(T)$	Total cost of the system per unit time for $T \leq N$.
$TVC(T)$	Total cost per unit time.

10.4 Mathematical Formulation

The model developed here is dependent upon demand function which follows innovation diffusion process where potential market size is dynamic varies with time.

The demand function $\lambda(t)$ is defined as follows:

$$\lambda(t) = p\big(\overline{N}(t) - N(t)\big), \text{ where, } \overline{N}(t) = N_0 e^{\alpha t}, \alpha > 0, N(0) = 0 \quad (10.1)$$

The differential equation defining the instantaneous state of the inventory level at any time t, $I(t)$ in the interval $(0, T)$ is given by

$$\frac{dI(t)}{dt} = -\lambda(t), \quad 0 \leq t \leq T \quad (10.2)$$

$$I(T) = 0 \quad (10.3)$$

$$\Rightarrow I(t) = \frac{pN_0}{(\alpha + p)}\big(e^{\alpha T} - e^{-pT}\big) - \frac{pN_0}{(\alpha + p)}\big(e^{\alpha t} - e^{-pt}\big), 0 \leq t \leq T \quad (10.4)$$

$$\int_0^T I(t)dt = \frac{TpN_0}{(\alpha + p)}\big(e^{\alpha T} - e^{-pT}\big) - \frac{pN_0}{(\alpha + p)}\left(\frac{1}{\alpha}\big(e^{\alpha T} - 1\big) + \frac{1}{p}\big(e^{-pT} - 1\big)\right) \quad (10.5)$$

Therefore, Inventory carrying cost per unit time (IHC) is

$$IHC = \frac{IC}{T}\int_0^T I(t)dt = \frac{ICpN_0}{(\alpha + p)}\big(e^{\alpha T} - e^{-pT}\big)$$

$$- \frac{ICpN_0}{T(\alpha + p)}\left(\frac{1}{\alpha}\big(e^{\alpha T} - 1\big) + \frac{1}{p}\big(e^{-pT} - 1\big)\right) \quad (10.6)$$

$$\text{Also, } I(0) = Q = \int_0^T \lambda(t)dt \quad (10.7)$$

$$\Rightarrow Q = \frac{pN_0}{(\alpha + p)}\big(e^{\alpha T} - e^{-pT}\big) \quad (10.8)$$

Therefore, Purchasing Cost per unit time (PC) is

$$PC = \frac{CQ}{T} = \frac{CpN_0}{T(\alpha + p)}\big(e^{\alpha T} - e^{-pT}\big) \quad (10.9)$$

Also, Ordering cost per unit time $(OC) = \dfrac{A}{T}$ (10.10)

Here, a case has been considered where Supplier allows a fixed credit period M to the Retailer and the Retailer also allows a fixed credit period N to the Customer. On the basis of this assumption the following three cases are discussed.

Case-(1) $M \leq T$

$$\text{Interest charged per unit time } (IC) = \dfrac{CI_c}{T} \int_M^T I(t)dt \qquad (10.11)$$

$$\Rightarrow IC = \dfrac{CI_c(T-M)pN_0}{T(\alpha+p)}(e^{\alpha T} - e^{-pT})$$
$$- \dfrac{CI_c pN_0}{T(\alpha+p)}\left(\dfrac{1}{\alpha}(e^{\alpha T} - e^{\alpha M}) + \dfrac{1}{p}(e^{-pT} - e^{-pM})\right) \qquad (10.12)$$

$$\text{Interest earned per unit time } (IE) = \dfrac{PI_e}{T}\int_N^M t\lambda(t)dt \qquad (10.13)$$

$$\Rightarrow IE = \dfrac{PI_e pN_0}{T(\alpha+p)}\left(e^{\alpha M}\left(M - \dfrac{1}{\alpha}\right) - e^{\alpha N}\left(N - \dfrac{1}{\alpha}\right)\right)$$
$$- \dfrac{PI_e pN_0}{T(\alpha+p)}\left(e^{-pM}\left(M + \dfrac{1}{p}\right) - e^{-pN}\left(N + \dfrac{1}{p}\right)\right) \qquad (10.14)$$

The total cost per unit time, $K_1(T) = IHC + PC + OC + IC - IE$ (10.15)

$$\Rightarrow K_1(T) = \dfrac{ICpN_0}{(\alpha+p)}(e^{\alpha T} - e^{-pT}) - \dfrac{ICpN_0}{T(\alpha+p)}\left(\dfrac{1}{\alpha}(e^{\alpha T} - 1) + \dfrac{1}{p}(e^{-pT} - 1)\right)$$
$$+ \dfrac{CpN_0}{T(\alpha+p)}(e^{\alpha T} - e^{-pT}) + \dfrac{A}{T} + \dfrac{CI_c(T-M)pN_0}{T(\alpha+p)}(e^{\alpha T} - e^{-pT})$$
$$- \dfrac{CI_c pN_0}{T(\alpha+p)}\left(\dfrac{1}{\alpha}(e^{\alpha T} - e^{\alpha M}) + \dfrac{1}{p}(e^{-pT} - e^{-pM})\right)$$
$$- \dfrac{PI_e pN_0}{T(\alpha+p)}\left(e^{\alpha M}\left(M - \dfrac{1}{\alpha}\right) - e^{\alpha N}\left(N - \dfrac{1}{\alpha}\right)\right)$$
$$+ \dfrac{PI_e pN_0}{T(\alpha+p)}\left(e^{-pM}\left(M + \dfrac{1}{p}\right) - e^{-pN}\left(N + \dfrac{1}{p}\right)\right) \qquad (10.16)$$

Case-(2) $N \leq T \leq M$.

Interest Earned per unit time $(IE_1) = \frac{PI_e}{T} \int_N^T t\lambda(t)dt + \frac{I_eP}{T} \int_T^M \left[\int_0^T \lambda(t)dt \right] dt$

$= \frac{PI_e}{T} \int_N^T t\lambda(t)dt = \frac{PI_e pN_0}{T(\alpha+p)} \left(e^{\alpha T} \left(T - \frac{1}{\alpha}\right) - e^{\alpha N}\left(N - \frac{1}{\alpha}\right) \right)$

$- \frac{PI_e pN_0}{T(\alpha+p)} \left(e^{-pT}\left(T + \frac{1}{p}\right) - e^{-pN}\left(N + \frac{1}{p}\right) \right)$ (10.17)

and $\frac{I_eP}{T} \int_T^M \left[\int_0^T \lambda(t)dt \right] dt = \frac{I_eP(M-T)pN_0}{T(\alpha+p)} (e^{\alpha T} - e^{-pT})$ (10.18)

$\Rightarrow IE_1 = \frac{PI_e pN_0}{T(\alpha+p)} \left(e^{\alpha T}\left(T - \frac{1}{\alpha}\right) - e^{\alpha N}\left(N - \frac{1}{\alpha}\right) \right)$

$- \frac{PI_e pN_0}{T(\alpha+p)} \left(e^{-pT}\left(T + \frac{1}{p}\right) - e^{-pN}\left(N + \frac{1}{p}\right) \right)$

$+ \frac{I_eP(M-T)pN_0}{T(\alpha+p)} (e^{\alpha T} - e^{-pT})$ (10.19)

The total cost per unit time, $K_2(T) = IHC + PC + OC - IE_1$ (10.20)

$\Rightarrow K_2(T) = \frac{ICpN_0}{(\alpha+p)} (e^{\alpha T} - e^{-pT}) - \frac{ICpN_0}{T(\alpha+p)} \left(\frac{1}{\alpha}(e^{\alpha T} - 1) + \frac{1}{p}(e^{-pT} - 1) \right)$

$+ \frac{CpN_0}{T(\alpha+p)} (e^{\alpha T} - e^{-pT}) + \frac{A}{T} - \frac{I_eP(M-T)pN_0}{T(\alpha+p)} (e^{\alpha T} - e^{-pT})$

$- \frac{PI_e pN_0}{T(\alpha+p)} \left(e^{\alpha T}\left(T - \frac{1}{\alpha}\right) - e^{\alpha N}\left(N - \frac{1}{\alpha}\right) \right)$

$+ \frac{PI_e pN_0}{T(\alpha+p)} \left(e^{-pT}\left(T + \frac{1}{p}\right) - e^{-pN}\left(N + \frac{1}{p}\right) \right)$ (10.21)

Case-(3) $T \leq N$.

Interest Earned per unit time

Interest Earned per unit time $(IE_2) = \frac{I_eP}{T} \int_N^M \left[\int_0^T \lambda(t)dt \right] dt$ (10.22)

$\Rightarrow IE_2 = \frac{I_eP(M-N)pN_0}{T(\alpha+p)} (e^{\alpha T} - e^{-pT})$ (10.23)

Table 10.1 Sensitivity analysis with respect to p

Changes in optimal values due to p
For $M = 0.041\ (15/365),\ N = 0.03, \alpha = 0.11$

p	T^*	$TVC(T*)$	$Q(T^*)$
0.005	0.544	16,370	14.0
0.006	0.499	19,218	15.36
0.007	0.463	22,036	16.61
0.008	0.434	24,829	17.78
0.009	0.411	27,602	18.89

The total cost per unit time, $K_3(T) = IHC + PC + OC - IE_2$ \hfill (10.24)

$$\Rightarrow K_3(T) = \frac{ICpN_0}{(\alpha+p)}\left(e^{\alpha T} - e^{-pT}\right) - \frac{ICpN_0}{T(\alpha+p)}\left(\frac{1}{\alpha}(e^{\alpha T}-1) + \frac{1}{p}(e^{-pT}-1)\right)$$
$$+ \frac{CpN_0}{T(\alpha+p)}\left(e^{\alpha T} - e^{-pT}\right) + \frac{A}{T} - \frac{I_e P(M-N)pN_0}{T(\alpha+p)}\left(e^{\alpha T} - e^{-pT}\right)$$

\hfill (10.25)

To obtain the total optimum cost the following relationship has been established

$$TVC(T) = \begin{cases} K_1(T) & \text{if } M \leq T \\ K_2(T) & \text{if } N \leq T \leq M \\ K_3(T) & \text{if } T \leq N \end{cases} \quad (10.26)$$

10.5 Numerical Illustrations

A numerical example based on some hypothetical data is illustrated as given below. The parameters used to do the sensitivity analysis are described as follows.

$A = \$1100/\text{order},\ C = \$500/\text{unit},\ I = 0.25/\text{unit time},\ N_0 = 5000,\ I_c = 0.20/\text{year},\ I_e = 0.15/\text{year},\ P = \$1200/\text{unit}$ (Tables 10.1, 10.2, 10.3 and 10.4).

10.6 Graphical Representations

In this section the graphical representations showing convexity of the cost function is given in Figs. 10.1 and 10.2.

Table 10.2 Sensitivity analysis with respect to M

Changes in optimal values due to M
For $p = 0.005$, $\alpha = 0.11$ $N = 0.03$

M	$T*$	$TVC(T*)$	$Q(T*)$
(15/365)	0.544	16,370	14.0
(30/365)	0.543	16,255	13.98
(45/365)	0.541	16,134	13.94
(60/365)	0.539	16,005	13.87
(90/365)	0.531	15,729	13.66

Table 10.3 Sensitivity analysis with respect to α

Changes in optimal values due to p
For $M = 0.041$ (15/365), $N = 0.03$, $p = 0.005$

α	$T*$	$TVC(T*)$	$Q(T*)$
0.12	0.538	16,411	13.88
0.13	0.532	16,451	13.76
0.14	0.526	16,491	13.64
0.15	0.520	16,531	13.52
0.16	0.515	16,570	13.41

Table 10.4 Sensitivity analysis with respect to N

Changes in optimal values due to p
For $M = 0.246$(90/365), $p = 0.005$, $\alpha = 0.11$

N	$T*$	$TVC(T*)$	$Q(T*)$
(15/365)	0.532	15,732	13.68
(30/365)	0.534	15,754	13.75
(45/365)	0.539	15,789	13.88
(60/365)	0.546	15,839	14.05
(90/365)	0.564	15,979	14.53

10.7 Conclusion and Managerial Implications

This paper develops an economic order quantity model where the demand behaviour is dynamic varies with time. The potential market size is the part of the demand behaviour which is also dynamic in nature and varies with time exponentially. The model incorporates the concept of two-stage trade credit policy in which the supplier offers a credit period to the retailer and the retailer also offers credit period to the customer. The sensitivity analysis with respect to the different parameters such as coefficient of innovation, coefficient of the potential market size (α), credit period

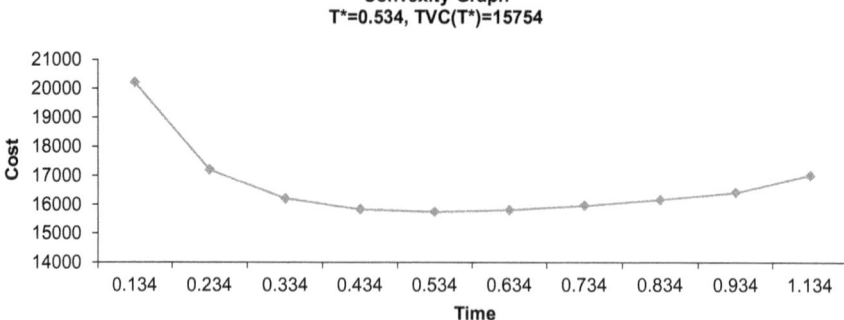

Fig. 10.1 Convexity Graph showing convexity of the cost function

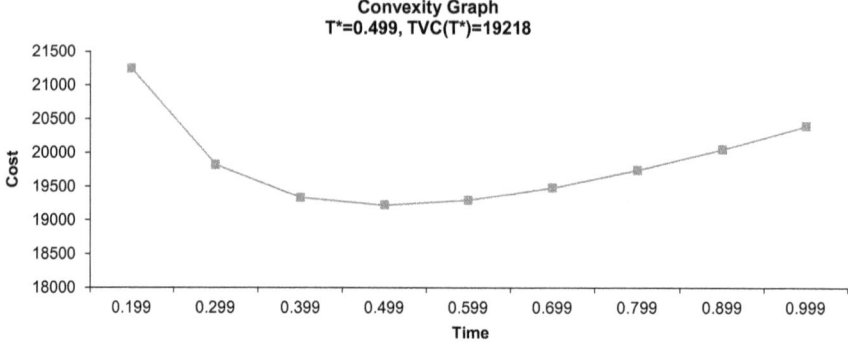

Fig. 10.2 Convexity Graph showing convexity of the cost function

offered to the retailer, credit period offered to the customer etc. has been performed to understand the behaviour of the model and it can also helps in the numerical validation. Through the sensitivity analysis with respect to the parameters as discussed above one can easily understand its impact over optimal replenishment policy. The manager can take the effective decisions which will be based on the optimal replenishment policy. The supply chain system can be well understood on the basis of decisions taken as discussed above. This paper can be extended by incorporating shortages, lost sales, deterioration etc. and it can also be done by explaining its closed-form solutions.

Acknowledgements The infrastructural support provided by FORE School of Management, New Delhi is gratefully appreciated.

References

1. Aggarwal SP, Jaggi CK (1995) Ordering policies of deteriorating items under permissible delay in payments. J Oper Res Soc 46(5):658–662
2. Aggarwal KK, Jaggi CK, Kumar A (2012) Economic order quantity model under fuzzy sense when demand follows innovation diffusion process having dynamic potential market size. Int J Serv Oper Manage 13(3):361–391
3. Aggarwal KK, Kumar A (2013) An inventory decision model for new products when demand depends on dynamic advertising expenditure. Int J Logist Syst Manage 15(4):424–444
4. Aggarwal KK, Kumar A (2014) Economic order quantity model with innovation diffusion criterion under influence of price dependent potential market size. Int J Innov Technol Manage 11(4):1450028-1–1450028-25
5. Banu A, Mondal SK (2018) Analyzing an inventory model with two-level trade credit period including the effect of customers' credit on the demand function using q-fuzzy number. Oper Res, 1–29
6. Bass FM (1969) A new-product growth model for consumer durables. Manage Sci 15(5):215–227
7. Chanda U, Kumar A (2011) Economic order quantity model with demand influenced by dynamic innovation effect. Int J Oper Res 11(2):193–215
8. Chanda U, Kumar A (2012) Economic order quantity model on inflationary conditions with demand influenced by innovation diffusion criterion. Int J Procurement Manage 5(2):160–177
9. Fourt LA, Woodlock JW (1960) Early prediction of market success for new grocery products. J Market 25(2):31–38
10. Goyal SK (1985) Economic order quantity under conditions of permissible delay in payments. J Oper Res Soc 36(4):335–338
11. Horsky D, Simon L (1983) Advertising and the diffusion of new products. Market Sci 2(1):1–17
12. Hwang H, Shinn SW (1997) Retailer's pricing and lot sizing policy for exponentially deteriorating product under the condition of permissible delay in payments. Comput Oper Res 24(6):539–547
13. Kumar A, Aggarwal KK (2012) Optimal replenishment policy for an inventory model with demand follows innovation diffusion process under permissible delay in payments. Int J Inf Manage Sci 23(2):177–197
14. Kumar A, Chanda U (2017) Economic order quantity under permissible delay in payments for new products in dynamic pricing-advertising condition. Int J Bus Innov Res 13(2):203–221
15. Kumar S, Handa N, Singh SR, Yadav D (2015) Production inventory model for two-level trade credit financing under the effect of preservation technology and learning in supply chain. Cogent Eng 2(1):1045221
16. Majumder P, Mondal SP, Bera UK, Maiti M (2016) Application of Generalized Hukuhara derivative approach in an economic production quantity model with partial trade credit policy under fuzzy environment. Oper Res Perspect 3:77–91
17. Qin J, Liu W (2014) The optimal replenishment policy under trade credit financing with ramp type demand and demand dependent production rate. Discrete Dynamics in Nature and Society
18. Saxena S, Singh V, Gupta RK, Mishra NK, Singh P, Green inventory supply chain model with inflation under permissible delay in finite planning horizon. Adv Sci Technol Eng Syst J 4(5):123–131
19. Sharif MN, Ramanathan K (1982) Polynomial innovation diffusion models. Technol Forecast Soc Chang 21:301–323
20. Teng JT, Yang HL, Chern MS (2013) An inventory model for increasing demand under two levels of trade credit linked to order quantity. Appl Math Model 37(14–15):7624–7632
21. Teng JT, Thompson GL (1996) Optimal strategies for general price-quality decision models of new products with learning production costs. Eur J Oper Res 93(3):476–489
22. Thangam A (2014) Retailer's inventory system in a two-level trade credit financing with selling price discount and partial order cancellations. J Ind Eng Int 10(1):3

23. Wu C, Zhao Q (2014) Supplier–buyer deterministic inventory coordination with trade credit and shelf-life constraint. Int J Syst Sci Oper Logist 1(1):36–46
24. Wu J, Al-Khateeb FB, Teng JT, Cárdenas-Barrón LE (2016) Inventory models for deteriorating items with maximum lifetime under downstream partial trade credits to credit-risk customers by discounted cash-flow analysis. Int J Prod Econ 171:105–115
25. Yang CT, Ouyang LY, Hsu CH, Lee KL (2014) Optimal replenishment decisions under two-level trade credit with partial upstream trade credit linked to order quantity and limited storage capacity. Mathematical Problems in Engineering

Chapter 11
Software Engineering Analytics—The Need of Post COVID-19 Business: An Academic Review

Somayya Madakam and Rajeev K. Revulagadda

Abstract "Man Proposes, God Disposes". COVID-19; the unexpected pandemic decease is a good testimony of the above proverb by exemplifying global business operations shutting down. Pre COVID-19, the business across the globe more or less are planned one; running smoothly; trying to reach their goals and objectives; satisfying the customers, finally leading to leapfrog the companies' profits. This was the scenario, before the coronavirus; the alias of COVID-19 pandemic, which was germinated in one of the most scientific cities in china; Wuhan. Entire world business collapsed; another global economic recession after 2008; touching all the continents, countries, business, tribes, religions, lifestyles, as well as professional lives. No medicine was unable to save the lakhs of human life from contamination. Even different country leaders practised various preventive methods in order to cure the decease to save human life. In this light, this book chapter will explore the state of the art of COVID-19 across the globe with respect to business. The entire data analysis is based on secondary on-line data and thematically narrated. The book chapter furtherly discussed in detail connotes, process and state of art of software engineering and software analytics in business for sustainability. The analysis reveals that software analytics technology is the only industry lightly affected, should grow rapidly and it is the only solution provider to improve the process, and sustain business via better-automated software. In long term incurs less cost, time management, less manual intervention, integration of enterprise departments, virtual meetings, and e-commerce are the beauty of software analytics.

Keywords Software engineering · Software analytics · Analytics · Software engineering analytics · Web of science · COVID-19 · Pandemic

---

S. Madakam (✉)
Information Technology, FORE School of Management, New Delhi, India
e-mail: somayya@fsm.ac.in

R. K. Revulagadda
Finance & Accounting, National Institute of Industrial Engineering (NITIE), Mumbai, India

11.1 Introduction

The five-letter word 'COVID-19' given by medical scientists to a deadly virus made the whole world chaos. It does not have any barrier either developed or developing, rich or poor. This twenty-first century never knows that it will face massive destruction, which cost lakhs of human lives. This did not stop just by affecting the health system of the nations but also economic, business and political stability. The coronavirus rippled in the entire business ecosystem; effected all the industries in the world. Organizations with the best performance in the pre-COVID era trembled down day by day and worsened the condition of its affiliates as well. The ripple thrashed one industry to another i.e. from transportation, service, manufacturing, tourism etc. Before knowing about how this virus affected industry to industry, one has to know about non-negligible information about this virus.

According to the findings of scientific study on COVID-19 by Chan et al. [1], the virus was first identified in Wuhan , China. This novel virus belongs to the family of coronavirus. Till date, a sea of studies were published continuously in the top journals including medical journals like lancet. However, the symptoms vary from person to person based on age and geographical locations i.e. from country to country, but common symptoms included pneumonia, high fever, dry cough etc. The rapid spread of the virus in population led to serious concern by governments. This initiated the imposing of lockdowns and shutting down the businesses slowly. The need for health care workers has risen rapidly and highlighted the lack of research in the medical field. Slowly, many countries have started to sense the joblessness of its citizens, which has affected the manufacturing, export etc. This eventually led to slow down the income growth and in worst conditions; some countries experienced a high rate of unemployment. As the ripple never gets back, it started continuously affecting with a negative effect on the financial statements of many companies. This phenomenon and different unknown circumstances led to the question of recession. Rather diving directly to the interpretation at a macro level, it will be beneficial to understand the impact of the pandemic on the industry after industry. Economic indicators of a country depend on a pool of sectors. For example, mostly developing countries like India are dependent on agriculture outcome. Though industries like service and banking are performing better, a high segment of the population is dependent on agriculture. Hence, it is evident that the impact on agriculture will show an effect on commodities export and imports. Lack of credit availability for purchasing the raw material, and increase in expenditure for seeds, fertilizers and unavailability of labor has increased the cost of production in the agriculture sector. On one side the countries experienced a fall of growth in income and wages and on the other side, prices are very odd. Essential commodities are available but the imposition of transport barriers made to purchase them at a high price. These issues are even worse for the people under the poverty line across nations. The other burdens like lack of medical staff and infrastructure further worsen the situation of the people. Hence, the negative impact on a high proportion of the population led to the downfall of sales in other industries. When compared with the agriculture sector, many sectors

11.2 COVID19-Effects in Business

11.2.1 *Tourism and Hospitality*

The impact of the pandemic on tourism is highly negative and the industry is in complete stress. When the lockdowns are imposed and travel restrictions are initiated in many countries, passenger commute was totally shut which led many aviation services to suspend. To reduce the outbreak caused by the pandemic, many tourists have cancelled their touring plans. Lockdown imposition has led to close down the hotels, restaurants, malls, museums, open-air theatres, archaeological sites etc. in many countries. UNWTO (UN-World Tourism Organization) raised the concern to all the countries that, governments are responsible to prioritize the health of the public and they should be responsible for protecting jobs.

In all the regions of the world as shown in Fig. 11.1, the tourism industry was worst hit by this pandemic. The revenues for Asia on tourism is decreasing at an increasing pace. This travel restriction has caused many jobs at risk. The demand for tourism and hospitality has to rise in due course, as this sector is the worst hit.

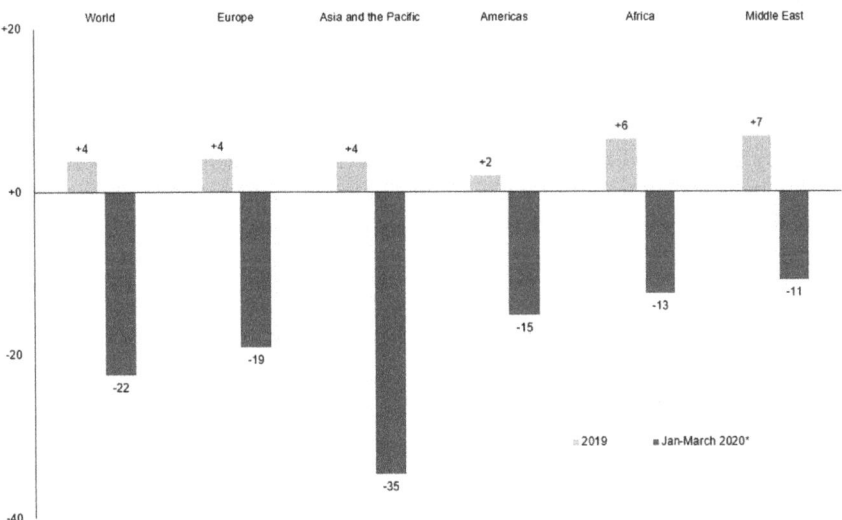

Fig. 11.1 International tourist arrivals, 2019 and Q1 2020 (% Change). *Source* UNWTO-World Tourism Organization (https://www.unwto.org/news/covid-19-international-tourist-numbers-could-fall-60-80-in-2020)

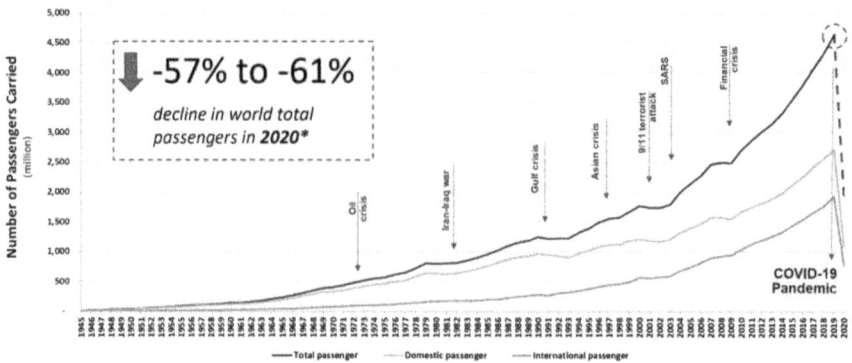

Fig. 11.2 World passenger traffic, 1945–2020*. *Source* ICAO, The International Civil Aviation Organization (https://www.icao.int/sustainability/Documents/COVID19/ICAO%20COVID%202020%2009%2023%20Economic%20Impact.pdf)

11.2.2 Airlines and Aviation

COVID-19 also affected the aviation sector with generating heavy losses and closing down of services. As the transmission rate is high from person to person through cough or sneeze, it is very reluctant for the aviation industry to continue the services. The potential loss of gross passenger operating revenues is approximately USD 375 to 395 billion. ICAO reveals that a reduction from 49 to 51% of seats offered by airlines and passenger overall reduction of 635 million passengers. The study reveals that this pandemic made the world passenger traffic with unprecedented decline ever. With the summer season missed, global travel got heavily impacted. The data in Fig. 11.2 from 1945 to 2020 shows that COVID pandemic crises made all other major crises of the world less impact on the world passenger traffic. The ripple effect caused by the demand of tourism has also affected the growth and survival of the aviation industry. If this severity level grows furthermore for international services, this will eventually affect the growth of the aviation sector. Thus due to COVID-19, this industry is heavily affected.

11.2.3 Oil and Gas

Oil and gas industry was also one of the hardest hit by the pandemic COVID-19. The imposition of lockdown and unavailability of the workforce, the industry has to stop either the operations or irregular in operating the physical operations. This has affected oil and gas production with a bad flow of continuity in operations. The industry has many problems in which unstable prices of oil barrels, doubt of employee safety and security in the pandemic environment and other crises that come from a different set. Those include supply chain risk and other unidentified risks. The depressed price continued for a longer time due to the crisis. The main cause for the reduction of barrels prices cannot be said unless identified all the possible impacts. The ongoing trade war between the countries has made sensitive prices to change continuously i.e. unstable. The companies, which are public sector, has an edge when compared to the companies that service the other industrial purposes. This COVID-19 has made the employees, and the management to think about the security and continual flow of business. The pandemic related issues but also sensitive towards the government's decisions not only affect this. Hence, with the ongoing challenges, this industry was hardly hit.

11.2.4 Automotive Manufacturing

The automotive manufacturing sector is a significant sector but it has the ripple effect caused by COVID-19; the fall in demand for products. The plant closures due to employee safety made furthermore worst on the performance of the automotive sector. The credit from banks and the burden of repayments to the credit lenders made some companies vulnerable. However, the demand has increased slowly, but it takes time to settle the ongoing issues of the automotive sector. The sales of the vehicle have rapidly fallen in all the countries. When compared with 2019, the industry is facing a turbulent environment Sales data of China, Europe and the US has sales down when compared with 2019 (Fig. 11.3). This shows the industries crucial issue in the pandemic season and considered as the worst hit.

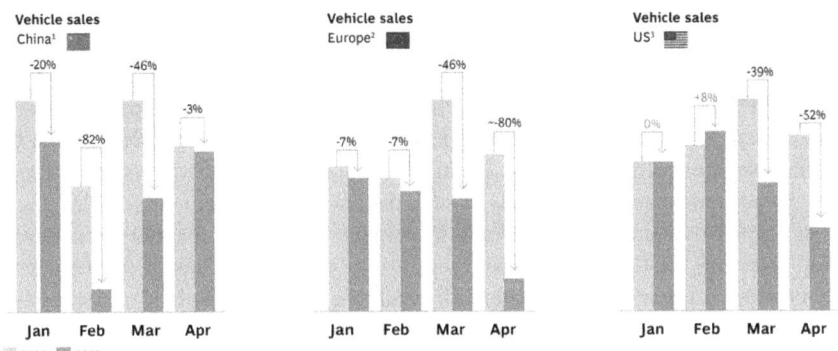

Fig. 11.3 Drop in auto sales. *Source* BCG, Boston Consulting Group (https://image-src.bcg.com/Images/Auto%20post%20COVID-19_052920_tcm9-249607.pdf)

11.2.5 Consumer Products

Consumer products are somehow good, but the revenues that rise from the consumption by the public will be high. The supply chain connectivity is also becoming a big problem, in the pandemic season. The consumption of alcohol and beverages will be hit a hard time in demand. However, daily essentials demand will be increased rapidly. The change in consumption will be seen in different types of different products. The things that changes in operations are very crucial in handling consumer products. Agility, portfolio evaluation, shopper insights, flexible production are some of the key elements that have to be considered for transformation in the industry. This not only helps the industry to survive from these pandemic environments but also helps to mitigate the risks from the unknown and uncertain barriers. Hence, this sector is not that worst hit but product segmentation plays a key role in overcoming the barriers. Along with these sectors, the pandemic environment also hit electronics but, after the movement and lifting up the travel restrictions and other key barriers, this helped to survive better than many worst hit business sectors. Hence, the COVID-19, most hits to better-performing sectors were explained. Coping up with the pandemic challenges and barriers by getting the customer insights, need and type of the product will help the industries to survive in these chaotic and unprecedented conditions. To overcome the ripples caused by the other business, the industries have to be agile and active.

11.2.6 Education

The education sector did not leave untouched by this pandemic. A long halt and uncertain timing have made the education sector unstable. Pandemic showed the impact on the nursery, primary, high school to professional educations. University management has made many efforts to limit the loss by pandemic lock-down by offering on-line teaching services. Though the internet has removed the boundaries between countries and helped many to seek the knowledge of top universities, there are still rigid barriers for many students. Mostly these issues are for students in developing countries. Students in rural and semi-urban explain the lack of infrastructure and monetary support to purchase the infrastructure. India is one of the examples that the education sector affected very badly. Still, the schools and colleges are closed due to the increasing number of corona cases. Nevertheless, software companies have tied relationships with academic institutes to offer the best online educational experience. Due to pandemic situations and physical classroom access barriers, the software giants are come up with video conferencing teaching tools or virtual classroom systems. Microsoft's teams; Cisco's Webex Meet; Zoom rooms; GoToWebinar, and Google talk etc. Depending on sub-scribed licenses, the number of students can get connection for online learning.

11.2.7 Software Industry

The impact on the software industry is also crucial for comprehensive development. This industry has an option to face the crisis by offering work from home options to the employees. This helped to cope up with the contracts and credit flow to the companies. The software sector helped to resist the unemployment to some extent. It helped the pharma companies and governments to track the real-time data of COVID-19 patients and tracking the victims on time. Understanding its characteristics and elements will further help the decision-makers to overcome these situations. The software industry got good opportunity to do the business from the home. However, the industry lost most of the business opportunities due to interdependency among all business. Software industry/engineering incorporates all aspects of software production from business strategy to designing and coding, testing, quality and management of large-scale complex software systems in a bigger way during post COVID-19; also to conduct business mostly on-line. Software industry pave the path for e-business and e-commerce round the clock to get competitive advantage as well as business sustain. Software engineering is about multi-person development of multi-version software that is large and complex based on the Windows, Linux and Android operating systems relating to office software, flight control systems, automation, education, and manufacturing applications. Software engineering is an old field emerging from the Computer Science and IT management; however now develop all kinds of business applications to run smoothly during post COCID-2019 or all kinds of situations.

11.3 Research Methodology

As the software analytics or software engineering analytics field is novel and much academic scribbling is not available, it is the need of the hour to do more rigorous research work and develop new applications and innovative findings for better software engineering process during post or in these pandemic days. Since, the concept is contemporary and not much empirical evidence is not available, the present manuscript graphed based on secondary data (articles). The google scholar and google databases are bases for searching the articles on this phenomenon. The writings carried out during 1-1-2019 to 30-6-2020. Since, almost one and half year duration, the authors got enough time to browse, came across and conceptualize the phenomenon in a better way, in which the scribblings are bringing the crystal clear ideas on the software engineering analytics connotes as well as on COVID-19 and its relationships. Besides the web of science articles are extracted for the composing this article. The author discovered 35 articles on software analytics. Besides, some of the articles and blogs considered with reference to COVID-19 phenomenon. The general awareness of academic and research database of the web of science is much more advanced in an Information Technology global world who are into the research field. The study by Bakkalbasi et al. [2] says that for year's researchers looking for this type of information had only one resource to consult: the Web of Science from Thomson Scientific (Clarivate Analytics). Therefore, the web of science is vital for any kind of research across the globe.

11.4 Software Development Life Cycle (SDLC)

The software is inevitable in our daily life. The entertainment devices like computers, mobiles, tablets, laptops, televisions, etc. run by different embedded software. Journeys are based on Geographical Information Systems (GIS), Geo-graphical Position System (GPS) and other remote sensing software etc. Besides, Management Information Systems (MIS), Enterprise Resource Planning (ERP), Computer Aided Design (CAD), Computer Aided Manufacturing (CAM) software ate backing organizations, manufacturing companies, and corporate business and so on. Government organizations, Academic Institutions, Research & Development (R&D) offices are handicapped without Microsoft office suite, International Business Machines (IBM)'s analytical Statistical Package for Social Sciences (SPSS), Atlas-ti, and N-Vivo many more application software suites. That means human life is completely dependent on the multiple software for more comfortable business life. Scripts, applications, programs and a set of instructions, all terms often used to describe software. In a nonprofessional point of view, the software is a collection of programs, the programs are nothing but a collection of logically, and sequential arranged instructions to do particular tasks or to resolve one issue. Different languages are using to develop the software including JavaScript, Python, R, Go, and many more. However, the base languages in decades back are FORTRAN (Formula Translation), COBOL (Common Business Oriented Languages), and, Pascal. Moreover, the C, C++, C#, PHP (Hypertext Preprocessor), HTML 5, Macromedia, Swift, Ruby are some of the best dynamic high-level languages using to develop many application software and utilities in these days. Currently the software development using high-level languages is becoming very easy in IDEs. By putting minimum efforts, anybody can write codes for application software development with the help of many available built-in libraries. Therefore, developing a software is very easy these days irrespective of application. Software are available in the market freely, so-called open source as well as other licensed software in which the user has to pay the amount for the software usage. Any software generally developed by Software Development Life Cycle (SDLC) process. Moreover, different companies practicing different development approaches including spiral, incremental, agile, RAD, DevOps to name a few.

11.4.1 Software Requirements Analysis

The first phase of software engineering or SDLC is Software Requirements Analysis (SRS). Here the actual product or service specifications will be taken from the end user or client. This business can be a kind of Business to Business (B2B), Business to Customer (B2C), Business to Government Business-to-Business (B2B), Consumer-to-Business (C2B), Business-to-Administration (B2A) and Consumer-to-Administration (C2A). All the software specifications are noted down very clearly including product details like media design, user interface, color, compatibility, language, operating system, and the number of workforce involvement, technical staff selection, project cost, and time of completion, payment options, and maintenance. The criticality of the Software Requirements Specifications phase of the software life cycle for the success of the whole software project is widely recognized and the attention played on it by software developers is more and more significant [3]. Early inspections of software requirements specifications are known to be an effective and cost-efficient quality assurance technique. However, inspections are often applied with the underlying assumption that they work equally well to assess all kinds of quality attributes [4]. Software Requirement Specifications are a key result of the requirement engineering process and an important basis for every large industrial software development project [5]. SRS document is the first deliverable product/milestone in the software development process and acts as a basis for the formal contract between the user and the developer of the software of an information system [6]. Moreover, the requirement specification is gaining increasing attention as a critical phase of software development [7].

11.4.2 Software Design

The concept of information hiding modularity is a cornerstone of modern software design thought, but its formulation remains casual and its emphasis on changeability imperfectly related to the goal of creating benefit in a given context [8]. The software design is the most critical part of the entire software engineering process. This is the abstract view or graphical form of any software product before transforming into reality. All the software requirement specifications logically converted into software design or model. The graphical representation of this phase is the medium from product idea to a physical/logical form of the software. The term "software architecture" typically refers to the bigger structures of a software system, whereas "software design" typically refers to the smaller structures [9]. However, software architecture and software design are two aspects of the same topic. Both are about how software structured in order to perform their tasks. Despite a diversity of software architectures supporting information visualization, it is often difficult to identify, evaluate, and re-apply the design solutions implemented within such frameworks

[10]. Even for software design, there are different new software products are available in the market including CAD, CAM, Microsoft project management and many more developed by Information Technology firms. Moreover, the software design-cognitive aspects cover a variety of areas including software analysis, design, coding and maintenance [11]. Apart from these, PLEASE is an executable specification language that supports software development by incremental refinement. PLEASE is part of the ENCOMPASS environment that provides automated support for all aspects of the development process [12]. Therefore, briefly, the software design is the blueprint for entire software product including time, cost, workforce, and logical relationships among submodules. That means the software design is the birds eye view of the software project. The tools and techniques using in software design are Flowchart, Data Flow Diagram (DFD), Data Dictionary, Structured English, Decision Table and Decision tree. The nest section will discuss in detail about the software implementation phase that means how to do the coding.

11.4.3 Software Implementation

This phase is a vital portion of all software engineering projects. Software implementation generally done at company level. It involves all the technical people including website designers, programmers, network administrators, systems analysts, functionalists, chief information officers, and database managers. Depending on the project type, the professionals will be involved in developing the software. The languages will be used based on the application either frontend-or-backend level. However, currently, the development of the app is becoming very easy due to the availability of built-in libraries, syntax prepared by regular English language and keywords well versed. This is the phase in which, the right programmer to identifying and assigning from the technical bench from the corporate. The functionalist always monitors the completion of the task, results checking, and modified inputs from the customer to developers based on the software model using. We knew that in today's software development environment, requirements often change during the product development life cycle to meet shifting business demands, creating endless headaches for development teams [13].

Moreover, properly managed, architecture-centric methods can be a cost-effective addition to the software development process and will increase system and product quality [9]. However, a model of software development maturity describes managerial processes that can be used to attack software development difficulties from the managerial control perspective at five maturity levels [14]. An optimal software development process regarded as being dependent on the situational characteristics of individual software development settings including the nature of the application(s) under development, team size, requirements volatility and personal experience [15]. The developing team should meet all the criteria as per Software Requirement Specifications. The next phase of software implementation is software testing; discussing in the next section.

11.4.4 Software Testing

The Software testing is also an important part of software engineering. Testing generally done whether the software requirement specifications are met or not. If not, the deviations can be incorporated to meet the exact specifications, otherwise convincing the client with the developed specifications. Therefore testing is the measurement of the exactly predefined list at the time of customer project specifications. In software engineering or development process, the product divided into sub-modules in which the developers can easily code and finish it as early as possible in terms of big projects. In this case, the submodules are not only divided but also tested individually and later by integrated as one. Software testing is a critical stage in entire software development used to ensure that a program meets required specifications and does not contain errors in programming code [16]. Exhaustive testing of computer software is intractable, but empirical studies of software failures suggest that in some cases testing can be effectively exhaustive [17]. Recent years have witnessed a surge of interest in symbolic execution for software testing, due to its ability to generate high-coverage test suites and find deep errors in complex software applications [18]. Moreover, in commercial software development organizations, increased the complexity of products, shortened development cycles, and higher customer expectations of quality have placed a major responsibility on the areas of software debugging, testing, and verification [19].

11.4.5 Software Deployment

At the customer's site, the technical team will install necessary application software along with necessary hardware and networking components based on the clients' requirement. Worldwide one of the most important parts of the software development life cycle (SDLC) is user acceptance. Software deployment is a post-production activity that performed for or by the customer of a piece of software [20]. Even though the product is not 100% as per the specifications, convincing the user to accept the software is a vital part. This completely depends on the situation, level of customer acceptance and skill of the functionalist to convince the customer/client. This happens at the time of deployment of the software in the customer premises. The functional analysts or coordinator of software development will assist. Software deployment process consists of all the preparing a software application to run and operate in a specific environment and encompasses right from installation, configuration, testing and making changes to optimize the performance. The process already discussed in the seminal paper by saying that the software deployment is an evolving collection of interrelated processes such as release, install, adapt, reconfigure, update, activate, deactivate, remove, and retire [21]. Moreover, software applications are no longer stand-alone systems. They are increasingly the result of integrating heterogeneous collections of components, both executable and data, possibly dispersed over a computer network [22]. Hence, software development should be taken care of by both parties to make the software project successfully. The deployment process includes installation of software and hardware in the premises of client along with documentation and user training. Moreover, the software conversion process done in one of the four approaches either by (1) Parallel Strategy (2) Direct Strategy (3) Pilot Study Strategy (4) Phased Approach Strategy. This process make sure that (1) How to execute the packages? (2) How to enter the data? (3) How to process the data (processing details)? (4) How to take out the reports?

11.4.6 Software Maintenance

Maintenance is the most important post activity of any business sales. After delivering the software product to the end customer or user, most of the products and services are needs technical support from the supplier of the product. Many times the user may not be aware of the small repairing issues also, which may occur at the time of using the software product. Even though the entire software usage and bug deductions are given in the manual at the time of software delivery, installation and, users are scared of how to use and debugging the errors. Since most of the software developed with the help of computing programming languages, it is better to provide maintenance support annually. Nowadays companies are coming with different maintenance packages like 1-year free service otherwise, if the user goes for extended maintenance—the charges will be very less cost sort of. However, software maintenance and evolution characterized by their huge cost and slow speed of implementation [23]. Management has turned to software engineering tools designed to support software maintenance as a potential solution to maintaining productivity and quality problems [24]. Software maintenance is the dominant factor contributing to the high cost of software [25]. Therefore, the bugs need to be analysed before shutting down the systems. In this scenario, if we develop automatic prior intimation systems or automatic maintenance systems or would be grateful to system people. Moreover, software engineering comprehends several disciplines devoted to prevent and remedy malfunctions and to warrant adequate behaviour [26]. Annual Maintenance Contract (AMC) stands for annual maintenance contract where the company charges some lumsum amount from their customer for specified product for a fixed period and fixed services. This AMC includes (1) Knowing the full capabilities of the system (2) Knowing the required changes or the additional requirements (3) Studying the performance.

All industries today require computer scientists with advanced skill sets and an ability to develop, test, and maintain software engineering applications to meet ever-evolving modern industry and commercial requirements. A systems analyst produces a software requirement specification that precisely describes the attributes of the software to be produced [10]. As we are aware that the general software development process involves software requirements analysis, design, implementation, testing, integration, deployment, and maintenance. However, in the business landscape, there are different software development models, i.e. water-fall, spiral, the process of prototype, incremental, phases of iterative development and the principles of agile, and dev-ops methods. Local and multinational companies like Mphasis, Hindustan Computers Limited (HCL) Technologies, Mind-tree, IBM, Rolta India Ltd, Oracle Financial Services Software, Wipro, and Tata Consultancy Services (TCS) are working so hard to develop software in different domains including healthcare, pharmaceutical, manufacturing, aviation, marine, entertainment since decades. The IT/ITes companies offering multiple services relating to custom business solutions, web branding, internet marketing, collaborative content management, database migration, customization, application development including outsourcing, ERP solutions, iPhone apps development, collaborative commerce, programming and quality assurance, and testing services, multimedia offering and consulting. We are all familiar with traditional software engineering research and practices including requirements engineering, software project management, design, development, testing, and implementation of software applications. This software engineering discipline developed as a practice which has been well established over the past 50 years. Software development is an art; it is a skill and complex group task too. Software development is a professional activity. It is completely programming by different software developers including computer engineers, functional analysts, and business tycoons.

11.5 Software Analytics

11.5.1 Literature Analysis

The following are the total articles published as on dated 30/06/2019
 Total Articles Published: 35
 Sum of the Times Cited: 316
 Average Citations per Item: 9.03
 H-index per these articles: 10
 Out of the 35-research manuscript, the software analytics/software engineering analytics connotes clearly defined including the physiognomies and importance of field. From the above data Table 11.1, it was understood that up to now the research is on software analytics in practice including understanding users' behavior with software operation data mining and recommendations for mass spectrometry data quality metrics for open access data. Besides, there are different practices including visual software analytics for the build optimization of large-scale software systems. Some of the academic studies are based on qualitative including a retrospective study of software analytics projects: in-depth interviews with practitioners. Besides, the papers are on continuously assessing and improving software quality with software analytic tools: a case study. Like that, many faces of software analytics were scribbled. Using Analytics to Guide Improvement during an Agile-DevOps Transformation another study which intensifying both Agile-DevOps models. However,

Table 11.1 Software analytics in practice (Articles Titles: 35)

Title of the articles
Software Analytics: So What?
On the automatic classification of app reviews
Tuning for software analytics: Is it really necessary?
Human–computer interaction in evolutionary visual software analytics
Understanding users' behavior with software operation data mining
Recommendations for mass spectrometry data quality metrics for open access data (corollary to the Amsterdam principles)
Behavioral Response to a Just-in-Time Adaptive Intervention (JITAI) to Reduce Sedentary Behavior in Obese Adults: Implications for JITAI Optimization
Student and Faculty Member Perspectives on Lecture Capture in Pharmacy Education
A Retrospective Study of Software Analytics Projects: In-Depth Interviews with Practitioners
Recommendations for Mass Spectrometry Data Quality Metrics for Open Access Data
Rapid Releases and Patch Blackouts A Software Analytics Approach
Are Software Analytics Efforts Worthwhile for Small Companies? The Case of Amisoft
Recommendations for Mass Spectrometry Data Quality Metrics for Open Access Data
Roundtable: What's Next in Software Analytics
Searching under the Streetlight for Useful Software Analytics
How Evolutionary Visual Software Analytics Supports Knowledge Discovery
Recommendations for mass spectrometry data quality metrics for open access data (corollary to the Amsterdam principles)
Visual software analytics for the build optimization of large-scale software systems
Knowledge discovery in software teams by means of evolutionary visual software analytics
Investigating and Projecting Population Structures in Open Source Software Projects: A Case Study of Projects in GitHub
The Unreasonable Effectiveness of Software Analytics
Implementation of a consumer-focused eHealth intervention for people with moderate-to-high cardiovascular disease risk: protocol for a mixed-methods process evaluation
IRISH: A Hidden Markov Model to detect coded information islands in free text
Addressing problems with replicability and validity of repository mining studies through a smart data platform
The Many Faces of Software Analytics Introduction
Using Analytics to Guide Improvement during an Agile-DevOps Transformation
Roundtable: Research Opportunities and Challenges for Emerging Software Systems
Understanding software artifact provenance
Towards base rates in software analytics Early results and challenges from studying Ohloh
Bad smells in software analytics papers
Software Engineering Data Analytics: A Framework Based on a Multi-Layered Abstraction Mechanism

(continued)

Table 11.1 (continued)

Title of the articles
Continuously Assessing and Improving Software Quality with Software Analytics Tools: A Case Study
Software Analytics: What's Next?
Experience report on applying software analytics in incident management of online service

some research papers intensify about bad smells in software analytics papers. Moreover, what is Next? In software, the Analytics buzz word in the future. From the articles it can be understood that software analytics is interdisciplinary and can bring applications in all fields.

11.5.2 Journal Publications with Frequency

The above Fig. 11.4 shows that the ascending order of Journal of publications. IEEE Software (11), Science of Computer Programming (4), Computers in Hu-man Behavior (2), IEICE Transactions on Information and Systems (2), Information and Software Technology (2), American Journal of Pharmaceutical Edu-cation (1), Automated Software Engineering (1), BMJ Open (1), Computational Statistics (1), Empirical Software Engineering (1), Health Psychology (1), IEEE Access (1), Journal of Computer Science and Technology (1),, Journal of Proteome Research (1), Molecular & Cellular Proteomics(1), Proteomics (1), Proteomics Clinical Applications (1), and Requirements Engineering (1).

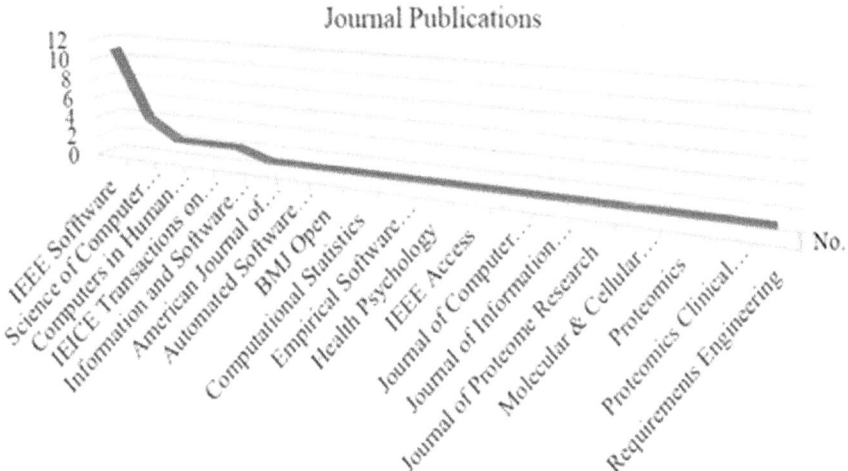

Fig. 11.4 The Journal wise publications

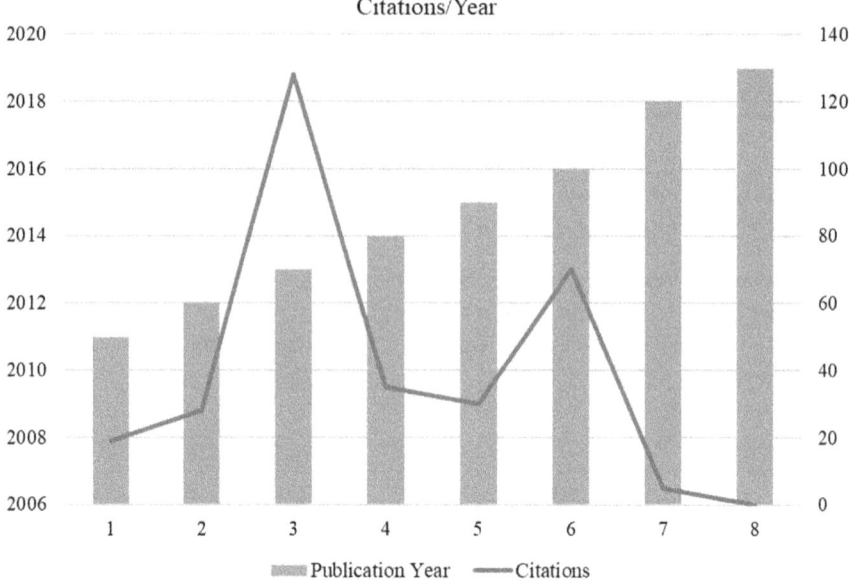

Fig. 11.5 Citations/year

11.5.3 Citations/Year

Less had been talked about the phenomenon of software analytics and software engineering analytics. We are getting this evidence even from Fig. 11.5. The graphs explore that the number of academic and research publication citations over a period is very less. That means the field is new and contemporary. Besides the phenomenon is well represented in the Graph 5. by year wise publication citations including 2011 (19), 2012 (28), 2013 (128), 2014 (35), 2015 (30), 2016 (70), 2018 (5), and up to now June 2019 it was only zero citations as per web of science database. However, there was ongoing research during 2011, 2012 and 2013. Later there was a sudden decrease in software analytics and software engineering analytics research, publications and its citations. Probably, the analytics field was expanding its applications in other fields like business analytics, big data analytics, marketing analytics, HR analytics, and social media analytics and so on. Hence, the researchers started working and exploring the above field, instead of working, publishing, and citing on software analytics phenomenon.

11.5.4 Software Engineering Analytics: Connotes

The software development is a data-rich activity with many sophisticated metrics. Software engineering data are in many forms such as user stories, use cases, requirements specifications, issue and bug reports, source code, test cases, execution logs, app reviews, user and develop mailing lists, discussion threads, and so on. For example, the well-known Web browser, Mozilla Firefox project, had over 1 million bug reports by 2014. Hidden in those big data are valuable insights about the quality of the development process, and the experience that software users receive. Despite large volumes of data and many types of metrics, software projects continue to be difficult to predict and risky to conduct [27]. We are well versed with 4–5 decades onwards on software development; it is a data-rich activity with many sophisticated metrics. Yet engineers often lack the tools and techniques necessary to leverage these potentially powerful information resources to-ward decision-making [28]. Hence, analytics is the need of hour across the globe in current scenario. The word "Analytics" is becoming a buzz in every domain all over the world. Analytics is now eye-catching from the public. We often come across the phrases with Analytics as suffix like 'Data Analytics'; 'Business Analytics'; 'Big Data Analytics'; 'Descriptive Analytics'; 'Diagnostic Analytics'; 'Prescriptive Analytics'; 'Predictive Analytics'; 'HR Analytics'; 'People Analytics'; 'Supply Chan Analytics', 'Marketing Analytics'; 'Consumer Analytics'; 'Fraud Analytics'; 'Financial Analytics'; Software Analytics'; and many more.

The word Analytics is not new and derived long back from the Greek word "Analytic', which means 'of or proceeding by analysis' or 'skilled in the careful study of something'. That means the Analysis is nothing but the methods to follow one particular task systematically. These analytics using for analysing the huge amount of data generated from different sources and trying to find out the patterns using the software. These patterns or insights are using for different purposes in all sorts of business, depending on the nature of business. Therefore, whatever, the sort of analytics, the objective of analytics is to understand the hidden patterns/insights in the organizational data? As technology continues to advance, we constantly generate an ever-increasing amount of data. Data analytics is the process of examining large data sets to uncover hidden patterns, unknown correlations, market trends, customer preferences, and other useful business information. In data analytics, the data is huge in volume in generation, processing and storage levels. This data is generating in the giga / tera / zetta / peta / yottabytes per second from different electronic computing and communication devices. Besides, presently even the data is germinating from the physical objects or things too because of their connectivity to the internet. We knew that social media sites like Skype, Facebook, WhatsApp, LinkedIn, Twitter, Instagram, Myspace, Telegram, Viber, etc. for every second producing official, personal, and social information in the form of text, pictures, audios, and videos. All this data is sometimes structured / unstructured and many times mixtures of these two. So analysing this complex data with Analytics tools for useful decisions of netizens is essential in this world.

In this line, the software analytics also entering as an emerging concept, in which all the traditional Software Engineering research and its practices include requirements engineering, software project management, design, development, testing and implementation of software applications merged with this new technology "Analytics". While coming to the software analytics is to enable software practitioners to perform data exploration and analysis to obtain insightful and actionable information for data-driven tasks around software and services [29]. Software analytics is the analytics specific to the domain of software systems taking into the account source code, static and dynamic characteristics as well as related processes of their development and evolution. Besides, software analytics is to enable software practitioners to perform data exploration and analysis to obtain insightful and actionable information for data-driven tasks around software and services. Nowadays, a huge amount of data continuously generated at a rapid rate during the engineering of software systems. In the last decade, modern data analytics technologies have enabled the creation of software analytics tools offering real-time visualization of various aspects related to software development and usage. These tools seem to be particularly attractive for companies doing agile software development [30]. With software analytics, software practitioners explore and analyze data to obtain insightful, actionable information for tasks regarding software development, systems, and users [31]. Abdellatif et al. [32] said that is concerned with the analysis of all software artifacts, not only source code, defines software analytics. Its importance comes from the need to extract support insights and facts from the available software artifacts to facilitate decision-making. Artefacts are available from all software development life cycle steps, beginning with the proposal and project initiation phases and ending with the project closure and customer satisfaction surveys. Yet engineers often lack the tools and techniques necessary to leverage these potentially powerful information resources toward decision-making. Despite large volumes of data and many types of metrics, software projects continue to be difficult to predict and risky to conduct [28]. Software engineering analytics (Shah [33]) is the process of development, and testing activities—uncovering quintessential insights and recommendations. It is a very complex and data-heavy activity, involving millions of lines of code, a huge bug database, and complex testing frameworks.

However, understanding and analyzing user needs plus using of core programming languages like C, C++, Java, etc. and testing, with the help of tools like maven, ant, gradle etc., configuration tools like chef, puppet, etc. play an important role in Software Engineering. Software analytics and the use of computational methods on "big data" in software engineering is transforming the ways software is developed, used, improved and deployed [34]. Developers sometimes develop their own ideas about good and bad software, based on just a few past projects. Using software analytics, we can correct those misconceptions. Software analytics lets software engineers learn about AI techniques [35] Off course, the sweeter fruit of software analytics is spoiling: data-mining techniques are oblivious to the software domain [36]. Besides, this literature, now the researchers and academicians, and software engineers set out scribbling on software analytics.

11.6 Conclusion

The Software Development Life Cycle (SDLC) is a systematic process for building software that ensures the quality and correctness of the software built. SDLC process aims to produce high-quality software, which meets customer expectations. Software engineering is never a dying product. As long as the human being exists on the earth, the software and their applications are inevitable in their day to life. Consequently, both local and multinational companies need to plan, design, develop and deploy the novel software time-to-time-based on user requirements. That is why the novel software analytics is very essential to predict better software solutions without any manual interventions. Therefore, the Software Engineering Analytics/ Software Analytics is the enriching and upcoming field across the global. Many multinational Information Technology /Information Technology enabled Services companies are trying to develop the new analytical software in which it can automatically identify the errors, auto-mate the software process and make sure integration of submodules. Later this will cause user-level acceptance of the software at the specified time within the budget. Hence, the dawn of software analytics or software engineering is emerging; the academicians, researchers, and corporates have to come together to reap the fruits across the media, banking, finance, entertainment, heavy industries, manufacturing and other IT related software analytics product and services. Hoping that the day is very soon, where we will have software analytics and will take care of our software products in the future for reducing the cost and time. Besides, reducing the complexities in implementing software and error-free. Therefore, it is the last minute all the corporate people should work seriously for the better software analytics. Therefore, in this post-COVID-19 scenario, all the global business sectors need to adopt the software deployment in full-pledged in all departments including human resource, training & development, production, manufacturing, projects, operations, systems, marketing, sales, accounting and finance to name a few for operational efficiency. Moreover, all levels in all industries including transportation, aviation, mining, media, entertainment, healthcare, governance, security, education and special applications. The software deployment process will smoothen the process even during COVID-19 (pandemic) situations like round the clock in all business disciplines. The developments of software analytics or software engineering analytics will definitely help business organizations to surpass any kind of pandemics including COVID-19 to sustain by proper deployment.

References

1. Chan JFW, Yuan S, Kok KH, To KKW, Chu H, Yang J, et al (2020) 333 RWS Poon, HW Tsoi, SKF Lo, KH Chan, VKM Poon, WM Chan, JD Ip, JP Cai, 334 VCC Cheng, H. Chen, CKM Hui, KY Yuen, A familial cluster of pneumonia associated 335 with the 2019 novel coronavirus indicating person-to-person transmission: a study of a 336 family cluster. Lancet 395:514–523
2. Bakkalbasi N, Bauer K, Glover J, Wang L (2006) Three options for citation tracking: Google Scholar, Scopus and Web of Science. Biomed Digit Libr 3(1):7
3. Fabbrini F, Fusani M, Gnesi S, Lami G (2000) Quality evaluation of software requirement specifications. In: Proceedings of the software and internet quality week 2000 conference, pp 1–18
4. Salger F, Engels G, Hofmann A (2009) Inspection effectiveness for different quality attributes of software requirement specifications: an industrial case study. In: 2009 ICSE workshop on software quality. IEEE, pp 15–21
5. Pekar V, Felderer M, Breu R (2014) Improvement methods for software requirement specifications: a mapping study. In: 2014 9th international conference on the quality of information and communications technology. IEEE, pp 242–245
6. Singh Y, Sabharwal S, Sood M (2004) A systematic approach to measure the problem complexity of software requirement specifications of an information system. Int J Inf Manage Sci 15(1):69–90
7. Hassine J, Dssouli R, Rilling J (2004) Applying reduction techniques to software functional requirement specifications. In: International workshop on system analysis and modeling. Springer, Berlin, Heidelberg, pp 138–153
8. Sullivan KJ, Griswold WG, Cai Y, Hallen B (2001) The structure and value of modularity in software design. ACM SIGSOFT Softw Eng Notes 26(5):99–108
9. Shaw M, Garlan D (1996) Software architecture, vol 101. Prentice Hall, Englewood Cliffs
10. Heer J, Agrawala M (2006) Software design patterns for information visualization. IEEE Trans Visual Comput Gr 12(5):853–860
11. Détienne F (2001) Software design–cognitive aspect. Springer Science & Business Media
12. Terwilliger RB, Campbell RH (1989) Please: executable specifications for incremental software development. J Syst Softw 10(2):97–112
13. Rising L, Janoff NS (2000) The Scrum software development process for small teams. IEEE Softw 17(4):26–32
14. Jiang JJ, Klein G, Hwang HG, Huang J, Hung SY (2004) An exploration of the relationship between software development process maturity and project performance. Inf Manag 41(3):279–288
15. Clarke P, O'Connor RV (2012) The situational factors that affect the software development process: towards a comprehensive reference framework. Inf Softw Technol 54(5):433–447
16. Kaner C, Bach J, Pettichord B (2008) Lessons learned in software testing. Wiley
17. Kuhn DR, Wallace DR, Gallo AM (2004) Software fault interactions and implications for software testing. IEEE Trans Software Eng 30(6):418–421
18. Cadar C, Sen K (2013) Symbolic execution for software testing: three decades later. Communications 56(2):82–90
19. Hailpern B, Santhanam P (2002) Software debugging, testing, and verification. IBM Syst J 41(1):4–12
20. Dearle A (2007) Software deployment, past, present and future. In: Future of software engineering (FOSE'07). IEEE, pp 269–284
21. Hall RS, Heimbigner D, Wolf AL (1999) A cooperative approach to support software deployment using the software dock. In: Proceedings of the 1999 international conference on software engineering (IEEE Cat. No. 99CB37002). IEEE, pp 174–183
22. Carzaniga A, Fuggetta A, Hall RS, Heimbigner D, Van Der Hoek A, Wolf AL (1998) A characterization framework for software deployment technologies. Colorado State University Fort Collins Department of Computer Science

23. Bennett KH, Rajlich VT (2000) Software maintenance and evolution: a roadmap. In: Proceedings of the conference on the future of software engineering. ACM, pp 73–87
24. Dishaw MT, Strong DM (1998) Supporting software maintenance with software engineering tools: a computed task–technology fit analysis. J Syst Softw 44(2):107–120
25. Yau SS, Collofello JS (1980) Some stability measures for software maintenance. IEEE Trans Software Eng 6:545–552
26. Bertolino A (2007) Software testing research: achievements, challenges, dreams. In: 2007 future of software engineering. IEEE Computer Society, pp 85–103
27. Buse RP, Zimmermann T (2010) Analytics for software development. In: Proceedings of the FSE/SDP workshop on future of software engineering research. ACM, pp 77–80
28. Buse RP, Zimmermann T (2012) Information needs for software development analytics. In: Proceedings of the 34th international conference on software engineering. IEEE Press, pp 987–996
29. Zhang D, Dang Y, Lou JG, Han S, Zhang H, Xie T (2011) Software analytics as a learning case in practice: Approaches and experiences. In: Proceedings of the international workshop on machine learning technologies in software engineering. ACM, pp 55–58
30. Martínez-Fernández, S., Vollmer, A. M., Jedlitschka A, Franch X, López L, Ram P, et al (2019) Continuously assessing and improving software quality with software analytics tools: a case study. IEEE access
31. Zhang D, Han S, Dang Y, Lou JG, Zhang H, Xie T (2013) Software analytics in practice. IEEE Softw 30(5):30–37.2
32. Abdellatif TM, Capretz LF, Ho D (2015) Software analytics to software practice: a systematic literature review. In: Proceedings of the first international workshop on BIG data software engineering. IEEE Press, pp 30–36
33. Shah A—Senior Management Trainee ERS Practice (2017) Engineering analytics–what is next in software engineering? Accessed dated on 20/6/2019. https://www.hcltech.com/blogs/engineering-analytics-what-next-software-engineering
34. Menzies T, Williams L, Zimmermann T (2016) Perspectives on data science for software engineering. Morgan Kaufmann
35. Menzies T, Zimmermann T (2018) Software analytics: what's next? IEEE Softw 35(5):64–70
36. Hassan AE, Hindle A, Runeson P, Shepperd M, Devanbu P, Kim S (2013) Roundtable: what's next in software analytics? IEEE Softw 30(4):53–56

Chapter 12
The Rise and Fall of the SCOR Model: What After the Pandemic?

Nteboheng Pamella Phadi and Sonali Das

Abstract Charged by the evolving industrial revolutions, contemporary supply chains are characterised by integrated Digital Supply Chain networks. Though the benefits of global supply chain are vast, the COVID-19 pandemic has exposed the vulnerability of these networks in managing supply chain risks. While the history of the Supply Chain Operations Reference (SCOR) model is characterised by both criticism and approval, in this paper we argue that it can still serve as an opportunity of edification to achieve supply chain agility and resilience via strategic and intelligent re-design. To achieve this, we propose the use of the SCOR model and its associated best practices, particularly focusing on the SCOR performance attributes, namely, Cost, Reliability, Agility, Responsiveness and Asset Management to motivate revisions to existing supply chain risk management approaches as a strategy to recover from the aftermath of the COVID-19 pandemic.

Keywords Digital supply chain · SCOR · Resilience · Supply chain performance · COVID-19

12.1 Introduction: The SCOR Model as a Supply Chain Management Foundation

Supply chain systems have existed in formal, as well as informal, ways as long as humans have traded goods and services. The nature of such supply chains has often evolved organically as a result of impetus from new technologies and other socio-economic-political challenges presented to the system. The 4th Industrial Revolution (4IR) particularly, has introduced positive disruptive digital technologies which have facilitated the interconnectedness of supply chain networks across the globe. The potential impact of supply chain disruptions has thus evolved from a localised to a global impact scale, and as a result, the need to focus on supply chain resilience has increased. A distressing consequence of this is the impact of the COVID-19 epidemic,

N. P. Phadi · S. Das (✉)
Department of Business Management, University of Pretoria, Pretoria 0083, South Africa
e-mail: sonali.das@up.ac.za

which has exposed the lack of resilience of global supply chains and highlighted the need for network analyses and advanced resilience analytics methodologies [33]. In addition, global supply chains are also susceptible to trade disruptions, decreased foreign investment and disrupted tourism sectors, resulting in damaged economies, especially in less industrialised countries, which has further exacerbated during the COVID-19 pandemic [72]. The COVID-19 pandemic has also exposed the exceptional uncertainty in world economies as indicated by the World Uncertainty Index (WUI)[1] [6]. The founders the WUI have further expanded their model to generate the World Pandemic Uncertainty Index (WPUI), which of course specifically focuses on the uncertainty caused by pandemics. According to the WPUI data, the COVID-19 pandemic has thus far generated the greatest amount of uncertainty compared to any previous pandemics, which include the SARS of 2002, the 2009 Swine Flu, and the 2014 Ebola outbreak, to name a few [7].

The aforementioned issues highlight that improving supply chain resilience is not only natural and necessary, but that the more urgent question is how to achieve this in a systematic manner. The purpose of this paper is two folds: first, we briefly discuss the emergence of the supply chain operations reference (SCOR) model in the backdrop of the various industrial revolutions and why it may have lost favour with practitioners; and second, we suggest how, in the event of a pervasive pandemic like the COVID-19, the SCOR model can still be adapted to make recovery both efficient and sustainable, at the local level as well at the inter-connected global scale, providing also an opportunity to less developed countries to reduce the supply chain divide across countries.

12.2 The Evolution of Supply Chain Management

Historically, emerging technologies have altered business models and changed the end-to-end view of supply chains. Industrial Revolutions (IR) have been described as an implied growth in an industry and its related sectors in the economy. Furthermore, they are a result of increased net capital accumulation for a country, increased ability to trade worldwide, technological innovations, as well as, the increase of goods and services as a result of increased populations [38, 63]. In the lead up to the 4IR, there has been three industrial revolutions. Originating in the Great Britain, the First Industrial Revolution spanned the period 1760–1840 [22], and was characterised by the discovery of steam as a source of energy for machinery. It initially impacted the textile industry, but quickly paved the way for various other inventions for other industries including the mining sector [23]. However, the most revolutionary and lasting impact of the first industrial revolution was in the rail transportation sector for both goods and passengers [80].

---

[1] WUI is calculated using text mining of reports generated by Economist Intelligence Units reports in 143 countries and considers both developed and developing countries [6]: 2).

The Second Industrial Revolution was characterised by the use of electrical energy to enable mass production using the assembly-line concept, and also referred to by some historians as the "The Technological Revolution" [80, 88]. Originating in the United States of America, it flourished during the period 1850–1970 and marked the completion of the world's first transcontinental railroad [22, 87]. The Third Industrial Revolution began in 1969 and was driven by the extensive use of digitalisation methods [53, 71]. It enabled the efficient use of electronics, telecommunication devices and the computers and consequently, increased the cost efficiency of digital manufacturing practices and commerce, further enabled by the increased access to the internet [86]. Some of the noteworthy technological innovations of the Third Industrial Revolution include the creation of the first intelligent humanoid robot developed in 1972 in Japan, and the inception of the Google search engine in 1998 [56]. The significance of the Third Industrial Revolution can be indicative from the observation of [29], who noted that a Google search of "the third industrial revolution" yields more than 150 million entries.

The current era is that of the Fourth Industrial Revolution, also referred to as Industry 4.0, and is characterised by technologies such as cloud computing, advanced robotics and autonomous systems [89]. Although the Industry 4.0 concept is believed to have begun in Germany. [55] and [84] are of the view that this concept bares similarities to technological developments that occurred in other European countries as well during this period. None-the-less, in its report, the [89] noted the unparalleled degree and rate at which contemporary global supply chains are being altered as a result of the 4IR.

12.3 The SCOR Model—A Foundation for Supply Chain Management

By definition, a supply chain is concerned with the integration of interrelated business processes comprising primarily of the following aspects: (i) acquisition of raw materials and parts, (ii) transformation of the raw materials and parts, and (ii) distribution of the finished products [30]. A more simple definition is offered by the Charted Institute of Procurement and Supply (CIPS), who a define a supply chain as the activities undertaken by an organisation to deliver goods and services to its consumers [19]. Figure 12.1 offers a traditional supply chain view and its associated key activities.

The management of supply chains is also of great importance to ensure that each activity is executed as efficiently and as optimal as possible. [25] define the management of such supply chains as: *the end-to-end business process integration that occurs between end user and suppliers of products, services and information that add value to customers,* and is recognised by the term *Supply Chain Management (SCM).* [62] further define SCM as the systematic and strategic coordination of business functions within a company and across business entities that form part of

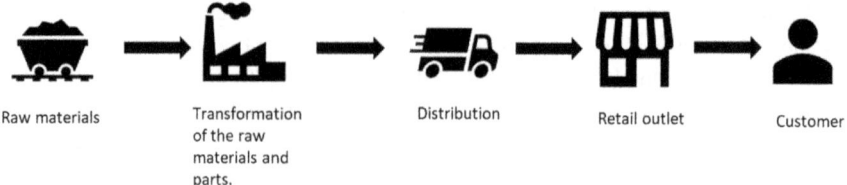

Fig. 12.1 A traditional supply chain view and its associated key activities (Icons used to construct Figs. 12.1, 12.3 and 12.4 were downloaded from Microsoft icons library.)

a company's supply chain. The authors further state that the objective of SCM is to improve the long-term performance of companies and its overall supply chain.

Originally introduced to the industry by consultants in the 1980s [51], SCM is a discipline and research area that has rightly received a great deal of attention over the years. Scholarly references that stand-out in this domain include [51]. Over the past several years the authors have reported in at least 30 publications, their efforts to propose and refine various SCM frameworks [52]. Both academia and industry have, since the 1990s, made attempts to give structure to the SCM discipline by introducing various frameworks [51] and [26], with the most widely recognised being the Supply Chain Operations Reference (SCOR)®[2] model [83]. The SCOR model was developed in 1996, and it serves as a guideline to the mapping, benchmarking, and improvement of supply chain operations, and is endorsed by the Supply Chain Council (SCC). The history of the SCC is in itself fascinating. It was founded in 1996 by a consulting company Pittiglio, Rabin, Todd and McGrath (PRTM) and comprised of 69 practitioner companies who initially met as an informal association to discuss the group's views on SCM. The group of companies later re-organised themselves as a non-profit organisation whose members represent various industries, including those involved in manufacturing, distribution, and retail [73]. The SCC later merged with the American Production and Inventory Control Society (APICS) in the year 2015 and has since published later versions of the SCOR model [69].

The SCOR model highlights six primary SCM components namely: *Plan, Source, Make, Deliver, Return* and *Enable* functions, and that these activities are all linked [37, 90] The model emphasises the business process activities which are linked to satisfying customer demand [2]. The Plan processes are associated with planning activities of supply chain operations which include the gathering of customer demands, the collection of information of resources being utilised in the supply chain, balancing customer demands to resources capabilities and subsequently identifying any gaps; the Source processes include the schedule development and ordering of goods and services of the specific supply chain; the Make processes describe the supply chain activities that are executed to convert sourced materials or are performed to create content for; the Deliver processes comprises of activities associated with the fulfilment and maintenance of customer orders, the Return processes which are

---

[2]Henceforth in this document, the use of "SCOR" will implicitly imply "SCOR®".

associated with the reverse flow of goods from the customer; and finally the Enable are descriptive of the management of the supply chain [74].

12.3.1 Supply Chain Optimisation in the Digital Supply Chain Era

The evolution of industrial technologies together with SCM best practices enabled organisations to review their supply chain process and optimise them. The history of SCM includes various attempts by scholars to use the SCOR model as a foundation to optimise supply chain processes.

Supply Chain Optimisation (SCO) entails the strategic planning and redesigning of supply chains in response to tangible or probable changes in the marketplace. Common practices in SCO include the use of data to run computational simulations for the analysis and optimisation of supply chain activities, such as, vendor selection, inventory and transportation, to name a few [73].

The application of Supply Chain Analytics (SCA) brings forth an opportunity for organisations to leverage the data-rich environment that they now find themselves in. SCA, also referred to in some literature as Business Analytics, comprises of the utilisation of data and quantitative tools to optimise supply chain performance by providing valuable insight to data [21]. The analytics tools that are applied in SCA are classified as either being descriptive, predictive, or prescriptive analytics in nature. Table 12.1 provides brief explanations for each of the SCA tools, as well as the supply chain-related questions that each of the tools answer when applied. While

Table 12.1 Explanation of supply chain analytics tools

	Descriptive analytics	Predictive analytics	Prescriptive analytics
Description	Descriptive analytics makes use of information that is flowing through the supply chain and provides insight about its current state [79] Mathematical summary statistics are common descriptive analytics techniques	Predictive analytics predicts the future behaviour of a system by analysing its data through the application of mathematical algorithms. Typical predictive analytics tools include data, text and web mining [77]	The prescriptive analytics process involves deriving optimal decision given a set of scenarios [39]
Insight derived from data	What is the current state of the organisation supply chain processes?	What trends are present in the organisation supply chain?	What will be the supply chain outcome should the supply chain processes be subjected to a particular scenario?

the concept of SCA is not new, what differentiates the current state of SCA to its historical applications is the increased volume of data flowing through these supply chains, as well as, the technological applications that are associated with analysing this data [66].

Data that is available for analysis in supply chains can be categorised as either being small data or big data. Data that is definite, comprises of predefined dataset attributes and can typically be accumulated to a spreadsheet for analysis is referred to as small data, while big data refers to high data volume that is both structured and unstructured which requires advanced analytical systems to access, organise, analyse and interpret [60]. The primary sources of supply chain small data include existing information system flows within the organisation and these typically include the limited information flowing through Enterprise Resource Planning (ERP) Systems and Transport Management Systems (TMS). ERP systems are useful tools in supply chains because they link information that is generated in various parts of the organisation by implementing various software modules [65]. The technology use in TMS on-the-other-hand, facilitates the planning, execution and supply chain optimisation of an organisation by providing visibility into operations, document control and customer service levels [68]. As adopted from the definition used in the field of Market Research, secondary data sources encompass the information that has been generated by other organisations or scholars but can be adopted and used to extrapolate useful insight for decision-making [8]. An example of such data is that supplied by government institutions depicting economic statistics such as Consumer Confidence Indicators (CCI) data, which refers to the extent of consumer optimism about the state of the economy [59]. A review of literature demonstrates how the historical data of CCI can be used as input in forecasting models to predict the future state of associated variables. An example in manufacturing is where a low consumer confidence can be associated with a decline in customer spending, which could signal the need to decrease inventory in advance as sales are anticipated to decline and a decision to invest in new project or facilities may be deferred due to low CCI [43].

Primary and secondary big data sources are defined similar to that of small data sources, where the former is found within the organisation and the latter is found external to the organisation, and both can be used to derive value-adding supply chain insights. An example of primary big data sources in organisations is that of data extracted from sensors, typically connected to more than one computer device, that can be viewed in real-time, while secondary big data, differently, includes data that is publicly available databases such as social media [12].

The output of both predictive and descriptive analytics using customer data ultimately inform decisions for achieving the ideal supply chain state. Using prescriptive analytics organisations can improve their supply chains while taking also taking consideration their current supply chain constraints [13]. The analysis of unstructured data in supply chains enables organisations to anticipate markets and this results in decision-making processes that address the supply chain changes and risks [81].

Supply chains are increasingly metamorphosing into digital networks that are interconnected and constantly generating data. This trend has resulted in interconnected and complex supply chains which have also become vulnerable to disruptions caused by variations such as pandemics, natural disasters, and political unrests, both local and global [42]. SCM scholars and practitioners thus need to review the traditional frameworks that inform the design and optimisation of supply chains and call into question their relevance and adaptability in the digital supply chain era, and to develop appropriate best practice guidelines for SCM using data.

12.4 Is the Scor Model Still Relevant?

The need for a structured approach to contemporary supply chain, and its associated risks in each level and process is not only necessary, but rather urgent. The emerging supply chain landscape is one that is highly characterised and influenced by disruptive technologies. It is thus important to define the characteristics of this new landscape and question whether traditional frameworks, such as the SCOR model, can still be used as a foundation to address the emerging challenges of Digital Supply Chain (DSC) networks. In the sub-sections below, we present a brief history of the SCOR model evolution in the past, discuss how it fell out of favour, and then proceed to present some adaptation opportunities within the SCOR models.

12.4.1 The Evolution of the Scor Model

In the past, changes to the SCOR model have been instituted as and when deemed necessary by the SCC in response to changes in the supply chain environment [70]. In Version 4.0 of the SCOR model, performance attributes were introduced, and process descriptions were also adjusted to reflect the defined process metrics to the appropriate supply chain object [70]: 58). Versions 5.0 and 6.0 saw the inclusion of additional supply chain processes, specifically, the "Return" process and the incorporation of processes related to retail, return and electronic business respectively [54, 70]. Published in 2015, Version 7.0 included two changes, namely, the simplification of SCOR performance indicators and the addition of new procedures to supply chain best practices. The fundamental revisions of Version 8.0 focused in the areas of performance indicators, best practices, and a system view of the supply chain database [70]. Version 9.0 of the SCOR model finally expanded to include risk management capabilities [17], while Versions 10.0 and 11.0 introduced a section addressing People Skills in supply chains and the introduction of the "Cost to Serve" metric respectively ([2, 73]). To answer the question of the relevance of the SCOR model in the contemporary supply chain, one must retrospectively analyse the fall of the SCOR model and then evaluate how the SCOR model can still remain a candidate for resilient adaptation.

12.4.2 The Fall of the Scor Model

The SCOR model is not without its flaws. [61] performed a critical analysis of the SCOR model and assessed its contribution to the integration of both business processes and information systems. In their study, the authors saw it fit to propose a SCOR-based reference model that would extend upon the SCOR reference model and argued that the SCOR model only enables the configuration of process maps based physical product flows, and does not facilitate the flow of information within the supply chain [61]. Reference [40] cited one of the shortfalls of the SCOR model as the process of reviewing the various proposed supply chain metrics, and were of the view that having to select the appropriate metrics and monitor them for a supply chain is rather tedious. Another shortfall of the SCOR model is derived from the implications of a study performed by [54] in which the authors investigated the relationship between supply chain management planning practices and supply chain performance based on the SCOR model processes, which they defined as key decision areas. The [54] study was based on SCOR version 4.0, and their findings insinuate that the application of best practice models such as SCOR do not necessarily provide the same perceived supply chain performance benefits in every supply chain. The aforementioned discussion lends that there is an opportunity the SCOR model to be further refined and be adapted, and adopted, to accommodate technological advances within the supply chain environment.

The introduction of advanced technologies has however altered the view of traditional supply chains. While the traditional supply chain comprises of physical locations that are linked through transportation lines, contemporary supply chain is one of a digital nature [5, 18]. The resulting dynamic set of supply networks that have emerged as a result of the 4IR, are known as a Digital Supply Chain (DSC) network [27]. According to [49], DSC's improve the communication between organisations that are part of this supply chain network through the strategic and operative exchange of information. The inter-organisational links within the DSC enable the linking of information systems between the actors from the point of sourcing until when the product or service is supplied to the customer [49]. The adaptation of 4IR technologies in supply chains has resulted in an increase of information that flows through the supply chain.

As practitioners and scholars are gaining knowledge about the DSCs, we see emerging trends in the contemporary SCM landscape. Digital Supply Chain Management (DSCM) is one of those emerging trends, which in simple terms, refers to the management of supply chains that have adopted technologies in their operations [44]. Other trends include opportunities which have been present to organisations to expand their supply chain networks into previously neglected regions, and we see this in third world countries such as South Africa (SA). Case in point is the investment made by South African municipalities who are reported to have spent on infrastructure that intends to provide free Wi-Fi connection to its township residents [57]. Furthermore, the competition of supply chains has also been accentuated by increased customer expectations and the growing trend towards the individualisation and customisation of products and services [58]. The mass customisation campaign by the German sports brand Adidas demonstrates this point well, when back in 2017 the brand launched a campaign that allowed their customers to order and customise products to their liking, which were delivered within the cycle time of four hours [24].

12.4.3 Is the Resurrection of the Scor Model Possible?

Despite the noted shortfalls, the SCOR model continues to serve as a model which many scholars and practitioners use as reference in the development of innovative supply chain improvement models. SCOR related literature published between the years of 2000–2005 focused mainly on the applications of the SCOR model, either in isolation or in combination with other methodologies such as Lean Manufacturing and Six Sigma, which aim to reduce process cycle time by eliminating process deficiencies, and help to improve the quality of process outputs by reducing process variation respectively [82]. This period also saw that authors were attentive to the topics of defining SCOR, its associated metrics and how it can be applied in practice [14–15]. Contemporary operational SCM literature themes focus on the optimisation of day-to-day supply chain activities using both mathematical models and simulation models that are reflective of the organisation's real-life supply chain configuration [64, 75] and [11]. These models are built using e-SCOR, a simulation software tool based on the SCOR model and comprised of drag-and-drop blocks that were used to assemble a particular virtual supply chain model and allows for a virtual environment to test and analyse various supply chain configurations before their implementation [11].

The latest version of the SCOR model is SCOR 12.0 [1] which has been updated to address several drivers of supply chain success in the current supply chain environment, including blockchain technology [3]. Blockchain technology is defined as a digital distribution ledger of transactions that cannot be altered as result of cryptographic methods [35], and is applied to prevent information that is shared from being compromised. In addition to blockchain technology, SCOR 12.0 aims to enable digital supply chain strategies of organisations by presenting modernised best practices and processes. It should be noted that SCOR 12.0, in its discussion, states that the model does not dictate how an organisation should execute its functions, nor does it dictate the tailoring of its information/system flows [1].

The SCOR model generically comprises of three hierarchical levels: Levels 1 and 2 provide standard supply chain architecture process, and level 3 describes the implementation of the architecture [1, 74]. Figure 12.2 provides a descriptive overview of the aforementioned SCOR model levels. Other noted outputs of the SCOR 12.0 model is in its ability to link supply chain business process, performance metrics, best practices and technologies into an integrated structure that supports communication among supply chain stakeholders, and also seeks to increase the effectiveness of supply chain management and related supply chain improvement activities [2].

Level		Description	Comment
1	 ▲	Major processes	Defines the scope, content and performance targets of the supply chain.
2	 ▲	Process categories	Defines the operations strategy; process capabilties are set.
3	 ▲	Process elements	Defines the configuration of individual processes. The ability to execute is set. Focus is on processes, inputs/outputs, skills, perfromance, best practices and capabilities.

Fig. 12.2 SCOR hierarchical levels as adopted from APICS [2]

The COVID-19 pandemic, particularly, has exposed the vulnerability of contemporary supply chains and the absence of resilience. Supply chain resilience refers to the ability of an organisation to respond and recover from supply chain disruptions by conserving its operating levels at the target level of connectedness, as well as, maintaining its control over the structure and function of its processes [46]). The pandemic has altered and disrupted global supply chains, yet the performance domains of the SCOR model are still relevant in guiding the required rapid response to the pandemic. In the following section we specifically aim to guide the re-application of the SCOR model's associated best practices in response to the COVID-19 uncertain landscape.

12.5 Re-designing the Future of Supply Chains Using Scor

The foundations of SCM that are presented by the SCOR model are still relevant in contemporary supply chains and as such, we are of the view that future attempts to construct a novel SCM framework should be based on the learnings of and from the SCOR model. The construction of a novel SCM framework needs to appreciate that the neoteric supply chain landscape comprises of global supply chains and that these are complex, and comprise of various organisations that work together, yet may be based in separate geographical locations [78]. These global supply chains are characterised by organisations that either distribute their products across international borders, have facilities in other countries or source supplies from other countries [48]. The networks comprise of both Business-to-Business (B2B) and Business-to-Customer (B2C) transactions, which ultimately translates to links between the global supply chain, local supply chains and customers, as illustrated by Fig. 12.3. According to [33], the nature of these Intertwined Supply Networks (ISN) is such that firms can play the roles of both competitors and suppliers within the same supply

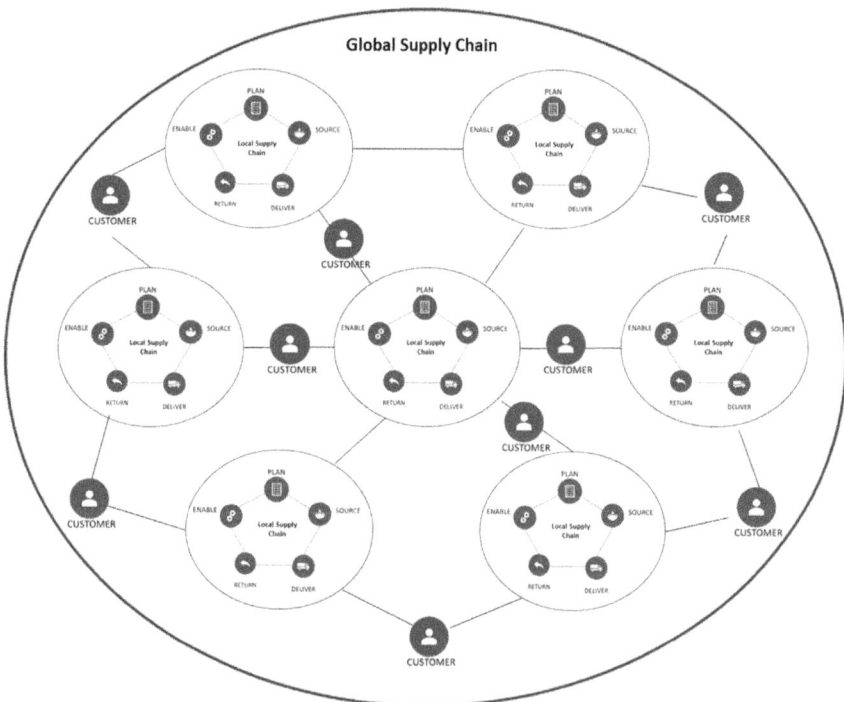

Fig. 12.3 The interconnected global supply chain, local supply chains and customers

chain network. Supply Chain Risk Management (SCRM) is necessary for every supply chain, regardless of whether they are integrated or independent. The nature of SCRM is such that strategies in supply chains are developed to identify, assess, treat and monitor possible supply chain risks. These supply chain risks may occur as a consequence of the acts of man or nature [31]. Thus, global supply chains are naturally vulnerable to disruptions caused by both natural as well as human-made disasters [42].

The COVID-19 pandemic presents an ideal case study of how organisations can continue to leverage the foundation provided by traditional frameworks such as the SCOR model. Supply chain performance measurement denotes how well the supply chain system is operating and the use of the SCOR model in this regard provides a systematic approach to the measurement thereof [28]. SCOR 11.0 links the measurement of supply chain performance monetary-related metrics and neglects the domains of Reliability, Responsiveness and Asset Management [73], see Fig. 12.4. The existing performance attributes of the SCOR model provide a strategic approach for the prioritisation and alignment of the supply chain performance to the business strategy [3].

The five main performance dimensions of the SCOR model include [91] and [3]:

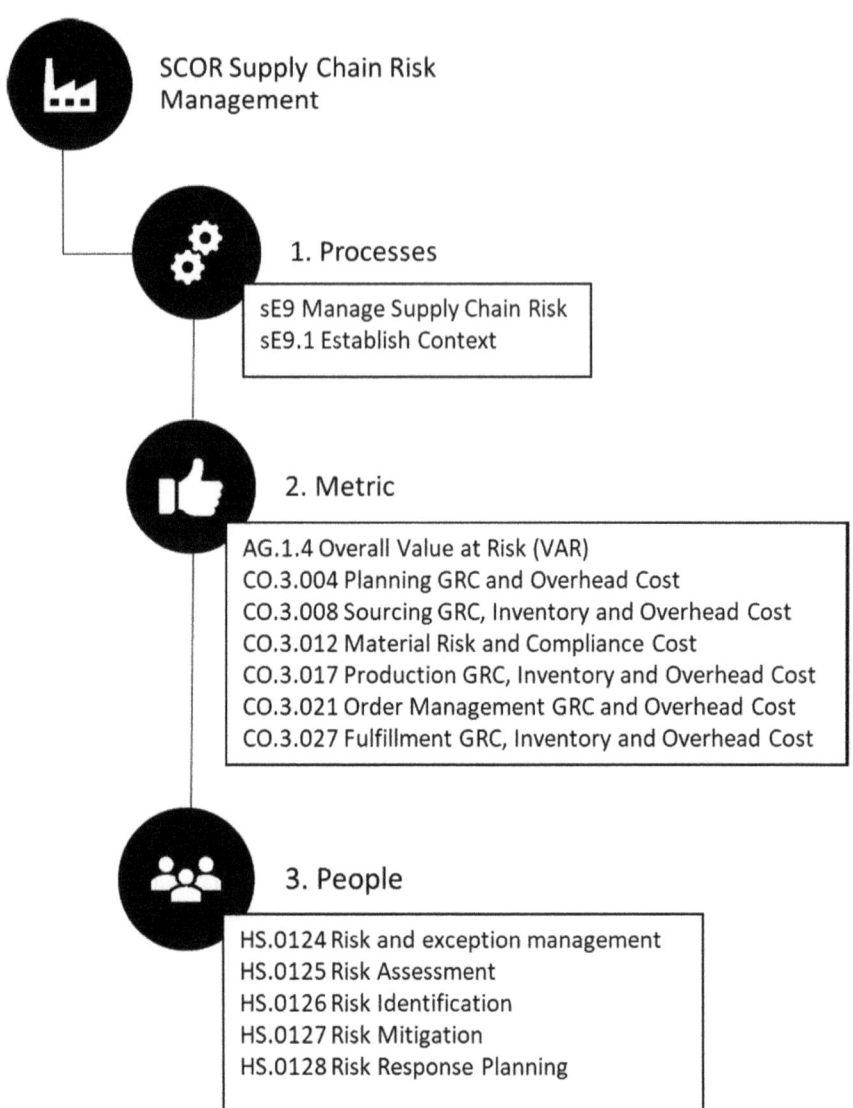

Fig. 12.4 Supply chain risk management processes, metric and people capabilities as adopted from SCOR 11.0

- Reliability—The reliability performance attribute measures the ability to fulfil supply chain tasks within the defined parameters and typically include metrics that measure the delivery of products at the right quality, place, and quantity.
- Responsiveness—This refers to the speed at which tasks are performed to fulfil customer demand and typical metrics include cycle-time.

- Agility—This performance domain refers to the supply chain response to marketplace dynamics while maintaining a competitive advantage. Adaptability and Overall Value at Risk are typical metrics for this performance domain.
- Cost—The costs associated with the operation of supply chain processes is measured by means of the cost performance dimensions. Cost of Goods Sold and Total Supply Chain Cost are typical metrics in this area.
- Asset Management—Lastly, the efficient utilisation of assets is measured through the Asset Management performance dimension, this influenced by strategic supply chain decisions such as the reduction of inventory and insourcing versus outsourcing of supply chain operations. Metrics for this domain include Cash-to-Cash Cycle Time, Return on Supply Chain Fixed Assets and Return on Working Capital.

The traditional SCOR performance view of SCRM does however need to be altered to focus on supply chain resilience rather than the minimisation of supply chain costs during times of disruption. The failure of supply chains during COVID-19 has been as a result of supply chain practices that accentuate efficiency instead of resilience [66]. The post COVID-19 supply chain design needs to measure all domains of the SCOR performance attributes in order to achieve resilience, as we propose in the sub-sections below.

12.5.1 Re-designing for Supply Chain Reliability

The COVID-19 pandemic has resulted in the disruption of material flow throughout the global supply chain. Worldwide, countries have closed their boarders to curb the spread of COVID-19 infections, and for countries that source their raw materials and products from international markets, this has subsequently meant supply shocks to their supply chains. Order fulfilment, which is reflective of supply chain reliability, was one of the significantly impacted supply chain aspects because of the pandemic disruption. This meant that products could not reach consumers, which ultimately resulted in lost profits.

Electronic Data Interchange (EDI), as identified by SCOR, is a best practice approach which implies automated transactions into vendor data bases, as well as, into their ordering systems [73]. DSC networks facilitate EDI transactions through blockchain technology, also referred to as extreme automation [32]. The cited benefits of EDI include supply chain cost and time reduction and improved customer satisfaction as supply chain lead time are ultimately reduced. The Africa Medical Supplies Platform (AMSP) provides an ideal example of how the medical supply chain has been reconfigured Africa's medical supply chain to achieve reliability through such a system. The online marketplace provides access to African Union Member States to vetted manufacturers and procurement strategic partners [85]. A prerequisite of such a strategy does of course require significant investment in Information Communication and Technology (ICT) infrastructure by the supply chain actors. The pandemic has demonstrated the importance of technology in the current supply chain landscape and has made the case for previously resistant or sluggish adopters to adopt new technologies or risk extinction. Organisations now need to establish a culture where new technologies are constantly explored and embraced [50].

While various supply chains have seen a significant decrease in demand for their goods and services, food and medical supply chains have experienced a surge in demand [76], and this has necessitated supply chain systems that optimally facilitate demand planning and forecasting. The SCOR model directs that Demand Planning and Forecast best practices provides an anticipated outlook on customer demand with the input of various stakeholders, including suppliers and supply chain planners, and makes use of customer data [73]. An immense opportunity to enhance Demand Planning lies in the use of secondary data sources that have proven to be useful in the development of forecasting models. Google Trends, for example, is a type of online big data that is representative of public sentiments. The trends provide unfiltered data samples and its results are representative of all Google search queries within a specified timeframe, and the data is also normalised according to time and location of query and the results are scaled between 0 and 100 [34]. The value of Google Trends has become apparent over the years and this has led to the application of online big data being used in several industries [45]. In emerging markets, empirical data demonstrates how Google Trends can be used to forecast the sales of automobiles [20]. In tourism, forecasting models based on such data has also proven to be of value in when planning for tourism demand in tourism-dependant economies such as the Caribbean [10]. Demand Planning and Forecasting Models post COVID-19 should capitalise on the use of big data and the insights it provides to improve supply chain optimisation.

12.5.2 Re-designing for Supply Chain Agility and Responsiveness

The COVID-19 pandemic has brought about a shift in the needs of consumers and has consequently rendered the strategic focus of some supply chains obsolete while asserting pressure on others. The reduced demand in supply chains during the pandemic have required Agility and Responsiveness in the form of redesigned business models and the optimal use of technologies. Some automotive manufacturers, for example, have experienced a sharp fall in car sales but have repurposed their manufacturing plants for the production of valves for respirators in response to the increased demand for these for ventilators [41]. Additionally, various manufacturers of fragrances have repurposed their factories to produce alcohol-based sanitizers. Moreover, there has been an increase of e-commerce activities and logistics services enabled by mobile technology [47].

Organisations who did not have to completely change their business models during the pandemic, on-the-other-hand, have also had to redesign their supply chain planning processes to achieve total order fulfilment. The total order fulfilment cycle time of activities that are executed to meet customer demand is a key metric for supply chain responsiveness [73]. Additionally, supply chain network planning is a SCOR best practice that enables responsiveness [73] and is achieved through the application of SCA techniques to test scenarios in a virtual supply chain environment. For organisations who have not already done so, there is a need for the investment in supply chain technology to facilitate agility and responsiveness [4]. The visualisation of material movement across the supply network enables the understanding of the physical material movements and this ultimately yields insight for decisions that result in supply chain resilience [33].

The resulting DSC networks are expected to facilitate this process by providing real-time data that will enable supply chain planners to make data-driven decisions, as this will result in agile and responsive supply chains. Even though the DSC network has a vast amount of technology to develop advanced forecast models, some disruptive events can unfortunately not be foreseen, and it is therefore imperative to incorporate agility in the design of supply chains. An agile supply chain is one that can proactively anticipate disruption and is prepared for sudden changes that impact the Plan, Source, Make, Deliver, Return and Enable processes [73]. While we have witnessed the consequences of the integrated nature of global supply chains, it does not however justify anti-globalism, but rather calls for sourcing from multiple countries to alleviate the excess dependency on a single country, like say China [9].

An agile supply chain strategy seeks ways in which flexibility can be incorporated into their business processes. As such, it is anticipated that in manufacturing, for example, organisations will move towards a China + 1 strategy [36], which implies that sourcing options will not be limited to only China, instead, organisations will invest in alternative sourcing strategies, and this approach of course has additional cost implications. One might argue that in this example the substitution for China is not so simple as 60% of global consumer goods exports are accounted for by the country. Nevertheless, the aftermath of the global supply chain disruption presents an opportunity for other countries to attract international business by investing in their manufacturing capabilities, and coupled attractive offerings for land, labour, and logistics.

12.5.3 Re-designing for Supply Chain Asset Management

SCOR11.0 highlights that SCRM applications to Asset Management should reduce inventory variability. To achieve this during times of demand uncertainty, organisations are to leverage insights obtained from data to inform their inventory models. The SCOR model identifies collaborative inventory planning as a best practice that is intended to include key customers in the planning process. The metrics in this sphere include Return on Working Capital, Supply Chain Revenue and Forecasting.

Return on working capital assesses the amount of investment relative to the organisation's working capital versus supply chain revenue; Supply Chain Revenue on the other hand refers to the revenue generated through supply chain operations, while Forecasting refers to the anticipated customer demand [73].

Asset Management optimisation post the COVID-19 pandemic should focus on generating optimal value from the organisation's assets and to achieve this, organisations are to enhance their supply chain capabilities across their networks. This would of course require investment on infrastructure by organisations and the countries within which they operate. The investment by countries in their own capabilities also implies that local firms would get an opportunity to shift their strategic sourcing processes from international markets to local sourcing, thus reducing supply lead time and Cost of Goods Sold (COGS). The post-COVID19 supply chain landscape will thus require collaborative integrated planning to be merged with SCA practices to achieve the organisational targets in these performance areas. While the Total Cost to Serve will initially increase, the prospect of a resilient supply chain is sufficient motivation for this trade-off.

12.6 Concluding Remarks

Supply chains have always been impacted by the revolutions of humankind. From the first steam engine to the Internet of Things, scholars and practitioners have leveraged these changes to improve supply chains. The pandemic has exposed the unpreparedness of the global supply chains to instantaneous and unforeseen change in the world. Contemporary global supply chains led supply networks to a place of comfort, where sourcing strategies comprised of a single country. We have observed that prior to the pandemic, many organisations have considered China as a single source for lost cost manufacturing, and with the closing of borders, this left their supply chains destitute.

Supply chain recovery beyond the COVID-19 pandemic will require organisations to revisit SCOR best practices and align these with their supply chain performance measures. Supply Chain Management as a practice has seen many changes over time, and in parallel to this evolution has been the evolution in the SCOR model. Though criticised by scholars, the model arguably remains the most relevant and structured framework that provides a common language for the management between supply chains.

Supply chain resilience cannot be sustained only at an organisational level but needs to have national policies in place to enable resilience. While the increased leveraging of technology might be an effortless action for some organisations, organisation which operate in certain countries are constrained by limited infrastructure and resources in their country of operation. The governments of nations also have a role to play in enabling the increased utilisation of assets of organisations operating within their borders by exploring alternate markets. In countries, whose businesses are driven by tourism for example, the pandemic has resulted in a distressed economy. Private–public partnerships, in this instance, should seek to leverage other resources within the country while also investing in technological applications that will facilitate their potential unique supply chains [67]. Governments of countries thus need to facilitate recovery through collaboration, and by investing in projects that will sustain supply chains post the COVID-19 era. The investment in technology requires subsequent investment in skills and development of local human resources as this will have a direct impact on the fostering of the required supply chain optimisation capabilities.

Developing countries specifically are to seek alternate ways in which they can strengthen their economic portfolios. The strategic response by supply chains does not call for the end of global supply chains, instead, it calls for new supply chain strategies that are informed by the SCOR performance domains and reviewed trade policies. The proposed SCOR developments provide an opportunity for even disadvantaged economies to re-invent their supply chains as they refocus the strategies of their value chains. Increased foreign investments, revived economies and local job creation are just some of the potential benefits of a vigorous value chain. The collective efforts by governments and organisations are to focus on the enhancement of both technological and human skill capabilities. In conclusion, the novel COVID-19 virus has had an impact on world economies like no other pandemic, nonetheless, supply chain scholars and practitioners have an opportunity review traditional supply chain approaches and adapt them to the current. The "new normal" will require a trade-off between Cost to Serve and supply chain resilience. To achieve supply chain Reliability, Responsiveness and Agility, SCOR model best practices for SCRM are to focus beyond supply chain costs. The ability to appreciate the measurement of these domains will be the determining which organisations will survive the pandemic wave and which ones may sink.

References

1. APICS (2017) SCOR 12.0. https://www.logsuper.com/ueditor/php/upload/file/20190530/1559181653829933.pdf. Accessed 09 March 2020
2. APICS (2019) Quick reference guide supply chain operations reference model. https://www.apics.org/docs/default-source/scor-ptoolkits/apics-scc-scor-quick-reference-guide.pdf. Accessed 25 May 2019
3. APICS (2020) SCOR 12.0. https://www.apics.org/apics-for-business/frameworks/scor12. Accessed 05 March 2020
4. Accenture (2020) COVID-19: boost agility with supply chain planning. https://www.accenture.com/gb-en/insights/consulting/coronavirus-boost-agility-responsiveness-supply-chain-planning. Accessed 08 June 2020
5. Agrawal P, Narain R (2018) Digital supply chain management: an overview. IOP Conf Ser Mater Sci Eng, 455(1)
6. Ahir H, Bloom N, Furceri D (2018) The world uncertainty index. Available at SSRN 3275033
7. Ahir H, Bloom N, Furceri D (2020) World pandemic uncertainty index. World Uncertainty Index. https://worlduncertaintyindex.com/data/. Accessed 27 July 2020
8. Andrei T (2018) The Internet-secondary data source in marketing research. Ann-Econ Ser 6:92–97

9. Baldwin R, Tomiura E (2020) Thinking ahead about the trade impact of COVID-19. Economics in the Time of COVID-19 59
10. Bangwayo-Skeete PF, Skeete RW (2015) Can Google data improve the forecasting performance of tourist arrivals? Mixed-data sampling approach. Tour Manage 46:454–464
11. Barnett MW, Miller CJ Analysis of the virtual enterprise using distributed supply chain modeling and simulation: an application of e-SCOR. In: 2000 winter simulation conference proceedings (Cat. No. 00CH37165), 2000. IEEE, pp 352–355
12. Bekker A (2017) Big Data: EXAMPLES, SOURCES AND TECHNOLOGIES EXPLAINED. ScienceSoft. https://www.scnsoft.com/blog/what-is-big-data. Accessed 11 June 2020
13. Biswas S, Sen J (2016) A proposed framework of next generation supply chain management using big data analytics. In: … of national conference on emerging trends in …
14. Bolstorff P (2002) How does SCOR measure up. Supply Chain Technol News 5:22–25
15. Bolstorff P, Rosenbaum R (2003) Supply chain excellence: a handbook for dramatic improvement using the SCOR model. J Supply Chain Manag 39(4):38
16. Bolstorff P (2001) How do I use SCOR? Supply chain world. Supply-Chain Council
17. BusinessWire (2008) New SCOR® 9.0 release includes expanded risk management capabilities and addresses sustainability efforts. https://www.businesswire.com/news/home/20080317005798/en/New-SCOR-9.0-Release-Includes-Expanded-Risk. Accessed 30 March 2020
18. Büyüközkan G, Göçer F (2018) Digital supply chain: literature review and a proposed framework for future research. Comput Ind 97:157–177
19. CIPS (2020) What is a supply chain? Charted Institute of procurement and supply https://www.cips.org/knowledge/procurement-topics-and-skills/supply-chain-management/what-is-a-supply-chain/. Accessed 13 Nov 2020
20. Carrière-Swallow Y, Labbé F (2013) Nowcasting with Google Trends in an emerging market. J Forecast 32(4):289–298
21. Chae B, Olson D, Sheu C (2014) The impact of supply chain analytics on operational performance: a resource-based view. Int J Prod Res 52(16):4695–4710
22. Chen J (2019) Industrial revolution. https://www.investopedia.com/terms/i/industrial-revolution.asp. Accessed 30 March 2020
23. Clark G (2007) The industrial revolution. 2020 (23 March)
24. Connelly T (2019) Adidas pursues rival Nike with in-store personalisation and manufacturing. https://www.thedrum.com/news/2017/03/21/adidas-pursues-rival-nike-with-store-personalisation-and-manufacturing. Accessed 05 June 2020
25. Cooper MC, Lambert DM, Pagh JD (1997) Supply chain management: more than a new name for logistics. The Int J Logist Manag 8(1):1–14
26. Croom S, Romano P, Giannakis M (2000) Supply chain management: an analytical framework for critical literature review. Eur J Purchas Supply Manag 6(1):67–83
27. Deloitte (2019). https://www2.deloitte.com/us/en/pages/about-deloitte/articles/press-releases/deloitte-ascm-unveil-digital-capabilities-model-for-supply-networks.html. Accessed 19 March 2020
28. Dissanayake CK, Cross JA (2018) Systematic mechanism for identifying the relative impact of supply chain performance areas on the overall supply chain performance using SCOR model and SEM. Int J Prod Econ 201:102–115
29. Dosi G, Galambos L, Orsanigo L (2013) The third industrial revolution in global business. Cambridge University Press
30. Fahimnia B, Farahani RZ, Marian R, Luong L (2013) A review and critique on integrated production–distribution planning models and techniques. J Manuf Syst 32(1):1–19
31. Fan Y, Stevenson M (2018) A review of supply chain risk management: definition, theory, and research agenda. Int J Phys Distrib Logist Manag
32. Fiaidhi J, Mohammed S, Mohammed S (2018) EDI with blockchain as an enabler for extreme automation. IT Prof 20(4):66–72
33. Golan MS, Jernegan LH, Linkov I (2020) Trends and applications of resilience analytics in supply chain modeling: systematic literature review in the context of the COVID-19 pandemic. Environ Syst Decis 1

34. Google (2020) FAQ about Google Trends data. https://support.google.com/trends/answer/4365533?hl=en. Accessed 18 May 2020
35. Hackius N, Petersen M (2017) Blockchain in logistics and supply chain: trick or treat? In: Digitalization in supply chain management and logistics: smart and digital solutions for an Industry 4.0 Environment. proceedings of the hamburg international conference of logistics (HICL), Vol. 23. Berlin: epubli GmbH, pp 3–18
36. Hedwall M (2020) The ongoing impact of COVID-19 on global supply chains. World Health Organisation. https://www.weforum.org/agenda/2020/06/ongoing-impact-covid-19-global-supply-chains/. Accessed 08 June 2020
37. Huan SH, Sheoran SK, Wang G (2004) A review and analysis of supply chain operations reference (SCOR) model. Supply Chain Manag: An Int J
38. Hudson P (2014) The industrial revolution. Bloomsbury Publishing
39. Huisman D (2015) To what extent do predictive, descriptive and prescriptive supply chain analytics affect organizational performance?
40. Hwang G, Han S, Jun S, Park J (2014) Operational performance metrics in manufacturing process: based on SCOR model and rfid technology. Int J Innov, Manag Technol 5(1):50–55
41. Ivanov D, Dolgui A (2020) Viability of intertwined supply networks: extending the supply chain resilience angles towards survivability. A position paper motivated by COVID-19 outbreak. Int J Prod Res
42. Jabbarzadeh A, Fahimnia B, Sabouhi F (2018) Resilient and sustainable supply chain design: sustainability analysis under disruption risks. Int J Prod Res 56(17):5945–5968
43. Jain S (2014) Why consumer confidence is an important economic indicator. https://marketrealist.com/2014/09/why-consumer-confidence-important-economic-indicator/.
44. Johnson K (2019) What is digital supply chain management? https://www.bitsight.com/blog/what-is-digital-supply-chain-management. Accessed 03 June 2020
45. Jun S-P, Yoo HS, Choi S (2018) Ten years of research change using Google Trends: from the perspective of big data utilizations and applications. Technol Forecast Soc Chang 130:69–87
46. Kamalahmadi M, Parast MM (2016) A review of the literature on the principles of enterprise and supply chain resilience: major findings and directions for future research. Int J Prod Econ 171:116–133
47. Khan SAR, Jabbour CJC, Mardani A, Wong C-Y (2020) Supply chain and technology innovation during COVID-19 outbreak
48. Koberg E, Longoni A (2019) A systematic review of sustainable supply chain management in global supply chains. J Clean Prod 207:1084–1098
49. Korpela K, Hallikas J, Dahlberg T (2017) Digital supply chain transformation toward blockchain integration. In: Proceedings of the 50th Hawaii international conference on system sciences
50. Laluyaux F (2020) Covid-19 crisis shows supply chains need to embrace new technologies. World Health Organisation. https://www.weforum.org/agenda/2020/04/covid-19-crisis-shows-supply-chains-need-to-embrace-new-technologies/. Accessed 08 June 2020
51. Lambert DM, Cooper MC (2000) Issues in supply chain management. Ind Mark Manage 29(1):65–83
52. Lambert DM, Enz MG (2017) Issues in supply chain management: progress and potential. Ind Mark Manage 62:1–16
53. Lasi H, Fettke P, Kemper H-G, Feld T, Hoffmann M (2014) Industry 4.0. Bus Inf Syst Eng 6(4):239–242
54. Lockamy A, McCormack K (2004) Linking SCOR planning practices to supply chain performance. Int J Oper Prod Manag
55. Lu Y (2017) Industry 4.0: a survey on technologies, applications and open research issues. J Ind Inf Integr 6:1–10
56. Magal P (2019) Timeline of revolutions. https://www.industrialiotseries.com/2019/02/18/timeline-of-revolutions/. Accessed 30 March 2020
57. Mahlangu I (2019) Tshwane brings back free Wi-Fi in the townships. https://www.sowetanlive.co.za/news/south-africa/2019-04-29-tshwane-brings-back-free-wi-fi-in-the-townships/. Accessed 05 June 2020

58. McKinsey (2020). https://www.mckinsey.com/business-functions/operations/our-insights/supply-chain-40--the-next-generation-digital-supply-chain. Accessed 03 June 2020
59. Mcwhinney J (2018) Understanding the consumer confidence index. https://www.investopedia.com/insights/understanding-consumer-confidence-index/. Accessed 18 March 2020
60. Miglani S (2016) Big Data and small data: what's the difference? Dataversity. https://www.dataversity.net/big-data-small-data/#. Accessed 10 June 2020
61. Millet P-A, Schmitt P, Botta-Genoulaz V (2009) The SCOR model for the alignment of business processes and information systems. Enterp Inf Syst 3(4):393–407
62. Min S, Zacharia ZG, Smith CD (2019) Defining supply chain management: in the past, present, and future. J Bus Logist 40(1):44–55
63. More C (2002) Understanding the industrial revolution. Routledge
64. Munoz A, Dunbar M (2015) On the quantification of operational supply chain resilience. Int J Prod Res 53(22):6736–6751
65. Oghazi P, Rad FF, Karlsson S, Haftor D (2018) RFID and ERP systems in supply chain management. Eur J Manag Bus Econ
66. Ojha R, Ghadge A, Tiwari MK, Bititci US (2018) Bayesian network modelling for supply chain risk propagation. Int J Prod Res 56(17):5795–5819
67. Omarjee L (2020) SA's deep recession is dragging down neighbouring states. News24. https://www.news24.com/fin24/Economy/South-Africa/sas-deep-recession-is-dragging-down-neighbouring-states-20200717. Accessed 17 June 2020
68. Oracle (2020) What is a transportation management system? https://www.oracle.com/scm/what-is-transportation-management-system/. Accessed 15 Oct 2020
69. Persson F (2011) SCOR template—a simulation based dynamic supply chain analysis tool. Int J Prod Econ 131(1):288–294
70. Poluha RG (2007) Application of the SCOR model in supply chain management. Cambria Press
71. Pouspourika K (2019) The 4 industrial revolutions. https://ied.eu/project-updates/the-4-industrial-revolutions/. Accessed 23 March 2020
72. Rajan R (2020) Pursue self-interest by helping other economies too - Industrialised Europe and Asia have contained the virus but elsewhere prospects are bleaker. The Financial Times Limited. https://faculty.chicagobooth.edu/-/media/faculty/raghuram-rajan/articles/pursue-self-interest-by-helping-other-economies-too-_-financial-times.pdf. Accessed 27 July 2020
73. SCC (2012) SCOR11.0. supply-chain.org. Accessed 5 March 2020
74. SCC (2019) SCOR 10.0. www.supply-chain.org. Accessed 23 March 2020
75. Schütz P, Tomasgard A (2011) The impact of flexibility on operational supply chain planning. Int J Prod Econ 134(2):300–311
76. Seifert MR (2020) Digesting the shocks: how supply chains are adapting to the COVID-19 lockdowns. Institute for Management Development. https://www.imd.org/research-knowledge/articles/supply-chains-adapting-to-covid-19/. Accessed 16 June 2020
77. Seyedan M, Mafakheri F (2020) Predictive big data analytics for supply chain demand forecasting: methods, applications, and research opportunities. J Big Data 7(1):1–22
78. Short JL, Toffel MW, Hugill AR (2016) Monitoring global supply chains. Strateg Manag J 37(9):1878–1897
79. Souza G (2014) Supply chain analytics. Bus Horiz
80. Supply Chain Game Changer (2020) The Industrial Revolution from Industry 1.0 to 5.0. https://supplychaingamechanger.com/the-industrial-revolution-from-industry-1-0-to-industry-5-0/. Accessed 25 March 2020
81. SupplyChainBrain (2015) Challenges of the digital supply chain. https://www.supplychainbrain.com/articles/22016-challenges-of-the-digital-supply-chain-1.
82. Swartwood D (2003) Using Lean, Six Sigma, and SCOR to improve competitiveness. Pragmatek Consulting Group
83. Theeranuphattana A, Tang JC (2008) A conceptual model of performance measurement for supply chains. J Manuf Technol Manag

84. Tjahjono B, Esplugues C, Ares E, Pelaez G (2017) What does industry 4.0 mean to supply chain? Procedia Manuf 13:1175–1182
85. Tralac (2020) COVID-19: African Union Chair, President Ramaphosa launches the Africa Medical Supplies Platform, Africa's Unified Continental Response to fight the pandemic. https://www.tralac.org/news/article/14674-covid-19-african-union-chair-president-ramaphosa-launches-the-africa-medical-supplies-platform-africa-s-unified-continental-response-to-fight-the-pandemic.html. Accessed 07 June 2020
86. Troxler P (2013) Making the 3rd industrial revolution. FabLabs: of machines, makers and inventors. Transcript Publishers, Bielefeld
87. Vale R (2016) Second industrial revolution: the technological revolution. htttps://richmondvale.org/en/blog/second-industrial-revolution-the-technological-revolution. Accessed 23 March 2020
88. Van Herreweghe M (2015) The 4th industrial revolution: opportunity and imperative: evolution for some, revolution for others. IDC Manufacturing Insights White paper, 1–11
89. WEF (2017) Impact of the fourth industrial revolution on supply chains. https://www3.weforum.org/docs/WEF_Impact_of_the_Fourth_Industrial_Revolution_on_Supply_Chains_.pdf. Accessed 23 Feb 2020
90. White SK (2018) What is SCOR? A model for improving supply chain management. CIO Africa. https://www.cio.com/article/3311516/what-is-scor-a-model-for-improving-supply-chain-management.html. Accessed 09 March 2020
91. Zanon LG, Arantes RFM, Calache LDDR, Carpinetti LCR (2019) A decision making model based on fuzzy inference to predict the impact of SCOR® indicators on customer perceived value. Int J Prod Econ 107520

Part II
Management Optimization

Chapter 13
A Comparative Study on Multi-objective Evolutionary Algorithms for Tri-objective Mean-Risk-Cardinality Portfolio Optimization Problems

Georgios Mamanis

Abstract In this research paper we experimentally investigate three state-of-the-art evolutionary multi-objective optimization algorithms and measure their efficiency and effectiveness in problems of multi-objective portfolio optimization. Especially we solve the mean-risk-cardinality portfolio optimization problem with six different measures of risk. Three different modern and state-of-the-art Multi-Objective Evolutionary Algorithms (MOEAs) are employed: Strength Pareto Evolutionary Algorithm (SPEA2), Multi-Objective Evolutionary Algorithm based on decomposition (MOEA/D) and S-Metric Selection Evolutionary Multi-Objective Algorithm (SMS-EMOA). Experimental results show that the best algorithm considering the C metric is MOEA/D while the best algorithm considering the hypervolume metric is SPEA2 while being the fastest approach. This suggests that the best approach for solving the problem is to run all the algorithms for a number of replicates and take the elite non-dominated solutions from the combined pool of solutions generated by the three algorithms.

Keywords Portfolio optimization · Portfolio selection · Multi-objective evolutionary algorithms · Multi-objective optimization

13.1 Introduction

Portfolio optimization constitutes one of the most important problems of financial economics. Actually it is one of the two important problems in financial economics. Financial economics is essentially concerned with two questions: how much to save and how to save, that is, how to invest income not consumed [18]. Generally, the problem consists of finding an optimal distribution of the available funds among various assets. A fundamental theory for solving this problem was given by Harry Markowitz who introduced the mean-variance model [40]. In this model, variance is used to define risk while mean is used to define the performance/reward of the

---

G. Mamanis (✉)
CGSoft, Athens, Greece
e-mail: gmamanis@pme.duth.gr

portfolio. The mean-variance model results in a bi-objective quadratic optimization problem. There are two conflicting objective functions, mean and variance. The expected return (mean) should be maximized while variance should be simultaneously minimized. The solution of this multi-objective optimization problem provides a special solution set which is called efficient in modern portfolio theory parlance and gives the best trade-off portfolios between mean and variance. The image of the efficient set in mean-variance space determines the efficient frontier [25]. The only constraints that are utilized in the Markowitz's mean-variance model are the budget constraint which guarantees that all the available capital is invested (i.e. the fractions invested in each asset must sum to one) and the non-negativity constraints which forbid short-selling (i.e. the fractions invested in each asset must be non-negative). For finding the efficient frontier with these constraints, efficient exact algorithms that rely on quadratic optimization exist. Markowitz himself proposed the critical line method.

The research on portfolio selection is now being focused on three directions: (i) the development of different measures of risk [50], (ii) the introduction of additional objectives (beyond mean and risk) and (iii) the introduction of additional real-world constraints (beyond budget and non-negativity constraints) [2].

Many additional constraints can be considered as objectives. In multi-objective optimization, as pointed out in [55], "we distinguish an objective from a constraint when it is not easy to fix a right-hand side value for the constraint without knowing the levels of the other objectives". In this study we solve the mean-risk-cardinality portfolio optimization problem with various measures of risk. Thus the cardinality constraint is considered as additional objective to the classical mean-risk portfolio model as it is exceptionally hard for the decision maker to know on beforehand the ideal number of assets that should be added in his/her portfolio without looking at all the tradeoffs between risk, return and the cardinality of the portfolio. Regardless, a financial investor may lose significant portfolios with substantial tradeoff between the objectives when he/she is compelled to fix the number of financial securities in the portfolio on in advance. Hence, we focus on the second category of including additional objectives in portfolio optimization problem. Multi-objective optimization is a natural and promising field of study for portfolio optimization.

This paper extends the work done by Anagnostopoulos and Mamanis [2], by considering different risk measures in the mean-risk-cardinality portfolio optimization model. Anagnostopoulos and Mamanis [2] consider only the mean-variance-cardinality portfolio optimization model. Furthermore, the paper of [2] compares only Pareto-based multi-objective evolutionary algorithms (MOEAs). This study compares three modern, state-of-the-art, representative MOEAs as identified in the recent paper of [26].

According to [26], there are right now three fundamental ideal models for MOEA designs. These are the (i) Pareto-based MOEAs, (ii) Indicator-based MOEAs and (iii) Decomposition-based MOEAs. The distinction between these algorithms is primarily because of differences in the selection operators and are these operators that are actually compared. From the first category we choose SPEA2 as it was performed best in the mean-variance-cardinality portfolio optimization problem of [2]. From the second category we choose as representative algorithm the SMS-EMOA (S Metric Selection Evolutionary Multi-objective Optimization Algorithm), [27] as suggested by the tutorial of [26] and for the same reason from the third category we choose MOEA/D (Multi-objective Evolutionary Algorithm based on Decomposition) [64].

The portfolio optimization models that will be studied have been proposed theoretically in the specialized literature but they have never been solved. Their solution consist a contribution to portfolio management. Furthermore, the algorithms have been applied, tested and compared for the first time in this problem thus making a contribution to methodology. The algorithms have proved their effectiveness in artificial test functions but they must also be tested in real-world problems.

The remainder of the paper is organized as follows. In Sect. 13.2 the three-objective portfolio selection problem considered in this research is described. Section 13.3 presents the three multiobjective evolutionary algorithms and how they were implemented in this particular problem. Section 13.4 presents a literature review on multi-objective evolutionary algorithms for portfolio optimization problems. Section 13.5 is devoted to numerical results, and some concluding remarks are presented in Sect. 13.6.

13.2 Portfolio Optimization

The problem of portfolio selection comprises of finding a best allocation of the available funds among various assets. There are two well established models for portfolio choice under risk and uncertainty: the expected utility maximization/ stochastic dominance approach and the reward-risk models [51].

The classical portfolio optimization model considers a one investment period and n available assets for investment. The investor ought to decide the proportion $x = (x_1, \ldots, x_n)$ of the primary funds to be invested in the available assets, where the decision variable w_i is the weight assigned in risky asset $i = 1, \ldots, n$. The return on each (risky) asset which is considered for consideration in the portfolio is a random variable R_i. The investor's objective is to maximize the random portfolio return $R(x) = \sum_{i=1}^{n} x_i R_i$ under the constraint that the sum of the weights, being proportions, must aggregate to one $\sum_{i=1}^{n} x_i = 1$ (i.e. it is assumed that the investor is fully invested).

Optimality between different random variables is not an obvious concept and the debate on the choice of a criterion with respect to which one should optimize the portfolio optimization is still open [20]. As stated above we will present the bi-objective mean-risk model for portfolio choice and its extension to the multi-objective portfolio optimization models.

Since Markowitz's fundamental paper, the issue of picking among various random variables R(x) is figured as a mean-risk bi-objective optimization problem, where the mean portfolio return is maximized, while a risk measure is minimized dependent upon a bunch of constraints that characterize the feasible set of portfolios. The problem is multi-objective in nature, actually bi-objective. There are two conflicting objective functions, mean and risk. The expected return (mean) should be maximized while risk should be at the same time minimized. A portfolio that simultaneously optimizes the two objective functions does not exist. Thus, the optimal trade-off portfolios between the two objectives, mean and risk are hunted. These trade-off portfolios form a special solution set which is called efficient in modern portfolio theory parlance. The image of the efficient set in mean-risk space defines the so called efficient frontier [25]. The intention of bi-objective optimization and in general multi-objective optimization is to find the efficient frontier and the set of efficient solutions. This is also in accordance with the Markowitz's approach for portfolio selection which suggests solving the portfolio choice problem using a two-step process. The first step requires the computation of the efficient frontier and the portfolios that define the efficient set. The second step involves the choice of a portfolio from this frontier that reflects best the investors' tolerance towards risk. The bi-objective portfolio optimization problem that must be solved is given below.

$$\max \mu(x)$$
$$\min \rho(x)$$
$$s.t.\ x \in X = \left\{ x \in R^n | \sum_{i=1}^{n} x_i = 1, x_i \geq 0 \right\} \tag{13.1}$$

The set of efficient portfolios is comprised by all feasible portfolios which are not dominated by any other portfolio in the feasible set.

$$E = \left\{ x^1 \in X | \nexists x^2 \in X : x^2 \succ x^1 \right\}. \tag{13.2}$$

The symbol $\succ$ stands for the Pareto dominance relation. In the mean-risk portfolio management context a solution x^2 is said to dominate a solution x^1 if $\mu(x^2) > \mu(x^1)$ and $\rho(x^2) \leq \rho(x^1)$ or $\mu(x^2) \geq \mu(x^1)$ and $\rho(x^2) < \rho(x^1)$.

The image of the efficient set in the objective space defines the efficient frontier (or non-dominated frontier).

$$EF = \{(\rho(x), \mu(x)),\ \ x \in E\}. \tag{13.3}$$

Harry Markowitz established the mean-variance model [40]. In this model, variance is used to measure the risk $\rho(x)$ while mean is used to define the return on the portfolio. The only constraints that are utilized in the Markowitz's mean-variance model are the budget constraint which guarantees that all the available funds are invested (i.e. the fractions invested in each asset must sum to one) and the non-negativity constraints which forbid short-selling (i.e. the fractions invested in each asset must be non-negative).

For computing the variance and other risk measures we need the following notation. It is assumed for every asset return a discrete probability distribution with S states of nature. The discrete probability distribution can be produced using any scenario generation technique or by using historical simulation. In this paper, we use the historical approach.

Let r_{js} be the return of asset j under scenario s. All scenarios are considered equally likely, thus $p_s = 1/S$. For any portfolio x its return under scenario s is given by the following equation:

$$z_s(x) = \sum_{j=1}^{n} r_{js} x_j, \quad s = 1, \ldots, S. \tag{13.4}$$

Thus we have a set of S returns equally likely.
The mean of the portfolio is therefore calculated using the formula

$$\mu(x) = \sum_{s=1}^{S} z_s(x) p_s. \tag{13.5}$$

It is known from probability theory that the expected return or mean is the sum of the possible values of the random variable times the probability each possible value has. All these theory applies to discrete random variables.

The variance of the portfolio is given by

$$V(x) = \frac{1}{S} \sum_{s=1}^{S} (z_s(x) - \mu(x))^2. \tag{13.6}$$

Variance measures the mean value of the squared distribution of each value of the discrete random variable from its mean or expected value.

Other than variance several risk measures has also been proposed thus defining different mean-risk models according to the risk measure used. The measures of risk that have been proposed are: the Mean Absolute Deviation (MAD), Semi-variance, Value-at-Risk, Expected Shortfall, Maximum Loss [50].

An advantage of these measures of risk is that their implementation does not need any distribution assumption of returns to be made. Accordingly, we count on the non-parametric methods to estimate these quantities [9].

The semi-variance of the portfolio is given by

$$SV(x) = \frac{1}{N} \sum_{s=1}^{S} \left[\max(0, \mu(x) - z_s(x))^2 \right], \tag{13.7}$$

where N is the total number of observations below the mean.

Semi-variance like variance measures the mean of the squared value of the values of the discrete random variable except that it counts only the values that are below the mean or expected return of the random variable.

The Value-at-Risk (*VaR*) at a given confidence level α is the maximum level of loss that the portfolio will not exceed with a probability α. Probability α is a user defined parameter and is usually set at a very small number (e.g. 0.01, 0.05 or 0.1) in order to account only for extreme losses. In this study we use $\alpha = 0.1$. The negative sign in the VaR equation shown below is used in order to describe loss since $z_s(w)$ describes return.

$$VaR_a(x) = -\inf\left\{z_{(s_a)}(x) | \sum_{j=1}^{s_a} p_{(j)} \geq a\right\}, \quad (13.8)$$

where $z_{(j)}$ are the ordered returns such that $z_{(1)}(x) \leq z_{(2)}(x) \leq \cdots \leq z_{(s)}(x)$ and $p_{(j)}$ their corresponding probabilities of occurrence.

Expected shortfall (*ES*) is the average loss conditioned that exceeds *VaR*.

$$ES_a(x) = -E\{z_s(x)|z_s(x) < -VaR_a(x)\} \quad (13.9)$$

$$ES_a(x) = -\frac{\sum_{s=1}^{S} z_s(x) \mathbf{1}_{\{z_s(x) < -VaR_a(x)\}}}{\sum_{s=1}^{S} \mathbf{1}_{\{z_s(x) < -VaR_a(x)\}}} \quad (13.10)$$

where $\mathbf{1}_{\{z_s(x) < -VaR_a(x)\}}$ is **1** if the expression in brackets is true and zero otherwise. Thus we consider only the returns of the discrete random variable that are below the Value-at-Risk described above and we divide it by the number of occurrences of such numbers. The minus sign is used to define loss since z defines return.

The maximum loss is equal to: $ML(x) = -z_{(1)}(x)$. (13.11)

Maximum loss is the minimum value of the possible values of the discrete random variable. The minus sign is used to define loss since z defines returns. For example a return of -2% is equal to a loss of 2%.

The mean absolute deviation is calculated using the following equation:

$$MAD(x) = \frac{1}{S}\sum_{s=1}^{S} |z_s(x) - \mu(x)|. \quad (13.12)$$

Recently, numerous researchers have perceived the practicality of incorporating extra objectives beyond mean and risk into the portfolio optimization model [6, 24, 62]. A very good theoretical work on multi-objective portfolio optimization models has been performed in [54–56]. In these studies, the authors defined the so-called suitable-portfolio investor who is additionally concerned with the cardinality of the portfolio, the maximum amount invested in any asset, the social responsibility, the amount invested in R&D and so forth. The additional objective functions converts the efficient frontier into a surface in a high dimensional space.

In this research, additionally to risk and return we take into account an extra discrete objective function which minimizes the cardinality of the portfolio. The objective function is included by summing up the number of non-negative proportions in the portfolio and should be minimized.

$$\min \quad card(x) = \sum_{i=1}^{n} \mathbf{1}_{\{x_i > 0\}} \tag{13.13}$$

Including a third objective (in addition to mean and risk) into the portfolio selection model the efficient frontier is transformed into a surface in the three-dimensional space and computing the exact efficient surface is very difficult if not impossible. However, a discrete approximation of the efficient surface is usually acceptable. The Pareto dominance relation described above for the mean-risk portfolio optimization problem should be transformed. In the mean-risk-cardinality portfolio optimization problem we say that a portfolio x^2 dominates another portfolio x^1 if $\mu(x^2) \geq \mu(x^1)$, $\rho(x^2) \leq \rho(x^1)$, $card(x^2) \leq card(x^1)$ with at least one strict inequality.

In sum, in this research we study the mean-risk-cardinality portfolio selection problem with non-negativity constraints. Except the added difficulty of the incorporation of a third criterion, there is a special one which is imposed by the non-smoothness of the extra objective function. These issues have led us to investigate the ability of the state-of-the-art evolutionary multi-objective algorithms in order to compute a good approximation of the true efficient surface.

13.3 Multi-objective Evolutionary Algorithms

Evolutionary algorithms (EAs) are population-based, random search heuristics that imitate the principles of Darwin's theory of evolution, and are appropriate for tackling optimization problems with tough search landscapes (e.g., large solution spaces, multimodal search spaces, constraints, nonlinear and non-differentiable functions, multiple objectives). The last capacity of EAs to take care of situations with multiple objectives has offered ascend to the field of evolutionary multi-objective optimization. The EAs intended for multi-objective optimization problems are called Multi-objective Evolutionary Algorithms (MOEAs) and they contrast from traditional EAs mainly in the selection operator. The main supremacy of MOEAs is that they produce a good approximation of the efficient frontier in a single run and within little computing time.

In this study we use three modern, representative MOEAs, as are identified in the recent paper of [26] namely SPEA2 [68], MOEA/D [64] and SMS-EMOA [27] to investigate the multi-objective portfolio selection model domain for the optimal trade-off solutions optimistically to give a good estimation of the (unknown) efficient set and its corresponding efficient surface. Furthermore, an additional goal of this study is to compare the three algorithms in the mean-risk-cardinality portfolio optimization problem making a contribution in methodology and showing a new application domain for intelligent algorithms.

SPEA2 is a prevalent and powerful Pareto-based MOEA. It adopts a typical set of possible solutions-individuals together with a repository-archive so as to guarantee the conservation of good non-dominated solutions. From the outset, the repository is set to the empty set and the population (of size N^{pop}) to a random sample of the solution space. At each iteration-generation, and while a halting criterion is not fulfilled, SPEA2 calculates the fitness of the solutions-individuals from both the repository and the normal population. SPEA2 uses a blended procedure to underline non-dominated individuals dependent on the dominance rank and dominance count method, and a grouping method to preserve diversity. To start with, every solution is appointed a strength value which is equal to the number of solutions that dominates. From that point, the fitness of an individual is basically the sum total of the strengths of its dominators. In this manner, the non-dominated individuals have zero fitness. Minimization of fitness is assumed. The density information is consolidated by adding to the fitness value of every solution a value that is equal to the inverse of the k-th smallest Euclidean distance (measured in objective space) plus two. Next, the repository is hopefully upgraded by the best solutions of the repository and the normal population. The solutions in the archive experience a reproduction scheme which is equivalent to traditional evolutionary algorithms and the outcome (the offspring population) comprises the population of the next generation. At the last iteration the best individuals, portfolios in our case, from both the repository and the final population is supplied by the algorithm.

MOEA/D is another popular and effective multi-objective evolutionary algorithm. In general, MOEA/D breaks down the multi-objective optimization problem into several sub-problems, every last one of them focusing on various parts of the efficient frontier. For decomposing the multi-objective optimization problem MOEA/D works either with Chebychev scalarizations, or other scalarization methods. MOEA/D manages a set of solutions, and every solution is associated with a particular sub-problem i.e. a weight vector. The weight vectors are chosen with such a way so that are equitably dispersed in the search space. For generating an offspring only the neighborhood of the parent solution is considered. Furthermore, to store all non-dominated solutions it produces during the search an unbounded external archive is maintained.

SMS-EMOA [27] is a classical paradigm of indicator-based multi-objective evolutionary algorithms. A performance indicator is a scalar measure that computes the quality of an efficient frontier. The SMS-EMOA utilizes the hypervolume indicator as a performance indicator. An algorithm which maximizes the hypervolume indicator yields efficient points with good proximity and diversification characteristics. The hypervolume indicator calculates the size the efficient frontier dominates bounded by a reference point. SMS-EMOA plans to maximize this indicator by advancing a population of solutions. This is accomplished by considering the contribution of solutions to the hypervolume indicator in the selection technique. Toward the begin the algorithm randomly produces a population of solutions. Next it creates an offspring solution utilizing recombination and mutation operators. At that point this offspring solution replaces that solution from the population which has the minor (exclusive) hypervolume contribution. The hypervolume contribution of a solution is the difference between the hypervolume of the population minus the hypervolume of the population without that particular solution. For applying SMS-EMOA the fast implementation described in [32] has been used as suggested by Emmerich and Deutz [26].

When evolutionary algorithms and of course multi-objective evolutionary algorithms are applied in practical optimization problems, a crucial issue for their performance is the solution representation (coding, chromosomal data structure). For the portfolio selection problem Streichert et al. [57, 58] introduced a hybrid representation, where a binary string is added to indicate the existence or not of a security in the solution portfolio leading in better algorithm performance. In the hybrid solution representation two arrays are used for characterizing a portfolio. A binary vector indicates if a particular financial security participates in the portfolio or not, and a real-valued vector is used to calculate the proportion weights of the budget invested in the available financial securities. Thus we have:

$$\Delta = \{\delta_1, \ldots, \delta_n\}, \quad \delta_i \in \{0, 1\}, i = 1, \ldots, n \quad (13.14)$$

$$W = \{w_1, \ldots, w_n\}, \quad 0 \leq w_i \leq 1, i = 1, \ldots, n \quad (13.15)$$

In order to compute the portfolio x associated with the above representation we make the following calculations: first, the weights of the financial securities that are not part of the portfolio are vanished (i.e. $w_i = 0$, if $\delta_i = 0$). From that point the excess weights are normalized to fulfill the budget constraint. Thus the proportion weight x_i is calculated by $x_i = \frac{w_i}{\sum_{j=1}^{n} w_j}$ for every $i = 1, \ldots, n$.

For generating the offspring population, the so-called uniform crossover operator in every array of the chromosome has been used. According to uniform crossover operator two selected individuals-solutions produce a single child. The value for each array is selected with equal probability from one or another parent. The offspring population was subject also for mutation. Distinctive mutation probabilities for each array have been utilized. In the real-valued array the Gaussian random mutation was applied with standard deviation 0.05, while in the binary string bit flip mutation in an arbitrarily characterized position was applied.

The algorithms stopped when 150,000 solutions-portfolios were produced. For crossover and mutation probabilities we have used the following values. Crossover probability was set at 0.9, Gaussian mutation probability at 1.0 and bit flip mutation probability at 0.01 for all algorithms. Population size was fixed to 500 individuals for every evolutionary procedure. The archive size for SPEA2 was set to 300 as well. MOEA/D utilizes an unbounded archive size.

13.4 Multi-objective Evolutionary Algorithms for Portfolio Optimization Problems—A Literature Review

The popularity of metaheuristics to support decision-making in finance is gaining momentum among researchers—which is reasonable due to the complexity of real-life financial problems [1].

A decision maker in practical portfolio management confronts problems of discrete and multi-objective aspects (with two or more objectives). Following the paradigms of multi-objective optimization and the Markowitz approach the investor desires to find the efficient sets of portfolios and their efficient frontiers (or at least a good approximation of them).

In general, the solution of a multi-objective problem can be partitioned in two distinct steps: the optimization of several objective functions and the decision of what kind of tradeoffs are relevant from the decisions maker perspective. Multi-objective optimization techniques are classified by most multi-objective optimization texts as a priori, a posteriori and interactive approaches [16]. The second step involves the selection of the appropriate optimization technique which will explore for the best solution or solutions of the specified optimization problem. Most MOEAs for portfolio selection problems have been embraced a posteriori approaches, i.e. all objectives are viewed as equivalent significant and the target is to compute the set of non-dominated solutions-portfolios from which the decision-maker-investor will choose the most appropriate. In this paper we consider these techniques.

Metaheuristics are a core topic for research in operations research and computer science the last decades [11]. They seem suitable for solving practical and complex portfolio selection models as these models have attributes such as non-convex objective functions and search spaces. Multi-objective evolutionary algorithms (MOEAs), on the other hand, while suitable to handle non-convex objective functions and search spaces are additionally capable to tackle problems with multiple objectives in a natural manner. MOEAs provide a natural tool for solving complex portfolio selection problems with additional objective functions and/or real-world constraints, however they have been less investigated in the specialized literature [36, 39]. The primary preferred position of MOEAs, particularly contrasting them with single-objective metaheuristics, is that they compute the efficient set and the respective efficient frontier in a single run of the algorithm. Single-objective metaheuristics require tackling a few optimization problems in order to produce an estimate of the true efficient frontier.

Multi-objective evolutionary algorithms are applied as early as 1997 in portfolio optimization problems [61]. The majority of papers on portfolio optimization with MOEAs solves bi-objective problems and considers only variance as a risk measure [41]. In this study, we solve the tri-objective mean-risk-cardinality portfolio selection problem with different measures of risk in addition to variance.

13.4.1 Mean-Variance Portfolio Optimization Using MOEAs

Vedarajan et al. [61] considered a mean-variance optimization model with a top weight bound for every financial security. For solving the problem they applied a multi-objective genetic algorithm namely Non-dominated Sorting Genetic Algorithm (NSGA) [53]. Diosan [22] compared three popular Pareto-based MOEAs, namely: NSGA-II [21], PESA [19] and SPEA2 [68] for solving the classical mean-variance portfolio selection model. Various computational experiments were conducted using real-world data. The outcome of the research shows that PESA outperforms NSGA-II and SPEA2 for the considered experiments. Mishra et al. [43] compared three multi-objective evolutionary techniques on the classical Markowitz mean-variance portfolio selection model. The algorithms compared were: PAES [34], APAES [47], and NSGA-II. They used only the smallest data set from the OR-Library with 31 assets to perform their experiments. NSGA-II was observed to be the best algorithm among the three while PAES was the worst. On the same standard mean-variance model and data set, [44], have also proposed and compared a multi-objective particle swarm optimization algorithm (MOPSO), PESA and microGA [17]. MOPSO was found to be the best algorithm based on a collective summary of quality metrics.

Radziukyniene and Zilinskas [49] experimentally investigated several multi-objective algorithms on the classical mean-variance problem. The multi-objective metaheuristics compared were: Fast Pareto genetic algorithm (FastPGA) [29], Multi-Objective Cellular genetic algorithm (MOCeLL) [45], Archive-based hybrid Scatter Search algorithm [46] and NSGA-II. The experiments have been performed using data from 10 Lithuanian stocks. The results were shown that MOCeLL outperforms the other algorithms.

Duran et al. [23] performed a comparison of various evolutionary multi-objective techniques, namely NSGA-II, SPEA2, and IBEA (Indicator-Based Evolutionary Algorithm) [67]. They constructed a data set using weekly returns of 26 mutual funds that are traded in the Caracas stock exchange. The research showed that NSGA-II performed better than SPEA2 while IBEA achieved a mixed performance. There were instances which IBEA provided the best results while in others the worst.

Streichert et al. [57, 58] raised an important issue in the solution of portfolio optimization problems with metaheuristics (that of solution representation) and showed that the solution representation and the variation operators may considerably affect the performance of the algorithm. In their first study, they compared several solution representations within the context of a MOEA (the authors applied an algorithm based on NSGA-II and its predecessor NSGA) on different portfolio optimization models with cardinality, buy-in and roundlot constraints. They conducted experiments using the smallest data set (with 31 assets) from the OR-Library. At first a binary representation with a 32-bit string for each decision variable x_i was used. This representation was compared with a real-valued representation where every variable x_i is encoded in a vector of real values between 0 and 1. In both representations, an additional binary string was introduced in order to specify whether a financial security participates in the portfolio (1) or not (0). This extension is called by the authors knapsack representation or hybrid encoding. They found that the hybrid representation outperforms the standard approaches. The best approach was the hybrid encoding with the real-valued vector. They solve a mean-variance portfolio selection model with buy-in thresholds, roundlots and cardinality constraints.

Armananzas and Lozano [5] compared and adapted three notable optimization algorithms, simulated annealing [7], greedy search, and ant colony optimization to the cardinality constrained mean-variance portfolio optimization model. They performed their experiments utilizing the five data sets from the OR-Library. The research has shown that the ACO and the SA heuristics perform the best as observed by visual inspections of the generated efficient frontiers. On the same mean-variance portfolio selection problem with quantity, cardinality and roundlot constraints. Chiam et al. [15] introduced an order-based solution representation and analyzed how the additional constraints affect the evolutionary search progress and the efficient frontier achievable. They compared the newly proposed solution representation with those of hybrid and real-valued solution representations of [57, 58] using three quality metrics. For the standard portfolio problem they observed that the order-based representation and the hybrid encoding were of the same quality considering generational distance. However, for three problem instances the order-based representation was better than the hybrid representation with respect to diversity.

Skolpadungket et al. [52] performed a comparative study of various multi-objective evolutionary algorithms on the mean-variance portfolio selection problem with cardinality constraints, floor constraints and roundlots. The algorithms tested were: Vector Evaluated Genetic Algorithm (VEGA), Fuzzy VEGA, Multiobjective Optimization Genetic Algorithm (MOGA), Strength Pareto Evolutionary Algorithm 2 (SPEA2), and Non-dominated sorting genetic algorithm II (NSGA-II). They based their analysis using data from the OR-Library of the Hang Seng data set which contains 31 assets. SPEA2 performed the best for both of the instances tested.

Chen et al. [14] proposed a novel technique for portfolio selection, namely multi-objective extremal optimization (MOEO) [13]. They utilized the cardinality constrained Markowitz model and they compared their approach to NSGA-II, SPEA2 and PAES. The authors test their proposed technique using the five data sets from the OR-Library. They use the front spread [10] and C (coverage) metrics to compare the performance of the algorithms. The results show that MOEO performs best considering the front spread metric. Concerning the C metric it was observed that MOEO performed better than SPEA2 and PAES and a little worse than NSGA-II.

On the mean-variance cardinality constrained portfolio selection problem [3] compared the effectiveness of five state-of-the-art MOEAs namely: Non-dominated Sorting Genetic Algorithm II (NSGA-II), Strength Pareto Evolutionary Algorithm 2 (SPEA2), Pareto Envelope-based Selection Algorithm (PESA), the Niched Pareto Genetic Algorithm 2 (NPGA2) [28], and e-Multi-objective Evolutionary Algorithm (e-MOEA) [33]. The experimental results demonstrated that SPEA2 is the best technique. Furthermore, NSGA-II and e-MOEA have shown similar performance.

Branke et al. [12] introduced a hybrid algorithm that merges a multi-objective evolutionary algorithm with the critical line algorithm for portfolio selection problems with complex constraints. Especially, they tackled a mean-variance portfolio optimization problem with buy-in threshold and cardinality constraints and a two-objective mean-variance model with the 5–10–40 constraint.

Suganya and Vijayalakshmi Pai [60] proposed a Pareto-archived evolutionary wavelet network (PEWN) to handle the mean-variance version of portfolio optimization models with bounding, class, shortsale and cardinality constraints. The wavelet coefficient shrinkage method was employed for the estimation of the input variables (covariance matrix and expected returns of the assets). Experimental studies have been performed using daily quoted prices from the Tokyo Stock Exchange (Nikkei225 index: March 2002 to March 2007) and Bombay Stock Exchange (BSE200 index: July 2001 to July 2006).

Mishra et al. [42] address a realistic mean-variance portfolio optimization problem considering cardinality, budget and quantity constraints. They propose a new multiobjective optimization technique, which they call non-dominated sorting multiobjective particle swarm optimization (NS-MOPSO) and they compared with four Multi-objective Evolutionary Algorithms based on non-dominated sorting (PESA-II, SPEA2, NSGA-II, 2 LB-MOPSO) and one based on decomposition (MOEA/D). The computational results got from the examination are additionally compared with those of single objective metaheuristics such as the simulated annealing, tabu search, genetic algorithm and particle swarm optimization. The results showed a superiority of (NS-MOPSO).

Lwin et al. [38] studied the Markowitz's mean-variance portfolio selection problem with quantity, cardinality, roundlot and pre-assignment constraints. An efficient learning-guided hybrid evolutionary multiobjective technique is proposed to handle the constrained portfolio selection model in the extended mean-variance framework. The suggested algorithm was compared against four state-of-the-art multiobjective evolutionary techniques, namely Strength Pareto Evolutionary Algorithm 2 (SPEA2), Pareto Archived Evolution Strategy (PAES), Non-dominated Sorting Genetic Algorithm II (NSGA-II) and Pareto Envelope-based Selection Algorithm II (PESA-II). Experimental results are outlined for openly accessible ORlibrary datasets from seven market indices including up to 1318 financial securities. Exploratory outcomes on the portfolio selection problem with the additional constraints exhibit that the proposed algorithm fundamentally beats the four notable multiobjective evolutionary algorithms concerning the quality of produced efficient frontier in the conducted experiments.

Zhou et al. [66] introduces a multi-objective genetic algorithm namely DEA-MOEA/D by integrating decomposition method and DEA (Data Envelopment Analysis) method for the mean-variance cardinality constrained portfolio model. The results show that DEA-MOEA/D is better than FDH-MOGA, MOEA/D and NSGA-II, not only for test functions, but also for the portfolio model.

Liagkouras and Metaxiotis [37], introduces a new multi-objective evolutionary Algorithm (MOEA) for the solution of the mean-variance cardinality constrained portfolio optimization problem (CCPOP). The suggested MOEA incorporates an efficient encoding scheme especially designed for the CCPOP. Furthermore, the submitted MOEA included a new mutation and crossover operator tailor-made to perform well with the new representation scheme. Seven different datasets from several stock markets are utilized for testing the proposed algorithm. The performance of the proposed efficiently encoded multiobjective portfolio optimization solver (EEMPOS) is contrasted with two popular and very efficient and effective MOEAs, namely NSGA-II and MOEA/D. The research results demonstrate that the proposed EEMPOS outperforms the two different MOEAs for all performance metrics considered for a fraction of computing time needed by the other algorithms.

13.4.2 Mean-Risk Portfolio Optimization Using MOEAs

Zeiaee and Jahed-Motlagh [63] apply NSGA-II in a mean-VaR portfolio optimization model. They use data from 12 stocks of the Tehran Stock exchange. They provide the efficient frontier using summary attainment surfaces. The efficient frontier showed an acceptable diversity of portfolios capturing different trade-offs between expected return and VaR. Krink and Paterlini [35] proposed a novel evolutionary multi-objective algorithm for portfolio optimization: DEMPO—Differential Evolution for Multi-objective Portfolio Optimization. They tested their technique with NSGA-II in a classic mean-variance, mean-VaR and mean-ES portfolio optimization problem using daily observations from 219 stocks of the Italian stock market. The results showed that the DEMPO outperformed NSGA-II.

Zhang et al. [65] applied a newly developed algorithm MOEA/D [64] on a portfolio selection model with minimum transaction units, transaction costs and cardinality constraints. Eight test instances with up to 150 decision variables were constructed based on data from the German stock index DAX. For comparison purposes they tested their new approach with NSGA-II having the same reproduction repair and solution representation with MOEA/D implementation. The experimental outcomes demonstrated that MOEA/D is better than NSGA-II in the majority of the problem instances considered.

Anagnostopoulos and Mamanis [4] investigated the ability and compared the effectiveness of NSGA-II, SPEA2 and PESA, on the mean-VaR and mean-ES portfolio selection models with quantity, cardinality and class constraints. To test the proposed algorithms they used daily returns from 96 stocks included in the US S&P 100 index. From the computational experiments it was not observed an apparent dominant technique. All algorithms have shown good quality and robust results as compared also with efficient points obtained using the exact method of CPLEX commercial package.

Baixauli-Soler et al. [8] tackled a mean-VaR portfolio optimization problem with minimum transaction lots and transaction costs and solved it using SPEA2. They used daily data of 50 stocks from the Eurostoxx 50 index. They managed to obtain reliable results from the solution of four models: the standard mean-VaR portfolio selection model, the mean-VaR portfolio optimization model with minimum transaction units, the mean-VaR portfolio selection model with transaction costs and the mean-VaR portfolio selection model including both minimum transaction lots and transaction costs. They also showed that by not including real constraints into the portfolio selection model this lead to inefficient solutions.

13.4.3 Three-Objective Portfolio Optimization Using MOEAs

Vedarajan et al. [61] proposed a three-objective portfolio selection problem, where the third additional objective measures the transaction cost due to changes in portfolio weights which is to be minimized. For solving the problem they applied the NSGA.

Ong et al. [48] suggested an algorithm which includes the grey and possibilistic regression models to forecast expected return and covariance matrix for a three-objective portfolio selection model with two sources of risk (an uncertainty risk function and the classic variance). In order to handle the multi-objective quadric optimization problem, a multi-objective evolution algorithm which transfers the vector objective function into a scalar was employed. A computational example with six stocks was built in order to compare the suggested method to the classical mean-variance model. The suggested technique has been shown to provide more workable and precise results than the Markowitz model.

Fieldsend et al. [31] formulated the cardinality constrained portfolio optimization problem as a tri-objective problem where they seek to compute all possible tradeoffs between return, risk and the cardinality of the portfolio. This is performed by formulating the number of assets in the portfolio as an extra objective to be minimized. Fieldsend search for the efficient frontiers which represent the trade-off among return, cardinality and variance and using a modified MOEA to compute the efficient surface in one execution of the algorithm. They base their solution approach in a simple (1 + 1)-evolution strategy [30]. They test their method using weekly asset returns from the US S&P 100 index and emerging stock markets for the period January 1992 to December 2003.

Anagnostopoulos and Mamanis [2] have performed a computational study with the state-of-the-art MOEA techniques, on the same three-criterion problem introducing however additional practical constraints (quantity and class constraints). They test the ability of MOEAs to solve large-scale instances with 200 and 300 assets generated randomly. The MOEAs tested and compared were: SPEA2, PESA and NSGA-II. The results revealed a clear win of SPEA2 with PESA coming next while being the fastest approach.

Subbu et al. [59] introduced a hybrid multiobjective optimization technique that merges evolutionary algorithm and linear optimization to simultaneously maximize a return measure and minimize two measures of risk. The return is described by portfolio's book yield while the two sources of risk are measured by value-at-risk and variance. The constraints are all linear functions of the portfolio weights expressing duration and convexity mismatches. For identifying the efficient frontier they employ Pareto Sorting Evolutionary Algorithm (PSEA).

Radziukyniene and Zilinskas [49] experimentally investigated several multiobjective algorithms on a three-objective mean-variance-annual-dividend-yield portfolio selection model. The multi-objective metaheuristics compared were: Multi-Objective Cellular genetic algorithm (MOCeLL) [45], Fast Pareto genetic algorithm (FastPGA) [29], NSGA-II and Archive-based hybrid Scatter Search algorithm [46]. The experiments have been performed using data from 10 Lithuanian stocks. The results were shown that MOCeLL performs best concerning generational distance and hypervolume while NSGA-II and FastPGA are the best algorithms considering the inverted generational distance.

Based on the above literature review on evolutionary multi-objective algorithms for portfolio optimization we aim to apply the state-of-the-art MOEAs in different mean-risk-cardinality portfolio optimization models. We make a contribution both in finance as we solve problems that have never been solved in the specialized literature and to computer science, comparing and identifying the best representative MOEAs for solving the problems and showing a new application domain for intelligent algorithms. More specifically: The corresponding MOEAs have never been compared in such a problem. Their effectiveness has been proven in other fields and artificial functions but they should be tried in various practical problems as well. Furthermore, as it is shown from the literature review, the majority of papers deal with mean-variance portfolio optimization problems, fewer studies consider different risk measures in a mean-risk framework and even fewer consider tri-objective portfolio optimization problems.

13.5 Experimental Results

We demonstrate here the computational outcomes acquired by applying MOEAs in the mean-risk-cardinality portfolio selection models just as a cross-algorithm performance comparison. The data required were collected from the yahoo finance webpage and they are referred to the S&P 100 index. Daily returns from 2 October 2012 to 2 October 2017 of 94 assets have been computed and each computed rate of return was considered to define a different scenario, thus the total number of scenarios were $T = 1257$.

All algorithms have utilized the same parameter settings. A population and archive size of 300 individuals have been used. MOEA/D uses an unbounded archive. A crossover probability of 0.9 was used for all the three algorithms. Mutation probabilities of 0.01 for the Δ array and 1.0 for the W set were used. The algorithms were stopped after 150,000 solutions were generated.

For comparing the three different MOEAs the Set Coverage (C-metric) and hypervolume metric have been employed. The C-metric is calculated as follows: Let A and B be two approximation sets of the efficient frontier. $C(A, B)$ is defined as the percentage of the solutions in B that are covered (it is dominated or it is equal) by at least one solution in A, i.e.,

$$C(A, B) = \frac{|\{u \in B\} | \exists v \in A : v \succeq u|}{|B|}. \tag{13.16}$$

$C(A, B) = 1$ means that all solutions in B are covered by some solutions in A, while $C(A, B) = 0$ implies that no solution in B is covered by a solution in A. According to [69], binary quality measures like C-metric overcome the limitations of unary measures like hypervolume and, if properly designed, are capable of indicating whether A is better than B.

Hypervolume metric can quantify how well the computational methods perform in computing solutions along the full extent of the efficient frontier. It essentially computes the volume of the objective space dominated by the solutions generated by the corresponding algorithm bounded by a reference point. Consequently, the higher the hypervolume metric value the better.

SMS-EMOA and MOEA/D are properly oriented so as to minimize the three objective functions. To express the expected return objective in minimization form, the expected return objective is transformed as $\mu(x)$. In order to compute the hypervolume metric SPEA2 is also operates in the minimization problem although this is not a requirement.

All algorithms have been run 10 times for every portfolio problem on identical computers (Intel(R) Core (TM) i5-7200U, 2.5 GHZ, 4.00 GB) and coded using Microsoft Visual C++.

Table 13.1 shows the means of the C-metric values for all portfolio models and Table 13.2 gives the mean computing time used by each algorithm for each problem.

With respect to computational time, as shown in Table 13.2, SPEA2 is the fastest approach followed by MOEA/D and SMS-EMOA. On average, SPEA2 requires about 13.5% of the CPU time that MOEA/D needs and 2.3% of SMS-EMOA. In addition, MOEA/D requires on average only about 17% the time that SMS-EMOA needs.

With respect to C-metric, as presented in Table 13.1, it is observed that SPEA2 is the worst algorithm. On average its solutions are covered by the other two algorithms in approximately more than 41% for all portfolio problems while it covers only approximately 4.7% of the solutions of the other two algorithms. MOEA/D is slightly better than SMS-EMOA since it covers on average roughly 14% of the generated solutions of SMS-EMOA and SMS-EMOA covers only 4.3% of the generated solutions of MOEA/D.

Concerning the hypervolume metric, as shown in Table 13.3 the results are completely different. SPEA2 wins in five of six risk measures as shown with bold font. This suggests that MOEA/D and SMS-EMOA provide solutions with good proximity (that is why they win considering C-metric) but with poor coverage and diversity of the efficient surface. This phenomenon can be seen in the following figures, where we see that SPEA2 approaches areas with more assets in the portfolio (but less risk) than SMS-EMOA and MOEA/D.

Thus we observe that although SMS-EMOA and MOEA/D covers approximately 50% of the solutions generated by SPEA2 this also shows that there are another 50% of solutions not covered by the two algorithms. This recommends that the best

Table 13.1 C metric for all mean-risk-cardinality portfolio optimization problems

	MOEA/D	SMS-EMOA	SPEA2
Variance			
MOEA/D	–	0.105	0.369
SMS-EMOA	0.007	–	0.479
SPEA2	0.004	0.044	–
Value at risk			
MOEA/D	–	0.154	0.295
SMS-EMOA	0.109	–	0.476
SPEA2	0.084	0.084	–
Expected shortfall			
MOEA/D	–	0.158	0.454
SMS-EMOA	0.047	–	0.475
SPEA2	0.030	0.036	–
Mean absolute deviation			
MOEA/D	–	0.166	0.630
SMS-EMOA	0.022	–	0.566
SPEA2	0.006	0.039	–
Maximum loss			
MOEA/D	–	0.146	0.224
SMS-EMOA	0.054	–	0.245
SPEA2	0.059	0.123	–
Semi-variance			
MOEA/D	–	0.112	0.319
SMS-EMOA	0.018	–	0.472
SPEA2	0.010	0.045	–

Table 13.2 CPU time in seconds for each algorithm and each mean-risk-cardinality portfolio problem

	Variance	VaR	ES	MAD	ML	SV
MOEA/D	7101.59	1383.01	5052.40	5392.13	4406.37	4077.06
SMS-EMOA	32,326.94	16,350.71	26,327.99	27,318.23	33,584.29	24,086.28
SPEA2	630.07	557.71	652.90	668.62	604.73	579.64

Table 13.3 Hypervolume for all algorithms and risk measures

	Variance	VaR	ES	MAD	ML	SV
MOEA/D	0.8976174	0.8572079	0.8585893	0.8747477	0.8733369	0.8974866
SMS-EMOA	**0.8981439**	0.8608451	0.8589865	0.8746798	0.8734708	0.8980951
SPEA2	0.8981417	**0.8625269**	**0.8593118**	**0.8749439**	**0.8745495**	**0.8981631**

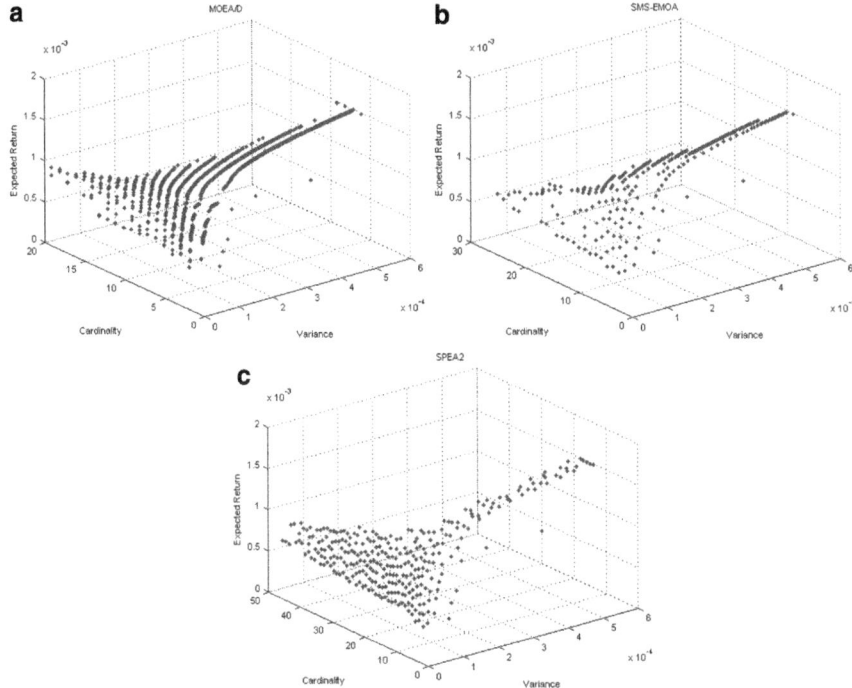

Fig. 13.1 Mean-variance-cardinality efficient surface, **a** MOEA/D, **b** SMS-EMOA, **c** SPEA2

technique for tackling the problem is to execute all the algorithms several times and take the efficient portfolios from the combined pool of solutions generated by the three algorithms.

In Fig. 13.1, 13.2, 13.3, 13.4, 13.5 and 13.6 we see the efficient surface for the Mean-Risk-Cardinality portfolio optimization problem generated by all algorithms for different risk measures. It is seen for each cardinality level the mean-risk efficient frontier for different risk measures generated by each algorithm. In general, for the problem, it is seen that as the number of assets in the portfolio increases the risk decreases but the expected return of the portfolio decreases as well. And this is observed independently for each risk measure.

Furthermore, for a fixed level of expected return, there are various portfolios with varying risk but generally portfolios with smaller risk contain more securities and this is certainly a tradeoff since investors prefer to have small portfolios. By examining the above surfaces decision makers can therefore find portfolios that suit to their preferences the best.

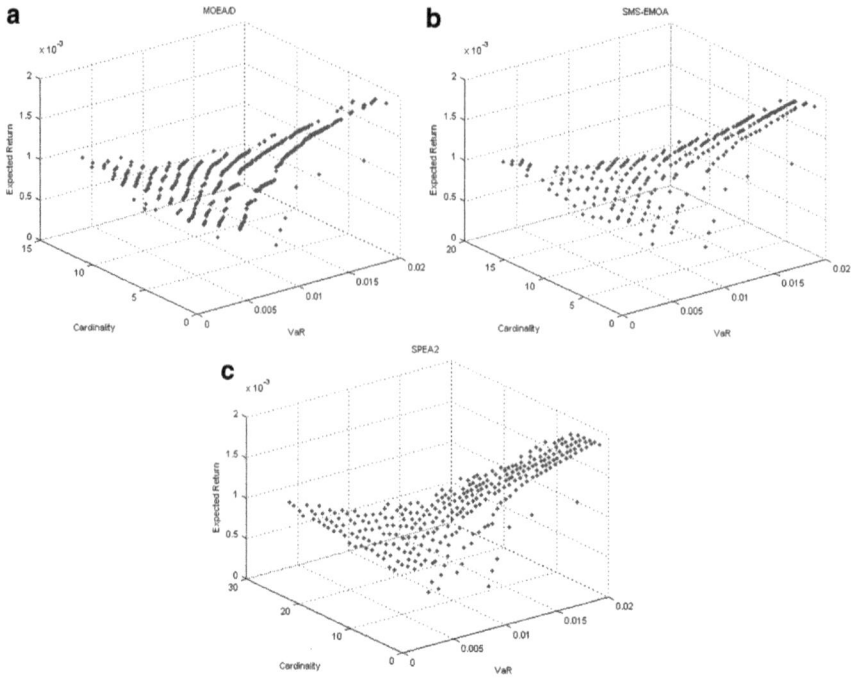

Fig. 13.2 Mean-VaR-cardinality efficient surface, **a** MOEA/D, **b** SMS-EMOA, **c** SPEA2

13.6 Conclusion

Multi-objective portfolio optimization is gaining momentum the last years. Portfolio optimization is an inherently multi-objective problem from its origin. However the last years additional criteria has been proposed in the classical mean-risk portfolio selection models. In this research we considered the mean-risk-cardinality portfolio selection model with six different risk measures. Three different state-of-the-art Multi-Objective Evolutionary Algorithms (MOEAs) were employed: Strength Pareto Evolutionary Algorithm (SPEA2), Multi-Objective Evolutionary Algorithm based on decomposition (MOEA/D) and S-Metric Selection-Evolutionary Multi-Objective Algorithm (SMS-EMOA). Experimental results demonstrated that the best algorithm considering the C metric was MOEA/D while the best algorithm considering the hypervolume metric was SPEA2 while being the fastest approach. This recommends that the best technique for tackling the problem is to execute all the algorithms several times and take the efficient portfolios from the combined pool of solutions generated by the three algorithms.

As future research other MOEAs can be used like memetic and convolution-based MOEAs which are alternative MOEAs frameworks in addition to Pareto-based, indicator-based and decomposition-based MOEAs which were considered in

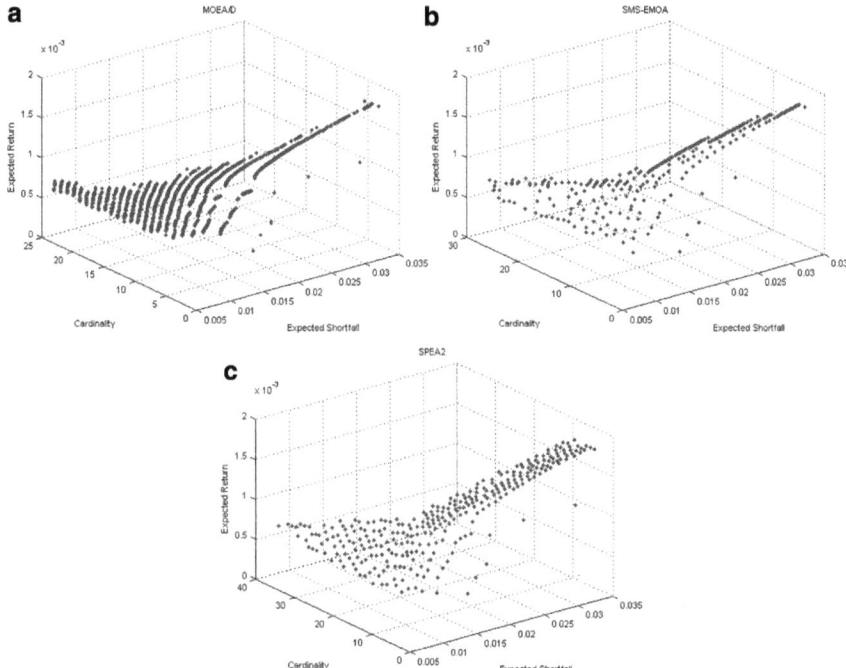

Fig. 13.3 Mean-ES-cardinality efficient surface, **a** MOEA/D, **b** SMS-EMOA, **c** SPEA2

this study. Furthermore there are also other less explored but important multiobjective algorithms (from the family of multi-objective metaheuristics and not only evolutionary algorithms) e.g., the multiobjective versions of ant colony optimization, particle swarm optimization, scatter search, simulated annealing, tabu search and GRASP. A comparative study of different approaches seems particularly interesting and necessary.

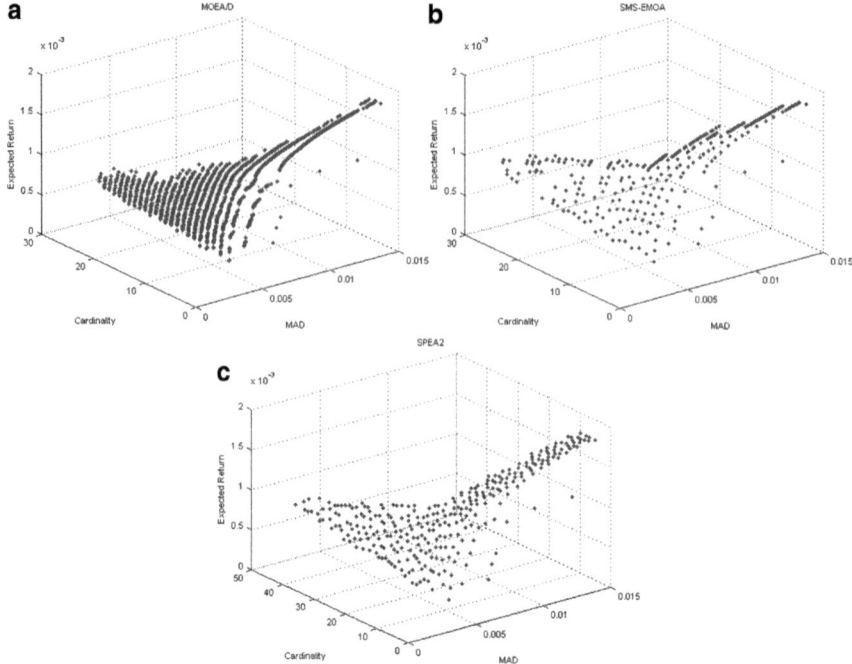

Fig. 13.4 Mean-MAD-cardinality efficient surface, **a** MOEA/D, **b** SMS-EMOA, **c** SPEA2

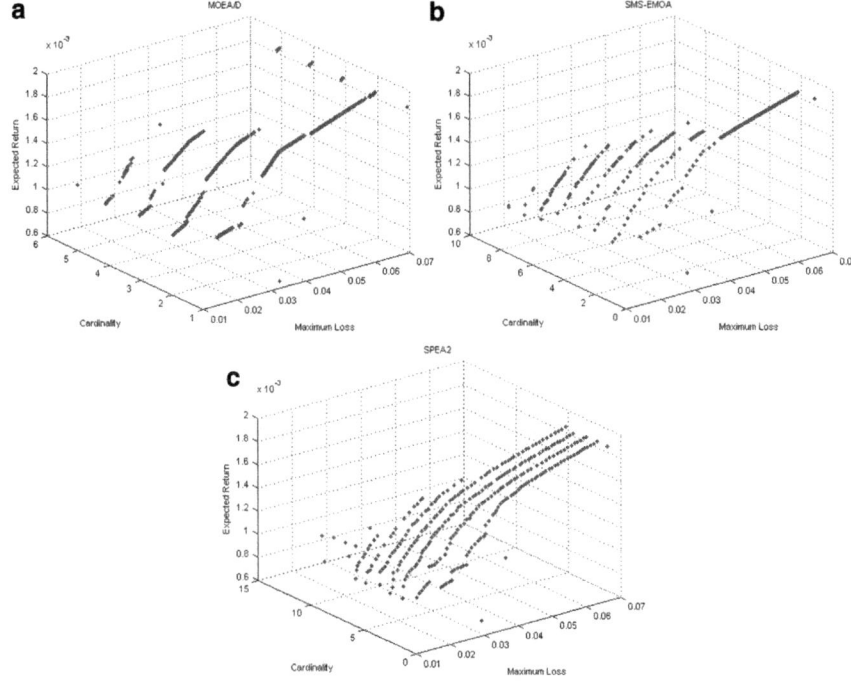

Fig. 13.5 Mean-ML-cardinality efficient surface, **a** MOEA/D, **b** SMS-EMOA, **c** SPEA2

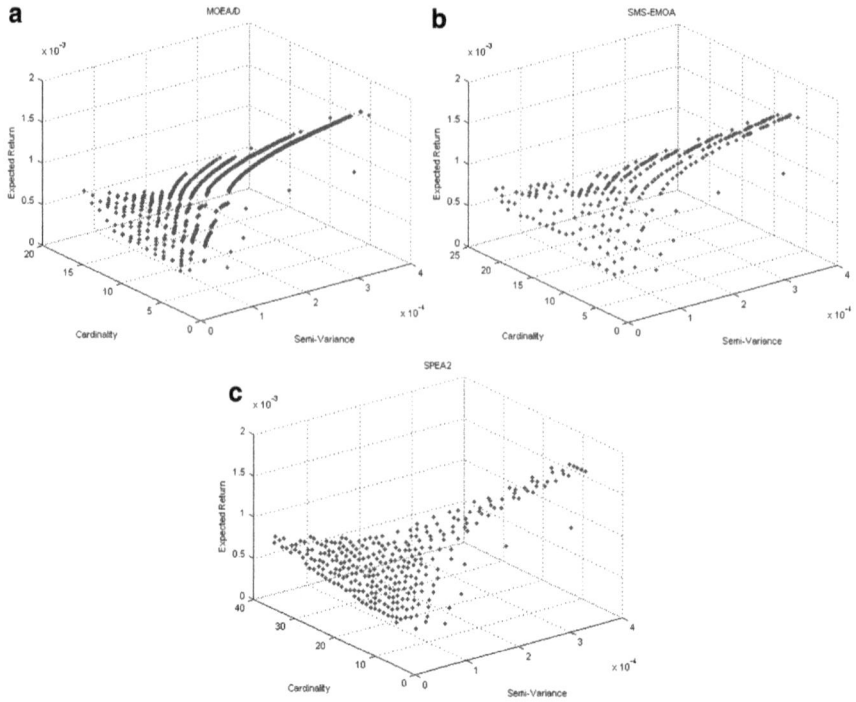

Fig. 13.6 Mean-SV-cardinality efficient surface, **a** MOEA/D, **b** SMS-EMOA, **c** SPEA2

References

1. Amparo SD, Angel AJ, Renatas K (2017) A survey on financial applications of metaheuristics. ACM Comp Surv 50(15). https://doi.org/10.1145/3054133
2. Anagnostopoulos KP, Mamanis G (2010) A portfolio optimization model with three objectives and discrete variables. Comput Oper Res 37:1285–1297
3. Anagnostopoulos KP, Mamanis G (2011) The mean-variance cardinality constrained portfolio optimization problem: an experimental evaluation of five multiobjective evolutionary algorithms. Exp Syst Appl 38:14208–14217
4. Anagnostopoulos KP, Mamanis G (2011) Multiobjective evolutionary algorithms for complex portfolio optimization problems. Comput Manag Sci 8:259–279
5. Armananzas R, Lozano JA (2005) A multiobjective approach to the portfolio optimization problem. In: IEEE congress on evolutionary computation, pp 1388–1395
6. Bana e Costa CA, Soares JO (2004) A multicriteria model for portfolio management. Eur J Fin 10:198–211
7. Bandyopadhyay S, Saha S, Maulik U, Deb K (2008) A simulated annealing-based multiobjective optimization algorithm: AMOSA. IEEE Trans Evol Comput 12:269–283
8. Baixauli-Soler JS, Alfaro-Cid E, Fernandez-Blanco MO (2011) Mean-VaR portfolio selection under real constraints. Comput Econ 37:113–213
9. Benati S, Rizzi R (2007) A mixed integer linear programming formulation of the optimal mean/value-at-risk portfolio problem. Eur J Oper Res 176:423–434
10. Bosman PAN, Thierens D (2003) The balance between proximity and diversity in multiobjective evolutionary algorithms. IEEE Trans Evol Comput 7:174–188

11. Boussaid I, Lepagnot J, Siarry P (2013) A survey on optimization metaheuristics. Inf Sci 237:82–117
12. Branke J, Scheckenbach B, Stein M, Deb K, Schmeck H (2009) Portfolio optimization with an envelope-based multi-objective evolutionary algorithm. Eur J Oper Res 199:684–693
13. Chen MR, Lu YZ (2008) A novel elitist multiobjective optimization algorithm: multiobjective extremal optimization. Eur J Oper Res 188:637–651
14. Chen MR, Weng J, Li X (2009) Multiobjective extremal optimization for portfolio optimization problem. In: IEEE international conference on intelligent computing and intelligent systems, pp 552–556
15. Chiam SC, Tan KC, Al MA (2008) Evolutionary multi-objective portfolio optimization in practical context. Int J Autom Comput 5:67–80
16. Coello CAC, Lamont GR, Van Veldhuizen DA (2007) Evolutionary algorithms for solving multi-objective problems, 2nd edn. Springer, Berlin
17. Coello CAC, Pulido GT (2001) A micro-genetic algorithm for multiobjective optimization. In: Zitzler E et al (eds) First international conference on evolutionary multi-criterion optimization. Lecture notes in computer science 1993, pp 126–140
18. Constantinides GM, Malliaris AG (1995) Portfolio theory. In: Jarrow RA, Maksimovic V, Ziemba WT (eds) Handbooks in operations research and management science, vol 9, pp 1–30
19. Corne DW, Knowles JD, Oates MJ (2005) The pareto envelope-based selection algorithm for multiobjective optimization. In: Schoenauer M et al (eds) Proceedings of the parallel problem solving from nature, VI conference, Lecture notes in computer science 1917. Springer, Paris, France, pp 839–848
20. De Giorgi E (2005) Reward-risk portfolio selection and stochastic dominance. J Bank Fin 29:895–926
21. Deb K, Pratap A, Agarwal S, Meyarivan T (2002) A fast and elitist multiobjective genetic algorithm: NSGA II. IEEE Trans Evol Comput 6:182–197
22. Diosan L (2005) A multi-objective evolutionary approach to the portfolio optimization problem. In: Proceedings of the international conference on computational intelligence for modeling control and automation, pp 183–188
23. Duran FC, Cotta C, Fernandez AJ (2009) Evolutionary optimization for multiobjective portfolio selection under Markowitz's model with application to the Caracas stock exchange. In: Chiong R (ed) Nature-inspired algorithms for optimisation, SCI 193, pp 489–509
24. Ehrgott M, Klamroth K, Schwehm C (2004) An MCDM approach to portfolio optimization. Eur J Oper Res 155:752–770
25. Elton EJ, Gruber MJ, Brown SJ (2014) Modern portfolio theory and investment analysis, 9th edn. Wiley, Hoboken
26. Emmerich MTM, Deutz AH (2018) A tutorial on multiobjective optimization: fundamentals and evolutionary methods. Nat Comput 17:585–609
27. Emmerich M, Beume N, Naujoks B (2005) An EMO algorithm using the hypervolume measure as selection criterion. In: Coello Coello CA et al (eds) EMO 2005, LNCS 3410, pp 62–76
28. Erikson M, Mayer A, Horn J (2001) The Niched Pareto genetic algorithm 2 applied to the design of groundwater remediation systems. In: Zitzler E et al (eds) Lecture notes in computer science, vol 1917, pp 681–695
29. Eskandari H, Geiger CD (2008) A fast Pareto genetic algorithm approach for solving expensive multiobjective optimization problems. J Heuristics 14:203–241
30. Fieldsend JE, Everson R, Singh S (2003) Using unconstrained elite archives for multi-objective optimization. IEEE Trans Evol Comput 7:305–323
31. Fieldsend JE, Matatko J, Peng M (2004) Cardinality constrained portfolio optimization. In: Lecture notes in computer science (including subseries lecture notes in artificial intelligence and lecture notes in bioinformatics), vol 3177, pp 788–793
32. Guerreiro AP, Fonseca CM (2018) Computing and updating hypervolume contributions in up to four dimensions. IEEE Trans Evol Comput 22:449–463
33. Hanne T (2007) A multiobjective evolutionary algorithm for approximating the efficient set. Eur J Oper Res 176:1723–1734

34. Knowles J, Corne D (2003) Approximating the nondominated front using the Pareto archived evolution strategy. Evol Comput 8:149–172
35. Krink T, Paterlini S (2011) Multiobjective optimization using differential evolution for real world portfolio optimization. Comput Manag Sci 8:157–179
36. Kumar D, Mishra KK (2017) Portfolio optimization using novel co-variance guided Artificial Bee Colony algorithm. Swarm Evol Comput 33:119–130
37. Liagkouras K, Metaxiotis K (2018) A new efficiently encoded multiobjective algorithm for the solution of the cardinality constrained portfolio optimization problem. Annals Oper Res 267:281–319
38. Lwin K, Qu R, Kendal G (2014) A learning-guided multi-objective evolutionary algorithm for constrained portfolio optimization. Appl Soft Comput 24:757–772
39. Macedo LL, Godinho P, Alves MJ (2017) Mean-semivariance portfolio optimization with multiobjective evolutionary algorithms and technical analysis rules. Exp Syst Appl 79:33–43
40. Markowitz HM (1952) Portfolio selection. J Fin 7:77–91
41. Metaxiotis K, Liagkouras K (2012) Multiobjective evolutionary algorithms for portfolio management: a comprehensive literature review. Exp Syst Appl 39:11685–11698
42. Mishra SK, Panda G, Majhi R (2014) A comparative performance assessment of a set of multiobjective algorithms for constrained portfolio assets selection. Swarm Evol Comput 16:38–51
43. Mishra SK, Meher S, Panda G, Panda A (2009) Comparative performance evaluation of multi-objective optimization algorithms. In: World congress on nature and biologically inspired computing, pp 1338–1342
44. Mishra SK, Panda G, Meher S, Majhi R (2010) Multiobjective evolutionary algorithms for financial portfolio design. Int J Comput Vis Rob 1:236–247
45. Nebro AJ, Durillo JJ, Luna F, Dorronsoro B, Alba E (2006) A cellular genetic algorithm for multiobjective optimization. In: Proceedings of NICSO, pp 25–36
46. Nebro AJ, Luna F, Alba E, Beham A, Dorronsoro B (2006b) AbYSS adapting scatter search for multiobjective optimization. Technical report ITI-2006-2. Departamento de Lenguajes y Ciencias de la Computación, University of Malaga
47. Oltean M, Grosan C, Abraham A, Koppen M (2005) Multiobjective optimization using adaptive Pareto archive evolution strategy. In: 5th international conference on intelligent systems design and applications (ISDA'05), Piscataway, NJ, IEEE Press, pp 558–563
48. Ong CS, Huang JJ, Tzeng GH (2005) A novel hybrid model for portfolio selection. Appl Math Comput 169:1195–1210
49. Radziukyniene I, Zilinskas A (2009) Approximation of Pareto set in multi objective portfolio optimization. In: Advances in electrical engineering and computational science series. Lecture notes in electrical engineering, vol 39, pp 551–562
50. Righi MB, Borenstein D (2017) A simulation comparison of risk measures for portfolio optimization. Fin Res Lett 24:105–112
51. Roman D, Mitra G (2009) Portfolio selection models: a review and new directions. Wilmott J 1:69–85
52. Skolpadungket P, Dahal K, Harnpornchai N (2007) Portfolio optimization using multi-objective genetic algorithms. In: IEEE congress on evolutionary computation, pp 516–523
53. Srinivas N, Deb K (1994) Multiobjective optimization using nondominated sorting in genetic algorithms. Evol Comput 2:221–248
54. Steuer RE, Qi Y, Hirschberger M (2005) Multiple objectives in portfolio selection. J Fin Decis Making 1:1–26
55. Steuer RE, Qi Y, Hirschberger M (2007) Suitable-portfolio investors, nondominated frontier sensitivity, and the effect of multiple objectives on standard portfolio selection. Annals Oper Res 152:297–317
56. Steuer RE, Qi Y, Hirschberger M (2008) Portfolio selection in the presence of multiple criteria. In: Zopounidis C et al (eds) Handbook of financial engineering. Springer, Berlin, pp 3–24
57. Streichert F, Ulmer H, Zell A (2004) Comparing discrete and continuous genotypes on the constrained portfolio selection problem. In: Proceedings of conference on genetic and evolutionary computation. Lecture notes in computer science, vol 3103, pp 1239–1250

58. Streichert F, Ulmer H, Zell A (2004) Evaluating a hybrid encoding and three crossover operators on the constrained portfolio selection problem. In: Proceedings of congress on evolutionary computation, vol 1. IEEE Press, USA, Portland, pp 932–939
59. Subbu R, Bonissone PP, Eklund N, Bollapragada S, Chalermkraivuth K (2005) Multiobjective financial portfolio design: a hybrid evolutionary approach. IEEE Congr Evol Comput 2:1722–1729
60. Suganya NC, Vijayalakshmi Pai GA (2010) Pareto-archived evolutionary wavelet network for financial constrained portfolio optimization. Intell Syst Account Fin Manag 17:59–90
61. Vedarajan G, Chan L, Goldberg D (1997) Investment portfolio optimization using genetic algorithms. In: Proceedings of the late breaking papers at the genetic programming conference, pp 255–263
62. Xidonas P, Mavrotas G, Psarras J (2010) Equity portfolio construction and selection using multiobjective mathematical programming. J Glob Opt 47:185–209
63. Zeiaee M, Jahed-Motlagh MR (2009) A heuristic approach for value at risk based portfolio optimization. In: Proceedings of the 14th international CSI computer conference (CSICC'09), pp 686–691
64. Zhang Q, Li H (2007) MOEA/D: a multiobjective evolutionary algorithm based on decomposition. IEEE Trans Evol Comput 11:712–731
65. Zhang Q, Li H, Maringer D, Tsang E (2010) MOEA/D with NBI-style Tchebycheff approach for portfolio management. IEEE Congr Evol Comput 1–8. https://doi.org/10.1109/CEC.2010.5586185
66. Zhou Z, Liu X, Xiao H, Wu S, Liu Y (2018) A DEA-based MOEA/D algorithm for portfolio optimization. Cluster Comput. https://doi.org/10.1007/s10586-018-2316-7
67. Zitzler E, Kunzli S (2004) Indicator-based selection in multiobjective search. In: Yao X et al (eds) PPSN 2004, vol 3242. LNCS, Springer, Heidelberg, pp 832–842
68. Zitzler E, Laumanns M, Thiele L (2001) SPEA2: improving the strength Pareto evolutionary algorithm, TIK-103. Department of Electrical Engineering, Swiss Federal Institute of Technology, Zurich, Switzerland
69. Zitzler E, Thiele L, Laumanns M, Fonseca CM, Grunert da Fonseca V (2003) Performance assessment of multiobjective optimizers: an analysis and review. IEEE Trans Evol Comput 7:117–132

Chapter 14
Portfolio Insurance and Intelligent Algorithms

Vasilios N. Katsikis and Spyridon D. Mourtas

Abstract Minimizing portfolio insurance (PI) costs is an investment strategy of great importance. In this chapter, by converting the classical minimum-cost PI (MCPI) problem to a multi-period MCPI (MPMCPI) problem, we define and investigate the MPMCPI under transaction costs (MPMCPITC) problem as a nonlinear programming (NLP) problem. The problem of MCPI gets more genuine in this way. Given the fact that such NLP problems are widely handled by intelligent algorithms, we are introducing a well-tuned approach that can solve the challenging MPMCPITC problem. In our portfolios' applications, we use real-world data and, along with some of the best memetic meta-heuristic and commercial methods, we provide a solution to the MPMCPITC problem, and we compare their solutions to each other.

Keywords Portfolio selection · Multi-period portfolio insurance · Transaction costs · Nonlinear programming · Meta-heuristic optimization

14.1 Introduction

In finance, a portfolio is defined as a group of financial assets that investors hold. Optimizing a portfolio is considered as the optimal allotment procedure of assets, according to specific objective parameters such as returns maximization and costs minimization. Portfolio optimization consequently plays a role of great importance in financial decisions.

In this chapter, we focus exclusively in reducing the insurance costs of a portfolio, which is a way of reducing the portfolio's costs. Portfolio Insurance (PI) refers to a hedging technique employed to mitigate portfolio risks with no need to sell off assets as their valuation falls. Also, PI is based on the risk transfer principle, i.e. protection of one person is the liability of another, and its cost is the mechanism to equilibrate its demand with supply. Because the reduction of costs in portfolios plays a significant role in the outcome, minimizing PI costs is an investment strategy of

---

V. N. Katsikis (✉) · S. D. Mourtas
Department of Economics, Division of Mathematics and Informatics, Kapodistrian University of Athens, Sofokleous 1 Street, 10559 Athens, Greece
e-mail: vaskatsikis@econ.uoa.gr

© The Author(s), under exclusive license to Springer Nature Switzerland AG 2021
S. Patnaik et al. (eds.), *Computational Management*, Modeling and Optimization in Science and Technologies 18, https://doi.org/10.1007/978-3-030-72929-5_14

great importance. Consequently, the minimum-cost PI (MCPI) problem relates to an investment strategy which solves an important cost minimization problem.

In the literature, one can find different approaches of the MCPI problem (see [1–12]). For example, by using quantile method along with expected shortfall criteria as presented in [2], the authors analyze the PI method of constant proportion if the multiple will differ over time and they demonstrate how the multiple can be selected to fulfill the guarantee term, at a given rate of possibility and for different financial market terms. In [9], the authors identify a permissible spectrum of multiplier rates by balancing the risk of deficiency, in which the ideal multiplier rate under the omega ratio lies. In this way, a two-step approach to the numerical optimization of the multiplier, of the constant proportion PI is introduced. In [10], an optimal investment problem under portfolio insurance and short-selling constraints, which is faced by a defined contribution pension fund manager who is averse to losses, is investigated. The task is to maximize the terminal wealth's expected S-shaped utility beyond a minimum guarantee. To solve that problem and derive the representations of the optimal trading strategies and wealth process, the authors apply the dual control method.

A common MCPI problem solving is focused on using linear programming methods. Linear programming is a process in a mathematical model whose requirements for achieving the best result, such as maximum return or lowest cost, are represented by linear relationships. Another common option is focused on using intelligent optimization algorithms. Intelligent optimization algorithms are practical alternative techniques to approach and solve a variety of challenging problems in many scientific fields. An intelligent optimization algorithm is a kind of widely used wavelength selection method that develops an algorithmic model, by mathematical abstraction from the context of biological behavior or material movement process, and then applies iterative calculation to solve problems of combinatorial optimization.

In this chapter, by defining a multi-period version of the MCPI (MPMPCI) problem, we also define and study the MPMPCI under transaction costs (MPMPCITC) problem. We approach the MPMPCITC problem with popular and recently developed meta-heuristic algorithms. More precisely, we compare the effectiveness of the Beetle Antennae Search (BAS) algorithm with two popular meta-heuristic algorithms, specifically the firefly algorithm (FA) and the bat algorithm (BA), and the GA MATLAB function. The advantages of using intelligent heuristic algorithms are that they are faster than the commercial MATLAB optimization functions (such as GA) with similar efficiency and, at the same time, they can be implemented easily in various programming languages. Notice that, finding a successful way to address multi-period financial problems is an ideal technical analysis instrument that can also help investors make smart choices.

The content in this chapter is distributable in 5 sections. Section 14.2 describes the problems of MPMCPI and MPMCPITC. Section 14.3 presents the memetic meta-heuristic algorithms. Section 14.4 includes applications with real-world data on the MPMCPITC problem. Conclusions remarks are provided in Sect. 14.5.

14.2 Portfolio Insurance

One way to decrease a portfolio's spending is to reduce its insurance costs (see [13–20]). For example, a novel strategy that generalizes the CPPI approach is presented in [17]. The main purpose of this strategy is to guarantee the investment target or floor while engaging in asset performance and, at the same time, minimizing the portfolio's downside risk. It is also shown that the strategy accounts for the investor's behavioral aspects, such as skewed likelihoods, risk-averse benefit behavior, and risk-seeking loss behavior. In [18], the authors analyze the efficiency of option-based and constant percentage PI strategies for a defined contribution fund targeting a minimum rate of retirement annuity income covered from inflation. They conclude that their option-based approach typically leads to higher cumulative retirement savings, whereas the constant ratio approach offers a more robust downside risk security to stock market jumps/volatilities.

In this section, we give a concise introduction to the MCPI problem and we define its multi-period version, named MPMCPI. Furthermore, we also define the MPMCPI under transaction costs (MPMCPITC) problem that minimizes as well the transaction costs and, simultaneously, seeks to reach full pay-off while keeping the pay-off over a price floor with respect to time-period which they belong. Note that our approach to the PI problem is resembling the problem described and investigated in several publications, such as [21–27].

The linear space $X = [x_1, x_2, \ldots, x_n] \in \mathbb{R}^{m \times n}$, is called the space of marketed securities, where $x_i \in \mathbb{R}^m$ denotes i-security, $i = 1, 2, \ldots, n$, with information from m successive price records. A vector $\theta = (\theta_1, \theta_2, \ldots, \theta_n)$ of $\mathbb{R}^n$, is called a portfolio, if θ_i implies the i-th security's number of shares. For the classical two period model, by considering θ, which is not zero, as the initial portfolio at the 1-st period, its pay-off at the 2-nd period is the outcome of the following formula:

$$G(\theta) = \sum_{i=1}^{n} \theta_i x_i \tag{14.1}$$

where $x_i \in \mathbb{R}^m$, $i = 1, 2, \ldots, n$, and G is the pay-off operator. Given a "floor" price $\phi \in \mathbb{R}$, the insured pay-off on the portfolio θ in the insurance price p and at the "floor" ϕ is the supremum $G(\theta) \vee \phi \cdot 1$, where $1 \in \mathbb{R}^m$ is a vector of ones, and "$\vee$" signifies the operator of supremum. Based on that, the minimum-cost insured portfolio q is the solution to the MCPI problem below:

$$\min_q p^T \cdot q \tag{14.2}$$

$$\text{subject to } X \cdot q \geq G(\theta) \vee \phi \cdot 1. \tag{14.3}$$

The problem of (14.2)–(14.3) can be formulated in a linear programming (LP) form as below:

$$\min_q p^T \cdot q \qquad (14.4)$$

$$\text{subject to } -X \cdot q \leq \min\{-G(\theta), -\phi \cdot 1\} \qquad (14.5)$$

$$0 \leq q \leq X(1,:) \cdot \theta \cdot \left[\frac{1}{x_1(1)}, \ldots, \frac{1}{x_n(1)}\right]^T, \qquad (14.6)$$

where $0 \in \mathbb{R}^n$ is a zero vector. In addition, the right part of (14.6) is the greatest quantity of each stock that an investor is able to hold by putting all the payout of the portfolio into each stock. Therein, $x_i(1)$ implies the first element of the vector x_i, $i = 1, \ldots, n$.

14.2.1 Multi-period Portfolio Insurance Under Transaction Costs

We consider the MPMCPI problem to be a multi-period version of the MCPI problem. Considering that m is the observations' number of one period then β denotes the observations' number of one sub-period. According to this, the following must hold, $m = \sum_{t=0}^{[m/\beta]} \beta t$, where t is the number of the sub-periods and $[x]$ denotes the integer part of x. Consequently, we transform the space $X \in \mathbb{R}^{m \times n}$ to $X(t) = [x_1(t), x_2(t), \ldots, x_n(t)] \in \mathbb{R}^{\beta \times n}$, where $x_i(t) \in \mathbb{R}^\beta$, comprises of the prices of the i-security, over time t.

Our model consists of $[m/\beta] + 1$ time-periods, and we consider $q(0) = \theta$, and $q(t)$, $t = 1, 2, \ldots, [m/\beta]$, as the requested ones, we have that $q(t) \in \mathbb{R}^n$ for $t = 0, 1, \ldots, [m/\beta]$ time-periods. Moreover, there exists two variants of operator G, the G_1, that works as the G operator, and the G_2, that excludes the previous period's insurance costs. Consequently, we have

$$G_1(q(t-1)) = \sum_{i=1}^n q_i(t-1)x_i(t)$$

and

$$G_2(q(t-1)) = \sum_{i=1}^n q_i(t-1)x_i(t) - \sum_{i=1}^n q_i(t-1)p_i(t-1),$$

where $p(t) = [p_1(t), p_2(t), \ldots, p_n(t)] \in \mathbb{R}^n$ is a vector of multi-period insurance prices and $\sum_{i=1}^n q_i(t-1)p_i(t-1)$ is the insurance cost for the preceding period. Notice that, for consistency reasons with Eq. (14.1), it must hold $\sum_{i=1}^n q_i(0)p_i(0) = 0$, hence, for $t = 1$ we have $G_1 = G_2 = G$.

Given a "floor" price $\phi(t) \in \mathbb{R}$, the insured pay-off on $q(t-1)$ in the insurance price $p(t)$ and at the floor $\phi(t)$ is the supremum $G_2(q(t-1)) \vee \phi(t)$. According to this, the multi-period minimum-cost insured portfolio $q(t)$ solves the following MPMCPI problem:

$$\min_{q(t)} p^T(t) \cdot q(t) \tag{14.7}$$

$$\text{subject to } G_1(q(t)) \geq G_2(q(t-1)) \vee \phi(t) \cdot 1. \tag{14.8}$$

The problem of (14.7)–(14.8) can be formulated in a LP form as below:

$$\min_{q(t)} p^T(t) \cdot q(t) \tag{14.9}$$

$$\text{subject to } -X(t) \cdot q(t) \leq \min\{-\text{pay-off}, -\phi(t) \cdot 1\} \tag{14.10}$$

$$0 \leq q(t) \leq \widehat{X}(t) \cdot q(t-1) \cdot \left[\frac{1}{\hat{x}_1(t)}, \ldots, \frac{1}{\hat{x}_n(t)}\right], \tag{14.11}$$

where pay-off $= \left(X(t) - p^T(t-1)\right) \cdot q(t-1)$ and $\widehat{X}(t) = X((t-1)\beta + 1, :)$. Note that $\widehat{X}(t) = [\hat{x}_1(t), \ldots, \hat{x}_n(t)] \in \mathbb{R}^n$.

Furthermore, we assume that the portfolio's costs are separable as proposed in [28] and we use their multi-period version as introduced in [27]. Thus, we define $\kappa(t)$ to be the sum of the transaction costs related to the atomic exchanges, that is,

$$\kappa(t) = \sum_{i=1}^{n} \kappa_i(t), \tag{14.12}$$

where $\kappa_i(t)$ denotes the i transaction cost in time period t. Our strategy is enhanced to manage composite transaction costs. Considering that ζ^-, ζ^+ are the cost prices associated with the sale and buy of assets i, and δ^-, δ^+, are the fixed costs associated with the sale and buy of assets i, the fixed along with the linear multi-period transaction cost is the following:

$$\kappa_i(t) = \begin{cases} 0, & q_i(t) = q_i(t-1) \\ \delta_i^+ + \zeta_i^+(q_i(t) - q_i(t-1))x_i(t), & q_i(t-1) < q_i(t) \\ \delta_i^- + \zeta_i^-(q_i(t-1) - q_i(t))x_i(t), & q_i(t) < q_i(t-1) \end{cases} \tag{14.13}$$

Notice that this function is explicitly non-convex, with the exception of zero fixed cost condition.

By adding the transaction cost function (14.13) into the problem of (14.9)–(14.11), the MPMCPITC problem can be formulated in a NLP form as follows:

$$\min_{q(t)} p^\mathrm{T}(t) \cdot q(t) + \kappa(t) \tag{14.14}$$

subject to $- X^\mathrm{T}(t) \cdot q(t) \leq \min\{\kappa(t) - \text{pay-off}, -\phi(t) \cdot 1\}$ (14.15)

$$0 \leq q(t) \leq \widehat{X}(t) \cdot q(t-1) \cdot \left[\frac{1}{\hat{x}_1(t)}, \ldots, \frac{1}{\hat{x}_n(t)}\right]. \tag{14.16}$$

Notice that we exclude from the portfolio's pay-off the prior insurance and transaction costs. In this way, the MPMCPITC problem becomes more genuine. Also, note that (14.14) is the objective function of the MPMCPITC problem.

14.3 Intelligent Algorithms

Intelligent algorithms are practical alternative techniques for tackling and solving a variety of challenging problems. Nature has been an inspiration for several meta-heuristic algorithms being introduced. Without being told, but by experience, it has managed to solve problems. The principal inspiration behind the early meta-heuristic algorithms was natural selection and survival of the fittest. Via various communication styles, different species interact with each other. In this section, we present the BAS algorithm in detail and we give some brief introduction to the firefly and bat algorithms. Furthermore, our approach on the MPMCPITC problem is presented.

14.3.1 Beetle Antennae Search Algorithm

The searching behavior of a beetle is described by a metaheuristic algorithm, namely BAS, see [29]. Such a technique permits novel optimisation algorithms to be developed (see [30–36]).

At i-th time, consider the location of beetle as a vector, x_i, $i = 1, 2, \ldots$, and signify the odour concentration at position x to be $f(x)$ defined as a fitness function. The maximum of $f(x)$ refers to the source odour point. The searching behavior model is described by the random path of the beetle searching as described in the following:

$$b = \frac{\mathrm{rnd}(g, 1)}{\|\mathrm{rnd}(g, 1)\|},$$

$$B = \frac{b}{2^{-52} + \|b\|}, \tag{14.17}$$

where rnd(·) implies a random function, and g denotes the dimensions of position. For imitating the movements of the beetle's antennae, the searching behaviors of right-hand (x_R) and left-hand (x_L) side are formulated as follows:

$$x_R = x_i + Bd_i, \tag{14.18}$$

$$x_L = x_i - Bd_i, \tag{14.19}$$

where d is the sensing diameter of antennae equivalent to the exploit ability. Furthermore, the behavior of detecting can be formulated as follows:

$$x_i = x_{i-1} + B\delta_i \text{sign}(f(x_R) - f(x_L)), \tag{14.20}$$

where δ is the search step size that accounts for convergence speed after a decreasing i function, and sign(·) represents a sign function. Lastly, the update rules of d and δ are the followings:

$$d_i = 0.988 d_{i-1} + 0.001, \tag{14.21}$$

$$\delta_i = 0.988 \delta_{i-1}. \tag{14.22}$$

Algorithm 1 Algorithm of Beetle Antennae Search (BAS).

Input: Write an objective function $f(x_i)$, where $x_i = [x_1, x_2, ..., x_n]$, and initialize the parameters x_0, d_0, δ_0, i.
1: **While** $i < K_{max}$ **OR** (stop criterion)
2:   Set the vector unit B in line with (17)
3:   Investigate in variable space with two antennae in line with (18) and (19)
4:   Update the state variable x_i in line with (20)
5:   **if** $f(x_i) < f_{best}$ **then**
6:     Set $f_{best} = f(x_i)$ and $x_{best} = x_i$
7:   **end if**
8:   Update d and δ in line with (21) and (22), respectively
9:   Set $i = i + 1$
10: **end while**
Output: x_{best}, f_{best}.

14.3.2 Popular Meta-Heuristic Algorithms

In this chapter, the effectiveness of BAS is compared with two well known meta-heuristic algorithms as well as the GA MATLAB function. These algorithms are the firefly algorithm (FA) (see [37]) and the bat algorithm (BA) (see [38]).

FA is a swarm-based meta-heuristic algorithm that imitates how fireflies communicate with their flashing lights. The algorithm considers that all fireflies are unisex. That is, any firefly can attract any other firefly. Furthermore, a firefly's attractiveness is directly proportional to its brightness, which rely on the objective function. That is, a firefly will be attracted to a firefly that is brighter.

BA is inspired by the echolocation characteristics of microbats and was introduced in [39]. The majority of microbats are insectivores. Echolocation, which is a type of sonar, is used by microbats to locate their roosting crevices, prevent obstacles, and detect prey in the dark. These bats send out a very loud pulse of sound and listen to the echo reflecting back from the objects around them. The echolocation behavior of Microbats can be structured in a way that it is associated with objective function optimization, and this allows for the formulation of new optimization algorithms.

Of course there are several other algorithm options in the literature we might have used but in this chapter we put the BA and FA against BAS because of their efficiency and originality.

14.3.3 Multi-objective Optimization

The MPMCPITC is an optimization problem which not has only a single objective, since we want to minimize a portfolio's insurance and transaction costs when seeking to reach the highest pay-off while at the same time maintaining the pay-off above a floor price. In this case, we are dealing with a multi-objective optimization problem. Note that all of the meta-heuristic algorithms discussed in this section are specifically applicable to unconstrained optimization as we use their regular form. For this purpose, it is important to use some external approaches that can maintain alternatives in a feasible area. Note that there are many methods to achieve that but, in this chapter, we use the penalty function method presented in [40].

The penalty function method alter the objective function of the problem by adding a penalty function s. Considering that $f(x) = p^T(t) \cdot q(t) + \kappa(t)$, as in (14.14), we optimize $Z(x, R)$ instead of $f(x)$, where

$$Z(x, R) = f(x) + s(R, r(x)). \tag{14.23}$$

In (14.23), R denotes a set of penalties and $r(x)$ represents the inequality constraint functions, (14.15) and (14.16). Since there is more than one penalty function presented in [40], we choose to use

$$s = R\langle r_j(x)\rangle^2, \quad (14.24)$$

where $\langle \cdot \rangle$ implies the bracket-operator with $\langle z \rangle = z$, if z is negative, else $\langle z \rangle = 0$. Furthermore, $r_j(x)$ is a penalty condition handling the j-th inequality constraint. As the infeasible signs are imported with a negative price, the bracket operator is an external penalty condition. This operator is specifically employed to resolve inequality constraints.

In Algorithm 2, the algorithmic process that manages MPMCPITC's equality/inequality constraint is shown. Note that this algorithm include MATLAB code.

Algorithm 2 Penalty function approach on the MPMCPITC problem.

Input: The penalty parameter R; f the objective function (14); A and b the left and right part of (15), respectively; the lower limit q^- and upper limit q^+ of (16).
1: Set $s = R(\text{sum}((A > b).(b - A)^2 + (q^- > q).(q^- - q)^2 + (q > q^+).(q - q^+)^2))$
2: Set $Z = f + s$

Output: The outcome of Z.

14.3.4 The Main Algorithm for the MPMCPITC Problem

First, we introduce the supplementary Algorithm 3 that creates the variables of the MPMCPITC problem of (14.14)–(14.16). Note that all the algorithms in this section include MATLAB code.

Algorithm 3 Algorithm for the construction of (14)-(16) problem's variables.

Input: The marketed space X; the insurance prices p; the time t; the delay parameter β; the floor ϕ; the previous portfolio $q_{-1} = q(t-1)$; and the previous insurance and transaction costs $y_{-1} = y(t-1)$.
1: **function** $[q^-, q^+, A, b, P]$ = problem$(X, p, \phi, q_{-1}, y_{-1}, t, \beta)$
2:   Set $n = \text{size}(X, 2)$ and $A = X((t-1)\beta + 1 : t\beta, :)$
3:   Set pay-off $= -Aq_{-1} + y_{-1}$
4:   Set $b = \min(\text{pay-off}, -\phi(t)\text{ones}(\beta, 1))$
5:   Set $P = p(t)$ and $q^- = \text{zeros}(n, 1)$
6:   Set $q^+ = A(1, :)q_{-1}./A(1, :)'$
7: **end function**

Output: The q^-, q^+, A, b, P at time t.

In the MPMCPITC, we presume that the costs of insurance of the portfolio include a standard price plus a risk rate of the assets. If the price rates related to the asset risk is denoted by ξ and the fixed price by λ, then the function of the fixed and linear insurance prices is composed as below:

$$p_i = \lambda + \xi \cdot \text{Var}\left[\frac{x_i}{\max(x_i)}\right], i = 1, 2, \ldots, n, \tag{14.25}$$

where $\text{Var}[Y]$ implies the variance of Y. Since $x_i \in \mathbb{R}^\beta$, it is the variance of its β normalized prices that measures the risk of the i asset.

The insurance price of any asset relies on the level of risk it bears, where there is a possibility that the expected return will be different from the real return. Hence, the prices of insurance are becoming more realistic.

The following Algorithm 4 is the main algorithm which includes the complete procedure for solving the MPMCPITC problem of (14.14)–(14.16) along with the construction of the insurance prices vector.

Algorithm 4 Main algorithm for the solution of MPMCPITC problem.

Input: The marketed space $X = [x_1, x_2, \ldots, x_n]$ which is a matrix of n time series as column vectors of m prices; the delay parameter $\beta \leq m$, $\beta \in \mathbb{N}$; the price rates ξ and the fixed price λ of (25); the portfolio $\theta \in \mathbb{R}^n$ and the floor vector $\phi \in \mathbb{R}^\beta$.

1: Set $[m, n] = \text{size}(X)$, $s = [m/\beta]$ and $p = \text{zeros}(s, n)$
2: **for** $i = 1:\beta:m$
3:     Set $Y = X(i:i + \beta - 1, :)$
4:     Set $p((i - 1)/\beta + 1, :) = \lambda + \xi \text{var}(Y./\max(Y))$
5: **end for**
6: Set $q = \text{zeros}(n, s)$ and $f_q = \text{zeros}(1, s)$
7: $[q^-, q^+, A, b, P] = \text{problem}(X, p, \phi, \theta, 0, 1, \beta)$
8: Set $q(:, 1)$ the optimal solution of the penalty function of Algorithm 2 by using the variables q^-, q^+, A, b, P of the previous step via any intelligent algorithm
9: Set $f_q(1)$ the outcome of (14) for $q(:, 1)$
10: **for** $t = 2:s$
11:     $[q^-, q^+, A, b, P] = \text{problem}(X, p, \phi, x(:, t - 1), y(:, t - 1), t, \beta)$
12:     Set $q(:, t)$ the optimal solution of the penalty function of Algorithm 2 by using the variables q^-, q^+, A, b, P of the previous step via any intelligent algorithm
13:     Set $f_q(t)$ the outcome of (14) for $q(:, t)$
14: **end for**

Output: The optimal solution of the MPMCPITC problem, $q(t)$ and $f_q(t)$.

14.4 Portfolios' Applications with Real-World Data

In this section, 4 applications are presented in portfolios comprise of 4, 8, 12 and 16 stocks, respectively. For all the applications, the results have been produced from BAS, BA, FA and GA, where we have set

- $\delta_i^+ = \delta_i^- = 0.1$, $\zeta_i^+ = 0.06$ and $\zeta_i^- = 0.04$ in (14.13),
- $R = 1e5$ in Algorithm 2,
- $d_0 = 2$, $\delta_0 = 2$ and $K_{max} = 1200$ in BAS,
- population size 100, max iterations 1000, parameter alpha 0.97, parameter gamma 0.1, initial loudness 1 and initial pulse rate 1 in BA,
- population size 100, randomness reduction factor 0.97, absorption coefficient 0.01, attractiveness constant 1, randomness strength 1 and maximum number of iterations 1000 in FA,
- GA MATLAB function in its default settings.

In addition, the portfolios stocks are listed in Table 14.1 by their ticker symbol.

14.4.1 Application in 4 Stocks' Portfolio

Assume that the marketed space is $X = [x_1, \ldots, x_4]$, according to the daily stocks close prices of Group A in Table 14.1 for the period 1/7/2020-4/8/2020. Because this period has 24 observations, $X \in \mathbb{R}^{24 \times 4}$. By setting $\beta = 6$, we have a 5 period model and, hence, $X(t) \in \mathbb{R}^{6 \times 4}$ with $t \in [1, 4]$. Moreover, given a portfolio $\theta = [10, 2, 2, 1]^T$, we set the floor $\phi = [580, 600, 620, 730]^T$ and the parameters $\xi = 10$, $\lambda = 7$ in (14.25). The results are presented in Fig. 14.1 where:

- Fig. 14.1a depicts the portfolio's $q(t)$ pay-off, which is $G_1(q(t)) = X(t) \cdot q(t)$, produced by BAS, BA, FA and GA, along with the floor price.
- Figure 14.1b displays the portfolio's $q(t)$ cost, which is the result of (14.14), produced by BAS, BA, FA and GA. Notice that the portfolio's cost is the sum of insurance and transaction cost.

Table 14.2 indicates the time usage of BAS, BA, FA and GA, in App. 4.1.

Table 14.1 Market vector stocks, $X(t)$, which are divided in four groups and used in the MPMCPI problems of Apps. 4.1-4.4

Group	Market			
A	AMD	AAL	F	GE
B	NIO	PINS	VALE	SRNE
C	DKNG	INO	UAL	WFC
D	CCL	HRB	INTC	UBER

Fig. 14.1 The pay-off, which is the outcome of $G_1(q(t))$, and the cost, which is the outcome of (14.14), of a 4 stocks' portfolio in App. 4.1

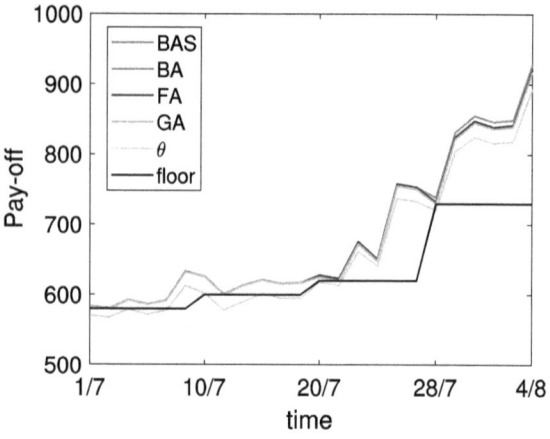

(a) Pay-off of Portfolios.

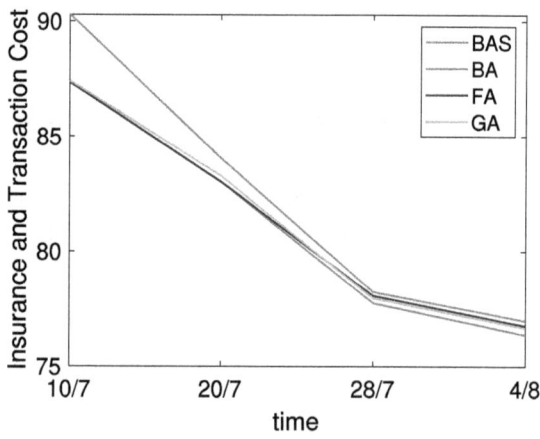

(b) Cost of Portfolios.

Table 14.2 Applications 4.1-4.4 execution time for solving the MPMCPI problem with BAS, BA, FA and GA

Application	BAS	BA	FA	GA
4.1	0.1 s	2.2 s	95 s	4 s
4.2	0.2 s	3.5 s	160 s	150 s
4.3	0.2 s	3.5 s	166 s	160 s
4.4	0.2 s	6 s	276 s	232 s

14.4.2 Application in 8 Stocks' Portfolio

Assume that the marketed space is $X = [x_1, \ldots, x_8]$, according to the daily stocks close prices of Group A and Group B in Table 14.1 for the period 7/5/2020-31/7/2020. Because this period has 60 observations, $X \in \mathbb{R}^{60 \times 8}$. By setting $\beta = 10$, we have a 7 period model and, hence, $X(t) \in \mathbb{R}^{10 \times 8}$ with $t \in [1, 6]$. Moreover, given a portfolio $\theta = [7, 2, 1, 1, 2, 4, 0, 1]^T$, we set the floor $\phi = [530, 570, 550, 570, 580, 600]^T$ and the parameters $\xi = 10$, $\lambda = 5$ in (14.25). The results are presented in Fig. 14.2 where:

- Fig. 14.2a depicts the portfolio's $q(t)$ pay-off, which is $G_1(q(t)) = X(t) \cdot q(t)$, produced by BAS, BA, FA and GA, along with the floor price.

Fig. 14.2 The pay-off, which is the outcome of $G_1(q(t))$, and the cost, which is the outcome of (14.14), of a 8 stocks' portfolio in App. 4.2

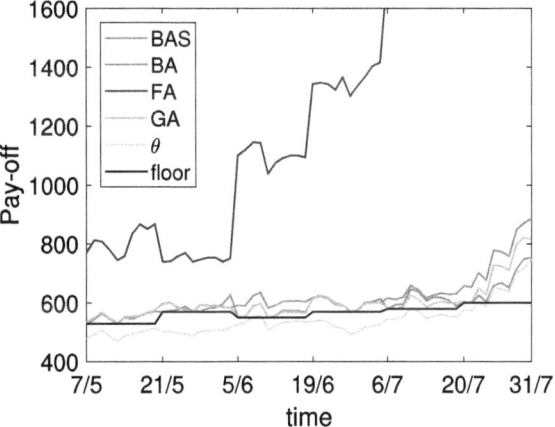

(a) Pay-off of Portfolios.

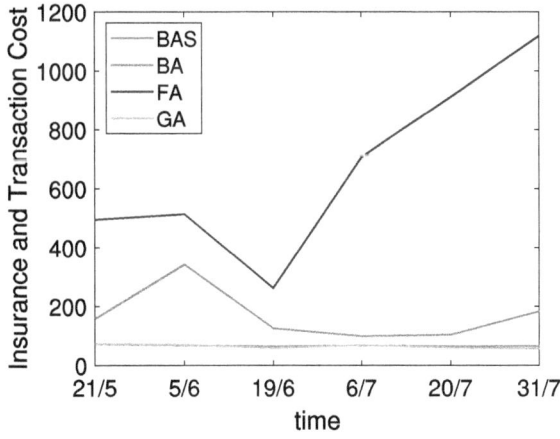

(b) Cost of Portfolios.

- Figure 14.2b displays the portfolio's $q(t)$ cost, which is the outcome of (14.14), produced by BAS, BA, FA and GA. Notice that the portfolio's cost is the sum of insurance and transaction cost.

Table 14.2 indicates the time usage of BAS, BA, FA and GA, in App. 4.2.

14.4.3 Application in 12 Stocks' Portfolio

Assume that the marketed space is $X = [x_1, \ldots, x_{12}]$, according to the daily stocks close prices of Group A, Group B and Group C, in Table 14.1 for the period 25/3/2020-31/7/2020. Because this period has 60 observations, $X \in \mathbb{R}^{90 \times 12}$. By setting $\beta = 15$, we have a 7 period model and, hence, $X(t) \in \mathbb{R}^{15 \times 12}$ with $t \in [1, 6]$. Moreover, given a portfolio $\theta = 3\text{ones}(1, 12)$, we set the floor $\phi = [700, 650, 680, 720, 750, 780]^T$ and the parameters $\xi = 10, \lambda = 5$ in (14.25). The results are presented in Fig. 14.3 where:

- Fig. 14.3a depicts the portfolio's $q(t)$ pay-off, which is $G_1(q(t)) = X(t) \cdot q(t)$, produced by BAS, BA, FA and GA, along with the floor price.
- Figure 14.3b displays the portfolio's $q(t)$ cost, which is the outcome of (14.14), produced by BAS, BA, FA and GA. Notice that the portfolio's cost is the sum of insurance and transaction cost.

Table 14.2 indicates the time usage of BAS, BA, FA and GA, in App. 4.3.

14.4.4 Application in 16 Stocks' Portfolio

Assume that the marketed space is $X = [x_1, \ldots, x_{16}]$, where X comprises of the daily stocks close prices of Group A, B, C and D in Table 14.1 for the period 10/3/2020-30/7/2020. Because this period has 100 observations, $X \in \mathbb{R}^{100 \times 16}$. By setting $\beta = 10$, we have a 11 period model and, hence, $X(t) \in \mathbb{R}^{10 \times 16}$ with $t \in [1, 10]$. Moreover, given a portfolio $\theta = \text{ones}(1, 16)$, we set the floor $\phi = [840 : 40 : 1200]^T$ and the parameters $\xi = 10, \lambda = 5$ in (14.25). The results are presented in Fig. 14.4 where:

- Fig. 14.4a depicts the portfolio's $q(t)$ pay-off, which is $G_1(q(t)) = X(t) \cdot q(t)$, produced by BAS, BA, FA and GA, along with the floor price.
- Figure 14.4b displays the portfolio's $q(t)$ cost, which is the outcome of (14.14), produced by BAS, BA, FA and GA. Notice that the portfolio's cost is the sum of insurance and transaction cost.

Table 14.2 indicates the time usage of BAS, BA, FA and GA, in App. 4.4.

Fig. 14.3 The pay-off, which is the outcome of $G_1(q(t))$, and the cost, which is the outcome of (14.14), of a 12 stocks' portfolio in App. 4.3

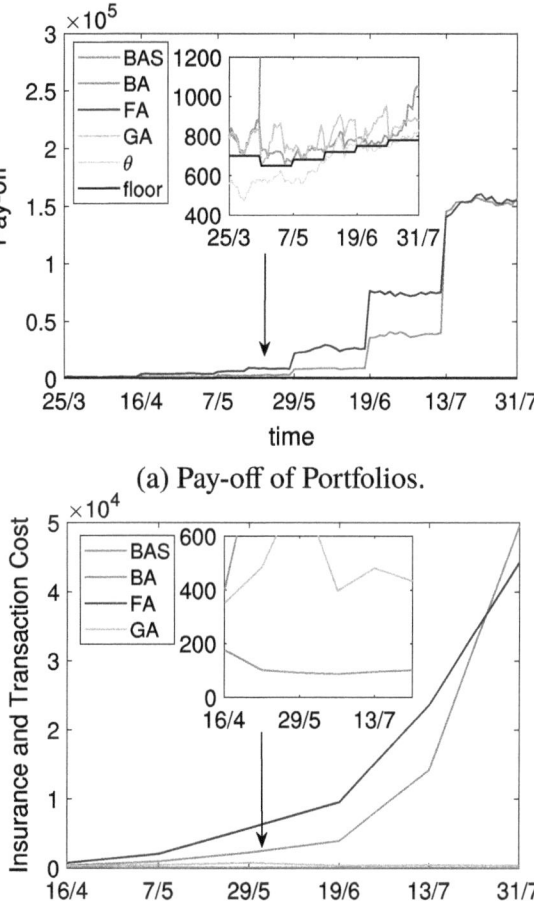

(a) Pay-off of Portfolios.

(b) Cost of Portfolios.

14.4.5 Results and Performance Comparison of BAS, BA, FA and GA

The MPMCPI problem minimizes (14.14) and, at the same time, keeps the portfolio's pay-off above a floor price. Based on that, in Figs. 14.1a–14.4a, we observe that the outcomes of BAS, BA, FA and GA keep the portfolio's pay-off, which is the outcome of $G_1(q(t))$, above the floor price in all the applications of this section. Furtremore, Figs. 14.1b-14.4b show the value of (14.14) in the Apps. 14.1-14.4, respectively.

More precisely, in Fig. 14.1a, b, we observe that the outcomes of BAS, BA, FA and GA, for the 4 stocks' portfolio described in App. 4.1, are pretty close in solving the MPMCPI problem. For the 8 stocks' portfolio described in App. 4.2, we note in

Fig. 14.4 The pay-off, which is the outcome of $G_1(q(t))$, and the cost, which is the outcome of (14.14), of a 16 stocks' portfolio in App. 4.4

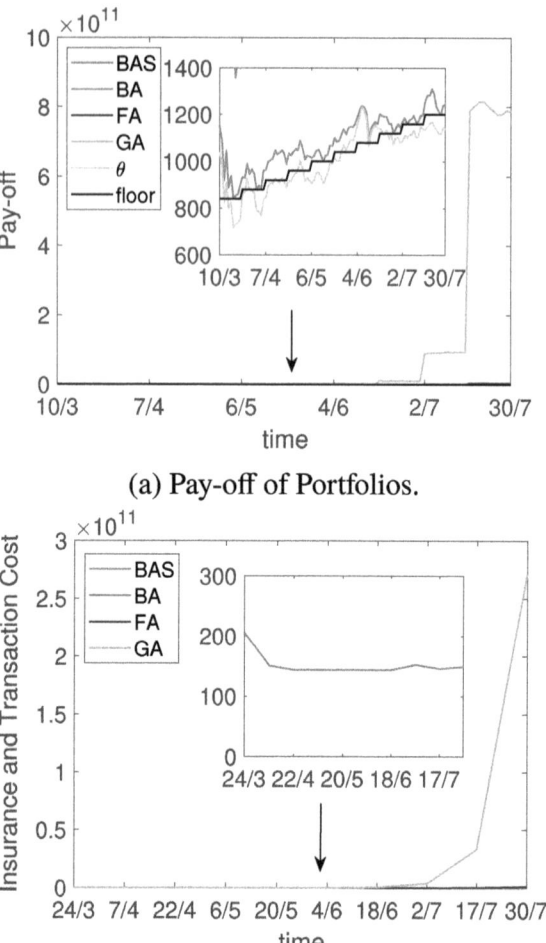

(a) Pay-off of Portfolios.

(b) Cost of Portfolios.

Figs. 14.2a and 14.2b that the outcomes of BAS and GA are reasonably similar and are also better than the outcome of BA. Moreover, the outcome of FA is the worst. For the 12 stocks' portfolio described in App. 4.3, we observe in Figs. 14.3a, b that the outcome of BAS is better than the outcome of GA. In addition, the outcomes of BA and FA are close and the worst. For the 16 stocks' portfolio described in App. 4.4, we observe in Figs. 14.4a, b that the outcome of BAS is much better than the outcomes of BA, FA and GA. Note that when the portfolios costs are high in Figs. 14.1b–14.4b, the portfolio's pay-off in Figs. 14.1a–14.4a, are high too. However, given that minimizing the portfolios costs is the main task of the MPMCPI problem, the higher the costs in Figs. 14.1b–14.4b, the worse the method's solution is.

The time usage of BAS, BA, FA and GA, in Apps. 4.1-4.4, is presented in Table 14.2, which includes their average execution time. Therein, we note that BAS is always close to 20 times faster than BA, when BA is always close to 45 times faster than FA. Furthermore, the time consumptions of FA and GA are always close, except from the App. 4.1, which deals with a 4 stocks' portfolio.

Overall, BAS produces the same or better results than BA, FA and GA in Apps. 4.1-4.4 and, at the same time, BAS is far less time-consuming. Based on that, we can conclude that in the MPMCPI problem, BAS is more efficient than BA, FA and GA.

14.5 Conclusion

This chapter introduces a multi-period version of the MCPITC problem, namely MPMCPITC, and shows its solution. The BAS efficacy in a financial NLP problem has been demonstrated in four applications. Each of these applications solve the MPMCPITC problem by using real-world data in different time periods. Furthermore, the portfolios that they optimize vary in size. We concluded, according to our portfolios' applications results and their time consumption, that the BAS approach yields an exceptional solution to the problem of MPMCPITC. Finally, the application of BAS to the MPMCPITC problem is also very antagonistic, or even better alternative, to the BA, FA and GA approaches.

References

1. Annaert J, Van Osselaer S, Verstraete B (2009) Performance evaluation of portfolio insurance strategies using stochastic dominance criteria. J Bank Finance 33(2):272–280
2. Ameur HB, Prigent JL (2014) Portfolio insurance: Gap risk under conditional multiples. Eur J Oper Res 236(1):238–253
3. Katsikis VN, Mourtas SD (2020) Optimal portfolio insurance under nonlinear transaction costs. Journal of Modeling and Optimization. 12(2):117–124
4. Barucci, E., Fontana, C.: Portfolio, Insurance and Saving Decisions. In: Financial Markets Theory. Springer, London. 55–121 (2017)
5. Heckel T, Soupé F, de Carvalho RL (2016) Portfolio insurance with adaptive protection. Journal of Investment Strategies. 5:1–15
6. Annaert J, Ceuster MD, Vandenbroucke J (2019) Mind the Floor: Enhance Portfolio Insurance without Borrowing. The Journal of Investing. 28:39–50
7. George, J., Trainor, W.: Portfolio insurance using leveraged ETFs. Available at SSRN 3055199 (2017)
8. Maalej H, Prigent JL (2016) On the stochastic dominance of portfolio insurance strategies. Journal of Mathematical Finance. 6(1):14
9. Biedova, O., Steblovskaya, V.: Multiplier optimization for constant proportion portfolio insurance (cppi) strategy. International Journal of Theoretical and Applied Finance (IJTAF), World Scientific Publishing Co. Pte. Ltd. **23**, 1–22 (2020)
10. Dong, Y., Zheng, H.: Optimal investment of DC pension plan under short-selling constraints and portfolio insurance. Insurance: Mathematics and Economics. **85**, 47–59 (2020)
11. Persio, L.D., Oliva, I., Wallbaum, K.: Options on constant proportion portfolio insurance with guaranteed minimum equity exposure. Appl Stochastic Models Bus Ind. 1–15 (2020)

12. Dichtl H, Drobetz W, Wambach M (2017) A bootstrap-based comparison of portfolio insurance strategies. The European Journal of Finance. 23(1):31–59
13. Baofeng, Y., Hailong, L.: Research of dynamical portfolio insurance strategies in shanghai security market. Management Review. **7** (2005)
14. El-Adaway IH, Kandil AA (2010) Construction risks: single versus portfolio insurance. Journal of Management in Engineering. 26(1):2–8
15. Hohmann, R.: Constant-proportion-portfolio-insurance, In: Portfolio Insurance reloaded. Springer. 9–23 (2018)
16. Hohmann, R.: Portfolio Insurance Reloaded. Springer (2018)
17. Escobar M, Lichtenstern A, Zagst R (2020) Behavioral portfolio insurance strategies. Fin Markets Portfolio Mgmt 34:353–399
18. Xu, M., Sherris, M., Shao, A.W.: Portfolio insurance strategies for a target annuitization fund. Available at SSRN 3417818 (2019)
19. Asano T, Osaki Y (2020) Portfolio allocation problems between risky and ambiguous assets. Ann Oper Res 284:63–79
20. Bertrand, P., luc Prigent, J.: Equilibrium of financial derivative markets under portfolio insurance constraints. Economic Modelling. **52**, 278–291 (2016)
21. Aliprantis CD, Brown DJ, Werner J (2000) Minimum-cost portfolio insurance. Journal of Economic Dynamics & Control. 24:1703–1719
22. Katsikis VN (2007) Computational methods in portfolio insurance. Appl Math Comput 189(1):9–22
23. Katsikis VN (2008) Computational methods in lattice-subspaces of C[a, b] with applications in portfolio insurance. Appl Math Comput 200:204–219
24. Katsikis, V.N.: Computational and mathematical methods in portfolio insurance - A MATLAB-based approach. In: Matlab - Modelling, Programming and Simulations. IntechOpen (2010)
25. Katsikis VN, Mourtas SD (2019) A heuristic process on the existence of positive bases with applications to minimum-cost portfolio insurance in C[a, b]. Appl Math Comput 349:221–244
26. Katsikis, V.N., Mourtas, S.D.: ORPIT: A Matlab Toolbox for Option Replication and Portfolio Insurance in Incomplete Markets. Computational Economics. (2019)
27. Katsikis VN, Mourtas SD, Stanimirović PS, Li S, Cao X (2020) Time-varying minimum-cost portfolio insurance under transaction costs problem via beetle antennae search algorithm (BAS). Appl Math Comput 385:
28. Lobo MS, Fazel M, Boyd S (2007) Portfolio optimization with linear and fixed transaction costs. Ann Oper Res 152:341–365
29. Jiang, X., Li, S.: BAS: Beetle Antennae Search Algorithm for Optimization Problems. arXiv preprint. abs/1710.10724 (2017)
30. Wu Q, Lin H, Jin Y, Chen Z, Li S, Chen D (2020) A new fallback beetle antennae search algorithm for path planning of mobile robots with collision-free capability. Soft Comput 24:2369–2380
31. Wu Q, Shen X, Jin Y, Chen Z, Li S, Khan AH, Chen D (2019) Intelligent Beetle Antennae Search for UAV Sensing and Avoidance of Obstacles. Sensors. 19:1758
32. Khan, A.T., Cao, X., Li, S., Hu, B., Katsikis, V.N.: Quantum beetle antennae search: A novel technique for the constrained portfolio optimization problem. SCIENCE CHINA Information Sciences (2020)
33. Khan AH, Cao X, Li S, Katsikis VN, Liao L (2020) BAS-ADAM: an ADAM based approach to improve the performance of beetle antennae search optimizer. IEEE/CAA Journal of Automatica Sinica. 7(2):461–471
34. Jiang X, Lin Z, He T, Ma X, Ma S, Li S (2020) Optimal path finding with beetle antennae search algorithm by using ant colony optimization initialization and different searching strategies. IEEE Access. 8:15459–15471
35. Medvedeva, M.A., Katsikis, V.N., Mourtas, S.D., Simos, T.E.: Randomized time-varying knapsack problems via binary beetle antennae search algorithm: Emphasis on applications in portfolio insurance. Math Meth Appl Sci. 1–11 (2020)

36. Khan, A.H., Cao, X., Katsikis, V.N., Stanimirovic, P., Brajevic, I., Li, S., Kadry, S., Nam, Y.: Optimal Portfolio Management for Engineering Problems Using Nonconvex Cardinality Constraint: A Computing Perspective. IEEE Access. 1–1 (2020)
37. Yang X (2010) Firefly algorithm, stochastic test functions and design optimisation. Int. J. Bio Inspired Comput. 2(2):78–84
38. Yang, X.: Nature-inspired optimization algorithms. Elsevier (2014)
39. Yang, X.: A new metaheuristic bat-inspired algorithm. In: González, J.R., Pelta, D.A., Cruz, C., Terrazas, G., Krasnogor, N. (eds) Nature Inspired Cooperative Strategies for Optimization, (NICSO 2010). Studies in Computational Intelligence. Springer, Berlin, Heidelberg. **284**, 65–74 (2010)
40. Deb, K.: Optimization for Engineering Design: Algorithms and Examples. PHI. second ed. (2013)

Chapter 15
On Interval-Valued Multiobjective Programming Problems and Vector Variational-Like Inequalities Using Limiting Subdifferential

B. B. Upadhyay and Priyanka Mishra

Abstract This article deals with the generalized vector variational-like inequalities, namely Minty and Stampacchia vector variational-like inequalities for interval-valued functions and interval-valued multiobjective programming problem. Under strong LU-invexity hypothesis, we establish equivalence among the solutions of considered vector variational-like inequalities and LU-efficient minimizers of order s of interval-valued multiobjective programming problem. Moreover, we also considered the weak forms of the considered vector variational-like inequalities and derive the equivalence between the solutions of the vector variational-like inequalities and the LU-strict minimizers of order s of interval-valued multiobjective programming problem. Furthermore, with the help of KKM-Fan theorem, we derive certain conditions under which the solutions of vector variational-like inequalities exist.

Keywords Multiobjective programming problems · Strong LU-invexity · LU-efficient minimizers

15.1 Introduction

The assumption of convexity is often too strong for the applications in optimization theory, economics, and probability theory, see [11, 31, 33]. Due to wider applications of convex function, many researchers have shown their interest in the study of the special class of functions, which possesses several properties of convex functions. In this aspect, Hanson [14] introduced a class of nonconvex functions, which was later named as invex functions by Craven [5]. Invex function has various applications in the field of variational inequality problems and nonlinear optimization, for more exposition, see [3, 25, 39]. The notion of strongly η-invex functions are introduced by Jeyakumar and Mond [17]. The concept of invexity was extended by Reiland [32] for nonsmooth functions.

B. B. Upadhyay (✉) · P. Mishra
Department of Mathematics, Indian Institute of Technology Patna, Patna, Bihar 801106, India
e-mail: bhooshan@iitp.ac.in

Interval-valued optimization is one of the deterministic optimization models to deal with the inexact, imprecise or uncertain data. In interval-valued multiobjective programming problem (for short (IVMPP)), the coefficients of objective and constraint functions are compact intervals. To deal with the functions with interval coefficients, Moore [26, 27] introduced the concept of interval analysis. In many real world problems, the algorithms for solving multiobjective programming problems (for short, (MPP)) terminate after finite steps and we get approximate solutions only. Therefore, in computational as well as analytical point of view, the study of approximate solutions becomes very useful. Many authors have been studied several variants of approximate efficient solution, see [7, 12, 34, 36, 38]. During the study of the convergence of numerical techniques, Cromme [6] defined the notion of strict local minimizers. The notion of strict local minimizer was extended by Auslender [2] to higher order strict local minimizer. Jiménez [18] gave the concept of strict minimizer of order s, for multiobjective programming problems.

For finite dimensional spaces, Giannessi [8] defined the concept of vector variational inequality. In literature, several researchers studied the applications of Minty vector variational-like inequality (for short (MVVLI)) and the Stampacchia vector variational-like inequality (for short (SVVLI)) for (IVMPP), for more details, we refer to [1, 4, 9, 10, 13, 15, 20, 22, 24, 37]. Li and Yu [21] showed the equivalence among the solutions of (MVVLI), (SVVLI) and multiobjective programming problem (for short (MPP)) for directionally differentiable invex functions. Oveisiha and Zafarani [30] showed the equivalence between the solution of (MVVLI) and (IVMPP) using η-invex functions with limiting subdifferential. Zhang et al. [41] established the relations between the solutions of (IVMPP), (GMVVI) and (GSVVI). Upadhyay and Mishra [35] proved the relationship among the solutions of (GMVVLI), (GSVVLI) and (IVMPP) using Clarke subdifferential.

Motivated by the works of [23, 34–36], we consider a class of (GMVVLI), (GSVVLI) with its weaker form (WGMVVLI), (WGSVVLI), respectively in terms of limiting subdifferentials for interval-valued functions and a class of (IVMPP). We showed the equivalence between the solutions of (GMVVLI), (GSVVLI) and LU-efficient minimizer of order s with respect to (for abbreviation w.r.t.) ϑ of (IVMPP). Furthermore, we obtain existence results for the solutions of (GMVVLI) and (GSVVLI).

The organization of this paper is given as: In Sect. 15.2, some basic definitions and preliminaries are given. We proved the equivalence between the solutions of (GMVVLI), (GSVVLI) and the LU-efficient minimizer of order s w.r.t. ϑ of (IVMPP) in Sect. 15.3. In Sect. 15.4, we showed the equivalence between the solutions of (WGMVVLI), (WGSVVLI) and LU-strict minimizer of order s w.r.t. ϑ of (IVMPP). In Sect. 15.5, we derive existence results for the solutions of (GMVVLI), (GSVVLI).

15.2 Definitions and Preliminaries

Let $\emptyset \neq \Omega \subseteq \mathbb{R}^n$ and $\vartheta : \Omega \times \Omega \to \mathbb{R}^n$ be a vector valued function. Let $\Psi : \Omega \to \mathbb{R}$ be lower semicontinuous function.

For $u, y \in \mathbb{R}^n$, following notions for equality and inequalities will be used throughout the sequel:

(i) $u = y$, $\iff u_k = y_k$, $\forall k = 1, 2, \ldots, n$;
(ii) $u < y$, $\iff u_k < y_k$, $\forall k = 1, 2, \ldots, n$;
(iii) $u \leqq y$, $\iff u_k \leq y_k$, $\forall k = 1, 2, \ldots, n$;
(iv) $u \leq y$, $\iff u_k \leq y_k$, $\forall k = 1, 2, \ldots, n$, $k \neq l$, and $u_l < y_l$ for some l.

The following notions of interval analysis from Moore [26].

Let $\mathcal{I}$ and $\mathcal{I}^+$ denotes the classes of all closed intervals in $\mathbb{R}$ and $\mathbb{R}^+$, respectively. Let $B = [b^L, b^U] \in \mathcal{I}$ be a closed interval, where b^L and b^U denotes the lower and upper bounds of B, respectively. If $b^L = b^U$ then B reduces to a real number.

For $B = [b^L, b^U]$, $D = [d^L, d^U] \in \mathcal{I}$, we have

(i) $B + D = [b^L + d^L, b^U + d^U]$;
(ii) $-B = [-b^U, -b^L]$;
(iii) $B \times D = [\min_{bd}, \max_{bd}]$, where $\min_{bd} = \min\{b^L d^L, b^L d^U, b^U d^L, b^U d^U\}$ and $\max_{bd} = \max\{b^L d^L, b^L d^U, b^U d^L, b^U d^U\}$.

Then, we can show that

$$B - D = B + (-D) = [b^L - d^U, b^U - d^L],$$

$$kB = \{kb : b \in B\} = \begin{cases} [kb^L, kb^U], & k \geq 0, \\ |k|[-b^U, -b^L], & k < 0, \end{cases} \quad (15.1)$$

where, $k \in \mathbb{R}$.

Let $B = [b^L, b^U]$, $D = [d^L, d^U] \in \mathcal{I}$, then we define

1. $B \preceq_{LU} D \iff b^L \leq d^L$ and $b^U \leq d^U$,
2. $B \prec_{LU} D \iff B \preceq_{LU} D$ and $B \neq D$.

Remark 1 $B = [b^L, b^U]$, $D = [d^L, d^U] \in \mathcal{I}$ are comparable iff $B \preceq_{LU} D$ or $B \succeq_{LU} D$. B and D are not comparable if one of the following holds:

$b^L \leq d^L$ and $b^U > d^U$; $b^L < d^L$ and $b^U \geq d^U$; $b^L < d^L$ and $b^U > d^U$;

$b^L \geq d^L$ and $b^U < d^U$; $b^L > d^L$ and $b^U \leq d^U$; $b^L > d^L$ and $b^U < d^U$.

Consider an interval-valued vector $\mathbf{B} = (B_1, \ldots, B_p)$, where each component $B_k = [b_k^L, b_k^U]$, $k = 1, 2, \ldots, p$ is a closed interval. Let two interval-valued vectors $\mathbf{B}$ and $\mathbf{D}$ be such that, B_k and D_k are comparable for each $k = 1, 2, \ldots, p$, then

1. $\mathbf{B} \preceq_{LU} \mathbf{D}$ iff $B_k \preceq_{LU} D_k$ for each $k = 1, \ldots, p$;
2. $\mathbf{B} \prec_{LU} \mathbf{D}$ iff $B_k \preceq_{LU} D_k$ for each $k = 1, \ldots, p$, and $B_r \prec_{LU} D_r$ for some r.

The function $g : \mathbb{R}^n \to \mathcal{I}$ is called an interval-valued function, if $g(z) = [g^L(z), g^U(z)]$, where $g^L, g^U : \mathbb{R}^n \to \mathbb{R}$ satisfying $g^L(z) \leq g^U(z)$, for all $z \in \mathbb{R}^n$.

Now, we recall the following concepts related with limiting subdifferential from [28].

Definition 1 The Fréchet subdifferential of Ψ at $y \in \Omega$, is defined as

$$\hat{\partial}\Psi(y) := \left\{ \xi \in \mathbb{R}^n : \liminf_{y' \to y} \frac{\Psi(y') - \Psi(y) - \langle \xi, y' - y \rangle}{\|y' - y\|} \geq 0 \right\}.$$

Definition 2 The limiting subdifferential of Ψ at $\tilde{w} \in \Omega$, is defined as

$$\partial_M \Psi(\tilde{w}) := \limsup_{y \to \tilde{w}} \hat{\partial}\Psi(y),$$

where, lim sup is the Painlevé-Kuratowski outer limit.

Definition 3 [25] The function $\Psi : \Omega \to \mathbb{R}$ is locally Lipschitz on Ω, if for each $\tilde{y} \in \Omega$, there exist constants $L, \delta > 0$, s.t. for every $y_1, y_2 \in B(\tilde{y}; \delta) \cap \Omega$, one has

$$|\Psi(y_1) - \Psi(y_2)| \leq L\|y_1 - y_2\|.$$

Definition 4 [25] The set Ω is an invex set w.r.t. ϑ, if for every $y_1, y_2 \in \Omega$, one has

$$y_2 + \mu\vartheta(y_1, y_2) \in \Omega, \quad \forall \mu \in [0, 1].$$

From now onwards, Ω is an invex set w.r.t. ϑ, unless otherwise specified.

Definition 5 The function $\Psi : \Omega \to \mathbb{R}$ is strongly preinvex of order s, if we can get a scalar $\beta > 0$, such that, for all $y_2, y_1 \in \Omega$ and $\mu \in [0, 1]$, one has

$$\Psi(y_2 + \mu\vartheta(y_1, y_2)) \leq (1 - \mu)\Psi(y_2) + \mu\Psi(y_1) - \mu(1 - \mu)\beta\|\vartheta(y_1, y_2)\|^s.$$

Definition 6 The function $\Psi : \Omega \to \mathbb{R}$ is strongly invex of order s, if we can get a scalar $\beta > 0$, such that, for all $y_2, y_1 \in \Omega$, one has

$$\Psi(y_2) - \Psi(y_1) \geq \langle \xi, \vartheta(y_2, y_1) \rangle + \beta\|\vartheta(y_2, y_1)\|^s, \quad \forall \xi \in \partial_M \Psi(y_1).$$

Definition 7 The interval-valued function $\Psi : \Omega \to \mathcal{I}$ is strongly LU-invex of order s, iff the real valued functions $\Psi^L(z)$ and $\Psi^U(z)$ are strongly invex of order s.

Definition 8 [19] The setvalued map $\Gamma : \Omega \to 2^\Omega$ is strongly invariant monotone of order s, if we can get a scalar $\beta > 0$, such that

$$\langle \xi, \vartheta(y_2, y_1)\rangle + \langle \zeta, \vartheta(y_1, y_2)\rangle \leq -\beta \{||\vartheta(y_2, y_1)||^s + ||\vartheta(y_1, y_2)||^s\},$$

for any $y_1, y_2 \in \Omega$, and any $\xi \in \Gamma(y_1), \zeta \in \Gamma(y_2)$.

Condition A. [40] Let $\Psi : \Omega \to \mathbb{R}$, then

$$\Psi(y_2 + \vartheta(y_1, y_2)) \leq \Psi(y_1), \quad \forall y_1, y_2 \in \Omega.$$

Condition C. [29] Let $\vartheta : \Omega \times \Omega \to \mathbb{R}^n$, then for all $y_2, y_1 \in \Omega$, $\mu \in [0, 1]$, one has

(i) $\vartheta(y_1, y_1 + \mu\vartheta(y_2, y_1)) = -\mu\vartheta(y_2, y_1),$
(ii) $\vartheta(y_2, y_1 + \mu\vartheta(y_2, y_1)) = (1 - \mu)\vartheta(y_2, y_1).$

Remark 2 Yang et al. [40] have proved that, for all $\mu_1, \mu_2 \in [0, 1]$, one has

$$\vartheta(y_1 + \mu_2\vartheta(y_2, y_1), y_1 + \mu_1\vartheta(y_2, y_1)) = (\mu_2 - \mu_1)\vartheta(y_2, y_1).$$

Lemma 1 [16, 35] Let Ψ be strongly invex of order s, then $\partial_M \Psi$ is strongly invariant monotone of order s, that is

$$\langle \xi, \vartheta(y_1, y_2)\rangle + \langle \zeta, \vartheta(y_2, y_1)\rangle \leq -\beta \{||\vartheta(y_1, y_2)||^s + ||\vartheta(y_2, y_1)||^s\},$$

for all $y_2, y_1 \in \Omega$, $\xi \in \partial_M \Psi(y_2)$ and $\zeta \in \partial_M \Psi(y_1)$.

Theorem 1 [28] Let Ψ be Lipschitz on an open set containing $[y_2, y_1]$ in Ω. Then, one has

$$\Psi(y_1) - \Psi(y_2) \leq \langle \xi, y_1 - y_2\rangle, \text{ for some } \xi \in \partial_M \Psi(u); \ u \in [y_2, y_1[.$$

Lemma 2 [35] Let Ψ be strongly invex of order s on Ω, such that ϑ satisfy the Condition C, then Ψ is strongly preinvex of order s on Ω.

We study the following interval-valued multiobjective programming problem:

$$\text{(IVMPP)} \quad \text{Minimize} \quad \boldsymbol{\Psi}(y) = (\Psi_1(y), \ldots, \Psi_p(y))$$
$$\text{subject to } y \in \Omega,$$

where $\Psi_k = [\Psi_k^L, \Psi_k^U] : \Omega \to \mathcal{I}$ and $\Psi_k^L, \Psi_k^U : \Omega \to \mathbb{R}$, $k \in \mathcal{K} := \{1, 2, \ldots, p\}$ are lower semicontinuous functions on Ω.

The following notions of LU-efficient minimizer and LU-strict minimizer are the adaptations of the notions of efficient minimizer and strict minimizer given in Upadhyay and Mishra [35].

Definition 9 The point $\tilde{w} \in \Omega$ is a LU-efficient minimizer of order s w.r.t. ϑ of (IVMPP), if for some interval-valued vector $\boldsymbol{\beta}$ with $\beta_k \in \mathcal{I}^+$, $k \in \mathcal{K}$, there exists no $y \in \Omega$, such that

$$\boldsymbol{\Psi}(y) - \boldsymbol{\Psi}(\tilde{w}) \prec_{LU} \boldsymbol{\beta}||\vartheta(y, \tilde{w})||^s.$$

Definition 10 The point $\tilde{w} \in \Omega$ is a LU-strict minimizer of order s w.r.t. ϑ of (IVMPP), if for some interval-valued vector $\boldsymbol{\beta}$ with $\beta_k \in \mathcal{I}^+$, $k \in \mathcal{K}$, there exists no $y \in \Omega$, such that

$$\Psi_k(y) - \Psi_k(\tilde{w}) \prec_{LU} \beta_k ||\vartheta(y, \tilde{w})||^s.$$

Now, we study the following generalized Minty and Stampacchia type vector variational-like inequalities in terms of limiting subdifferentials:

(GMVVLI) To find $\tilde{w} \in \Omega$, such that for all $\xi_k^L \in \partial_M \Psi_k^L(y)$ and $\xi_k^U \in \partial_M \Psi_k^U(y)$, $k \in \mathcal{K}$ one has

$$(\langle \xi_1^L, \vartheta(y, \tilde{w}) \rangle, \ldots, \langle \xi_p^L, \vartheta(y, \tilde{w}) \rangle) \not< 0,$$
$$(\langle \xi_1^U, \vartheta(y, \tilde{w}) \rangle, \ldots, \langle \xi_p^U, \vartheta(y, \tilde{w}) \rangle) \not< 0, \; \forall y \in \Omega.$$

(GSVVLI) To find $\tilde{w} \in \Omega$, such that, for any $y \in \Omega$, we can get $\zeta_k^L \in \partial_M \Psi_k^L(\tilde{w})$ and $\zeta_k^U \in \partial_M \Psi_k^U(\tilde{w})$, $k \in \mathcal{K}$, such that

$$(\langle \zeta_1^L, \vartheta(y, \tilde{w}) \rangle, \ldots, \langle \zeta_p^L, \vartheta(y, \tilde{w}) \rangle) \not< 0,$$
$$(\langle \zeta_1^U, \vartheta(y, \tilde{w}) \rangle, \ldots, \langle \zeta_p^U, \vartheta(y, \tilde{w}) \rangle) \not< 0.$$

(WGMVVLI) To find $\tilde{w} \in \Omega$, such that for all $\xi_k^L \in \partial_M \Psi_k^L(y)$ and $\xi_k^U \in \partial_M \Psi_k^U(y)$, $k \in \mathcal{K}$, one has

$$(\langle \xi_1^L, \vartheta(y, \tilde{w}) \rangle, \ldots, \langle \xi_p^L, \vartheta(y, \tilde{w}) \rangle) \not\prec 0,$$
$$(\langle \xi_1^U, \vartheta(y, \tilde{w}) \rangle, \ldots, \langle \xi_p^U, \vartheta(y, \tilde{w}) \rangle) \not\prec 0, \; \forall y \in \Omega.$$

(WGSVVLI) To find $\tilde{w} \in \Omega$, such that, for any $y \in \Omega$, we can get $\zeta_k^L \in \partial_M \Psi_k^L(\tilde{w})$ and $\zeta_k^U \in \partial_M \Psi_k^U(\tilde{w})$, $k \in \mathcal{K}$, such that

$$(\langle \zeta_1^L, \vartheta(y, \tilde{w}) \rangle, \ldots, \langle \zeta_p^L, \vartheta(y, \tilde{w}) \rangle) \not\prec 0,$$
$$(\langle \zeta_1^U, \vartheta(y, \tilde{w}) \rangle, \ldots, \langle \zeta_p^U, \vartheta(y, \tilde{w}) \rangle) \not\prec 0.$$

15.3 Relationship Between (GMVVLI), (GSVVLI) and (IVMPP)

In this section, using the powerful tool of limiting subdifferential, we establish certain relations among the solutions of (GMVVLI), (GSVVLI) and LU-efficient minimizer of order s w.r.t. ϑ of (IVMPP).

Theorem 2 *Let each Ψ_k, $k \in \mathcal{K}$ be strongly LU-invex of order s, ϑ is skew and $\tilde{w}$ is an LU-efficient minimizer of order s of (IVMPP) w.r.t. ϑ, then $\tilde{w}$ solves (GMVVLI). Furthermore, let ϑ satisfy the Condition C, each Ψ_k^L, Ψ_k^U, $k \in \mathcal{K}$ be locally Lipschitz* on Ω and satisfy the Condition A. Let $\tilde{w}$ solves (GMVVLI), then $\tilde{w} \in \Omega$ is an LU-efficient minimizer of order s w.r.t. ϑ of (IVMPP).

Proof Assume that $\tilde{w} \in \Omega$ be an LU-efficient minimizer of order s w.r.t. ϑ of (IVMPP). Therefore, we can get an interval-valued vector $\boldsymbol{\beta}$ with $\beta_k \in \mathcal{I}^+$, $k \in \mathcal{K}$, such that there exists no $y \in \Omega$ satisfying

$$\boldsymbol{\Psi}(y) - \boldsymbol{\Psi}(\tilde{w}) \prec_{LU} \boldsymbol{\beta} \|\vartheta(y, \tilde{w})\|^s. \tag{15.2}$$

From strong LU-invexity of order s of Ψ_k, $k \in \mathcal{K}$, for any $y \in \Omega$, we get

$$\begin{aligned}\Psi_k^L(\tilde{w}) - \Psi_k^L(y) &\geq \langle \xi_k^L, \vartheta(\tilde{w}, y) \rangle + \beta_k^L \|\vartheta(\tilde{w}, y)\|^s, \ \forall \xi_k^L \in \partial_M \Psi_k^L(y), k \in \mathcal{K}, \\ \Psi_k^U(\tilde{w}) - \Psi_k^U(y) &\geq \langle \xi_k^U, \vartheta(\tilde{w}, y) \rangle + \beta_k^U \|\vartheta(\tilde{w}, y)\|^s, \ \forall \xi_k^U \in \partial_M \Psi_k^U(y), k \in \mathcal{K}. \end{aligned} \tag{15.3}$$

Since ϑ is skew and $\beta_k^L, \beta_k^U > 0$, from (15.3), we get

$$\begin{aligned}\Psi_k^L(y) - \Psi_k^L(\tilde{w}) - \beta_k^L \|\vartheta(y, \tilde{w})\|^s &\leq \langle \xi_k^L, \vartheta(y, \tilde{w}) \rangle - 2\beta_k^L \|\vartheta(y, \tilde{w})\|^s, \ \forall k \in \mathcal{K}, \\ \Psi_k^U(y) - \Psi_k^U(\tilde{w}) - \beta_k^U \|\vartheta(y, \tilde{w})\|^s &\leq \langle \xi_k^U, \vartheta(y, \tilde{w}) \rangle - 2\beta_k^U \|\vartheta(y, \tilde{w})\|^s, \ \forall k \in \mathcal{K}. \end{aligned} \tag{15.4}$$

From (15.2) and (15.4), for any $y \in \Omega$, we have

$$\begin{aligned}(\langle \xi_1^L, \vartheta(y, \tilde{w}) \rangle, \ldots, \langle \xi_p^L, \vartheta(y, \tilde{w}) \rangle) &\not< 0, \ \forall \xi_k^L \in \partial_M \Psi_k^L(y), k \in \mathcal{K}, \\ (\langle \xi_1^U, \vartheta(y, \tilde{w}) \rangle, \ldots, \langle \xi_p^U, \vartheta(y, \tilde{w}) \rangle) &\not< 0, \ \forall \xi_k^U \in \partial_M \Psi_k^U(y), k \in \mathcal{K}. \end{aligned}$$

Therefore, $\tilde{w}$ solves (GMVVLI).

Conversely, let $\tilde{w} \in \Omega$ solves (GMVVLI), but not LU-efficient minimizer of order s w.r.t. ϑ of (IVMPP). Then, for some $y \in \Omega$, and for all interval-valued vector $\boldsymbol{\beta}$ with $\beta_k \in \mathcal{I}^+$, $k \in \mathcal{K}$, we have

$$\boldsymbol{\Psi}(y) - \boldsymbol{\Psi}(\tilde{w}) \prec_{LU} \boldsymbol{\beta} \|\vartheta(y, \tilde{w})\|^s. \tag{15.5}$$

Let $y(\mu) := \tilde{w} + \mu \vartheta(y, \tilde{w})$ for all $\mu \in [0, 1]$. Let $\mu' \in (0, 1)$, then from Theorem 1, we may get $\mu_k \in]0, \mu']$ and $\xi_k^L \in \partial_M \Psi_k^L(y(\mu_k))$, $\xi_k^U \in \partial_M \Psi_k^U(y(\mu_k))$, such that

$$\begin{aligned}\mu' \langle \xi_k^L, \vartheta(y, \tilde{w}) \rangle &\leq \Psi_k^L(\tilde{w} + \mu' \vartheta(y, \tilde{w})) - \Psi_k^L(\tilde{w}), \ \forall k \in \mathcal{K}, \\ \mu' \langle \xi_k^U, \vartheta(y, \tilde{w}) \rangle &\leq \Psi_k^U(\tilde{w} + \mu' \vartheta(y, \tilde{w})) - \Psi_k^U(\tilde{w}), \ \forall k \in \mathcal{K}. \end{aligned} \tag{15.6}$$

From Lemma 2, each Ψ_k, $k \in \mathcal{K}$ is strongly preinvex of order s. Therefore, we may get $\beta_k \in \mathcal{I}^+$, such that

$$\begin{aligned}\Psi_k^L(y(\mu')) &\leq (1 - \mu') \Psi_k^L(\tilde{w}) + \mu' \Psi_k^L(y) - \mu'(1 - \mu') \beta_k^L \|\vartheta(y, \tilde{w})\|^s, \ \forall k \in \mathcal{K}, \\ \Psi_k^U(y(\mu')) &\leq (1 - \mu') \Psi_k^U(\tilde{w}) + \mu' \Psi_k^U(y) - \mu'(1 - \mu') \beta_k^U \|\vartheta(y, \tilde{w})\|^s, \ \forall k \in \mathcal{K}. \end{aligned} \tag{15.7}$$

From (15.6) and (15.7), we get

$$\begin{aligned}\langle \xi_k^L, \vartheta(y, \tilde{w})\rangle &\leq \Psi_k^L(y) - \Psi_k^L(\tilde{w}) - (1-\mu')\beta_k^L\|\vartheta(y,\tilde{w})\|^s, \ \forall k \in \mathcal{K},\\ \langle \xi_k^U, \vartheta(y, \tilde{w})\rangle &\leq \Psi_k^U(y) - \Psi_k^U(\tilde{w}) - (1-\mu')\beta_k^U\|\vartheta(y,\tilde{w})\|^s, \ \forall k \in \mathcal{K}.\end{aligned} \quad (15.8)$$

From (15.8), we get

$$\begin{aligned}\langle \xi_k^L, \vartheta(y, \tilde{w})\rangle &\leq \Psi_k^L(y) - \Psi_k^L(\tilde{w}), \ \forall k \in \mathcal{K},\\ \langle \xi_k^U, \vartheta(y, \tilde{w})\rangle &\leq \Psi_k^U(y) - \Psi_k^U(\tilde{w}), \ \forall k \in \mathcal{K}.\end{aligned} \quad (15.9)$$

Let $\mu^* < \min\{\mu_1, \mu_2, \ldots, \mu_p\}$. From Lemma 1, each $\partial_M \Psi_k^L$ and $\partial_M \Psi_k^U$, $k \in \mathcal{K}$ are strongly invariant monotone of order s. Therefore, for every $\xi_k^L \in \partial_M \Psi_k^L(y(\mu_k))$ and $\xi_k^{*L} \in \partial_M \Psi_k^L(y(\mu^*))$, we get

$$\langle \xi_k^L, \vartheta(y(\mu^*), y(\mu_k))\rangle + \langle \xi_k^{*L}, \vartheta(y(\mu_k), y(\mu^*))\rangle \leq -\beta_k^L[\|\vartheta(y(\mu_k), y(\mu^*))\|^s +$$
$$\|\vartheta(y(\mu^*), y(\mu_k))\|^s], \ \forall k \in \mathcal{K}. \quad (15.10)$$

Since ϑ is skew, from (15.10), we get

$$\langle \xi_k^L - \xi_k^{*L}, \vartheta(y(\mu_k), y(\mu^*))\rangle \geq \beta_k^L \|\vartheta(y(\mu_k), y(\mu^*))\|^s, \ \forall k \in \mathcal{K}. \quad (15.11)$$

From (15.11) and Remark 2, we have

$$\bar{\beta}_k^L \|\vartheta(y, \tilde{w})\|^s \leq \langle \xi_k^L - \xi_k^{*L}, \vartheta(y, \tilde{w})\rangle, \text{ where } \bar{\beta}_k^L = (\mu_k - \mu^*)^{s-1}, \ \forall k \in \mathcal{K}. \quad (15.12)$$

From (15.12), we get

$$\langle \xi_k^L, \vartheta(y, \tilde{w})\rangle \geq \langle \xi_k^{*L}, \vartheta(y, \tilde{w})\rangle + \bar{\beta}_k^L \|\vartheta(y, \tilde{w})\|^s, \ \forall k \in \mathcal{K}. \quad (15.13)$$

Similarly,

$$\langle \xi_k^U, \vartheta(y, \tilde{w})\rangle \geq \langle \xi_k^{*U}, \vartheta(y, \tilde{w})\rangle + \bar{\beta}_k^U \|\vartheta(y, \tilde{w})\|^s, \ \forall k \in \mathcal{K}. \quad (15.14)$$

From (15.9), (15.13) and (15.14), we get

$$\begin{aligned}\Psi_k^L(y) - \Psi_k^L(\tilde{w}) &\geq \langle \xi_k^{*L}, \vartheta(y, \tilde{w})\rangle + \bar{\beta}_k^L \|\vartheta(y, \tilde{w})\|^s, \ \forall k \in \mathcal{K},\\ \Psi_k^U(y) - \Psi_k^U(\tilde{w}) &\geq \langle \xi_k^{*U}, \vartheta(y, \tilde{w})\rangle + \bar{\beta}_k^U \|\vartheta(y, \tilde{w})\|^s, \ \forall k \in \mathcal{K}.\end{aligned} \quad (15.15)$$

From (15.15), we get

$$\begin{aligned}\langle \xi_k^{*L}, \vartheta(y, \tilde{w})\rangle &\leq \Psi_k^L(y) - \Psi_k^L(\tilde{w}) - \bar{\beta}_k^L \|\vartheta(y, \tilde{w})\|^s, \ \forall k \in \mathcal{K},\\ \langle \xi_k^{*U}, \vartheta(y, \tilde{w})\rangle &\leq \Psi_k^U(y) - \Psi_k^U(\tilde{w}) - \bar{\beta}_k^U \|\vartheta(y, \tilde{w})\|^s, \ \forall k \in \mathcal{K}.\end{aligned} \quad (15.16)$$

From (15.5) and (15.16), we have

$$\begin{aligned}(\langle \xi_1^{*L}, \vartheta(y, \tilde{w})\rangle, \ldots, \langle \xi_p^{*L}, \vartheta(y, \tilde{w})\rangle) \leq 0, \ \forall \xi_k^{*L} \in \partial_M \Psi_k^L(y(\mu^*)), k \in \mathcal{K},\\(\langle \xi_1^{*U}, \vartheta(y, \tilde{w})\rangle, \ldots, \langle \xi_p^{*U}, \vartheta(y, \tilde{w})\rangle) \leq 0, \ \forall \xi_k^{*U} \in \partial_M \Psi_k^U(y(\mu^*)), k \in \mathcal{K}.\end{aligned}$$
(15.17)

Multiplying (15.17) by $-\mu^*$ and from Remark 2, we have

$$\begin{aligned}(\langle \xi_1^{*L}, \vartheta(y(\mu^*), \tilde{w})\rangle, \ldots, \langle \xi_p^{*L}, \vartheta(y(\mu^*), \tilde{w})\rangle) \leq 0, \ \forall \xi_k^{*L} \in \partial_M \Psi_k^L(y(\mu^*)), k \in \mathcal{K},\\(\langle \xi_1^{*U}, \vartheta(y(\mu^*), \tilde{w})\rangle, \ldots, \langle \xi_p^{*U}, \vartheta(y(\mu^*), \tilde{w})\rangle) \leq 0, \ \forall \xi_k^{*U} \in \partial_M \Psi_k^U(y(\mu^*)), k \in \mathcal{K},\end{aligned}$$

which contradicts our assumption.

Theorem 3 *Let each Ψ_k, $k \in \mathcal{K}$ be strongly LU-invex function of order s on Ω and $\tilde{w} \in \Omega$ solves (GSVVLI), then $\tilde{w}$ is an LU-efficient minimizer of order s w.r.t. ϑ of (IVMPP).*

Proof Let $\tilde{w} \in \Omega$ solves (GSVVLI), then for each $y \in \Omega$, we can get $\zeta_k^L \in \partial_M \Psi_k^L(\tilde{w})$ and $\zeta_k^U \in \partial_M \Psi_k^U(\tilde{w})$, $k \in \mathcal{K}$, such that

$$\begin{aligned}(\langle \zeta_1^L, \vartheta(y, \tilde{w})\rangle, \ldots, \langle \zeta_p^L, \vartheta(y, \tilde{w})\rangle) \not\leq 0,\\(\langle \zeta_1^U, \vartheta(y, \tilde{w})\rangle, \ldots, \langle \zeta_p^U, \vartheta(y, \tilde{w})\rangle) \not\leq 0.\end{aligned}$$
(15.18)

From strong LU-invexity of order s of Ψ_k, $k \in \mathcal{K}$, we may get $\beta_k \in \mathcal{I}^+$, $k \in \mathcal{K}$, such that

$$\begin{aligned}\langle \zeta_k^L, \vartheta(y, \tilde{w})\rangle + \beta_k^L \|\vartheta(y, \tilde{w})\|^s \leq \Psi_k^L(y) - \Psi_k^L(\tilde{w}),\\\langle \zeta_k^U, \vartheta(y, \tilde{w})\rangle + \beta_k^U \|\vartheta(y, \tilde{w})\|^s \leq \Psi_k^U(y) - \Psi_k^L(\tilde{w}), \ \forall y \in \Omega.\end{aligned}$$
(15.19)

From (15.18) and (15.19), we can say there exists no $y \in \Omega$, such that

$$\Psi(y) - \Psi(\tilde{w}) \prec_{LU} \beta \|\vartheta(y, \tilde{w})\|^s.$$

Hence, $\tilde{w} \in \Omega$ is LU-efficient minimizer of order s w.r.t. ϑ of (IVMPP).

From Theorems 2 and 3, we conclude the following relation between the solutions of (GSVVLI) and (GMVVLI).

Corollary 1 *Let each Ψ_k, $k \in \mathcal{K}$ be strongly LU-invex function of order s and ϑ is skew. Let ϑ and Ψ_k^L, Ψ_k^U, $k \in \mathcal{K}$ satisfy the Condition C and Condition A, respectively. If $\tilde{w} \in \Omega$ solves (GSVVLI), then $\tilde{w}$ also solves (GMVVLI).*

15.4 Relationship Between (WGMVVLI), (WGSVVLI) and (IVMPP)

In this section, we showed the equivalence between the solutions of (WGMVVLI), (WGSVVLI) and LU-strict minimizer of order s w.r.t. ϑ of (IVMPP).

Proposition 1 Let each Ψ_k, $k \in \mathcal{K}$ be strongly LU-invex of order s and $\tilde{w}$ solves (WGSVVLI), then $\tilde{w}$ also solves (WGMVVLI).

Proof Assume that $\tilde{w} \in \Omega$ solves (WGSVVLI), then we can get $\zeta_k^L \in \partial_M \Psi_k^L(\tilde{w})$ and $\zeta_k^U \in \partial_M \Psi_k^U(\tilde{w})$, such that for any $y \in \Omega$, we get

$$\begin{aligned}(\langle \zeta_1^L, \vartheta(y, \tilde{w}) \rangle, \ldots, \langle \zeta_p^L, \vartheta(y, \tilde{w}) \rangle) \not< 0, \\ (\langle \zeta_1^U, \vartheta(y, \tilde{w}) \rangle, \ldots, \langle \zeta_p^U, \vartheta(y, \tilde{w}) \rangle) \not< 0.\end{aligned} \quad (15.20)$$

From Lemma 1, $\partial_M \Psi_k^L$ and $\partial_M \Psi_k^U$ are strongly invariant monotone of order s. Therefore, we can get $\beta_k \in \mathcal{I}^+$, $k \in \mathcal{K}$, such that

$$\begin{aligned}\langle \xi_k^L - \zeta_k^L, \vartheta(y, \tilde{w}) \rangle \geq \beta_k^L (\|\vartheta(y, \tilde{w})\|^s + \|\vartheta(\tilde{w}, y)\|^s), \; \forall k \in \mathcal{K}, \\ \langle \xi_k^U - \zeta_k^U, \vartheta(y, \tilde{w}) \rangle \geq \beta_k^U (\|\vartheta(y, \tilde{w})\|^s + \|\vartheta(\tilde{w}, y)\|^s), \; \forall k \in \mathcal{K}.\end{aligned} \quad (15.21)$$

for all $\zeta_k^L \in \partial_M \Psi_k^L(\tilde{w})$, $\zeta_k^U \in \partial_M \Psi_k^U(\tilde{w})$, $\xi_k^L \in \partial_M \Psi_k^L(y)$, $\xi_k^U \in \partial_M \Psi_k^U(y)$ and $y \in \Omega$. From (15.20) and (15.21), we get

$$\begin{aligned}(\langle \xi_1^L, \vartheta(y, \tilde{w}) \rangle, \ldots, \langle \xi_p^L, \vartheta(y, \tilde{w}) \rangle) \not< 0, \; \forall \xi_k^L \in \partial_M \Psi_k^L(y), k \in \mathcal{K}, \\ (\langle \xi_1^U, \vartheta(y, \tilde{w}) \rangle, \ldots, \langle \xi_p^U, \vartheta(y, \tilde{w}) \rangle) \not< 0, \; \forall \xi_k^U \in \partial_M \Psi_k^U(y), k \in \mathcal{K}.\end{aligned}$$

Hence, $\tilde{w} \in \Omega$ solves (WGMVVLI).

Proposition 2 Let each Ψ_k^L and Ψ_k^U, $k \in \mathcal{K}$ be locally Lipschitz, ϑ is skew and affine in first argument. If $\tilde{w} \in \Omega$ solves (WGMVVLI), then $\tilde{w}$ also solves (WGSVVLI).

Proof Assume that $\tilde{w}$ solves (WGMVVLI), then for each $y \in \Omega$, we have

$$\begin{aligned}(\langle \xi_1^L, \vartheta(y, \tilde{w}) \rangle, \ldots, \langle \xi_p^L, \vartheta(y, \tilde{w}) \rangle) \not< 0, \; \forall \xi_k^L \in \partial_M \Psi_k^L(y), k \in \mathcal{K}, \\ (\langle \xi_1^U, \vartheta(y, \tilde{w}) \rangle, \ldots, \langle \xi_p^U, \vartheta(y, \tilde{w}) \rangle) \not< 0, \; \forall \xi_k^U \in \partial_M \Psi_k^U(y), k \in \mathcal{K}.\end{aligned}$$

Let $y(\mu_n) = \tilde{w} + \mu_n \vartheta(y, \tilde{w})$, where $\mu_n \downarrow 0$. Since $\tilde{w}$ solves (WGMVVLI), then

$$\begin{aligned}(\langle \xi_{1_n}^L, \vartheta(y(\mu_n), \tilde{w}) \rangle, \ldots, \langle \xi_{p_n}^L, \vartheta(y(\mu_n), \tilde{w}) \rangle) \not< 0, \; \xi_{k_n}^L \in \partial_M \Psi_k^L(y(\mu_n)), k \in \mathcal{K}, \\ (\langle \xi_{1_n}^U, \vartheta(y(\mu_n), \tilde{w}) \rangle, \ldots, \langle \xi_{p_n}^U, \vartheta(y(\mu_n), \tilde{w}) \rangle) \not< 0, \; \xi_{k_n}^U \in \partial_M \Psi_k^U(y(\mu_n)), k \in \mathcal{K}.\end{aligned} \quad (15.22)$$

Since, ϑ is skew and affine in first argument, we get

$$\begin{aligned}0 &= \langle \xi_{k_n}^L, \vartheta(y(\mu_n), y(\mu_n)) \rangle \\ &= \mu_n \langle \xi_{k_n}^L, \vartheta(y, y(\mu_n)) \rangle + (1 - \mu_n) \langle \xi_{k_n}^L, \vartheta(\tilde{w}, y(\mu_n)) \rangle, \; \forall \xi_{k_n}^L \in \partial_M \Psi_k^L(y(\mu_n)).\end{aligned} \quad (15.23)$$

Similarly,

$$\mu_n \langle \xi_{k_n}^U, \vartheta(y, y(\mu_n)) \rangle + (1 - \mu_n) \langle \xi_{k_n}^U, \vartheta(\tilde{w}, y(\mu_n)) \rangle = 0, \; \forall \xi_{k_n}^U \in \partial_M \Psi_k^U(y(\mu_n)). \tag{15.24}$$

From (15.22), (15.23) and (15.24), we get

$$\begin{aligned} (\langle \xi_{1_n}^L, \vartheta(y, y(\mu_n)) \rangle, \ldots, \langle \xi_{p_n}^L, \vartheta(y, y(\mu_n)) \rangle) \not< 0, \\ (\langle \xi_{1_n}^U, \vartheta(y, y(\mu_n)) \rangle, \ldots, \langle \xi_{p_n}^U, \vartheta(y, y(\mu_n)) \rangle) \not< 0. \end{aligned} \tag{15.25}$$

Since $\partial_M \Psi_k^L$ and $\partial_M \Psi_k^U$, $k \in \mathcal{K}$ are compact, $\xi_{k_n}^L \in \partial_M \Psi_k^L(y(\mu_n))$, $\xi_{k_n}^U \in \partial_M \Psi_k^U(y(\mu_n))$, $\xi_{k_n}^L \to \zeta_k^L$, $\xi_{k_n}^U \to \zeta_k^U$ and $y(\mu_n) \to \tilde{w}$ as $n \to \infty$, we have $\zeta_k^L \in \partial_M \Psi_k^L(\tilde{w})$ and $\zeta_k^U \in \partial_M \Psi_k^U(\tilde{w})$.

Therefore, for all $y \in \Omega$, we can get $\zeta_k^L \in \partial_M \Psi_k^L(\tilde{w})$ and $\zeta_k^U \in \partial_M \Psi_k^U(\tilde{w})$, such that

$$\begin{aligned} (\langle \zeta_1^L, \vartheta(y, \tilde{w}) \rangle, \ldots, \langle \zeta_p^L, \vartheta(y, \tilde{w}) \rangle) \not< 0, \\ (\langle \zeta_1^U, \vartheta(y, \tilde{w}) \rangle, \ldots, \langle \zeta_p^U, \vartheta(y, \tilde{w}) \rangle) \not< 0. \end{aligned}$$

We conclude the equivalence between the solutions of (WGSVVLI) and (WGMVVLI), from Propositions 1 and 2.

Theorem 4 *Let each Ψ_k^L and Ψ_k^U, $k \in \mathcal{K}$ be locally Lipschitz and Ψ_k be strongly LU-invex of order s. Let ϑ be skew and affine in first argument, then $\tilde{w}$ is a solution of (WGSVVLI) iff $\tilde{w}$ solves (WGMVVLI).*

Proposition 3 *Let each Ψ_k, $k \in \mathcal{K}$ be strongly LU-invex of order s and $\tilde{w} \in \Omega$ solves (WGSVVLI), then $\tilde{w}$ is LU-strict minimizer of order s w.r.t. ϑ of (IVMPP).*

Proof On contrary assume that $\tilde{w} \in \Omega$ solves (WGSVVLI), but not LU-strict minimizer of order s w.r.t. ϑ of (IVMPP). Therefore, for some $y \in \Omega$ and for all interval-valued vector $\boldsymbol{\beta}$ with $\beta_k \in \mathcal{I}^+$, $k \in \mathcal{K}$, we get

$$\boldsymbol{\Psi}(y) - \boldsymbol{\Psi}(\tilde{w}) \prec_{LU} \boldsymbol{\beta} \|\vartheta(y, \tilde{w})\|^s. \tag{15.26}$$

From strong LU-invexity of order s of Ψ_k, $k \in \mathcal{K}$, we can get $\beta_k \in \mathcal{I}^+$, $k \in \mathcal{K}$, such that

$$\begin{aligned} \langle \zeta_k^L, \vartheta(y, \tilde{w}) \rangle \leq \Psi_k^L(y) - \Psi_k^L(\tilde{w}) - \beta_k^L \|\vartheta(y, \tilde{w})\|^s, \forall \zeta_k^L \in \partial_M \Psi_k^L(\tilde{w}), k \in \mathcal{K}, \\ \langle \zeta_k^U, \vartheta(y, \tilde{w}) \rangle \leq \Psi_k^U(y) - \Psi_k^U(\tilde{w}) - \beta_k^U \|\vartheta(y, \tilde{w})\|^s, \forall \zeta_k^U \in \partial_M \Psi_k^U(\tilde{w}), k \in \mathcal{K}. \end{aligned} \tag{15.27}$$

From (15.26) and (15.27), we get

$$\begin{aligned} (\langle \zeta_1^L, \vartheta(y, \tilde{w}) \rangle, \ldots, \langle \zeta_p^L, \vartheta(y, \tilde{w}) \rangle) < 0, \; \forall \zeta_k^L \in \partial_M \Psi_k^L(\tilde{w}), k \in \mathcal{K}, \\ (\langle \zeta_1^U, \vartheta(y, \tilde{w}) \rangle, \ldots, \langle \zeta_p^U, \vartheta(y, \tilde{w}) \rangle) < 0, \; \forall \zeta_k^U \in \partial_M \Psi_k^U(\tilde{w}), k \in \mathcal{K}, \end{aligned}$$

which is a contradiction.

Proposition 4 *Let each Ψ_k, $k \in \mathcal{K}$ be strongly LU-invex of order s and ϑ is skew. Let $\tilde{w} \in \Omega$ be LU-strict minimizer of order s w.r.t. ϑ of (IVMPP), then $\tilde{w}$ solves (WGMVVLI).*

Proof On contrary assume that $\tilde{w} \in \Omega$ is a LU-strict minimizer of order s w.r.t. ϑ of (IVMPP), but does not solves (WGMVVLI). Then, we can get $y \in \Omega$, such that

$$\begin{aligned}(\langle \xi_1^L, \vartheta(y,\tilde{w})\rangle, \ldots, \langle \xi_p^L, \vartheta(y,\tilde{w})\rangle) < 0, \quad \forall \xi_k^L \in \partial_M \Psi_k^L(y), k \in \mathcal{K},\\ (\langle \xi_1^U, \vartheta(y,\tilde{w})\rangle, \ldots, \langle \xi_p^U, \vartheta(y,\tilde{w})\rangle) < 0, \quad \forall \xi_k^U \in \partial_M \Psi_k^U(y), k \in \mathcal{K}.\end{aligned} \quad (15.28)$$

From strong LU-invexity of order s of Ψ_k, $k \in \mathcal{K}$ and using skew property of ϑ, we can get $\beta_k \in \mathcal{I}^+$, $k \in \mathcal{K}$, such that

$$\begin{aligned}\Psi_k^L(y) - \Psi_k^L(\tilde{w}) \le \langle \xi_k^L, \vartheta(y,\tilde{w})\rangle - \beta_k^L \|\vartheta(y,\tilde{w})\|^s, \quad \forall \xi_k^L \in \partial_M \Psi_k^L(y), k \in \mathcal{K},\\ \Psi_k^U(y) - \Psi_k^U(\tilde{w}) \le \langle \xi_k^U, \vartheta(y,\tilde{w})\rangle - \beta_k^U \|\vartheta(y,\tilde{w})\|^s, \quad \forall \xi_k^U \in \partial_M \Psi_k^U(y), k \in \mathcal{K}.\end{aligned} \quad (15.29)$$

Since $\beta_k^L, \beta_k^U > 0$, $\forall k \in \mathcal{K}$, from (15.29), we have

$$\begin{aligned}\Psi_k^L(y) - \Psi_k^L(\tilde{w}) - \beta_k^L \|\vartheta(y,\tilde{w})\|^s \le \langle \xi_k^L, \vartheta(y,\tilde{w})\rangle - 2\beta_k^L \|\vartheta(y,\tilde{w})\|^s \le \langle \xi_k^L, \vartheta(y,\tilde{w})\rangle,\\ \Psi_k^U(y) - \Psi_k^U(\tilde{w}) - \beta_k^U \|\vartheta(y,\tilde{w})\|^s \le \langle \xi_k^U, \vartheta(y,\tilde{w})\rangle - 2\beta_k^U \|\vartheta(y,\tilde{w})\|^s \le \langle \xi_k^U, \vartheta(y,\tilde{w})\rangle.\end{aligned} \quad (15.30)$$

From (15.28) and (15.30), we have

$$\Psi(y) - \Psi(\tilde{w}) \prec_{LU} \beta \|\vartheta(\tilde{w},y)\|^s, \quad \forall y \in \Omega,$$

which contradicts our assumption.

Theorem 5 *Let each Ψ_k^L and Ψ_k^U, $k \in \mathcal{K}$ be locally Lipschitz and Ψ_k be strongly LU-invex of order s. Let ϑ be skew and affine in first argument, then $\tilde{w} \in \Omega$ solves (WGSVVLI) iff $\tilde{w}$ is a LU-strict minimizer of order s w.r.t. ϑ of (IVMPP).*

15.5 Existence Results for (GMVVLI) and (GSVVLI)

Definition 11 [21] A map $\Gamma : \Omega \to 2^\Omega$ is said to be a KKM map if for each $\{z_1, \ldots, z_l\} \subseteq \Omega$, one has

$$\mathrm{co}\{z_1, \ldots, z_l\} \subseteq \bigcup_{k=1}^{l} \Gamma(z_k),$$

where $\mathrm{co}\{z_1, \ldots, z_l\}$ is the convex hull of $\{z_1, \ldots, z_l\}$.

Lemma 3 *[21] Let Ω be convex set. Let $\Gamma : \Omega \to 2^\Omega$ be KKM map, such that $\Gamma(y)$ is closed for each $y \in \Omega$. If for some point $\bar{y} \in \Omega$, $\Gamma(\bar{y})$ is compact, then*

$$\bigcap_{y\in\Omega} \Gamma(y) \neq \emptyset.$$

Theorem 6 *Let Ψ_k^L and Ψ_k^U, $k \in \mathcal{K}$ be lower semicontinuous functions and ϑ is affine in second argument. Let the following conditions hold:*

1. *For every $y_1, y_2 \in \Omega$,*

$$\langle \xi_k^L, \vartheta(y_2, y_1)\rangle + \langle \zeta_k^L, \vartheta(y_1, y_2)\rangle \geq 0,$$
$$\langle \xi_k^U, \vartheta(y_2, y_1)\rangle + \langle \zeta_k^U, \vartheta(y_1, y_2)\rangle \geq 0,$$

   for each $\xi_k^L \in \partial_M \Psi_k^L(y_2)$, $\zeta_k^L \in \partial_M \Psi_k^L(y_1)$, $\xi_k^U \in \partial_M \Psi_k^U(y_2)$ and $\zeta_k^U \in \partial_M \Psi_k^U(y_1)$, $k \in \mathcal{K}$.

2. *For all $y \in \Omega$,*

$$(\langle \xi_1^L, \vartheta(y, y)\rangle, \ldots, \langle \xi_p^L, \vartheta(y, y)\rangle) \not\geq 0, \; \forall \xi_k^L \in \partial_M \Psi_k^L(y), k \in \mathcal{K},$$
$$(\langle \xi_1^U, \vartheta(y, y)\rangle, \ldots, \langle \xi_p^U, \vartheta(y, y)\rangle) \not\geq 0, \; \forall \xi_k^U \in \partial_M \Psi_k^U(y), k \in \mathcal{K}.$$

3. *The set valued map*

$$\Gamma(y_2) := \left\{ y \in \Omega \;\middle|\; \begin{array}{l} (\langle \xi_1^L, \vartheta(y_2, y_1)\rangle, \ldots, \langle \xi_p^L, \vartheta(y_2, y_1)\rangle) \not\leq 0, \; \forall \xi_k^L \in \partial_M \Psi_k^L(y_2), k \in \mathcal{K}, \\ (\langle \xi_1^U, \vartheta(y_2, y_1)\rangle, \ldots, \langle \xi_p^U, \vartheta(y_2, y_1)\rangle) \not\leq 0, \; \forall \xi_k^U \in \partial_M \Psi_k^U(y_2), k \in \mathcal{K} \end{array} \right\}$$

   is closed for all $y_2 \in \Omega$.

4. *There exist a compact set $\emptyset \neq G \subseteq \Omega$ and a compact convex set $\emptyset \neq M \subseteq \Omega$ and for each $y_1 \in \Omega \setminus G$, we can get $y_2 \in M$, such that $y_1 \notin \Gamma(y_2)$.*

Then (GMVVLI) is solvable on Ω.

Proof For all $y_2 \in \Omega$, we define a map

$$\hat{\Gamma}(y_2) := \left\{ y \in \Omega \;\middle|\; \begin{array}{l} (\langle \zeta_1^L, \vartheta(y_1, y_2)\rangle, \ldots, \langle \zeta_p^L, \vartheta(y_1, y_2)\rangle) \not\geq 0, \; \forall \zeta_k^L \in \partial_M \Psi_k^L(y_1) \\ (\langle \zeta_1^U, \vartheta(y_1, y_2)\rangle, \ldots, \langle \zeta_p^U, \vartheta(y_1, y_2)\rangle) \not\geq 0, \; \forall \zeta_k^U \in \partial_M \Psi_k^U(y), k \in \mathcal{K} \end{array} \right\}.$$

Evidently, $y \in \hat{\Gamma}(y)$.

Next, we prove that $\hat{\Gamma}(y)$ is KKM map on Ω. On contrary, let $\{u_1, \ldots, u_l\} \subseteq \Omega$, $\mu_k \geq 0$, $k = 1, \ldots, l$, with $\sum_{k=1}^{l} \mu_k = 1$, such that

$$\tilde{w} = \sum_{k=1}^{l} \mu_k u_k \notin \bigcup_{k=1}^{l} \hat{\Gamma}(u_k). \tag{15.31}$$

Therefore, for each u_k, $k = 1, \ldots, l$, we have

$$\begin{array}{l} (\langle \zeta_1^L, \vartheta(\tilde{w}, u_k)\rangle, \ldots, \langle \zeta_p^L, \vartheta(\tilde{w}, u_k)\rangle) \geq 0, \; \forall \zeta_k^L \in \partial_M \Psi_k^L(\tilde{w}), k \in \mathcal{K}, \\ (\langle \zeta_1^U, \vartheta(\tilde{w}, u_k)\rangle, \ldots, \langle \zeta_p^U, \vartheta(\tilde{w}, u_k)\rangle) \geq 0, \; \forall \zeta_k^U \in \partial_M \Psi_k^U(\tilde{w}), k \in \mathcal{K}. \end{array} \tag{15.32}$$

Since ϑ is affine in second argument, one has

$$(\langle \zeta_1^L, \vartheta(\tilde{w}, \sum_{k=1}^l \mu_k u_k)\rangle, \ldots, \langle \zeta_p^L, \vartheta(\tilde{w}, \sum_{k=1}^l \mu_k u_k)\rangle) \geq 0, \ \forall \zeta_k^L \in \partial_M \Psi_k^L(\tilde{w}), k \in \mathcal{K},$$

$$(\langle \zeta_1^U, \vartheta(\tilde{w}, \sum_{k=1}^l \mu_k u_k)\rangle, \ldots, \langle \zeta_p^U, \vartheta(\tilde{w}, \sum_{k=1}^l \mu_k u_k)\rangle) \geq 0, \ \forall \zeta_k^U \in \partial_M \Psi_k^U(\tilde{w}), k \in \mathcal{K},$$

that is

$$(\langle \zeta_1^L, \vartheta(\tilde{w}, \tilde{w})\rangle, \ldots, \langle \zeta_p^L, \vartheta(\tilde{w}, \tilde{w})\rangle) \geq 0, \ \forall \zeta_k^L \in \partial_M \Psi_k^L(\tilde{w}), k \in \mathcal{K},$$
$$(\langle \zeta_1^U, \vartheta(\tilde{w}, \tilde{w})\rangle, \ldots, \langle \zeta_p^U, \vartheta(\tilde{w}, \tilde{w})\rangle) \geq 0, \ \forall \zeta_k^U \in \partial_M \Psi_k^U(\tilde{w}), k \in \mathcal{K},$$

which is a contradiction. Therefore, $\hat{\Gamma}(y)$ is KKM map on Ω. Now, we prove that $\hat{\Gamma}(y) \subseteq \Gamma(y), \forall y \in \Omega$.

Let $\tilde{w} \notin \Gamma(y)$, for some $y \in \Omega$, then from the definition of Γ, we get

$$\begin{aligned}(\langle \xi_1^L, \vartheta(y, \tilde{w})\rangle, \ldots, \langle \xi_p^L, \vartheta(y, \tilde{w})\rangle) \leq 0, \forall \xi_k^L \in \partial_M \Psi_k^L(y), k \in \mathcal{K},\\ (\langle \xi_1^U, \vartheta(y, \tilde{w})\rangle, \ldots, \langle \xi_p^U, \vartheta(y, \tilde{w})\rangle) \leq 0, \forall \xi_k^U \in \partial_M \Psi_k^U(y), k \in \mathcal{K}.\end{aligned} \quad (15.33)$$

Using hypothesis (1), we get-

$$\begin{aligned}\langle \xi_k^L, \vartheta(y, \tilde{w})\rangle + \langle \zeta_k^L, \vartheta(\tilde{w}, y)\rangle \geq 0,\\ \langle \xi_k^U, \vartheta(y, \tilde{w})\rangle + \langle \zeta_k^U, \vartheta(\tilde{w}, y)\rangle \geq 0,\end{aligned} \quad (15.34)$$

for all $\xi_k^L \in \partial_M \Psi_k^L(y), \xi_k^U \in \partial_M \Psi_k^U(y), \zeta_k^L \in \partial_M \Psi_k^L(\tilde{w})$ and $\zeta_k^U \in \partial_M \Psi_k^U(\tilde{w})$, $k \in \mathcal{K}$. From (15.33) and (15.34), we get

$$\begin{aligned}(\langle \zeta_1^L, \vartheta(\tilde{w}, y)\rangle, \ldots, \langle \zeta_p^L, \vartheta(\tilde{w}, y)\rangle) \geq 0, \ \forall \zeta_k^L \in \partial_M \Psi_k^L(\tilde{w}), k \in \mathcal{K},\\ (\langle \zeta_1^U, \vartheta(\tilde{w}, y)\rangle, \ldots, \langle \zeta_p^U, \vartheta(\tilde{w}, y)\rangle) \geq 0, \ \forall \zeta_k^U \in \partial_M \Psi_k^U(\tilde{w}), k \in \mathcal{K}.\end{aligned} \quad (15.35)$$

Therefore, $\tilde{w} \notin \hat{\Gamma}(y)$. Hence, Γ is a KKM map. From hypotheses, $\Gamma(y)$ is a compact set. Using the KKM-Fan Theorem

$$\bigcap_{y \in \Omega} \Gamma(y) \neq \emptyset.$$

Therefore, we can get a $\tilde{w} \in \Omega$, such that

$$\begin{aligned}(\langle \xi_1^L, \vartheta(y, \tilde{w})\rangle, \ldots, \langle \xi_p^L, \vartheta(y, \tilde{w})\rangle) \nleq 0, \ \forall \xi_k^L \in \partial_M \Psi_k^L(y), k \in \mathcal{K},\\ (\langle \xi_1^U, \vartheta(y, \tilde{w})\rangle, \ldots, \langle \xi_p^U, \vartheta(y, \tilde{w})\rangle) \nleq 0, \ \forall \xi_k^U \in \partial_M \Psi_k^U(y), k \in \mathcal{K}.\end{aligned}$$

Hence, (GMVVLI) has a solution on Ω.

Example 1 Let $\Psi_1, \Psi_2 : [-\frac{1}{2}, \frac{1}{2}] \to \mathcal{I}$ and $\vartheta : [-\frac{1}{2}, \frac{1}{2}] \times [-\frac{1}{2}, \frac{1}{2}] \to \mathbb{R}$ be defined as:

$$\Psi_1^L(y) = \begin{cases} y+1, & y > 0, \\ 2y, & y \le 0, \end{cases} \quad \Psi_1^U(y) = \begin{cases} y+1, & y > 0, \\ 2y + e^y, & y \le 0, \end{cases}$$

$$\Psi_2^L(y) = \begin{cases} e^y - 1, & y > 0, \\ y^2 + 2y, & y \le 0, \end{cases} \quad \Psi_2^U(y) = \begin{cases} e^y, & y > 0, \\ y^2 + y, & y \le 0, \end{cases}$$

and $\vartheta(w, y) = \begin{cases} w - y, & w > 0 \text{ and } y > 0, \text{ or } w < 0 \text{ and } y < 0, \\ 0, & w = 0 \text{ and } y = 0, \\ 1 - y, & \text{elsewhere.} \end{cases}$

Now, we can evaluate that

$$\partial_M \Psi_1^L(y) = \begin{cases} 1, & y > 0, \\ [2, \infty), & y = 0, \\ 2, & y < 0, \end{cases} \quad \partial_M \Psi_1^U(y) = \begin{cases} 1, & y > 0, \\ \{1, 3\}, & y = 0, \\ 2 + e^y, & y < 0, \end{cases}$$

$$\partial_M \Psi_2^L(y) = \begin{cases} e^y, & y > 0, \\ \{1, 2\}, & y = 0, \\ 2y + 2, & y < 0, \end{cases} \quad \partial_M \Psi_2^U(y) = \begin{cases} e^y, & y > 0, \\ [1, \infty), & y = 0, \\ 2y + 1, & y < 0. \end{cases}$$

For all $\xi_1^L \in \partial_M \Psi_1^L(y)$ and $\zeta_1^L \in \partial_M \Psi_1^L(w)$, we get

$$\langle \xi_1^L, \vartheta(y, w) \rangle + \langle \zeta_1^L, \vartheta(w, y) \rangle = \begin{cases} \langle 1, y-w \rangle + \langle 1, w-y \rangle, & y > 0, w > 0; \\ \langle 2, y-w \rangle + \langle 2, w-y \rangle, & y < 0, w < 0; \\ \langle 1, 1-w \rangle + \langle 2, 1-y \rangle = (1-w) + 2(1-y), & y > 0, w < 0; \\ \langle 2, 1-w \rangle + \langle 1, 1-y \rangle = 2(1-w) + (1-y), & y < 0, w > 0, \end{cases} \ge 0.$$

For all $\xi_1^U \in \partial_M \Psi_1^U(y)$ and $\zeta_1^U \in \partial_M \Psi_1^U(w)$, we get

$$\langle \xi_1^U, \vartheta(y, w) \rangle + \langle \zeta_1^U, \vartheta(w, y) \rangle =$$

$$\begin{cases} \langle 1, y-w \rangle + \langle 1, w-y \rangle, & y > 0, w > 0; \\ \langle 2+e^y, y-w \rangle + \langle 2+e^w, w-y \rangle, & y < 0, w < 0; \\ \langle 1, 1-w \rangle + \langle 2+e^w, 1-y \rangle, & y > 0, w < 0; \\ \langle 2+e^y, 1-w \rangle + \langle 1, 1-y \rangle, & y < 0, w > 0, \end{cases}$$

$$\ge 0.$$

For all $\xi_2^L \in \partial_M \Psi_2^L(y)$ and $\zeta_2^L \in \partial_M \Psi_2^L(w)$, we get

$$\langle \xi_2^L, \vartheta(y,w) \rangle + \langle \zeta_2^L, \vartheta(w,y) \rangle =$$
$$\begin{cases} \langle e^y, y-w \rangle + \langle e^w, w-y \rangle = (e^y - e^w)(y-w), & y > 0, w > 0; \\ \langle 2y+2, y-w \rangle + \langle 2w+2, w-y \rangle = 2(y-w)^2, & y < 0, w < 0; \\ \langle e^y, 1-w \rangle + \langle 2w+2, 1-y \rangle, & y > 0, w < 0; \\ \langle 2y+2, 1-w \rangle + \langle e^w, 1-y \rangle, & y < 0, w > 0, \end{cases}$$
$$\geq 0.$$

For all $\xi_2^U \in \partial_M \Psi_2^U(y)$ and $\zeta_2^U \in \partial_M \Psi_2^U(w)$, we get

$$\langle \xi_2^U, \vartheta(y,w) \rangle + \langle \zeta_2^U, \vartheta(w,y) \rangle =$$
$$\begin{cases} \langle e^y, y-w \rangle + \langle e^w, w-y \rangle = (e^y - e^w)(y-w), & y > 0, w > 0; \\ \langle 2y+1, y-w \rangle + \langle 2w+1, w-y \rangle = 2(y-w)^2, & y < 0, w < 0; \\ \langle e^y, 1-w \rangle + \langle 2w+1, 1-y \rangle, & y > 0, w < 0; \\ \langle 2y+1, 1-w \rangle + \langle e^w, 1-y \rangle, & y < 0, y > 0, \end{cases}$$
$$\geq 0.$$

Furthermore, $\vartheta(y,y) = 0$, for all $y \in [-\frac{1}{2}, \frac{1}{2}]$. Hence,

$$(\langle \xi_1^L, \vartheta(y,y) \rangle, \langle \xi_2^L, \vartheta(y,y) \rangle) \not\geq 0,$$
$$(\langle \xi_1^U, \vartheta(y,y) \rangle, \langle \xi_2^U, \vartheta(y,y) \rangle) \not\geq 0,$$

for all $\xi_1^L \in \partial_M \Psi_1^L(y)$, $\xi_1^U \in \partial_M \Psi_1^U(y)$, $\xi_2^U \in \partial_M \Psi_2^U(y)$ and $\xi_2^U \in \partial_M \Psi_2^U(y)$.
For all $y \in [-\frac{1}{2}, \frac{1}{2}]$,

$$\Gamma(y) = \begin{cases} [-\frac{1}{2}, y], & y > 0, \\ [-\frac{1}{2}, \frac{1}{2}], & y = 0, \\ [-\frac{1}{2}, y] \cup [0, \frac{1}{2}], & y < 0. \end{cases}$$

Let $G = [0, \frac{1}{2}]$ and $M = [-\frac{1}{2}, -\frac{1}{4}]$. Clearly, M is convex and for any $y \in [-\frac{1}{2}, \frac{1}{2}] \setminus G$, we can get a $w < y$, such that $y \notin \Gamma(w)$.

Moreover, we can verify that $\tilde{w} = 0$ solves (GMVVLI).

In the same way as Theorem 6, we conclude the following result for the existence of solution of (GSVVLI).

Theorem 7 *Let $\Psi_k^L, \Psi_k^U : \Omega \to \mathbb{R}$, $k \in \mathcal{K}$ be lower semicontinuous functions and ϑ is affine in second argument. Let the following conditions hold:*

1. For every $y, w \in \Omega$,

$$\langle \xi_k^L, \vartheta(y,w) \rangle + \langle \zeta_k^L, (w,y) \rangle \geq 0,$$
$$\langle \xi_k^U, \vartheta(y,w) \rangle + \langle \zeta_k^U, (w,y) \rangle \geq 0,$$

for each $\xi_k^L \in \partial_M \Psi_k^L(w)$, $\zeta_k^L \in \partial_M \Psi_k^L(y)$, $\xi_k^U \in \partial_M \Psi_k^U(w)$ and $\zeta_k^U \in \partial_M \Psi_k^U(y)$, $k \in \mathcal{K}$.

2. For any $y \in \Omega$,

$$(\langle \xi_1^L, \vartheta(y,y) \rangle, \ldots, \langle \xi_p^L, \vartheta(y,y) \rangle) \not\geq 0, \; \forall \xi_k^L \in \partial_M \Psi_k^L(y),$$
$$(\langle \xi_1^U, \vartheta(y,y) \rangle, \ldots, \langle \xi_p^U, \vartheta(y,y) \rangle) \not\geq 0, \; \forall \xi_k^U \in \partial_M \Psi_k^U(y), k \in \mathcal{K}.$$

3. The set valued map

$$\Gamma(w) = \left\{ y \in \Omega \; \middle| \; \begin{array}{l} (\langle \zeta_1^L, \vartheta(w,y) \rangle, \ldots, \langle \zeta_p^L, \vartheta(w,y) \rangle) \not\leq 0, \; \forall \zeta_k^L \in \partial_M \Psi_k^L(y), \\ (\langle \zeta_1^U, \vartheta(w,y) \rangle, \ldots, \langle \zeta_p^U, \vartheta(w,y) \rangle) \not\leq 0, \; \forall \zeta_k^U \in \partial_M \Psi_k^U(y), k \in \mathcal{K}. \end{array} \right\}$$

is closed for all $w \in \Omega$.

4. There exist a compact set $\emptyset \neq G \subseteq \Omega$ and a compact convex set $\emptyset \neq M \subseteq \Omega$, such that for any $y \in \Omega \setminus G$, we can get a $w \in M$, such that $y \notin \Gamma(w)$.

Then (GSVVLI) is solvable on Ω.

15.6 Conclusions

In this article, we have considered generalized vector variational-like inequalities (GMVVLI) and (GSVVLI) with their weaker forms (WGMVVLI) and (WGSVVLI), respectively in terms of limiting subdifferentials and a class of (IVMPP). Under strong LU-invexity of order s hypothesis, we have showed equivalence between the solutions of (GMVVLI), (GSVVLI) and LU-efficient minimizers of order s of (IVMPP). We also establish the equivalence among the solutions (WGMVVLI), (WGSVVLI) and LU-strict minimizers of order s of (IVMPP). Moreover, we have obtained some existence results for the solutions of (GMVVLI), (GSVVLI), with the help of KKM-Fan theorem. Suitable numerical example has been given to justify the significance of obtained results. The results presented in this paper generalize, unify and sharpens the works of Li and Yu [21], Upadhyay et al. [34] and Upadhyay and Mishra [35, 36].

Acknowledgements The first author is supported by SERB, DST, Government of India, through grant number "ECR/2016/001961".

References

1. Al-Homidan S, Ansari QH (2010) Generalized Minty vector variational-like inequalities and vector optimization problems. J Optim Theory Appl 144:1–11
2. Auslender A (1984) Stability in mathematical programming with nondifferentiable data. SIAM J Control Optim 22:239–254
3. Ben-Israel A, Mond B (1986) What is invexity? J Austral Math Soc Ser B 28:1–9
4. Chen SI (2020) Existence results for vector variational inequality problems on hadamard manifolds. Optim Lett 14:2395–2411
5. Craven BD (1981) Invex functions and constrained local minima. Bull Austral Math Soc 24:357–366
6. Cromme L (1978) Strong uniqueness: a far-reaching criterion for the convergence of iterative numerical procedures. Numer Math 29:179–193
7. Deng S (1997) On approximate solutions in convex vector optimization. SIAM J Control Optim 35:2128–2136
8. Giannessi F (1980) Theorems of the alternative quadratic programs and complementarity problems. In: Cottle RW, Giannessi F, Lions JL (eds) Variational inequalities and complementarity problems (1980)
9. Giannessi F (2000) Vector variational inequalities and vector equilibria. Kluwer Academic, Mathematical Theories, Dordrecht
10. Golestani M, Sadeghi H, Tavan Y (2018) Nonsmooth multiobjective problems and generalized vector variational inequalities using quasi-efficiency. J Optim Theory Appl 179:896–916
11. Green J, Heller WP (1981) Mathematical analysis and convexity with applications to economics, vol 1, Handbook of Mathematical Economics, pp 15–52
12. Gupta D, Mehra A (2008) Two types of approximate saddle points. Numer Funct Anal Optim 29:532–550
13. Gupta P, Mishra SK (2018) On Minty variational principle for nonsmooth vector optimization problems with generalized approximate convexity. Optimization 67:1157–1167
14. Hanson MA (1981) On sufficiency of the Kuhn-Tucker conditions. J Math Anal Appl 80:545–550
15. Hartman P, Stampacchia G (1980) On some nonlinear elliptic differential functional equations. Acta Math 115:153–188
16. Jabarootian T, Zafarani J (2006) Generalized invariant monotonicity and invexity of nondifferentiable functions. J Global Optim 36:537–564
17. Jeyakumar V, Mond B (1992) On generalized convex mathematical programming. J Austral Math Soc Ser B 34:43–53
18. Jiménez B (2002) Strict efficiency in vector optimization. J Math Anal Appl 265:264–284
19. Karamardian S, Schaible S (1990) Seven kinds of monotone maps. J Optim Theo Appl 66:37–46
20. Lee GM (2000) On relations between vector variational inequality and vector optimization problem. In: Yang XQ, Mees AI, Fisher ME, Jennings LS (eds) Progress in optimization II: contributions from Australia
21. Li R, Yu G (2017) A class of generalized invex functions and vector variational-like inequalities. J Inequal Appl pp 2–14
22. Mishra SK, Laha V (2016) On Minty variational principle for nonsmooth vector optimization problems with approximate convexity. Optim Lett 10:577–589
23. Mishra SK, Wang SY (2006) Vector variational-like inequalities and nonsmooth vector optimization problems. Nonlinear Anal pp 1939–1945
24. Mishra SK, Upadhyay BB (2013) Some relations between vector variational inequality problems and nonsmooth vector optimization problems using quasi efficiency. Positivity 17:1071–1083
25. Mishra SK, Upadhyay BB (2015) Pseudolinear functions and optimization. CRC Press, Chapman and Hall
26. Moore RE (1966) Interval analysis. Prentice-Hall, Englewood Cliffs, New Jersey

27. Moore RE (1979) Methods and applications of interval analysis. SIAM Studies in Applied Mathematics, Philadelphia
28. Mordukhovich BS (2006) Variational analysis and generalized differentiation I: basic theory. Springer, Berlin
29. Mohan SR, Neogy SK (1995) On invex sets and preinvex functions. J Math Anal Appl 189:901–908
30. Oveisiha M, Zafarani J (2013) Generalized Minty vector variational-like inequalities and vector optimization problems in Asplund spaces. Optim Lett 7:709–721
31. Rahtu E, Salo M, Heikkilä J (2006) A new convexity measure based on a probabilistic interpretation of images. IEEE Trans Pattern Anal Mach Intell 28:1501–1512
32. Reiland TW (1990) Nonsmooth invexity. Bull Austral Math Soc 42:437–446
33. Smith P (1985) Convexity methods in variational calculus. Research Studies Press, Letchworth
34. Upadhyay BB, Mohapatra RN, Mishra SK (2017) On relationships between vector variational inequality and nonsmooth vector optimization problems via strict minimizers. Adv Nonlinear Var Inequal 20:1–12
35. Upadhyay BB, Mishra P (2020) On generalized Minty and Stampacchia vector variational-like inequalities and nonsmooth vector optimization problem involving higher order strong invexity. J Sci Res 64:182–191
36. Upadhyay BB, Mishra P (2020) On vector variational inequalities and vector optimization problems. Soft computing: theories and applications. Springer, Singapore, pp 257–267
37. Upadhyay BB, Mishra P, Mohapatra RN, Mishra SK (2019) On the applications of nonsmooth vector optimization problems to solve generalized vector variational inequalities using convexificators. Adv Intell Sys Comput. https://doi.org/10.1007/978-3-030-21803-4_66
38. Ward DE (1994) Characterizations of strict local minima and necessary conditions for weak sharp minima. J Optim Theo Appl 80:551–557
39. Weir T, Mond B (1988) Pre-invex functions in multiple objective optimization. J Math Anal Appl 136:29–38
40. Yang XM, Yang XQ, Teo KL (2003) Generalized invexity and generalized invariant monotonicity. J Optim Theo Appl 117:607–625
41. Zhang J, Zheng Q, Ma X, Li L (2016) Relationships between interval-valued vector optimization problems and variational inequalities. Fuzzy Optim Decis Mak 15:33–55

Chapter 16
Portfolio Optimization Using Multi Criteria Decision Making

Namita Srivastava

Abstract Finance represents "money" and the process of acquiring needed funds. The basic idea of financial mathematics is making appropriate decisions in the face of uncertainty. Fuzzy set theory, soft computing, optimization are some of the tools for handling the uncertainty in finance. One of the key objectives of financial mathematics is to construct the best investment strategy that minimizes risks and maximizes the return in the real world. The portfolio optimization model is a mathematical model in which return, risk, dividend and liquidity are objective functions. The main objective is to develop optimal portfolio in the increasingly important areas of Multi Criteria Decision Making (MCDM).

Keywords MCDM · MODM · Mathematical programming · Goal programming · SWOT analysis

16.1 Introduction

1.1 Globalization and integration of economies have made financial decision making a complex process involving a number of variables and constraints. Various tools have been used by the decision makers to overcome these complexities. Traditional investigative approaches like econometric and statistical analysis are based on presentation of structured and clean data, which is not the case with the data available to the financial analyst or the decision maker.

1.2 With all the constraints and complexities researchers and decision makers have sought to address the problem of financial decision making by adopting an integrated approach with realistic optimizations, using the most sophisticated and quantitative analysis tools. This now makes the problem of financial decision making a mathematical problem whose solution aids the decision maker to make the best possible choice amongst available alternatives.

1.3 1950 onwards, mathematics and operation research has been widely used as a tool to address the problems of financial management and decision making. Over

N. Srivastava (✉)
Department of Mathematics, MANIT Bhopal, Bhopal, India

the last three decades "Multi Criteria Decision Making Technique" has evolved as a major tool for addressing this problem. Roy, 1988, [11] used a multi criteria approach to deal with such ill structured and unclean data to overcome the various problems faced by the financial decision makers. The Multi Criteria Decision Making Model is now widely used by the decision makers when faced with multiple and conflicting decision variables (such as goal, constraints, criteria and objectives).

1.4 A portfolio for any business is much wider and complex in connotation than for an individual. The business portfolio will start from the sources of funding and will include all the current and planned business activities in the future, along with the investments of surpluses while maintaining adequate liquidity at all times for smooth conduct of business activities. Hence the Financial decision making for businesses will be much more complex and will acquire the colour of organizational values and culture.

1.5 To explain the problem of financial decision making and the complexities involved in the same we may consider the example of Company XYZ, which has a cash surplus of a few million at the closure of the financial year. The management of the company is considering the option to invest these surplus funds in various available mutual funds so that investments are made with requisite liquidity. The mutual funds under consideration have been grouped into three categories, by the investment consultants according to their ratings and performances, as high-risk high returns, medium risk medium returns and low risk low returns. The economic data of the country available do not show a very vibrant economy at this point of time and there is a factor of uncertainty in the government policies in near future. Let's assume there are five people, A, B, C, D, E on the board of directors each with their own opinions and preferences for making such investments. A prefers high returns at any cost, B who is a safe player prefers low returns with high security, C prefers both security and high returns, D does not like to invest in mutual funds but would like to invest these funds in the human capital of the company, and E would like to invest in the expansion projects of the company. The Chief Executive Officer has given the job of devising a suitable investment proposal which should be acceptable to all the directors, to financial analyst Mr. FA in the company, who is having sleepless nights as the date of board meeting wherein this investment proposal has to be finalized is approaching fast. If we analyze this problem then we would be able to make out the complexities and uncertainties which shroud the financial decision making, but these decisions cannot wait as any delay in making such decisions is not only the financial loss but also the loss of opportunity etc. To handle the situations as described here Analysts and decision makers would employ multi criteria decision making algorithms to arrive at best possible options and make the decision accordingly.

16.2 An Overview of Multi Criteria Decision Making (MCDM) Process

2.1 Multi Criteria Decision Making process is a complex intermixture involving both quantitative and qualitative analysis and interpretation of available data. Over the years researchers and analysts have adopted various methods and tools for selecting the probable and possible solutions. For a mathematician, this problem has been considered as a problem in Operation Research, for designing optimal computational tools for decision making in the maze of available financial data. These techniques have been used in various other fields like energy, environment, sustainability, supply chain management, quality management, construction and project management, production management etc.,

2.2 Like Multi Attribute Utility Theory, MCDM too involves choice and selection of elective ratings and their weight by the decision maker. These ratings and weights represent the values determined as per predefined utility functions. The complete picture is interdependent on the suggested alternatives available to the analyst or decision maker. Choices with higher utility become a natural preference for decision makers.

2.3 In case of probabilistic obscure payoffs, the decision making is not only complex but also quite vague and vulnerable. Abstraction and vagueness in available data and its conversion into reliable information will not be possible for a number of reasons such as fractional ignorance, unclean data and unavailable information. Even future events which are beyond the control of the decision maker can have influence on the decision. The situation is like that of a black box thus traditional MCDM techniques fail to provide reliable and optimistic solutions in such situations.

2.4 Decision making can be simply understood as making the most logical and best choices from the available alternatives. Depending on the problem at hand, MCDM techniques provide a structured and logical approach for coming at available alternatives and then choosing the best. MCDM process has special qualities, and is able to deal with the numerous and clashing criteria, various estimation models and discrete alternatives.

2.5 The aim here is to study various MCDM techniques which have been adopted by the decision makers and to test the different accessible methodologies for the expansion of MCDM into group decision-making circumstances for the treatment of uncertainty.

2.6 Depending on the type of problem, various steps and elements involved in the MCDM process of decision making are illustrated as follows.

Let the set of decision variables selected for making the decision by the decision maker be represented as "x", then the various steps leading to the final decision are represented by the following equations.

Let the set of alternatives available be "A", the function of these decision variables can be represented as

$$A = f(x) \tag{16.1}$$

Set of criteria for suitable outcomes if represented as "y", the set of outcomes "B" on the basis of alternatives determined at (Eq. 16.1) will be:

$$B = g(A, y) = g((f(x), y) \qquad (16.2)$$

If the preferences which are there are represented as a set "Z" then the final decision "C" on the basis of outcomes and preferences can be represented as:

$$C = h(B, z) = h(g((A, y), z) = h(g(f(x), y, z) \qquad (16.3)$$

2.7 Thus it is seen that suitability of the decision taken is a function of the decision variables, criteria and preference simpliciter. These three variables "x", "y" and "z" taken together can be said to be the decision framework. The interplay between the three variables is dependent on the functions "f", "g" and "h" which are to be determined by the application of quantitative and qualitative techniques. With the advances in computation technologies both digital and analog, it is now possible to determine these functions quite precisely and accurately. Though we have represented the entire process in form of three simple equations but still the computation of the functions is not a simple and straightforward task and no unique solution suitable in all situations can be determined as each problem of decision making is unique with its own set unstructured and unreliable data.

16.3 Classification of Multi Criteria Decision Making Processes

3.1 The first step towards decision making is identification of decision variables and alternatives. These decision variables and alternatives can be discrete or continuous. In continuous choice space the available decision variables and alternatives are infinite but not very well defined whereas in the discrete choice space they are limited and well defined. Broadly depending on choice space, whether continuous or discrete, the MCDM process can be classified as "Multi objective decision-making process (MODM)" and "Multi Attribute Decision Making Process (MADM)".

3.2 Irrespective of the fact whether the choice space is continuous or discrete, the problem of decision in a multi criteria environment requires proper identification and aggregation of the decision variables in the form which they can be compared and evaluated. This aggregation can be represented by the function g_{ij} in a two-dimensional choice space as follows:

$$g_{ij} = g_{ij}(x_i, y_j)$$
$$\text{such that } g_{ij} > g_{i-1, j-1} \text{ and so on} \qquad (16.4)$$

Table 16.1 Known methods of Quantitative Evaluation of Decision Variables

Weighted methods	Sequential elimination	Mathematical programming	Spatial proximity analysis
Deduced preferences	Option verses standard examination crosswise over attributes-disjunction and conjunctive constraints	Global objective—linear programming	Iso preference graphs—indifference graphs
Directly assessed preferences: trade–offs (general aggregation)	Alternative versus alternative: comparison across attributes	Goal constraints—goal programming	Multi-dimensional, non metric scaling
Directly assessed preferences-maximum (specialized aggregation)		Immediate or local objective—interactive multi criteria programming	Graphical preferences

3.3 Such aggregation of the decision variables or factors into comparable alternatives is crucial and the starting point of the decision-making process. Quantitative evaluation of the decision variables, alternatives against the known and unknown constraints defining the criterion for decision making, is an essential and integral part of the MCDM process. Since the techniques which are employed are well known to mathematicians and analysts without going into unnecessary details of the same, we highlight those which have been employed as per the available literature in Table 16.1.

3.4 MODM is employed for planning the best choice with various goals depending on the consistent decision variables as per the requirements, whereas MADM is used for selecting the best available options amongst the available options.

3.5 A comparison between the two techniques on the basis of features and applications of each of each is given in Table 16.2.

16.4 Quantitative Techniques Applied for the MCDM

4.1 **Mathematical Programming**: Traditionally the problem of mathematical programming in a two-dimensional space is represented by the equation

$$u_{ij} = \text{Min Max}\big(g_{ij}(x, y)\big) \tag{16.5}$$

where (x, y) belong to the set representing choice space, and u is the most optimum alternative.

Table 16.2 Comparison of MADM and MODM techniques

Feature	MADM	MODM
Basis of criteria	Attributes	Objectives
Basis of model	Alternatives identification	Targets and their relationship with the criterion
Choice space	Discrete	Continuous
Objectives	Implied	Well defined and explicit
Alternatives available	Finite	Infinite
Control	Very limited	Significant
Basis of decision	Outcome oriented	Process oriented
Applications	Defined practical problems	Undefined suitable for research and design

However, such a simplistic optimization may not be true in the case of decision-making processes for the reason of uncertainties, conflicts and preferences. Therefore, the most optimal solution may not truly represent the most optimum decision. In the case of decision making it is not the optimal solution that is acceptable but the compromise solution which suits all criteria, uncertainties and preferences will be acceptable one. So, the equation at 5, will accordingly be modified as follows:

$$\{u_{ij}\} = (g_{ij}(x, y)) \qquad (16.6)$$

where $\{u_{ij}\}$ represents the set of all the alternatives based on the aggregation of decision variables in the choice space. Choice of the most suitable alternative is then worked out on the basis of various methodologies both iterative and interactive depending on the factors of uncertainties, conflicts and preferences. These methodologies are based on the improvements and tradeoffs for such improvements. Several such methodologies have been suggested in the available literature by Benayoun et al. (1971) [17], Zions and Wallenius (1976) [8], Wierzbicki (1980) [6], Steuer and Choo (1983) [16], Korhonen (1988) (Boran, Göztepe et al. 2008), Korhonen and Wallenius (1988) [9], Siskos and Despotis (1989) [4], Lofti et al. (1992) [4].

4.2 **Goal Programming**: This approach has been developed by Charnes and Cooper (1961) [15]. This approach is suitable in cases where the objective is not well defined but is set of many objectives that are desirable. In case of the single objective such as profit maximization the problem can be defined in a single dimensional space. But as the number of objectives sought to be achieved multiply the problem assumes the character of a multidimensional problem and the solution needs to be determined in multi-dimensional space, for optimal allocation to achieve the most optimal solution which satisfies all the objectives aimed at. Such multi-dimensional problems have been dealt by adopting the procedure of goal programming, which can be represented in following form:

$$u_{ij} = (g_{ij}(x, y)) + \text{Min Max } h_{ij}\left(d_{ij}^+, d_{ij}^-\right) \qquad (16.7)$$

where u_{ij} is targeted goal value, g_{ij} is the function for aggregation of the decision variables in the decision space, h_{ij} is the function of deviation variables and d_{ij}^+, d_{ij}^- are deviations from the targeted value.

The above formulation shows that in the Goal Programming approach, the objective function has now been used as a constraint in the goal optimization. The simplicity of this approach makes it the most popular choice amongst the researchers and analysts.

4.3 Outranking relation theory: As per this theory developed by Roy [10, 13]], the outranking relation is a binary relation enabling the analyst to determine the behavior of alternative u_i over the alternative u_j on the basis of some outranking feature. The selection is more appropriate if it can be shown that $u_i \geq u_j$ and there is no contrary possibility or evidence. This scheme involves a two stage processes. In the first stage, the set of suitable alternatives $\{u_i, u_j\}$ is determined and the second stage is determination of outranking character, on the basis of which suitable alternatives are ranked against each other. ELECTRE methods [12] and PROMETHEE methods [2] are the most widely used methods. These methods are widely discussed in literature.

4.4 Utility function Methods: These are based on the multi-attribute utility theory which extends the utility theory to the multidimensional space. The utility function is defined on the basis marginal utility of alternative against the predefined criterion and the corresponding trade off.

$$U(u_{ij}) = \sum_{m=1}^{n} p_m h_m g_{ij}(x, y) \qquad (16.8)$$

where $U(u_{ij})$ is a linear additive function for alternative u_{ij}.
p_m is constant representing the trade-off (generally $\Sigma\, p_m = 1$).
h_m is the marginal utility function for alternative u_{ij}

$$\text{If}, U(u_{ij}) \geq U(u_{kl}), then u_{ij} > u_{kl} \qquad (16.9)$$

Then, u_{ij} becomes the preferred choice over u_{kl}.

The additive utility function is developed on the basis of information in respect of the criteria trade-offs and the form of the marginal utility functions. This information is to be made available to the analyst by the decision makers through interactive procedures which makes this method too cumbersome and time consuming. To overcome these problems, a preference disaggregation approach (PDA) [2] has been suggested in the literature. The repetitive nature of the financial decisions and real time support which is necessarily makes PDA as one of the most preferred methods for financial decision making.

4.5 Several variations of utility function methods have been discussed in the literature under the category of Utility Additive Method (UTA) [14] and Utilities

Additive and Discriminate Methods (UTADIS) [18] methods. These methods are more refinements of utility function methods and provide for more determination of the utility functions based on the past preferences and practices of the decision maker and his judgement policies. By following these methods, the interactive processes are made iterative in nature and tend to reduce the time spent in getting the preference and judgement policies interactively of the decision maker each time.

4.6 Multi Group Hierarchical Discrimination Method extends the PDA framework of UTADIS methods to complex sorting and classification problems involving multiple groups. This method addresses the sorting problem, through a hierarchical procedure, wherein the groups are distinguished progressively starting from the group of most preferred alternatives and then discriminating between the alternatives in the other groups. A detailed discussion on this method is available in Zopounidis and Doumpos (2000a) [5].

4.7 Rough Set Theory proposed by Palwak (1982) [7] has been used as a tool to determine the dependencies amongst the attributes and determine their significance. It has been found very effective in case of unclean and ill structured data. This theory has been applied to choose and ranking problems also. The rationale behind the application of this theory to decision making is ascribed to the fact that some information is associated with every alternative which will aid the decision-making process. This information can be in relation to condition (criteria and alternatives) and the decision. The traditional framework of the rough set theory has evolved as a preference modelling framework. The rough set approach addresses the criteria, i.e. attributes with preference ordered domains, and groups. The rough approximations are defined as per the dominant relationship and not on the basis of indiscernibility relationships. The decision rules so developed form the basis of a preferential model.

16.5 Financial Decision Problem

5.1 We have briefly discussed the various quantitative methods and models that have been developed and are available in literature to handle the problem of financial decision making. In general, the entire process of financial decision making can be described in a basic eight step process as depicted sequentially in Fig. 16.1.

5.2 The financial decision-making problem in case of investment to be made is akin to portfolio optimization in terms of criteria identified and references. As stated earlier the solution to such a problem is not the most optimal but a compromised solution.

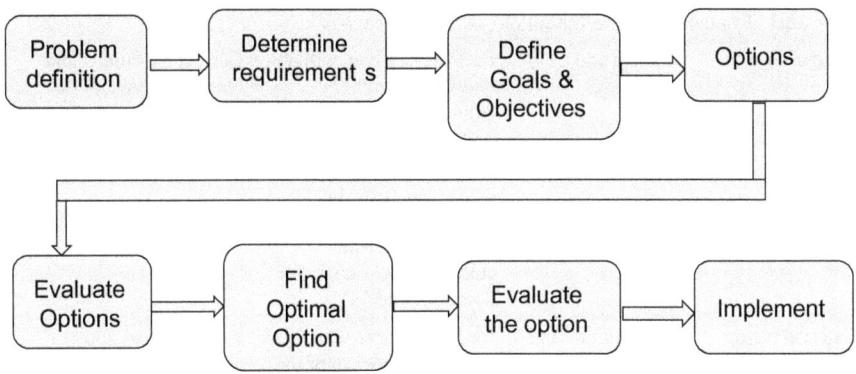

Fig. 16.1 Process of financial decision making

16.6 Risk Modeling and SWOT Analysis

6.1 Czars in business prefer using SWOT (Acronym for Strength, Weakness, Opportunity and Threat) Analysis to evaluate the available options for positioning themselves while finalizing their strategies optimally. SWOT analysis an analytical and qualitative tool has been used for all types of decision making and involves logical evaluation of all the available options by the decision makers.

6.2 Modern SWOT analysis will be presented as a 2 × 2 matrix in two pair of dimensions i.e. Strength and Weakness being mapped against opportunities and threats. The Template is normally a four quadrant template with each quadrant identified as Strength, Weakness, Opportunity and Threats. Thus logically mapping the internal factors within the business organization i.e. Strength and Weakness against the external factors i.e. Opportunities and Threats. The entire paradigm is as indicated in Table 16.3.

6.3 Thus the SWOT matrix that would be derived in the process will be represented graphically in four quadrants of the two dimensional chart in the following manner (Fig. 16.2).

6.4 By application of the SWOT analysis of the available options, the decision maker can determine all the options vis a vis the internal and external factors and also in terms of long and short terms goals and objectives he has set for himself. Such analysis will aid in the risk modelling and assignment of risk scores qualitatively which can then be compared and evaluated against those obtained quantitatively by the application of MCDM methods. Such an analysis will provide a framework whereby quantitative and qualitative process will be used simultaneously to evaluate the available portfolio and suggest the long and short terms modifications in the portfolio continually on real time basis.

6.5 Jernej Prisenk and Andreja Borec, applied such an approach for evaluation of production and marketing of the food products. Srdjevic, Z., Bajcetic, R. & Srdjevic, B. used similar approach for identifying the decision making criteria

Table 16.3 Evaluation criteria for options in SWOT analysis

Internal factors	Strength and weaknesses	Factors clearly identifiable and which can be controlled by the decision maker. Internal factors which would influence the decisions. These factors can be identified exactly and modeled with sufficient accuracy in the decision process	Current Situations and conditions. Available resources and investment opportunities
External factors	Opportunities and threats	External factors which are beyond the control of the decision maker. Create uncertainties and risks to the decision making process. These factors are not easy to identify and will be based on probabilistic evaluation of future trends and need to be modelled with the factor of uncertainty associated with them	Future factors and uncertainties, dependent on market and economic parameters on a later date

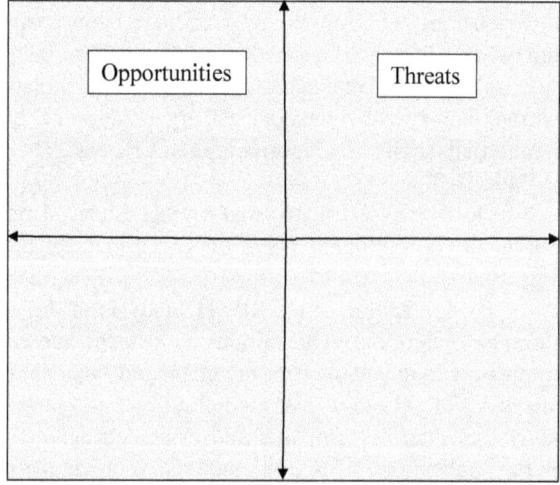

Fig. 16.2. SWOT 2 × 2 matrix for evaluation of options

on the basis of SWOT/ PESTLE analysis. Iraj Mahdavi, Hamed Fazlollahtabar, Mohammad Mahdi Paydar and Armaghan Heidarzade analyzed the role of information technology in Industrial Development in Iran, by applying the MCDM and SWOT factors. Aikaterini Papapostolou, Charikleia Karakosta, Georgios Apostolidis and Haris Doukas applied a combination of SWOT analysis with multi criteria methods viz, AHP and fuzzy TOPSIS, to assess crucial factors in the implementation of cross border cooperation in the field of RES and prescribe strategy for its successful implementation.

6.6 Ample literature is available where the qualitative tools in form of SWOT analysis have been used along with the quantitative approach of Multi Criteria Decision Making models for evaluation of available options. The literature very clearly suggests that both qualitative and quantitative approaches can be used in tandem for portfolio management in complex environments.

16.7 Advantages of MCDM Methods

In decision making for conflicting objectives in different units MCDM provides a good help. The main benefit of MCDM is that it is very lucid, the choices created provide better clarity to other people, providing methods in organizing and working through the data, and helping individuals better perceive the problem from their own and others' viewpoints. Following are the main advantages of MCDM.

- It is open and clear.
- The selection of any decision problem is available for study and change, if it is felt to be improper.
- Various goals can be considered simultaneously.
- As it is a multi-disciplinary method, it can be very well applied to a wide field of decision-making problems in real life.
- It gives a complete approach that shows trade-offs between the conflicting issues.
- It is a significant way of communication within the decision-making group and also for a wider community.
- Its scores and weights when utilized are likewise defined and are created by built-up procedures. Those can likewise be cross-referred to various origins of data information on relative qualities and rectified important if necessary.
- It gives a functional decision process by encouraging to fuse genuine subjectivity and experience of the decision-makers.
- It assists with seeing what might be the outcomes of providing distinctive requests of significance to various goals or making various appraisals to the presentation of the accessible alternatives against various destinations.
- It is simple to adopt appraisal led design, which can be refined with the development process.
- Management problems can be structured properly by MCDM.
- The model can be used to serve as focus for conversation.
- It offers a technique that prompts rational, legitimate and logical decisions.
- It speeds up the problem-solving system in any association.
- It deals with blended sets of data including qualitative and quantitative measurements.
- It is designed to empower participative planning and decision making.

- It enables the DM to understand the situation and arrive at the best solution as decision is very transparent and constructive.
- It viably eliminates personal favors in the decision of a specific other alternative.
- Goals and objectives can be better understood by MCDM.
- Selected decision criteria are properly evaluated.
- The best alternative on which the consensus is developed is identified and promoted in MCDM.
- It provides the most cost-effective decision to the Decision maker for each decision criteria between the conflicting criteria.

16.8 Conclusion and Future Course of Action

8.1 In this paper we have reviewed the existing literature on the applications of Multi Criteria Decision Making processes in the Financial Decision making. Reviewing the literature provides interesting insight into application of these methods for portfolio optimization, in complex business environments. The complexities multiply substantially in the globalized environments as the factor impacting the decisions have multiplied substantially. The modeling of decision variables have become as complex as the global environment with cross border factors impacting the fiscal health of the businesses and corresponding investment opportunities. Market integration and cross capital flows create uncertainties and complicate the decision making process.

8.2 Number of factors which are uncertain are not amenable to quantitative analysis for the reason of non-availability of reasonably clean and tested data. These factors need to be evaluated by application of the qualitative techniques and applied in the decision making process. Though the quantitative techniques have been developed from 1960 onwards and we have referred to a lot of published literature available on the subject, the integrated approach making use of both qualitative and quantitative techniques in emerging uncertain fiscal environments is a recent phenomenon and is in a very nascent stage. Research in future needed to bridge this gap and qualitative methods need to be applied by the decision makers in tandem with the quantitative multi criteria decision making methods for better optimization of the portfolios.

References

1. Boran S et al (2008) A study on election of personnel based on performance measurement by using analytic network process (ANP). Int J Comput Sci Netw Secur 8(4):333–338
2. Brans J-P, Vincke P (1985) Note—a preference ranking organisation method: (The PROMETHEE Method for Multiple Criteria Decision-Making). Manage Sci 31(6):647–656
3. Brans J-P et al (1986) How to select and how to rank projects: the PROMETHEE method. Eur J Oper Res 24(2):228–238

4. Cavalcante C, De Almeida A (2007) A multi-criteria decision-aiding model using PROMETHEE III for preventive maintenance planning under uncertain conditions. J Qual Maintenance Eng
5. Cheng, E. W. and H. Li (2001) Analytic hierarchy process. Measur Bus Excellence
6. Hwang C-L Yoon K (1981) Methods for multiple attribute decision making. In: Multiple attribute decision making. Springer, pp 58–191
7. Kumar NV, Ganesh L (1996) A simulation-based evaluation of the approximate and the exact eigenvector methods employed in AHP. Eur J Oper Res 95(3):656–662
8. Lin C, Wu W-W (2004) A fuzzy extension of the DEMATEL method for group decision making. Eur J Oper Res 156(1):445–455
9. Meade L, Sarkis J (1999) Analyzing organizational project alternatives for agile manufacturing processes: an analytical network approach. Int J Prod Res 37(2):241–261
10. Roy B (1968) Classement et choix en présence de points de vue multiples. Revue française d'informatique et de recherche opérationnelle 2(8):57–75
11. Roy B (1987) Des critères multiples en recherche opérationnelle: pourquoi? Université Paris-Dauphine
12. Roy B (1991) The outranking approach and the foundations of ELECTRE methods. Theory Decis 31(1):49–73
13. Roy B, Bouyssou D (1993) Aide multicritère à la décision: méthodes et cas. Economica Paris
14. Saaty TL (2000) Fundamentals of decision making and priority theory with the analytic hierarchy process. RWS Publications
15. Saaty TL, Vargas LG (2013) The analytic network process. In: Decision making with the analytic network process. Springer, pp 1–40
16. Satty T (2001) Desion making with dependence and feedback: the analytic network process. RWS Publications, Pittburgh
17. Triantaphyllou E et al (1997) Determining the most important criteria in maintenance decision making. J Qual Maintenance Eng
18. Zadeh LA (1996) Fuzzy sets. In: Fuzzy sets, fuzzy logic, and fuzzy systems: selected papers by Lotfi A Zadeh. World Scientific, pp 394–432.

Chapter 17
Uncertain Multi-objective Transportation Problems and Their Solution

Vandana Y. Kakran and Jayesh M. Dhodiya

Abstract Nowadays, it is seen that the transportation of goods is a crucial part in each and every sector of our Indian economy. The parameters like transportation cost, supply and demand etc. involved in the Transportation problem (TP) cannot be always precisely defined and can vary due to certain reasons like weather conditions or incomplete information. This chapter focusses on studying the multi-objective TP (MOTP) in the uncertain domain area. To solve the uncertain MOTP model, Expected value and Dependent chance constraint models have been developed using the basic concepts of uncertainty theory which are further converted into their deterministic forms for computational work. The deterministic form of DCCM model is further converted into linear model with the help of Charnes and Cooper's transformations. The deterministic models are then solved using the fuzzy programming technique and weighted sum methodology to obtain the compromise solution. At last, numerical illustration is also given to demonstrate the working of the proposed models.

Keywords Multi objective transportation problem · Uncertain programming · Dependent chance programming · Fuzzy programming technique

17.1 Introduction

Transportation problem (TP) is an essential segment in this economic and competitive world. Minimizing the total amount for shipping a product from various origins to destinations is the principal objective in traditional TPs. The basic model of TP was first introduced by Hitchcock [15] to obtain minimum value of the shipping cost/amount for supplying the products from several sources to destinations. As per the model given by Hitchcock [15], TP essentially manages with two dimensions i.e. available sources (supply) and available destinations (demand). In the existing economic environment, single objective TP fails to satisfy the requirements of the

V. Y. Kakran (✉) · J. M. Dhodiya
Department of Applied Mathematics and Humanities, S.V. National Institute of Technology, Surat, India

decision makers for the complex real world problems which encourages the introduction of multi objective TPs (MOTPs). Various problems in this current era can be expressed as MOTPs and their solution needs the attention of conflicting objectives which can be observed in the MOTPs. In MOTPs, the notion of optimal solution turns insignificant and a new concept of pareto optimal (non-dominated) solution is introduced here which states that there can be no improvement in any of the objective function without compromising in the optimal value of at least one objective function. The MOTPs have a number of alternative Pareto-optimal solutions.

MOTPs have been investigated by several researchers in the past years. Lee and Moore [21] focussed on optimizing TP with multiple objective functions. Delphi [10] proposed three heuristic algorithms to solve MOTPs. Zimmerman [42] introduced fuzzy programming technique (FPT) for solving linear problems with multiple objectives. Bit et al. [5] developed an algorithm using FPT to obtain the solution of MOTPs in linear form. Verma et al. [35] utilized the membership functions of non-linear type in FPT for solving MOTPs. Zangiabadi and Maleki [39, 40] solved MOTPs using fuzzy goal programming approach and later, they applied the same approach with non-linear membership functions for obtaining the solutions. Kakran and Dhodiya [17] have also utilized fuzzy programming technique for solving MOTP. A MOTP model having non linear shipping cost with multi-choice demand has been studied by Maity and Kumar Roy [25]. Nasseri and Bavandi [29] presented MOTP with multi-choice parameters and applied fuzzy approach to find the compromise solution.

Upto now, assuming that the parameters involved in MOTP are some fixed quantities, numerous effective algorithms were presented by many researchers for solving MOTPs assuming that the parameters involved in it are some fixed quantities. However, defining these parameters accurately is not always possible in real life problems. Therefore, to handle imprecise environments, Zadeh [38] firstly presented the theory of fuzzy sets and Moore [28] gave the concept of interval numbers. Liu [22] developed another way of handling indeterminacy called uncertainty theory and enhanced it later [23]. Chanas and Kuchta [6] defined TP assuming the cost co-efficients as fuzzy numbers and proposed a solution algorithm for it. Gupta and Kumar [13] proposed a method to solve linear MOTP with parameters considered as fuzzy numbers. Vinoliah and Ganesan [37] studied the fuzzy MOTP to obtain the properly efficient solutions. Christi and Kalpana [8] obtained the solution of MOTP in the fuzzy environment with non linear membership functions. Biswas and Modak [4] developed a fuzzy goal programming model to obtain the solution for unbalanced MOTP under the fuzzy random environment. Salama et al. [33] wrote an article specialized multi-objective unbalanced transportation data problems in the fuzzy random environment by using fuzzy programming technique. Bagheri et al. [1, 2] focussed on fully fuzzy and fuzzy MOTP and obtained the solution using DEA approach. Khalifa [19] utilized a signed distance method for MOTP under fuzzy environment. Mahajan and Gupta [24] studied a balanced MOTP assuming all the involved variables as intutionistic fuzzy numbers. Gupta et al. [14] studied multi objective capacited TP in fuzzy environment with mixed type of constraints. Jana et al. [16] gave the application of FPT to solve solid TP with additional constraints.

Latpate and Kurade [20] studied MOTP for crude oil with multi index using fuzzy NSGA II. Vijayalakshmi and Vinotha [36] explored ecological intuitionistic fuzzy MOTP with non linear cost. Das et al. [9] studied MOTPs by considering the objective function co-efficients and other parameters as interval numbers and obtained the solution for it. Porchelvi and Anitha [32] developed an algorithm to solve MOTP where all the parameters were assumed as interval numbers. Osman et al. [30] presented a FPT for solving MOTP in the rough interval environment. Patel and Dhodiya [31] used grey situation decision theory based on grey number to obtain a compromise solution of interval MOTP. Midya and Roy [27] analyzed MOTP with fixed charges using rough programming. Bera and Mondal [3] studied a credit linked two-stage MOTP in rough and bi-rough environments.

Various researchers have also studied the transportation model under the uncertain environment introduced by Liu [22]. Sheng and Yao [34] used unit costs, supplies and demands as uncertain numbers rather than random or fuzzy variables to formulate the TP model. Kakran and Dhodiya [18] have studied the extended forms of transportation problem in the uncertain environment. Guo et al. [12] studied the TP assuming the costs and random supplies as uncertain variables. Maity et al. [26] studied MOTP under the uncertain environment with cost reliability. Zhao and Pan [41] proposed an uncertain TP model with transfer costs in which all the parameters are regarded as uncertain numbers. Deyi et al. [11] also studied TP in which the transportation cost and the truck times were assumed as uncertain variables.

In this chapter, we have studied MOTP under the uncertain environment introduced by Liu [22]. As the uncertain model of MOTP is difficult to solve, the Expected Value Model (EVM) and Dependent Chance constrained model (DCCM) are utilized for solving the uncertain MOTP which are further converted into their deterministic forms using basic fundamentals of uncertainty. The deterministic form of DCCM model consists of fractional objectives which are converted into the linear objectives with the help of Charnes and Cooper's transformations [7]. These deterministic formulations are then solved using the FPT and weighted sum approach. A numerical example is illustrated at the end and the results obtained for both the models with given methodologies are compared with each other.

17.2 Preliminaries

This section introduces the basic theory of uncertainty.

Definition 1 (Liu [22]) An uncertain measure $\mathcal{M}$ is a function $\mathcal{M} : \mathcal{L} \to [0, 1]$ such that it obeys the stated following axioms:

1. $\mathcal{M}\{\Gamma\} = 1$.
2. $\mathcal{M}\{A\} + \mathcal{M}\{A^c\} = 1$ for any event $A \in \mathcal{L}$.
3. $\mathcal{M}\{\bigcup_{k=1}^{\infty} A_k\} \leq \sum_{k=1}^{\infty} \mathcal{M}\{A_k\}$, for each countable sequence of events $\{A_k\}$.

Here, $\mathcal{L}$ is a σ algebra on a non empty set Γ and the triplet $(\Gamma, \mathcal{L}, \mathcal{M})$ is known as the uncertainty space.

Definition 2 (Liu [22]) For an uncertain variable denoted by η, the uncertainty distribution $\Phi : \mathcal{R} \to [0, 1]$ is defined as

$$\Phi(z) = \mathcal{M}\{\eta \leq z\}, z \in \mathcal{R}.$$

For a linear uncertain variable denoted by $\mathcal{L}(c, d)$, we have

$$\Phi(z) = \frac{z - c}{d - c}, \quad c \leq z \leq d,$$

otherwise it is 0 and 1 if $z \leq c$ and $z \geq d$ respectively. Also $c, d \in \mathcal{R}$ and $c < d$.

Definition 3 (Liu [22]) For any linear uncertain variable $\mathcal{L}(c, d)$, the inverse uncertainty function denoted by Φ^{-1} is given by

$$\Phi^{-1}(\beta) = (1 - \beta)c + \beta d.$$

Theorem 1 (Liu [22]) *For any uncertain variable η, we have*

$$\mathcal{M}\{\eta \leq y\} = \Phi(y), \quad \mathcal{M}\{\eta \geq y\} = 1 - \Phi(y), \quad y \in \mathcal{R}.$$

Theorem 2 (Liu [22]) *If b and s are positive real numbers, then $\mathcal{M}\{b - \eta \geq s\} \geq \beta$ holds if and only if $b - \Phi^{-1}(\beta) \geq s$, where β is the chance level.*

Theorem 3 (Liu [22]) *If the expected value exists for an uncertain variable η, it is given by*

$$E[\eta] = \int_0^1 \Phi^{-1}(\beta) d\beta.$$

For linear uncertain variable $\mathcal{L}(c, d)$, $E[\eta] = \frac{c+d}{2}$.

Theorem 4 (Liu [22]) *Let $\eta_1, \eta_2, \ldots, \eta_n$ be independent linear uncertain variables $\mathcal{L}(c_1, d_1), \mathcal{L}(c_2, d_2), \ldots, \mathcal{L}(c_n, d_n)$ with decision variables $x_1, x_2, \ldots, x_n$. So, when $\overline{f} \in \left[\sum_{j=1}^{n} c_j x_j, \sum_{j=1}^{n} d_j x_j\right]$ then $\mathcal{M}\left\{\sum_{j=1}^{n} \xi_j x_j \leq \overline{f}\right\} = \frac{\overline{f} - \sum_{j=1}^{n} c_j x_j}{\sum_{j=1}^{n} (d_j - c_j) x_j}$, otherwise it is 0 when $\overline{f}$ lies to the left of interval $\left[\sum_{j=1}^{n} c_j x_j, \sum_{j=1}^{n} d_j x_j\right]$ and 1 if $\overline{f}$ lies to the right of interval $\left[\sum_{j=1}^{n} c_j x_j, \sum_{j=1}^{n} d_j x_j\right]$.*

Since the uncertain variables do not follow any justified ordership in the uncertain environment, the critical problem arising in uncertain models is ranking of the uncertain variables. Liu [22] introduced four ranking criterias to overcome this problem. This chapter focuses here only on the Expected value criterion (EVC) and chance criterion given by Liu [22]. For any two uncertain variables η and ξ, EVC says that $\eta < \xi$ iff $E[\eta] < E[\xi]$ and the chance criterion says that $\eta < \xi$ iff $\mathcal{M}\{\eta \geq \bar{y}\} < \mathcal{M}\{\xi \geq \bar{y}\}$ for some predefined level $\bar{y}$.

17.3 Mathematical Model of MOTP

The basic formulation of the mathematical model for MOTP is shown in the model (1).

Model 1:

$$\min Z^k(x) = \sum_{i=1}^{m} \sum_{j=1}^{n} C_{ij}^k x_{ij}, \quad \forall k$$

subject to the given const:

$$\sum_{j=1}^{n} x_{ij} \leq a_i, \quad \forall i,$$

$$\sum_{i=1}^{m} x_{ij} \geq b_j, \quad \forall j,$$

$$x_{ij} \geq 0, \forall i, \quad \forall j$$

Here, C_{ij}^k is the notation used for transportation cost/penalty for shipping products from source i to destination j for the objective function k and a_i, b_j are the notations used to denote the capacity of the supplier i and demand at the destination j respectively. x_{ij} denotes the number of products getting supplied from the source-destination pair (i, j). Here, we have used the notations $\forall i$ to represent $i = 1$ to m, $\forall j$ to represent $j = 1$ to n and $\forall k$ to represent $k = 1$ to p. The above model (1) is formulated by considering all the parameters as crisp numbers. So, to involve indeterminacy of the parameters, the variables C_{ij}^k, a_i and b_j in the MOTP are replaced by uncertain parameters $\xi_{ij}^k, \tilde{a}_i$ and $\tilde{b}_j$ respectively. The MOTP model formulated with uncertain parameters is known as uncertain MOTP model, denoted by UMOTP.

17.4 Uncertain Model of MOTP

Replacing all the crisp parameters C_{ij}^k, a_i and b_j as uncertain variables $\xi_{ij}^k, \tilde{a}_i$ and $\tilde{b}_j$ in the mathematical model (1), we get the uncertain model (2).

Model 2:

$$\min Z^k(x, \xi) = \sum_{i=1}^{m} \sum_{j=1}^{n} \xi_{ij}^k x_{ij}, \quad \forall k,$$

subject to the given const:

$$\sum_{j=1}^{n} x_{ij} \leq \tilde{a}_i, \quad \forall i,$$

$$\sum_{i=1}^{m} x_{ij} \geq \tilde{b}_j, \quad \forall j,$$

$$x_{ij} \geq 0, \forall i, \quad \forall j$$

The model (2) mentioned here is known as an Uncertain MOTP (UMOTP) and does not have any natural ordership in the uncertain variables. So, to rank the uncertain variables, EVC and chance criterion is utilized and such models are stated as EVM and DCCM respectively.

17.4.1 Expected Value Model

The basic concept in EVM is to solve the objective function subject to the given constraints by applying the expected criterion on all the involved uncertain variables in the objectives and the constraints. The formulation of the EVM is given in model (3):

Model 3:

$$\min E\left[Z^k(x, \xi)\right] = E\left[\sum_{i=1}^{m} \sum_{j=1}^{n} \xi_{ij}^k x_{ij}\right] \quad \forall k,$$

subject to the given const:

$$E\left[\sum_{j=1}^{n} x_{ij} \leq \tilde{a}_i\right], \quad \forall i,$$

$$E\left[\sum_{i=1}^{m} x_{ij} \geq \tilde{b}_j\right], \quad \forall j,$$

$$x_{ij} \geq 0, \forall i, \quad \forall j.$$

17.4.2 Dependent Chance Constrained Model

In DCCM model, the objective turns out to maximize the chance of meeting these objectives instead of optimizing the uncertain objective function directly. The DCCM model may be formulated as follows, so that the chance of meeting these objectives in an uncertain environment is maximized:

Model 4:

$$\max \mathcal{M}\{Z_k\} = \mathcal{M}\left\{\sum_{i=1}^{m}\sum_{j=1}^{n} \xi_{ij}^k x_{ij} \leq \overline{f}_k\right\} \quad \forall k,$$

subject to the given const:

$$\mathcal{M}\left\{\sum_{j=1}^{n} x_{ij} \leq \tilde{a}_i\right\} \geq \beta_i, \quad \forall i,$$

$$\mathcal{M}\left\{\sum_{i=1}^{m} x_{ij} \geq \tilde{b}_j\right\} \geq \gamma_j, \quad \forall j,$$

$$x_{ij} \geq 0, \forall i, \quad \forall j.$$

where $\beta_i, \forall i$ and $\gamma_j, \forall j$ are predetermined confidence levels.

17.5 Deterministic Models

17.5.1 EVM Model

The deterministic form equivalent to the uncertain EVM is given in the model (5).

Model 5:

$$\min E\left[Z^k(x,\xi)\right] = \sum_{i=1}^{m}\sum_{j=1}^{n} x_{ij} \int_{0}^{1} \Phi_{\xi_{ij}^k}^{-1}(\alpha) d\alpha \quad \forall k,$$

subject to the given const:

$$\sum_{j=1}^{n} x_{ij} - \int_{0}^{1} \Phi_{\tilde{a}_i}^{-1}(\beta_i) d\beta_i \leq 0, \quad \forall i,$$

$$\int_0^1 \Phi_{\tilde{b}_j}^{-1}(\gamma_j) d\gamma_j - \sum_{i=1}^m x_{ij} \leq 0, \quad \forall j,$$

$$x_{ij} \geq 0, \forall i, \quad \forall j.$$

where $\beta_i, \forall i$ and $\gamma_j, \forall j$ are predetermined confidence levels.

17.5.2 DCCM Model

The DCCM model may be formulated as follows, so that the chance of meeting these objectives in an uncertain environment is maximized:

Model 6:

$$\max \mathcal{M}\{Z_k\} = \mathcal{M} \left\{ \sum_{i=1}^m \sum_{j=1}^n \xi_{ij}^k x_{ij} \leq \overline{f}_k \right\} \quad \forall k,$$

subject to the given const:

$$\sum_{j=1}^n x_{ij} - \Phi_{\tilde{a}_i}^{-1}(1 - \beta_i) \leq 0, \quad \forall i,$$

$$\Phi_{\tilde{b}_j}^{-1}(\gamma_j) - \sum_{i=1}^m x_{ij} \leq 0, \quad \forall j$$

$$x_{ij} \geq 0, \forall i, \quad \forall j.$$

where $\beta_i, \forall i$ and $\gamma_j, \forall j$ are predetermined confidence levels.

17.6 Charnes and Cooper's Transformation for DCCM Model

In this section, we discuss the Charnes and Cooper's transformations [7] used for converting linear fractional problem (LFP) to the Linear problem. Consider a fractional problem:

$$\min = \frac{P'x + \alpha}{Q'x + \beta}$$

Subject to the given constraints:

$$Ax \leq b, \quad x \geq 0,$$

Suppose that $S = \{x : Ax \leq b, x \geq 0\}$ is compact set and $Q'x + \beta > 0$ for each $x \in S$. Assume $t = \frac{1}{Q'x+\beta}$ and $y = tx$, then LFP in the above model converts to the following linear model:

$$\min = P'y + \alpha t$$

subject to the given const:

$$Q'y + \beta t = 1,$$
$$Ay - bt \leq 0, \quad y \geq 0, t \geq 0.$$

This transformation gives a feasible solution (y, t) to the above linear model, when $t > 0$ and if solution $(\bar{y}, \bar{t})$ is optimal to the above linear program, then $\bar{x} = \frac{\bar{y}}{\bar{t}}$ is also optimal to the fractional model.

Now if, $Q'x + \beta < 0, \forall x \in S$, then the assumption changes to $-t = \frac{1}{Q'x+\beta}$, $y = tx$ and accordingly the linear form of the fractional model can be obtained.

17.7 Methods for Solving MOTP

Various solution methods available in the literature can be used to attain the non dominant solution for the optimization problems with multiple objectives (MOOPs). In this chapter, we have utilized two methods: weighted sum method (WSM) and fuzzy programming technique (FPT) for obtaining the non-dominated solutions.

17.7.1 Weighted Sum Method

This method aims in converting the MOOPs into a scalar problem by considering a weighted sum of all the objectives. By multiplying each objective with a user specified weight, it converts multiple objectives into a single objective which can be optimized easily under the given problem constraints. In this method, the objective function is formulated as:

$$\min \sum_{k=1}^{p} w_p Z_p(x)$$

where w_p denotes the weight vector used to give the preference to the individual objective functions. The weight vector can be chosen such that $\sum_{k=1}^{p} w_p = 1$.

17.7.2 Fuzzy Programming Technique for Solving MOTP

In this section, fuzzy approach given by Zimmermann [42] to obtain the solution for MOOPs is discussed. The sequential steps of the FPT for obtaining the solution for MOTP are mentioned as below:

Step 1: Initially, solve each objective of MOTP individually as a single objective problem w.r.t the given constraints of the problem and avoiding all the other objective functions.

Step 2: Obtain the minimum value of Z_k denoted by l_k and maximum value of Z_k denoted by u_k for every p objectives.

Step 3: Formulate the linear membership function $\mu_k(Z_k)$ for all the k objective functions if the objective functions are of minimization type:

$$\mu_k(Z_k) = \begin{cases} 1, & \text{if } Z_k \leq l_k. \\ \frac{u_k - Z_k}{u_k - l_k}, & \text{if } l_k < Z_k < u_k \\ 0, & \text{if } Z_k \geq u_k, \forall k \end{cases}$$

and if the objective functions are of maximization type formulate the linear membership function $\mu_k(Z_k)$ as:

$$\mu_k(Z_k) = \begin{cases} 0, & if\ Z_k \leq l_k. \\ \frac{Z_k - l_k}{u_k - l_k}, & if\ l_k < Z_k < u_k \\ 1, & if\ Z_k \geq u_k, \forall k \end{cases}$$

Step 4: Now, formulate the fuzzy mathematical model as:

$$maximize\ \lambda$$

Subject to the given const:

$$\mu_k(Z_k) \geq \lambda,$$
constraints in the model (5) or (6),
with $\lambda = \min\{\mu_k(Z_k(x))\} \geq 0$

Step 5: Solve the single objective model formed in Step 4. The obtained solution will be a compromise solution for the MOTP.

17.8 Numerical Illustration

Consider the MOTP with two objective functions in the uncertain environment with three sources and four destinations. One aims to evaluate the number of products shipped from available sources to the desired destinations such that both the objective functions are optimized. This problem deals with linear uncertain variables $\mathcal{L}(c, d)$ only.

Using the above numerical data given in Tables 17.1, 17.2, 17.3 and 17.4, the uncertain model (7) can be formulated as:

Model 7:

$$\min \tilde{Z}_1 = (2,4)x_{11} + (4,6)x_{12} + (3,5)x_{13} + (5,7)x_{14} + (2,5)x_{21} + (5,6)x_{22}$$
$$+ (2,8)x_{23} + (3,4)x_{24} + (3,4)x_{31} + (2,7)x_{32} + (2,4)x_{33} + (1,5)x_{34};$$
$$\min \tilde{Z}_2 = (4,5)x_{11} + (3,5)x_{12} + (4,6)x_{13} + (2,4)x_{14} + (5,7)x_{21} + (3,4)x_{22}$$
$$+ (5,7)x_{23} + (2,5)x_{24} + (6,7)x_{31} + (2,5)x_{32} + (6,8)x_{33} + (3,6)x_{34};$$

Table 17.1 Capacity of the sources $\tilde{a}_i$

$\tilde{a}_1$	$\tilde{a}_2$	$\tilde{a}_3$
$\mathcal{L}$ (100,200)	$\mathcal{L}$ (200,300)	$\mathcal{L}$ (150,200)

Table 17.2 Demands at destinations $\tilde{b}_j$

$\tilde{b}_1$	$\tilde{b}_2$	$\tilde{b}_3$	$\tilde{b}_4$
$\mathcal{L}$ (100,120)	$\mathcal{L}$ (90,110)	$\mathcal{L}$ (80,100)	$\mathcal{L}$ (100,120)

Table 17.3 The transportation cost ξ_{ij}^1 for the 1st objective

Mines/Cities	1	2	3	4
1	$\mathcal{L}$ (2,4)	$\mathcal{L}$ (4,6)	$\mathcal{L}$ (3,5)	$\mathcal{L}$ (5,7)
2	$\mathcal{L}$ (2,5)	$\mathcal{L}$ (5,6)	$\mathcal{L}$ (2,8)	$\mathcal{L}$ (3,4)
3	$\mathcal{L}$ (3,4)	$\mathcal{L}$ (2,7)	$\mathcal{L}$ (2,4)	$\mathcal{L}$ (1,5)

Table 17.4 The transportation cost ξ_{ij}^2 for the 2nd objective

Mines/Cities	1	2	3	4
1	$\mathcal{L}$ (4,5)	$\mathcal{L}$ (3,5)	$\mathcal{L}$ (4,6)	$\mathcal{L}$ (2,4)
2	$\mathcal{L}$ (5,7)	$\mathcal{L}$ (3,4)	$\mathcal{L}$ (5,7)	$\mathcal{L}$ (2,5)
3	$\mathcal{L}$ (6,7)	$\mathcal{L}$ (2,5)	$\mathcal{L}$ (6,8)	$\mathcal{L}$ (3,6)

subject to the given const:

$$x_{11} + x_{12} + x_{13} + x_{14} \leq (100, 200);$$
$$x_{21} + x_{22} + x_{23} + x_{24} \leq (200, 300);$$
$$x_{31} + x_{32} + x_{33} + x_{34} \leq (150, 200);$$
$$x_{11} + x_{21} + x_{31} \geq (100, 120);$$
$$x_{12} + x_{22} + x_{32} \geq (90, 110);$$
$$x_{13} + x_{23} + x_{33} \geq (80, 100);$$
$$x_{14} + x_{24} + x_{34} \geq (100, 120);$$

The above uncertain model (7) can solved using the EVM and the DCCM which are obtained as below:

1. *Expected value model*

Using the data values of the numerical given above, the corresponding model (8) equivalent to model (5) is obtained by taking the expected values of the uncertain parameters is given as follows:

Model 8:

$$\min E[Z_1] = 3x_{11} + 5x_{12} + 4x_{13} + 6x_{14} + 3.5x_{21} + 5.5x_{22} + 5x_{23} + 3.5x_{24}$$
$$+ 3.5x_{31} + 4.5x_{32} + 3x_{33} + 3x_{34};$$
$$\min E[Z_2] = 4.5x_{11} + 4x_{12} + 5x_{13} + 3x_{14} + 6x_{21} + 3.5x_{22} + 6x_{23}$$
$$+ 3.5x_{24} + 6.5x_{31} + 3.5x_{32} + 7x_{33} + 4.5x_{34};$$

subject to the given const:

$$x_{11} + x_{12} + x_{13} + x_{14} \leq 150;$$
$$x_{21} + x_{22} + x_{23} + x_{24} \leq 250;$$
$$x_{31} + x_{32} + x_{33} + x_{34} \leq 175;$$
$$x_{11} + x_{21} + x_{31} \geq 110;$$
$$x_{12} + x_{22} + x_{32} \geq 100;$$
$$x_{13} + x_{23} + x_{33} \geq 90;$$
$$x_{14} + x_{24} + x_{34} \geq 110;$$

In order to achieve the solution of the above model (8) using weighted sum approach, we have the formulation as:

$$\min (w_1 E[Z_1] + w_2 E[Z_2]);$$

subject to the given constraints of model (8).

17 Uncertain Multi-objective Transportation Problems and Their Solution

Table 17.5 Solution for EVM with various weight vectors in weight sum method

w_1	w_2	$E[Z_1]$	$E[Z_2]$	Solution
1	0	1442.5	1905	$x_{11} = 110, x_{12} = 40, x_{24} = 85, x_{32} = 60, x_{33} = 90, x_{34} = 25$
0.8	0.2	1450	1830	$x_{11} = 110, x_{13} = 15, x_{24} = 110, x_{32} = 100, x_{33} = 75$
0.6	0.4	1475	1780	$x_{11} = 110, x_{13} = 40, x_{24} = 110, x_{32} = 100, x_{33} = 50$
0.5	0.5	1475	1780	$x_{11} = 110, x_{13} = 40, x_{24} = 110, x_{32} = 100, x_{33} = 50$
0.4	0.6	1475	1780	$x_{11} = 110, x_{13} = 40, x_{24} = 110, x_{32} = 100, x_{33} = 50$
0.2	0.8	1575	1730	$x_{11} = 110, x_{13} = 40, x_{23} = 50, x_{24} = 110, x_{32} = 100$
0	1	1665	1730	$x_{11} = 110, x_{13} = 40, x_{22} = 90, x_{23} = 50, x_{24} = 110, x_{32} = 10$

Where w_1 and w_2 denote the weight parameters used to measure the preference given to the objective functions. The different pareto optimal solutions can be obtained for this transformed single objective problem by considering various combinations of weight vectors such that $w_1 + w_2 = 1$. The obtained results are displayed in Table 17.5.

Now, we apply Fuzzy technique to get the non dominated solution of the model (8) with linear membership function by transforming model (8) into the following model using the steps mentioned in Sect. 17.7.2:

$$maximize\ \lambda$$

subject to the given const:

$$\frac{u_k - Z_k}{u_k - l_k} \geq \lambda, k = 1, 2,$$

and the constraint of the model (8).

where $l_1 = 1442.5, l_2 = 1730$ and $u_1 = 2907.5, u_2 = 3125$ are the minimum and maximum values for objective functions Z_1 and Z_2 respectively. This fuzzy programming model is solved with the help of Lingo optimizing tool to obtain the solution set as given below:

$$\lambda = 0.9688602, x_{11} = 110, x_{13} = 40, x_{23} = 6.56,$$
$$x_{24} = 110, x_{32} = 100, x_{33} = 43.44$$

and remaining all values are zero in this solution vector. The optimal solutions for the objective functions are obtained as $E[Z_1] = 1488.12$ and $E[Z_2] = 1773.440$.

λ gives the minimum value of the two membership functions in the solution set i.e. $\lambda_t = \min\{\mu_t(Z_t)\}$. The individual membership function values λ_1 and λ_2 are obtained as $\mu_1(Z_1) = 0.9688602$ and $\mu_2(Z_2) = 0.9688602$ for the objective functions Z_1 and Z_2 respectively. So, we have $\lambda = 0.9688602 = \min(0.9688602, 0.9688602)$. With the increasing value of λ, the solution is found to be improving for the problem approaching to the individual optimal values of the objectives.

It can be seen from Fig. 17.1 that the speckled line in the two graphs (a) and (b) show the grade of membership function on the vertical line and the corresponding

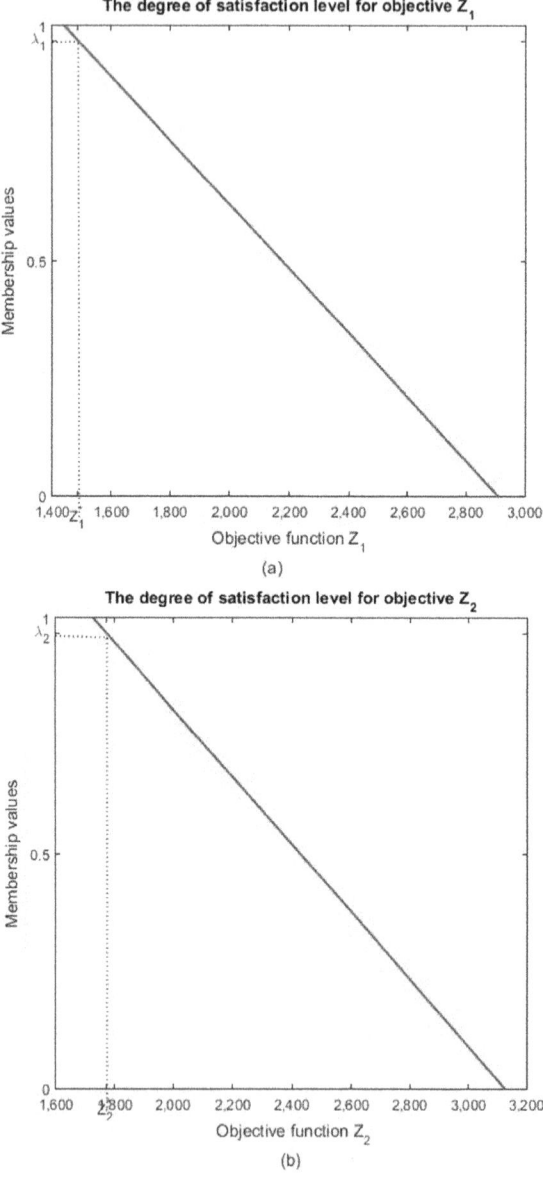

Fig. 17.1 Graph for membership function values versus objective function values

17 Uncertain Multi-objective Transportation Problems and Their Solution

objective value achieved at λ is shown on the horizontal line. For example, the non-dominated value achieved is $(Z_1^* = 1488.12)$ at $\lambda = 0.9688602$ grade of membership in 1st objective whereas the non-dominated value achieved is $(Z_2^* = 1773.440)$ at $\lambda = 0.9688602$ membership value in the 2nd objective.

2. *Dependent Chance constrained model*

The deterministic model of DCCM is obtained with the help of Theorem 2 with values $\overline{f}_1$ and $\overline{f}_2$ the mentioned in Sect. 17.2. The preassumed values of $\overline{f}_1$ and $\overline{f}_2$ are 1950 and 2400 respectively which represent the predetermined maximal costs for the objective functions. Also, the values of all the confidence levels are assumed as 0.9. Thus, the deterministic DCCM model is stated in model (9):

Model 9:

$$\max \mathcal{M}\{Z_1\} = \frac{(1950 - (2x_{11} + 4x_{12} + 3x_{13} + 5x_{14} + 2x_{21} + 5x_{22} + 2x_{23} + 3x_{24} + 3x_{31} + 2x_{32} + 2x_{33} + 1x_{34})}{(2x_{11} + 2x_{12} + 2x_{13} + 2x_{14} + 3x_{21} + 1x_{22} + 6x_{23} + 1x_{24} + 1x_{31} + 5x_{32} + 2x_{33} + 4x_{34})}$$

$$\max \mathcal{M}\{Z_2\} = \frac{(2400 - (4x_{11} + 3x_{12} + 4x_{13} + 2x_{14} + 5x_{21} + 3x_{22} + 5x_{23} + 2x_{24} + 6x_{31} + 2x_{32} + 6x_{33} + 3x_{34})}{(1x_{11} + 2x_{12} + 2x_{13} + 2x_{14} + 2x_{21} + 1x_{22} + 2x_{23} + 3x_{24} + 1x_{31} + 3x_{32} + 2x_{33} + 3x_{34})}$$

subject to the given const:

$$x_{11} + x_{12} + x_{13} + x_{14} \leq 0.9 * 100 + 0.1 * 200;$$
$$x_{21} + x_{22} + x_{23} + x_{24} \leq 0.9 * 200 + 0.1 * 300;$$
$$x_{31} + x_{32} + x_{33} + x_{34} \leq 0.9 * 150 + 0.1 * 200;$$
$$x_{11} + x_{21} + x_{31} \geq 0.1 * 100 + 0.9 * 120;$$
$$x_{12} + x_{22} + x_{32} \geq 0.1 * 90 + 0.9 * 110;$$
$$x_{13} + x_{23} + x_{33} \geq 0.1 * 80 + 0.9 * 100;$$
$$x_{14} + x_{24} + x_{34} \geq 0.1 * 100 + 0.9 * 120;$$

Converting the above model (9) into linear form using the Charnes and Coopers transformations [7], we get:

Model 10:

$$\max \mathcal{M}\{Z_1\} = 1950t - (2y_{11} + 4y_{12} + 3y_{13} + 5y_{14} + 2y_{21} + 5y_{22} + 2y_{23}$$
$$+ 3y_{24} + 3y_{31} + 2y_{32} + 2y_{33} + 1y_{34});$$
$$\max \mathcal{M}\{Z_2\} = 2400t - (4y_{11} + 3y_{12} + 4y_{13} + 2y_{14} + 5y_{21} + 3y_{22} + 5y_{23}$$
$$+ 2y_{24} + 6y_{31} + 2y_{32} + 6y_{33} + 3y_{34});$$

subject to the given const:

$$2y_{11} + 2y_{12} + 2y_{13} + 2y_{14} + 3y_{21} + 1y_{22} + 6y_{23}$$
$$+ 1y_{24} + 1y_{31} + 5y_{32} + 2y_{33} + 4y_{34} = 1;$$
$$1y_{11} + 2y_{12} + 2y_{13} + 2y_{14} + 2y_{21} + 1y_{22} + 2y_{23}$$

$$+ 3y_{24} + 1y_{31} + 3y_{32} + 2y_{33} + 3y_{34} = 1;$$

$$y_{11} + y_{12} + y_{13} + y_{14} \leq (0.9 * 100 + 0.1 * 200)t;$$
$$y_{21} + y_{22} + y_{23} + y_{24} \leq (0.9 * 200 + 0.1 * 300)t;$$
$$y_{31} + y_{32} + y_{33} + y_{34} \leq (0.9 * 150 + 0.1 * 200)t;$$
$$y_{11} + y_{21} + y_{31} \geq (0.1 * 100 + 0.9 * 120)t;$$
$$y_{12} + y_{22} + y_{32} \geq (0.1 * 90 + 0.9 * 110)t;$$
$$y_{13} + y_{23} + y_{33} \geq (0.1 * 80 + 0.9 * 100)t;$$
$$y_{14} + y_{24} + y_{34} \geq (0.1 * 100 + 0.9 * 120)t;$$

The results for the above model (10) using weighted sum method are shown in Table 17.6.

The results shown in Table 17.6 are in terms of variables y and t which are to be retransformed back in terms of original decision variables x_{ij}'s using the transformation $x = y/t$. Say for the case $w_1 = w_2 = 0.5$ the obtained solution set in terms of X decision vector is $x_{11} = 110, x_{22} = 108, x_{24} = 76, x_{31} = 8, x_{33} = 98, x_{34} = 42$ with objective values $Z_1 = 0.902061856$ and $Z_2 = 0.9304124$. Likewise, the X decision vector can be obtained for the other cases of the weighted sum method.

Now, we apply FPT to get the non dominant solution of the model (10) with linear membership function by transforming the model (10) into following single objective problem using the steps described in Sect. 17.7.2:

$$maximize\ \lambda$$

subject to the given const:

$$\frac{Z_k - L_k}{U_k - L_k} \geq \lambda, k = 1, 2,$$

and constraints in the model (10).

where $l_1 = 0.1305286, l_2 = 0.4127580$ and $u_1 = 0.9476923, u_2 = 0.9717742$ are the minimum and maximum values for objective functions Z_1 and Z_2 respectively. This fuzzy programming model can be solved in the Lingo software to get the solution as:

$$\lambda = 0.9361960, t = 0.001288660, y_{11} = 0.1417526,$$
$$y_{22} = 0.1391753, y_{23} = 0.00244044, y_{24} = 0.1011921,$$
$$y_{31} = 0.01030928, y_{33} = 0.1238482, y_{34} = 0.05086979$$

and remaining all values are zero in this solution vector. The optimal solutions for the objective functions are obtained as $\mathcal{M}\{Z_1^*\} = 0.8955540$ and $\mathcal{M}\{Z_2^*\} = 0.9361067$.

Table 17.6 Solution table for DCCM model with various weight vectors in weight sum method

w_1	w_2	$\mathcal{M}\{Z_1\}$	$\mathcal{M}\{Z_2\}$	Solution
1	0	0.9020619	0.9304124	$t = 0.001288660, y_{11} = 0.1417526, y_{22} = 0.1391753, y_{24} = 0.09793814,$ $y_{31} = 0.01030928, y_{33} = 0.1262887, y_{34} = 0.05412371$
0.8	0.2	0.9020619	0.9304124	$t = 0.001288660, y_{11} = 0.1417526, y_{22} = 0.1391753, y_{24} = 0.09793814,$ $y_{31} = 0.01030928, y_{33} = 0.1262887, y_{34} = 0.05412371$
0.6	0.4	0.9020619	0.9304124	$t = 0.001288660, y_{11} = 0.1417526, y_{22} = 0.1391753, y_{24} = 0.09793814,$ $y_{31} = 0.01030928, y_{33} = 0.1262887, y_{34} = 0.05412371$
0.5	0.5	0.9020619	0.9304124	$t = 0.001288660, y_{11} = 0.1417526, y_{22} = 0.1391753, y_{24} = 0.09793814,$ $y_{31} = 0.01030928, y_{33} = 0.1262887, y_{34} = 0.05412371$
0.4	0.6	0.8637703	0.9639175	$t = 0.001288660, y_{11} = 0.1417526, y_{22} = 0.1391753, y_{23} = 0.01435935,$ $y_{24} = 0.1170839, y_{31} = 0.01030928, y_{33} = 0.1119293, y_{34} = 0.03497791$
0.2	0.8	0.8637703	0.9639175	$t = 0.001288660, y_{11} = 0.1417526, y_{22} = 0.1391753, y_{23} = 0.01435935,$ $y_{24} = 0.1170839, y_{31} = 0.01030928, y_{33} = 0.1119293, y_{34} = 0.03497791$
0	1	0.7013245	0.9675497	$t = 0.001158940, y_{11} = 0.04569536, y_{13} = 0.08178808, y_{22} = 0.07483444,$ $y_{23} = 0.03178808, y_{24} = 0.1367550, y_{31} = 0.09105960, y_{32} = 0.05033113$

This solution set can be retransformed back in the terms of original decision variables using the transformation $x = y/t$ which are obtained as follows:

$$x_{11} = 110, x_{22} = 108, x_{23} = 1.893784, x_{24} = 78.52506,$$
$$x_{31} = 8, x_{33} = 96.10619, x_{34} = 39.47495.$$

λ gives the minimum value of the membership functions i.e. $\lambda_t = \min\{\mu_t(Z_t)\}$ in the solution set. The individual membership values λ_1 and λ_2 are obtained as $\mu_1(Z_1) = 0.9361960$ and $\mu_2(Z_2) = 0.9361960$ for the objective functions Z_1 and Z_2 respectively. So, we have $\lambda = 0.9361960 = \min(0.9361960, 0.9361960)$. With the increasing value of λ, the solution is found to be improving for the problem approaching to the individual optimal values of the objectives.

It can be seen from Fig. 17.2 that the speckled line in the plotted graphs (a) and (b) show the grade of membership function on the vertical line and the non-dominant value obtained corresponding to this membership value is shown on the horizontal line. For example, the objective value achieved is $\mathcal{M}\{Z_1^*\} = 0.8955540$ at $\lambda = 0.9361960$ membership value in the 1st objective and the objective value achieved is $\mathcal{M}\{Z_2^*\} = 0.9361067$ at $\lambda = 0.9361960$ membership value in the 2nd objective function.

17.9 Results and Comparison

Tables 17.7 and 17.8 gives the comparison of the results for both the models obtained using Weighted sum approach and Fuzzy programing technique. We can note from Table 17.7 that the expected value of the 1st objective obtained using the weighted sum approach is better than that of the FPT whereas, for the second objective function FPT gives a better solution than weighted sum approach. Similarly, we can see from Table 17.8 that the measure value of the 1st objective function obtained using weighted sum is better than FPT but for 2nd objective function FPT gives a better objective value. So, we can say that the results obtained for the EVM and DCCM models using the WSM and FPT are non-dominating to each other.

But, the selection of weights in weighted sum method does not always guarantee that the obtained solution will be acceptable. So, changing the weight vectors randomly does not give us appropriate solution whereas, in FPT we can have more than one solution with the exponential membership function than linear membership function. So, FPT is a better approach for solving uncertain MOTP and obtaining their solutions.

17 Uncertain Multi-objective Transportation Problems and Their Solution

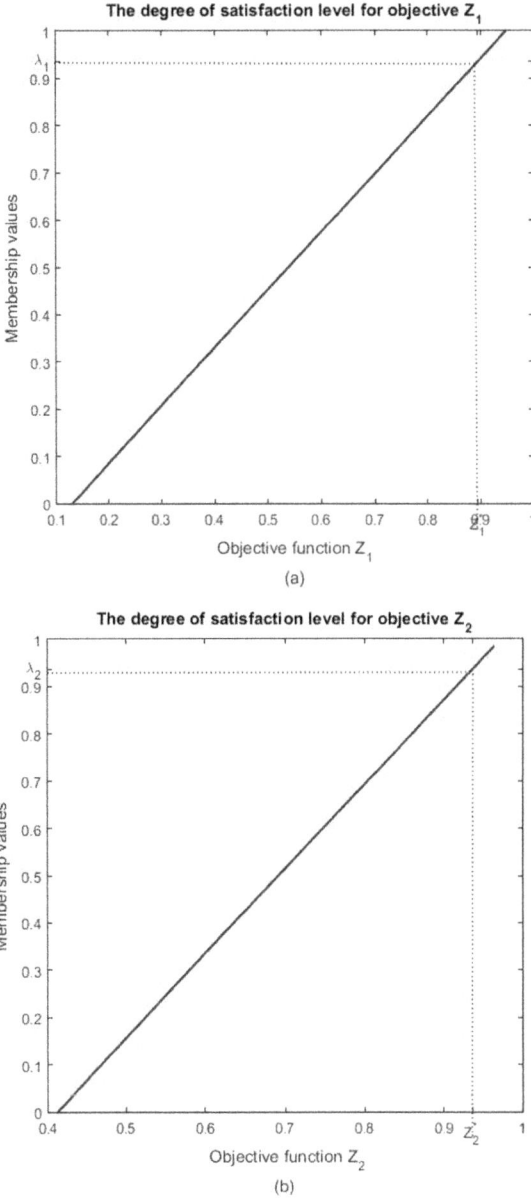

Fig. 17.2 Graph for membership function values versus objective function values

Table 17.7 Comparison table for EVM model using weighted sum approach and FPT

WSM with equal weights	Fuzzy programming technique (FPT)
$E[Z_1^*] = 1475$	$E[Z_1^*] = 1488.120$
$E[Z_2^*] = 1780$	$E[Z_2^*] = 1773.440$

Table 17.8 Comparison table for DCCM model using weighted sum approach and FPT

WSM with equal weights	Fuzzy programming technique (FPT)
$\mathcal{M}\{Z_1^*\} = 0.9020619$	$\mathcal{M}\{Z_1^*\} = 0.8955540$
$\mathcal{M}\{Z_2^*\} = 0.9304124$	$\mathcal{M}\{Z_2^*\} = 0.9361067$

17.10 Conclusion

In this chapter, a MOTP involving uncertain variables is inspected. Initially, using the basic definitions of uncertainty, the uncertain mathematical model of MOTP is converted into its deterministic form with EVM and DCCM. In addition, the deterministic DCCM model is converted into its linear form using Charnes and Cooper's transformations. Further, both of these deterministic models are solved using FPT and weighted sum method. The numerical illustration is also provided to explain the application of each method and the obtained results were compared for both the methods. Finally, non dominated solutions of the problem are obtained with EVM and DCCM using LINGO 18.0 software. In future, the work can be extended by considering more complex imprecise environments like random uncertain or fuzzy uncertain environments.

References

1. Bagheri M, Ebrahimnejad A, Razavyan S, Hosseinzadeh Lotfi F, Malekmohammadi N (2020) Solving the fully fuzzy multi-objective transportation problem based on the common set of weights in DEA. J Intell Fuzzy Syst 39(3):3099–3124
2. Bagheri M, Ebrahimnejad A, Razavyan S, Lotfi FH, Malekmohammadi N (2020) Fuzzy arithmetic DEA approach for fuzzy multi-objective transportation problem. Oper Res 1–31 (2020)
3. Bera RK, Mondal SK (2020) Credit linked two-stage multi-objective transportation problem in rough and bi-rough environments. Soft Comput 1–26 (2020)
4. Biswas A, Modak N (2017) On solving multiobjective transportation problems with fuzzy random supply and demand using fuzzy goal programming. Int J Oper Res Inf Syst (IJORIS) 8(3):54–81
5. Bit A, Biswal M, Alam S (1992) Fuzzy programming approach to multicriteria decision making transportation problem. Fuzzy Sets Syst 50(2):135–141
6. Chanas S, Kuchta D (1996) A concept of the optimal solution of the transportation problem with fuzzy cost coefficients. Fuzzy Sets Syst 82(3):299–305
7. Charnes A, Cooper WW (1962) Programming with linear fractional functionals. Naval Res Logistics Q 9(3–4):181–186

8. Christi MA, Kalpana I (2016) Solutions of multi objective fuzzy transportation problems with non-linear membership functions. Int J Eng Res Appl 6(11):52–57
9. Das S, Goswami A, Alam S (1999) Multiobjective transportation problem with interval cost, source and destination parameters. Eur J Oper Res 117(1):100–112
10. Delphi AM (2016) Heuristic algorithms for solving multiobjective transportation problems. J Math Res 8(3) (2016)
11. Deyi M, Wanlin Z, Xiaoding C (2013) A transportation problem with uncertain truck times and unit costs. Ind Eng Manag Syst 12(1):30–35
12. Guo H, Wang X, Zhou S (2015) A transportation problem with uncertain costs and random supplies. Int J e-Navig Marit Econ 2:1–11
13. Gupta A, Kumar A (2012) A new method for solving linear multi-objective transportation problems with fuzzy parameters. Appl Math Model 36(4):1421–1430
14. Gupta S, Ali I, Ahmed A (2020) An extended multi-objective capacitated transportation problem with mixed constraints in fuzzy environment. Int J Oper Res 37(3):345–376
15. Hitchcock F (1941) The distribution of a product from several sources to numerous locations. J Math Phys 20:224–230
16. Jana B et al (2020) Application of fuzzy programming techniques to solve solid transportation problem with additional constraints. Oper Res Decis 30 (2020)
17. Kakran VY, Dhodiya JM (2020) Fuzzy programming technique for solving uncertain multi-objective, multi-item solid transportation problem with linear membership function. In: Advanced engineering optimization through intelligent techniques. Springer, Berlin, pp 575–588 (2020)
18. Kakran V, Dhodiya JM (2020) Charnes and Cooper's transformations for solving uncertain multi-item fixed charge solid transportation problem. Sci Technol Asia 142–156 (2020)
19. Khalifa HAE (2019) A signed distance method for solving multi-objective transportation problems in fuzzy environment. Int J Res Ind Eng 8(3):274–282
20. Latpate R, Kurade SS (2020) Multi-objective multi-index transportation model for crude oil using fuzzy NSGA-II. IEEE Trans Intell Transp Syst (2020)
21. Lee SM, Moore LJ (1973) Optimizing transportation problems with multiple objectives. AIIE Trans 5(4):333–338
22. Liu B (2007) Uncertainty theory. In: Uncertainty theory. Springer, Berlin, pp 205–234 (2007)
23. Liu, B.: Uncertainty theory. In: Uncertainty theory. Springer, Berlin, pp 1–79 (2010)
24. Mahajan S, Gupta S (2019) On fully intuitionistic fuzzy multiobjective transportation problems using different membership functions. Ann Oper Res 1–31 (2019)
25. Maity G, Kumar Roy S (2016) Solving a multi-objective transportation problem with nonlinear cost and multi-choice demand. Int J Manag Sci Eng Manag 11(1):62–70
26. Maity G, Roy SK, Verdegay JL (2016) Multi-objective transportation problem with cost reliability under uncertain environment. Int J Comput Intell Syst 9(5):839–849
27. Midya S, Roy SK (2020) Multi-objective fixed-charge transportation problem using rough programming. Int J Oper Res 37(3):377–395
28. Moore RE (1966) Interval analysis 4
29. Nasseri SH, Bavandi S (2020) Solving multi-objective multi-choice stochastic transportation problem with fuzzy programming approach. In: 2020 8th Iranian joint congress on fuzzy and intelligent systems (CFIS). IEEE, pp 207–210
30. Osman MS, El-Sherbiny MM, Khalifa HA, Farag HH (2016) A fuzzy technique for solving rough interval multiobjective transportation problem. Int J Comput Appl 147(10)
31. Patel J, Dhodiya J (2017) Solving multi-objective interval transportation problem using grey situation decision-making theory based on grey numbers. Int J Pure Appl Math 113(2):219–233
32. Porchelvi RS, Anitha M (2018) On solving multi objective interval transportation problem using fuzzy programming technique. Int J Pure Appl Math 118(6):483–491
33. Salama A, Wahed ME, Yousif E, A multi-objective transportation data problems and their based on fuzzy random variables
34. Sheng Y, Yao K (2012) A transportation model with uncertain costs and demands. Inf Int Interdisc J 15(8):3179–3186

35. Verma R, Biswal M, Biswas A (1997) Fuzzy programming technique to solve multi-objective transportation problems with some non-linear membership functions. Fuzzy Sets Syst 91(1):37–43
36. Vijayalakshmi P, Vinotha JM (2020) Multi choice ecological intuitionistic fuzzy multi objective transportation problem with non linear cost. Solid State Technol 2675–2689
37. Vinoliah EM, Ganesan K (2018) Solution to a multi-objective fuzzy transportation problem-a new approach. Int J Pure Appl Math 119(9):385–393
38. Zadeh LA (1965) Fuzzy sets. Inf Control 8(3):338–353
39. Zangiabadi M, Maleki HR (2013) Fuzzy goal programming technique to solve multiobjective transportation problems with some non-linear membership functions
40. Zangiabadi M, Maleki H (2007) Fuzzy goal programming for multiobjective transportation problems J Appl Math Comput 24(1–2):449–460
41. Zhao G, Pan D (2020) A transportation planning problem with transfer costs in uncertain environment. Soft Comput 24(4):2647–2653
42. Zimmermann HJ (1978) Fuzzy programming and linear programming with several objective functions. Fuzzy Sets Syst 1(1):45–55

Part III
Computational Intelligence

Chapter 18
Agile Computational Intelligence for Supporting Hospital Logistics During the COVID-19 Crisis

Rafael D. Tordecilla, Leandro do C. Martins, Miguel Saiz, Pedro J. Copado-Mendez, Javier Panadero, and Angel A. Juan

Abstract This chapter describes a case study regarding the use of 'agile' computational intelligence for supporting logistics in Barcelona's hospitals during the COVID-19 crisis in 2020. Due to the lack of sanitary protection equipment, hundreds of volunteers, the so-called "Coronavirus Makers" community, used their home 3D printers to produce sanitary components, such as face covers and masks, which protect doctors, nurses, patients, and other civil servants from the virus. However, an important challenge arose: how to organize the daily collection of these items from individual homes, so they could be transported to the assembling centers and, later, distributed to the different hospitals in the area. For over one month, we have designed daily routing plans to pick up the maximum number of items in a limited time—thus reducing the drivers' exposure to the virus. Since the problem characteristics were different each day, a series of computational intelligence algorithms was employed. Most of them included flexible heuristic-based approaches and biased-randomized

---

R. D. Tordecilla (✉) · L. C. Martins · M. Saiz · P. J. Copado-Mendez · J. Panadero · A. A. Juan
IN3—Computer Science Department, Universitat Oberta de Catalunya, 08018 Barcelona, Spain
e-mail: rtordecilla@uoc.edu

L. C. Martins
e-mail: leandrocm@uoc.edu

M. Saiz
e-mail: msaiz@weoptimize.net

P. J. Copado-Mendez
e-mail: pcopadom@uoc.edu

J. Panadero
e-mail: jpanaderom@uoc.edu

A. A. Juan
e-mail: ajuanp@uoc.edu

R. D. Tordecilla
Universidad de La Sabana, Chía, Cundinamarca, Colombia

M. Saiz
weOptimize, 43700 El Vendrell, Spain

P. J. Copado-Mendez · J. Panadero · A. A. Juan
Euncet Business School, Terrassa 08225, Spain

© The Author(s), under exclusive license to Springer Nature Switzerland AG 2021
S. Patnaik et al. (eds.), *Computational Management*, Modeling and Optimization in Science and Technologies 18, https://doi.org/10.1007/978-3-030-72929-5_18

algorithms, which were capable of generating, in a few minutes, feasible and high-quality solutions to quite complex and realistic optimization problems. This chapter describes the process of adapting several of our 'heavy' route-optimization algorithms from the scientific literature into 'agile' ones, which were able to cope with the dynamic daily conditions of real-life routing problems. Moreover, it also discusses some of the computational aspects of the employed algorithms along with several computational experiments and presents a series of best practices that we were able to learn during this intensive experience.

Keywords Computational intelligence · Operations management · Hospital logistics · Vehicle routing problems · Biased-randomized algorithms

18.1 Introduction

The COVID-19 pandemic crisis is one of the toughest global challenges we have faced in decades. Several leaders have even stated it as "the biggest one since the Second World War". The exponential growth of cases that needed medical attention led to a sudden shortage of protective material, so that the medical and support staff were subject to higher risk to also become infected, endangering the needed level of attention in hospitals and also a faster spread of COVID-19. By March 2020, the pandemic had a strong impact in countries like Italy and Spain. As it happened in other regions, in the metropolitan area of Barcelona, a community of volunteers, the so-called "Coronavirus Makers" Community, arose with the aim to supply protective material to the staff working in the hospitals, nursing homes, and emergency medical attention. The main tool used were home 3D-printers, which helped to iterate the design very fast in order to reach, within a few days, the design level that was considered acceptable by the staff in charge of guaranteeing safety and quality of the produced items (Fig. 18.1). It soon became noticeable that the bottleneck was the logistic side of this endeavor, due to the fact that the lock-down situation meant that each 3D-printer was located at each individual home, and route planning needed technological support, in order not to expose the drivers to more risk than strictly necessary.

This chapter discusses the experience of matching several professional and personal profiles that typically work with very different approaches, because of the nature of their work: scientists, volunteers, makers and entrepreneurs. In this case, it was needed to find a fast way of applying the knowledge gathered within years' of research to an urgent need, where every day counts. The target was to support the makers community (with each maker located in his/her individual home) on their voluntary initiative to supply the sanitary staff with as much protective material as possible and with a limited time to avoid unnecessary exposure for the drivers.

 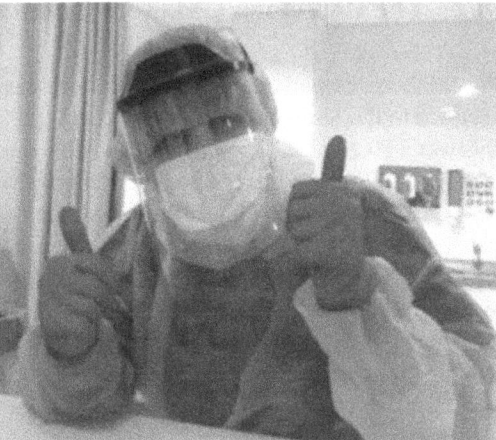

Fig. 18.1 The main goal of the Makers' community was to supply protective items to hospitals and healthcare centers

The main contributions of our work are: *(i)* it describes a real-life case in which computational intelligence was used to support hospital logistics during the COVID-19 pandemic crisis; *(ii)* it illustrates how real-life logistics might be rich in the sense that they combine multiple routing problems with dynamic characteristics and constrains—which might vary from day to day; *(iii)* it provides an example of how 'agile' optimization can be applied—in combination with other technologies—to support decision making in scenarios under stress; and *(iv)* it discusses how to develop new agile-optimization tools that can efficiently cope with the aforementioned scenarios. The rest of the chapter is distributed as follows: Sect. 18.2 introduces the Barcelona's "makers" community, which is at the heart of this project. Section 18.3 provides an overview of the methodological approach that we developed to generate the daily routing plans. Section 18.4 describes our addressed problems from an academic point of view, and provides a literature review regarding these problems. Additionally, it reviews the concept of biased-randomized algorithms, which are used here to develop 'agile' (flexible and fast) approaches that allow to deal efficiently with the daily challenges in a few minutes of computation. Section 18.5 explains how we adapted previously developed and published algorithms for different variants of the vehicle routing and team orienteering problems into faster and more flexible algorithms. Section 18.6 describes the heuristic-based approach used to solve our problem. Section 18.7 presents several examples of daily solutions that were generated during the one and a half month that this project was operative. Section 18.8 examines managerial insights and best practices derived from the project. Finally, Sect. 18.9 summarizes the main contributions of this work.

18.2 The Barcelona's Makers Community

The 'Makers' community was born to contribute with creative capacity and offer a service to the healthcare system, the geriatric staff, and home-support personnel. This was a 100% altruistic and non-profitable initiative. The aim was to alleviate the need for additional protective material in hospitals and health centers derived from the scarcity of resources due to the unprecedented level of demand worldwide generated by the COVID-19 outbreak. The initiative was conceived in less than 48 hours between March 12th and 14th, and grew at an average rate of 1, 000 new volunteers per day during the first two weeks. The makers community from Barcelona and the surrounding provinces adhered soon to this initiative, and the community was already handing out material to the hospitals on March 16th, 2020. By the end of that same week, the Barcelona community had grown significantly, leading to a communicate to their community on March 21st informing that the next routes will only collect material for makers that have a minimum amount of parts manufactured—despite the fact that, in those critical times, every shield counted in order to protect the sanitary staff. Figure 18.2 provides an example of the problem magnitude in the area of Barcelona, for a specific crisis period. As we can notice, several pickup and delivery points are geographically distributed, being the coordination of both loading and unloading activities the next challenge to be faced. This was the first sign that the logistics were becoming a bottleneck and further help was needed.

Figure 18.3 depicts the overall process. Firstly, every maker is printing the cap through a 3D printer, and then the caps are disinfected by the maker. Due to the lockdown restrictions, the makers cannot (and must not) transport the caps themselves, so there is a transport arranged by the node coordinators to pick up all material and hand it out to the person being responsible of that area. Then, the node team is assembling the finished product (3D-printed cap with an acetate shield and a head-adjusting elastic band). Afterwards, the finished product is delivered to the medical centers, which complete a second disinfection round, typically via ozone sterilization.

The proposal to collaborate to the research team was made on March 23rd. The work was done in one-day sprints, setting the highly ambitious target to be able to have route proposals following the adapted algorithms by the next day at 9 a.m. or before. The first sprint was almost successful. For a few minutes it could not be implemented due to several challenges in the database interfaces and synchronization between all planners involved. It became successful the day after, and from that day on it was possible to propose near-optimal routes to the drivers. The route proposal is handed out to the drivers via a native file that can be opened in a popular mapping application of common use in cell phones.

A process was established in order to improve the synchronization between all the volunteers involved in the node planning and the research team (Fig. 18.4). In this specific case, transport 1 (pick up at makers' homes) and transport 2 (delivery at sanitary centers) were planned and executed together in order to gain further advantage from the algorithms already in place. The makers community continued to

Fig. 18.2 Some of the pickup locations geographically distributed in the area of Barcelona

work on this initiative for over two months (March, April, and part of May), reaching a total of 75, 000 or more face shields, above 11, 000 door openers, and over 53, 000 ear savers through all nodes established in the metropolitan area of Barcelona.

18.3 Routes Generation Process

The "*Proposal with optimized routes*" step in Fig. 18.4 is the process in which routes are generated through the use of agile computational algorithms. Such process is showed extensively in Fig. 18.5. Initially, details about each point to visit are provided in a spreadsheet by the Route coordinator, namely: quantity of medical supplies to pick up, municipality, address, postal code, geographic coordinates and type of node (origin, destination or mandatory node). As showed in Fig. 18.3, demand nodes can either offer (makers and preparation nodes) or consume (hospitals and preparation nodes) medical supplies. Nevertheless, consumption points do not have an established demand, since they consume all supplies offered by the makers. Therefore, data for demand are only given by pickup points.

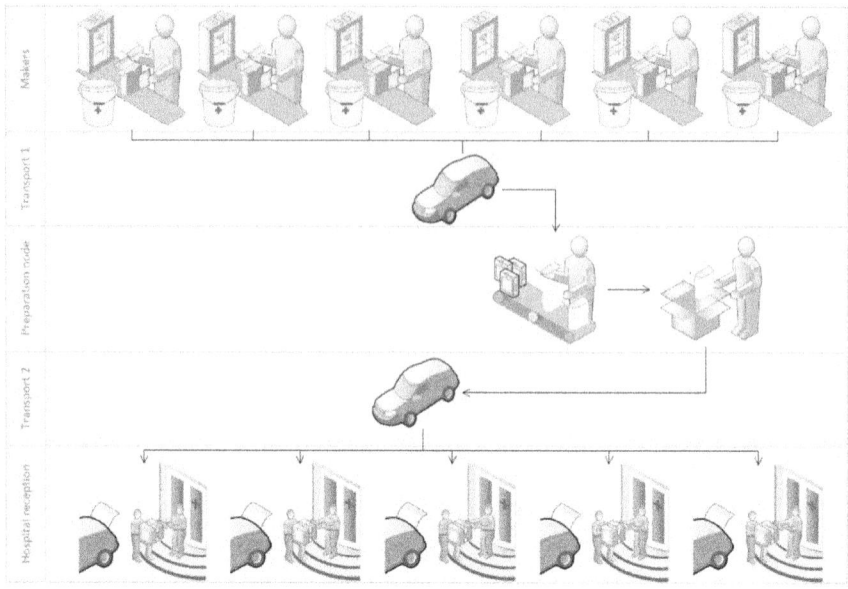

Fig. 18.3 Overview of the material flow

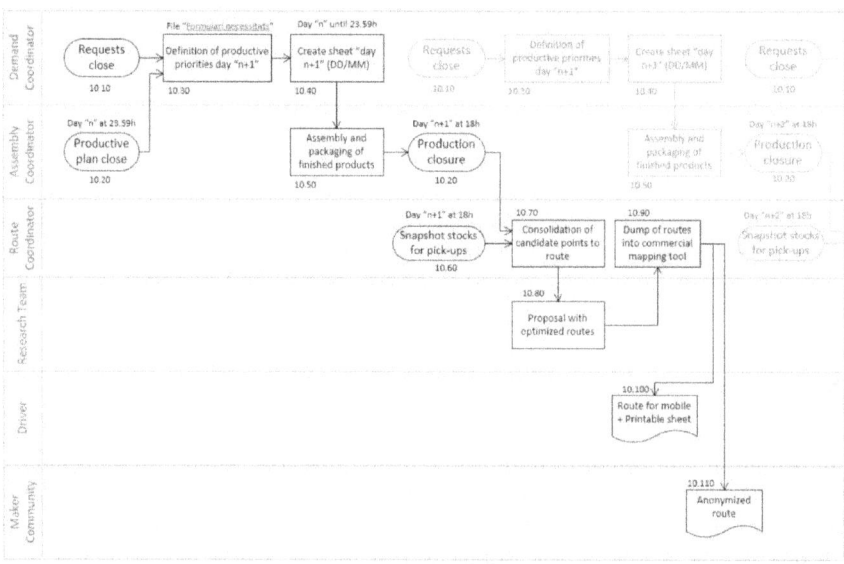

Fig. 18.4 Daily node planning and interaction with research team

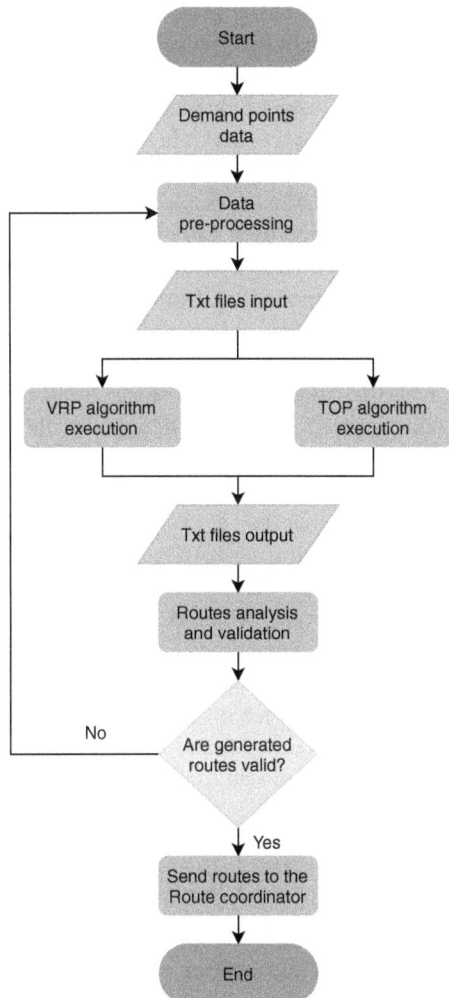

Fig. 18.5 Flowchart of the routes generation process

Input data is then analyzed by the research team to identify characteristics of each instance. This step is necessary since such requirements are set every day, posing a new challenge for the team. Details about these requirements are provided in Sect. 18.7. In general, our hospital-logistics problem shows characteristics of several rich variants of the vehicle routing problem (VRP). Nevertheless, we address all daily challenges as variants of two big groups: *(i)* a VRP-related challenge, in which all customers (makers' houses) are visited and the objective is to minimize total time requested in completing the collection; and *(ii)* a rich team orienteering problem (TOP), in which a time limit must be met and, hence, not all customers can be visited –since the number of drivers and pick-up vehicles is also limited. Therefore, some customers are not visited, seeking a maximum reward while satisfying the constraints.

Notice that, since the problem characteristics change every day, the challenge we faced every night was typically not a pure VRP or a pure TOP, but a hybrid between them.

Once the instance characteristics have been identified, the spreadsheet is adjusted and processed to convert it in standardized *txt* files through a Python code. This code generates two files: *(i)* a nodes-file with details about demand, coordinates, and type of node; and *(ii)* an edges-file with information about the estimated travel time and the distance between each pair of nodes in the instance. A web mapping service is called by the Python code to estimate these parameters. Then, VRP and TOP algorithms are executed in parallel. Both algorithms were coded as Java applications. They are able of generating near-optimal solutions in a matter of seconds. Each algorithm outputs a standardized *txt* file with the next variables for each designed route: total travel time, total traveled distance, total collected elements, total visited nodes and nodes sequence to visit. Additionally, the output file highlights the route with the highest travel time, since this variable represents one of the main objectives to be minimized by the algorithms. Next, the research team analyzes the designed routes to guarantee that they are valid according to the instance requirements. Also, routes are depicted using a VBA/Excel application. Both travel times and routes depiction are used for the validation process. For example, some instances require a node-clustering process, which is performed by the research team in the pre-processing step. However, sometimes generated routes are quite different in terms of total travel time, which should be avoided so that volunteer drivers have similar travel times in each route. Hence, if non-balanced routes are generated by the algorithms, data is pre-processed again and the process is repeated. Finally, when valid routes are obtained, they are sent to the coordinator, who provides them to the drivers.

18.4 Addressed Problems and Related Works

The first academic mention of a vehicle routing problem was made by [12], as a generalization of the traveling salesman problem. More than 60 years have passed since then, in which a large set of variants has been identified. Each variant of this problem has shown its relevance both academically—given the algorithmic challenges that they pose—and economically—given their high applicability in real-world problems. A thorough study of these variants, solving methods, and applications is carried out by [52]. More recently, a concise research is showed by [49]. Also, a review centered on VRP-related themes instead of traditional variants is made by [53]. Similar to our work, the urgency and worrying caused by the spread of COVID-19 around the world have encouraged researchers to recognize the potential of solving these classical combinatorial optimization problems to overcome its related challenges, e.g., [11], and [34]. For instance, the pandemic has increased travel and processing times in vehicle routing activities, given the introduction of labors such as cleaning, disinfection and wearing personal protection equipment [33]. An Indonesian real-case study addressed by [39] shows that the COVID-19 pandemic has also increased uncertainty and complexity in food supply chains.

18.4.1 The Vehicle Routing Problem

The VRP is a classical *NP-hard* problem [29] in which a set of customers has a known demand that must be satisfied by a single depot. This depot has a virtually unlimited capacity, and a fleet of homogeneous capacitated vehicles are employed to deliver the demanded units of product. Each customer must be visited only once. Due to the limited capacity of the cargo vehicles, this situation forces the design of several routes in order to satisfy all the customers' demands (Fig. 18.6). Finally, once the vehicle has delivered all its assigned load, it must return to the depot. The objective is to minimize transportation costs, usually measured in terms of total traveled distance or time [40].

Real-world problems hardly show characteristics of a pure VRP as described above, since real-cases are usually much more complex. This is the case of our hospital-logistics problem. For instance, the most common faced variant is related to

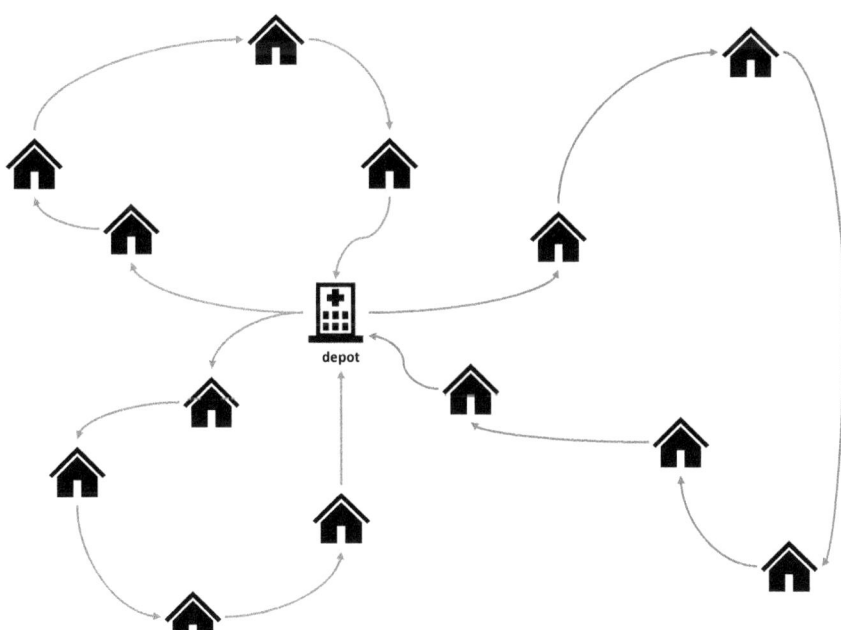

Fig. 18.6 An example of a VRP solution with a single central depot

Fig. 18.7 An example of a OVRP solution with different start and end depots

the possibility that vehicles start and end their routes in different nodes. The relaxation of this constraint defines what we call *open* VRP (OVRP), which, apart from minimizing the travel costs, also aims to minimize the number of used vehicles [40]. From the earlier definition stated by [48], the OVRP has been studied and enriched by many other constraints [7]. Among them, we can highlight the rich variants with multiple depots [8, 28], heterogeneous fleet of vehicles [54] or even a conjunction of them [25, 51]. Although most of these works address a single-objective, there exist multi-objective studies whose aim, for instance, to reduce the total number of routes, the total travel cost, and the longest route altogether [46]. According to [30], the range of applications that ends up in OVRPs is commonly found in contexts where contractors—who are not employees of the delivery company—use their vehicles and do not return to the depot, such as home delivery of packages and newspapers. In our application context, volunteers start their daily routes from an origin depot (hospitals, health centers, or even their homes) and finish at the end depot (usually, the hospitals or health care centers), where the unloading of the collected goods is performed. Figure 18.7 presents a feasible solution for this variant, in which four routes are designed to serve all customers in order to minimize transportation costs.

18.4.2 The Team Orienteering Problem

In contrast with classical VRPs, which require visiting all customers, there is no obligation regarding this constraint when solving a Team Orienteering Problem (TOP).

Besides, the objective is to maximize the collected reward by visiting a set of nodes subject to a set of constraints. It might be the case of many situations in which there are no vehicles (or volunteers) enough for visiting all the pick-up locations from the daily problem data, as faced in this hospital-logistics project. By being an extension of the *NP-hard* orienteering problem (OP) [17], the TOP is also an *NP-Hard* problem. Similarly to the VRP, this problem aims to define a set of vehicle routes in order to optimize one specific objective. In the case of the TOP, the objective is to define a set of vehicle routes such that the total reward collected from visiting each of their assigned nodes, or targets, is maximized [4]. Each target must be visited once and only by a single vehicle, being each characterized by a positive reward that is collected as soon as it is visited. Given this particularity, the TOP has been ultimately linked to rescue operations [45], in which this set of targets—places or tasks—must be chosen to be visited or performed according to their relevance. In our case, the reward is represented by the number of medical supplies to be collected, weighted by their importance at each period from the pandemic crisis. Examples of those materials are face shields, surgical masks, ventilators, handles for opening doors, etc. Therefore, for each pick-up location, its reward is measured by the sum of the weighted available material. Also, we assume that it is not possible to get a partial reward from nodes. The complete route must be performed before a maximum time limit is achieved. Otherwise, the route cannot be properly performed, and all the collected rewards are discarded. Some of its variants still consider precedence constraints [23], time-windows [26], stochastic travel times [36] and dynamic rewards [43], which transforms the TOP into an even harder problem to solve. Figure 18.8 represents a solution example composed of three routes, being each one of a different color—blue, red, and green—and visiting 2, 3, and 4 collection points, respectively. Besides, notice that the origin and end depots are different, and volunteers do not visit 5 locations due to the limited length of the routes. The nodes which are not considered are more likely to be those ones with lower reward values, i.e., those with lower weighted available material, in which visiting them does not compensate for the effort to get there.

18.4.3 Rich Pickup and Delivery Routing Problems

The general pickup and delivery problem (GPDP) combines the OVRP and the TOP. This problem was first addressed in the literature by [47], and concerns with defining a set of optimal routes to satisfy a set of transportation requests—each requiring both pickup and delivery under capacity and precedence constraints–, in order to minimize transportation costs. The transportation is performed from a set of origins to a set of destinations, without any transshipment at other locations. The origins represent the pickup locations, while the destinations are the delivery points.

The classical PDP has been widely extended by incorporating several characteristics into the problem in order to bring it as closest as possible to real-life environments. Constraints such as time-windows to visit costumers [3, 14], heterogeneous

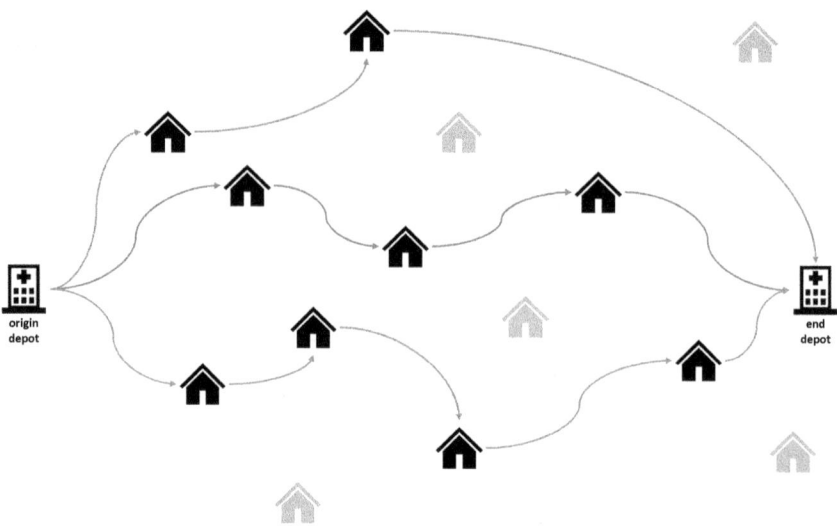

Fig. 18.8 An example of a TOP solution with different start and end depots

fleet of vehicles [2], multi-depot [50], a combination of them [6], and multi-echelon distribution [5, 32], or even uncertainties [22] have been addressed into the classical problem, resulting in a large number of variants of this classical problem, frequently called as *rich* pickup and delivery problems (RPDPs). According to [38], the term 'rich' refers to the additional features that are incorporated in order to accommodate the different characteristics of the various problem types. By incorporating those features, the RPDP has become even harder to be solved to optimality, which results in the proposal of many approximated approaches for solving them. In our case, the problem addressed in this work can be described as a PDP by considering the last node to be visited (end depot) as the delivery node, while the remaining ones, except from the first node (start depot), are pickup points.

18.4.4 Biased-Randomized Algorithms

Heuristic and metaheuristic algorithms have become the default standard when dealing with rich and realistic vehicle routing problems [16]. Biased randomized (BR) optimization algorithms make use of Monte Carlo simulation and skewed probability distributions to introduce a non-uniform random behavior into a constructive heuristic. Thus, the heuristic is transformed into a more powerful probabilistic algorithm, which can be run in virtually the same wall-clock time as the original heuristic if parallelization techniques are employed [15]. One of the main advantages of BR algorithms is their ability to generate multiple promising solutions that still follow the logic behind the original heuristic.

BR algorithms have been successfully used during the last years to solve different rich and realistic variants of vehicle routing problems [13], permutation flow-shop problems [20], location routing problems [41], facility location problems [35], waste collection problems [21], arc routing problems [19], horizontal cooperation problems [42], constrained portfolio optimization problems [27], and e-marketing problems [31].

A different class of BR algorithms was introduced by [18] for solving combinatorial optimization problems. Because its core is a genetic algorithm, the biased random–key genetic algorithms (BRKGA) aim to bias the selection of parents for generating new solutions. Recently, this solving methodology has been developed for solving an OVRP with capacity and distance constraints [44]. In literature, the BRKGA has been also largely applied for solving different scheduling problems [1, 9, 10, 24].

18.5 Adapting Heavy Methods into Agile Algorithms

Metaheuristic frameworks are robust tools that provide optimal or pseudo-optimal solutions to optimization problems, which are typically large-scale and *NP-hard*, in a reasonable computing time. In order to obtain near-optimal solutions, these frameworks are composed of complex operators which are designed specifically to deal with the optimization problem to solve. Although these methods have been widely applied in the context of computational intelligence to cope with traditional optimization problems, they are not very useful when dealing with real-time and dynamic optimization problems that vary each day, as the ones faced in our hospital-logistics project. Hence, the concept of 'agile' optimization has arisen to deal with this new kind of problems, which are highly dynamics and require flexible and even 'light' algorithms able to provide good solutions very fast without having to complete time-consuming set-up processes. This refers to a new optimization paradigm for *NP-hard* and large-scale optimization problems that need to be solved in 'real time'. In this context, 'agile' algorithms follows the next principles: *(i)* extremely fast execution (i.e., seconds or even milliseconds); *(ii)* easy to implement and modify; *(iii)* flexible enough to deal with different problems and variants; *(iv)* parameter-less, avoiding complex and time-costly fine-tuning processes; and *(v)* specifically designed to run iteratively every few seconds or minutes, as new data is available.

In particular, agile optimization is based on the hybridization of biased-randomized algorithms and parallel computing. The intrinsic characteristics of biased-randomized algorithms make them a perfect candidate to be massively parallelized. In effect, they are good candidates to be executed using massive parallel processing architectures [37]. Usually, biased-randomized heuristics are wrapped into a multi-start framework, which is a sequential and iterative approach. Hence, a different solution is generated in each iteration. Typically, the multi-start methods are composed of two

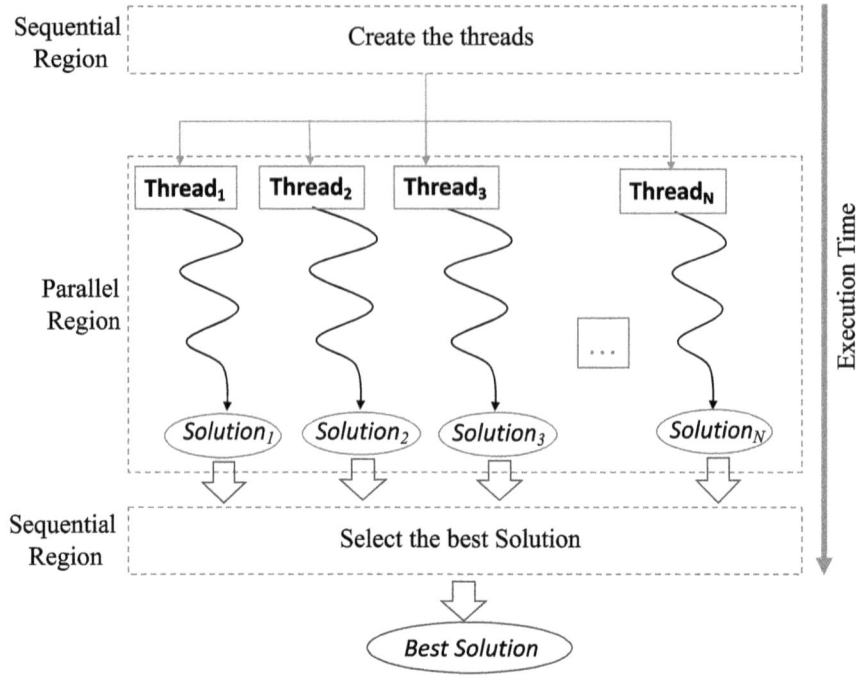

Fig. 18.9 Parallel execution of a biased-randomized algorithm

phases: a first one in which a new solution is generated using a deterministic constructive heuristic, and a second one in which the deterministic heuristic is turned into a probabilistic one, applying a skewed probability distribution to induce an oriented (biased) random behavior, without losing the logic behind of the original heuristic. Further, in this second step the algorithm compares the newly generated solution with the best solution obtained so far—updating the latter whenever appropriate. The main idea behind agile optimization is as follows: instead of using a multi-start sequential approach, many threads of the biased-randomized algorithm can be run in parallel (e.g., using different computers or GPUs), as shown in Fig. 18.9. As a result, several alternative solutions are generated in the same wall-clock time as the one employed by the original heuristic—that is, milliseconds in most cases. Then, different solutions are provided. Some of these solutions outperform the one generated by the original heuristic, while others show different characteristics that might be interesting for the decision maker.

18.6 Solving Approach

Algorithm 1 shows our TOP-like algorithm used to solve the hospital-logistics problem. It works as follows: firstly, a dummy solution is generated (line 1), being composed of one route per location (house). For each route, a vehicle departs from the origin depot, visits the location, and then returns to the destination depot. In case this route cannot be performed within the maximum driving time, its respective location is discarded from the problem, since visiting this location is not possible. The second stage regards the computation of savings that are associated with each edge connecting two different locations (line 2). This computation considers both the travel time required to traverse that edge and the aggregated reward generated by visiting both locations. Since each edge is associated with two directions (arcs) in which it is traversed, individual savings are calculated for each arc. Then, these lists are sorted from higher to lower savings. Next, based on this sorted savings list, a route-merging process is started. The arc with the highest saving, i.e., that one at the top of the sorted list, is selected in each iteration (line 4). By using the selected arc, its two corresponding routes are merged into a new one, as far as this new route does not violate the driving-range constraint (line 9). The selected edge is later removed from the savings list and this process is repeated until the list is empty. As a result, a list of routes is generated and sorted according to the total collected reward (line 15). Finally, from the sorted list of routes, the first n routes are selected, where n is the size of vehicles fleet. This heuristic is later extended into a probabilistic algorithm by introducing a biased-randomization behavior, which smooths the original greedy behavior of the heuristic. As explained in Sect. 18.4.4, biased-randomization techniques employ skewed probability distributions to induce an 'oriented' (non-uniform) random behavior into deterministic procedures, consequently transforming them into randomized algorithms while preserving the logic behind the original greedy heuristics. For doing so, we employ a geometric probability distribution with a single parameter β ($0 < \beta < 1$), which controls the relative level of greediness present in the randomized behavior of the algorithm. After a fine-tuning process, $\beta = 0.3$ has presented a good performance, being this value selected to be used in our computations.

18.7 Examples of Daily Routing Plans

Many variants of classical problems show increasing complexity due to the incorporation of extra decisions and operational constraints in order to reflect real-life cases. In our context, the problems are notably dynamic since they must be frequently modified, even on a daily basis. These daily challenges are determined by limited resources and a variety of operational decisions, such as:

1. *A maximum tour length for drivers*: as they are volunteers (and also to avoid excessive exposition to risk), the total time that drivers can dedicate to pick up

Algorithm 1 Example of one heuristic-based approach employed.

```
 1: sol ← generateDummySolution(Inputs)
 2: savingList ← computeSortedSavingList(Inputs)
 3: while (savingList is not empty) do
 4:     arc ← selectNextArc(savingList, β)
 5:     iRoute ← getStartingRoute(arc)
 6:     jRoute ← getClosingRoute(arc)
 7:     newRoute ← mergeRoutes(iRoute, jRoute)
 8:     timeNewRoute ← calcRouteTravelTime(newRoute)
 9:     isMergeValid ← validateMergeDrivingConsts(timeNewRoute, drivingRange)
10:     if (isMergeValid) then
11:         sol ← updateSolution(newRoute, iRoute, jRoute, sol)
12:     end if
13:     deleteEdgeFromSavingList(arc)
14: end while
15: sortRoutesByProfit(sol)
16: deleteRoutesByProfit(sol, maxVehicles)
17: return sol
```

elements is limited. Routes must be as balanced as possible, so that the work time of each driver is similar. Additionally, this total length must include a service time per visited node, which is assumed to be constant.

2. *A limited number of vehicles*: the number of volunteers are variable each day, which limits the quantity of routes that can be designed. This condition, jointly with the limit in the drivers' work time, makes that a few nodes must be skipped some days. These instances are then solved preferentially using a TOP-like algorithm.

3. *Origin and arrival nodes are the same or different*: most instances require an OVRP-like solution, in which drivers depart from a point that is different to the final destination. However, sometimes this constraint can be relaxed, as we will explain later.

4. *Mandatory nodes to visit*: the three previous constraints rely on a TOP-like algorithm, resulting in locations that are not visited. However, there are some cases in which specific nodes must be mandatorily visited. These mandatory nodes to visit are more likely to be the preparation nodes, the medical centers, or some makers who offer a large number of medical supplies.

5. *Segmentation of nodes*: drivers are more willing to visit the makers and medical centers depending on the geographical zone where they are located. Hence, a segmentation process is required, in which nodes are grouped in clusters. Besides, some instances include multiple origin or arrival depots, and each cluster must contain only one pair origin-arrival. Whenever these conditions show, the solving process is semi-automatic in order to create properly the clusters.

6. *Precedence constraints*: sometimes it is mandatory to pass through a specific node to pick up supplies before making any delivery at the medical centers, e.g., to visit a preparation node before a hospital. Hence, this situation imposes a mandatory precedence in a specific group of nodes.

7. *Pickups and deliveries*: despite most nodes are pickup points, the loaded freight must be unloaded in somewhere, either into an intermediate location during the routing operation, or, more commonly, at the end of this process. Therefore, routes are frequently characterized by both operations.

As we can notice, all this dynamism, in terms of daily-based problem conceptualization, enlarges the problems' complexity. Therefore, our solving methodologies must be flexible enough to deal with them smartly, quickly, and efficiently. Below, we re-describe the most common problem variants we faced during the development of this project, now based on their definitions and particularities (1)–(7), previously defined:

- **VRP**: As described, the consideration of this problem relies on operating scenarios in which the fleet of vehicles is large enough to visit all the problem locations without disrespecting any constraint regarding vehicle capacity and/or tour length. Moreover, the routes must necessarily start and end at the same point. By being the simplest problem, it is barely considered in its classical definition, resulting in frequent adaptations, which lead us to the following variants.
- **OVRP**: Unlike classical VRPs, which impose the routes starting and ending their trajectories at the same depot, OVRPs allow different points for starting and concluding the operation of a route. It consists of another typical situation since the drivers usually start their routes from home—or any location—and finish them at hospitals or health care centers, where the collected loads must be delivered.
- **PDP**: Several requests must be processed daily, requiring both picking up and delivering of goods. This might be the case in which all cases rely on, where a set of pickups locations are visited for loading and, later, this freight is unloaded at the delivery points, usually represented by the end depot.
- **TOP**: As described, a common application context that relies on the definition of TOP regards the situation where a limited number of vehicles are available to process a large number of visits. Therefore, as part of the decision to be made, a set of nodes must be discarded during the routing in order to satisfy the constraint related to the maximum tour length, which, implicitly, defines the number of needed vehicles.

Based on these VRP variants, Fig. 18.10 presents a visual schema for classifying the studied problems according to their main particularities (1) to (7). Notice that the main core of each problem limits its definition. In other words, for instance, the OVRP and VRP are defined by use or not of constraint (3), respectively. The TOP is mainly defined by constraint (2), which might result in non-visited nodes. Finally, the PDP is defined by constraint (7). The remaining characteristics are considered, in some way, as minor, but no less important, part of the problem definition. By visualizing this classification schema, we can notice that these problems are commonly characterized by not only distinct singularities but also, in some cases, common components. Such particularities define which problem is the most suitable one for representing the real scenario.

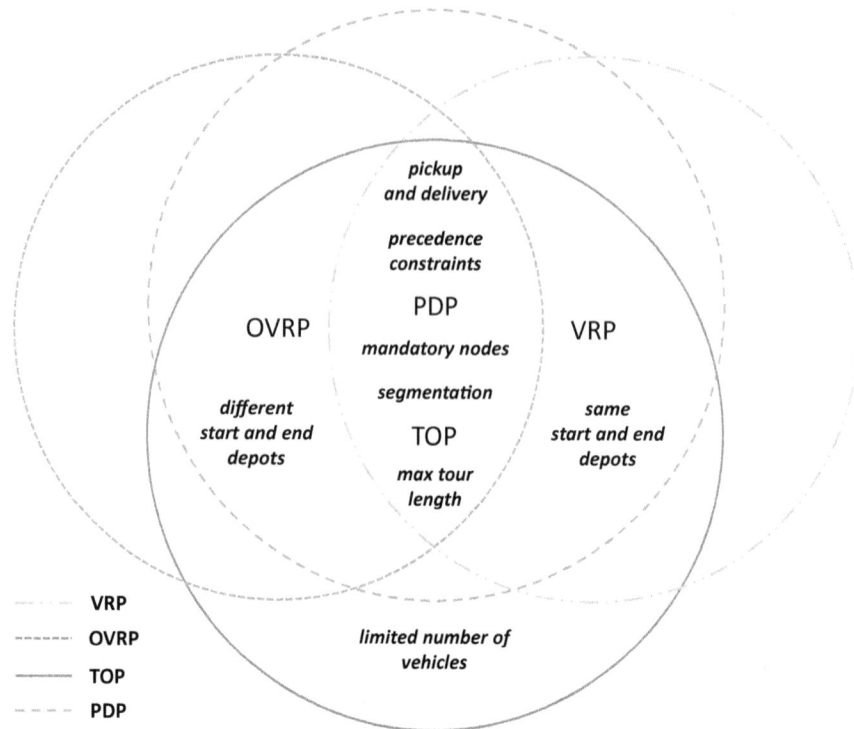

Fig. 18.10 The visual classification of the problems regarding their specifications

Taking into account that there were not service during most weekends—specially as the urgency for the new material was lower after the first weeks–, a total of 29 instances (days) were solved during this COVID-19 crisis. Table 18.1 displays both the known characteristics of each instance (input columns) and obtained results (output columns). Additionally, the instance name is shown, which corresponds to the date for which the instance was solved. Notice that the service time decreases to 4 minutes from the instance *apr-04*. Initially, coordinators estimated a constant service time per node of 7 minutes, however, drivers suggested a shorter time given the experience acquired in previous days. In general, the origin node is not the same as the arrival node. Nevertheless, some instances allow to relax this constraint. They are marked with an asterisk in Table 18.1. Strictly speaking, these instances correspond to an OVRP. However, since the arrival node must also be visited before starting the route, i.e., there are *mandatory nodes* and *precedence constraints*, it was possible to adjust these instances in the pre-processing step, so that a VRP-like problem was solved.

The output columns in Table 18.1 show the maximum tour length (MTL) and the number of visited nodes according to the results obtained by each algorithm. As the VRP algorithm is designed to visit always all points, the number of nodes in the

Table 18.1 Instances' inputs and outputs

Instance	Input							Output				Best strategy
	Number of clusters	Maximum tour length (h)	Service time (min)	Number of vehicles	Mandatory nodes	Precedence constraints	Same origin-arrival node	VRP algorithm		TOP algorithm		
								MTL (hh:mm)	Visited nodes	MTL (hh:mm)	Visited nodes	
Mar-25	1	5	7	6				3:50	95	3:52	95	VRP
Mar-26	1	5	7	4				4:38	77	4:59	77	VRP
Mar-27	2	5	7	6				3:36	62	3:43	62	VRP
Mar-28	1	5	7	2				4:59	48	5:09	48	VRP
Mar-30	1	5	7	2	✓			4:01	32	4:15	32	VRP
Mar-31	2	7	7	2				6:32	53	6:36	53	VRP
Apr-01	2	7	7	2	✓	✓		6:39	31	6:41	31	VRP
Apr-02	2	5	7	2				4:58	29	5:02	29	VRP
Apr-03	1	5	7	2				4:05	29	4:14	29	VRP
Apr-04	1	4	4	1	✓			5:56	22	4:07	14	TOP
Apr-06a	2	6	4	2	✓			5:58	46	6:07	46	VRP
Apr-06b	1	5	4	1	✓			3:12	15	3:17	15	VRP
Apr-07	1	5	4	1	✓			4:35	17	4:36	17	VRP
Apr-08	1	5	4	1				5:28	19	4:53	16	TOP
Apr-09a	2	6	4	2	✓	✓	*	6:15	51	5:54	49	TOP
Apr-09b	1	6	4	1	✓	✓	*	5:43	22	5:48	22	TOP
Apr-10	1	6	4	1				5:21	25	5:31	25	VRP
Apr-13a	2	5	4	2				4:59	39	5:05	39	VRP
Apr-13b	1	6	4	1				6:34	24	5:54	21	TOP

(continued)

Table 18.1 (continued)

Instance	Input							Output				
	Number of clusters	Maximum tour length (h)	Service time (min)	Number of vehicles	Mandatory nodes	Precedence constraints	Same origin-arrival node	VRP algorithm		TOP algorithm		Best strategy
								MTL (hh:mm)	Visited nodes	MTL (hh:mm)	Visited nodes	
Apr-14	1	6	4	1				5:30	21	5:33	21	VRP
Apr-15	1	7	4	1	✓	✓	*	8:38	39	6:32	28	TOP
Apr-17	1	6	4	1	✓	✓	*	6:53	33	6:22	29	TOP
Apr-18	2	5	4	2	✓	✓		4:55	38	4:54	38	TOP
Apr-20	1	5	4	1	✓	✓	*	5:32	21	4:48	17	TOP
Apr-22	1	5	4	1	✓	✓	*	6:33	21	4:46	16	TOP
Apr-25	1	5	4	1	✓	✓	*	5:37	23	4:59	20	TOP
Apr-28	1	5	4	1	✓	✓	*	6:06	23	4:54	19	TOP
Apr-30	1	6	4	1				5:34	14	5:40	14	VRP
May-02	1	5	4	1				3:30	13	3:36	13	VRP

corresponding column represents the total input nodes in the instance. Conversely, the number yielded by the TOP algorithm is less than or equal to the total nodes. Skipping nodes is necessary in some instances given the limitations in both tour lengths and available vehicles. For example, the instance *apr-13b* imposes that the single available vehicle must not take more than 6 hours in completing its tour. The VRP algorithm yields a total travel time of 6 hours and 34 minutes to visit 24 nodes, which violates such constraint. Alternatively, the TOP algorithm designs a 21-node route that takes 5 hours and 54 minutes. Therefore, the TOP algorithm is the best strategy to solve this instance.

The hardness of the travel time constraint depends on the problem instance, since drivers are not the same every day. Hence, the travel time is a soft constraint in instances *apr-04* and *apr-17*. Anyway, the TOP algorithm is the best strategy in these 2 instances since the time yielded by the VRP algorithm is prohibitively high. The rest of the instances have the total travel time as a hard constraint. For most of them the VRP algorithm is the best strategy, because it yields a shorter time than the TOP algorithm, guaranteeing complete routes visiting all nodes. Figure 18.11 shows an example of the best routes obtained by VRP and TOP algorithms for the instance *apr-08*. Three nodes are skipped in the second case to meet the time constraint of 5 hours, which generates savings of 35 minutes with respect to the first case. This example shows the advantages of using a TOP algorithm when the time is limited, since it finds a good balance between the reward offered by each node and the cost of service.

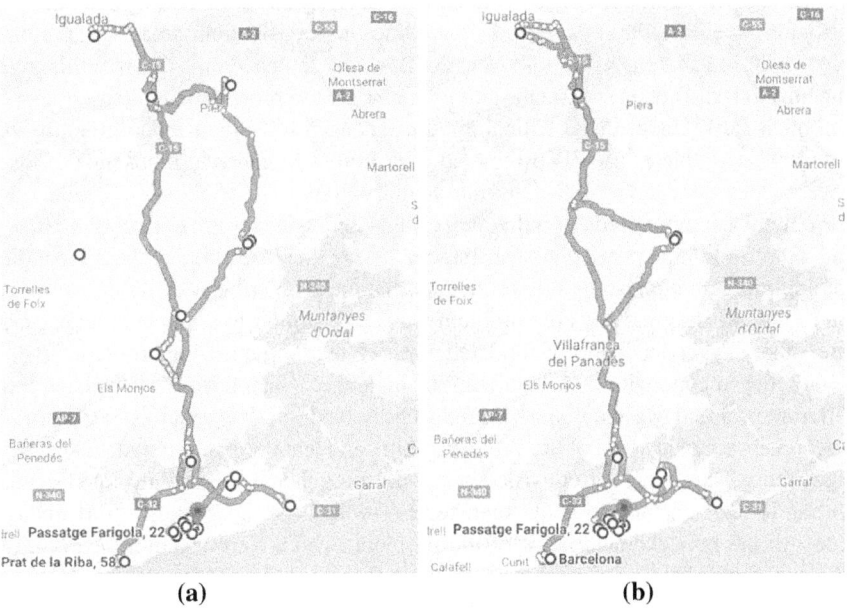

Fig. 18.11 Routes generated for the instance *apr-08* by VRP (**a**) and TOP (**b**) algorithms

18.8 Managerial Insights

This experience shows how synergies appear when there is a diverse team working together towards a challenging and common goal. Firstly, through the combination of the know-how from research and scientific field and the one from business and entrepreneurship contexts. Secondly, through the collaboration between companies (logistics), institutions (the ICSO@IN3 research group) and volunteers (the "coronavirus makers" community), with very different organizational conditions that provide a broader scope of the project benefits. For example, the fast experiential feedback from the field creates a rich feedback loop in order to tailor further the process followed at all levels.

We have also seen how a new restriction appearing in the real-use cases may need a completely different approach to its solution. Such approaches have been evolved through years of research. However, once the knowledge exists, even in a more theoretical context, it can be quickly applied to the real use case by means of the know-how and experience from the researchers and a clear explanation of the need from the field side.

18.9 Conclusions and Future Work

This chapter has discussed how computational intelligence has been a useful support for dealing with complex logistics challenges during the COVID-19 pandemic. In particular, we describe a case study regarding the metropolitan area of Barcelona (Spain) during March, April, and May 2020, where the pandemics were at his high and hospitals did not have enough sanitary material as to protect their nurses, doctors, and other staff. Under those critical circumstances, a self-organized community of "makers" was able to use 3D printers at their homes to generate thousands of face covers, open-door devices, and similar sanitary items. The challenge of collecting these items from hundreds of individual houses and using a limited fleet of vehicles and a threshold time per service was huge, not only due to the size of the collection-routing problem but also mainly to the fact that the problem was evolving day after day. Thus, while some days the problem was more similar to a rich vehicle routing one, other days it needed to be modeled as a rich team orienteering problem.

In order to cope with this optimization challenge, which basically consisted in a different problem every day that needed to be solved in a few minutes, our team of researchers adapted some of the vehicle routing and team orienteering metaheuristic algorithms. The adaptation consisted in transforming 'heavy' algorithms into flexible and agile ones capable to adapt themselves—with little or no extra effort on our side—to the new characteristics of the problem, which were changing every day. The experience has shown us that: *(i)* in crisis scenarios like the one described in the paper, a more 'agile' optimization paradigm is requested –in contrast to the use of complex algorithms that focus on the solving of a single optimization problem, more flexible and fast algorithms are needed; and *(ii)* rapid development environments, such as those provided by the Python programming language, might significantly reduce the effort and time required to adapt algorithms to new variants of routing optimization problems.

As future work, we plan to develop an agile algorithm—based on the use of biased-randomization and parallelization techniques–, which hybridizes aspects from rich vehicle routing problems and team orienteering problems. Such an algorithm, which also should not require from time-consuming fine-tuning processes, might be very valuable if a new wave of this pandemic (or a similar one) hits us again in the future.

Acknowledgements We would like to thank the "makers" community for the effort they put in helping other people during this crisis. We also appreciate the technical support received from the firms FHIOS Smart Knowledge and ACT OR.

References

1. Andrade CE, Silva T, Pessoa LS (2019) Minimizing flowtime in a flowshop scheduling problem with a biased random-key genetic algorithm. Expert Syst Appl 128:67–80
2. Avci M, Topaloglu S (2016) A hybrid metaheuristic algorithm for heterogeneous vehicle routing problem with simultaneous pickup and delivery. Expert Syst Appl 53:160–171
3. Aziez I, Côté JF, Coelho LC (2020) Exact algorithms for the multi-pickup and delivery problem with time windows. Eur J Oper Res
4. Bayliss C, Juan AA, Currie CS, Panadero J (2020a) A learnheuristic approach for the team orienteering problem with aerial drone motion constraints. Appl Soft Comput 106280
5. Bayliss C, Martins LDC, Juan AA (2020b) A two-phase local search with a discrete-event heuristic for the omnichannel vehicle routing problem. In: Computers and industrial engineering , 106695
6. Bettinelli A, Ceselli A, Righini G (2014) A branch-and-price algorithm for the multi-depot heterogeneous-fleet pickup and delivery problem with soft time windows. Math Programming Comput 6:171–197
7. Braekers K, Ramaekers K, Van Nieuwenhuyse I (2016) The vehicle routing problem: state of the art classification and review. Comput Industrial Eng 99:300–313
8. Brandão J (2020) A memory-based iterated local search algorithm for the multi-depot open vehicle routing problem. Eur J Oper Res 284:559–571
9. Brandão JS, Noronha TF, Resende MG, Ribeiro CC (2015) A biased random-key genetic algorithm for single-round divisible load scheduling. Int Trans Oper Res 22:823–839
10. Brandão JS, Noronha TF, Resende MG, Ribeiro CC (2017) A biased random-key genetic algorithm for scheduling heterogeneous multi-round systems. Int Trans Oper Res 24:1061–1077
11. Chen D, Pan S, Chen Q, Liu J (2020) Vehicle routing problem of contactless joint distribution service during covid-19 pandemic. Transp Res Interdisc Perspect 8
12. Dantzig GB, Ramser JH (1959) The truck dispatching problem. Manage Sci 6:80–91
13. Dominguez O, Juan AA, de la Nuez I, Ouelhadj D (2016) An ils-biased randomization algorithm for the two-dimensional loading hfvrp with sequential loading and items rotation. J Oper Res Soc 67:37–53
14. Dumas Y, Desrosiers J, Soumis F (1991) The pickup and delivery problem with time windows. Eur J Oper Res 54:7–22
15. Ferone D, Gruler A, Festa P, Juan AA (2019) Enhancing and extending the classical grasp framework with biased randomisation and simulation. J Oper Res Soc 70:1362–1375

16. Fikar C, Juan AA, Martinez E, Hirsch P (2016) A discrete-event driven metaheuristic for dynamic home service routing with synchronised trip sharing. Eur J Industrial Eng 10:323–340
17. Golden BL, Levy L, Vohra R (1987) The orienteering problem. Naval Res Logistics (NRL) 34:307–318
18. Gonçalves JF, Resende MG (2011) Biased random-key genetic algorithms for combinatorial optimization. J Heuristics 17:487–525
19. Gonzalez-Martin S, Juan AA, Riera D, Castella Q, Muñoz R, Perez A (2012) Development and assessment of the sharp and randsharp algorithms for the arc routing problem. AI Commun 25:173–189
20. Gonzalez-Neira EM, Ferone D, Hatami S, Juan AA (2017) A biased-randomized simheuristic for the distributed assembly permutation flowshop problem with stochastic processing times. Simul Modell Pract Theor 79:23–36
21. Gruler A, Fikar C, Juan AA, Hirsch P, Contreras-Bolton C (2017) Supporting multi-depot and stochastic waste collection management in clustered urban areas via simulation-optimization. J Simul 11:11–19
22. Györgyi P, Kis T (2019) A probabilistic approach to pickup and delivery problems with time window uncertainty. Eur J Oper Res 274:909–923
23. Hanafi S, Mansini R, Zanotti R (2020) The multi-visit team orienteering problem with precedence constraints. Eur J Oper Res 282:515–529
24. Homayouni SM, Fontes DB, Gonçalves JF (2020) A multistart biased random key genetic algorithm for the flexible job shop scheduling problem with transportation. Int Trans Oper Res
25. Husakou A, Hvattum LM, Danielsen K, Hoff A (2020) An application of the multi-depot heterogeneous fixed fleet open vehicle routing problem. Int J Adv Oper Manage 12:142–155
26. Karabulut K, Tasgetiren MF (2020) An evolution strategy approach to the team orienteering problem with time windows. Comput Industrial Eng 139
27. Kizys R, Juan AA, Sawik B, Calvet L (2019) A biased-randomized iterated local search algorithm for rich portfolio optimization. Appl Sci 9:3509
28. Lahyani R, Gouguenheim AL, Coelho LC (2019) A hybrid adaptive large neighbourhood search for multi-depot open vehicle routing problems. Int J Prod Res 57:6963–6976
29. Lenstra JK, Kan AR (1981) Complexity of vehicle routing and scheduling problems. Networks 11:221–227
30. Li F, Golden B, Wasil E (2007) The open vehicle routing problem: algorithms, large-scale test problems, and computational results. Comput Oper Res 34:2918–2930
31. Marmol M, Martins, LDC, Hatami S, Juan AA, Fernandez V (2020) Using biased-randomized algorithms for the multi-period product display problem with dynamic attractiveness. Algorithms 13:34
32. Martins LDC, Hirsch P, Juan AA (2020) Agile optimization of a two-echelon vehicle routing problem with pickup and delivery. Int Trans Oper Res
33. Nucci F (2021) Multi-shift single-vehicle routing problem under fuzzy uncertainty. In Intelligent and fuzzy techniques: smart and innovative solutions. Springer, Heidelberg, pp 1620–1627
34. Pacheco J, Laguna M (2020) Vehicle routing for the urgent delivery of face shields during the covid-19 pandemic. J Heuristics 26:619–635
35. Pagès-Bernaus A, Ramalhinho H, Juan AA, Calvet L (2019) Designing e-commerce supply chains: a stochastic facility-location approach. Int Trans Oper Res 26:507–528
36. Panadero J, Juan AA, Bayliss C, Currie C (2020) Maximising reward from a team of surveillance drones: a simheuristic approach to the stochastic team orienteering problem. Eur J Industrial Eng 14:485–516
37. Parhami B (2006) Introduction to parallel processing: algorithms and architectures. Springer Science & Business Media
38. Parragh SN, Doerner KF, Hartl RF (2008) A survey on pickup and delivery problems. J für Betriebswirtschaft 58:21–51
39. Perdana T, Chaerani D, Achmad ALH, Hermiatin FR (2020) Scenarios for handling the impact of covid-19 based on food supply network through regional food hubs under uncertainty. Heliyon 6

40. Pisinger D, Ropke S (2007) A general heuristic for vehicle routing problems. Comput Oper Res 34:2403–2435
41. Quintero-Araujo CL, Caballero-Villalobos JP, Juan AA, Montoya-Torres JR (2017) A biased-randomized metaheuristic for the capacitated location routing problem. Int Trans Oper Res 24:1079–1098
42. Quintero-Araujo CL, Gruler A, Juan AA, Faulin J (2019) Using horizontal cooperation concepts in integrated routing and facility-location decisions. Int Trans Oper Res 26:551–576
43. Reyes-Rubiano L, Juan A, Bayliss C, Panadero J, Faulin J, Copado P (2020) A biased-randomized learnheuristic for solving the team orienteering problem with dynamic rewards. Transp Res Proc 47:680–687
44. Ruiz E, Soto-Mendoza V, Barbosa AER, Reyes R (2019) Solving the open vehicle routing problem with capacity and distance constraints with a biased random key genetic algorithm. Comput Industrial Eng 133:207–219
45. Saeedvand S, Aghdasi HS, Baltes J (2020) Novel hybrid algorithm for team orienteering problem with time windows for rescue applications. Appl Soft Comput 96
46. Sánchez-Oro J, López-Sánchez AD, Colmenar JM (2020) A general variable neighborhood search for solving the multi-objective open vehicle routing problem. J Heuristics 26:423–452
47. Savelsbergh MW, Sol M (1995) The general pickup and delivery problem. Transp Sci 29:17–29
48. Schrage L (1981) Formulation and structure of more complex/realistic routing and scheduling problems. Networks 11:229–232
49. Sharma SK, Routroy S, Yadav U (2018) Vehicle routing problem: recent literature review of its variants. Int J Oper Res 33:1–31
50. Sombuntham P, Kachitvichayanukul V (2010) A particle swarm optimization algorithm for multi-depot vehicle routing problem with pickup and delivery requests. In: World Congress on Engineering 2012 (July), pp 4–6 (2012) London, UK, Citeseer, pp 1998–2003
51. Tavakkoli-Moghaddam R, Meskini M, Nasseri H, Tavakkoli-Moghaddam H (2019) A multi-depot close and open vehicle routing problem with heterogeneous vehicles. In: 2019 International Conference on Industrial Engineering and Systems Management (IESM). IEEE, pp 1–6
52. Toth P, Vigo D (2014) Vehicle routing: problems, methods, and applications. SIAM
53. Vidal T, Laporte G, Matl P (2019) A concise guide to existing and emerging vehicle routing problem variants. Eur J Oper Res
54. Yousefikhoshbakht M, Dolatnejad A (2017) A column generation for the heterogeneous fixed fleet open vehicle routing problem. Int J Prod Manage Eng 5:55–71

Chapter 19
Multi-objective Assignment Problems and Their Solutions by Genetic Algorithm

Anita R. Tailor and Jayesh M. Dhodiya

Abstract Assignment problem (AP) has been usually used to solve decision-making problems in the industrial organization, manufacturing system, developing service system, etc., is to optimally resolve the problem of n-activities to n-devices such that total cost/time can be minimized or total profit/sales can be maximized. In today's optimization problems, the single objective optimization problems (SOOPs) are not more sufficient to hold the problem facts, hence multi-objective optimization problems (MOOPs) are considered. The purpose of MOOPs in the mathematical programming (MP) structure is to optimize several objective functions under some constraints. In research, the multi-objective field does not give a single optimal solution, but a set of efficient solutions because there are frequently conflicts between the various objectives. In the multi-objective assignment problem (MOAP), significant research concerns related to the study of effective solutions. The current chapter focuses on a genetic algorithm (GA) based approach to find a solution to MOAP. In order to achieve an effective allocation plan, the decision-maker (DM) must specify different aspiration levels (ALs) according to his/her preferences and different shape parameters (SPs) in the exponential membership function (EMF) to show the effect of integration on the effective solution of MOAP. Numerical illustrations are provided to express the usefulness of a specific approach related to the data set from realistic circumstances. This research turned out to be a GA based approach provides effective output based on analysis to take the decision regarding the situation.

Keywords Multi-objective assignment problem · Genetic algorithm · Shape parameter · Aspiration level

---

A. R. Tailor (✉)
Navyug Science College, Surat, India

J. M. Dhodiya
S. V. National Institute of Technology, Surat, India

19.1 Introduction

In an optimization problem, MOOP copes the optimization of multiple incompatible objective functions. The intention of MOOPs in the MP agenda is to optimize several objective functions under some constraints. The MOOPs are more practical but more tricky to solve than single-objective version. When considering numerous objective functions, there usually does not available single optimal solution. The conception of optimality is definitely derived from the Pareto dominance that stimulates a partial order on the solutions. The Pareto best possible solutions are feasible solutions in parallel a way one cannot recover the recital of one objective with no deterioration of another objective. As they stand for all possible negotiations between the objectives, a DM would hence choose an efficient solution (ES). However, when a large numeral of efficient solutions are available, it is difficult for the DM to select his/her most favored solution from the entire efficient ones. So, to determine an "optimal" solution in such models, more information is required and it can be gained from the subjective preferences of the DM. This preferred solution is measured as the "Best Compromise" solution as the trade-offs are made between the various objectives [1].

This chapter deals with an alternative approach to solve a special case of the MOOP as a MOAP. In classical assignment problem, the function is to be optimized with their decision variables while MOAP involves numerous objectives like minimization of time, distance, cost; maximization of the profit, quality, sales etc., [2]. MOAPs can be found in various essential application areas like facility layout problem, manpower planning, planning and capital budgeting, an arrangement of project networks, loading and grouping for manufacturing systems, scheduling of timetable problem, assigning tasks to computers in computer networks, nurses scheduling at a hospital, payments scheduling on accounts, scheduling variable extent of television marketable, etc. [1]. Here, an important research issue in MOAPs is to find efficient solutions as per the requirement of DM.

19.2 Literature Review

An optimal solution of MOAP in the real-world situation need not be deterministic because many parameters are responsible for it. Lots of researchers provided attention on the solution of MOAP and give the best solution. Tsai et al. [3] have extended the conventional minimize SOOP to the multi-objective decision-making problem of three objectives like cost, time, and quality using a fuzzy concept by maximization of the membership function. Tuyttens et al. [4] provided a 'multi-objective simulated annealing' (MOSA) method to improve the efficiency of bi-criteria AP with a greedy approach and also its results are compared with an exact method. Ibrahim et al. [5] proposed a 0–1 linear goal programming model to convert the multi-criteria assignment problem (MCAP) into a classical AP and its solution obtained by the Hungarian algorithm. Przybylski et al. [6] presented a synthesis of the two-phase method (TPM) for the MOAP to find all ESs with an improved upper bound using a ranking approach. Bao et al. [7] used the 0-1 programming method to formulate a MOAP into a linear programming problem by converting all the objectives in normal form and applied the different weights to each objective. Bufardi [1] addressed the efficiency of feasible solutions of an MCAP in two steps as in the step-1, determined a feasible solution of an MCAP is efficient or not, and if not then it is obtained by the branch and bound algorithm in step-2. Kagade and Bajaj [8] have utilized linear and non-linear membership functions to solve the MOAP and also compared the results which are obtained by both membership functions.

Odior et al. [9] provided a technique to determine all real ESs of a general MOAP. Przybylski et al. [10] have solved tri-objective assignment problem using the TPM to find all supported ESs and also found non-supported ESs using enumerative methods. Li et al. [11] have discussed an effective TPM with distributed auction algorithm is used for finding supported Pareto optimal solution to solve bi-objective assignment problem. Adiche and Aïder [12] have developed a metaheuristic method for solving MOAPs which are derived from the dominance cost variant of the MOSA and hybridizes neighborhood search procedures that involve of multi-objective branch and bound search. Ozlen et al. [13] have initiated an improved recursive algorithm to produce the set of every nondominated objective vector for the MOAP.

Przybylski et al. [10] generalized the TPM to find solutions of tri-objective AP; in phase-I, all supported ESs are computed and non-supported ESs are found by enumerative methods in phase-II. De and Yadav [14] accessible an interactive fuzzy goal programming approach for solving MOAP which is characterized by a EMF. Ratli et al. [15] presented a multi-objective GA to solve MAOP. Gupta et al. [16] have used fuzzy approach with GA to solve a cost-time-quality objectives AP with many realistic resource constraints. Basirzadeh et al. [17] have proposed a technique to derive the best non-dominated point which possess the smallest distance to the ideal point for the MOAPs. Hassan Shirdel and Mortezaee [18] have proposed a 'Data Envelopment Analysis' model to evaluate arc efficiency for MCAP when multiple weights on arcs in a network are given. Tiwari et al. [19] have proposed an algorithm to solve of MOAP with different membership functions.

Jayalakshmi and Sujatha [20] have expressed a new method to solve MOAP using the optimal solution of single AP and obtained the ideal solution and the set of all Pareto optimal solution. Medvedeva and Medvedev [21] have introduced two approximation algorithm to solve MOAP via Lagrangian relaxation of many constraints. Hammadi [22] provided the multi-objective tabu search method for MOAP to generate non-dominated alternatives. Belhoul et al. [23] have resolute an efficient method to find the best compromise solutions to MOAP. Huang and Lim [24] have provided a hybrid Genetic algorithm to solve Three index AP.

To find the solution of MOAP, the MOAP is transformed into a single objective nonlinear optimization problem (SONOP) with many pragmatic constraints and consequently, this particular problem turn into an "NP-hard" problem. To solve such problems, GA is a suitable course of action. Thus, this chapter proposes a GA based approach to ascertain the solution of MOAP. This chapter also discusses effects of the convergence to the ESs of MOAP when DM gives a distinct aspiration level (AL).

In the end, the chapter includes a comparison of the developed approach with other approaches.

19.3 Multi-objective Assignment Problem Formulation

The general mathematical formulation of MOAP is as follows [2, 3, 4, 5, 6, 8, 9, 10, 11, 12, 13, 14, 16, 20, 21, 22, 25]:

$$(\mathbf{P}-1) \min Z^k(x) = \sum_{i=1}^{n} \sum_{j=1}^{n} c_{ij}^k x_{ij}, \quad k = 1, 2, \ldots, m$$

Subject to the constraints:

$$\sum_{i=1}^{n} x_{ij} = 1 \quad \text{for all } j \tag{19.1}$$

$$\sum_{j=1}^{n} x_{ij} = 1 \quad \text{for all } i \tag{19.2}$$

$$x_{ij} = \begin{cases} 1; & \text{if ith worker perform to jth job} \\ 0; & \text{else} \end{cases} \tag{19.3}$$

where $i, j = 1, 2, \ldots, n$. $x_{ij} = (x_{11}, \ldots, x_{nn})$ is the matrix of decision variable and c_{ij} is the cost of conveying the jth job to the ith worker.

19.4 Some Preliminaries

Aspiration level

In many MOOPs, it is imperative to obtain a solution that faithfully reflects the DM's judgment. In order to make decisions based on the diversity of value-judgment and complicated modification in the nature of decision making, the aspiration level has provided the satisfactory trading method that can work well not simply with the quantitative DM judgment but also with the dynamics of significance opinion of them. Aspiration level does not necessitate any uniformity of judgment of DM because DM often changes his/her attitude even throughout the decision-making process [26, 27].

Positive Ideal Solution (PIS) and Negative Ideal Solution (NIS)

The minimum and the maximum value of an objective function is defined by PIS and NIS respectively. They are required to compute the membership value for all objective functions [28–30].

Exponential membership Function

In MOOPs, the different ALs of DM are distinguished by fuzzy membership functions for the objective functions. Moreover, the membership function is used for describing the performance of the vague information, utilization of the fuzzy numbers of DM, preferences towards uncertainty, etc. The EMF gives healthier demonstration then others and provides the flexibility to express grade of exactness in parameter values. It also reflects reality better than the linear membership function [16, 28, 29, 30].

If Z_k^{PIS} and Z_k^{NIS} are PIS and NIS of objective Z_k, the EMF $\mu_{Z_k}^E$ is defined by

$$\mu_{Z_k}^E(x) = \begin{cases} 1; & \text{if } Z_k \leq Z_k^{PIS} \\ \frac{e^{-S\psi_k(x)} - e^{-S}}{1 - e^{-S}}, & \text{if } Z_k^{PIS} < Z_k < Z_k^{NIS} \\ 0; & \text{if } Z_k \geq Z_k^{NIS} \end{cases} \quad (19.4)$$

where, $\psi_k(x) = \frac{Z_k - Z_k^{PIS}}{Z_k^{NIS} - Z_k^{PIS}}$ and $S \neq 0$ is shape parameter (SP) which specified by DM such that $\mu_{Z_k}(x) \in [0, 1]$. The membership function is strictly convex (concave) for $S < 0$ ($S > 0$) in $[Z_k^{PIS}, Z_k^{NIS}]$.

Genetic algorithm

Considering the adaptive evolution and usual selection of biological systems, GA is a distinguished arbitrary search and global optimization method. This is also a suitable method to solve discrete, non-linear, and non-convex global optimization problems rather than any traditional method because it seeks for the optimal solution by simulating the natural evolutionary process, and is performed by imitating the evaluation principle and chromosome processing work in usual genetics [25, 31, 32, 33, 34, 35, 36, 37]. It has proven many key advantages like as strong robustness, convergence to large-scale most favorable and analogous search competence, etc. GA has high-quality acknowledgment in determining a variety of NP-hard problems.

In GAs, chromosomes are coded along with the problem, and fitness function is used to measure chromosomes. Apply gene operators such as selection, crossover, and mutation to produce the new population [28–30]. Thus, the newly generated population included some of the parent population and child population, which can be used to find an ES.

Convergence criteria

If after getting some particular value i.e. optimum value, we say GA is converged. For an NP-hard problem, GA has converged at a large-scale optimum is impractical, unless you already have the test data set with the best solution known. Moreover, the dimension of the problem so affects the convergence of GA. One can define the size of the chromosome (number of solutions) based on the problem's parameter. Also, one should note that increasing the size of the chromosome affects the GA's rate of convergence [28, 29].

19.5 Solution Methodology of MOAP Using GA Based Approach

This section introduces a GA based approach for MOAP to find out the finest ESs. As well as this approach provides better litheness to solve the MOAP in terms of different variety of ALs for all objective functions.

19.5.1 Steps to Find the Efficient Solution of MOAP

The following is a step-by-step description of a GA based approach to find efficient solutions to the MOAP.

Step-1: Consider the MOAP (P − 1).
Step-2: Convert the maximum objective function of the problem into minimum form.
Step-3: Find out the PIS and NIS of each Z_K for all k.
Step-4: Find μ_{Z_k} for each objective function.
Step-5: In this step, convert the MOAP into the SOOP which define as follows.

$$(P-2) \max W = \prod_{k=1}^{m} \mu_{Z_k}$$

Subject to the constraints: (1)–(3)

$$\mu_{Z_k}(x) - \overline{\mu_{Z_k}}(x) \geq 0; \quad k = 1, 2, \ldots, m \tag{19.5}$$

where, $\overline{\mu_{Z_k}}(x)$ is the preferred AL of fuzzy goals in proportion to all objectives.
Step-6: To find different assignment plans for P − 2 which is developed in step-5 using GA with verities of the SPs.

1. **Encoding of Chromosomes**

Encoding of chromosomes is the primary task to solve any problem with GA. To create the solution for the SOOP through GA, the data structure of chromosomes in the encoding space must be considered, which indicates the solution to the problem. In the data structure of chromosomes, put all 0s to $n \times n$ genes on a chromosome, and thereafter arbitrarily select gene of the chromosome, put 1s in every rows and columns of chromosomes precisely one time such that it satisfies constraints (1)–(3). In chromosome structure, each string can express individually in 2^r; $r = 0 : N - 1$ way.

2. **Evolution of fitness function**

In GA, the fitness function is a key factor to solve any optimization problem. With satisfaction of constraints (1)–(3) and (5), the objective function of P – 2 is evaluated to check the fitness of chromosomes.

3. **Selection**

The Selection operator is utilized to decide how chromosomes are preferred from the present population and will treat as parents in favor of the next genetic operations, viz., crossover/mutation/next population which will have the highest fitness. To find the solution of MOAP, tournament selection is used by reason of its competence and simple execution. In this selection, n-chromosomes are arbitrarily preferred from the population and judge against each other. The maximum fit chromosome (winner) is preferred for the subsequent generation, whereas others are ineligible. This process is unremitting until the population size and the number of winners become same [16, 28, 29, 30, 38].

4. **Crossover**

Crossover is a course of action of obtaining two parent solutions and producing a child from them. For MOAP, partially matched crossover (PMX) is used to generate new offspring. In PMX, two chromosomes are allied and two crossover points are chosen at random. These two crossover points provide a matching option, that is used to change a cross by swapping positions one by one.

5. **Threshold construction**

In order to preserve the population's diversity later than the crossover, a threshold is built to produce MOAP's solution. In such step, select some chromosomes from the parent and child population sets in favor of the new iterations. To make the threshold, first, the whole population must be sorting in an ascending/descending order according to its objective function values and then select the encoded entity strings from all groups. Rest on these objective function values, the population is separated into four groups: above $\mu + 3 * \sigma$, among $\mu + 3 * \sigma$ and μ, among μ and $\mu - 3 * \sigma$, and below $\mu - 3 * \sigma$. As a result, the capable string cannot be missed where σ and μ are standard deviation and expected value of the objective function values of parenthood and childhood population respectively [16, 28, 29, 30, 38].

6. **Mutation**

Here, we have used a swap mutation operator among the available numerous mutation operators to improve the missing genetic materials. It is also used to arbitrarily allocate genetic information. In this mutation, two arbitrary spots are chosen in a string and the consequent values are swapped among spots.

7. Termination criteria

A genetic algorithm is run over a given number of iterations until a termination condition has been reached, that is, the population's best fitness has not changed for a convinced number of generations.

As a result, two cases are executed to find the SOOP's solution using GA: (i) Without mutation (ii) With mutation.

In both cases, an algorithm converged to an optimal solution of SOOP, and at the end ES of MOAP with different combinations of SP and AL are obtained.

If DM approved the achieved solution, considers it an ideal solution and terminate the solution procedure otherwise change the SP, AL and reiterate the step-1 to 6 till an agreeable ES is achieved.

19.6 Problems and Result Analysis

This section present the MOAPs and their solutions using GA based approach. To evaluate MOAP, the model is coded.

19.6.1 MOAP-1

Consider the two objective assignment problem to minimize operation cost and operation time as described below [3, 7, 14, 39].

$$\min C = 10x_{11} + 8x_{12} + 15x_{13} + 13x_{21} + 12x_{22} + 13x_{23} + 8x_{31} + 10x_{32} + 9x_{33}$$

$$\min T = 13x_{11} + 15x_{12} + 8x_{13} + 10x_{21} + 20x_{22} + 12x_{23} + 15x_{31} + 10x_{32} + 12x_{33}$$

Subject to the constraints: (1)–(3); $i, j = 1, 2, 3$

In MOAP-1, number of objectives, tasks and workers are 2, 3, and 3 respectively. Table 19.1 gives PIS and NIS of Z_1 and Z_2.

Using PIS and NIS value for Z_1 and Z_2 in P – 1, an equivalent crisp model can be formulated for MOAP-1 is as follows:

Objective function

Table 19.1 PIS and NIS values of Z_1 and Z_2 of MOAP-1

Objective	Cost (Z_1)	Time (Z_2)
PIS	29	28
NIS	38	45

19 Multi-objective Assignment Problems and Their Solutions ...

Table 19.2 Different combinations of shape parameter and aspiration level

Case	Shape parameters	Aspiration levels
1	$(-5, -5)$	$(0.95, 0.75)$
2	$(-5, -5)$	$(0.7, 0.95)$
3	$(-3, -3)$	$(0.9, 0.65)$
4	$(-2, -2)$	$(0.75, 0.7)$
5	$(-1, -1)$	$(0.65, 0.6)$
6	$(1, 1)$	$(0.4, 0.45)$

$$\max w = \mu_{Z_1} \cdot \mu_{Z_2}$$

Subject to the constraints: (1)–(3) and (5); $i, j = 1, 2, 3$ and $k = 1, 2$ where,

$$\mu_{Z_1} = \frac{\exp\left(-S\left(\begin{array}{c}10x_{11} + 8x_{12} + 15x_{13} + 13x_{21} + 12x_{22} + 13x_{23} \\ +8x_{31} + 10x_{32} + 9x_{33} - 29\end{array}\right)/9\right) - \exp(-S)}{(1 - \exp(-S))}$$

$$\mu_{Z_2} = \frac{\exp\left(-S\left(\begin{array}{c}13x_{11} + 15x_{12} + 8x_{13} + 10x_{21} + 20x_{22} \\ +12x_{23} + 15x_{31} + 10x_{32} + 12x_{33} - 28\end{array}\right)/17\right) - \exp(-S)}{(1 - \exp(-S))}$$

Here, we obtain ESs for the P − 2 by taking the different combinations of SP and AL which are given by the DM.

Different combinations of SP and AL are shown in Table 19.2 which is given by DM.

Table 19.3 describes the computational results and its corresponding optimal allocations according to distinct combinations which shown in Table 19.2.

According to different combinations of SP and AL, Table 19.3 shows the assignment plans for cost and time objectives. It also shows that the change in SP, influences the degree of satisfaction level for each objective function and all achieved solutions are consistent with the DM's desire.

Convergence rate of GA for MOAP-1

The convergence rate of GA for MOAP-1 is obtained for case-1 of Table 19.2 as shown in Fig. 19.1. For the convergence rate, we used different number of iterations and populations with the values of max $W = \prod_{i=1}^{2} Z_i$.

In the both cases, the GA based algorithm converges after 10 populations and 5 iterations for case-1 of Table 19.2. For other cases of Table 19.2, the convergence rate of GA approximately remain the same.

Figure 19.2 shows the ESs of cost and time objective values at different combinations of SP and AL. Figure 19.2 indicates the solutions of cost and time objectives

Table 19.3 Summary results for MOAP-1

Case	Degree of satisfaction	Membership values		Objective values		Optimal allocation plans
		μ_{Z_1}	μ_{Z_2}	Z_1	Z_2	
1	0.9442	0.9442	0.9536	30	37	$x_{12} = x_{21} = x_{33} = 1$
2	0.9110	0.9950	0.9110	33	35	$x_{11} = x_{23} = x_{32} = 1$
3	0.7959	0.9793	0.7959	30	37	$x_{12} = x_{21} = x_{33} = 1$
4	0.7053	0.9611	0.7053	30	37	$x_{12} = x_{21} = x_{33} = 1$
5	0.6743	0.6743	0.7035	33	35	$x_{11} = x_{23} = x_{32} = 1$
6	0.4324	0.4324	0.4661	33	35	$x_{11} = x_{23} = x_{32} = 1$

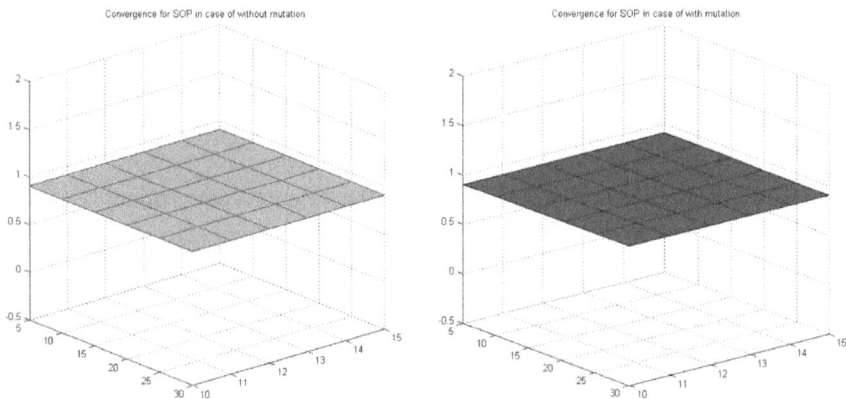

Fig. 19.1 Convergence rate of GA for case-1

as (30, 37), (30, 37), (30, 37), (33, 35) and (33, 35) at $(-5, -5), (-3, -3), (-2, -2), (-1, -1)$ and $(1, 1)$ SPs and their consequent (0.95, 0.75), (0.9, 0.65), (0.75, 0.7), (0.65, 0.6) and (0.4, 0.45) ALs, respectively.

Figure 19.3 indicates the degree of satisfactions as (0.9442, 0.9536), (0.9950, 0.9110), (0.9793, 0.7959), (0.9611, 0.7053), (0.6743, 0.7053) and (0.4324, 0.4661) of the cost and time objectives at $(-5, -5), (-5, -5), (-3, -3), (-2, -2), (-1, -1)$ and $(1, 1)$ SPs and their consequent (0.95, 0.75), (0.7, 0.95), (0.9, 0.65), (0.75, 0.7), (0.65, 0.6) and (0.4, 0.45) ALs respectively. Figure 19.3 also shows that the changes in SP values straightforwardly affect to all objectives.

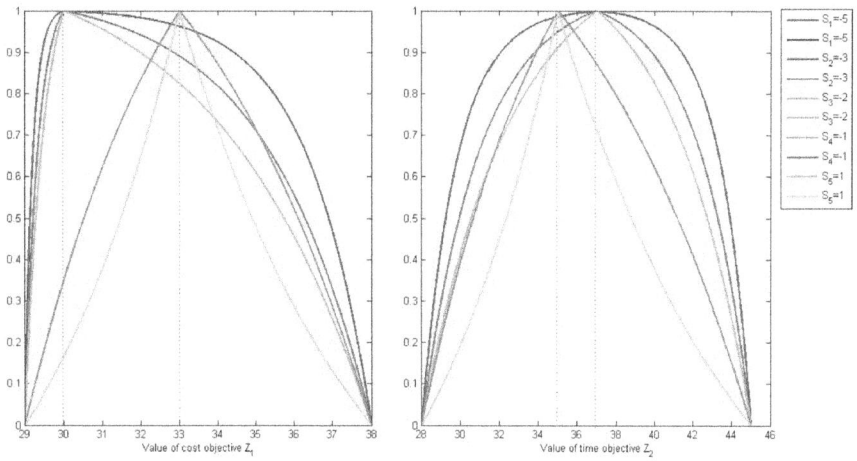

Fig. 19.2 Efficient solutions of MOAP-1 at different combinations of SP and AL

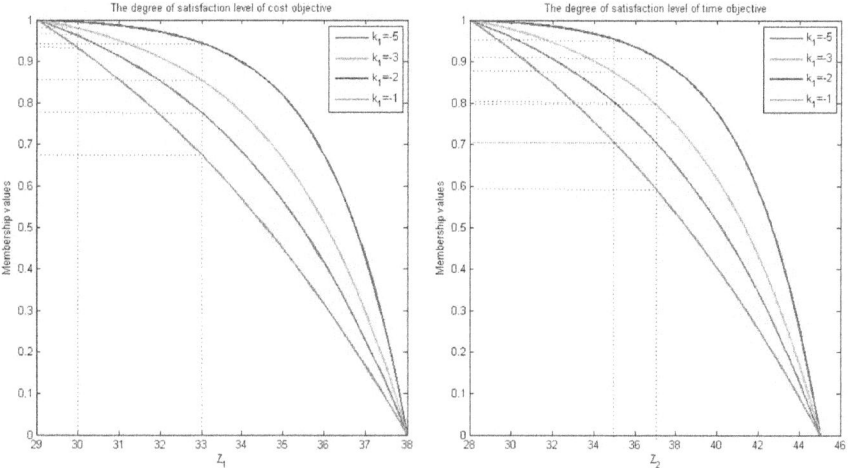

Fig. 19.3 The degree of satisfaction of the MOAP-1

19.6.2 MOAP-2

The cost-time and quality objective assignment problem with 6 doers and 6 tasks has been preferred from the article of Bao et al. and Tsai et al. [3, 7, 39] which is given as follows:

$$\min Z_1 = 6x_{11} + 3x_{12} + 5x_{13} + 8x_{14} + 10x_{15} + 6x_{16} + 6x_{21} + 4x_{22} \\ + 6x_{23} + 5x_{24} + 9x_{25} + 8x_{26} + 11x_{31} + 7x_{32} + 4x_{33} + 8x_{34}$$

$$+ 3x_{35} + 2x_{36} + 9x_{41} + 10x_{42} + 8x_{43} + 6x_{44} + 10x_{45} + 4x_{46}$$
$$+ 4x_{51} + 6x_{52} + 7x_{53} + 9x_{54} + 8x_{55} + 7x_{56} + 3x_{61} + 5x_{62}$$
$$+ 11x_{63} + 10x_{64} + 12x_{65} + 8x_{66}$$

$$\min Z_2 = 4x_{11} + 20x_{12} + 9x_{13} + 3x_{14} + 8x_{15} + 9x_{16} + 6x_{21}$$
$$+ 18x_{22} + 8x_{23} + 7x_{24} + 17x_{25} + 8x_{26} + 2x_{31} + 8x_{32}$$
$$+ 20x_{33} + 7x_{34} + 15x_{35} + 7x_{36} + 12x_{41} + 13x_{42} + 14x_{43}$$
$$+ 6x_{44} + 9x_{45} + 10x_{46} + 9x_{51} + 8x_{52} + 7x_{53} + 14x_{54} + 5x_{55}$$
$$+ 9x_{56} + 17x_{61} + 13x_{62} + 3x_{63} + 4x_{64} + 13x_{65} + 7x_{66}$$

$$\min Z_3 = 1x_{11} + 3x_{12} + 1x_{13} + 1x_{14} + 1x_{15} + 5x_{16} + 3x_{21}$$
$$+ 5x_{22} + 3x_{23} + 5x_{24} + 7x_{25} + 5x_{26} + 1x_{31} + 7x_{32} + 5x_{33}$$
$$+ 3x_{34} + 3x_{35} + 7x_{36} + 5x_{41} + 9x_{42} + 3x_{43} + 5x_{44} + 7x_{45}$$
$$+ 3x_{46} + 3x_{51} + 9x_{52} + 7x_{53} + 5x_{54} + 3x_{55} + 3x_{56} + 3x_{61}$$
$$+ 3x_{62} + 5x_{63} + 7x_{64} + 5x_{65} + 7x_{66}$$

Subject to the constraints: (1)–(3); $i, j = 1, 2, \ldots, 6$

Where cost unit is defined in thousands, time unit is defined in weeks and quality levels are defined as 1, 3, 5, 7 and 9.

In MOAP-2, number of objectives, tasks and workers are 3, 6, and 6 respectively. For MOAP-2, the PIS and NIS of Z_1, Z_2 and Z_3 are shown in Table 19.4 as well as

Table 19.4 PIS and NIS of Z_1, Z_2 and Z_3 of MOAP-2

Objective	Z_1	Z_2	Z_3
PIS	25	32	14
NIS	59	98	38

Table 19.5 Different combinations of shape parameter and aspiration level

Case	Shape parameters	Aspiration levels
1	(−5, −5, −5)	(0.9, 0.85, 0.7)
2	(−5, −3, −1)	(0.8, 0.75, 0.85)
3	(−3, −3, −3)	(0.7, 0.8, 0.9)
4	(−3, −1, −2)	(0.65, 0.7, 0.75)
5	(−2, −2, −2)	(0.65, 0.8, 0.75)
6	(−2, −5, −3)	(0.75, 0.85, 0.8)
7	(−1, −1, −1)	(0.55, 0.6, 0.6)

different combinations of the SP and AL are shown in Table 19.5 which are given by DM.

Using PIS and NIS value for Z_1, Z_2 and Z_3 in P − 1, an equivalent crisp model can be formulated for MOAP-2 is as follows:

Objective function

$$\max w = \mu_{Z_1} \cdot \mu_{Z_2} \cdot \mu_{Z_3}$$

Subject to the constraints: (1)–(3) and (5); i, $j = 1, 2, \ldots, 6$ and $k = 1, 2, 3$ where,

$$\mu_{Z_1} = \frac{\exp\left[-S\left(\begin{array}{l} 6x_{11} + 3x_{12} + 5x_{13} + 8x_{14} + 10x_{15} + 6x_{16} + 6x_{21} \\ +4x_{22} + 6x_{23} + 5x_{24} + 9x_{25} + 8x_{26} + 11x_{31} + 7x_{32} \\ +4x_{33} + 8x_{34} + 3x_{35} + 2x_{36} + 9x_{41} + 10x_{42} + 8x_{43} \\ +6x_{44} + 10x_{45} + 4x_{46} + 4x_{51} + 6x_{52} + 7x_{53} + 9x_{54} \\ +8x_{55} + 7x_{56} + 3x_{61} + 5x_{62} + 11x_{63} + 10x_{64} \\ +12x_{65} + 8x_{66} - 25 \end{array}\right)/34\right] - \exp(-S)}{(1 - \exp(-S))}$$

$$\mu_{Z_2} = \frac{\exp\left[-S\left(\begin{array}{l} 4x_{11} + 20x_{12} + 9x_{13} + 3x_{14} + 8x_{15} + 9x_{16} \\ +6x_{21} + 18x_{22} + 8x_{23} + 7x_{24} + 17x_{25} + 8x_{26} \\ +2x_{31} + 8x_{32} + 20x_{33} + 7x_{34} + 15x_{35} + 7x_{36} \\ +12x_{41} + 13x_{42} + 14x_{43} + 6x_{44} + 9x_{45} + 10x_{46} + \\ 9x_{51} + 8x_{52} + 7x_{53} + 14x_{54} + 5x_{55} + 9x_{56} + 17x_{61} \\ +13x_{62} + 3x_{63} + 4x_{64} + 13x_{65} + 7x_{66} - 32 \end{array}\right)/66\right] - \exp(-S)}{(1 - \exp(-S))}$$

$$\mu_{Z_3} = \frac{\exp\left[-S\left(\begin{array}{l} 1x_{11} + 3x_{12} + 1x_{13} + 1x_{14} + 1x_{15} + 5x_{16} \\ +3x_{21} + 5x_{22} + 3x_{23} + 5x_{24} + 7x_{25} + 5x_{26} \\ +1x_{31} + 7x_{32} + 5x_{33} + 3x_{34} + 3x_{35} + 7x_{36} \\ +5x_{41} + 9x_{42} + 3x_{43} + 5x_{44} + 7x_{45} + 3x_{46} \\ +3x_{51} + 9x_{52} + 7x_{53} + 5x_{54} + 3x_{55} + 3x_{56} \\ +3x_{61} + 3x_{62} + 5x_{63} + 7x_{64} + 5x_{65} + 7x_{66} - 14 \end{array}\right)/24\right] - \exp(-S)}{(1 - \exp(-S))}.$$

The ESs and their consequent allocation plans of MOAP-2 are given in the Table 19.6 for distinct combinations of SP and AL which are precise by the DM.

Table 19.6 shows the assignment plans for Z_1, Z_2 and Z_3 of MOAP-2 according to distinct combinations of SP and AL.

Convergence rate of GA for MOAP-2

Figure 19.4a, b show the ESs of SOOP and MOAP in the cases of mutation and without the mutation operator for case-1 of Table 19.5 respectively. The GA based algorithm converges in the case of with mutation operator, solutions of the SOOP

Table 19.6 Summary results for MAOP-2

Case	Degree of satisfaction	Membership values			Objective values			Optimal allocations
		μ_{Z_1}	μ_{Z_2}	μ_{Z_3}	Z_1	Z_2	Z_3	
1	0.9672	0.9726	0.9803	0.9965	36	50	16	$x_{13} = x_{21} = x_{34} = x_{46} = x_{55} = x_{62} = 1$
2	0.9241	0.9241	0.9735	1.0000	42	41	14	$x_{14} = x_{23} = x_{31} = x_{46} = x_{55} = x_{62} = 1$
3	0.9013	0.9013	0.9448	0.9851	37	47	16	$x_{11} = x_{23} = x_{34} = x_{46} = x_{55} = x_{62} = 1$
4	0.8176	0.8176	0.9150	1.0000	42	41	14	$x_{14} = x_{23} = x_{31} = x_{46} = x_{55} = x_{62} = 1$
5	0.8395	0.8395	0.9099	0.9716	37	47	16	$x_{11} = x_{23} = x_{34} = x_{46} = x_{55} = x_{62} = 1$
6	0.9465	0.9465	0.9582	0.9660	30	58	18	$x_{13} = x_{24} = x_{35} = x_{46} = x_{51} = x_{62} = 1$
7	0.7537	0.7537	0.8515	0.9494	37	47	16	$x_{11} = x_{23} = x_{34} = x_{46} = x_{55} = x_{62} = 1$

and MOAP are converging after 300 iterations with 50 populations else solutions of the SOOP and MOAP are converging after 650 iterations with 200 populations. These figures also provided alternative solutions to DM as per his/her prerequisite. For cases of Table 19.5, the convergence criteria of GA approximately remain the same.

Figure 19.5 represents the ESs of cost, time and quality objectives and its allocation plans at different combinations of SP and AL. It indicates the ESs (36, 50, 16), (42, 41, 14), (37, 47, 16), (42, 41, 14), (37, 47, 16), (30, 58, 18) and (37, 47, 16) at (−5, −5, −5), (−5, −3, −1), (−3, −3, −3), (−3, −1, −2), (−2, −2, −2), (−2, −5, −3) and (−1, −1, −1) SPs and their corresponding (0.9, 0.85, 0.7), (0.8, 0.75, 0.85), (0.7, 0.8, 0.9), (0.65, 0.7, 0.75), (0.65, 0.8, 0.75), (0.75, 0.85, 0.8) and (0.55, 0.6, 0.6) ALs, respectively.

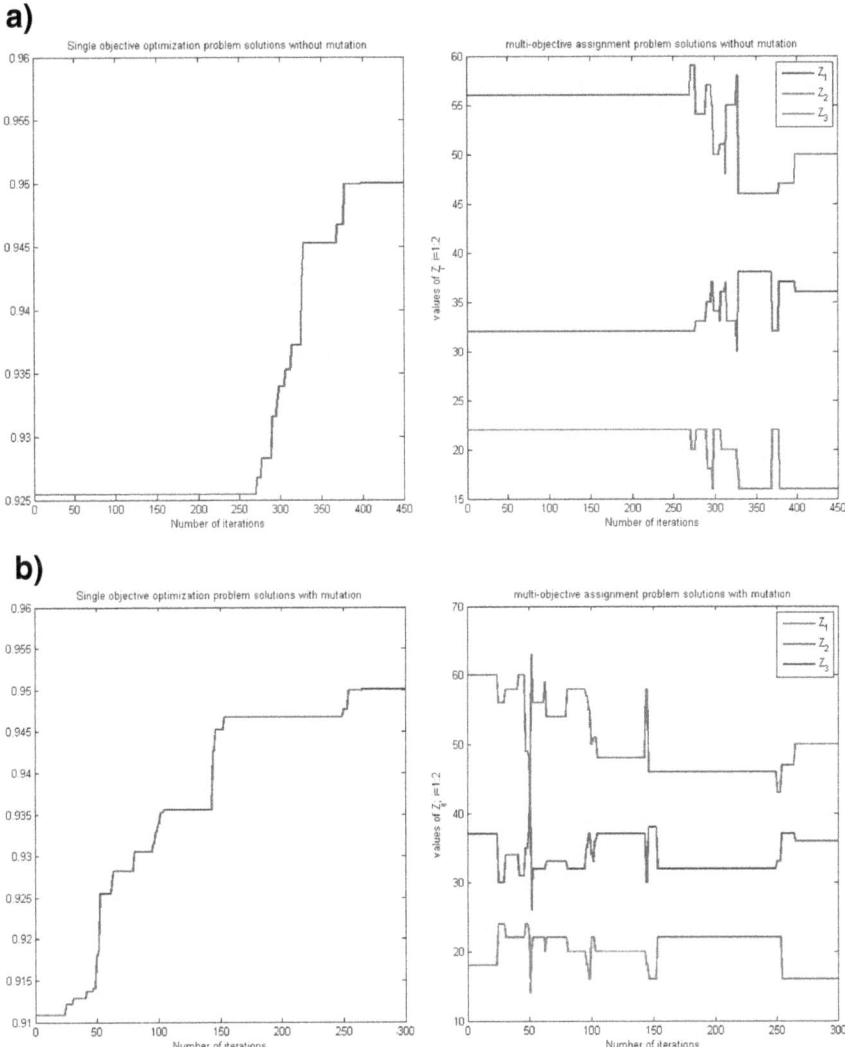

Fig. 19.4 a Convergence rate of GA in case of with mutation for case-1. **b** Convergence rate of GA in case of with mutation for case-1

Figure 19.6 shows the degree of satisfaction as (0.9726, 0.9803, 0.9965), (0.9241, 0.9735, 1), (0.9013, 0.9448, 0.9851), (0.8176, 0.9150, 1), (0.8395, 0.9099, 0.9716), (0.9465, 0.9582, 0.9660) and (0.7537, 0.8515, 0.9494) of cost, time and quality objectives at (−5, −5, −5), (−5, −3, −1), (−3, −3, −3), (−3, −1, −2), (−2, −2, −2), (−2, −5, −3) and (−1, −1, −1) SPs and their consequent (0.9, 0.85, 0.7), (0.8, 0.75, 0.85), (0.7, 0.8, 0.9), (0.65, 0.7, 0.75), (0.65, 0.8, 0.75), (0.75, 0.85, 0.8) and (0.55, 0.6, 0.6) ALs respectively.

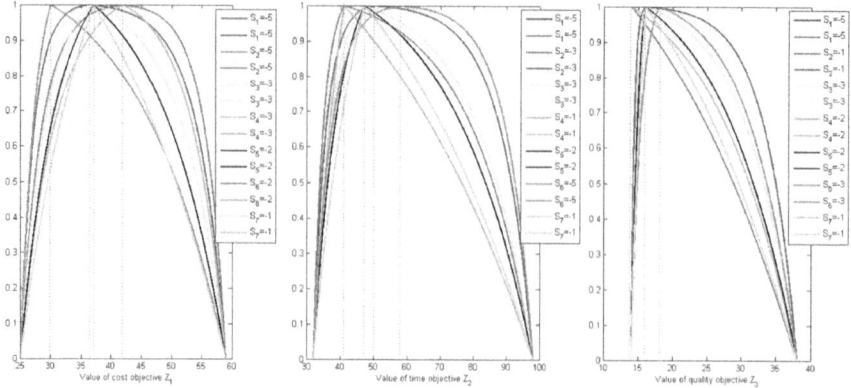

Fig. 19.5 Efficient solutions of MOAP-2 at different combinations of SP and AL

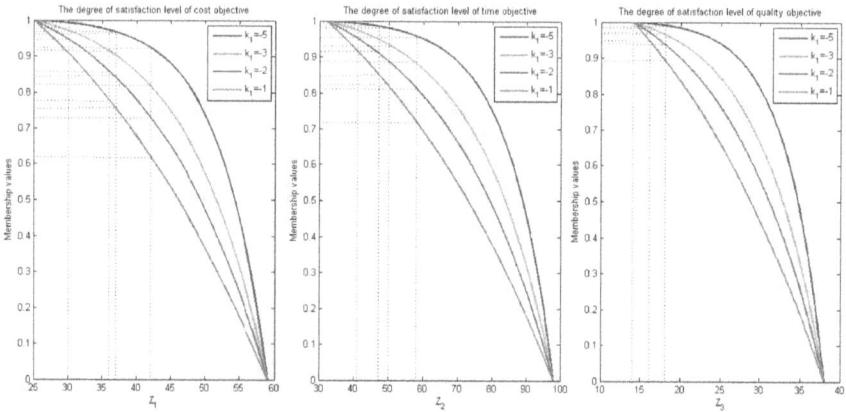

Fig. 19.6 The degree of satisfaction of goal for MOAP-2

Figures 19.3 and 19.6 also show the best advantage to use the EMF with a variety of SPs in MOAP-1 and MOAP-2 and also show that the alterations in SP values directly affect to all objectives. In the sense of changing SPs, the GA based approach confers the flexibility and lots of information are shown in Tables 19.3 and 19.6 as well as Figs. 19.3 and 19.6. The GA based approach also provides the different schemes for allocation plan to DM.

The GA based approach treats each objective consistently. So, different solutions are chosen by DM in different situations according to DM's necessity. Depending upon the predilections of the DM for the all objectives, the preferred compromise solution also can be adapted by alternating the SP's values as well as the AL's values.

19.7 Comparison of GA Based Approach with Other Approaches

To further assessment of the GA based approach, we compared the obtained results of MOAP-1 and MOAP-2 by other approaches which are expressed in Tables 19.7 and 19.8.

Tables 19.7 and 19.8 indicate the GA based approach can provide an alternative approach to find the MOAP's solutions with more influences. Form these tables, we can say that GA based approach provides a set of solutions according to DM AL and his/her choices SP that are beneficial in today's world rather than other approaches.

Table 19.7 Comparison of GA based approach with other approaches of MOAP-1

Bao et al. approach [7]	Interactive Fuzzy goal programming approach [14]	Kagade [8]	Atul Tiwari [19]	Shape parameter	Aspiration level	Developed hybrid approach
$Z_1 = 30$, $Z_2 = 37$ $x_{12} =$ $x_{21} = x_{33}$	$Z_1 = 30$, $Z_2 = 37$ $x_{12} = x_{21} =$ x_{33}	$Z_1 = 30$, $Z_2 = 37$ $x_{12} =$ $x_{21} = x_{33}$	$Z_1 = 30$, $Z_2 = 37$ $x_{12} =$ $x_{21} = x_{33}$	$(-5, -5)$	$(0.95, 0.75)$	$Z_1 = 30$, $Z_2 = 37$ $x_{12} =$ $x_{21} = x_{33}$
				$(-5, -5)$	$(0.7, 0.95)$	$Z_1 = 33$, $Z_2 = 35$ $x_{11} =$ $x_{23} = x_{32}$
				$(-3, -3)$	$(0.9, 0.65)$	$Z_1 = 30$, $Z_2 = 37$ $x_{12} =$ $x_{21} = x_{33}$
				$(-2, -2)$	$(0.75, 0.7)$	$Z_1 = 30$, $Z_2 = 37$ $x_{12} =$ $x_{21} = x_{33}$
				$(-1, -1)$	$(0.65, 0.6)$	$Z_1 = 33$, $Z_2 = 35$ $x_{11} =$ $x_{23} = x_{32}$
				$(1, 1)$	$(0.4, 0.45)$	$Z_1 = 33$, $Z_2 = 35$ $x_{11} =$ $x_{23} = x_{32}$

Table 19.8 Comparison of GA based approach with other approach of MOAP-2

Bao approach [7]	Shape parameter	Aspiration level	Developed hybrid approach
$Z_1 = 42$, $Z_2 = 41$, $Z_3 = 14$ $x_{14} = x_{23} = x_{31} = x_{46} = x_{55} = x_{62}$	$(-5, -5, -5)$	$(0.9, 0.85, 0.7)$	$Z_1 = 36$, $Z_2 = 50$, $Z_3 = 16$ $x_{13} = x_{21} = x_{34} = x_{46} = x_{55} = x_{62}$
	$(-5, -3, -1)$	$(0.8, 0.75, 0.85)$	$Z_1 = 42$, $Z_2 = 41$, $Z_3 = 14$ $x_{14} = x_{23} = x_{31} = x_{46} = x_{55} = x_{62}$
	$(-3, -3, -3)$	$(0.7, 0.8, 0.9)$	$Z_1 = 37$, $Z_2 = 47$, $Z_3 = 16$ $x_{11} = x_{23} = x_{34} = x_{46} = x_{55} = x_{62}$
	$(-3, -1, -2)$	$(0.65, 0.7, 0.75)$	$Z_1 = 42$, $Z_2 = 41$, $Z_3 = 14$ $x_{14} = x_{23} = x_{31} = x_{46} = x_{55} = x_{62}$
	$(-2, -2, -2)$	$(0.65, 0.8, 0.75)$	$Z_1 = 37$, $Z_2 = 47$, $Z_3 = 16$ $x_{11} = x_{23} = x_{34} = x_{46} = x_{55} = x_{62}$
	$(-2, -5, -3)$	$(0.75, 0.85, 0.8)$	$Z_1 = 30$, $Z_2 = 58$, $Z_3 = 18$ $x_{13} = x_{24} = x_{35} = x_{46} = x_{51} = x_{62}$
	$(-1, -1, -1)$	$(0.55, 0.6, 0.6)$	$Z_1 = 37$, $Z_2 = 47$, $Z_3 = 16$ $x_{11} = x_{23} = x_{34} = x_{46} = x_{55} = x_{62}$

19.8 Conclusion

GA based approach rendered the solution of the MAOP with respect to many choices of SPs and their corresponding ALs. The EMF used to effectively design the SOOP with the product operator. The SPs in the EMFs are used to depicted classes of accuracy in objective functions. The consequential SONOP with some resource constraint is "NP-hard" and the various alternatives of ALs desired for the objectives specified by the DM, so it is work out with GA. This approach describes how the different parameters affect the algorithm and their effect on convergence to concluding ES. This approach also provided situation wise all possible solution to take the decision for DM, which is highly beneficial for DM to take decision according to the situation.

References

1. Bufardi A (2008) On the efficiency of feasible solutions of a multicriteria assignment problem. Open Oper Res J 2:25–28
2. White DJ (1984) A special multi-objective assignment problem. J Oper Res Soc 759–767.
3. Tsai C-H, Wei C-C, Cheng C-L et al (1999) Multi-objective fuzzy deployment of manpower. Int J Comput Internet Manag 7(2):1–7

4. Tuyttens D, Teghem J, Fortemps P, Nieuwenhuyze KV (2000) Performance of the mosa method for the bicriteria assignment problem. J Heuristics 6(3):295–310
5. Gungor I, Gunes M (2000) Fuzzy multiple criteria assignment problems for fusion: the case of Hungarian algorithm. In: Proceedings of the third international conference on information fusion, 2000. FUSION 2000, vol 1, IEEE, pp TUD4–8
6. Przybylski A, Gandibleux X, Ehrgott M (2005) The biobjective assignment problem. Technical report. Research report
7. Bao C-P, Tsai M, Tsai M-I (2007) A new approach to study the multi-objective assignment problem. WHAMPOA Interdisc J 53:123–132
8. Kagade K, Bajaj V (2009) Fuzzy approach with linear and some non-linear membership functions for solving multi-objective assignment problems. J Adv Comput Res 1:14–17
9. Odior A, Charles-Owaba O, Oyawale F (2010) Determining feasible solutions of a multicriteria assignment problem. J Appl Sci Environ Manag 14:1
10. Przybylski A, Gandibleux X, Ehrgott M (2010) A two phase method for multi-objective integer programming and its application to the assignment problem with three objectives. Discrete Optim 7(3):149–165
11. Li C, Park C, Pattipati KR, Kleinman DL (2011) Distributed algorithms for biobjective assignment problems. In: 2011 50th IEEE conference on decision and control and European control conference (CDC-ECC). IEEE, pp 5893–5898
12. Adiche C, Aïder M (2010) A hybrid method for solving the multi-objective assignment problem. J Math Modell Algorithms 9(2):149–164
13. Ozlen M, Burton BA, Macrae CA (2014) Multi-objective integer programming: an improved recursive algorithm. J Optim Theory Appl 160(2):470–482
14. De P, Yadav B (2011) An algorithm to solve multi-objective assignment problem using interactive fuzzy goal programming approach. Int J Contemp Math Sci 6(34):1651–1662
15. Ratli M, Eddaly M, Jarboui B, Lecomte S, Hanafi S (2013) Hybrid genetic algorithm for bi-objective assignment problem. In: Proceedings of 2013 international conference on industrial engineering and systems management (IESM). IEEE, pp 1–6
16. Gupta P, Mehlawat MK, Mittal G (2013) A fuzzy approach to multicriteria assignment problem using exponential membership functions. Int J Mach Learn Cybern 4(6):647–657
17. Basirzadeh H, Morovati V, Sayadi A (2014) A quick method to calculate the super-efficient point in multi-objective assignment problems. J Math Comput Sci 10:157–162
18. Hassan Shirdel G, Mortezaee A (2015) A dea-based approach for the multicriteria assignment problem. Croatian Oper Res Rev 6(1):145–154
19. Tiwari AK, Tiwari A, Samuel C, Pandey SK (2013) Flexibility in assignment problem using fuzzy numbers with nonlinear membership functions. Int J Ind Eng Technol 3(2):1–10
20. Jayalakshmi M, Sujatha V (2018) A new algorithm to solve multi-objective assignment problem. Int J Pure Appl Math 119(16):719–724
21. Medvedeva OA, Medvedev SN (2018) A dual approach to solving a multiobjective assignment problem. IOP Conf Ser J Phys Conf Ser (973)
22. Hammadi AMK (2017) Solving multi objective assignment problem using Tabu search algorithm. Global J Pure Appl Math 13(9):4747–4764
23. Belhoul L, Lucie G, Daniel V (2014) An efficient procedure for finding best compromise solutions to the multi-objective assignment problem. Comput Oper Res 49:97–106
24. Huang G, Lim A (2006) A hybrid genetic algorithm for the three-index assignment problem. Eur J Oper Res 172(1):249–257
25. Toroslu IH, Arslanoglu Y (2007) Genetic algorithm for the personnel assignment problem with multiple objectives. Inf Sci 177(3):787–803
26. Nakayama H (1995) Aspiration level approach to interactive multi-objective programming and its applications. In: Advances in multicriteria analysis. Springer, Berlin, pp 147–174
27. Nakayama H, Yun Y, Yoon M (2009) Interactive programming methods for multi-objective optimization. In: Sequential approximate multiobjective optimization using computational intelligence, pp 17–43

28. Dhodiya JM, Tailor AR (2016) Genetic algorithm based hybrid approach to solve fuzzy multi-objective assignment problem using exponential membership function. Springerplus 5(1):2028
29. Dhodiya JM, Tailor AR (2018) Genetic algorithm based hybrid approach to solve uncertain multi-objective COTS selection problem for modular software system. J Intell Fuzzy Syst 34(4):2103–2120
30. Tailor AR, Dhodiya JM (2016) Genetic algorithm based hybrid approach to solve optimistic, most-likely and pessimistic scenarios of fuzzy multi-objective assignment problem using exponential membership function. J Adv Math Comput Sci 1–19
31. Rajan K (2013) Adaptive techniques in genetic algorithm and its applications. Ph.D. thesis, Kottayam
32. Sahu A, Tapadar R (2006) Solving the assignment problem using genetic algorithm and simulated annealing. In: IMECS, pp 762–765
33. Sani H, Yabo M (2016) Solving timetabling problems using genetic algorithm technique. Int J Comput Appl 134:15
34. Sivanandam S, Deepa S (2007) Introduction to genetic algorithms. Springer Science & Business Media, Berlin
35. Tosun U (2014) A new recombination operator for the genetic algorithm solution of the quadratic assignment problem. Procedia Comput Sci 32:29–36
36. Wu B, Tu X, Wu J (2000) Generalized self-adaptive genetic algorithms. J Univ Sci Technol Beijing Eng Ed 7(1):72–75
37. Younas I (2014) Using genetic algorithms for large scale optimization of assignment, planning and rescheduling problems. Ph.D. thesis, KTH Royal Institute of Technology
38. Tailor AR, Dhodiya JM (2016) A genetic algorithm based hybrid approach to solve multi-objective interval assignment problem by estimation theory. Indian J Sci Technol 9(35):0974–5645
39. Tailor AR, Dhodiya JM (2016) Genetic algorithm based hybrid approach to solve multi-objective assignment problem. Int J Innov Res Sci Eng Technol 5(1):524–535

Chapter 20
Role of Evolutionary Approaches to Solving Multi-objective Optimization Problems

Surbhi Tilva and Jayesh M. Dhodiya

Abstract This Chapter aims to provide the fundamental knowledge for finding the solution of the multi-objective optimization problem (MOOP). Here, the main concentration is on the intelligent meta-heuristic approaches, especially evolutionary approaches. Since the mid-1980s, the Evolutionary approaches are in trend for solving MOOP and developed a very substantial work in the past two decades. Moreover, and despite the maturity of this field, there are yet various essential demands situated in front of the existent-world. This chapter gives a brief note about them and describes the multiple approaches for the effective formulation of the compromise solutions. The essential generation of the MOOP is provided in this chapter, here the concepts of Pareto-optimality, Pareto-front, and many others are introduced, with a specific concentration on open research problems or topics, despite particular research areas. The supreme goal of this chapter is to actuate students and researchers and for developing the new theories in this field, as that will bring the maintainable in this field work for an upcoming couple of decades.

Keywords Evolutionary approaches · Multi-objective optimization problem · Pareto front · Pareto optimal

20.1 Introduction

The process of optimization is a basic cycle in numerous businesses, executives, and designing tools and applications. An optimization problem generally manages to look through the best solution set from all the attainable solution sets, and it is known as an optimal solution. A significant piece of exploration and application in the region of enhancement is consider as a solitary goal while practically viable tasks think about at least two or more than two destinations. The presence of a few opposing or conflicting destinations is ordinary in bunches of issues and which makes the advancement issues further captivating to illuminate. Most genuine choice and arranging circumstances include numerous conflicting measures that ought to be considered all the while. In

---

S. Tilva (✉) · J. M. Dhodiya
S. V. National Institute of Technology, Surat 395007, India

these fields, numerous and frequently conflicting targets should be satisfied. They are known as multi-objective optimization problems (MOOPs), and their answer has pulled in consideration of analysts for a long time.

There are mainly three goals to seek after for solving a MOOP: (1) the act of convergent, (2) the assorted variety in the solution set, and (3) the solution set's dissemination consistency. Moreover, the received non-dominated solution sets must be as close as conceivable to the Pareto optimal front of the MOOP. For the single-objective optimization problem (SOOP), this objective is like the interest to convergent at global optimum. Many times, there are uncountable solution sets which are Pareto optimal. Normally, just a finite number of solution sets is created throughout the process of optimization. Moreover, for reducing the computational expense, the quantity of created solution sets must be restricted. In any case, the biggest conceivable opportunity of decision ought to be put up on a Decision Maker (DM). Hence, a very much dispersed approximation set is requested, which is an objective that comprises itself of two demands: (1) the set that is as extensive as could be expected under the circumstances and (2) a distribution that is as uniformly divided as could reasonably be expected. Pareto optimal fronts may not be continuous, so all the things considered, a precisely uniform circulation of solution sets, is preposterous. In any case, the non-dominated solution sets must spread over the complete area of the Pareto front and reconstruct the fundamental curve of Pareto front as effectively as could be expected under the circumstances. These requirements don't have a counterpart in SOOP since all things considered, just a single optimum solution is produced.

MOOPs are solved with two types of approaches, specifically known as classical and evolutionary approaches. In classical approaches, we substitute overall goals into an equivalent solitary goal that means the MOOP converts into SOOP. The final objective is to search the solution set that maximizes or minimizes this single objective while keeping up the framework's physical constraints or cycle. The optimization solution set brings about a solitary worth that returns a trade-off between all goals. The craftsmanship in this cycle is to generate the function to accomplish this ideal trade-off solution.

Transformation of the MOOP into a SOOP is typically done by amassing every goal in a weighted function or converting everything except one of the destinations into constraints. This way to deal with unraveling MOOP has many restrictions: (1) it requires from the earlier information about the overall significance of the destinations, and the restrictions on the destinations that are transformed over into constraints (2) from the amassed function derives just a single solution set; (3) compromise solutions among destinations can't be search out effortlessly, (4) the solution set can't be feasible except for the convex space of search.

This basic optimization theory is not, at this point, worthy of frameworks with numerous conflicting destinations. Simultaneously, specialists may want to know every potential solution set of all the destinations. For the real world, it is called trade-off analysis. In the real world scenario, there are many instances of the demand to execute trade-off analysis. For instance, structuring appropriated regulators while decreasing expenses are two conflicting goals. Correspondingly, to put more utilitarian squares on a chip while limiting that chip region or potentially power dissemination are conflicting goals. To search the automobile that travels the largest length in a particular fixed time, although expecting the minimum energy, is a MOOP. Limiting the working expense of business, although keeping up a steady work power, is conflicting. Thus it is a MOOP [1–3]. There are so numerous standard life's issues that go under the MOOP, for example, CPM, assignment problem, transportation problem, COTS selection problem, and so forth [4–7].

20.2 Multi-objective Optimization Problem

The goal of MOOP is to search the arrangement of satisfactory solution sets and provide them to the DM, which will pick among them at that point. Extra requirements or standards are prescribed either previously or after the DM's search process can help direct, refine, or limit the pursuit. Yet, we will take a gander at the conventional situation where there is no earlier DM data.

20.2.1 Model of MOOP

Think about the issue of buying the most proficient merchandise. Assume here we have accepted it as we need to purchase the couch. Expect two rules are utilized to decide this effectiveness: (a) cost secured to make a specific couch, and (b) nature of the material that is utilized simultaneously. The presence of the mind can be utilized to get every likely arrangement. For instance, if we need a couch with too quality and remarkable look, at that point our spending plan must be high, and if we have an essential spotlight on spending plan and our financial plan is low, at that point we need to compromise with the nature of the couch.

20.2.2 Formulation of MOOP

The general formulation of MOOP which has K objective functions is described as below:

$$\text{Min } Z(x) = [z_1(x), z_2(x), \ldots, z_K(x)]^T$$

Subject to constraints,

$$g_j(x) \geq 0, \quad j = 1, 2, \ldots m.$$
$$h_k(x) = 0, \quad k = 1, 2, \ldots o.$$

$$x_i^L \le x_i \le x_i^U, \quad i = 1, 2, \ldots n.$$

where x_i^U, and x_i^L represents the upper and lower bounds for the i-th decision variable. m stands for total number of inequality constraint; o stands for total number of equality constraint. The solution set x_i that satisfies the $(m + o)$ constraints is called the feasible solution set, and a collection of every feasible solution sets represents a feasible region and symbolized as Ω. For the general formulation of MOOP, we consider the minimize MOOP as the maximize problem can be easily transferred to minimize according to the principle of duality through multiply every goal by -1 and constraints are changed according to the rules of duality.

20.2.3 Basic Definitions

- Take a pair of solution sets $x, y \in \mathbb{R}^k$, then x **dominates** y and symbolized as $x \prec y$, when $z_i(x) \le z_i(y) \ \forall \ i = 1, \ldots, K$, and $z_i(x) < z_i(y)$ for at least one i.
- The solution set $x \in X \subset \mathbb{R}^k$ is **non-dominate** wrt X, when $\nexists$ another x' in X such that $z(x') \prec z(x)$.
- The solution set $x \in X \subset \mathbb{R}^k$ is **weakly non-dominate** wrt X, when $\nexists$ another x' in X such that $z_i(x') < z_i(x) \ \forall \ i = 1, \ldots, K$.
- The solution sets $x^* \in F \subset \mathbb{R}^k$ is **Pareto-optimal** when x^* is non-dominate wrt F, where F is feasible region.
- The **Pareto optimal set** P^* is expressed as: $P^* = \{x^* \in F | x^* \text{ is Pareto-optimal}\}$
- The **Pareto front** pf^* is expressed as: $pf^* = \{f(x^*) \in \mathbb{R}^k | x^* \in P^*\}$.

20.3 Evolutionary Approaches (EAs)

20.3.1 Non-Pareto EAs

Vector Evaluate Genetic Algorithm: VEGA

In 1985, Schaffer [8] developed one of the first choices to conform EAs to deal with MOOPs known as VEGA. The fundamental thought is to subdivide the population into n equal parts as there are n goals, and it is called n sub-populations. After that, the selection operator performed on each of the solutions set by considering the related goals. Then, to perform the other evolutionary operators, the whole population is gathered when the selection mechanism was completed. This procedure is replicated in every iteration. The discernible VEGA issue is that it inclines toward the best solution sets of every goal independently; however, it doesn't advance the endurance of good trade-off solution sets. This issue is called speciation (by its similarity in hereditary qualities). Schaffer has searched this issue and attempted to solve by utilizing the mating limitations (i.e., not permitting recombination between solution sets of a similar sub-population). Also, some more heuristic rules were applied to the mechanism of selection. Despite VEGA's restrictions, a few analysts, in the long run, discovered applications in which such a plan could be more powerful (see, for instance, [9]). In the work of Richardson et al. [10], it was also exhibited that the VEGAs strategy is similar to the linear combination of goals if you utilized the proportional selection that implies VEGA has restrictions concerning non-convex Pareto fronts.

Vector Optimized Evolutionary Strategy: VOES

After VEGA proposed, the Kursawe [11] developed the VOES for handling the MOOPs in 1990. The assignment mechanism of fitness in VOES is similar to VEGA. Additionally, Kursawe utilized some genetic facts with the help of nature. The solution set is presented through the diploid chromosome, and every chromosome has two strings: one is dominant, and the other is recessive in VOES. Two distinct solution sets (both have a design variable y and the related procedure vector δ) are utilized, particularly in a population. Subsequently, the solution set y is measured through computing: (1) Z^d dependent on the genotype of dominant string and (2) Z^r dependent on the genotype of recessive string. Here, we have given the mechanisms of evolution and selection. The selection procedure is carried out in N levels. The probability vector provided by the client is utilized for choosing a goal at each level. This vector can be varied or fixed throughout the iterations. Suppose the n-th goal is chosen, the fitness of individual solution set y is counted as the weighted sum of recessive and dominant goal measure as below:

$$Z(y) = \frac{2}{3} Z_n^d(y) + \frac{1}{3} Z_n^r(y) \tag{20.1}$$

The population is sorted according to the goal and $(\frac{N-1}{N})$-th part is chosen as parent from the population at each selection level. Every time we utilized the survived population from past sorting, this process replicated N times. Hence, the connection among the no. of children γ and no. of parents η can be represented below:

$$\eta = \left(\frac{N-1}{N}\right)^N \gamma \tag{20.2}$$

For example, we get $\eta = 0.25\gamma$ for the bi-objective OP. An external archive that keep a non-dominate solution sets received from the starting of the simulation run and in which every new η solution sets are copied. The verification of non-domination is created on the archive after adding the new solution sets, and just the new non-dominated solution sets are held. The mechanism of niching is utilized to uplift the diversity among solution sets by eliminating the solution sets from the crowded region, and it is performed when the external archive size becomes more than the size of the archive.

VOES utilizes the non-domination verification for ensuring the convergence of solution sets and niching for supporting the diversity of solution sets. These highlights are fundamental to the structure of great MOEA. Sadly, Kursawe evaluated the exhibition of his approach on a solitary test issue, and then further no test appraisals were sought after since Kursawe's unique examination.

Weight-Based GA: WBGA

WBGA is also known as HLGA, which means Hajela and Lin Genetic Algorithm. In 1992, Hajela and Lin presented the WBGA [12]. For every goal, the coefficient of weighted is allocated. Every population's solution set has its vector of weight coefficient coded in its string concatenate with its decision variables, not like the classical weight sum method. Thus, the WBGA was capable of searching the multiple non-dominate solution sets in a single simulation run. This approach's main problem is how to preserve the variety of weight coefficients for the solution sets of population. Two methods were proposed for this purpose. One method utilized the mechanism of niching over the substring, representing the vector of the weight coefficient. Whereas other methods picked, sub-populations are assessed for distinct pre-specified vectors of weight likewise to VEGA. As WBGA is the method depends on weight, subsequently, the solution sets lying on the non-convex portion of the Pareto front cannot search through this approach.

20.3.2 Pareto EAs

20.3.2.1 Non-elitist Approaches

Multi-objective GA: MOGA

In 1993, Fonseca and Fleming developed a first Pareto-based evolutionary approach, and it is known as MOGA [13]. To motivate the search in the direction of real Pareto front and keep the diversity among the population, MOGA unitedly utilized the concept of Pareto-based position and niching mechanism. Every solution set has provided a position that is represented as a function of the no. of solution sets dominated by it. Suppose the nd^s is the no. of solution sets dominating a specific solution sett y at the iteration s, then the position at s of y solution set is defined as follow:

$$pos^s(y) = 1 + nd^s \qquad (20.3)$$

By the mechanism of position, all the non-dominated solution sets are assigned at position 1 see the Fig. 20.1. The technique of MOGA is according to the position of the solution set and an average fitness value of the population. The procedure for calculating the fitness value is described below. Initially, the population is arranged by position. After that, the fitness value is provided to every solution set according

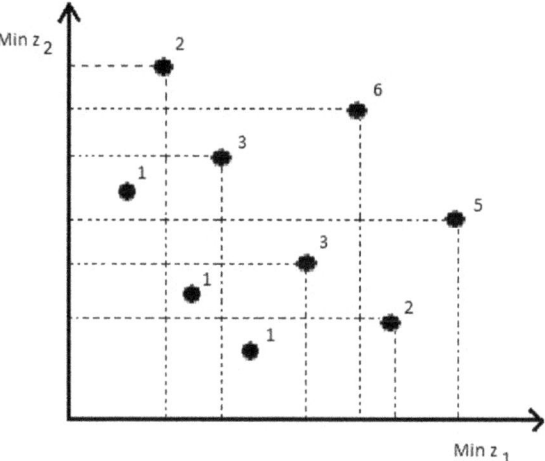

Fig. 20.1 Ranking process of MOGA

to the interpolation of best to the worst position by some predefined function. At last, the solution sets which provided the same position got the average value of the fitness. That guarantees the every solution sets which has the same position is examined with an indistinguishable frequency. This data is utilized to keep up consistent global population fitness with a suitable measure of specific weight. Moreover, MOGA applies the niching concept and utilizes the niche radius parameter, and here it is defined by σ_{rd} that should be assigned carefully. For generating the uniform distribution of the approximate Pareto front, the mechanism of niching is carried out on the objective space. Figure 20.2 gives the illustration of the niching mechanism. The solution sets occupying under the area of niching radius are given a penalty in

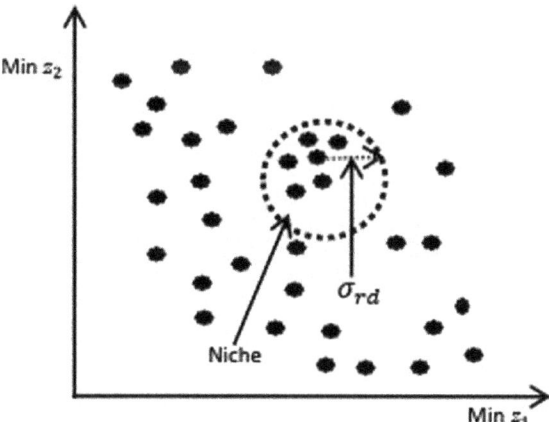

Fig. 20.2 Mechanism of niching

their fitness value. Even though in MOGA, the assigned fitness value is depends on the Pareto dominant concept, it is possible that the solution sets having the same position may not have the same fitness value. That may create an undesirable inclination for the specific area of search space. Especially, MOGA might be effective with the Pareto front geometry, along with the solution sets density on the search space. Besides, the niching mechanism shows kindness towards the solution sets with lower positions over the solution sets with higher positions when these last are more crowded.

Niche Pareto GA: NPGA

Horn et al. [14] developed NPGA that contrasts from the earlier proposed MOEAs in the operator of selection. The VEGA and MOGA utilized the proportional selection method rather than this approach utilizes the binary tournament selection. In the tournament selection, the randomly two solution sets y, and z are chosen among the parent population J. Afterward, both the solution sets are examined dependent on Pareto dominance with every solution set of an arbitrarily chosen sub-population I whose size is id, where $id \ll |J|$. When anyone from the two solution sets is non-dominate for all solution sets of sub-population and another solution set is dominant through at least one solution set, then the non-dominated solution set is kept. The mechanism of niching is applied to choose the solution set from y and z that lies in the least crowded region in the scenarios where both or neither solution sets are dominated through the solution sets of sub-population I.

The process of NPGA is related to the value of σ_{rd} along with the id. From the numerical results reported in the article of Horn et al. [14], we can conclude that the size of the population is greater than the id. Whereas, when id is excessively enormous, the non-dominate solution set is well-emphasized, yet its complexity will be its top level. Then again, when id is excessively little, then the verification of non-domination can boisterous that it can't focus on the non-dominate solution set sufficiently. Furthermore, jd relies upon the number of goals that to be optimizing.

Non-dominate Sorting GA: NSGA

NSGA [15] depends on the technique of non-dominated sorting, which is demonstrated in Fig. 20.3. This technique characterizes the solution sets of the population into many positions. The technique of non-dominated sorting starts through searching the non-dominate solution sets among the population. All these solution sets are assigned at position one and provide the largest dummy value of fitness. These solution sets are then deleted among the population, then again search the non-dominate solution sets among the remaining population. Moreover, the non-dominate solution sets of this time are assigned at position two, and the dummy value of fitness is assigned smaller than the previous one. This procedure is repeated until all the solution sets of the population are positioned. For maintaining the diversity of solution sets, the mechanism of niching is applied to a decision space rather than an objective space to reduce the value of fitness according to the value of σ_{rd}. The sharing in each position is accomplished by counting the value of sharing function among two solution sets, m and n, in a similar position as below:

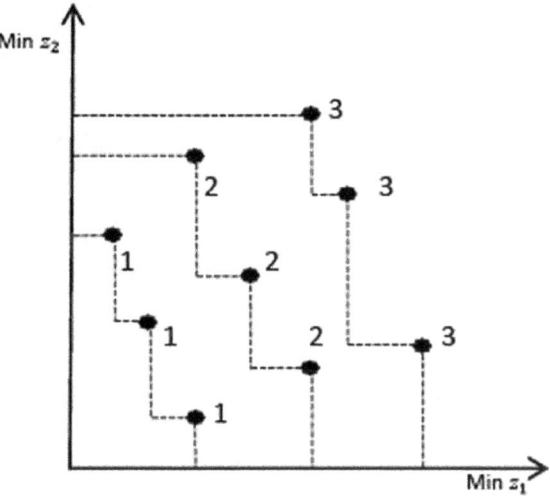

Fig. 20.3 Mechanism of non-dominate sorting

$$\begin{cases} Sh_{e_{mn}} = 1 - \left(\frac{e_{mn}}{\sigma_{rd}}\right)^2, & \text{if } e_{mn} < \sigma_{rd}, \\ 0, & \text{otherwise.} \end{cases} \quad (20.4)$$

where e_{mn} denotes the Euclidean distance connecting m and n. Afterward, the above value of sharing function for every solution sets is added in their respective position to calculate the parameter of niche count. Finally, the value of shared fitness for every solution set is calculated via divide the value of dummy fitness through niche count. Best solution sets are prioritized ever above the other solution sets. Thus, the new solution sets that are nearer to the non-dominate solution sets are more preferred. The mechanism of niching provides the approach to spread the non-dominate solution sets over the Pareto front. At the same time, the more affectability regarding the parameter σ_{rd} provides the lesser productive execution of NSGA.

20.3.2.2 Elitist Approaches

Elitism implies that the elite solution sets can't be removed from the population's archive gene pool in favor of worse solution sets [16]. In the accompanying, we survey the most popular elitist MOEAs [17].

Strength Pareto EA: SPEA/SPEA2

In 1999, Zitzler and Thiele [18] developed the SPEA that utilizes two populations: (1) the principle populace J, and (2) an archive population C, that is consist of the non-dominated solution sets in the whole process. At first, randomly, the population J is produced, and the archive population C is vacant. Afterward, the C is loaded up by the non-dominate solution sets from J. Then, the solution sets from C, that dominate through any of the solution set from C are erased. Furthermore, if the quantity of non-dominate solution sets surpasses the size of archive $|C|$, C is rationalizing with the help of cluster technique that will be discussed later on. When every population and the archive's solution sets have allocated the value of fitness, the selection of binary tournament and substitution put for satisfying pool mating. Subsequently, apply the genetic operators and formulate the new population J. When the stopping criterion is achieved, then procedure will be stopped, otherwise, the non-dominate solution sets from J are transferred to the archive C, and the whole procedure will be replicated.

The assigning of fitness value in SPEA has a two-phase procedure. Initially, the non-dominate solution sets from the archive C are positioned. Afterward, the solution sets from the J population is evaluated. Moreover, each solution set m from archive C has provided the strength value $g_m \in [0, 1[$, that is relative to number of solution sets in J, that dominate through m. The strength g_m is defined as below:

$$g_m = \frac{I}{|J|+1} \qquad (20.5)$$

where I signifies the no. of solution sets in J that are cover through m and $|J|$ is the population's total size. The fitness value of solution set $n \in J$ is received by adding the strength of every non-dominate solution sets $m \in C$ that dominates n. The received amount is added by 1 to ensure that archive solution sets have preferred execution over J solution sets. Minimize the fitness value, and it is defined as follow:

$$Z_n = 1 + \sum_{m, m \leq n} g_m \qquad (20.6)$$

The technique of cluster is applied to decrease the archive size by preserving its features. The basic concept is to divide the archive into A clusters, where $A < |C|$ and every solution sets of the same cluster have similar features. The clusters' process starts with generating the cluster of every component of the initial non-dominate solution set of the archive. Following this, the two groups are picked through the measurement of the distance to be consolidated in single group. A distance is counted as the mean Euclidean distance among the two solution sets over the groups. At the time of accomplishing the cluster technique, the new non-dominate solution sets from the archive contain the centroid solution sets for every group. The authors have also provided the outcomes in favor of SPEA as compared to other MOEAs.

Zitzler et al. [19] have studied and distinguished the three shortcomings for SPEA. The first one is occurred at assigning the fitness value. The solution sets that dominate through the same solution sets of the archive have the same fitness value. Thus, when the archive comprised just one solution set, then every solution set of the population has a similar position that doesn't depend on whether they dominate one another. Therefore, the selection pressure is diminished considerably, and thus the SPEA executes like the random search method. The second is for the estimation of density.

When the numerous solution sets of the recent iteration are Pareto equal, then none or almost no data can be acquired depending on the partial order characterized by dominance. In this circumstance, which will probably happen when the quantity of destinations surpasses two, the information of density must be utilized to direct the search the more efficaciously. Grouping utilizes this knowledge, yet just concerning with the archive, not with primary population-the third for archive truncation scheme. Even though grouping technique utilized in SPEA can decrease the non-dominate solution set avoid destruction in features, it might lose outrageous (external) solution sets. Moreover, these solution sets must be placed with the archive for receiving the well-distributed non-dominate solution sets. To deal with the SPEA's shortcomings, Zitzler et al. [19] have developed the SPEA2, which is the updated version of SPEA. Rather than SPEA, SPEA2 utilizes the fine-grain fitness assigning scheme, that integrates with data of density. Moreover, the size of archive is constant, which means When the quantity of non-dominate solution set is smaller over the archive's predefined size, then dominant solution sets fill up the archive; with SPEA, the size of the archive might be shift after some time. Moreover, mechanism of cluster, that is summoned for non-dominate solution sets, surpasses the size of the archive. It is changed by the substitutive method of truncation, which has the same characteristics yet maintain the boundary solution sets. In SPEA2, just the solution sets from the archive are performing the process of mating selection not like SPEA.

The SPEA2 assignment of fitness for a specific solution set m considers the number of solution sets that dominate the m; moreover, the number of solution sets dominate through m. Every solution set m of the population J. The archive C has provided the strength value g_m defines the number of solution sets is dominant through m:

$$g_m = |n|n \in J \cup C \wedge m \leq n| \tag{20.7}$$

Then raw fitness R_m is calculated as below:

$$R_m = \sum_{n \in J+C, n \leq m} g_n \tag{20.8}$$

The R_m is found out through its dominators' strengths from the main population and archive instead of SPEA, where just solution sets from the archive are taken in this specific situation. Note that, here, we have to minimize the fitness, i.e., $R_m = 0$ related with solution set which is non-dominate. Simultaneously, the higher worth of R_m implies that numerous solution sets dominate m. Figure 20.4b represents this technique.

The scheme of raw fitness provides the niching sort dependents on concept of Pareto dominant. Moreover, this technique turns ineffective if a major amount of solution sets are non-commanded with one another. Consequently, more data regarding density is consolidated to segregate among the solution sets with indistinguishable raw fitness values. SPEA2 adapted the method of l-th nearest neighbor in technique of density estimation. Moreover, for every solution set m, calculated the distances of each solution set n from the $J \cup C$ in the space of objective. Then, arranged in

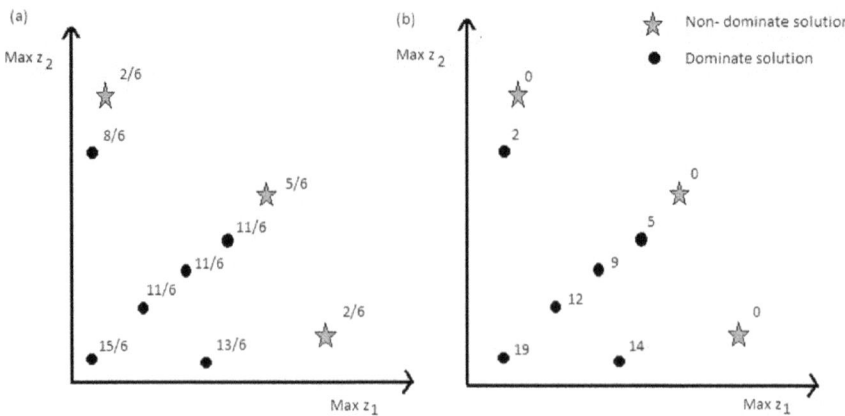

Fig. 20.4 Comparison between SPEA and SPEA2 for the technique of fitness assignment [19]

increasing order and stored. Then, the l-th closest neighbor provides the looked for distance and indicated by μ_m^l. The l parameter is typically defined as $\sqrt{|J|+|C|}$. The density δ_m of the solution set m is:

$$\delta_m = \frac{1}{\mu_m^l + 2} \tag{20.9}$$

Here, two is add in denominator, for guarantee its worth is more prominent than 0 and $\delta_m < 1$. At last, the fitness of a specific solution set m is acquired by adding density information and raw fitness as below:

$$T_m = R_m + \delta_m \tag{20.10}$$

The environmental selection mechanism for SPEA2 is different from SPEA one by maintaining the solution sets of a boundary. Also, the number of solution sets is stored in external is fixed throughout the time.

Non-dominate Sorting GA II: NSGA-II

The updated version of NSGA is known as NSGA-II [20, 21]. The less computation complexity, elitist method, and a strategy for diversity which doesn't require any extra parameter are the noticeable highlights of NSGA-II. The basic concept of NSGA-II is defined as below. In NSGA-II, first, produce the offspring population O_o through utilizing the genetic operators on an arbitrarily formulated parent population J_o. The fundamental iteration of NSGA-II is distinct from the first generation award. Initially, both the populations J_s and O_s are consolidated to shape a population T_s whose size $2 * M$, where $|J_s| = |O_s| = M$. After that, the sorting of non-dominate solution sets is executed to characterize whole population T_s. When sorting of non-dominate solution set is accomplished, population T_s becomes similarly partitioned into a few classes as NSGA. Then, the new parent populace J_{s+1} is formulated by the solution sets of the best non-dominate fronts, each in turn. As the size of the total population is $2 * M$, so it might not be every front lie in M spaces as it is the size of the new population J_{s+1}. When the final permitted front is taken, it might consist of more solution sets then the leftover spaces accessible in J_{s+1}. Rather than deleting the random solution sets from the last front, NSGA-II utilizes the mechanism of niching to pick solution set from the last front, that lie in the least crowded area. Moreover, for every positioning stage, the value of crowding distance is calculated by adding Euclidean distance among both neighborhood solution sets from either side of the solution set on every objective function, as represented by Fig. 20.5. For maintaining the boundary solution sets, these latter provide the infinite crowding distance. The concept of crowding distance value discussed later on along with the concept of non-dominate sorting and many more.

Pareto Archived ES/Pareto Envelope based Selection Algorithm: PAES/PESA

Knowles and Corne [22, 23] developed the $a(1 + 1) - ES$ $((1 + 1) - ES)$, known as PAES, for estimate entire Pareto front. The basic idea of PAES is taken from the achievement of $(1 + 1) - ES$ for solving SOOP. This is the fact that authors have conformed this approach of the search for MOOPs. Initially, in PAES, the child ch_o is formulated from the arbitrarily generated parent J_0. At every iteration i, the obtained non-dominated solution sets are kept in a pre-defined archive size. Initially, both the solution sets, J_i and ch_i, are analyzed. When one solution set is non-dominate, and the other solution set is the dominant one. The dominant one is deleted, and non-dominate solution set is kept as a parent for the next iteration. Suppose both the solution sets J_i and ch_i are non-dominate. In that case, the new solution set is compared with the

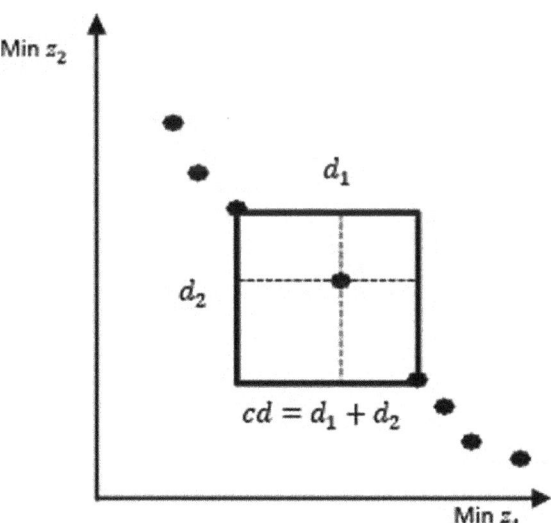

Fig. 20.5 Crowding distance

reference population of previous archive non-dominate solution sets, i.e., solution sets from the archive. If the comparison with the solution set from the archive is failed to provide a better solution set, then the solution set, which lies in the least crowd area, is preferred. DM has defined the maximum size of the archive that provides the wanted number of final solution sets. Every child ch_i that isn't dominant through its parent J_i, then compared with every solution set from the archive. After that, the solution set, which dominates the archive's solution sets, is always kept as parents for the next iteration and put in the archive. If the solution set is dominant through the archive's solution sets, are eliminated, and those who are non-dominate are kept and/or archived according to the CDV. The significant characteristic of PAES is its methodology for advancing diversity in the approximation set. PAES utilizes a versatile hyper-gridding framework in the objective space to partition it into O non-covering hyper-boxes. Having a place of a specific solution set for a specific area in the hyper-box is dictated through the value of objectives that determines a coordinates of solution sets. For the situation where the solution set is non-dominate according to the solution sets of an archive, then the concept of crowding distance applied on the number of solution sets that are lying in a specific hyper-box for deciding the solution set is rejected or not.

The specific benefit of a technique of diversity preservation is that it doesn't need any additional parameters like niche radius σ_{rd}. Yet, the principle core of PAES is the sensitiveness of the execution of this approach to the O parameter of the hyper-gridding framework Fig. 20.6.

The modified version of PAES was developed by the same authors [24], which is named as PESA. It has the same archive and technique of diversity maintenance as the PAES. In PESA, only the solution sets of the archive are performing the genetic operations such as SPEA2. Initially, generate the arbitrarily little internal population

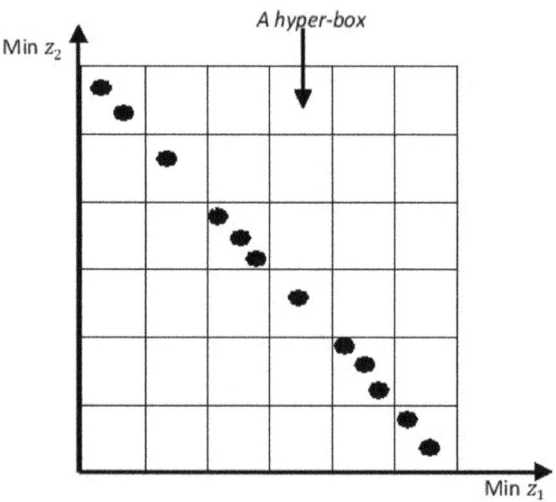

Fig. 20.6 PAES hyper-gridding system with d = 6

IJ. It utilizes a huge external population EJ. Initially, EJ is vacant. Afterward, archive EJ is improved by the elite solution sets in a similar way as performed in PAES. Suppose that the stopping criterion is achieved, then the approach gives back the solution sets of EJ; otherwise, IJ is occupied by the new solution sets through the following operations. By using the probability b, both parents are chosen from EJ. Subsequently, a single child is generated through a crossover. The mutation is applied to this child. By using the probability $(1 - b)$, the chosen parent among EJ is mutated. Afterward, archive EJ is improved, and the complete procedure is replicated.

The PAES, PESA, requires adjusting the archive's size and the parameter O of the gridding framework. By modifying the value of O, exponentially, the hyper-boxes total number varies that effects the distribution of the final population. PESA-II is the updated version of PESA. It was developed by Corne et al. [25], in which selection is according to area, and the selection subject is currently a hyper-box, not just the solution set, which means 1st choose the hyper-box, afterward the solution set browsed among a chosen hyper-box. An inspiration driving PESA is to decrease a total expense of Pareto positioning.

Indicator-Based EA: IBEA

Zitzler and Künzli [26] developed the IBEA in which selection according to the contribution of a solution set for a specific quality indicator. Thus, IBEA can be view as the 3rd evaluation of MOEA. In IBEA, first, generate a random population J. Then, every solution t of the population J is consecutively removed from the population J. The approach calculates the fitness value of t related to the loss in quality. Then, the solution set, which has the poorest fitness value, is deleted from the population. Then fitness value of each solution set is again calculate as population is shortened. This scheme of selection is utilized for generating the mating pool and in the selection of the environment. The most important part of IBEA is its sensitiveness towards the parameter l, which is utilized for scaling the value of fitness function as the execution of this approach is very much related to this parameter, which is studied to relate with the taken MOOP. Some other IB selection approach is named as S Metric Selection-based EMOA: SMS-EMOA [27], that is a collaboration of sorting of non-dominate and the mechanism of indicator-based selection. The crucial factor in this kind of approaches is effort in computation to calculate the values of a quality indicator for the specific non-dominate solution set [28].

Multi-objective EA Based on Decomposition: MOEAD

Zhang and Li [29] developed a most famous approach depends on decomposition. The essential concept of MOEAD is to break down the MOOP into M sub-problems, where M is the size of population. All of these sub-problems are solved turn by turn. To perform the MOEAD, the set of weight vectors w_i is needed. Thus, the w_i is formulated with the aim of solution sets that will be spread over the complete Pareto front. In MOEAD, Euclidean distance between weighted vectors is utilized for deciding the neighborhood of S weight vectors for every vector w_i. Afterward, every solution set of the population has provided a weight vector, and the dependent sub-problem is optimized according to the function of scalarizing. After that, both the solution sets among the neighborhood weighted vectors are mate and generate the offspring solution set. Then, with the help of a scalarizing function, an offspring solution set is evaluated. When the created new solution set was performed, it could also take over the many present solution sets of its neighboring sub-problems. There are three variants of scalarizing functions that are received for MOEAD: (1) weight sum method, (2) weight Tchebycheff method, and (3) boundary intersection method. The diversity of MOEAD has been tackled according to the similarity among the set of weighted vectors, i.e., depends on the related solution sets of a neighborhood. MOEAD is a very good approach for searching a few consistently distribute Pareto solution sets with a little expense of computation. Also, MOEAD has exhibited intriguing outcomes on many issues that have a large number of goals. Yet, its primary weakness is the decomposition of diversity and the distribution of solution sets when the problem deal with the bad scale.

20.3.3 New EAs

All the swarm intelligence based approaches and evolutionary approaches required for solving the SOOPs and MOOPs are probabilistic. They need standard controlling parameters such as no. of iterations, population size, elite size, etc. Other than the standard control parameters, distinct approaches need their self approach specific control parameters. For instance, GA utilizes selection operator, crossover probability, mutation probability; the PSO utilizes cognitive and social parameters, inertia weight; ABC utilizes the no. of scout bees, onlooker bees, employed bees, and breaking point; HS utilizes rate of pitch adjusting, rate of harmony memory consideration, and the no. of improvisations. Correspondingly, different approaches, for example, EP, DE, ACO, FF, CSA, GSA, BBO, FPA, ALO, IWO, and so forth require the adjustment of their algorithm-specific parameters. The best possible adjustment of the algorithm-specific parameters is a pivotal part that influences the approaches' execution, as referenced previously. The inappropriate adjustment of parameters further expands a computation attempts or gives the local optimal solution set. Thinking about this reality, in 2011, Rao et al. [30] presented the teaching learning-based optimization (TLBO) algorithm, that doesn't require even single algorithm-specific parameter. The TLBO requires standard control parameters such as population size and no. of iterations for its functioning. The TLBO has increased broad acknowledgment surrounded by the optimization researchers [31]. By analyzing the achievement of the TLBO, one more approach that is independent of the algorithm-specific parameter was developed by Rao [32] and named the Jaya algorithm (JA). Although, in contrast to two stages means learner stage and teacher stage of the TLBO, the JA has just a single stage. The JA is basic in an idea and has indicated better execution when contrasted with other optimization approaches.

Teaching–Learning-Based Optimization Algorithm: TLBO

In 2011, Rao et al. [30] built up the TLBO, that doesn't need adjustment of any algorithm-specific parameters for its functioning. An approach expresses two essential ways of updating: (i) by the educator (called a teacher stage) and (ii) by the association with different students (called a student stage). In this optimization approach, student's clustering is taken as population, and distinct subjects suggested to the students are taken as distinct decision variables of the SOOP or MOOP. A student's outcome is corresponding to fitness estimation of a considered problem. The best solution set in whole population is assigned as instructor. Each decision variable is a parameter associated with the goals of the provided problem, and best solution set is best estimation of a goal work.

The functioning of TLBO is separated by two sections, the Learner stage and the Teacher stage. The process of both these stages is clarified below [31].

- **Teacher stage**: The first part of the TLBO is the Teacher stage, where students learn through the educator. Along this stage, an educator attempts to expand the class's overall outcome in the subject taught by that person is dependent upon their potential. For any iteration t, expect that there is n no. of sub. (i.e., decision variables), m no. of students (i.e., population size, $l = 1, 2, \cdots, m$) and $M_{s,t}$ is average outcome of the students in a specific sub. s ($s = 1, 2, \cdots, n$). A best general outcome $Y_{total-lbest,t}$ taken every subject to combine acquired in a whole population of students, is taken as outcome of best student $lbest$. Although, as an instructor is typically assumed as an exceptionally learn individual who trains students to have better outcomes, the best student searched taken through the approach as the educator. The variation among the recent mean outcome of each subject and the instructor's parallel outcome for each subject is provided as follows:

$$Var M_{s,l,t} = r_t(Y_{s,lbest,t} - T_f M_{s,t}) \quad (20.11)$$

where, $Y_{s,lbest,t}$ is the outcome of the best student in sub. s. T_f stands for teaching factor that chooses the estimation of the average to be modified, and r_t stands for arbitrary no. lies in between 0 and 1. The estimation of T_f could be 1 or 2. The estimation of T_f is chosen haphazardly by equivalent likelihood as,

$$T_f = round[1 + rand(0, 1)\{2 - 1\}] \quad (20.12)$$

The estimation of T_f isn't provided as an input to an approach, and the approach arbitrarily chooses its worth from the Eq. (20.12). After directing several examinations on numerous benchmark functions, it is observed that the approach achieves good outcomes if the estimation of T_f is somewhere in the range of 1 and 2. In any case, the approach is established to execute much better if estimation of T_f can be 1 or 2. Consequently, to make an easier approach, the T_f is recommended to consider as 1 or 2, relying upon the measures provided from the Eq. (20.12).

Because of a $VarM_{s,l,t}$, the current solution set is updated in the teacher stage as per the accompanying expression.

$$Y'_{s,l,t} = Y_{s,l,t} + VarM_{s,l,t} \qquad (20.13)$$

where, $Y'_{s,l,t}$ represents updated estimation of $Y_{s,l,t}$. $Y'_{s,l,t}$ is considered whether it provides better esteem. At the end of the teacher stage, every considered esteem is kept up because they are the input for the learner stage. The learner stage relies on the educator stage.

- **Learner stage**: The second aspect of the TLBO is the learner stage, where students improve their skills through communication among themselves. A student associates haphazardly with different students to improve their skills. A student learns more when other student has more information over her/him. Taking m as the population size, the learning process of this stage is clarified below.

Arbitrarily select two students A and B with the end goal that $Y'_{total-A,t} \neq Y'_{total-B,t}$ Where, $Y'_{total-A,t}$ and $Y'_{total-B,t}$ represents updated estimations of $Y_{total-A,t}$ and $Y_{total-B,t}$ of A and B individually toward the accomplished of educator stage.

$$\begin{aligned} &\text{If } \quad Y'_{total-A,t} < Y'_{total-B,t} \\ &\text{then} \\ &\quad Y''_{s,A,t} = Y'_{s,A,t} + r_t(Y'_{s,A,t} - Y'_{s,B,t}), \end{aligned} \qquad (20.14)$$

$$\begin{aligned} &\text{If } \quad Y'_{total-B,t} < Y'_{total-A,t} \\ &\text{then} \\ &\quad Y''_{s,A,t} = Y'_{s,A,t} + r_t(Y'_{s,B,t} - Y'_{s,A,t}). \end{aligned} \qquad (20.15)$$

$Y''_{s,A,t}$ is considered whether it provides a better esteem.

Equations (20.14) and (20.15) are for minimizing problem. In the case of maximizing problem, Eqs. (20.16) and (20.17) are utilized.

$$\begin{aligned} &\text{If } \quad Y'_{total-B,t} < Y'_{total-A,t} \\ &\text{then} \\ &\quad Y''_{s,A,t} = Y'_{s,A,t} + r_t(Y'_{s,A,t} - Y'_{s,B,t}), \end{aligned} \qquad (20.16)$$

$$\begin{aligned} &\text{If } \quad Y'_{total-A,t} < Y'_{total-B,t} \\ &\text{then} \\ &\quad Y''_{s,A,t} = Y'_{s,A,t} + r_t(Y'_{s,B,t} - Y'_{s,A,t}). \end{aligned} \qquad (20.17)$$

Non-dominate Sorting TLBO: NSTLBO

NSTLBO is produced to deal with MOOPs. The extension of TLBO is known as NSTLBO. The NSTLBO comes under the posterior approach to deal with MOOPs and keeps up a various solution sets. NSTLBO contains the teacher stage and learner stage, likewise the TLBO. Moreover, to deal with various goals efficaciously and expeditiously, the NSTLBO is integrated by a crowding distance computations and non-dominated sorting approach.

Initially, start with the random initial population's generation as the J number of solution sets. Then sorted the initial population and positioned depends on concept of non-dominant and constraint dominance. The prevalence between the solution sets is decided as follows: The priority given to the concept of constraint dominance (CD) and afterward concept of non-dominance (ND), and afterward on the crowding distance value (CDV) of the solution sets. The student with a most elevated position (rank = 1) is chosen as a class educator. By chance that there be two or more than two students with a similar position, at that point, the student with the most elevated CDV is chosen as an instructor for a class. That provides guarantees which an educator is chosen among a scanty area of the search space (SS).

When an educator is chosen, students are updated as described in teacher stage of TLBO, i.e., as per Eqs. (20.11)–(20.13). Then, the solution sets of updated students (new students) are combined with the initial population to acquire $2J$ solution sets (students). Again, these students are arranged and positioned dependent on the concept of CD, ND, and CDV for every student is calculated. The student with a higher position is viewed as better than the other students. If both the students hold a similar position, at that point, the student with a higher CDV is viewed as better than the other. Established on the concept of new positioning and value of crowding distance, the J no. of best students are chosen. These students are additionally update through student stage of TLBO, i.e., as indicated by Eqs. (20.14) and (20.15) or (20.16) and (20.17).

The student's priority is decided according to the concept of CD, the concept of ND, and the CDV of the students. The student with the most elevated position is viewed as better than the other student. On the off chance that both the students hold a similar position, at that point, the student with a larger CDV is viewed as better than the other. At accomplishing the learner stage, all new students are joined with an old students and repeatedly sorted and positioned. Because of the new positioning and CDV, J no. of best students are chosen, and these students are straightforwardly update dependent on a teacher stage for next iteration [31, 33].

Non-dominated Sorting of the Population

In this methodology, population is arranged at many positions according to the concept of dominance given as bellow: the y_s solution set is called to dominate y_t solution set provided the y_s solution set is no longer poor than y_t solution set regarding every goal and besides the y_s solution set is strictly good over y_t solution set for at least single goal. When one of the above two terms are break, at that point y_s solution set doesn't dominate y_t solution set.

For the J solution sets, the solution sets that ain't dominated through any of the J solution sets are known as the non-dominate solution sets. Every non-dominate solution set recognized in a first sorting run is defined as position one and is erased from the set J. The rest of the solution sets in J solution sets are again sorted, and the strategy is rehashed until every solution sets of the J solution sets are sorted and positioned. For constrained MOOPs, the concept of constrained-dominance is utilized. The total no. of function evaluation for the NSTLBO is the two times of total no. of iterations multiply with size of population.

Crowding Distance Value (CDV)

The destination is to calculate density of solution sets around a specified solution set m is the basic concept of CDV. CDV defined for every solution set of the population. Subsequently, the crowding distance (cd_m) is defined as the average distance of a pair of solution sets on either side of the solution m is calculated along with all N goals. For computing the CDV of the solution m in the front f, the accompanying procedure is followed: Step 1: Determine the no. of solution sets lying on front f as $k = |f|$. For every solution set m has provide the $cd_m = 0$. Step 2: For every goal $n = 1, 2, \ldots, N$, sort the set in the worst order of z_n. Step 3: For $n = 1, 2, \ldots, N$, provide biggest CDV to limit solution sets $(cd_1 = cd_k = \infty)$, and for the remaining solution sets $s = 2$ to $(k - 1)$, CDV is calculated as below:

$$cd_s = cd_s + \frac{z_n^{s+1} - z_n^{s-1}}{z_n^{max} - z_n^{min}} \qquad (20.18)$$

where, s stands for the order of solution set, z_n provides the value of n-th goal, z_n^{min} and z_n^{max} are the values of population minimum and maximum for n-th goal.

Crowding-Comparison Operator

For finding the solution which is better than the other, the crowding comparison operator is utilized. It is established on the two precious factors: (1) Non-domination position, and (2) CDV of each solution s of the population. It is denoted by this $\prec_n$ symbol and it is described as below: $\prec_n$, if $(Rank_s < Rank_t)$ or $[(Rank_s = Rank_t)$ and $(cd_s > cd_t)]$. i.e., For the s and t two solution sets, which has contrasting non-domination positions, the solution set, which has a better or lower position, is prioritized. Whereas, suppose that the two solution sets have the same position $(Rank_s = Rank_t)$, at that point, the solution set situated in the lesser crowded area $(cd_s > cd_t)$ is chosen.

Constraint-Dominance Concept

The concept of constraint dominance is defined as follows: Consider the two solution sets s and t, the solution set s is called to constraint dominate solution set t, only when the prescribed events occur. The solution set t is not feasible and the solution set s is feasible. Both the solution sets s and t are infeasible, yet the solution set s has the total less constraint violation than solution set t. Both the solution sets s and t are feasible and moreover the solution set t is dominated by the solution set s. Thus, we can say that the concept of constraint dominance guarantees that the infeasible solution sets achieved a lower position than the feasible solution sets. For the infeasible solution sets, the solution sets with a total more constraint violation are defined at the lower position. For the feasible solution sets, the dominated feasible solution sets are defined at a lower position as compared to the non-dominated feasible solution sets.

Jaya Algorithm: JA

Recently, Rao [32] proposed the JA in 2016, which is easy to execute and doesn't need adjustment of any particular algorithm parameters. In JA, randomly J initial solution sets are formulated between decision variable's lower and upper limits. After that, every decision variables of each solution set is updated by utilizing Eq. (20.19). Suppose z be a goal that is optimized. Assume that v no. of decision variables. The value of objective function related to best solution set is denoted as z_{best}, and similarly, the value of objective function related to the worst solution set is denoted as z_{worst}.

$$V_{s+1,t,u} = V_{s,t,u} + R_{s,t,1}\left(V_{s,t,B} - |V_{s,t,u}|\right) - R_{s,t,2}\left(V_{s,t,W} - |V_{s,t,u}|\right) \quad (20.19)$$

where W and B defines the index of worst and best solution sets from the present population. The index of decision variable, iteration, and the solution set is represented as t, s, u, respectively. $V_{s,t,u}$ implies the tth decision variable of uth solution set at sth iteration. The random numbers are denoted as $R_{s,t,1}$ and $R_{s,t,2}$. They are lying in a range of [0, 1]. They work as the factor of scaling and provides better diversity. The primary goal of JA is to increase the value of fitness for every solution set of the population. Thus, JA attempts to transfer the value of an objective function of every solution set in the direction of the best solution set through uplifting the decision variables. When the decision variables are updated, there is a comparison between the updated one and the corresponding old one. Whichever provides the better value carried forward, i.e., that solution sets are taking part in the next iteration. In each iteration, the solution sets move closer to a better solution set by the JA, and also the solution set moves far from the worst solution set. In this way, a decent strengthening and broadening of the search space are accomplished. The JA consistently attempts to move nearer to progress (i.e., arriving at the best solution set) and attempts to dodge disappointment (i.e., transferring ceaselessly from the worst solution set). The JA endeavors to get successful through arriving at the best solution set, and thus it is named as Jaya which is a Sanskrit word whose meaning is triumph or win.

There are also a few variations of JA that are found in the literature. These variations are Self-Adaptive JA (SAJA) [34], Quasi-oppositional JA (QOJA) [35], Self-Adaptive multi-population JA (SAMPJA) [36], Self-Adaptive multi-population elitist JA (SAMPEJA) [37].

Multi-objective Jaya Algorithm: MOJA

In 2017, Rao et al. [38], has produced MOJA for tackling the MOOPs. The MOJA is a posteriori form of the JA for tackling MOOPs. According to the Eq. (20.19), the MOJA solution sets are uplifted likewise as in the JA. Yet, to deal with numerous goals efficaciously and expeditiously, the MOJA is integrated with the concept of non-dominated sorting and mechanism of crowding distance.

In SOOP, it is simple to search which one solution set is good than another solution set related to the objective function's corresponding value. On the other hand, it is a very difficult task to search the solution sets, which is the best and worst among the whole population in the case of MOOP. In the MOJA, the worst and best solution sets find out through the comparison of a position that is provided to the solution sets. The position of the solution sets depends on the concept of CD, ND, and CDV.

Initially, the random initial population is formulated J number of solution sets. This initial population is then sorted and assigned the position to each solution set according to the concept of CD and ND. The prevalence between solution sets is basically decided according to the concept of constraint-dominance. After that, the concept of ND and CDV of solution sets. The solution set, which has a greater position (position = 1), is viewed as better than the other solution set. However, if both the solution sets have the same position, then the solution set, which has a greater CDV, is prior to other solution set. This guarantees the solution set is chosen from the scanty area of the SS. A solution set which has most lower position is chosen as worst solution set. Similarly, a solution set which has most elevated position (position = 1) is chosen as best solution set. When the worst and best solution sets are chosen, then the solution sets are updated according to the Eq. (20.19).

When every solution set is updated, then the updated solution set is combined with the initial population, and we receive $2J$ solution sets. Again, these solution sets are sorted and provided the position according to the concept of CD, ND, and CDV. With the help of a new position and CDV, the J number of better solution sets are chosen. The solution set's prevalence between the solution sets is resolved according to their position and the CDV. The solution set, which has a greater position, is taken prior to the other solution sets. If some solution sets have the same position, then the solution set, which has a higher value of crowding distance, is preferred over the other solution sets. For each solution set, the MOJA finds the objective function's value just a single time at every iteration. Consequently, the total no. of function evaluates needed through the MOJA = Size of population * no. of iterations. Moreover, computationally if the approach is run more than a single time, at that point, by multiplying the population size with total number of iterations and runs, we could obtain the total number of function evaluations.

There are also a few variations of MOJA that are found in the literature. These variations are elitist JA (EJA) [39], Binary JA (BJA) [40], Improved JA (IJA) [41] and Multi-objective Quasi-oppositional JA (MOQOJA) [42].

Like wise the TLBO and JA, Rao has also develop the Rao algorithms which are three metaphor-less simple algorithms for solving unconstrained and constrained optimization problems [43, 44].

20.4 Conclusion

In this chapter, the fundamentals of MOOP are characterized. We divided MOEAs dependent on two fundamental scenario: (1) Utilization of the Pareto dominant for the selection operator, (2) elitism. First evaluation of MOEAs is viewed as the Non-Elitist methods, while the subsequent second evaluation relates to the elitist approaches. The utilization of performance indicators for the selection operator and scalarizing functions for breaking the first MOOP into the collection of sub-problems can be viewed as the third evaluation of MOEA. After that, the algorithms work independently of an algorithm-specific parameter is the forth evaluation of MOEA. Nowadays, there has been more concern for hybridization evolutionary approaches, which combines the original optimization approach with some different ideas to enhance its quality. The result of the acquired methodologies generally consistently gives preferred outcomes over the first one.

References

1. Stadler W (1988) Multicriteria optimization in engineering & in the sciences. Plenum Press, New York
2. Tabucanon M (1988) Multiple criteria decision making industry. Elsevier Science Publishers, Amsterdam
3. Coello C, VanVeldhuizen D, Lamont G (2002) Evolutionary algorithms for solving multi-objective problems. Kluwer Academic Publishers, New York
4. Tilva S, Dhodiya J (2019) Hybrid Jaya algorithm for solving multi-objective 0-1 integer programming problem. Int J Eng Adv Tech 9(2):4867–4871
5. Dhodiya J, Tailor A (2016) Genetic algorithm based hybrid approach to solve fuzzy multi-objective assignment problem using exponential membership function. Springerplus 5(1):20–28
6. Gen M, Li Y, Ida K (1999) Solving multi-objective transportation problem by spanning tree-based genetic algorithm. IEICE Trans Fundam Electron Commun Comput Sci 82(12):2802–2810
7. Dhodiya J, Tailor A (2018) Genetic algorithm based hybrid approach to solve uncertain multi-objective COTS selection problem for modular software system. J Int Fuzzy Syst 34:2103–2120
8. Schaffer J (1985) Multiple objective optimization with vector evaluated genetic algorithms. In: Proceedings of the 1st international conference on genetic algorithms. L. Erlbaum Associates Inc., pp 93–100
9. Coello C (2000) Treating constraints as objectives for single-objective evolutionary optimization. Eng Optim 32(3):275–308
10. Richardson J, Palmer M, Liepins G, Hilliard M (1989) Some guidelines for genetic algorithms with penalty functions. In: Proceedings of the third international conference on genetic algorithms. Morgan Kaufmann Publishers Inc., pp 191–197
11. Kursawe F (1990) A variant of evolution strategies for vector optimization. In: Parallel problem solving from nature. Springer, Berlin, pp 193–197
12. Hajela P, Lin C (1992) Genetic search strategies in multi criterion optimal design. Struct Optim 4(2):99–107
13. Fonseca C, Fleming P (1993) Genetic algorithms for multi-objective optimization: formulation discussion and generalization. In: ICGA, vol 93. Citeseer, pp 416–423

14. Horn J, Nafpliotis N, Goldberg D (1994) A niched Pareto genetic algorithm for multi-objective optimization. In: Proceedings of the first IEEE conference on evolutionary computation. IEEE World Congress on computational intelligence. IEEE, pp 82–87
15. Srinivas N, Deb K (1994) Muilti-objective optimization using non-dominated sorting in genetic algorithms. Evol Comput 2(3):221–248
16. Holland J (1975) Adaptation in natural and artificial systems: an introductory analysis with applications to biology, control, and artificial intelligence. University of Michigan Press, Michigan
17. Bechikh S, Chaabani A, Said L (2015) An efficient chemical reaction optimization algorithm for multi-objective optimization. IEEE Trans Cybern 45(10):2051–2064
18. Zitzler E, Thiele L (1999) Multi-objective evolutionary algorithms: a comparative case study and the strength Pareto approach. Evol Comput Trans IEEE 3(4):257–271
19. Zitzler E, Laumanns M, Thiele L (2001) Spea2: improving the strength Pareto evolutionary algorithm. TIK report, vol 103
20. Deb K, Agrawal S, Pratap A, Meyarivan T (2000) A fast elitist non-dominated sorting genetic algorithm for multi-objective optimization: Nsga-ii. In: Parallel problem solving from nature PPSN VI. Springer, New York, pp 849–858
21. Deb K, Pratap A, Agarwal S, Meyarivan T (2002) A fast and elitist multi-objective genetic algorithm: Nsga-ii. IEEE Trans Evol Comput 6(2):182–197
22. Knowles J, Corne D (1999) The Pareto archived evolution strategy: a new baseline algorithm for Pareto multi-objective optimisation. In: Proceedings of the 1999 Congress on evolutionary computation, CEC 99, vol 1. IEEE
23. Knowles J, Corne D (2000) Approximating the non dominated front using the Pareto archived evolution strategy. Evol Comput 8(2):149–172
24. Corne D, Knowles J, Oates M (2000) The Pareto envelope-based selection algorithm for multi-objective optimization. In: Parallel problem solving from nature PPSNVI. Springer, Berlin, pp 839–848
25. Corne D, Jerram N, Knowles J, Oates M et al (2001) Pesa-ii: region-based selection in evolutionary multi-objective optimization. In: Proceedings of the genetic and evolutionary computation conference (GECCO 2001). Citeseer
26. Zitzler E, Künzli S (2004) Indicator-based selection in multi-objective search. In: Parallel problem solving from nature-PPSN VIII. Springer, Berlin, pp 832–842
27. Beume N, Naujoks B, Emmerich M (2007) Sms-emoa: multi-objective selection based on dominated hypervolume. Eur J Oper Res 181(3):1653–1669
28. Azzouz N, Bechikh S, Said L (2014) Steady state IBEA assisted by MLP neural networks for expensive multi-objective optimization problems. In: Proceedings of the 2014 conference on genetic and evolutionary computation. ACM, pp 581–588
29. Zhang Q, Li H (2007) Moea/d: a multi-objective evolutionary algorithm based on decomposition. IEEE Trans Evol Comput 11(6):712–731
30. Rao R, Savsani V, Vakharia D (2011) Teaching-learning-based optimization: a novel method for constrained mechanical design optimization problems. Comput Aided Des 43:303–315
31. Rao R (2016) Teaching learning based optimization algorithm and its engineering applications. Springer International Publishing, Switzerland
32. Rao R (2016) Jaya: a simple and new optimization algorithm for solving constrained and unconstrained optimization problems. Int J Ind Eng Comput 7:19–34
33. Rao R, Rai D, Balic J (2016) Multi-objective optimization of machining and micro-machining processes using non-dominated sorting teaching-learning-based optimization algorithm. J Intell Manuf 2016
34. Rao R, More K (2017) Design optimization and analysis of selected thermal devices using self-adaptive Jaya algorithm. Energy Convers Manage 140:24–35
35. Rao R, Rai D (2017) Optimization of welding processes using quasi oppositional based Jaya algorithm. J Exp Theor Artif Intell 29(5):1099–1117
36. Rao R, Saroj A (2017) A self-adaptive multi-population based Jaya algorithm for engineering optimization. Swarm Evol Comput 37:1–26

37. Rao R, Saroj A (2019) An elitism-based self-adaptive multi-population Jaya algorithm and its applications. Soft Comput 23(12):4383–4406
38. Rao R, Rai D, Balic J (2017) A multi-objective algorithm for optimization of modern machining processes. Eng Appl Artif Intell 61:103–125
39. Rao R, Saroj A (2018) Multi-objective design optimization of heat exchangers using elitist-Jaya algorithm. Energy Syst 9(2):305–341
40. Prakash T, Singh V, Singh S, Mohanty S (2017) Binary Jaya algorithm based optimal placement of phasor measurement units for power system observability. Energy Convers Manage 140:34–35
41. Abarghooee R, Dehghanian P, Terzija V (2016) Practical multi-area bi-objective environmental economic dispatch equipped with a hybrid gradient search method and improved Jaya algorithm. IET Gener Trans Distrib 10(14):3580–3596
42. Warid W, Hizam H, Mariun N, Wahab N (2018) A novel quasi-oppositional modified Jaya algorithm for multi-objective optimal power flow solution. Appl Soft Comput 65:360–373
43. Rao R (2020) Rao algorithms: Three metaphor-less simple algorithms for solving optimization problems. Int J Ind Eng Comput 11(1):107–130
44. Rao R, Pawar R (2020) Constrained design optimization of selected mechanical system components using Rao algorithms. Appl Soft Comput 89:106–141

Chapter 21
Improving Financial Bankruptcy Prediction Using Oversampling Followed by Fuzzy Rough Feature Selection via Evolutionary Search

Pankhuri Jain, Anoop Kumar Tiwari, and Tanmoy Som

Abstract Recently, bankruptcy prediction has been addressed as one of the most interesting as well as challenging issues for financial institutions and business. For creditors and investors, bankruptcy prediction plays very interesting role by facilitating decision-making ability in various areas such as business, accounting, and finance etc. Due to assemblage of large volume of inconsistent, highly imbalanced, irrelevant and redundant data from companies and other creditors, its always a very challenging and complex task to handle financial risk associated with company by developing an effective prediction model. This chapter presents a new methodology for improving the bankruptcy prediction performance of various machine learning algorithms. Firstly, we convert imbalanced dataset consisting of bankrupt and non-bankrupt into balanced dataset by applying oversampling technique. Then, relevant and non-redundant features are generated based on fuzzy rough feature selection technique via evolutionary search. Furthermore, performance of various machine learning algorithms are fully analysed by applying them on this highly balanced reduced dataset. Moreover, discriminating ability of different features are analysed based on feature ranking algorithm. Finally, we present a comparative study of our best results with already existing results. From different experimental results, we can conclude that bankruptcy prediction can be enhanced by suitably adjusting the class distribution followed by fuzzy rough set based feature selection via evolutionary search.

Keywords Feature selection · Imbalanced dataset · SMOTE · Fuzzy rough set

---

P. Jain · T. Som (✉)
Department of Mathematical Sciences, IIT (BHU), Varanasi 221005, India
e-mail: tsom.apm@itbhu.ac.in

A. K. Tiwari
Department of Computer Science, Maulana Azad National Institute of Technology, Bhopal 462003, India

21.1 Introduction

Bankruptcy prediction [1–3] is an interesting research area for business as well as financial institutions in the recent decades, causing a substantial impact on stockholders, employees, people managing funds, academic community, financial market players, financial industry, customers and nation [4–6]. Bankruptcy is an critical business issue, which results in inability of the individual or the company to repay dues to creditors [2]. As a result of this issue, all the characteristics and bank accounts of the company need to liquidated to pay off the utmost amount of the pledges [7]. Moreover, bankruptcy has various charges on companies and/or individuals like access to credits in future, including reputation, and possible loss of valuable customers.

Due to the modern changes of the increasingly accelerated globalization, world economy, and more firms appear to collapse now more than ever [8, 9]. Therefore, it is growing importance to efficiently evaluate the bankruptcy. In the last few years, various computational intelligence and machine learning techniques have been presented to rectify the issue of bankruptcy prediction, such as artificial neural networks, support vector machines (SVMs), decision trees, and many more [3, 10–17].

21.2 Related Work

The problem of bankruptcy prediction have been basically considered as a supervised classification issue in these studies, where the decision class are supplied for all training samples. These classification techniques can achieve satisfying results to a certain degree based on huge number of labelled training instances or data points. Altman, Ohlson, and Zmijewski [2, 7, 18] presented that accounting ratios and stock market data comprise necessary information for evaluating financial status of a business and/or company. Chen et al. [19–21] and Goldstein et al. [22] showed that various evaluation models can be presented based on the state-of-the-art computational techniques, which can produce default prediction with great precision. These include the random forest, artificial neural networks, genetic algorithm, and decision trees as presented by Chandra et al. [23], Wilson and Sharda [14], Amjadian et al. [24] and Aoki and Hosonuma [25]. Moreover, Tian et al. [26], Cerchiello et al. [27], and Liang et al. [28] discussed the logit model, the deep neural network, and the support vector machine model, etc respectively. Tian et al. [26] suggested that bankruptcy prediction performance can be enhanced by selection of relevant and non-redundant features. The large amount of variation in number of normal and bankrupt cases makes bankruptcy datasets imbalanced. This imbalanced class distribution [29] negatively affects the generalization ability of the prediction models as standard machine learning algorithms concentrate towards the majority class to increase the average accuracy. In this problem, rare class distributions are neglected, which is more essential to determine. Faris et al. [30] proposed a hybrid technique to deal with bankruptcy problem with highly imbalanced class distribution. Bagged-pSVM and Boosted-pSVM methods were introduced and successfully implemented by Chen et al. [16] for bankruptcy prediction. Smiti and Soui [31] proposed BSM-SAES approach based on deep learning concept and presented effective results.

In this chapter, a new methodology is presented to cope with all the above mentioned issues. By fittingly modifying the class distribution based on SMOTE (Synthetic Minority Oversampling Technique), we obtain the optimally balanced datasets. Then, fuzzy rough set based feature selection with evolutionary search is utilized for selecting relevant and non-redundant features. Further, various machine learning algorithms are applied on optimally balanced reduced datasets. Moreover, ranks of the different features are presented by applying fuzzy rough attribute evaluator with ranker search that denotes the features ability to discriminate among classes. The schematic representation of entire methodology is given in Figure 21.1.

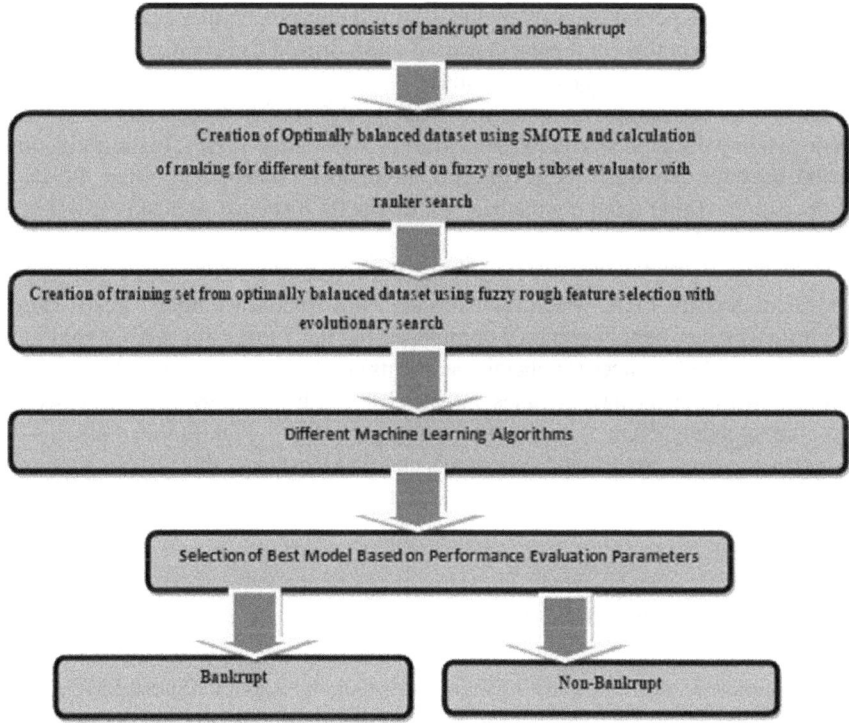

Fig. 21.1 Flowchart of the proposed work

21.3 Material and Methods

21.3.1 Dataset

We have employed the dataset from UCI repository [32] to conduct our entire experiments. This dataset was based on financial state of Polish companies and consisted of bankrupt and non-bankrupt. Entire data was distinguished in five classification cases as follows:

In the first case, the data comprises financial rates from first year of the forecasting period and equivalent decision class denotes the status after 5 years. This dataset consists of 7027 instances (financial statements). Here, 271 instances depicts bankrupted companies, while others did not bankrupt in the prediction period.

Second case consists financial rates from second year of the forecasting period with corresponding decision class representing the status of bankruptcy after 4 years. In the forecasting period, the data includes 10,173 data points (financial statements), where 9773 data points are representative of those firms did not bankrupt, while 400 instances indicates bankrupted companies.

Third case incorporates financial rates from third year of the forecasting period and corresponding decision that denotes bankruptcy status after 3 years. In the prediction period, the data consists of 10,503 data points (financial statements), where 495 data points express bankrupted companies, while 10,008 instances indicate those firms that did not bankrupt.

Fourth case includes financial rates from fourth year of the forecasting period and equivalent decision class, which indicates status of bankruptcy after 2 years. Here, 515 instances represent bankrupted companies and 9277 firms that did not bankrupt out of total 9792 instances (financial statements) in the prediction period.

In the fifth case, the data includes the rates from fifth year of the prediction period and corresponding decision represents bankruptcy after 1 year. In this dataset, 410 instances represent bankrupted companies and 5500 instances that did not bankrupt out of total 5910 instances (financial statements) in the forecasting period.

21.3.2 Classification Protocol

In this chapter, six machine learning algorithms namely: Random Forest, SMO, IBK, Rotation Forest, PART, and J48 are utilized to perform the experimental work. Experimental results demonstrates that Random Forest is the best performing technique. A brief description of Random Forest can be given as follows:

Random Forest: It is an ensemble method developed by Breiman [33]. This technique applies recursive partitioning to create numerous trees. Then, the outcomes are combined, which include an additional layer of randomness to bagging. The trees are generated by employing a bootstrap sample of the data, and the collection of candidate variables is defined as a subset of the variables chosen randomly at every step. For each tree, this method uses two-third of the training samples to construct tree and the remaining instances are applied to test the tree. This one third of data sample is defined as "Out of Bag" and utilized to standardize the performance of every tree. The generalization error of a forest is computed based on the information from the strength of individual trees and the correlation between them. Here, each tree is found to be unpruned, to attain low-bias trees; in the meantime, selection of random variable and bagging produces low correlation of the trees individually. Hence, this approach results in an ensemble technique that can produce both low bias as well as low variance. Finally, this technique provides overall model with better accuracy, improved general performance, and robust against overfitting, which results in performing well when compared to many other machine learning algorithms such as discriminant analysis, neural networks, naive bayes and support vector machines etc.

21.3.3 Optimal Balancing Protocol

Class imbalance appears in classification problems where the "normal" cases, or data points, differ considerably in number from "abnormal" cases. In class imbalance, the core issue is that learning techniques incited on imbalanced training sets demonstrate a bias affiliated with low overall performance on the minority class. Moreover, the less appearing classes are usually more influential when compared to frequent one, for example system failure, which is not adequate as it produces in more specificity value and less sensitivity value while forecasting the minority samples [34]. In the current chapter, we have balanced all the five years datasets as an ideal balancing ratio of 1:1 by employing SMOTE. A brief introduction of SMOTE can be depicted as follows:

SMOTE: SMOTE [35] is the frequently used over-sampling technique due to its computational efficiency, simplicity, and excellent performance [36]. However, SMOTE artificially interpolates new samples in minority class without investigating the majority instances, especially in locality regions with majority class. It produces artificial data point which lie between nearest minority class tuple's neighbour in the training set. Thus, an artificial instance is a combination of the seed sample's feature and randomly chosen k-nearest neighbours. The k parameter must be decided by the user. Here, each minority sample x is set to 5 (in general) via random variables, and there after the nearest neighbour is selected in the feature space to obtain the shortest path between the data points, prior to decide the Euclidean distance. Subsequently, the simulated data are produced amidst the nearest data. The SMOTE samples can be specified as the combinations of two similar minority class samples (m and m^k) associated linearly and is given by

$$n = m + z * (m^k - m) \tag{21.1}$$

where z changes from 0 to 1 and m^k is randomly chosen among the 5 minority class nearest neighbours of m. In last few years, SMOTE has been successfully applied to tackle class imbalance issues. In WEKA [37], the default value of nearest neighbours for SMOTE is taken as 5.

21.3.4 Feature Selection Protocol

Feature selection or attribute reduction technique addresses the dimensionality reduction problem by governing a subset of original features and removing irrelevant and redundant features to construct a good model for classification or prediction task. The classical rough set model, established by Pawlak [38] has been efficiently implemented as an interesting attribute reduction and rule learning tool. In the last twenty years, rough set theory has acquired considerable interest and has been successfully implemented to various fields. Rough set theory is an interesting mathematical tool that can be applied to reduce the irrelevant and redundant features using intrinsic information within the data without involving any additional information. For a given dataset with discrete-valued features, it is promising to obtain the most informative features or attributes among set of available features using rough set theory. However, rough set theory can be applied only on nominal data and does not consider the degree of overlapping available in the data. Fuzzy rough sets, as established by Dubois and Prade [39] can instantly and effectively cope with continuous data sets. The similarity between data samples is characterized by fuzzy rough set based techniques based on the information from fuzzy binary relations, which provides a way in which numerical feature values are no longer needed to be discretized. They are transformed to the membership degrees of instances corresponding to decisions; hence, more information of continuous feature values can be maintained. It can be observed from various research articles that fuzzy similarity relations has played a vital role in the successful applications of fuzzy rough set models. Combination of fuzzy and rough sets provides a key step in reasoning with uncertainty for continuous valued data. Fuzzy rough set concept has been efficiently and successfully carried out to eliminate few interesting shortcomings of the classical rough set based technique in various aspects. The notion of a dependency function in a conventional rough set model into the fuzzy occurrence was established by Jensen and Shen [40] and presented a feature selection algorithm based on fuzzy rough set concept. This concept was further extended by different researcher articles such as [41–47].

In this chapter, fuzzy rough set based feature selection with evolutionary search is utilized to select relevant as well as non-redundant features for improving the prediction of bankrupt.

21.3.5 Performance Evaluation Metrics

For evaluating the relative prediction performance of all the six machine learning algorithms, we have recorded the experimental results based on multiple performance evaluation metrics. These evaluation metrics are identified from the values of true positives (*TP*), false negatives (*FN*), true negatives (*TN*), and false positives (*FP*). *TP* denotes overall correctly predicted non-bankrupt, *FN* indicates the overall incorrectly predicted non-bankrupt, *TN* represents the number of correctly predicted bankrupt and *FP* is the number of incorrectly predicted bankrupt.

Sensitivity: This parameter produces the percentage of correctly predicted non-bankrupt and is specified by:

$$Sensitivity = \frac{TP}{TP + FN} \times 100 \tag{21.2}$$

Specificity: This evaluation metric represents the percentage of correctly predicted bankrupt and is calculated as follows:

$$Specificity = \frac{TN}{TN + FP} \times 100 \tag{21.3}$$

Accuracy: It is defined as the percentage of correctly predicted bankrupt and non-bankrupt and is represented as below:

$$Accuracy = \frac{TP + TN}{TP + FP + TN + FN} \times 100 \tag{21.4}$$

AUC: This parameter indicates the area under the receiver operating characteristic curve (ROC), the predictor is said to be superior as its value approaches to 1. It can be observed as one of the parameters which are robust to the imbalance nature of the information systems [48].

MCC: Mathew's correlation coefficient can be computed by using the following equation:

$$MCC = \frac{TP \times TN - FP \times FN}{\sqrt{(TP + FP)(TP + FN)(TN + FP)(TN + FN)}} \tag{21.5}$$

It is extensively employed evaluation metric for binary classifications. A predictor is considered as an excellent performing as it produces MCC value closer to 1.

In this chapter, entire experiments are performed using WEKA [37] which is a well-known open source Java based machine learning platform.

21.4 Experiments and Results

In this section, extensive experiments have been conducted for evaluating the effectiveness of our methodology. We have used six machine learning algorithms namely: Random Forest, Sequential minimization optimization (SMO) with puk kernel, Nearest Neighbor method (IBK), JRip, PART, and J48 to perform the experimental results.

The experiments have been carried out by using K-fold cross-validation approach with $K = 10$. This method is based on splitting the entire dataset into K mutually exclusive subsets and repeating the experiments K times. Further, one of the subsets is kept for testing and the remaining $K - 1$ subsets are used for training each time. Firstly, six classifiers have been employed on the original datasets and the results are recorded in Tables 21.2, 21.3, 21.4, 21.5 and 21.6. Secondly, SMOTE has been applied for all the five datasets to produce optimally balanced datasets. Thirdly, fuzzy rough subset selection with evolutionary search has been utilized for all the five datasets to produce reduced datasets. The reduct size are varied from 43 to 48 as recorded in Table 21.1. Then, all the six classifiers are applied on the optimally balanced reduced datasets. Table 21.7, 21.8, 21.9, 21.10 and 21.11 demonstrates the performances of different machine learning algorithms. From the experimental results, it can be concluded that better performance is obtained for classifiers applied on optimally balanced reduced datasets when compared to the results for original datasets. Sensitivity values are less for all the machine learning algorithms in case of original datasets due to their class imbalance nature, while all the algorithms (except IBK) have achieved better sensitivity values in case of optimally balanced reduced datasets. The best results have been reported for Random Forest with sensitivity of 96.4%, specificity of 96.9%, accuracy of 96.7%, AUC of 0.990, and MCC of 0.934 for optimally balanced reduced second year dataset. Features ranking based on fuzzy rough attribute evaluator with ranker search are recorded in Table 21.12. Receiver Operating Characteristic curve (ROC) is used to present the performance of different machine learning algorithms in an effective as well as convenient way. Figures 21.2, 21.3, 21.4, 21.5 and 21.6 are plots of ROC for all the five years optimally balanced reduced datasets.

Table 21.1 Dataset characteristics and reduct size

Dataset	Number of instances	Number of attributes	Reduct size
First year	7027	64	47
Second year	10,173	64	48
Third year	10,503	64	46
Fourth year	9792	64	48
Fifth year	5910	64	43

Table 21.2 Performance evaluation metrics for the machine learning algorithms on original first year dataset

Learning algorithms	Sensitivity	Specificity	Accuracy	AUC	MCC
Random Forest	1.5	99.8	96.0	0.819	0.048
SMO	0.0	100.0	96.1	0.500	−0.010
IBK	0.0	99.6	95.8	0.587	−0.012
JRip	5.5	99.7	96.1	0.523	0.148
PART	14.0	99.2	95.9	0.725	0.224
J48	21.0	99.3	96.2	0.694	0.319

Table 21.3 Performance evaluation metrics for the machine learning algorithms on original second year dataset

Learning algorithms	Sensitivity	Specificity	Accuracy	AUC	MCC
Random Forest	1.0	99.8	95.9	0.990	0.824
SMO	0.0	100.0	96.1	0.550	0.000
IBK	5.0	99.2	95.4	0.497	−0.006
JRip	8.3	99.8	96.2	0.543	0.230
PART	9.5	99.7	96.2	0.732	0.223
J48	15.3	99.4	96.1	0.640	0.262

Table 21.4 Performance evaluation metrics for the machine learning algorithms on original third year dataset

Learning algorithms	Sensitivity	Specificity	Accuracy	AUC	MCC
Random Forest	8.0	99.7	95.0	0.821	0.017
SMO	0.0	100.0	95.3	0.500	0.002
IBK	4.0	95.4	91.1	0.498	−0.005
JRip	9.1	99.8	95.5	0.545	0.232
PART	7.9	99.5	95.2	0.750	0.175
J48	12.9	99.2	95.1	0.715	0.220

Table 21.5 Performance evaluation metrics for the machine learning algorithms on original fourth year dataset

Learning algorithms	Sensitivity	Specificity	Accuracy	AUC	MCC
Random Forest	1.2	99.6	94.5	0.839	0.029
SMO	0.0	100.0	94.7	0.500	0.000
IBK	1.9	98.7	93.6	0.506	0.011
JRip	7.6	99.6	94.7	0.539	0.180
PART	9.3	99.4	94.7	0.769	0.194
J48	18.1	98.8	94.5	0.710	0.263

Table 21.6 Performance evaluation metrics for the machine learning algorithms on original fifth year dataset

Learning algorithms	Sensitivity	Specificity	Accuracy	AUC	MCC
Random Forest	24.4	98.8	93.6	0.890	0.355
SMO	5.0	100.0	93.1	0.502	0.053
IBK	4.4	98.9	92.4	0.519	0.074
JRip	27.6	98.5	93.6	0.640	0.372
PART	28.8	97.5	92.7	0.812	0.327
J48	30.0	97.7	92.7	0.744	0.351

Table 21.7 Performance evaluation metrics for the machine learning algorithms on optimally balanced reduced first year dataset

Learning algorithms	Sensitivity	Specificity	Accuracy	AUC	MCC
Random Forest	96.7	96.5	96.6	0.991	0.931
SMO	92.6	33.6	63.1	0.631	0.324
IBK	8.1	96.6	52.4	0.621	0.101
JRip	94.4	96.0	95.2	0.970	0.904
PART	92.7	94.0	93.7	0.965	0.875
J48	90.5	92.4	91.5	0.928	0.829

Table 21.8 Performance evaluation metrics for the machine learning algorithms on optimally balanced reduced second year dataset

Learning algorithms	Sensitivity	Specificity	Accuracy	AUC	MCC
Random Forest	96.4	96.9	96.7	0.990	0.934
SMO	77.6	46.7	62.2	0.622	0.256
IBK	21.3	92.2	56.8	0.569	0.192
JRip	92.8	95.4	94.1	0.958	0.882
PART	92.5	95.5	94.0	0.971	0.880
J48	91.9	93.4	92.7	0.931	0.854

Table 21.9 Performance evaluation metrics for the machine learning algorithms on optimally balanced reduced third year dataset

Learning algorithms	Sensitivity	Specificity	Accuracy	AUC	MCC
Random Forest	96.1	95.6	95.9	0.989	0.917
SMO	75.4	67.6	71.5	0.715	0.431
IBK	33.0	87.8	60.4	0.608	0.249
JRip	91.2	94.0	92.1	0.949	0.853
PART	91.1	94.1	92.6	0.960	0.853
J48	90.3	91.3	90.8	0.911	0.815

Table 21.10 Performance evaluation metrics for the machine learning algorithms on optimally balanced reduced fourth year dataset

Learning algorithms	Sensitivity	Specificity	Accuracy	AUC	MCC
Random Forest	95.1	94.6	94.9	0.984	0.898
SMO	79.5	60.1	69.8	0.698	0.403
IBK	18.0	93.1	55.6	0.557	0.168
JRip	90.5	93.0	91.8	0.942	0.835
PART	89.7	94.6	92.1	0.965	0.844
J48	89.2	91.3	90.2	0.917	0.805

Table 21.11 Performance evaluation metrics for the machine learning algorithms on optimally balanced reduced fifth year dataset

Learning algorithms	Sensitivity	Specificity	Accuracy	AUC	MCC
Random Forest	96.0	94.0	95.1	0.988	0.901
SMO	79.9	79.0	79.4	0.794	0.589
IBK	23.3	93.9	58.6	0.588	0.242
JRip	92.6	93.0	92.8	0.952	0.856
PART	92.6	92.9	92.7	0.966	0.855
J48	91.3	92.2	91.8	0.929	0.836

21.5 Conclusion

Bankruptcy prediction is an extensively considered topic due to its significance for the banking sector, especially credit risk management, which has been thrust into the spotlight as a result of the recent financial crisis. Recently, machine learning models have been successfully implemented in finance applications, and various researches investigated their use in bankruptcy prediction. However, various factors influence the real performance of machine learning algorithms, such as class imbalance, irrelevant and/or redundant features, and selection of adequate learning algorithm. This chapter considered all the above mentioned aspects. We converted the imbalanced datasets into optimally balanced dataset by suitably adjusting the class distribution using SMOTE. Later, irrelevant and redundant features were removed by fuzzy rough set based feature selection using evolutionary search. Thereafter, performances of different classifiers were evaluated on the basis of values of different performance evaluation parameters. By observing the experimental results, we can easily conclude that Random Forest algorithm produced the best results on the reduced optimally balanced dataset with sensitivity of 96.4%, specificity of 96.9%, accuracy of 96.7%, AUC of 0.990, and MCC of 0.934. Finally, we presented the ranks of different features according to their differentiating capability by using fuzzy rough attribute evaluator based on ranker search. In the future, we intend to improve our proposed methodology by using more accurate versions of fuzzy rough feature selection. Moreover, more

Table 21.12 Ranking of features for the dataset

Rank	Rank value	Actual ID	Rank	Rank value	Actual ID	Rank	Rank value	Actual ID
1	0.0275	Attr55	23	0.0034	Attr24	45	0.0006	Attr49
2	0.0174	Attr29	24	0.0034	Attr38	46	0.0004	Attr62
3	0.0161	Attr63	25	0.0033	Attr18	47	0.0004	Attr30
4	0.0147	Attr50	26	0.0033	Attr14	48	0.0004	Attr44
5	0.012	Attr16	27	0.0033	Attr7	49	0.0003	Attr43
6	0.0119	Attr33	28	0.0033	Attr25	50	0.0002	Attr20
7	0.01	Attr12	29	0.0031	Attr9	51	0.0002	Attr56
8	0.0084	Attr26	30	0.0031	Attr36	52	0.0002	Attr58
9	0.0082	Attr59	31	0.0028	Attr61	53	0	Attr11
10	0.0078	Attr34	32	0.0026	Attr22	54	0	Attr64
11	0.0078	Attr4	33	0.0026	Attr6	55	0	Attr21
12	0.0069	Attr41	34	0.0025	Attr35	56	0	Attr54
13	0.0067	Attr8	35	0.0016	Attr1	57	0	Attr60
14	0.0065	Attr17	36	0.0014	Attr31	58	0	Attr27
15	0.0059	Attr40	37	0.0013	Attr19	59	0	Attr53
16	0.0058	Attr57	38	0.0013	Attr23	60	0	Attr46
17	0.0052	Attr5	39	0.0013	Attr42	61	0	Attr45
18	0.0049	Attr48	40	0.0012	Attr39	62	0	Attr37
19	0.0044	Attr52	41	0.0011	Attr2	63	0	Attr28
20	0.004	Attr15	42	0.001	Attr3	64	0	Attr32
21	0.0037	Attr47	43	0.0009	Attr51			
22	0.0034	Attr10	44	0.0006	Attr13			

interesting experimental results can be recorded with the specific bankruptcy datasets for learning with label proportions (LLP). More accurate versions of SMOTE can be utilized to produce effective experimental results.

21 Improving Financial Bankruptcy Prediction Using Oversampling ... 467

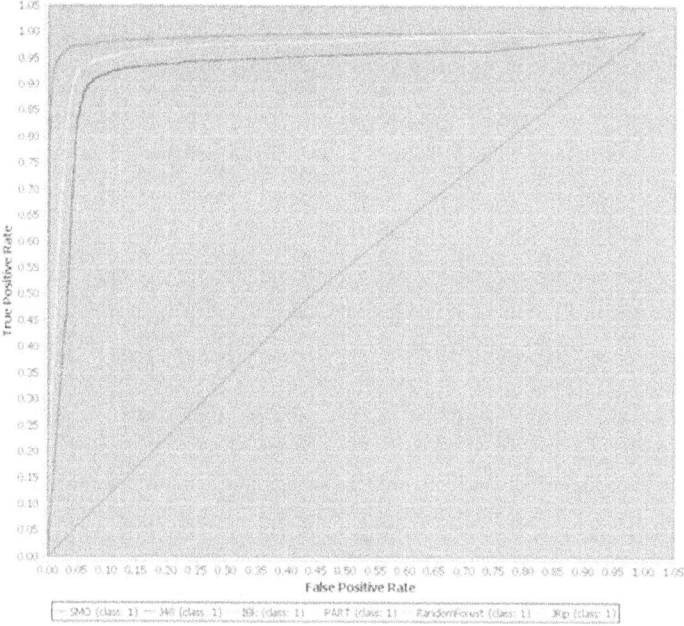

Fig. 21.2 AUC for six machine learning algorithms on optimally balanced first year dataset

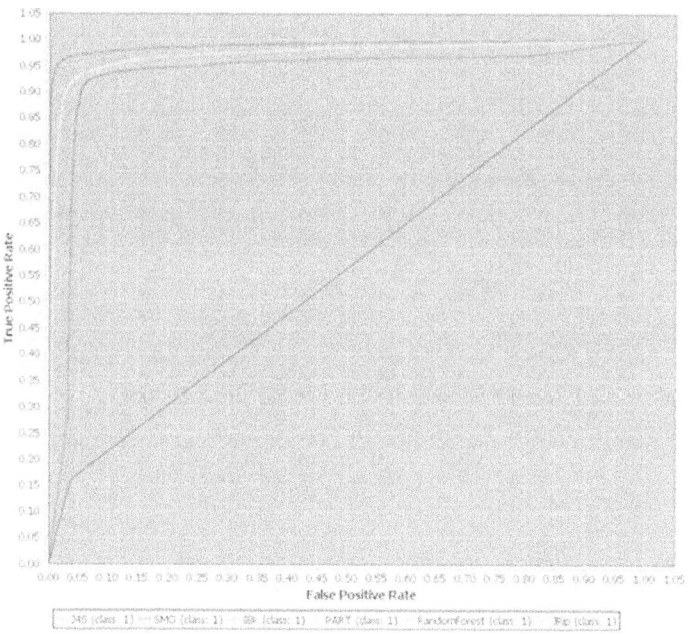

Fig. 21.3 AUC for six machine learning algorithms on optimally balanced second year dataset

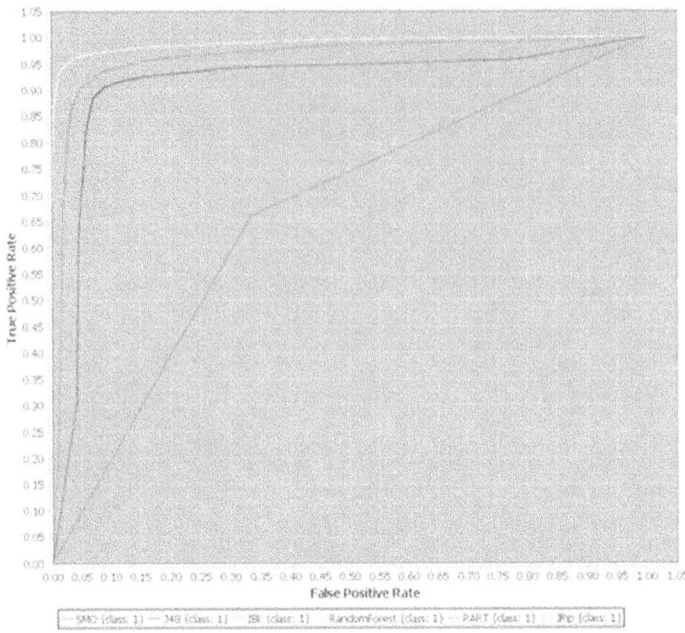

Fig. 21.4 AUC for six machine learning algorithms on optimally balanced third year dataset

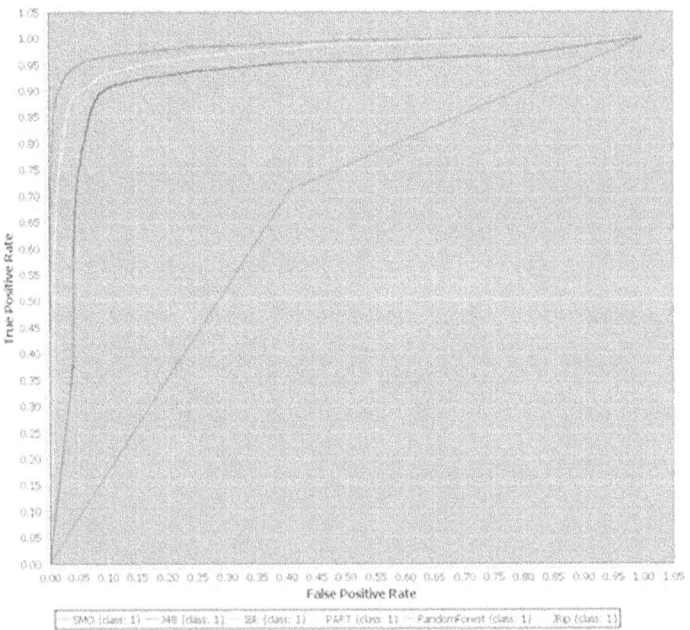

Fig. 21.5 AUC for six machine learning algorithms on optimally balanced fourth year dataset

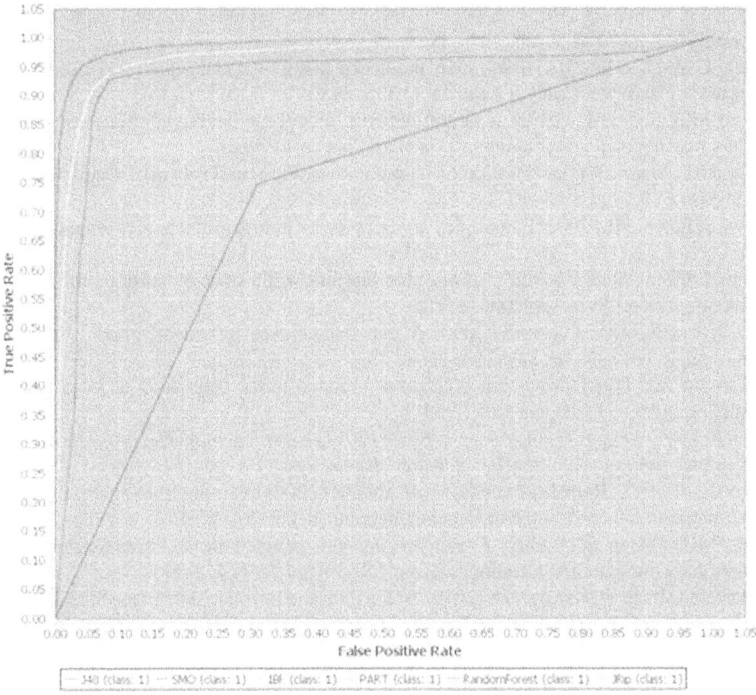

Fig. 21.6 AUC for six machine learning algorithms on optimally balanced fifth year dataset

References

1. Alaka HA, Oyedele LO, Owolabi HA, Kumar V, Ajayi SO, Akinade OO et al (2018) Systematic review of bankruptcy prediction models: towards a framework for tool selection. Expert Syst Appl 94:164–184
2. Altman EI (1968) Financial ratios, discriminant analysis and the prediction of corporate bankruptcy. J Finance 23(4):589–609
3. Zięba M, Tomczak SK, Tomczak JM (2016) Ensemble boosted trees with synthetic features generation in application to bankruptcy prediction. Expert Syst Appl 58:93–101
4. Bellovary JL, Giacomino DE, Akers MD (2007) A review of bankruptcy prediction studies: 1930 to present. J Finan Educ 1–42
5. Altman EI, Iwanicz-Drozdowska M, Laitinen EK, Suvas A (2020) A race for long horizon bankruptcy prediction. Appl Econ 52(37):4092–4111
6. Hu YC (2019) A multivariate grey prediction model with grey relational analysis for bankruptcy prediction problems. Soft Comput 24(6):4259–4268
7. Ohlson JA (1980) Financial ratios and the probabilistic prediction of bankruptcy. J Account Res 18(1):109
8. Shumway T (2001) Forecasting bankruptcy more accurately: a simple hazard model. J Bus 74(1):101–124
9. Jones S (2017) Corporate bankruptcy prediction: a high dimensional analysis. Rev Account Stud 22(3):1366–1422
10. Atiya AF (2001) Bankruptcy prediction for credit risk using neural networks: a survey and new results. IEEE Trans Neural Netw 12(4):929–935

11. Barboza F, Kimura H, Altman E (2017) Machine learning models and bankruptcy prediction. Expert Syst Appl 83:405–417
12. Bateni L, Asghari F (2016) Bankruptcy prediction using logit and genetic algorithm models: a comparative analysis. Comput Econ 55(1):335–348
13. Odom MD, Sharda R (1990) A neural network model for bankruptcy prediction. In: 1990 IJCNN international joint conference on neural networks. IEEE
14. Wilson RL, Sharda R (1994) Bankruptcy prediction using neural networks. Decis Support Syst 11(5):545–557
15. Chava S, Jarrow RA (2004) Bankruptcy prediction with industry effects. Rev Finance 8(4):537–569
16. Chen Z, Chen W, Shi Y (2020) Ensemble learning with label proportions for bankruptcy prediction. Expert Syst Appl 146:113155
17. Cao Y, Liu X, Zhai J, Hua S (2020) A two-stage Bayesian network model for corporate bankruptcy prediction. Int J Finance Econ
18. Zmijewski ME (1984) Methodological issues related to the estimation of financial distress prediction models. J Account Res 22:59
19. Chen Z, Qi Z, Wang B, Cui L, Meng F, Shi Y (2017) Learning with label proportions based on nonparallel support vector machines. Knowl Based Syst 119:126–141
20. Chen BC, Chen L, Ramakrishnan R, Musicant DR (2006) Learning from aggregate views. In: 22nd international conference on data engineering (ICDE'06). IEEE
21. Chen MC, Huang SH (2003) Credit scoring and rejected instances reassigning through evolutionary computation techniques. Expert Syst Appl 24(4):433–441
22. Goldstein I, Jiang W, Karolyi GA (2019) To FinTech and beyond. Rev Financ Stud 32(5):1647–1661
23. Chandra DK, Ravi V, Bose I (2009) Failure prediction of dotcom companies using hybrid intelligent techniques. Expert Syst Appl 36(3):4830–4837
24. Ahmadi F, Amjadian S, Pardegi K (2012) New approach to bankruptcy prediction using genetic algorithm. Int J Comput Appl 44(4):34–38
25. Aoki S, Hosonuma Y (2004) Bankruptcy prediction using decision tree. In: The application of econophysics. Springer, Japan, pp 299–302
26. Tian S, Yu Y, Guo H (2015) Variable selection and corporate bankruptcy forecasts. J Bank Finance 52:89–100
27. Cerchiello P, Nicola G, Ronnqvist S, Sarlin P (2017) Deep learning bank distress from news and numerical financial data. arXiv preprint arXiv:170609627
28. Liang D, Lu CC, Tsai CF, Shih GA (2016) Financial ratios and corporate governance indicators in bankruptcy prediction: a comprehensive study. Eur J Oper Res 252(2):561–572
29. Zoričák M, Gnip P, Drotár P, Gazda V (2020) Bankruptcy prediction for small- and medium-sized companies using severely imbalanced datasets. Econ Model 84:165–176
30. Faris H, Abukhurma R, Almanaseer W, Saadeh M, Mora AM, Castillo PA et al (2019) Improving financial bankruptcy prediction in a highly imbalanced class distribution using oversampling and ensemble learning: a case from the Spanish market. Prog Artif Intell 9(1):31–53
31. Smiti S, Soui M (2020) Bankruptcy prediction using deep learning approach based on borderline SMOTE. Inf Syst Front 22(5):1067–1083
32. Asuncion A, Newman D (2007) UCI machine learning repository
33. Breiman L (2001) Random forests. Mach Learn 45(1):5–32
34. Li H, Pi D, Wang C (2014) The prediction of protein-protein interaction sites based on RBF classifier improved by SMOTE. Math Probl Eng 2014
35. Chawla NV, Bowyer KW, Hall LO, Kegelmeyer WP (2002) SMOTE: synthetic minority over-sampling technique. J Artif Intell Res 16:321–357
36. He H, Garcia EA (2009) Learning from imbalanced data. IEEE Trans Knowl Data Eng 21(9):1263–1284
37. Hall M, Frank E, Holmes G, Pfahringer B, Reutemann P, Witten IH (2009) The WEKA data mining software: an update. ACM SIGKDD Explor Newsl 11(1):10–18
38. Pawlak Z (1982) Rough sets. Int J Comput Inf Sci 11(5):341–356

39. Dubois D, Prade H (1992) Putting rough sets and fuzzy sets together. In: Intelligent decision support. Springer, Berlin, pp 203–232
40. Jensen R, Shen Q (2008) New approaches to fuzzy-rough feature selection. IEEE Trans Fuzzy Syst 17(4):824–838
41. Qian Y, Wang Q, Cheng H, Liang J, Dang C (2015) Fuzzy-rough feature selection accelerator. Fuzzy Sets Syst 258:61–78
42. Chen J, Mi J, Lin Y (2020) A graph approach for fuzzy-rough feature selection. Fuzzy Sets Syst 391:96–116
43. Jensen R, Mac PN (2015) Towards scalable fuzzy–rough feature selection. Inf Sci 323:1–15
44. Jensen R, Shen Q (2007) Tolerance-based and fuzzy-rough feature selection. In: 2007 IEEE international fuzzy systems conference. IEEE, pp 1–6
45. Jain P, Tiwari AK, Som T (2020) A fitting model based intuitionistic fuzzy rough feature selection. Eng Appl Artif Intell 89:103421
46. Jain P, Tiwari AK, Som T (2019) Enhanced prediction of plant virus-encoded RNA silencing suppressors by incorporating reduced set of sequence features using SMOTE followed by fuzzy-rough feature selection technique. In: 2019 10th international conference on computing, communication and networking technologies (ICCCNT), pp 1–7
47. Jain P, Tiwari AK, Som T (2020) Enhanced prediction of anti-tubercular peptides from sequence information using divergence measure-based intuitionistic fuzzy-rough feature selection. Soft Comput 1–22
48. Myerson J, Green L, Warusawitharana M (2001) Area under the curve as a measure of discounting. J Exp Anal Behav 76(2):235–243

Chapter 22
An Integrated Fuzzy MCDM Approach to Supplier Selection—Indian Automotive Industry Case

Vijaya Kumar Manupati, G. Rajya Lakshmi, M. Ramkumar, and M. L. R. Varela

Abstract Multi-criteria decision-making (MCDM) techniques are being adopted in the supplier selection in the automotive industry. Due to their flexibility to interact with the factors needed in determining the supplier selection. The concept of a balanced scorecard is proposed for rating the supplier while evaluating their performance. Due to limitations in each MCDM technique, it is difficult to find a potential supplier using specific methodologies. Therefore, this paper proposes the integration of Fuzzy TOPSIS, Fuzzy AHP, and Fuzzy VIKOR in determining the potential supplier for a bus body building unit in the Indian context. After finding the rankings using each MCDM technique, spearman rank correlation coefficient is used to test the reliability in the ranks generated by each technique. The output results of this study help the company in establishing a robust supplier selection framework to select the best supplier.

Keywords Supplier selection · MCDM · Balanced score card · Fuzzy AHP · Fuzzy TOPSIS · Fuzzy VIKOR

V. K. Manupati (✉)
Department of Mechanical Engineering, National Institute of Technology Warangal, Warangal, India
e-mail: manupati.vijay@nitw.ac.in

G. Rajya Lakshmi
School of Mechanical Engineering, Vellore Institute of Technology, Vellore, India
e-mail: rajyalakshmi@vit.ac.in

M. Ramkumar
Operations and Quantity Methods, Indian Institute of Management Raipur, Raipur, India
e-mail: mramkumar@iimraipur.ac.in

M. L. R. Varela
Department of Production Engineering, University of Minho, Guimaraes, Portugal
e-mail: leonilde@dps.uminho.pt

22.1 Introduction

With the advent of technology, the needs of customers have been changed over the period and business with customer satisfaction as their top priority. To achieve the required amount of customer satisfaction, companies should consider several factors while manufacturing the required product. To manufacture a product with the necessary quality at a given time, a company should have an adequate supply of raw materials required for manufacturing the same. Literature reveals that supplier's efficiency is directly related to their performances to carry out their activities successfully. In the contemporary competing world, supplier selection plays a vital role in growing businesses to get ahead of their competitors [1].

The supplier selection process helps in identifying the most suitable suppliers who can meet the requirements of the buyer and provide products at the right price, quantity, quality, and time [2]. A proper supplier selection framework brings a significant difference in the organization's operational costs [3]. Later et al. [4] stated that many MCDM tools like ANP, DEA, TOPSIS, and AHP have been used vigorously in supplier selection problems. Parthiban et al. [5] studied various MCDM approaches used for supplier selection problems. Therefore, companies should have a proper supplier selection framework to select potential suppliers and build long-term and profitable relationships with the suppliers. A supplier selection framework is designed by integrating the three MCDM techniques such as Fuzzy TOPSIS, Fuzzy AHP, and Fuzzy VIKOR to find the potential supplier by choosing the alternative with better rankings in all the three MCDM techniques. The objective of this paper is to find a potential supplier of spare parts for a bus body-building unit located in south India. To accomplish the objective mentioned above, the deciding criteria for rating alternatives are decided in the initial step. Next, the alternatives are given ratings by the decision-makers with the help of linguistic variables. After that, using the ratings provided by the decision-makers, three MCDM techniques are implemented to find the rankings of alternatives in each technique. Then consistency of the results obtained by the proposed method is validated by using Spearman-Rank Correlation Coefficient. Finally, with the help of rankings obtained from these techniques, a potential alternative with a better ranking in all the three techniques is selected. Using this integrated method one can easily obtain a potential supplier and can also be confident with the rankings generated as the rankings are consistent. This framework can be applied to any industry with any number of criteria and alternatives. The limitations of each MCDM technique can be easily overcome by implementing integrated MCDM methodology.

22.2 Case Study

The company in this study is a bus body building unit located in the southern part of India which was established three years ago to manufacture luxury buses and intercity buses. The main focus of the company is on intercity buses with 15,000 employees currently. The current demand for the company's intercity buses has observed a significant increase due to its location advantage and also with orders from clients in public sectors. To match the increasing demand, it is required to maintain proper inventory to reduce the overhead costs and also the on-time delivery of the order to the clients. The company has been facing several issues with the suppliers of parts due to their inexperience in selecting proper supplier using a robust selection framework. To overcome these issues, the company in this study wants to replace its current set of suppliers with the potential supplier of parts in the market to reduce their current inventory costs and on-time delivery of the end products meeting the requirements of the clients. To achieve these objectives, the company has started looking for a better supplier selection framework to determine the potential supplier critically. Given the current scenario, company ABC wants to implement a robust multi-criteria decision-making framework to choose a potential supplier among the available pool of suppliers in the market.

22.3 Methodology

In this paper, we have used a fuzzy set theory to model the selection of a supplier for the automobile industry. This is solved as an FMCDM problem in which the attributes such as preferences of criteria and alternatives are given by the decision-makers. The attributes are taken in the form of linguistic variables and then converted to triangular fuzzy numbers (TFN) for further computation as articulated in Table 22.1. Triangular fuzzy numbers are used in this problem for two reasons, its computational simplicity [6] and its effectiveness in decision-making problems with imprecise and subjective information.

22.3.1 Proposed BSC Criteria

Table 22.2 shows the proposed BSC which the managers of the company has finally agreed upon which would provide us the required information while selecting a

Table 22.1 Fuzzy Linguistic variables and expression [7]

Linguistic expression	Linguistic expression	Triangular fuzzy numbers
Equally preferred (EP)	Very Low (VL)	(1, 1, 1)
Weakly preferred (WP)	Low (L)	(1, 3, 5)
Fairly strongly preferred (FSP)	Good (G)	(3, 5, 7)
Very strongly preferred (VSP)	High (H)	(5, 7, 9)
Absolutely preferred (AP)	Very High (HX)	(7, 9, 9)

Table 22.2 Proposed balanced score card criteria

Financial	Customer
Product price	Service and delivery
Product quality	Reputation
Distance to manufacturer	Level of Supply chain collaboration
Internal business	*Learning and growth*
Technical capability	Competitiveness
Production capacity	Employee satisfaction
Productivity	Standards consideration

supplier. The proposed BSC is used to assist the company to select its supplier from a pool of suppliers. With the help of this sub-criteria, the decision-makers have given their respective rating for the main criteria for each alternative supplier. Further, using these four criteria's potential suppliers are selected using Fuzzy AHP, Fuzzy TOPSIS, and Fuzzy VIKOR.

22.3.2 BSC-Fuzzy AHP

For the supplier selection, we have 3 decision-makers, 4 criteria, and alternatives. Based on the survey conducted and consultations with the three decision-makers they have given their ratings on the various criteria and alternatives that have been considered.

Step 1: In the Table 22.3 Comparative judgments of the weights of the criteria realized by decision makers are presented.

After the pairwise comparison matrix is obtained then the aggregation of the above fuzzy values is done by the geometric mean calculation. After aggregating the fuzzy number of the three decision-makers into one value than the synthetic values of each criterion. The obtained synthetic values of the criteria are shown in Table 22.4.

Using the above synthetic values, the V values are calculated as shown in the Table 22.5.

The similar calculations are followed the remaining pairwise comparison matrices shown in Tables 22.6, 22.7, 22.8 and 22.9.

On performing similar calculations to the remaining data the normalized weight vectors are obtained and are shown in Table 22.10.

The global performance values are calculated for the data shown in Table 22.10 and further are ranked based on their values. Table 22.11 shows the rankings obtained through Fuzzy AHP process.

Table 22.3 Comparative judgments of the weights

	CR1	CR2	CR3	CR4
DM1				
CR1	(1, 1, 1)	(3, 5, 7)	(5, 7, 9)	(0.11, 0.11, 0.14)
CR2	(0.14, 0.20, 0.33)	(1, 1, 1)	(0.20, 0.33, 1)	(3, 5, 7)
CR3	(0.11, 0.14, 0.20)	(1, 3, 5)	(1, 1, 1)	(1, 3, 5)
CR4	(7, 9, 9)	(0.14, 0.20, 0.33)	(0.20, 0.33, 1)	(1, 1, 1)
DM2				
CR1	(1, 1, 1)	(5, 7, 9)	(5, 7, 9)	(3, 5, 7)
CR2	(0.11, 0.14, 0.20)	(1, 1, 1)	(0.14, 0.20, 0.33)	(5, 7, 9)
CR3	(0.11, 0.14, 0.20)	(3, 5, 7)	(1, 1, 1)	(5, 7, 9)
CR4	(0.20, 0.33, 1)	(0.11, 0.14, 0.20)	(0.11, 0.14, 0.20)	(1, 1, 1)
DM3				
CR1	(1, 1, 1)	(3, 5, 7)	(3, 5, 7)	(5, 7, 9)
CR2	(0.14, 0.20, 0.33)	(1, 1, 1)	(0.11, 0.14, 0.20)	(7, 9, 9)
CR3	(0.14, 0.20, 0.33)	(5, 7, 9)	(1, 1, 1)	(5, 7, 9)
CR4	(0.11, 0.14, 0.20)	(0.11, 0.11, 0.14)	(0.11, 0.14, 0.20)	(1, 1, 1)

Table 22.4 Fuzzy Synthetic Values of each criteria

Sc1 = (7.11, 14.4, 28)	1/61.99, 1/40.43, 1/18.66 =	(0.11, 0.35, 1.50)
Sc2 = (4.22, 8.17, 11.33)		(0.06, 0.20, 0.60)
Sc3 = (3.11, 11.13, 19.33)		(0.05, 0.27, 1.03)
Sc4 = (1.33, 2.06, 11.33)		(0.02, 0.05, 0.60)

Table 22.5 Normalised Weight Vectors of the Criteria Matrix

Weight vector	(1, 0.76, 0.92, 0.62)
Normalized weight vector	(0.30, 0.23, 0.27, 0.18)

22.4 Integration of Fuzzy TOPSIS and Fuzzy AHP

In the order of above predefined steps fuzzy TOPSIS is carried out to find the ranking of suppliers.

Step 1: Creating the decision matrix

To determine the weights of the criteria used and to evaluate the alternatives of the questionnaire, three decision makers, who work and are involved in selecting the suitable supplier of the company are invited and asked to complete the questionnaires using the linguistic terms. Table 22.12 shows the input of decision makers for each alternatives with respect to each criteria.

Table 22.6 Fuzzy numbers of alternative suppliers related to criteria 1

Finance

	ALT1	ALT2	ALT3	ALT4
DM1				
ALT1	(1, 1, 1)	(0.11, 0.14, 0.20)	(0.14, 0.20, 0.33)	(1, 1, 3)
ALT2	(5, 7, 9)	(1, 1, 1)	(1, 3, 5)	(5, 7, 9)
ALT3	(3, 5, 7)	(0.20, 0.33, 1)	(1, 1, 1)	(1, 3, 5)
ALT4	(0.33, 1.00, 1.00)	(0.11, 0.14, 0.20)	(0.20, 0.33, 1.00)	(1, 1, 1)
DM2				
ALT1	(1, 1, 1)	(0.20, 0.33, 1)	(5, 7, 9)	(3, 5, 7)
ALT2	(1, 3, 5)	(1, 1, 1)	(1, 1, 3)	(1, 3, 5)
ALT3	(1, 0.33, 0.2)	(0.33, 1, 1)	(1, 1, 1)	(1, 3, 5)
ALT4	(0.14, 0.20, 0.33)	(0.20, 0.33, 1)	(0.20, 0.33, 1)	(1, 1, 1)
DM3				
ALT1	(1, 1, 1)	(0.14, 0.20, 0.33)	(5, 7, 9)	(3, 5, 7)
ALT2	(3, 5, 7)	(1, 1, 1)	(0.20, 0.33, 1.00)	(3, 5, 7)
ALT3	(0.11, 0.14, 0.20)	(1, 3, 5)	(1, 1, 1)	(0.11, 0.14, 0.20)
ALT4	(0.14, 0.20, 0.33)	(0.14, 0.20, 0.33)	(5, 7, 9)	(1, 1, 1)

Table 22.7 Fuzzy numbers of alternative suppliers related to criteria 2

Customer

	ALT1	ALT2	ALT3	ALT4
DM1				
ALT1	(1, 1, 1)	(7, 9, 9)	(5, 7, 9)	(3, 5, 7)
ALT2	(0.11, 0.11, 0.14)	(1, 1, 1)	(0.20, 0.33, 1.0)	(0.11, 0.11, 0.14)
ALT3	(0.11, 0.14, 0.20)	(1, 3, 5)	(1, 1, 1)	(0.14, 0.20, 0.33)
ALT4	(0.14, 0.20, 0.33)	(7, 9, 9)	(3, 5, 7)	(1, 1, 1)
DM2				
ALT1	(1, 1, 1)	(3, 5, 7)	(0.20, 0.33, 1)	(3, 5, 7)
ALT2	(0.14, 0.20, 0.33)	(1, 1, 1)	(7, 9, 9)	(1, 3, 5)
ALT3	(1, 3, 5)	(0.11, 0.11, 0.14)	(1, 1, 1)	(0.20, 0.33, 1)
ALT4	(0.14, 0.20, 0.33)	(0.20, 0.33, 1)	(1, 3, 5)	(1, 1, 1)
DM3				
ALT1	(1, 1, 1)	(1, 3, 5)	(0.11, 0.14, 0.20)	(3, 5, 7)
ALT2	(0.20, 0.33, 1)	(1, 1, 1)	(0.14, 0.20, 0.33)	(0.20, 0.33, 1)
ALT3	(5, 7, 9)	(3, 5, 7)	(1, 1, 1)	(5, 7, 9)
ALT4	(0.14, 0.20, 0.33)	(1, 3, 5)	(0.11, 0.14, 0.20)	(1, 1, 1)

Table 22.8 Fuzzy numbers of alternative suppliers related to criteria 3

Internal business

	ALT1	ALT2	ALT3	ALT4
DM1				
ALT1	(1, 1, 1)	(7, 9, 9)	(3, 5, 7)	(1, 3, 5)
ALT2	(0.11, 0.11, 0.14)	(1, 1, 1)	(0.20, 0.33, 1)	(0.11, 0.14, 0.20)
ALT3	(0.14, 0.20, 0.33)	(1, 3, 5)	(1, 1, 1)	(0.14, 0.20, 0.33)
ALT4	(0.20, 0.33, 1)	(5, 7, 9)	(3, 5, 7)	(1, 1, 1)
DM2				
ALT1	(1, 1, 1)	(5, 7, 9)	(5, 7, 9)	(3, 5, 7)
ALT2	(0.11, 0.14, 0.20)	(1, 1, 1)	(1, 3, 5)	(0.14, 0.20, 0.33)
ALT3	(0.11, 0.14, 0.20)	(0.20, 0.33, 1)	(1, 1, 1)	(1, 3, 5)
ALT4	(0.14, 0.20, 0.33)	(3, 5, 7)	(0.20, 0.33, 1)	(1, 1, 1)
DM3				
ALT1	(1, 1, 1)	(1, 0.33, 0.2)	(5, 7, 9)	(3, 5, 7)
ALT2	(1, 3, 5)	(1, 1, 1)	(3, 5, 7)	(1, 3, 5)
ALT3	(0.11, 0.14, 0.20)	(0.14, 0.20, 0.33)	(1, 1, 1)	(1, 1, 3)
ALT4	(0.14, 0.20, 0.33)	(0.20, 0.33, 1)	(0.33, 1, 1)	(1, 1, 1)

Table 22.9 Fuzzy numbers of alternative suppliers related to criteria 4

Learning and development

	ALT1	ALT2	ALT3	ALT4
DM1				
ALT1	(1, 1, 1)	(0.33, 1, 1)	(7, 9, 9)	(5, 7, 9)
ALT2	(1, 1, 3)	(1, 1, 1)	(3, 5, 7)	(1, 3, 5)
ALT3	(0.11, 0.11, 0.14)	(0.33, 0.2, 0.14)	(1, 1, 1)	(1, 1, 3)
ALT4	(0.11, 0.14, 0.20)	(0.20, 0.33, 1)	(0.33, 1, 1)	(1, 1, 1)
DM2				
ALT1	(1, 1, 1)	(3, 5, 7)	(1, 3, 5)	(1, 1, 3)
ALT2	(0.14, 0.2, 0.33)	(1, 1, 1)	(0.33, 1, 1)	(0.14, 0.20, 0.33)
ALT3	(0.20, 0.33, 1.00)	(1, 1, 3)	(1, 1, 1)	(0.20, 0.33, 1)
ALT4	(0.33, 1, 1)	(3, 5, 7)	(1, 3, 5)	(1, 1, 1)
DM3				
ALT1	(1, 1, 1)	(0.33, 1, 1)	(0.20, 0.33, 1)	(0.14, 0.20, 0.33)
ALT2	(1, 1, 3)	(1, 1, 1)	(0.14, 0.20, 0.33)	(1, 1, 3)
ALT3	(1, 3, 5)	(3, 5, 7)	(1, 1, 1)	(1, 3, 5)
ALT4	(3, 5, 7)	(0.33, 1, 1)	(0.20, 0.33, 1)	(1, 1, 1)

Table 22.10 Normalized weight vectors of all the pairwise comparison matrices

	CR1	CR2	CR3	CR4
ALT1	0.25	0.28	0.3	0.25
ALT2	0.27	0.22	0.24	0.24
ALT3	0.24	0.25	0.2	0.24
ALT4	0.22	0.24	0.24	0.25
W OF C	0.3	0.23	0.27	0.18

Table 22.11 Global performance values of alternative suppliers and rankings

	Global performance	Rankings
ALT1	0.2654	1
ALT2	0.2396	2
ALT3	0.2267	4
ALT4	0.231	3

Table 22.12 Rating of alternatives by decision maker 1, 2, and 3 (DM1,2,3)

	Decision maker 1				Decision maker 2				Decision maker 3			
	CR1	CR2	CR3	CR4	CR1	CR2	CR3	CR4	CR1	CR2	CR3	CR4
ALT1	G	H	H	HX	G	H	HX	HX	H	G	H	H
ALT2	H	HX	G	VL	H	HX	G	VL	H	X	G	L
ALT3	G	G	L	L	G	G	L	G	G	L	L	G
ALT4	HX	L	HX	G	H	G	G	H	HX	L	HX	G

Step 2: *Aggregate the evaluation of decision makers.*

From the fuzzy decision matrix, we can find the aggregated ratings of each alternative (R_{ij}) with respect to each criteria

Step 3: *Creating the normalized decision matrix*

Using the aggregated table of alternatives normalised decision matrix is obtained.

Inputs from questionnaire.

Step 4: *Creating the weighted normalized decision matrix.*

The weighted normalized decision matrix is obtained by multiplying the normalized matrix to the normalized aggregate weights of the criteria. Weights of criteria are obtained from Fuzzy AHP and are used here in Fuzzy TOPSIS so that the decision maker's perception is not changed with respect to the other MCDM technique and provides a consistency between rankings obtained by both Fuzzy TOPSIS and Fuzzy AHP. The ideal positive solution is obtained by taking the max of normalised weights

Table 22.13 Weighted normalised decision matrix

	CR1	CR2	CR3	CR4
ALT1	0.130	0.120	0.169	0.133
ALT2	0.160	0.164	0.112	0.031
ALT3	0.114	0.082	0.067	0.071
ALT4	0.186	0.069	0.166	0.093
A+	0.186	0.164	0.169	0.133
A-	0.114	0.069	0.067	0.031

of each criteria and for ideal negative solution minimum of normalised weights of each criteria among the alternatives is considered and depicted in Table 22.13.

Step 5: *Calculating the distance of each alternative from A^+ and A^-*

Table 22.14 depicts the distance calculated for each normalised weights of every alternative from A^+ and A^-.

Step 6: *Calculating the closeness coefficient and Ranking*

A closeness coefficient is to decide the ranking order of all alternatives. Ranking is given to the alternative with highest closeness coefficient as shown in Table 22.15.

Table 22.14 Distance of every alternative from A+ and A−

	D+	D−
ALT1	0.071	0.153
ALT2	0.120	0.114
ALT3	0.161	0.042
ALT4	0.103	0.137

Table 22.15 Closeness coefficient and ranks

Alternatives	CCi	Ranks
ALT1	0.683	1
ALT2	0.488	3
ALT3	0.207	4
ALT4	0.571	2

22.4.1 Integration of Fuzzy AHP and Fuzzy VIKOR

In the order of above predefined steps fuzzy VIKOR is carried out to find the ranking of suppliers. Using the same ratings of alternatives used in Fuzzy TOPSIS as given by decision makers, Fuzzy VIKOR is solved to obtain the rankings of supplier.

Step 1: After the ratings of alternatives are obtained from decision makers, they are converted into aggregated triangular fuzzy numbers.
Step 2: The best non-fuzzy performance value is calculated.
Step 3: Conversion of aggregated matrix to normalised matrix is illustrated in Table 22.16.
Step 4: Determine the distance rate of the alternative to the positive ideal solution and negative ideal solution and also Q_i as shown in Table 22.17.
Step 5: The final rankings are obtained by arranging in the ascending order.
Step 6: Finalise the rankings according based on Q are shown in Table 22.18.

Table 22.16 Normalised matrix

	CR1	CR2	CR3	CR4
ALT1	0.232	0.272	0.331	0.372
ALT2	0.276	0.371	0.229	0.127
ALT3	0.197	0.183	0.137	0.205
ALT4	0.293	0.173	0.301	0.294
f_j^*	0.293	0.371	0.331	0.372
f_j^-	0.197	0.173	0.137	0.127

Table 22.17 Distance rate of alternatives

	Si	Ri	S-	S*	R-	R*	Qi
ALT1	0.305	0.191	0.911	0.306	0.3	0.18	0.045
ALT2	0.376	0.18	0.911	0.306	0.3	0.18	0.058
ALT3	0.911	0.3	0.911	0.306	0.3	0.18	1
ALT4	0.329	0.23	0.911	0.306	0.3	0.18	0.228

Table 22.18 Final rankings based on value of Q

Final ranking		
Alternatives	Q	Rank
ALT1	0.0454	1
ALT2	0.0585	2
ALT3	1	4
ALT4	0.2282	3

Table 22.19 Spearman rank correlation coefficient results

	FTOPSIS	FAHP	FVIKOR
TOPSIS	1	0.8	1
AHP		1	0.8
VIKOR			1

22.4.2 Spearman Rank Correlation Coefficient

After finding the rankings of alternatives by using Fuzzy AHP, Fuzzy TOPSIS, Fuzzy VIKOR, to test the reliability in the rankings generated we have used spearman rank correlation coefficient using the Eq. (22.1) (D_a is the difference in rankings and A is no. of alternatives) to test the statistical significance of the difference in ranking obtained between the three MCDM techniques and are tabulated in Table 22.19.

$$R = 1 - \frac{6\sum_{a=1}^{A} D_a^2}{A(A^2 - 1)} \tag{22.1}$$

22.5 Results and Managerial Implications

Thus, by integrating Fuzzy AHP, Fuzzy VIKOR and Fuzzy TOPSIS rankings of the alternatives are obtained. The rankings obtained are found to be consistent by using the Spearman rank correlation coefficient and indicated a perfect association between them. The weights of criteria obtained also indicated that Finance (C1) plays a vital role in deciding the potential supplier as per this case study. In Table 22.20 we can see the final rankings of the alternatives obtained by considering the alternative with better rank in all the three MCDM technique. It is found that alternative 1 is the potential supplier for the company to achieving their objective of efficient supply chain.

Table 22.20 Rankings of alternatives

	FTOPSIS	FAHP	FVIKOR	Final Ranking
ALT1	1	1	1	1
ALT2	2	3	2	2
ALT3	4	4	4	4
ALT4	3	2	3	3

22.6 Conclusion

Newly established Indian companies are facing issues in maintaining long term relations with the supplier due to their inexperience in selecting a potential supplier. But few companies have taken the initiative to look out for a robust supplier selection process to enhance their organization's performance. This case study aims to present a solution methodology for an Indian Bus body building unit that needs to select a potential supplier for a certain part and maintain an efficient supply chain. This case models a real-world MCDM problem in a linguistic environment for supplier selection. Here, the criteria for the MCDM problem is proposed by using the concept of Balanced Scorecard to evaluate the performance of the supplier and later MCDM technique such as Fuzzy AHP, Fuzzy TOPSIS, Fuzzy VIKOR is used to find the rankings of alternatives. Results obtained through these methodologies are consistent which has been validated by using the Spearman Rank Correlation Coefficient indicating their reliability. Thus the best supplier can be selected by the company to achieve their operational goals.

References

1. Pan R, Zhang W, Yang S, Xiao Y (2014) A state entropy model integrated with BSC and ANP for supplier evaluation and selection. Int J Simulation Model 13(3):348–363
2. Yu C, Wong TN (2015) An agent-based negotiation model for supplier selection of multiple products with synergy effect. Expert Syst Appl 42(1):223–237
3. Zeydan M, Çolpan C, Çobanoğlu C (2011) A combined methodology for supplier selection and performance evaluation. Expert Syst Appl 38(3):2741–2751
4. Dweiri F, Kumar S, Khan SA, Jain V (2016) Designing an integrated AHP based decision support system for supplier selection in automotive industry. Expert Syst Appl 62:273–283
5. Parthiban P, Zubar HA, Garge CP (2012) A multi criteria decision making approach for suppliers selection. Procedia Eng 38:2312–2328
6. Kannan G, Pokharel S, Kumar PS (2009) A hybrid approach using ISM and fuzzy TOPSIS for the selection of reverse logistics provider. Resour Conserv Recycl 54(1):28–36
7. Kabir G, Hasin M (2012) Multiple criteria inventory classification using fuzzy analytic hierarchy process. Int J Ind Eng Comput 3(2):123–132

Part IV
Computational Management Applications

Chapter 23
Decision Support System to Assign Price Rebates of Fresh Horticultural Products Based on Quality Decay

Cláudia Matos, Vinicius Maciel, Carlos M. Fernandez, Tânia M. Lima, and Pedro D. Gaspar

Abstract Horticultural products ripeness brings out features like flavor, texture, aroma, skin changes and finally, generates waste due to its spoilage. To avoid or minimize it, many traders as supermarkets, mini-markets and groceries make changes in their fruit's prices just before expiration date. However, customers' acceptability changes during the products shelf life, which leads to selling decrease along products quality decay and, consequently, profit decrease. This behavior establishes a challenging scenario to manage stock replenishment and pricing strategies. Many studies present inventory management model for perishable food products but considering only physical quantity deterioration whereas some few authors discuss dynamic pricing, considering quantity and quality deterioration simultaneously. Aiming the optimization of profit in traders, this work introduces a decision support system to assign price rebates of fresh horticultural products based on quality decay. To achieve this goal, two methodologies were followed. The first one consists in using experimental test results for modeling purposes, based on Pontryagin's maximum principle, using apple, banana and strawberry. The former consists in using questionnaire as sensitivity analysis of quality from customers' perspective, bringing more reliability and criteria for modeling, since quality could be subjective. The result is a computational decision support system to predict the optimum price for a specific fruit during shelf life. The main objective is to extend the applicability of the computational tool in order to overcome challenges related to limitations of logistics, allowing mini-markets and groceries use this software.

Keywords Food waste · Mathematical modelling · Perishable products · Decision support system · Price rebates · Quality decay

---

C. Matos · T. M. Lima · P. D. Gaspar (✉)
Department of Electromechanical Engineering, University of Beira Interior, Convento de Santo António, 6201-001 Covilhã, Portugal
e-mail: dinis@ubi.pt

V. Maciel · C. M. Fernandez · P. D. Gaspar
C-MAST—Centre for Mechanical and Aerospace Science and Technologies, University of Beira Interior Covilhã, Covilhã, Portugal

23.1 Introduction

Nowadays, when it comes to food, it is common sense that is very important to avoid losses in all supply chain from production to consumer. In every point of view that waste is analyzed, it always must be avoided or reduced. According to United Nations, the 2nd objective for sustainable development of mankind is zero hunger [1] and 12th goal is sustainable consumption and production [2]. Is not about food waste itself, but to the complete supply chain as waste of limited resources as soil, water, fertilizers, human labor, oil, CO_2 emissions, electricity, among others. On the other hand, losses reduction brings cost reduction, less emissions, optimized consumption of resources, among others benefits [3].

Fruits, as perishable food product, travel a long journey before reaching the final consumer. The report of Food Waste (FW) [4] shows the waste in all stages of the Food Supply Chain (FSC). The data reveals that total fruit available in 2011 in Europe was 67.9 Mt and total of FW was 28.1 Mt, around 41%. The several stages of FSC are compared. Agricultural production or primary production is responsible for 11.1 Mt, 39.5%, processing and manufacturing for 6.1 Mt, 21.7%, retail and distribution for 0.8 Mt, 2.8%, and consumption in households and food services represents 10.1 Mt, 35.9%.

In agricultural production and processing and manufacturing stages there is a lot of food loss related to the agricultural and industrial processes, and product handling. In households and food services, the food waste is attributed to the lack of attention and awareness of the impact of buying more than needed, or leaving the food products either in the chiller or ambient air too long. In retail and distribution segment, as hyper and supermarkets, minimarkets and groceries, the main causes of food waste are related to extrinsic parameters of food such as air temperature and humidity during the shelf life. In this sector, consumers play a very important role in respect of FW. Their behavior choosing products reflects in whole FSC. The buying decision is made considering quality and price, which are always associated. At a lower or higher level, changes in the first lead to changes in the second, in a form of compensation. However, in the perishable food market, this relationship becomes even more evident and serious regarding to the losses described before. Therefore, minimizing this number through market strategies with appropriate pricing models is imperative [5].

Feng [6] proposed a model for dynamic pricing, quality investment, and replenishment of perishable items through the maximum principle of Pontryagin. The optimum point in dynamic systems as well as the use of data involves considering market size, product sensitivity to the price and quality, as well as the effort and the quality of the investment in the products acquisition. In contrast, Rabbani et al. [5, 7], proposed a model in which the number of promotions, relative to the change in prices, affects the profit maximization, considering the periods of the product purchase cycle and the sensitivity of the product to the variation of prices.

Wu et al. [8] proposed a new profit maximization model based on a dynamic pricing system. Only the inventory quality decay data was considered according to the product's price sensitivity, considering that products deterioration occurred before the end of the purchase cycle for a new batch. These works have in common the use of basic knowledge about the relationship between pricing and the deterioration rate initially presented by [8].

The current work evaluates the behavior of the retailers and distributors (supermarkets, mini-markets and groceries) and consumers considering the binomial quality and price, as well as the time since the harvest (quantitative) and visual aspect (qualitative) as quality variables, translating the concept of "all the characteristics of a product or service in meeting the requirements" according to the ISO 22000: 2018 [9] standard, and the variable sales price (quantitative), adopted as a variable in several branches of activities.

After analyzing the results of the study, a stock and sales management model for each group of perishable products was evaluated, which through a pricing strategy allows to increase the demand for products that are near end of validity, and leading simultaneously to reduce the FW [6]. This model aims to maximize profit by minimizing FW, also considering the distance and time between the distribution location and the end customers as variables.

23.2 Methodology

The study was divided into two phases. In the first phase, primary data acquisition was carried out by applying questionnaires to the target audience. The second phase consisted in performing mathematical modeling aiming profit´s maximization by minimizing waste.

23.2.1 Questionnaire

In the primary data acquisition phase, the questionnaires aimed to determine the best price or discount to be applied in the different stages of fruit´s ripeness: green, optimal maturation and overripe. Three different types of fruit were chosen, with different perishability [10], with objective to achieve the price/discount variations to be applied according to the fruit's features.

Factors such as harvest, perishability grade and sales volume were considered for the selection of fruits, resulting in strawberries as the most perishable product, banana with medium perishability and apple with less perishability, according to the data shown in Table 23.1.

The shelf life, mentioned on Table 23.1, depends of storage conditions, i.e., of the recommended air temperature and relative humidity [11].

Table 23.1 Relationship between shelf life and coefficient of deterioration (θ) [11]

Fruit	Shelf life [11]	Coefficient of deterioration (θ)
Apple	10–12 months	0.8
Strawberry	7–10 weeks	0.5
Banana	1–4 weeks	0.3

To increase confidence, the questionnaire was constituted by 2 inquiries. Both used the same question: "Which fruits would you like to buy?". The first part of the survey, without prices, shown in Fig. 23.1a used only photographs of the various ripening stages. The second part of the survey, shown in Fig. 23.1b, used the same photographs of the fruit, in a different order, with the price assigned to each of them. Note that, the order of ripening of the fruits was placed randomly to ensure that people would not be influenced by the answer to the first question. Thus, it is intended to confirm that the sensitivity to the change in price versus quality exists and it is important to be considered in the buying decision. The questionnaires were completed by people aged between 15 and 70 years old and residing in the central interior region of Portugal.

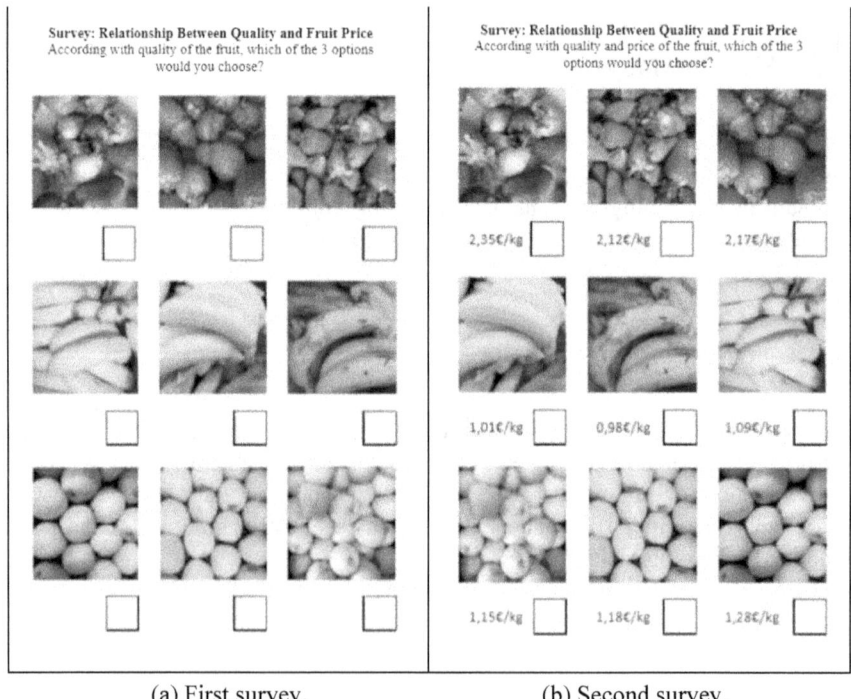

(a) First survey.   (b) Second survey.

Fig. 23.1 Questionnaire carried out. Left side, first inquiry. Right side, second inquiry

Table 23.2 Decision coefficient versus ripeness quality class

Ripeness quality class	Decision coefficient
M	2.00
P	2.66
V	1.33

After collecting and processing questionnaire data, it is possible to identify that the quality of the fruits, green (V), optimal maturation (M) and overripe (P) [11], influences the discount behavior, allowing the association of this behavior with the decision coefficient, shown in Table 23.2. Thus, the qualitative sensation was translated to a quantitative scale.

23.2.2 Mathematical Modeling

The use of mathematical modeling aimed to maximize profit by minimizing FW, assigning sales price of each fruit according to the customer's behavior, using a modeling adapted from [8]. The model is based on variables such as acquisition cost, stock maintenance, life cycle time, order quantity, deterioration coefficients and product sensitivity to price variation. In order to adapt the proposed model, the deterioration coefficient (θ) was related to the validity of each fruit [11], as mentioned in Table 1. Summarized equations are described in Eqs. 23.1 and 23.2 [8]. Thus, the total profit per unit time, denoted by $Z_1(p, T)$ is given by:

$$Z_1(p, T) = \frac{\{\text{Sales Revenue} - (\text{Order cost} + \text{Inventory Cost} + \text{Purchase Cost})\}}{T} \quad (23.1)$$

$$Z_1(p, T) = (p - c)D(p) - \frac{D(p)}{T} \left\{ \frac{ht_d}{\theta} \left[e^{\theta(T - t_d)} - 1 \right] + \frac{ht_d^2}{2} \right.$$
$$\left. + \frac{(h + \theta c)}{\theta^2} \left[e^{\theta(T - t_d)} - \theta(T - t_d) - 1 \right] \right\} - \frac{A}{T} \quad (23.2)$$

The model is based on variables such as acquisition cost, stock maintenance, cycle time, order quantity, deterioration coefficients and product sensitivity in price variation, as described below:

θ: constant of the deterioration rate.
α: demand parameter in relation to the market size.
β: product price sensitivity coefficient.
A: order-per-order cost.
c: purchase cost per unit.
h: storage cost per unit.

Fig. 23.2 Graphical representation of quality versus time decay [5]

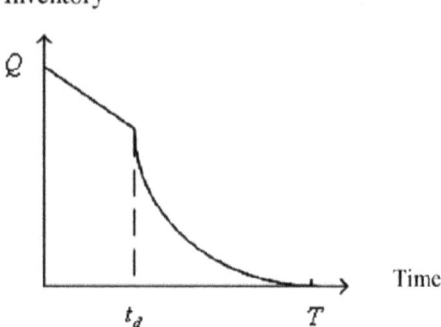

t_d: start of deterioration.
D: demand function ($\alpha \times p^{-\beta}$).
p: selling price per unit (or p_j).
Q: number of units per cycle.
Z: maximum profit.

In order to adapt the model proposed by [8], the deterioration coefficient, θ, was related to the validity of each fruit, as shown in Table 23.1. It is also assumed that the fruit products have a deterioration time less than the cycle time ($t_d < T$), represented by Fig. 23.2 [5].

After inserting data related to the user and the chosen product, initially the ideal cycle time of the fruit is calculated. Then, the suggested sale price and the optimum quantity to be acquired in the cycle is determined.

Finally, using the objective function of maximizing profit with the values suggested by the model, the profit forecast is given at the end of the period.

23.3 Results and Discussions

In view of the results obtained through the questionnaire, for the construction of an adapted model, the sensitivity coefficient in the price variation (β) was set to $\beta = 3$, which is higher than the highest decision coefficient obtained in the research. This procedure was required to apply the model to a range of fruit quality until the end (improper for consumption).

Using the proposed adaptation of the mathematical model [8], it was possible to create an application in Microsoft Excel using VBA mechanics as shown in Fig. 23.2. This worksheets software was selected due to its large usage by supermarkets, mini-markets and groceries around the world. It is possible to work on the above variables to find better scenarios with relation to pricing to obtain the maximum profit.

After software development using data from questionnaire and modeling, the computational decision support system was tested for different fruits in different scenarios. The first case study is related to strawberries due to its high perishability.

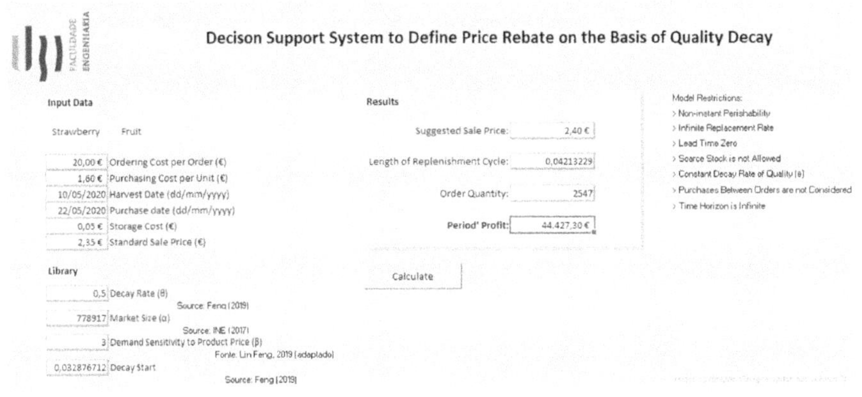

23 Decision Support System to Assign Price Rebates of Fresh ...

Fig. 23.3 Strawberry fruit. Simulation with the marketing price, €2.35 (first scenario)

The second case study is related to banana due to its medium perishability. Only high and medium perishability fruits were compared due to need to assign different marketing prices in a short time in order to sell the products and avoid food waste. The profit optimization scenarios tested for this fruit were related to different marketing prices assigned.

23.3.1 Case Study 1—Strawberry Profit Optimization

The first case study predicts the optimum profit for strawberries. In the first scenario, Fig. 23.3, a marketing price of €2.35 (real value) was used for the strawberries. Based in this input parameters, the computational tool predicts a profit of €44,427.30 in the period.

In the second scenario shown in Fig. 23.4, the marketing price was changed to the value proposed by the model (€2.40). The computational decision support system predicts an increase of profit, €44,496.64, in the same period of time. Thus, according the computational tool, an increase of 2% in the marketing price allowed to increase the profit €69.34.

23.3.2 Case Study 2—Banana Profit Optimization

This case study is applied to the profit optimization of banana, a medium perishable fruit. In the first scenario, shown in Fig. 23.5, the real value of marketing price was used, €1.50. In this case, a total profit of €431,955.61 was determined in the period.

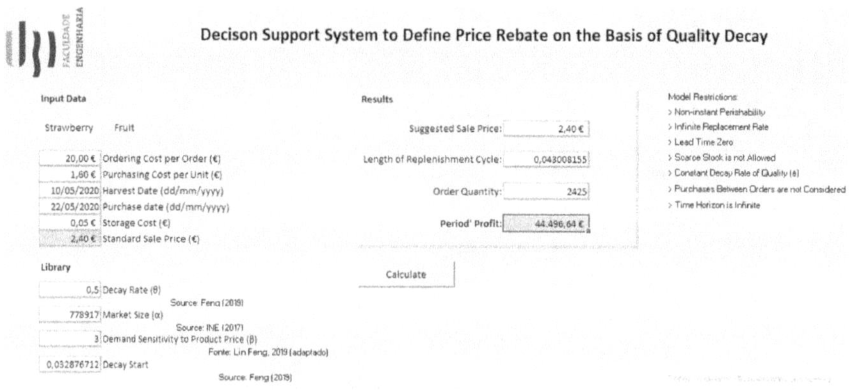

Fig. 23.4 Strawberry fruit. Simulation using the suggested price, €2.40 (second scenario)

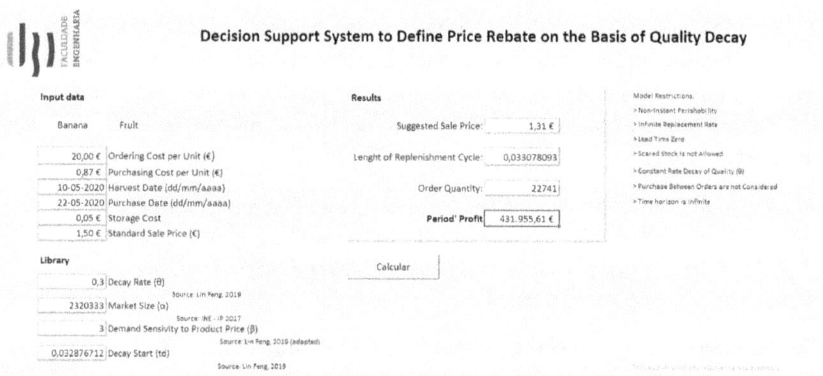

Fig. 23.5 Banana fruit. Simulation with the marketing price, €1.50 (first scenario)

After running the computational decision support system, a marketing price of €1.31 is proposed by the application. The total profit is predicted using the suggested sales price, as shown in Fig. 23.6.

In this case is possible to optimize the profit by reducing the sales price (−12.7%). While in the first scenario the total profit was €431,955.61, in the second one, even using a lower sales price, the total profit is higher, totaled €452,681.06 (+€20,725.45).

According to the results, it is possible to infer about the usefulness of the model. Due to the formulation restrictions, such as: non-instant perishability, infinite replacement rate, lead time being zero, scarce stock allowed, product deterioration at a constant rate, purchases between orders not considered as well as an infinite time horizon; the model will provide better results in large hypermarket chains than in minimarkets and groceries.

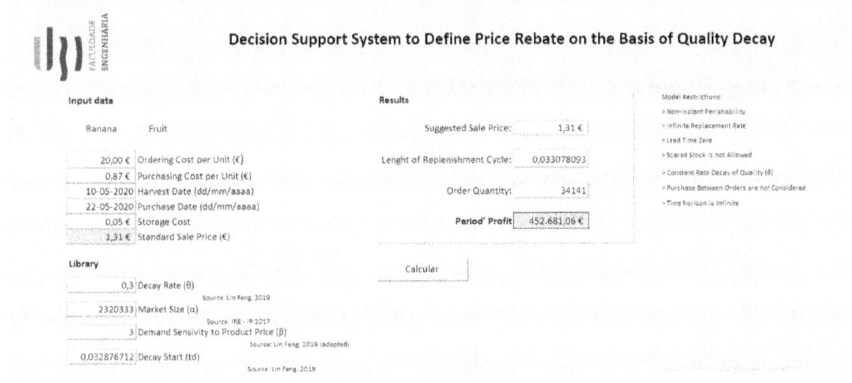

Fig. 23.6 Banana fruit. Simulation using the suggested price, €1.31 (second scenario)

In order to guarantee a larger reduction of waste, the transport must be carried out under controlled conditions and the travel time between the place where the product is supplied to the place where it will be marketed must be reduced to the minimum possible value.

23.4 Conclusions

Perishable food has a very long supply chain. Since agricultural production or primary production, processing and manufacturing, retail and distribution, to consumption in households and food services. Along this journey, in every stage, there is food waste. The quantity may vary in FSC, but at the end, totals around 41% of losses, if fruits are considered. The second objective of sustainable development proposed by the United Nations aims to zero hungry. It is imperative minimizing food waste to achieve it.

Different approaches in recent papers have been used to optimize stock replenishment for perishable products. From retail and distribution perspective, a model based in dynamic pricing for perishable products [8] was adapted for fruits in a specific market. This process demanded experimental tests and a formulation of a questionnaire to provide reliability to the model. Transforming qualitative in quantitative data, according to ISO 22000: 2018 standard requirements, enabled to adapt and change the model according to specific needs in a practical environment. After modeling, it was developed a simple and expedite computational decision support system using Microsoft Excel software, where it is possible to calculate maximum profit, using a very accurate approximation of reality, considering fruits quality decay and, also, quality from de customers' perspective and minimize food waste in this segment.

Must be highlighted the relationship between sales price and profit. For strawberry, increasing the sales price results in a profit increase, which is a logical behavior.

However, using the specific data for banana, increasing the sales price results in a profit reduction. It shows how this model is powerful and how mistakes could be done pricing without a very good knowledge.

However, despites of good results with modeling, which can solve the main problem proposed in this work, some limitations are revealed. For a good performance, the restrictions must be respected. These are non-instant perishability, infinite replacement rate, zero lead time, scarce stock allowed, product deterioration at a constant rate, purchases between orders not considered as well as infinite time horizon. Thus, although it can be applied in minimarkets and groceries, it will provide better results in large hypermarket chains that can achieved the above restriction conditions.

In future work, the model will be adapted and improved to provide better predictions in small size market stores that are more susceptible to the model restrictions. Although, the proposed model can be useful for sellers, as it ensures the best pricing of the fruit aiming also to reduce the food waste.

Acknowledgements This study is within the activities of project "PrunusPós—Optimization of processes for the storage, cold conservation, active and/or intelligent packaging and food quality traceability in post-harvested fruit products", project n.º PDR2020-101-031695, Partnership n.º 87, initiative n.º 175, promoted by PDR 2020 and co-funded by FEADER within Portugal 2020.

This work was supported in part by Fundação para a Ciência e Tecnologia (FCT) and C-MAST-Centre for Mechanical and Aerospace Science and Technologies, under project UIDB/00151/2020.

References

1. UN (2020) Goal 2 | Department of Economic and Social Affairs. In: United Nations. https://sdgs.un.org/goals/goal2. Accessed 21 Nov 2020
2. UN (2020) Goal 12 | Department of Economic and Social Affairs. In: United Nations. https://sdgs.un.org/goals/goal12. Accessed 21 Nov 2020
3. Bellina L (2016) 16. Feeding cities sustainably: the contribution of a 'zerofoodwaste-city' to sustainable development goal 2, 'zero hunger.' pp 113–118. https://doi.org/10.3920/978-90-8686-834-6_16
4. Caldeira C, De Laurentiis V, Corrado S et al (2019) Quantification of food waste per product group along the food supply chain in the European Union: a mass flow analysis. Resour Conserv Recycl 149:479–488. https://doi.org/10.1016/j.resconrec.2019.06.011
5. Rabbani M, Pourmohammad Zia N, Rafiei H (2016) Joint optimal dynamic pricing and replenishment policies for items with simultaneous quality and physical quantity deterioration. Appl Math Comput 287–288:149–160. https://doi.org/10.1016/j.amc.2016.04.016
6. Feng L (2019) Dynamic pricing, quality investment, and replenishment model for perishable items. Int Trans Oper Res 26:1558–1575. https://doi.org/10.1111/itor.12505
7. Rabbani M, Zia NP, Rafiei H (2014) Optimal dynamic pricing and replenishment policies for deteriorating items. Int J Ind Eng Comput 5:621–630. https://doi.org/10.5267/j.ijiec.2014.6.002
8. Wu KS, Ouyang LY, Te YC (2009) Coordinating replenishment and pricing policies for non-instantaneous deteriorating items with price-sensitive demand. Int J Syst Sci 40:1273–1281. https://doi.org/10.1080/00207720903038093

9. ISO (2018) Food safety management systems—requirements for any organization in the food chain. En Iso 220002018(3):32
10. INE (2018) Estatísticas Agrícolas 2017—Annual
11. Maiti R, Mandal D (2019) Post Harvest Management. Res Trends Bioresour Manag Technol, pp 137–166

Chapter 24
Recommendation Engine for Stock Market Trading

S. Sundarakamatchi and M. S. Gajanand

Abstract Investment management is a business problem that is applicable to individuals as well as businesses. Stock market trading can be daunting without the right strategy and tools. Decision making in the stock market is complex as it requires at least some basic understanding of economics, statistics and behavioral science. Due to this, traders face inertia and friction. Automation of the selection of stocks to be considered for trading is an application of computational intelligence, aimed at better discovery of opportunities for buying and selling in a stock market. The automatic selection process to pick stocks from the broad market is something which every trader would desire. In this chapter, a basic stock picking model based on technical analysis strategies is presented. This recommendation engine is proposed with an aim to shift the way one trades by removing some of the biases inherent in the manual process. By employing a few basic technical analysis strategies, the proposed model helps investors discover scrips without performing a detailed study. The resulting baseline list of stocks become choices from which traders can evaluate and determine the actual trades to make. The proposed recommendation engine yields encouraging results, and is practical and easy to implement.

Keywords Stock market · Portfolio optimization · Technical analysis · Stock trading

24.1 Introduction to Trading Systems

Trading systems are of a wide variety. At the simplest level are charting systems that represent historical data of stock prices and volumes (histogram and candle-stick pattern). Over these, indicators such as MACD, Fibonacci Retracement, Bollinger

---

S. Sundarakamatchi
MBA Graduate, Post Graduate Programme in Business Management, Indian Institute of Management Tiruchirappalli, Tiruchirappalli, Tamil Nadu, India

M. S. Gajanand (✉)
Faculty of Operations Management and Decision Sciences, Indian Institute of Management Tiruchirappalli, Tiruchirappalli, Tamil Nadu, India
e-mail: gajanand@iimtrichy.ac.in

Bands, RSI, etc. can be overlaid to bring out signals. With the advent of technology, these are easier to create, more accurate and near instantaneous with the most recent trading data delivered straight from the stock exchanges to our computers. The creation and implementation of trading strategies can be now aided by software. Achelis [1] highlighted that trading systems which can exploit whether a stock is trending or trading are very powerful. This is accomplished by using lagging indicators in a trending market and leading indicators in a trading market (also, called bull/bear or kangaroo market).

Electronic trading platforms are computer systems that execute trades. These connect stock exchanges with traders via a financial intermediary (broker). Typically, these platforms offer trading tools such as charting software and news feeds. In the recent days, these trading platforms have become the first touch point for an investor's foray into stock analysis and trading. Platforms such as ICICI Direct, Zerodha and Share Khan are some of the popular trading platforms in India. These brokers offer many trading features as a service to investors. There are other resources on the internet that offer comprehensive charting solutions along with options to integrate with brokers to execute trades. Examples include tradingview.com, stockcharts.com, etc. They can be used to study charts and indicators as well as to simulate trades.

Algorithmic trading is becoming quite prevalent and is used to run trades that simply meet a certain set of criteria. Many strategies can be implemented after testing them over past data and verifying their efficacy minimising the risk of losing one's capital. Such a process is called back-testing. Most of these platforms also come with an active investor community that exchanges ideas and writes scripts for stock screeners and performing strategy back-testing.

Automatic trading systems fall within algorithmic trading. The main advantage of these automated systems is that they are not prone to delays or biases inherent in human action. We can begin to see how automated trading lends itself so naturally to implement technical strategies. Our proposed system falls in this category.

24.2 Related Literature

24.2.1 Trading Analysis

Brown and Jennings [2] made a case for technical analysis used by rational investors to inform demand from historical prices. Neftci [3] created algorithms for technical analysis to exploit the properties of stock prices left out by linear models of Wiener-Kolmogorov theory. In the foreign exchange market, Taylor and Allen [4] observed a skew towards technical analysis for the shorter time ranges and fundamental analysis for the longer ones and noted complementarity between the two methods. Blume et al. [5] made a strong case for momentum indicators and the role volume plays by studying information quality outside of price action. Lo et al. [6] applied technical pattern recognition over a 31-year sample period on US stocks to counter the lack of theoretical checks in technical analysis. Ausloos and Ivanova [7] outline a mechanistic approach to technical analysis—particularly momentum indicators using price and volume drawing on physical principles. Griffioen [8] made a strong case for technical trading by applying the strategies to different data sets after correcting for transaction costs, risk and data snooping and found encouraging results. Menkhoff and Taylor [9] established the prevalent use of technical strategies among forex traders due to how it informed the pricing on non-fundamental factors. Tripathi [10] surveyed 93 investment analysis in the Indian stock market to establish that combination strategies with both fundamental and technical analysis were prevalent and time horizon of investment had also shortened.

24.2.2 Computational Methods for Stock Market Trading

Leigh et al. [11] attributed the prevalence of decision support systems to the development of high-performance desktop computing, machine learning, soft computing (as with neural networks and genetic algorithms), and combining diverse classification and forecasting systems. Eric et al. [12] determined trends in the stock market applying indicators like the MACD (Moving Average Convergence Divergence) and the RVI (Relative Volatility Index). Nair et al. [13] saw encouraging results with a neuro-fuzzy system that used technical analysis in feature extraction followed by decision tree in feature selection, which are then subjected to dimensionality reduction with the reduced dataset being applied back into the system. Rodríguez-González et al. [14] created a powerful neural network system using Relative Strength Index (RSI). Brown [15] showed that candlestick trading strategies did not show value for DJIA (Dow Jones Industrial Average) stocks. Ko et al. [16] argued for a timing strategy in the Taiwan stock market as opposed to the earlier buy and hold value investing strategy popularized by Fama and French [17].

Stanković et al. [18] showed machine learning techniques capturing the non-linear financial market models using MACD and RSI. de Souza et al. [19] built a trading system that attempted to simulate transactions in a portfolio of stocks from BRICS nations using technical analysis techniques. The returns were promising in Russia and India. They also established that a combination strategy proved more effective with technical analysis helping fundamental analysis to beat the buy and hold strategy. Song [20] predicts stock trends using machine learning methods and technical indicators such as RSI, ADX (Average Directional Movement Index), and SAR (Parabolic Stop and Reverse). Lutey and Rayome [21] showed that current earnings and prices at or near new highs can be combined with the Average Directional Index technical indicator to generate excess return in the Tehran stock market. Excess returns were statistically significant based on the difference in means tests (robust to recession only periods, and additional timeframes) and CAPM, Fama and French 3 and 5 factor models.

24.3 Technical Analysis

Great technical traders have been known to study stock prices and bet that they can profit off them without even knowing the name of the stock, its fundamentals or its industry. They rely on discipline and methods rather than getting swayed by emotions and market action. They set stop-losses, targets and ensure that they follow the market and reap their rewards. Towards the latter half of the 1800s, an American journalist—Charles Dow proposed a new way of analysing the stock market by using the average of a few representative stocks. He reasoned that the price of a stock already takes into account the appraisal (encompassing the knowledge, hopes and fears) of all the traders in a free market. He went on to develop the DJIA and established the Wall Street Journal publication (WSJ)—one of the most respected financial publications in the world.

After his death, from hundreds of Charles Dow's editorials in the WSJ, William Peter Hamilton published these ideas in his book, The Stock Market Barometer. Dow and Hamilton saw the measurement of the stock market as an economic measurement for the entire economy. These ideas became popular as the Dow Theory. Those principles of analysing the stock market was derived to create a different form of trading analysis. Its use as a tool for investing in the market came much later. Today, it forms the basis for the technical analysis method and has retained its value over time.

In the 1920s and 1930s, Richard Schabacker systematized the technical method—by inferring that the signals that were significant in the averages also applied equally in the charts of individual stocks. He also discovered new technical indicators and laid the groundwork for the trading system as we know it today. Seminal work by Robert Edwards and John Magee in 1948 in their book titled Technical Analysis of Stock Trends, established technical analysis as a scientific way to study stocks. Edwards et al. [22] outlined the important principles given below:

1. Averages discount everything (even unpredictable natural calamities with a slight time lag).
2. There are three trends.

    (a) Major or primary trends, that last a year or more—but no set time limits. Principle of Confirmation: Two averages must confirm in order for a primary trend to emerge (Here they meant the railway and overall industrial average created at that time. But it broadly means no single signal should be taken at face value without the confirmation of other market indicators).

    (b) Secondary or intermediate trends that form corrections in the primary trend.
    A loose rule to define a secondary trend is the price movement in opposition to the Primary Trend must last for at least 2–3 weeks and retrace at least one third of the preceding net move in the Primary direction.

(c) Minor or day to day fluctuations (that do not affect the Dow Theory)—usually the weekly swings.

3. Bull market is a type of Primary Trend where each price advance is higher than the previous high.

   (a) Accumulation phase—far-sighted investors start some activity even as volumes are subdued coming only from distressed sellers.
   (b) Steady price advance and increased activity—technical traders are active here.
   (c) Spectacular price advances cause public to flock to the market as volumes are very high and price bubbles start forming.
   (d) So, volume increases when prices rise and dwindles as prices fall.

4. Bear market is a type of Secondary Trend where each price decline is successively lower than the previous high and price advances fail to lift it above the top price of the previous rally.

   (a) Distribution phase—far-sighted investors book profits, as volumes are still high—fading with each rally.
   (b) Panic phase where prices mark a vertical drop and volumes rise.
   (c) Discouraged selling happens next with slower price declines driven by reluctant sellers making distress sales.
   (d) So, volume increases as prices fall and dries up when prices pick up.

In summary, technical analysis in investment valuation analyses historic prices to predict trends in price. The trader is urged to better understand the overarching trends to make trading decisions at the stock level. The movements in stock value and volume already factor in the underlying market sentiments.

1. Stock prices tend to move in trends. Market discounts everything.
2. Volume follows the trends.
3. A trend once set will continue until a reversal occurs.
4. As a corollary, a reversal can occur any time after a trend has been confirmed. So, an investor must always observe the general market trend and be ready to reverse positions if the signals suggest course correction.

24.4 Technical Strategies

Charting in and of itself is extensive. Reversal patterns are essential to the technical trader's methods. Charting and reading chart patterns to interpret future price movement is a popular technical analysis strategy. Reversals, consolidation patterns and gaps form the crux of chart reading. Candle stick patterns were made popular by Japanese commodity traders in rice trading and have led to the development of many types of charts and patterns—notably, Heikin Ashi, Kagi, Renko, etc. Usually this is the starting point for technical traders. The reversal of trends suggested in the chart patterns themselves lead to further scrutiny of the oscillators of the individual stocks to confirm a trading decision. This strategy is particularly popular in intra-day trading—more so with commodity and forex traders.

Indicators are many in number and each come with ardent practitioners who vouch for their efficacy. Applying them correctly involves patience and sometimes intuition. Technical analysis and trading are not so much limited by the lack of indicators but by the lack of correct understanding and application of them. With stocks screened using a multitude of technical indicators, it is possible to study the fundamentals of specific stocks and the macroeconomic conditions to determine the next course of action. This method of combining technical analysis with fundamental analysis is called a combination strategy and benefits traders by profiting from the best of both worlds. It is also a risk-averse strategy as we do not entirely depend on a fixed algorithm to make a trade for us.

Interpretation of chart pattern is subjective and not uniform across time periods. This is a contentious topic across traders. Two technical traders could study the same chart and come up with completely different conclusions. For this reason, as far as our application goes, we choose to avoid reading charts and instead focus on movement in indicators that can be captured more easily by computerised algorithms and requires lesser human scrutiny. Multiple confirmations by use of various indicators, eliminates upfront work by humans and instead helps us build an automatic screener that becomes the backbone of our recommendation engine for choosing the stocks.

24.4.1 Price Action

Price Action, as a trading technique, utilizes actual movements in prices in order to make trading decisions. This is the most basic technical trading strategy. Candlestick patterns are the most popular of price action strategies.

Price action becomes the data source from which all indicators are built. It is not an indicator by itself. We can take the last 12 or 18 months as the baseline on which to base our actions.

24.4.2 Reversal Strategies

Whenever there is a divergence in indicators and price action, the time is ripe for reversals. In a downtrend, if there is high volume and price is trending sideways or down, there is a possibility of course correction and an uptick in prices. Likewise, in an uptrend, with a fall in volume, there is pressure on the demand side signalling a fall in price and it is a signal to time the exit.

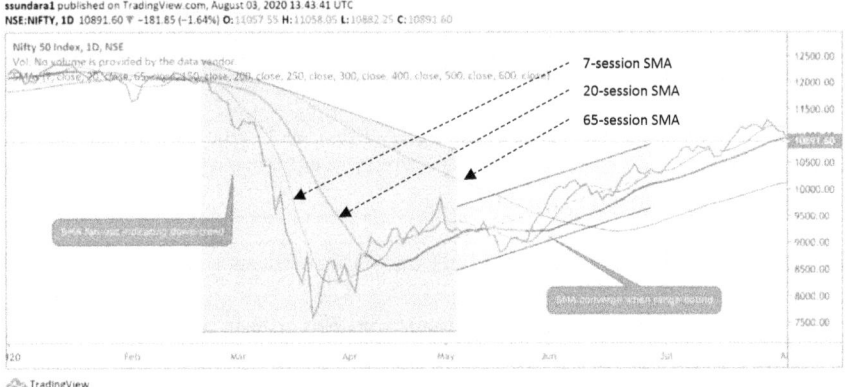

Fig. 24.1 Illustration of a trending and trading market using SMAs on NIFTY index chart

24.4.3 Trending and Trading

Depending on whether a market is trending or trading, one can determine the right strategies to pursue. In a trading market, some indicators are more effective than others. For example, Stochastic Oscillator, Klinger Volume Oscillator are more accurate in a trading market. Similarly, RSI (Relative Strength Index) and MFI (Money Flow Index) are reliable in a trending market. This is important to remember while implementing stock screening and trading strategies.

Many indicators can give clues on what type of a market it is. SMA (Simple Moving Averages) are easy to implement. A short-term, a medium-term and a long-term SMA are plotted over the index. When they fan-out, it is a trending market and when they converge it is a trading market. When the shorter-term SMA flares over the rest, it is an upwards trending market and vice versa. In the example presented in Fig. 24.1, NIFTY shows a serious downtrend and a fall of over 4000 points between February and March 2020. The use of 7-session SMA, 20-session SMA and 65-session SMA clearly illustrates the fan-out and the trending market. Similarly, when the fall ended, the SMAs converged and the market made a gradual recovery even as it remained range-bound, pointing to a trading market. Whenever the SMA lines cut the index line, we have used a different shade to bring attention to convergence or divergence in the trend.

24.4.4 Average Directional Movement Index (ADX)

To determine if a market is trending, Wilder [23] suggested using the ADX. It is a non-directional lagging indicator whose value ranges between 0 and 100. Values over 25 indicate the market is trending. The ADX requires a sequence of calculations based on Directional Movement (DM) and Average True Range (ATR).

Positive Directional Movement,

$$DM^+ = \text{Current High} - \text{Previous High} \qquad (24.1)$$

Negative Directional Movement,

$$DM^- = \text{Previous Low} - \text{Current Low} \qquad (24.2)$$

The greater of the two values obtained from Eqs. (24.1) and (24.2) gives the Current Directional Movement (CDM).

$$CDM = \text{Max}(DM^+, DM^-) \qquad (24.3)$$

Wilder [23] suggested the use of 14-sessions to obtain the Smoothened Directional Movements, which has been in use by traders even today.

Positive Smoothened Directional Movement,

$$SDM^+ = \sum_{14 \text{ Sessions}} DM^+ - \frac{\sum_{14 \text{ Sessions}} DM^+}{14} + CDM \qquad (24.4)$$

Negative Smoothened Directional Movement,

$$SDM^- = \sum_{14 \text{ Sessions}} DM^- - \frac{\sum_{14 \text{ Sessions}} DM^-}{14} + CDM \qquad (24.5)$$

ATR is a market volatility measure that was developed by Wilder. It is the average of the true range (TR).

$$TR = \text{Max}[(H - L), |H - C|, |L - C|] \qquad (24.6)$$

where,

H   is the Session High,
L   is the Session Low, and
C   is the Closing Price.

$$ATR = \frac{(6 \times ATR_{Previous}) + TR}{7} \qquad (24.7)$$

From these, we arrive at the Directional Index DI^+ or DI^-.

$$DI^+ = \frac{SDM^+}{ATR} \times 100 \qquad (24.8)$$

$$DI^- = \frac{SDM^-}{ATR} \times 100 \qquad (24.9)$$

Directional Movement Index,

$$DX = \frac{DI^+ - DI^-}{\sum |DI^+ + DI^-|} \times 100 \qquad (24.10)$$

Smoothing the results of the Directional Movement yields ADX.

$$ADX = \frac{(13 \times ADX_{Previous}) + DX}{14} \qquad (24.11)$$

Crossovers of DI^+ and DI^- offers entry and exit points. DI^- crossing above the DI^+ at ADX > 25 is a sign to exit the trade. In the same example used before, NIFTY's downtrend and range-boundedness are both confirmed by the ADX being over 25 for the former and lower than 25 for the latter Fig. 24.2. The ADX helps to capture what the SMA fan-out highlights. It is also a sell signal as the DI^- line has pulled up over the DI^+ line, indicating that the market is on a downward pull.

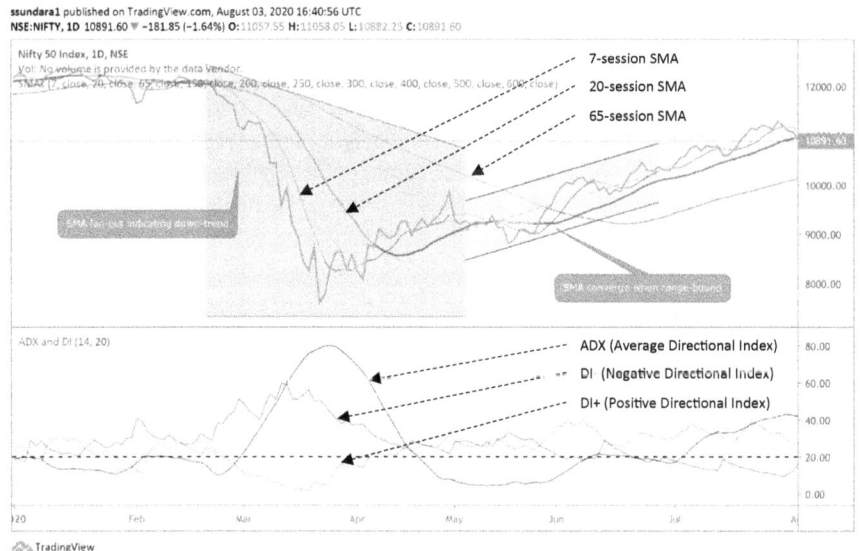

Fig. 24.2 Illustration of ADX, DI^+ and DI^- for the NIFTY index

24.4.5 Support and Resistance

While observing gaps in the charts, a technical analyst can start to notice price levels where reversals in trend will usually occur [22]. If a price level blocks off a downtrend or an uptrend and holds the prices even momentarily, and potentially starts to reverse course, it becomes either a Support level or a Resistance Level. In other words, a Support Level is a price point at which sufficient demand for a stock makes prices to move back up. Similarly, at the Resistance Level over supply of a stock is able to disrupt its upward movement. In very simple terms, at Support Levels, there is a concentration of demand whereas at Resistance Levels, there is a concentration of supply. Volumes will determine the strength of both zones. Breaking of these levels signifies a change in trend and for this reason, they are a useful basis for setting stop-loss orders.

Figure 24.3 shows an illustration of the various Support Levels for Reliance Industries (NSE) over the February–July 2020 time period.

Figure 24.4 shows the various Resistance Levels for Reliance Industries (NSE) over the April–July 2020 time period. In Figs. 24.3 and 24.4, the labels E, S and D indicate the events that affect prices.

E—Earnings (quarterly earnings released by an organization affects the prospects for a stock and so the price may rise or fall as per the expectation).

S—Stock-split (as of record date, typically stock value falls by half if it is a 1:2 split).

D—Dividends (typically price rises up to record date and then falls after dividend distribution happens).

Support and Resistance are important because they can be used in trading decisions to identify reversal in trend. We do this by identifying price levels at reversals from the past. Prior price behaviour are clues for future price behaviour. When price levels

Fig. 24.3 Illustration of support levels over the Reliance industries (NSE) stock chart

Fig. 24.4 Illustration of resistance levels over the Reliance industries (NSE) stock chart

flip between support and resistance, that behaviour determines the range of a market and new reversal levels, also called as bounces or breakouts.

When a stock hovers around its Support Level or its Resistance Level, there is a possibility of reversing course. Such points are ripe for investor action. In an uptrend, if the stock's momentum has run its course and the trend is not sustainable, then the gains will get erased. In such cases, the investors must be ready to book some gains. If the uptrend continues, then the current Resistance Level will be broken and a higher price level will be reached. In many cases, previous Resistance Levels turn out to be new Support Levels—further confirming the optimism in the uptrend. These moves supported by volume action act as a confirmation for the trader. In this case, an investor takes a stronger position in the stock and increases their exposure. Similarly, if a Support Level is broken, then downtrend is confirmed and it would be time to trigger those stop-loss orders and avoid further capital erosion.

If one were to connect all the Support Levels and form a line and then separately connect all the Resistance Levels, we end up with what is called a Price Channel. This could be a straight line or a wave pattern. This forms the foundation for charting patterns. Price Channels show the range of price movement that the stock has experienced as it moved over time. It also visually expresses the price trend. Figure 24.5 presents an illustration of Price Channels of Britannia Industries (NSE). The highlighted areas, bounded by the parallel lines indicate the price channels. We can see that Britannia has been on a gradual uptrend since the beginning of the six-month period charted, despite a steep drop at the start of the period.

24.5 Simple Strategy to Screen Stocks

Typically, technical analysis involves deciphering chart patterns followed by overlaying indicators/oscillators. The goal here is to build a stock screener that does not

Fig. 24.5 Price channels over the Britannia industries (NSE) stock chart

require human intervention. This means chart pattern reading is not a part of our approach. From all the technical analysis concepts outlined above, we highlight one basic strategy to show a possible implementation. The idea is to demonstrate the ease of building such a trading system as a computer application using easily available tools. The focus of our approach is not much on the robustness of the strategy or its success in trading.

Even for a beginner who is new to stock market trading, any personalised trading strategy can be implemented and tested quickly using a computer, depending on the requirements and risk preferences of the decision maker. This can be achieved by determining if a market is trending or trading and then by using 2 or 3 indicators that confirm one another, we can generate a list of stocks that are predicted to see an upward or downward pressure. Stocks facing upward pressure can be buy recommendations and stocks facing downward pressure are sell or short recommendations.

24.5.1 Proposed Approach

We use support levels and resistance levels that are readily available, say from an analyst or other sources. When the stock prices hover around these levels, they are ripe for reversals and all we need to do would be to pick out stocks whose current prices are around those levels. The computer requires a set of clear instructions to follow in order to create such a system. The computer application that we illustrate is a stock recommendation engine that serves as a screening tool to pick out specific scrips from the broad market. These scrips are trading at levels closer to their support level or resistance level. The proposed approach for building a computational application is broadly divided into the three stages. The flowchart of the proposed algorithm is presented in Fig. 24.6.

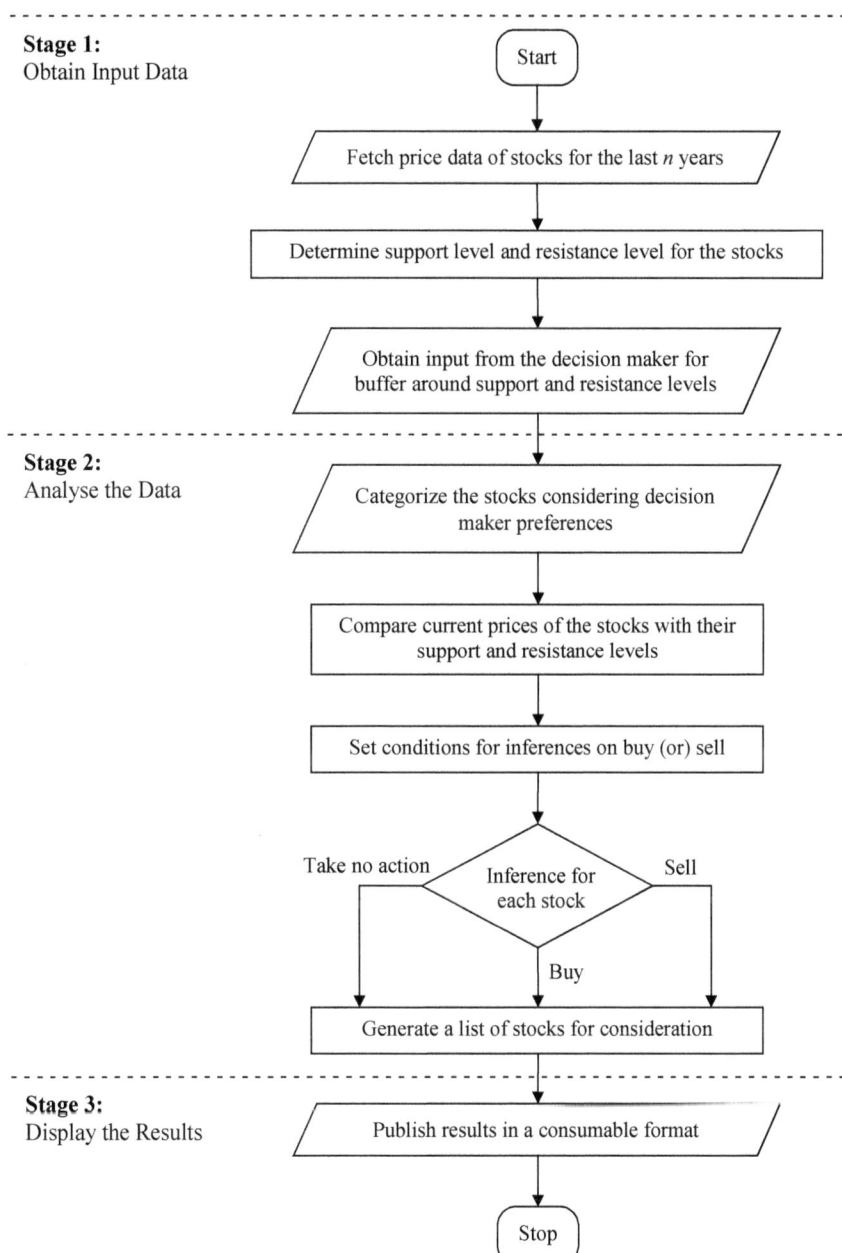

Fig. 24.6 Flowchart of the proposed approach

24.6 Illustration, Analysis and Discussion

24.6.1 Data Collection

Technical analysts believe that past price behaviour has a huge role in future price action. Technical strategies are tested against historic data and this limits the risk in making real trades using a personalised strategy. The free availability of historic data becomes important for this reason. Paid data is quite cost-prohibitive for individuals and may be acquired via secondary sources using a search engine like yahoo finance website or a charting and trader community like the tradingview website. For users of the python programming language, help comes in the form of the many python libraries that can download historic stock prices from yahoo finance quite reliably. The STOCKHISTORY function in Microsoft Excel (released in June 2020) can also pull up stock data easily. It can be used to fetch data directly from the trading exchanges.

The trading data is called "Bhav Copy" by traders. This Bhav Copy can be downloaded from both the NSE (National Stock Exchange of India Ltd.) and the BSE (Bombay Stock Exchange Ltd.) websites. NSE website also provides scrip-wise data download for up to 24 months with both price and volume data. Table 24.1 presents the compiled data for 20 well known stocks. The data of the stock prices were extracted (as on 25 September, 2020) from NSE, and then compiled to determine the support level, resistance level and last traded price.

24.6.2 Demonstration of the Proposed Approach

We developed a simple spreadsheet application to execute the proposed approach. We used Microsoft Excel to gather, compile store and retrieve the required data for the analysis. Using the Developer options in Microsoft Excel, we created a simple user-interface like the one in Fig. 24.7 by using buttons and labels. Error handling and built-in functions are an integral part of computer programming. One should ensure that every function written to execute the approach has options to handle various possibilities in input, output as well as errors,and especially a provision to handle blanks (no value) since the data that we fetch from websites may have missing values.

We present some details about the working of the proposed tool to highlight how a trader can use the tool Fig. 24.7 for screening the stocks. The Refresh button gathers the price data of NIFTY stocks for the last 5 years and computes the support and resistance levels. Then the trader can enter the Buy Range and Sell Range. Based on the relative value of the last traded price (LTP) to the support or resistance level, we can set a buy (or) sell flag on the said scrip. For setting a sell trade, we use a condition that if the last traded price is just below the resistance level, it is likely to reverse trend and fall and so we signal a sell. We input a range for this rather than

24 Recommendation Engine for Stock Market Trading 513

Table 24.1 Sample of the data extracted and compiled for 20 well known stocks

S. No.	Name	Support level	Resistance level	Last traded price (15 min delayed)
1	ACC	1253	1369.15	1303.65
2	Ashok Leyland	64.15	72.25	67.4
3	Axis Bank	384	426.65	402.45
4	Bombay Dyeing	58.3	64.35	60.8
5	BEML	580	631.3	605.6
6	BIOCON	379	431.65	408.25
7	Avenue Supermarts Limited	1995.1	2104.85	2047.35
8	GAIL	80.5	86.35	83
9	HAVELLS	645	688.7	664.2
10	HCL TECH	744.15	811.45	788.3
11	HDFC Bank	1003	1071.35	1030.4
12	HEG	696.25	828.9	741.95
13	HINDALCO	154.4	167.45	159.1
14	INDIAN Hotels	87.75	95.1	90.9
15	INFOSYS	948.3	1009	975.4
16	ITC	161.15	171.75	166.55
17	LUPIN	958	1019	979.6
18	MRF	54,500	57,098.1	55,618.85
19	NTPC	80.3	85.65	82.65
20	SBI	163.35	181.8	176.35

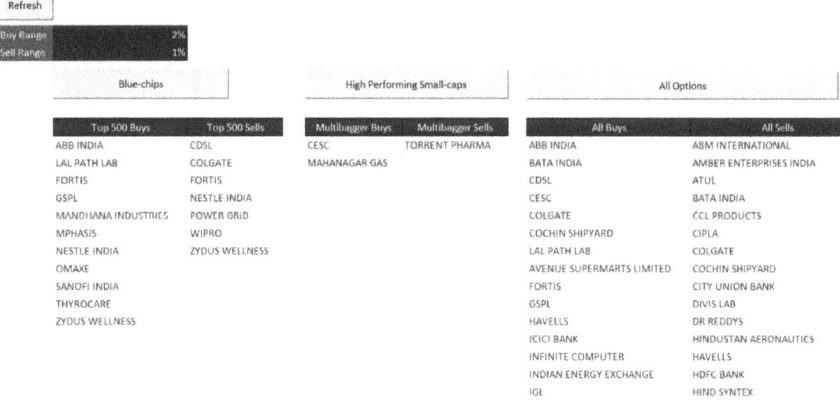

Fig. 24.7 Snapshot of the proposed tool

using the exact resistance level. Similarly, for setting a buy trade, we use a condition that if the last traded price is right above the support level, it may have bottomed out and so we signal a buy. Here 2% buy range means that a last traded price below 2% the resistance level becomes a buy signal. Similarly, the last traded price less than 1% above the support level becomes a sell signal.

Using the proposed algorithm presented in Fig. 24.6, the tool helps to bring up the potential buy and sell options. We have also included additional options like Safe Bets (blue-chip stocks) and Risky Options (high-performing small-cap stocks) to categorize the stocks. The trader may also identify and use a specific shortlist of stocks for categorizing them into more manageable chunks, according to one's risk appetite. The recommendations from the tool will serve as a good starting point for the trader to further optimize his/her investment portfolio.

24.7 Practical Implications

Wealth-creation in the stock market works best for actively-managed, long term investments. Picking the right scrips at the right price point and at the right time is very important to win in the stock market. Passively trading via index funds may cap the returns at the benchmark. Similarly, trading through Exchange-Traded Funds (ETFs) and Mutual Funds limit the earning potential to that of the alpha of the fund manager, not to mention the fees paid eating into the returns made. It is easy for an average investor to miss out on a bull run or stay put too long during a down-turn. Also, risk-averse investors might fail to identify potential opportunities that arise in the market. Too much focus on playing safe in blue-chip scrips limits the upside to their investments. This is because stocks that are normally not in a trader's radar might have improved their prospects. These are simply lost opportunities. Similarly, investors could misread a slump and stay invested too long or even get attached to scrips that are on a downward trend. This causes profit erosion or even erosion of the invested capital.

The movements in stock value and volume already factor in the underlying market sentiments. A possible methodology for using an automatic selection process to pick stocks from the broad market based on technical analysis, similar to the one proposed here, is something which every trader would desire.

24.8 Conclusion and Scope for Future Work

Automation of the selection of stocks to be considered for trading is an application of computational intelligence, aimed at better discovery of trading opportunities in a stock market. It reduces some of the inertia and friction faced by traders, and also avoids costly and unnecessary portfolio rejig. With a system to select stocks that meet a predetermined criterion, one would be left with an objectively curated list. This list could act as a starting point from which one may perform further analysis and actual trade. In this study, a stock picking model based on technical analysis strategies is presented. A recommendation engine, that is practical and easy to implement, is proposed for shortlisting stock choices for traders who may not be well versed with the stock market.

The focus of this study is to demonstrate the computer application of technical analysis trading strategies for stock trading and not to prove the efficacy of the strategy itself. Hence, thorough back-testing has not been made on the proposed strategy for profitability. This can be taken up for further analysis and has a scope for improvement in terms of testing the boundaries of the proposed approach and its effectiveness. We have used support levels for 'buy' signals and resistance levels for 'sell' signals, since it is a safe strategy considering a risk averse decision maker. The reverse could be true in a trending market. So effective strategies according to the market type—trending or trading could be created.

References

1. Achelis SB (2000) Technical analysis from A to Z: covers every trading tool.... from the absolute breadth Index to the zig zag. McGraw-Hill Education, New York
2. Brown DP, Jennings RH (1989) On technical analysis. Rev Financ Stud 2(4):527–551
3. Neftci SN (1991) Naive trading rules in financial markets and wiener-kolmogorov prediction theory: a study of "technical analysis". J Bus 549–571
4. Taylor MP, Allen H (1992) The use of technical analysis in the foreign exchange market. J Int Money Financ 11(3):304–314
5. Blume L, Easley D, O'hara M (1994) Market statistics and technical analysis: the role of volume. J Finance 49(1):153–181
6. Lo AW, Mamaysky H, Wang J (2000) Foundations of technical analysis: computational algorithms, statistical inference, and empirical implementation. J Financ 55(4):1705–1765
7. Ausloos M, Ivanova K (2002) Mechanistic approach to generalized technical analysis of share prices and stock market indices. Eur Phys J B Condens Matter Complex Syst 27(2):177–187
8. Griffioen GA (2003) Technical analysis in financial markets
9. Menkhoff L, Taylor MP (2007) The obstinate passion of foreign exchange professionals: technical analysis. J Econ Lit 45(4):936–972
10. Tripathi V (2008) Investment strategies in Indian stock market: a survey. Available at SSRN 1134668.
11. Leigh W, Purvis R, Ragusa JM (2002) Forecasting the NYSE composite index with technical analysis, pattern recognizer, neural network, and genetic algorithm: a case study in romantic decision support. Decis Support Syst 32(4):361–377
12. Eric D, Andjelic G, Redzepagic S (2009) Application of MACD and RVI indicators as functions of investment strategy optimization on the financial market. Zbornik Radova Ekonomskog Fakulteta U Rijeci: Časopis Za Ekonomsku Teoriju I Praksu 27(1):171–196
13. Nair BB, Dharini NM, Mohandas VP (2010) A stock market trend prediction system using a hybrid decision tree-neuro-fuzzy system. In: 2010 international conference on advances in recent technologies in communication and computing. IEEE, pp 381–385
14. Rodríguez-González A, García-Crespo Á, Colomo-Palacios R, Iglesias FG, Gómez-Berbís JM (2011) CAST: using neural networks to improve trading systems based on technical analysis by means of the RSI financial indicator. Expert Syst Appl 38(9):11489–11500
15. Brown C (2012) Technical analysis for the trading professional, strategies and techniques for today's turbulent global financial markets. McGraw-Hill Education, New York

16. Ko KC, Lin SJ, Su HJ, Chang HH (2014) Value investing and technical analysis in Taiwan stock market. Pac Basin Financ J 26:14–36
17. Fama EF, French KR (1992) The cross-section of expected stock returns. J Financ 47(2):427–465
18. Stanković J, Marković I, Stojanović M (2015) Investment strategy optimization using technical analysis and predictive modeling in emerging markets. Procedia Econ Finance 19:51–62
19. de Souza MJS, Ramos DGF, Pena MG, Sobreiro VA, Kimura H (2018) Examination of the profitability of technical analysis based on moving average strategies in BRICS. Financ Innovation 4(1):3
20. Song Y (2018) Stock trend prediction: based on machine learning methods. Doctoral dissertation, UCLA
21. Lutey M, Rayome D (2020) Portfolio management of high growth firms and technical buy points. J Appl Bus Econ 22(4)
22. Edwards RD, Magee J, Bassetti WHC (2007) Technical analysis of stock trends, 9th edn. CRC Press, Boca Raton; AMACOM, American Management Association, USA
23. Wilder JW (1978) New concepts in technical trading systems. Trend Research

Chapter 25
Cross-Listing Effect and Domestic Stock Returns: Some Empirical Evidence

Naliniprava Tripathy, Amit Tripathy, and Deepak Tandon

Abstract The present paper explored the influence of ADRs listing on the stock returns using event study methodology. The study also employed variance ratio and GARCH model to assess the influence of cross listing of ADRs on the volatility of underlying domestic stocks. A sample of eight companies considered which issued ADRs and listed in stock market of India during the period ranging from 1998 to 2017. The sample firms present significant positive abnormal domestic stock returns during the listing day and insignificant cumulative abnormal returns after the listing day. Results show that cross listing fails to bring any investment benefits to the shareholders. The findings of the study also demonstrating the response of market to the cross listing and volatility of equity returns of shareholders. The overall inferences suggested that issuing of ADRs by Indian firms do not have a substantial influence on the underlying local stock returns and shareholders value. The outcomes of the study are pertinent to the investor community, issuing companies and regulatory decision makers of the country.

Keywords Cross listing · American Depository Receipts (ADRs) · Event Study · Abnormal Returns

JEL Classification F30 · F65 · G15

N. Tripathy (✉)
Indian Institute of Management Shillong, Meghalaya, India
e-mail: nt@iimshillong.ac.in

A. Tripathy
National Institute of Technology, Rourkela, India

D. Tandon
International Management Institute, New Delhi, India
e-mail: deepaktandon@imi.edu

25.1 Introduction

The cross listing of shares engendered extensive debate among academicians and practitioners in the economic literature. A firm is concerned to cross list shares more than one stock exchange with the aim of increasing their stock valuation. Cross-listing eases the internationalizing of investor base and enables them to access the global equity capital to finance their investment opportunities [6]. It also facilitates to enhance visibility, access to more significant investors' base, fair valuation of cross-listed firms and trading volume. It also facilitated the firms to increases the firm's liquidity, reduces the transaction costs and explicit risks allied with macroeconomic shocks. Cross listing facilitated higher savings, development of markets, investments, risk diversification, better allocation of capital and economic progress of the country.

On the other hand, cross-listing bearings many risk, contagion, and distraction of economic activities. An international cross listing is often deliberated as a company's global business strategy [14]. Market segmentation theory proposed that cross listing elude segmentation of market and facilitate company's stock reachable to investors. Nevertheless, it increases the shareholder value, sharing the risk, lowering the cost of capital and upsurge the market valuation. Liquidity theory suggested that cross listing reduces the trading cost of investors and enhance the valuation of companies. Bonding theory proposed that cross listing imposed the company to support the enrichment of corporate governance practices, investor protection and minority shareholders wellbeing. Correspondingly, the signaling theory indicating that cross listing enabled the managers to disseminate the prospect and value information of the firm. Business strategy theory indicated that company's global strategy assists to increase the shareholders' value. However, the market response to cross listing have received considerably less attention in the literature in the overseas market. Charitou et al. [9] deliberated the improvement of company's corporate governance structure in the US stock exchange after listing. Lel and Miller [16] indicating the termination of poorly performing CEOs in the company who fails to safeguard the investor protection. Fresard and Salva [11] specifying that cross listing in the US make it difficult for the insiders to transmute the company's cash holdings into private benefits. On the contrary, Admati and Pfleiderer [2], Amihud and Mandelson [3], Domowitz et al. [10], postulating that cross listing of shares reduced the trading activity in the local market. Empirical evidence of the existing studies provides mixed findings on cross listing of shares. Hence, it motivates to exploring further research to shed some light on increasing of domestic returns after listing of American Depository Receipts (ADRs). However, there is no such sufficient number of studies in this domain found in India. Keeping on view, the present research study intends to observe the influence of international cross-listing events on shareholders' value employing the event study methodology. Further, the study tries to determine the influence of cross listing of ADRs on the volatility of underlying domestic stocks by using variance test ratio and GARCH model. Therefore, the study attempts to add value to the existing research by examining the promising influence of cross-listings on stock return, volatility, and increasing the value of the firm which will useful to the investors, traders, researchers as well as to regulators.

The remaining of the paper planned as follows: section two outlines the past reviews then the methodology and data explained in the third section. The empirical findings and analysis illuminate in the section four and the section five deals with concluding observation and managerial implications.

25.2 Literature Review

Cross listing decision and its market reaction is a significant area of research today. Howe and Kelm [13] observed that U.S. firms listed in the foreign stock exchange are making significantly negative abnormal returns in equally pre and post listing periods. Alexander et al. [1] observed the Canadian and non-Canadian stocks that listed in New York Stock Exchange (NYSE), American Express Company (AMEX) or National Association of Securities Dealers Automated Quotations (NASDAQ) and demonstrating that the return of both the stocks are truncated after cross listing. Chi [8] examined the week announcement force in the Japanese firms cross-listed in the United States and showing the positive announcement effects. Reilly et al. [21] tested the listing impact of US firms on Tokyo stock exchange and indicated the absence of cross listing result on firms' value and return. The US firms cross-listed in London stock exchange and Toronto stock exchange also indicated nonappearance of impact on shareholders wealth [15]. Domowitz et al. (1995) indicated a substantial growth of Mexican firms' value after international listing. Foerster and Karolyi [12] studied 11 countries' 153 foreign companies who listed ADRs in the US stock market and showing that these countries earned abnormal returns during the listing week but return declines after trading in the market. However, the study concluded that cross-listing ADRs are not able to add persistent values to their shareholders. Their evidence supported to market segmentation hypothesis. Miller [18] discussed the impact of announcement effect on share values for 181 foreign firms listed in NYSE or NASDAQ and reported that those who are raising capital through DR offerings earning positive abnormal returns whereas other raising capital from other sources providing negative returns. The study indicated that ADRs is positively influencing the shareholders' value. This finding is consistent with liquidity hypotheses.

Smirnova [24] examined the abnormal return generated in listing date for 16 Russian companies that are listing ADRs. However, all the companies earned negative abnormal return in the listing day. Silva and Chávez [25] examined ADRs issued by the firms in Latin American and found that ADRs do not demonstrate a liquidity advantage in the domestic market. Sarkissian and Schill [23] examined monthly stock returns around foreign listings of 25 host countries. The study showed that long-term gains of the firms are high for those who are cross-listed and geographically closer to their home market. The concluding observation stated that positive abnormal returns are achieving by the companies before cross listing but negative abnormal returns after cross listings.

This is consistent with market timing theory. The study suggested that firms those listed on American stock exchanges do not leverage any benefit on shareholders' value. Roosenboom and Dijk [22] made a comparative analysis of stock price response to the cross listing of eight major stock exchange and document significant abnormal returns found around announcement day of cross listing in all stock markets excluding Europe and Tokyo stock market. The study concluded that the effect of cross listing on shareholders wealth is positive in UK, and US but similar results are not found in Europe and Japan. Yong-Chen Su et al. [27] examined the causal relationship between ADRs and underlying stocks of Taiwan Companies. The study found that there is a unidirectional causality exist between Nasdaq ADRs and underlying domestic stocks of Taiwan and indicated that the quality of information improved after cross-listings of shares. Patharla [19] examined the influence of the cross-border listing of stocks by Wipro Ltd on its stock price in India. The study exhibited the discount of volatility of underlying stocks after ADR issues and signposted that domestic stock returns are not much influenced by cross-border listing.

The findings contradict with market segmentation theory. The study concluded that only volume of transaction increased after cross listing. Patel [20] examined the relationship between ADRs and their respective underlying stocks of Indian Stock Market. The study found a long-term relationship between ADRs and underlying stocks and stated that arrival of new information spreads more rapidly in ADR market and discovery takes place in ADR market. Maaji and Abdullah [17] examined the market reaction to international cross listing in Nigeria and indicating positive response on it. The study documented that market retorts positively to the announcement of cross listing and suggesting the value of firm increase in the long run. Justin and Prosper [7] examined semi-strong form of market efficiency and its response to major national news events of six frontier stock markets using event study methodology. The study indicated that six stock markets are least active in the world but their stock prices are relatively stable and their markets are not adequately delivering all pertinent accessible information, which lead to reduce their levels of trading activity. The study indicated that the six stock markets are not semi-strong form market efficient. Singh and Chakraborty [26] investigated the market efficiency of US, Indian stock markets, ADRs and their underlying stocks. The study shown the presence of inefficiency of Indian ADRs return and indicated that US stock market is more competent than Indian stock market.

25.3 Objectives of the Study

Despite the strong support for the presence of cross-listing premium, empirical evidence to date has been mixed in documenting the considerable benefits of international listings. In summary, the existing literature on market reactions to the cross-listing event remains inconclusive, with conflicting evidence. This motivates for further investigation of the cross-listing effect of ADRs in India. The objective of the present study is to investigate whether any abnormal returns generated by the underlying domestic stocks around the cross listing of ADRs and create value for shareholders.

25.4 Data and Methodology

ADRs are two types such as sponsored and unsponsored ADR. Unsponsored ADRs are issued without participation of the company trading in the over-the-counter exchange. On the other hand, sponsored ADRs are issued with the accord of the company categorized into three level. Level 1 ADRs are traded in over the counter exchange. Level II ADRs are listed on American stock exchange, AMEX or NASDAQ. These ADRs must follow the US General Accepted Accounting Principles or International Financial Reporting Standards. In Level III ADRs, an issuing company can raise capital as well as traded in NYSE, AMEX or NASDAQ. ADRs enables the companies to raise capital in US public markets. Most of the Indian company's ADRs are either Rule 144a program or Regulations depositary receipts. In our study, we have taken only those ADRs that fulfill the criteria of Rule 144a program or Regulations depositary receipts. Therefore, we have taken only level-II, level-III ADRs listed on New York Stock Exchange/NASDAQ, and its underlying stocks listed on Bombay Stock Exchange. These criteria are only complied by eight companies such as TATA Motors Ltd, ICICI Bank Limited, Dr. Reddy's Laboratories Limited, HDFC Bank Limited, Mahanagar Telephone Nigam Ltd, Infosys Limited, WIPRO Limited, Sesa Sterlite Limited (Vedanta Limited). So, our sample restricted to eight companies out of sixteen Indian ADRs.

The depository receipts introduced in India in the early 1990s. However, Reliance Industries and Grasim Industries were first issued GDRs/ ADRs in 1992. In the early stage, Indian companies traded GDRs/ ADRs in London and Luxembourg Stock Exchange. However, many of the companies in the year 1999 onwards listing their ADRs in the main U.S. stock exchanges through public issues. The domestic volatility of stocks after Rule 144A/Reg S issues were histrionically lower during the period 1990–1997 in India. In the past few years, there has been a steep decline in ADRs issues in India. In fact, in the year 2016–2017, not a single Indian company opted for equity offerings. The reason is that conversion of ADRs/GDRs into equity attract capital gains tax, as per the guidelines proposed in 2015, which was not previously prevailed in India. Hence, the period 1998 to 2017 period chosen to avoid the unique issues associated with it.

The year of data chosen based on three criteria such as Availability of effective date of listing, Availability of underlying stock price information for a period of 150 days before and 150 days after the listing date. Thirdly, only the maiden listing is considered, instead of any subsequent listing, as maiden listing is supposed to have maximum impact on domestic share prices in reducing information asymmetries. Due to data availability issue, missing data points and need of maintaining a balanced data, final data sample year restricted to 1998–2017. The sample in this study consists of firms listed on Bombay Stock Exchange (India), which issued ADRs during the period ranging from 1998 to 2017. The criteria for selection of the companies are the availability of historical daily prices of the underlying stocks 150 days before and after the event date. Companies those are not with such history dropped from our sample. Secondly, the firms considered which are operated in the top tier Over-The-Counter marketplace, i.e., Over-The-Counter Stocks (OTCQX). These are firms which cross-list their stocks more than once; our sample does not include the case of any such subsequent dual listing. Hence, we have taken only eight companies, which are meeting this criterion and seven companies out of eight sample listed in the New York Stock Exchange and traded in OTCQX. Two firms with ADRs trading in OTC dropped because of unavailability and reliability issues of data sources. OTCQX is the most prestigious option of trading in three-tiered over-the-counter marketplaces, followed by the other two: OTCQB (OTCQB, called "The Venture Market," is the middle tier) and OTC. Trading in the OTCQX marketplace is more expensive as compared to the latter two; however, it reduces costs of different listing as compared with the exchange listing process. We have chosen OTCQX since the listing demands a high level of financial disclosure requirements, disclosures, and sponsorship from a professional third-party advisor. To be included in the sample, a firm which have listing date, precise listing method and availability of daily local stock price in the Bombay Stock Exchange are considered.

The historical adjusted daily closing prices of the stocks in the sample collected from Bloomberg Database. The effective dates of cross listing are obtained from Citi Bank's ADR Depositary Database. BSE Sensex is taken for study, which is collected from www.bse.com. The reason for choosing Bombay Stock Exchange is that it is the oldest in Asia and fastest stock exchange in the world today. It also occupies the place in the top ten of global exchanges regarding market capitalization of its listed companies. It has a global reach with investors around the world. Therefore, BSE Sensex daily price employed as a proxy of the market index. Further, the reason for choosing Sensex as the market index in our study is two-fold. Firstly, this index is comparatively more accessible and connected with the sentiments of the investors, which are going to reflect the impact of any event reliably. Secondly, many of the Indian firms, which have issued ADRs in the past, have been included in this index from time to time.

The sample of the study consists of eight Indian firms. The Daily returns of the stocks calculated using the adjusted closing price: $R_t = \ln\left(\frac{P_t}{P_{t-1}}\right) * 100$ Where, P_t indicates current price and P_{t-1} denotes the previous day's price.

The listing date set for the event date (t = 0). The market coefficients assessed in the pre-listing period day -150 to day -21 and after listing period day + 21 to day + 150. The abnormal returns from Day −20 to Day + 20 is calculated employing coefficients from the pre-listing model. The study has adopted the event window of 41 trading days.

Figure 25.1 indicates the time line for the cross-listing.

Generally, event study methodology is used extensively in finance to empirically observe the market response to the international cross listings of stocks and its influence on firm value around event period. The assumption of the methodology is that

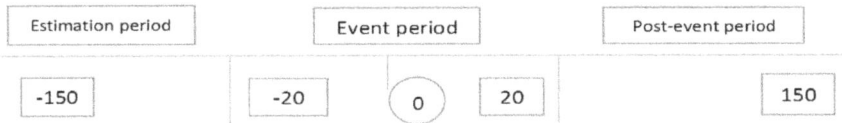

Fig. 25.1 Time-line for the cross-listing

stock market is efficient and the influence of an event will be replicated in the stock price. The underlying assumption is that if an event contains any information, it will cause the firms' future share price and this may lead to modify the current price. The event study measures the individual stock level affects. Hence, event study methodology is applied in economy wide events.. Event study measures the effects of the particular event on a specific stock level rather than the market level. The study has followed the event study methodology used by Miller [18]. The model describes that there is linear relationship exist between security return and market return.

The market model is expressed as follows:

$$R_{i,t} = \alpha_i + \beta_i R_{M,t} + e_{i,t} \tag{25.1}$$

R_{it} denotes return of the stock i on the day t; R_{mt} represents market return at time t; αi indicates intercept βi symbolizes coefficient and εit refers error term.

The abnormal returns calculated by the residuals of the Ordinary least square (OLS) market model as follows:

$$\varepsilon_{it} = R_{it} - [\alpha_i + \beta_i * R_{mt}] \tag{25.2}$$

The difference between the realized return and the expected return is the abnormal return and event date of firms represented as follows:

$$AR_{i,t} = R_{i,t} - E(R_{i,t}|\Omega_{i,t}) \tag{25.3}$$

AAR$_t$ denotes Average Abnormal Return for each period as follows:

$$AAR_t = \sum_{j=1}^{N} \frac{AR_{jt}}{N} \tag{25.4}$$

N indicates Number of securities where trading days relative to event day denotes t and ARR$_t$ represents the average of the abnormal returns of crossing events.

CAARs is calculated to determine the cumulative effect of the event over a specified number of days

$$CAAR_{t1,t2} = \frac{1}{N} \sum_{i=1}^{N} CAR_{i(t1,t2)} \tag{25.5}$$

25.5 Test Statistics

25.5.1 Cross-Sectional T-test

The cross-sectional t-test is robust to an event enlightened by [4]. Hence, it persuaded to upsurge the variance. The cross-sectional t-test is as:

$$T_{cross} = \frac{CAAR_{(t1,t2)}}{\hat{\sigma} CAAR_{(t1,t2)}} \quad (25.6)$$

The null hypothesis is that Cumulative Average Abnormal Return is equal to zero. The estimation of variance is based on the cross-section of Abnormal Returns.

$$\widehat{\sigma^2} CAAR(t1,t2) = \frac{1}{N(N-d)} \sum_{i=1}^{N} [CAR_i(t1,t2) - CAAR_i(t1,t2)^2] \quad (25.7)$$

25.5.2 Generalized Sign Test

The null hypothesis of generalized sign test is that the abnormal returns in the event window and the estimation window are not significantly different. The assessment window of 150 days calculated as:

$$\hat{p} = \frac{1}{n} \sum_{j=1}^{n} \frac{1}{150} \sum_{t=1}^{150} S_{jt} \quad (25.8)$$

where,

$$S_{jt} = \begin{cases} 1 \text{ if } AR_{jt} > 0 \text{ } if AR_{jt} > 0 \\ 0 \quad\quad\quad\quad\quad \text{otherwise} \end{cases}$$

$\hat{p}$ defined as:

$$Z_G = \frac{(w - n\hat{p})}{\sqrt{n\hat{p}(1-\hat{p})}} \quad (25.9)$$

25.5.3 Variance Ratio Test

The stock returns volatility before and after ADR is estimated using variance ratio test. The ratio expressed as follows

$$\text{Variance Ratio} = \frac{\text{Variance before cross-listing}}{\text{Variance after cross-listing}} \qquad (25.10)$$

If the ratio is higher than 1 and statistically significant, stock returns' variability improved after ADRs listing and vice versa.

25.6 The GARCH (1, 1) Model

The Generalized ARCH (GARCH) model developed by Bollerslev [5] stating that forecasts of time varying variance depend on the lagged variance of securities. The effect of volatility spillover is gauged by employing GARCH (1, 1) model. The GARCH (1, 1) model consists of conditional mean and variance equation. This model stated that conditional variance is depended on its past conditional variance and past innovations.

The GARCH model is expressed as follows:

$$R_t = R_{t-1} * \beta + \varepsilon_t \qquad (25.11)$$

R_t denotes daily return, R_{t-1} indicates lagged return, ε_t represents residual error term for the day t. β is the fixed parameter.

The conditional variance is represented as follow:

$$h_t = \omega + \sum_{i=1}^{q} \alpha i \varepsilon_{t-1}^2 + \sum_{j=1}^{p} \beta j h_{t-j} \qquad (25.12)$$

ω signifies mean, ε_{t-1}^2 is the ARCH term, h_{t-j} refers to last periods forecast variance (GARCH term), α and β indicates ARCH and GARCH coefficients. If α is high and β is low, then volatilities tend to be 'spiky'. If ($\alpha + \beta$) is close to unity indicating the volatility shock is persistent which in turn raises the stock price volatility.

25.7 Empirical Analysis

The descriptive statistics of all variables are shown in Table 25.1.

It can be seen from the Table 25.1 that the mean return of HDFC Bank and WIPRO are negative. The mean return of Infosys Limited is highest followed by

Table 25.1 Descriptive statistics

variable	Mean	Median	Std. Dev	Skewness	Kurtosis	Jarque–Bera
Sensex	0.000802	0.001063	0.016163	−0.136361	8.865292	6076.400 (0.000000)
TATA Motors	0.000801	0.000643	0.024016	−0.068015	5.091982	91.56028 (0.000000)
ICICI	0.003807	0.000403	0.044125	−0.112505	3.726180	12.04098 (0.000000)
DR Reddy's	0.000808	0.000112	0.027114	−0.021843	6.132038	204.4078 (0.000000)
HDFC Bank	−0.000277	0.000001	0.021290	−0.102781	8.153694	554.2254 (0.000000)
Mahanagar Telephone Nigam Ltd	5.30E-05	−0.001061	0.033325	−0.206003	7.241909	378.4070 (0.000000)
Infosys Limited	0.006398	0.003379	0.036294	−0.049134	3.422886	13.926864 (0.000000)
WIPRO	−0.000174	−3.70E-05	0.056231	−0.098494	3.412059	14.345767 (0.000000)
Sesa Sterlite Limited (Present name Vedanta Limited)	0.001108	−0.000816	0.025043	0.149498	8.061317	643.7983 (0.000000)

ICICI, Sesa Sterlite Limited, DR Reddy's. TATA Motors.and Mahanagar Telephone Nigam Ltd. All these companies' mean return is higher than the market return except HDFC Bank and WIPRO. The standard deviation of WIPRO is higher than all the Sample Company and stock market indicating that WIPRO is taking higher risk in the market but generating negative returns. All variables have negative skewness and excess kurtosis. High JB statistics indicating that the distribution is skewed to the right and showing leptokurtic.

The Jarque–Bera statistic is a measure of normality. Based on the result, the JB statistic indicates that all the variables exist non-normality and indicating the presence of Heteroscedasticity.

Table 25.2 presents the correlations among variables and indicating that the correlation coefficients value of all variables is smaller than 0.6. It observes from the Table 25.2 that correlation values in-between independent variables and dependent variables are small. It observed from table-2 that DR Reddy's, HDFC Bank and Infosys Limited are negatively correlated with stock market.

The multicollinearity problem of variables is determined using Variance Inflation Factor to know the degree of explaining factor of each independent variable by the other independent variables. VIF test results signposts that there is no multicollinearity problem found among variables since all variables are less than 10.

Table 25.2 Correlation coefficient of variables

	Sensex	TATA Motors	ICICI	DR Reddy's	HDFC Bank	Mahanagar Telephone Nigam Ltd	Infosys Limited	WIPRO	Sesa Sterlite Limited (Present name Vedanta Limited)	VIF
Sensex	1									
TATA Motors	**0.020**	1								1.019
ICICI	0.078	−0.015	1							1.008
DR Reddy's	−0.060	−0.040	−0.002	1						1.013
HDFC Bank	−0.001	−0.050	−0.019	0.049	1					1.012
Mahanagar Telephone Nigam Ltd	0.054	0.009	0.029	0.015	−0.003	1				1.008
Infosys Limited	−0.007	0.051	0.030	−0.027	−0.005	−0.068	1			1.010
WIPRO	0.115	0.096	0.065	−0.038	−0.081	0.016	−0.010	1		1.021
Sesa Sterlite Limited (Present name Vedanta Limited)	0.037	−0.050	−0.038	−0.081	0.013	−0.043	−0.036	0.008	1	1.015

The results of CAARs and AARs of the underlying stocks of all the eight ADRs presented in Table 25.3. Figure 25.2 shows the plot of CAARs in the event window. Table 25.3 shows CAARs for different time intervals along with the results of Crude Dependence Adjustment Test and the Generalized Sign Test (Fig. 25.3).

The output is presented in Table 25.3, which shows the AARs and CAARs for each day in the event window, around the listing date with their statistical significance levels. The daily average abnormal returns, and the cumulative abnormal returns for Day -20 through Day $+ 20$ around the listing date are documented in Table 25.3 demonstrating the significant positive cumulative average abnormal return over the interval $[-1, + 1]$ days around the first trading day. It is observed that out of 20 days 19 days shows positive abnormal returns before the event date. The analysis also demonstrates that AARs of eight stocks in each day displays positive mean abnormal returns also statistically significant at 1%, 5%, 10% level. The output exhibits that there is a significant increase of AARs in the day before the listing event i.e. 5.6 percent and decreases to 3.1 percent subsequently in the event date. However, in the listing day, Day 0, the domestic market exhibits positive statistically significant average abnormal return. The cumulative abnormal return between day $+ 2$ (T- 0.066) and day $+ 20$ is .069 (T-1.017). After the cross-listing, the stock value decreases gradually and its effects dissipate after the event day. From Table 25.3 and Fig. 25.2, it is noticed that before-issue period; the cumulative average abnormal return shows positive return however, after cross listing it drips slowly and obstinately and ends up with 6.9% at non-significance level. Eight out of 20 days' abnormal returns are negative after the international cross listing and statistically insignificant except one. In between day -7 and day -1, the stocks are earning a significant positive abnormal return. (T-statistic 6.641). In the post-listing period, the AARs again show uptrend immediately after the event date. Even though there exists a fluctuation in the AAR trend in the post-listing days, it shows an overall increase in the event window on 19 days. (T-statistic 7.118). In the event window of 41 days, 20 days AARs are significant. Throughout the event window, the CAAR is found positive. During the period between -20 to -1, the CAAR increases abruptly by 8.5 percent. (T-statistic of 1.793 and G Sign 1.693). On a listing day, there is an increase of 2% with significant T-statistic of 1.887. The cumulative abnormal returns are almost positive but statistically significant at 1% level in two cases only. Figure 25.2 shows that there is a frequent and quick reversal of directions in the movements of CAARs.

It is observed that the average cumulative abnormal returns exhibit positive returns but statistically insignificant indicating that cross listing fails to bring any investment benefits to the domestic country. The study indicates that the listing premium behaves very differently in the next day adjacent the events. The present is also in line with previous studies. The outcome of the study indicates that most of the firms show positive abnormal returns in the event date but firms fails to generate positive abnormal returns after cross listing. Therefore, the null hypothesis is rejected postulating that that cross listing does not increases the shareholders' value. The study exhibited that the market retorts to the cross listing which significantly encouraging to the volume

Table 25.3 AARs and CAARS in the event window and their significance (*, ** and *** Denote significance at 10%, 5% and 1%, respectively)

Day	AAR	T-statistic	CAAR	T-statistic
−20	0.009	0.871	0.009	0.871
−19	0.013	1.272	0.023	1.515
−18	0.01	0.959	0.024	1.288
−17	0.029	2.692**	0.039	1.825*
−16	0.007	0.674	0.036	1.505
−15	0.013	1.217	0.02	0.772
−14	0.011	1.055	0.024	0.859
−13	0.024	2.281**	0.035	1.179
−12	0.002	0.221	0.027	0.834
−11	0.01	0.977	0.013	0.379
−10	0.015	1.458	0.026	0.734
−9	0.012	1.137	0.028	0.749
−8	−0.01	−0.957	0.002	0.05
−7	0.043	4.098***	0.033	0.839
−6	0.007	0.646	0.05	1.225
−5	0.046	4.328***	0.053	1.243
−4	0.02	1.900*	0.066	1.511
−3	0.07	6.641***	0.091	2.013*
−2	0.029	2.721**	0.099	2.148*
−1	0.056	5.294***	0.085	1.792
0	0.031	2.910***	0.087	1.79
1	0.053	4.966***	0.083	1.679
2	0.013	1.262	0.066	1.299
3	0.042	4.005***	0.056	1.075
4	−0.008	−0.747	0.035	0.652
5	0.049	4.592***	0.041	0.754
6	−0.01	−0.927	0.039	0.705
7	0.036	3.351***	0.026	0.458
8	−0.02	−1.898*	0.015	0.27
9	0.043	4.059***	0.023	0.395
10	−0.018	−1.674	0.025	0.428
11	0.056	5.314***	0.039	0.643
12	−0.007	−0.613	0.05	0.818
13	0.066	6.266***	0.06	0.969
14	−0.014	−1.284	0.053	0.842

(continued)

Table 25.3 (continued)

Day	AAR	T-statistic	CAAR	T-statistic
15	0.063	5.982***	0.05	0.783
16	−0.004	−0.424	0.059	0.914
17	0.068	6.428***	0.064	0.974
18	0	0.029	0.068	1.034
19	0.075	7.118***	0.076	1.13
20	−0.006	−0.603	0.069	1.017

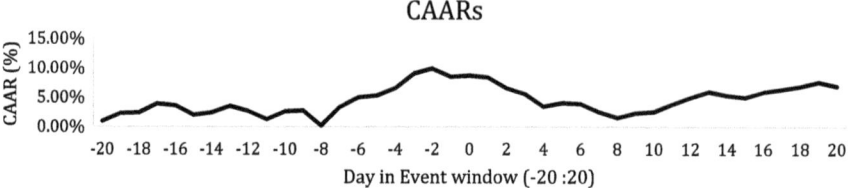

Fig. 25.2 CAARs around Listing Date

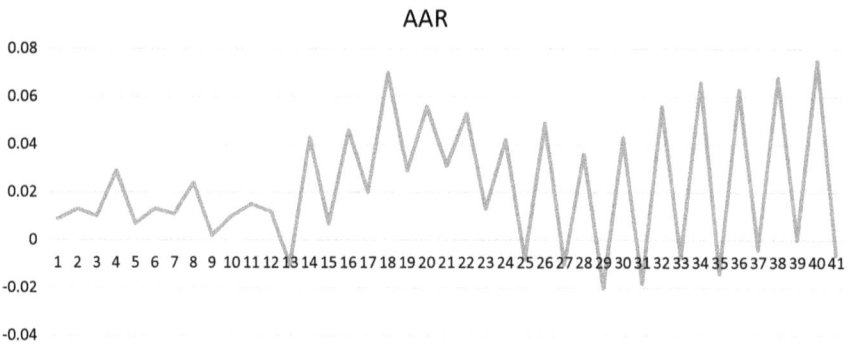

Fig. 25.3 AARs around Listing Date

of transactions but fails to influence the domestic stock returns. This phenomenon appears to be not consistent with the market segmentation hypothesis.

Results of variance analysis for each underlying stock presented in Table 25.4 and Fig. 25.4. It is observed that five out of eight sample stocks show variance ratio higher than one. All these five companies (Tata Motors, ICICI, Mahanagar Telephone Nigam Ltd, HDFC Bank, and Wipro) are statistically significant based on F-test. Therefore, it specifies that after the ADR listing, most of the companies are experiencing substantial volatility of returns. Thus, the investors are exposed to a higher amount of risk but unable to reap reasonable amount of return. The low degree of information transparency is seen between the local and the US markets in the study (Table 25.5).

25 Cross-Listing Effect and Domestic Stock Returns …

Table 25.4 Variance ratio for sample stocks

Company	Variance ratio
TATA Motors	1.363815*
ICICI	1.093162*
DR Reddy's	0.7437732
HDFC BANK	1.385986*
Mahanagar Telephone Nigam Ltd	1.238664*
Infosys Limited	0.9892758
WIPRO	1.363815*
Sesa Sterlite Limited (Present name Vedanta Limited)	0.8877031

*Denotes significance at 5 percent

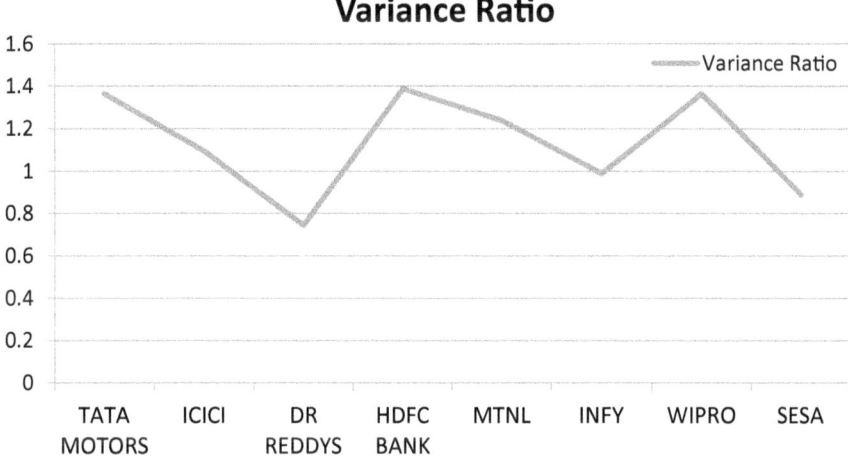

Fig. 25.4 Variance ratio

The results of GARCH model of all variables is presented in table-5. The findings illustrate that the arrival of new information caused the stock markets variations. The α and β are news coefficient. α indicates latest news whereas β specifies past news which is positive and statistical significance indicating that both recent and past news are influencing the variability of stock returns. Since the β value is very high, the past news is influencing more than recent news on the volatility of stock returns and indicating that the movement of stock price is very random. The output of table-5 postulates that the sum of $(\alpha + \beta)$ is very close '1. Hence volatility shocks are quite persistence. The low value of AIC and SIC indicates quite reasonable and fit for the model.

Table 25.5 GARCH (1, 1) model estimation

Variable	Coefficient	Std. error	t-statistics	Prob
Tata Motors	0.013698	0.035270	0.388368	0.6977
ICICI Bank	0.028770	0.020876	1.378136	0.1682
Dr.Reddy	−0.028235	0.032261	−0.875208	0.3815
HDFC	0.016261	0.035970	0.452076	0.6512
MTNL	0.028123	0.027359	1.027911	0.3040
INFY	0.003986	0.024620	0.161896	0.8714
WIPRO	0.030411⋆	0.015100	2.013991	0.0440
SESA	0.026156	0.031645	0.826528	0.4085
C	0.000836	0.000870	0.960763	0.3367
Variance equation				
ω	7.66E-05	3.83E-05	1.999319	0.0456
α	0.094783⋆	0.036932	2.566440	0.0103
β	0.687946⋆	0.136409	5.043274	0.0000
Diagnostic statistics				
Log likelihood 1285.363 Akaike info criterion −5.103659 Schwarz info criterion −5.002354 Hannan-Quinn criter −5.063904 Durbin-Watson stat 1.993343				

* Denotes significance at 5 percent

25.8 Concluding Observations

The present study aims to examine whether ADR issues increase the domestic stock returns after the listing event date and create any value for the shareholders using event study methodology. The study has used daily data from 1998 to 2017. Daily prices of BSE Sensex used as proxy of the market index. The findings of the study indicate that the mean return of HDFC Bank and WIPRO are negative. WIPRO is taking higher risk in the market but producing negative returns, also its share price is highly volatile in the market in comparison to Sample Company. The results of the study reveal that DR Reddy's, HDFC Bank and Infosys Limited are correlated negatively with the stock market. The study also finds CAARs show pre-listing run-up and post-listing run down, with occasional reversal of trends in both the pre- and post-listing periods. The outcome of the study demonstrating that international cross-listings caused the Indian stock returns in the first trading day and shortly thereafter which was experiencing high returns prior to foreign listing. The findings of the study witnessed that cross listing fails to bring any investment benefits to the shareholders since most of the cumulative abnormal returns are not significant statistically. However, the results postulate that the market reacted to the cross listing and increased the volume of transactions. Hence, the volatility of local stock returns increases during pre- and post-listing period. The result of the analysis also suggested that no Indian companies show positive and long-term abnormal returns. This may be attributed to time-specific variation effects of cross-listings and recent developments in the international capital markets. The results of the study also demonstrating that there is a considerable volatility of returns after the ADR listing for most of the companies in the sample. Hence, the investors exposed to increased expected risk premium. This may be attributed to lack of integrated capital market, lack of information and transparency escorted with public offerings. Therefore, it appears from the study that the ADR listing does not create tangible financial benefits for the shareholders. Further, the outcome of the study postulates that past news as well past news influence the stock return and indicating that volatility of shocks is persistence. Since stock market developments lead to economic development, policymakers are required to take necessary steps and right policy to conceptualize, facilitate and steer cross-listing efforts by firms.

25.9 Managerial Implication

The results of our study have significant implications for equity investors, policymakers, firms, and traders. It is important for the firms because the study indicates that cross listing does not increase the firm value but market reacts to cross listing and increasing the volume of transactions. Therefore, firms need to improve corporate governance structure, abate information irregularity, and match their accounting and reporting format with international guidelines to increase their firms' value. For equity investors and traders, cross listing has a temporary assenting effect on the underlying stock price but usually does not materialize a long time after listing date. Therefore, market regulators can take required steps for improvement in these areas. For policymakers, the findings of the study contribute to understanding the factors that assess the stock exchange's competitive position. Precisely, the results of this study add to the knowledge of the Indian cross-listing distinctiveness. To foster development in cross-listings, the suitable and harmonizing action required by both firms and policymakers. However, the future research study can be undertaken by the announcement date, which could have a substantial effect on the findings.

References

1. Alexander GJ, Eun CS, Janakiramanan S (1988) International listings and stock returns: some empirical evidence. J Financial Quantitative Anal 23(2):135–151
2. Admati A, Pfleiderer P (1988) A theory of intraday patterns: volume and price variability. Rev Financial Stud 1(1):3–40
3. Amihud Y, Mendelson H (1991) Liquidity, maturity and the yields on U.S. Treasury Securities. J Finance 46(4):1411–1425

4. Brown SJ, Warner JB (1980) Measuring security price performance. J Financ Econ 8(3):205–258
5. Bollerslev T (1986) Generalized autoregressive conditional heteroskedasticity. J Econometrics 13(1):307–327
6. Bancel F, Mittoo CR (2001) European managerial perceptions of the net benefits of foreign stock listings. Eur Financial Manage 7(2):213–236
7. Justin, CR, Prosper B-S (2017) Semi-strong form market efficiency in stock markets with low levels of trading activity: evidence from stock price reaction to major national and international events. Global Bus Rev 18(6):1447–1464
8. Chi CH (1988) American depository receipts issues by Japanese corporations: the information value on the US market (United States). PhD Dissertations, George Washington University
9. Charitou A, Lambertides N, Trigeorgis L (2007) Managerial discretion in distressed firms. Br Account Rev 39(4):323–346
10. Domowitz I, Glen G, Madhavan A (1997) Market segmentation and stock prices: evidence from an emerging market. J Finance 52(3):1059–1085
11. Fresard L, Salva C (2010) The value of excess cash and corporate governance: evidence from U.S. cross-listings. J Financial Econ 98(2):359–384
12. Foerster SR, Karolyi GA (1999) The effects of market segmentation and investor recognition on asset prices: evidence from foreign stocks listing in the United States. J Finance 54(3):981–1013
13. Howe JS, Kelm K (1987) The stock price impacts of overseas listings. Financ Manage 16(3):51–56
14. King MR, Mittoo UR (2007) What companies need to know about international cross-listing? J Appl Corp Financ 19(4):60–74
15. Lee I (1991) The impact of overseas listings on shareholder wealth: the case of the London and Toronto stock exchanges. J Bus Financ Acc 18(4):583–592
16. Lel U, Miller DP (2008) International cross-listing, firm performance, and top management turnover: a test of the bonding hypothesis. J Financ 63(4):1897–1937
17. Maaji MM, Abdullah SR (2014) Market reaction to international cross-listing: evidence from Nigeria. Int J Information Technol Bus Manage 27(1):13–25
18. Miller DP (1999) The market reaction to international cross-listings: evidence from depositary receipts. J Financ Econ 51(1):103–123
19. Patharla S (2013) Does ADR Issue influence return and volume of underlying domestic stocks? An evidence from the ADRs issue made by Wipro Ltd. Pacific Bus Rev Int 6(3):83–90
20. Patel SA (2014) ADRs and underlying stock returns: empirical evidence from India. AI Soc 29(3):44–52
21. Reilly W, Wagasuki (1990) A dual overseas listing: the impact on returns, risk and trading volume. Working paper (University of Notre Dame)
22. Roosenboom P, Dijk MA (2009) The market reaction to cross-listings: Does the estimation market matter? J Bank Finance 33(10):1898–1908
23. Sarkissian S, Schill MJ (2009) Are there permanent valuation gains to overseas listings? Rev Financial Stud 22(1):371–412
24. Smirnova E (2004) Impact on international exchanges: the case of emerging market's stocks. Eur Financ Manag 5(2):165–202
25. Silva AC, Chávez GA (2008) Cross-listing and liquidity in emerging market stocks. J Banking Finance 32(3):420–433
26. Singh A, Chakraborty M (2017) Examining efficiencies of Indian ADRs and their underlying stocks. Glob Bus Rev 18(1):144–162
27. Su Y-C, Huang H, In-Mancheng (2011) Dynamic causality relation between ADR and underlying stock. Int Res J Finance Econ 77(9):6–77

Chapter 26
Investigation on Supply Chain Vulnerabilities and Risk Management Practices in Indian Manufacturing Industries

Nikhil Gupta, R. Rajesh, and Yash Daultani

Abstract Supply chain plays a prominent role in any business activity. In order to make supply chains more responsive and efficient, we need to identify the vulnerabilities and risk involved in supply chains. The supply chain of an organisation depends upon six driving factors; information, inventory, sourcing, pricing, facility and transportation. We classify various risks in supply chains that influence the six major drivers of it; say for example under inventory; the fluctuation of demand, capacity limitations, stock out of products etc. Similarly, under information; the risks can be legal and regulatory risk, government laws or new tax implementation, lack of point of sale (POS) data, IT cyber security, customer data mishandling, anticipating bullwhip effect etc. And under facility drivers, the associated risks can include centralised or localised production, availability for stock keeping unit and lack of proper data storage, etc. Likewise, under transportation; the major risk factors includes poor logistic supply at supplier end and increased demand for responsiveness, shorter lead times, and shorter window for on time delivery of products, etc. Also, we observe that each driver in supply chain can represent a cluster of several risks, of which some of them are listed above. We study the risk factors using failure mode and effect analysis (FMEA) the causal relations among risk clusters were studied using DEMATEL analysis. We also use the fish bone diagram for representation of cause and effect relationship between the risk factors within a particular cluster. The whole exercise is to prioritise the risk factors based on their causal importance and the study is based on a questionnaire based survey. Overall, managers are benefitted, as the risk factors can be prioritised and the principal causal drivers under each cluster of risks can be identified using FMEA and DEMATEL for the case of Indian manufacturing industries.

Keywords FMEA · Supply chain vulnerability · Risk management · Causal analysis · DEMATEL

N. Gupta · R. Rajesh (✉)
ABV-Indian Institute of Information Technology and Management, Gwalior, India
e-mail: rajesh@iiitm.ac.in

Y. Daultani
Indian Institute of Management Lucknow, Lucknow, India

26.1 Introduction

Supply chain plays a prominent role in any business activity and it accounts for the highest share in terms of investment after raw material, machinery and technology. In today's world of competitiveness, manufacturing companies are giving emphasis on either supply chain efficiency or responsiveness and finding ways to minimise the cost of maintaining their supply chains. In order to make the supply chain more responsive and efficient, we need to identify the risk involved that can hinder the responsiveness and efficiency [10]. From this context, supply chain risk, vulnerability and disruption emerges as important topics in supply chain management. The sudden disturbance that affects the supply chain and seriously affecting its function is termed as supply chain vulnerability.

The study is indented to turn around the supply chain of a manufacturing company primarily focusing on the inbound logistics, intra-firm logistics and outbound logistics. Also, the study covers several risks associated with the failure of any of the drivers of the supply chain and its consequences over the entire supply chain. The risk factors in a supply chain are shown in Fig. 26.1.

The literature suggests four major categories of risks: supply, demand, operational, and security risks [1]. Borrowing from the definition of supply risk provided by Zsidisin [9], supply risk is the distribution of outcomes related to adverse events in inbound supply that affect the ability of the local firm to meet customer demand (in terms of both quantity and quality) within expected costs and time, or causes threats to customer life and safety. Operations risk is the distribution of outcomes related to adverse events within the firm that affect a firm's internal ability to produce goods and services, quality and timeliness of production, and/or profitability. Demand risk is the distribution of outcomes related to adverse events in the outbound flows that affect the likelihood of customers placing orders with the focal firm, and/or variance in the volume and assortment desired by the customer. Security risk is the distribution of outcomes related to adverse events that threaten human resources, operations

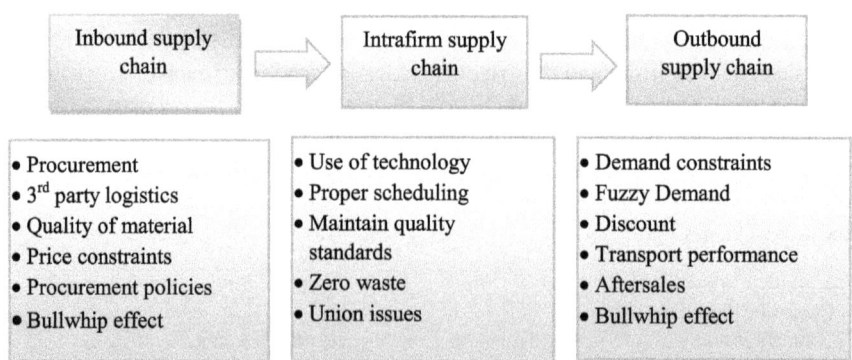

Fig. 26.1 Risk factors in supply chain

integrity, and information systems; and may lead to outcomes such as freight frauds, stolen data or proprietary knowledge, vandalism, crime, and sabotage.

26.1.1 Demand Side Risk

Demand-side risks result from disruptions emerging from downstream supply chain operations [2]. This includes the physical distribution of products to the end-customer with particular issues being transportation operations (e.g., a truck driver strike) and the distribution network (e.g., a fire in a warehouse). On the other hand, demand-side risks can originate from the uncertainty surrounding the random demands of the customers.

26.1.2 Supply Side Risk

Organisations are always associated in risk from supplier's network. Firms should proactively assess and manage the uncertainties in their supplier portfolio in order to guard it against costly supply disruptions. The threat of financial instability of suppliers and the consequences of supplier default, insolvency, or bankruptcy [3].

Poor logistics performance of logistics service providers

- Capacity fluctuations or shortages on the supply markets
- Opportunist behaviour of the supplier.

26.1.3 Catastrophic Risk

- Political instability, war, civil unrest, or other socio-political crises, trade war
- International terror attacks (9/11, 26/11 terror attacks)
- Diseases or epidemics (e.g., SARS, COVID-19)
- Natural disasters (e.g., earthquake, flooding, extreme climate, tsunami).

26.1.4 Regulatory, Legal and Bureaucratic Risk

- Union strike (worker union strike, sales tam strike, transporters strike etc.)
- Government laws (subsidies)
- New Tax implications (GST, Export and Import Tariffs).

26.2 Literature Review

26.2.1 Drivers of Supply Chain

After further study of supply chain disruption, we found more factors that are directly or indirectly causing defects in the supply chain and i have categorised it into disruptions in 6 drivers. These drivers are presented in Fig. 26.2 and the classification is shown in Fig. 26.3.

26.2.1.1 Inventory

"Inventory" means physical stock of goods, which is kept in hands for smooth and efficient running of future affairs of an organization at the minimum cost of funds blocked in inventories. The fundamental reason for carrying inventory is that it is physically impossible and economically impractical for each stock item to arrive exactly where it is needed, exactly when it is needed.

1. **Demand is higher than supply**
   When demand goes higher then supply, it results into frequent stock out, duplicity increase in prices and results into supply chain disruption.

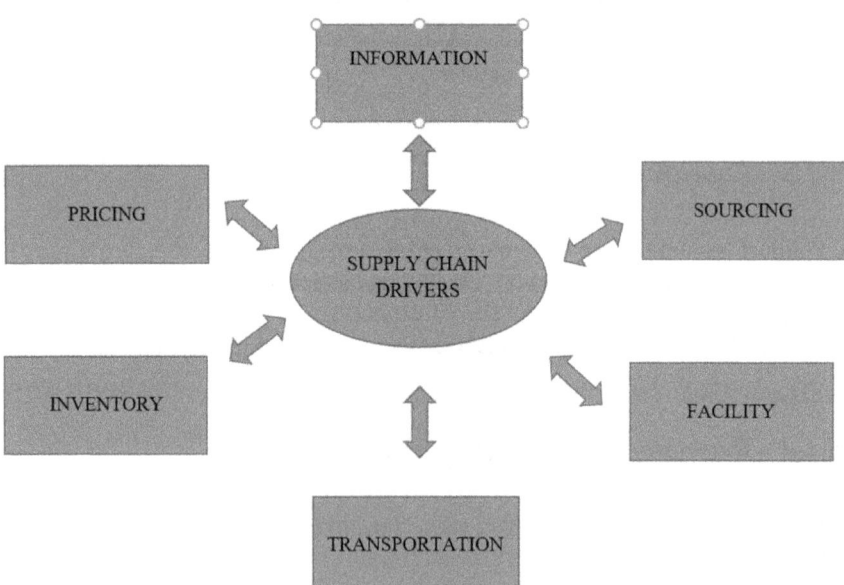

Fig. 26.2 Supply chain drivers

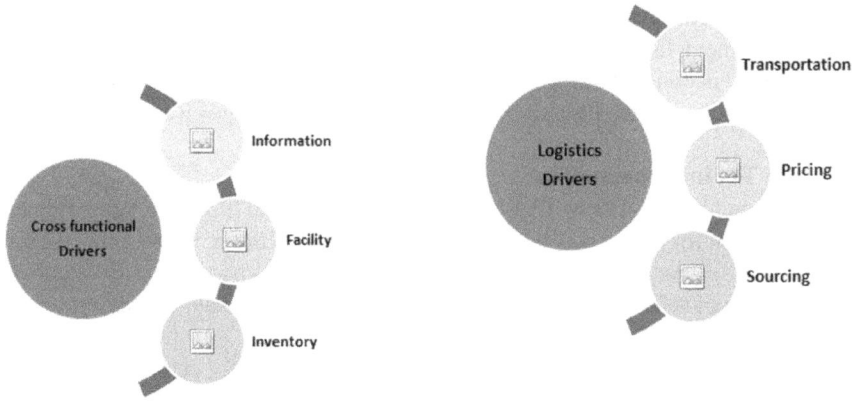

Fig. 26.3 Cross-functional and logistics drivers

2. **Supply is more than demand**
   When supply goes far beyond the actual demand, it results into piling of goods in warehouses, SKU etc. Which lead to locking up of money that ultimately leads to supply chain disruption.
3. **Fuzzy demand from customer side**
   False demand from the customer/competitor side to misguide the marketing/production strategy of our company leads to either stock out or high inventory which itself leads to supply chain disruption.
4. **Stock out of product**
   Stock out of product gives competitor the chance to sell more if the product is staple item, consumer may not wait for single minute and will buy product of competitor.
5. **Capacity limitations**
   Capacity limitation sometimes leads to supply chain disruption when demand is more than supply and the manufacturer/facilitator is not able to fulfil the demands because of limited supply, limited infrastructure, and lack of adequate forecasting measures.

26.2.1.2 Information

Flow of information is very important in managing supply chain. Organizations need to adopt it to support their supply chains and increase their efficiency by achieving tighter cooperation over the supply-chain [5].

1. **Legal and regulatory risk**
   Legal and regulatory risk is always something that cannot be avoided and that could certainly lead to supply chain disruption. Any nonconformity to the standards or violation to the government norms can lead to serious disruption in supply chain.

2. **Government laws**
   Government laws to protect environment like BS6, BS4 norms has led to decrease in demand of various automobile vehicles more over government decision to subsidise the Electric Vehicles in the country has given a boast in supply chain.
3. **New Tax implementation**
   Implementation of new TAX system GST had caused a sudden supply chain disruption for many industries in India, since most of the economy runs on cash in India so unavailability of cash led to difficulty in purchase and sales of goods, payment of salaries etc. which ultimately led to supply chain disruption.
4. **Lack of POS data**
   Point of scale data is very crucial in designing the supply chain and enabling supply chain visibility across the chain [4]; it gives an idea on

   How much to produce?
   How much to stock?
   How much to procure?

5. **IT/cyber security**
   IT/cyber security plays an important role in delivery of right information to the right channel and simultaneously safeguard against threats.
6. **False claims/propaganda**
   False claims/propaganda of poor quality of goods, adulteration, malpractices in any form leads to earning bad name for the company and eventually lead to supply chain disruption. E.g. Nestle was forced to stop production of instant noodles "Maggi" which there were allegations of lead contamination in it.
7. **Bullwhip effect**
   The bullwhip effect is a concept for explaining inventory fluctuations or inefficient asset allocation as a result of demand changes as we move further up the supply chain. The accumulation of inventory in anticipation of demand from the market sometimes leads to leads to accumulation of inventories where it is not as much needed and shortage where it is actually required which further leads to Supply Chain Disruptions.
8. **Customer data mishandling**
   Mishandling of customer data may sometimes take a negative turn against the firm as customer might feel that their privacy is at threatened and can lead to supply chain disruption.

26.2.1.3 Facility

Placement of production units, warehouses etc. in such a manner that they are viable to operate and at the same time at more reach to the customer [6].

1. **Centralised production**
   Centralised production involves production of goods at a single place; all the raw material, technology is brought at one single unit and production of goods are carried out there.
2. **Out-dated technology/inadequate technology**
   No proper devices for monitoring, recording, entry and exit of goods may lead to serious issues in forecasting, warehousing and capacity estimation of the actual data and may lead to supply chain disruption.
3. **Inadequate SKU for storage**
   Lack of proper stock keeping units for storage makes it very difficult for the supply chain to maintain a balance between demand and supply.
4. **Lack of proper data storage facility**
   Insufficiency of proper data storage is another factor included in the facility.

26.2.1.4 Transportation

Transportation is the heart of Supply chain which takes share of around 60% of total investment [8]. Driving transport is another name to driving the whole supply chain.

1. **Poor logistics at supplier end**
   Poor logistic services at supplier end will lead to delay in material arrival at manufacturer and will gradually lead to late goods production and will lead delay good arrival in the market. This will indirectly lead to supply chain disruption.
2. **Increased Demand for responsiveness**
   This results in shorter lead time, shorter window for on time delivery of product.

26.2.1.5 Sourcing

Sourcing can be of many things like raw materials, it can be technology items, sourcing can be of technology as well. Sourcing of right quantity and right quality of material is very important for a supply chain to work properly [7].

1. **Supplier quality problem**
   Quality of a raw material in the production cycle is the utmost thing which is needed for making standardised product. The raw material of sub-standard quality is detected at the production level and this lead to either rejection of the lot or the change of the vendor that indirectly delays all the process related to supply chain and this ultimately leads to supply chain disruption.
2. **Bankruptcy of supplier**
   Bankruptcy of the supplier leads to immediate shortage of the material at the manufacturing end especially when the supplier is a single hand-supplier. This can lead to serious supply chain disruption.
3. **Opportunistic behaviour of supplier**
   Supplier sometime shows an opportunistic face towards the manufacturer when the supply of raw material with the same quality is to be ensured or there may be

shortage of material. This lack of coordination in negotiation between supplier and manufacturer can lead to supply chain disruption.

4. **Reduction in supplier base**
   Purchase managers in big organisation tend to shrink their supplier base so as to make supply chain more rigid but supply chain demands more visibility.
5. **Product recall**
   Product recall is another supply chain disruption that occurs due to mismatch of product Quality or environmental regulation etc. If a company recalls its product, the entire cost is borne by the manufacturer, recalls creates tension in all sections of the supply chain, its tests consumer relationship with the brand, affects sales and even gives a negative image to the brand reputation.
6. **Lack of alternative supplier**
   Lack of alternative supplier leads to supply chain disruption when existing supplier suddenly stops supplying goods or there is sudden disruption form supplier side. There may be any reason for the disruption e.g. Fire, floods etc.

26.2.2 Research Gap

- Till now studies have included identification of the potential risk factors and then their prioritisation through any MCDM method.
- Our study includes identification and well as categorisation on the bases of supply chain drivers/clusters and then understanding which driver or cluster has highest risk factors under them.

26.2.3 Key Research Objective

- To identify the potential risk factors that is responsible for supply chain disruption and prioritises them.
- Finding Cause and Effect relationship between potential risk factors in each driver.

26.3 Research Methodology

26.3.1 Failure Modes and Effects Analysis (FMEA)

It is a systematic, proactive method for evaluating a process to identify where and how it might fail and to assess the relative impact of different failures, in order to identify the parts of the process that are most in need of change.

FMEA includes review of the following:

- Steps in the process
- Failure modes (What would go wrong?)
- Failure causes (Why would the failure happen?)
- Failure effects (What would be the consequences of each failure?).

Teams use FMEA to evaluate processes for possible failures and to prevent them by correcting the processes proactively rather than reacting to adverse events after failures have occurred. This emphasis on prevention may reduce risk of harm to both patients and staff. FMEA is particularly useful in evaluating a new process prior to implementation and in assessing the impact of a proposed change to an existing process.

The primary objective of an FMEA is to improve the design.

- For System FMEAs, the objective is to improve the design of the system
- For Design FMEAs, the objective is to improve the design of the subsystem or component.
- For Process FMEAs, the objective is to improve the design of the manufacturing process.

26.3.1.1 Steps to Failure Mode and Effect Analysis

Select a System to analyse.

1. Identify individuals from all departments with specific knowledge of processes, products and client needs to brainstorm potential failure modes.
2. Describe the process and/or product in detail.
3. Identify all potential failures. This includes all of the components, systems, processes and functions that could potentially fail to meet the quality or reliability standard and the potential causes.
4. Identify all the potential consequences of each failure.
5. Assign a severity rating (S) to each failure according to the significance of the impact it has. Severity is often ranked on a scale from 1 to 5, one being insignificant and 5 being catastrophic.
   Identify all possible root cause of each failure. Some companies use cause analysis tools in addition to the knowledge and experience of their staff.
6. Assign each cause an occurrence rating (O). This is often rated on a scale of 1 to 10, with 1 being rare and 5 being inevitable.
7. For each cause, identify current process controls that are in place to prevent these failures from impacting customers.
8. For each control, assign a detection rating (D) to determine how well the controls are able to detect the cause or failure mode once they have occurred, but before a customer is affected. This is typically rated on a scale of 1 to 5, with 1 meaning the problem will be detected with absolute certainty and meaning the control will most likely never detect the problem.

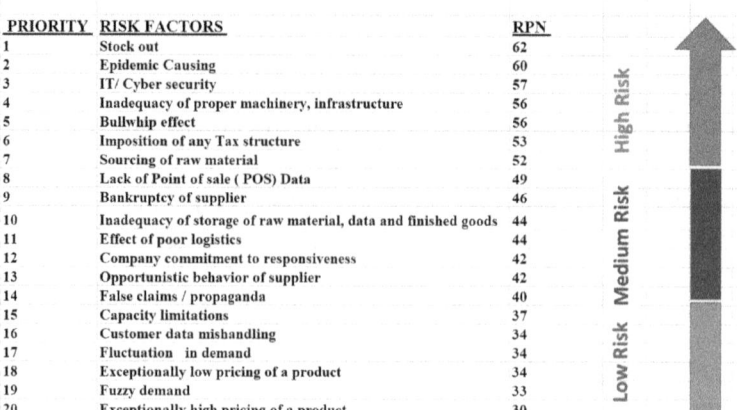

Fig. 26.4 FMEA—priority list

9. Determine a risk priority number (RPN) based on the rankings (S × O × D) for each potential failure and rank them.
10. Plan and implement changes to address the failures based on the RPN identifications.
11. Measure and document the success of each process change.

26.3.1.2 Questionnaire

For applying FMEA analysis, data is collected through a Questionnaire of 20 Questions respondents were required to give feedback on 3 parameters Severity, Occurrence, and Detection and 20 Responses were collected and Based on this the RPN (Risk Priority Number) is calculated. Then, risk factors were arranged from high to low priority. The risk factors observed in the order of priority is indicated in Fig. 26.4.

26.4 Results and Findings

With the above results from Failure mode and effect analysis we found the following,

1. Risk Factors like stock out of Product, IT/Cyber-security, Inadequacy of proper Machinery and Infrastructure, Bullwhip Effect, sudden Imposition of Tax structure and Sourcing of raw Material are kept under High Risk Priority
2. Risk factors like Lack of POS data, bankruptcy of supplier, inadequacy of storage of raw material, data and finished goods, effect of poor logistics, company commitment to responsiveness, opportunistic behaviour of supplier and false claims/propaganda regarding the product are kept under medium Risk Priority.

3. Risk factors like capacity limitations, customer data mishandling, fluctuations in demand, exceptionally low pricing of product, fuzzy demand, and exceptionally high pricing of a product are kept under Low Risk priority.
4. Stock out scores the highest RPN score, hence being the most important factor to be prioritised to avoid disruption in Supply chain in an Organization.
5. Epidemic or catastrophic event is found as the second most concerned factor to be prioritised to avoid supply chain disruption.
6. It also found exceptionally high pricing of a product being the least prioritised factor to avoid disruption in Supply chain in an Organization.

26.4.1 DEMATEL

Decision making trial and evaluation laboratory (DEMATEL) is considered as an effective method for the identification of cause-effect chain components of a complex system. It deals with evaluating interdependent relationships among factors and finding the critical ones through a visual structural model. For accurate data in each Driver, responses were taken from the industry Experts in the concerned domain. The DEMATEL analysis is shown in Fig. 26.5.

The influence ratings, total relation matrix, cause and effect relations and the digraphs are represented in Tables 26.1, 26.2, 26.3, 26.4, 26.5, 26.6, 26.7, 26.8, 26.9, 26.10, 26.11 and Figs. 26.6, 26.7, 26.8, 26.9, 26.10.

26.5 Conclusion

The above study concludes the following outcomes:
1. For efficient and responsive supply chain, an organisation should prioritise all the risk factors associated with it. In our study we found based on different respondents of varied expertise that organisation should avoid top priority Risk as mentioned in Fig. 26.4 (Priority list of FMEA).
2. Prioritisation of risk could significantly reduce the amount to efforts required to balance efficiency and responsiveness in a supply chain further avoiding disruptions.
3. Organisation can categorise their risk into 3 zones: Top priority, Medium priority and Low priority based on their nature of business (Risk factors for business can be different from each other).
4. There can be different" cause and effect" for particular Supply Chain Driver as can be seen the DEMATEL analysis for each driver in the study.
5. In the Inventory Driver, according to the analysis, capacity limitation, demand fluctuation and stock out are the cause behind disruption in supply chain while fuzzy demand serves as the effect of it (Ref. Table 26.3 and Fig. 26.6—Cause and Effect Matrix for Inventory).

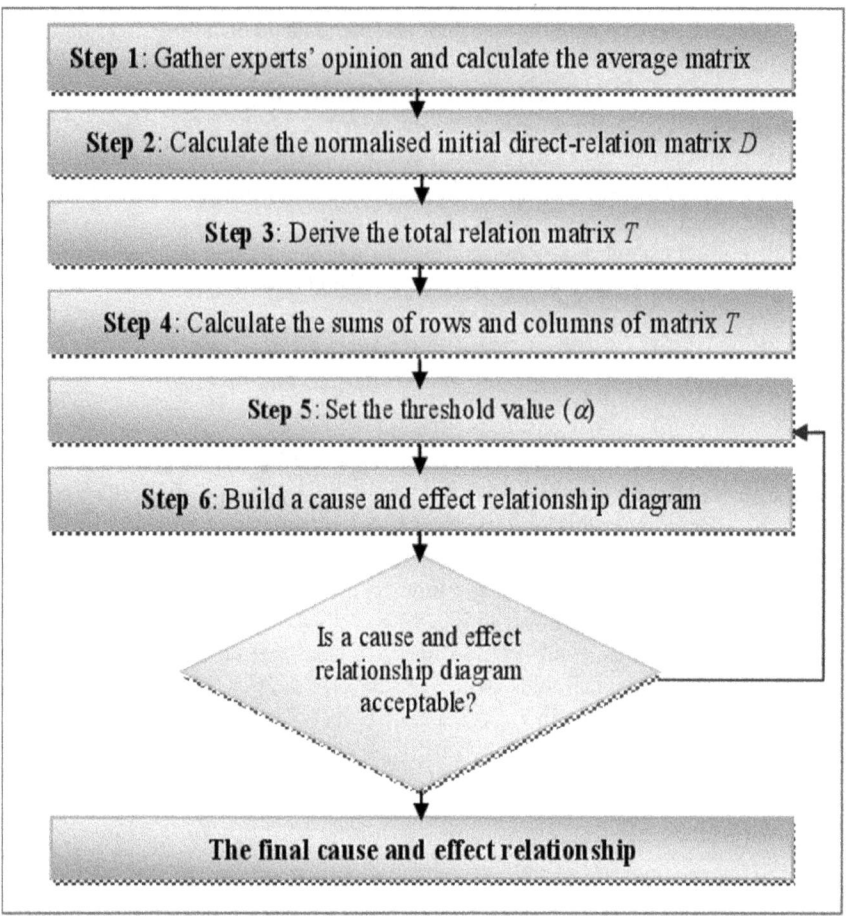

Fig. 26.5 DEMATEL analysis

Table 26.1 Comparison scale of DEMATEL

Numerical	Definition
0	No influence
1	Low influence
2	Medium influence
3	High influence
4	Very high influence

26 Investigation on Supply Chain Vulnerabilities and Risk …

Table 26.2 Total relation matrix for the driver inventory

	Capacity limitations	Fuzzy demand	Demand fluctuation	Stock out	D
CL	1.8	2.8	2.2	1.4	8.2
FD	2.1	2.5	2.3	1.4	8.3
DF	2.3	3.1	2.2	1.7	9.2
SO	2.0	2.7	2.1	1.3	8.1
R	8.2	11.0	8.7	5.8	

Table 26.3 Cause and effect matrix for the driver inventory

	D	R	D + R	D − R	
C.L	8.2	8.2	8.6	0.0	Cause
F. D	8.3	11.0	10.5	−2.7	Effect
D.F	9.2	8.7	9.2	0.4	Cause
S. O	8.1	5.8	8.1	2.3	Cause

Table 26.4 Total relation matrix for the driver information

	Gov. law	New tax implement	Bullwhip effect	Lack of POS data	IT/cyber security	False claims and propaganda	D
Government law	0.90	0.69	1.02	0.66	0.93	1.00	5.20
New tax implementation	0.86	0.37	0.81	0.47	0.62	0.79	3.92
Bullwhip effect	1.00	0.56	0.81	0.79	0.78	0.94	4.88
lack of POS Data	0.65	0.37	0.79	0.42	0.53	0.66	3.42
IT/cyber security	0.78	0.33	0.70	0.48	0.50	0.73	3.52
False claims and propaganda	1.12	0.48	1.00	0.76	0.87	0.80	5.03
R	5.31	2.80	5.13	3.58	4.24	4.91	

Table 26.5 Cause and effect matrix for information driver

	D	R	D + R	D − R	
Government law	5.2	5.3	10.5	−0.1	Effect
New tax implementation	3.9	2.8	6.7	1.1	Cause
Bullwhip effect	4.9	5.1	10.0	−0.2	Effect
Lack of POS data	3.4	3.5	6.9	0.0	Effect
IT/cyber security	3.5	4.2	7.8	−0.7	Effect
False claims and propaganda	5.0	4.9	9.9	0.1	Cause

Table 26.6 Total relation matrix for the driver sourcing

	Supplier quality problem	Bankruptcy of supplier	Opportunistic behaviour of supplier	Reduction in supplier base	Product recall	Lack of alternative supplier	D
Supplier quality problem	0.64	0.80	0.50	0.65	0.72	0.67	**3.97**
Bankruptcy of supplier	0.53	0.34	0.27	0.38	0.37	0.28	**2.17**
Opportunistic behaviour of supplier	0.78	0.71	0.41	0.61	0.77	0.68	**3.96**
Reduction in supplier base	0.69	0.67	0.40	0.43	0.67	0.51	**3.38**
Product recall	0.90	0.87	0.63	0.71	0.65	0.69	**4.45**
Lack of alternative supplier	0.76	0.74	0.49	0.63	0.75	0.47	**3.85**
R	**4.30**	**4.14**	**2.69**	**3.40**	**3.93**	**3.31**	

Table 26.7 Cause and effect matrix for the driver sourcing

	D	R	D + R	D − R	
Supplier quality problem	3.97	4.3	8.27	−0.33	**Effect**
Bankruptcy of supplier	2.16	4.136	6.296	−1.976	**Effect**
Opportunistic behaviour of supplier	3.95	2.69	6.64	1.26	**Cause**
Reduction in supplier base	3.38	3.4	6.78	−0.02	**Effect**
Product recall	4.44	3.92	8.36	0.52	**Cause**
Lack of alternative supplier	3.84	3.3	7.14	0.54	**Cause**

Table 26.8 Total relation matrix for transportation driver

	Poor logistics at supplier's ends	Delivery on time	Increased demand for responsiveness	Shorter window	D
Poor logistics at supplier's ends	1.2	2.1	1.9	1.2	**6.4**
Delivery on time	1.3	1.7	1.8	1.3	**6.1**
Increased demand for responsiveness	1.4	2.1	1.7	1.5	**6.6**
Shorter window	0.9	1.4	1.4	0.9	**4.6**
R	**4.8**	**7.3**	**6.8**	**5.0**	

26 Investigation on Supply Chain Vulnerabilities and Risk ...

Table 26.9 Cause and effect matrix for transportation driver

	D	R	D + R	D − R	
Poor logistics at supplier's ends	6.42	4.76	11.18	1.66	Cause
Delivery on time	6.12	7.28	13.4	−1.16	Effect
Increased demand for responsiveness	6.63	6.76	13.39	−0.13	Effect
Shorter window	4.61	4.98	9.59	−0.37	Effect

Table 26.10 Total relation matrix for facility driver

	Centralised production	Outdated technology	Inadequate Infra	lack of storage data facility	D
Centralised production	3.0	3.5	3.0	3.5	13
Outdated technology	3.3	3.2	3.0	3.5	13
Inadequate infrastructure	3.5	3.7	3.0	3.8	14
Lack of storage data facility	3.3	3.5	3.0	3.2	13
R	13.1	13.9	12.0	14.0	

Table 26.11 Cause and effect matrix for facility driver

	D	R	D + R	D − R	
Centralised production	13	13.07	26.07	−0.07	Effect
Outdated technology	13	13.93	26.93	−0.93	Effect
Inadequate infrastructure	14	12	26	2	Cause
Lack of storage data facility	13	13.99	26.99	−0.99	Effect

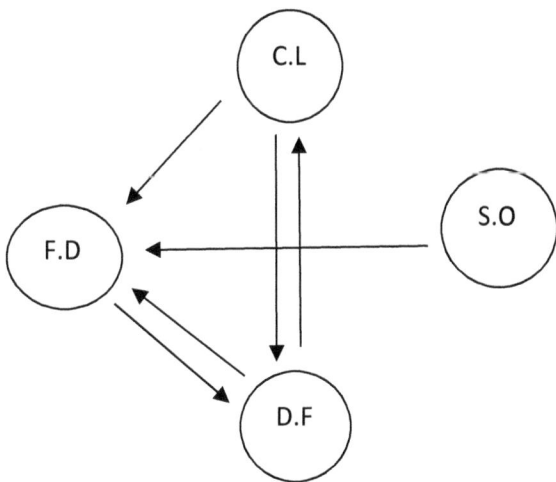

Fig. 26.6 Digraph for inventory driver

Fig. 26.7 Digraph for information driver

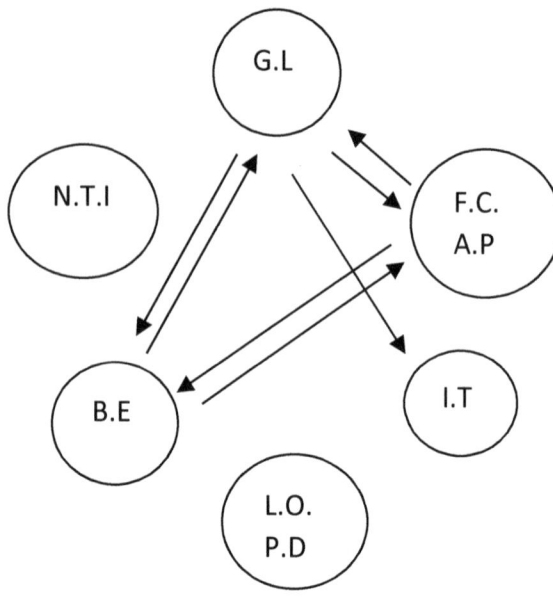

Fig. 26.8 Digraph for sourcing driver

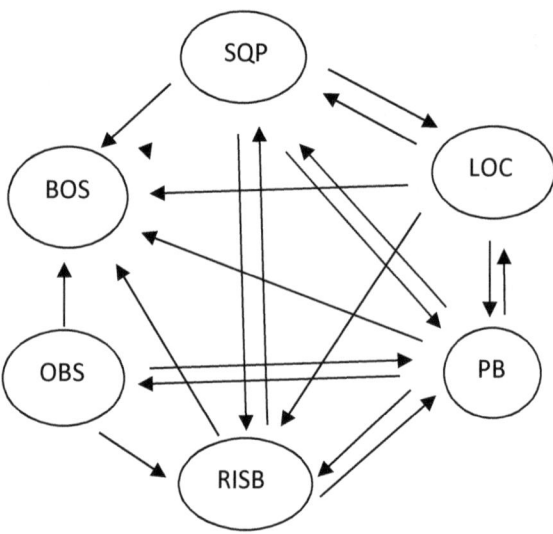

6. In the Information Driver, according to the analysis, new tax implementation and false claim propaganda are the cause behind disruption in supply chain, while Government laws, bullwhip effect, lack of POS data and IT/Cyber security serves as the effect of it (Ref. Table 26.5 and Fig. 26.7—Cause and Effect Matrix for Information Driver).

Fig. 26.9 Digraph for transportation driver

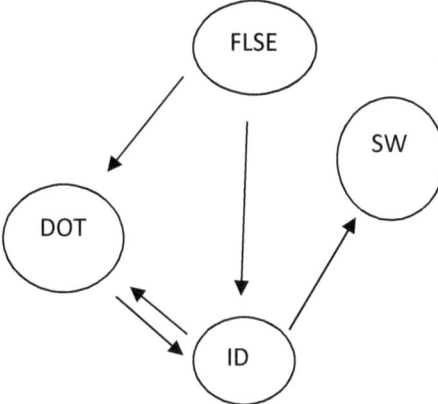

Fig. 26.10 Digraph for facility driver

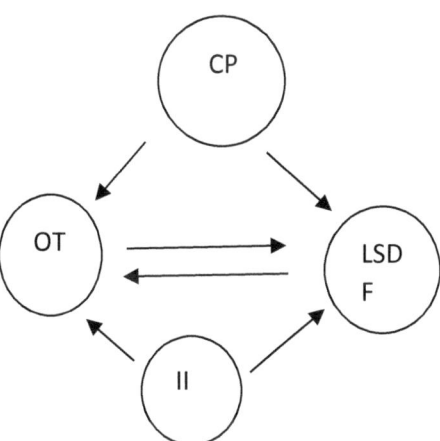

7. In the Sourcing Driver, according to the analysis, Opportunistic behaviour of supplier and product recall is the cause behind disruption in supply chain, while Supplier quality problem, bankruptcy of supplier, reduction in supplier base serves as the effect of it (Ref. Table 26.7 and Fig. 26.8—Cause and Effect Matrix for Sourcing Driver).
8. In the Transportation Driver, according to the analysis, Poor logistic at suppliers end is the cause behind disruption in supply chain while, delivery on time, increased demand for responsiveness, shorter window serves as the effect of it (Ref. Table 26.9 and Fig. 26.9—Cause and Effect Matrix for Transportation Driver).
9. In the Facility Driver, according to the analysis, inadequate infrastructure is the cause behind disruption in supply chain while, Centralised production, outdated technology, lack of storage data facility serves as the effect of it (Ref. Table 26.11 and Fig. 26.10—Cause and Effect Matrix for Facility Driver).

10. Next approach is to work on the financial aspect of supply chain and to mitigate the risk associate with it. For example: impact of inventory turnover (INVT), account receivables turnover (ART), property, plant and equipment turnover (PPPT) on Supply chain.

References

1. Anupindi R, Akella R (1993) Diversification under supply uncertainty. Manage Sci 39(8):944–963
2. Christopher M, Lee HL (2001) Supply chain confidence: the key to effective supply chains through improved visibility and reliability. Global Trade Manag 1–10
3. Sheffi Y (2001) Supply chain management under the threat of international terrorism. Int J Logist Manag 12(2):1–11
4. Tang CS (1999) Supplier relationship map. Int J Log Res Appl 2(1):39–56
5. Tang CS (2006) Robust strategies for mitigating supply chain disruptions. Int J Log Res Appl 9(1):33–45
6. Tsiakouri M (2008) Managing disruptions proactively in the supply chain: the approach in an auto-manufacturing production line. In: Production and operations management society's 19th annual conference, La Jolla, CA
7. Wasti SN, Liker JK (1997) Risky business or competitive power? Supplier involvement in Japanese product design. J Prod Innov Manag 14(5):337–355
8. Wilson MC (2007) The impact of transportation disruptions on supply chain performance. Transp Res Part E Logistics Transp Rev 43(4):295–320
9. Zsidisin GA (2003) A grounded definition of supply risk. J Purch Supply Manag 9(5/6):217–224
10. Zsidisin GA, Ragatz GL, Melnyk SA (2005) The dark side of supply chain management. Supply Chain Manag Rev 9(2):46–52

Chapter 27
Forecasting Long-term Electricity Demand: Evolution from Experience-Based Techniques to Sophisticated Artificial Intelligence (AI) Models

Abhishek Das and Somen Dey

Abstract Demand forecasting is one of the primary activities in the planning phase of any business and a key input to many crucial business decisions. It forms the basis of answering managerial questions such as how much raw materials to procure, how much resources to allocate, or how much to invest. Typically, demand for any product or service is stochastic and is linked to various extrinsic factors ranging from socio-economic conditions of the economy to consumers' taste and perception, and intrinsic factors like quality and value proposition of the product or service itself. The earliest attempts for electricity demand forecasting were of short-term and, based on wisdom, experience, and speculations of the vertically integrated electricity utilities, even before separate system operators came into practice. The first set of formalized methods made use of Trend Analysis, Econometrics, and End-Use Approaches. These methods, when applied individually, are prone to inherent errors; attempts were made to improve them or combine two or more of them to design hybrid techniques. With the realization that these methods are incapable of adequately capturing the variabilities of demand over time, emphasis began to be given on computationally intelligent methods. Subsequently, with advances in scientific knowledge and computational capabilities, smarter and more intelligent algorithms started being progressively used in this field. This chapter presents a systematic evolution of these methods with their merits, demerits and applicability criteria along with an outline on the shift from stand-alone demand forecasting to an integrated energy system modelling approach.

Keywords Long-term electricity demand forecasting · Trend analysis · Econometrics · Computationally intelligent methods · Energy system modelling

---

A. Das (✉)
Department of Management Studies, Indian Institute of Science, Bengaluru, India
e-mail: abhishekdas1@iisc.ac.in

S. Dey
School of Management Studies, Motilal Nehru National Institute of Technology Allahabad, Prayagraj, India
e-mail: somen@mnnit.ac.in

27.1 Introduction

Humans have always been fascinated about predicting the future, and prediction backed by logic is termed as forecast. In the world of business management, projections serve as a critical input for decision making. Be it an inventory manager's decision on the quantum of raw materials to be kept in stock, or a production manager's decision to have an additional production line or the decision of the board of directors to infuse additional capital or even, an HR manager's decision to hire fresh executives; all depend primarily on one input i.e. sales forecast. Therefore, the efficacy of these decisions, to a considerable extent, depends on the accuracy of sales forecasts.

Electricity is one of the critical drivers of economic growth, and investments in the electricity sector are capital intensive and long-term in nature. The majority of the projects have long gestation periods of almost a decade, which implies that investment decisions made today are purely based on speculations of how the markets would be ten years down the time frame. Thus, only a robust long-term forecast would ensure that these projects generate profits after being commissioned. Under-estimation of demand would lead to lower capacity additions and energy deficits in the country, impairing its development. On the other hand, over-estimation would result in excess capacity additions leading to the creation of non-performing assets (NPAs), which is again an impediment to a nation's economic growth. Thus, it is of utmost importance to have highly accurate forecasts of electricity demand for a long-term horizon of ten years or more.

27.2 Fundamentals of Electricity Demand

The operational value chain of the electricity sector can be summarised, as shown in Fig. 27.1.

Electricity is generated using a variety of technologies like coal-based thermal power plants, solar PV plants, wind generators, hydroelectric power stations, and nuclear power plants etc. The role of transmission utilities is to provide the necessary infrastructure for transporting the generated electricity to distribution centres. Distribution utilities are responsible for setting up and upkeep of infrastructure for the supply of electricity to each consumer premises. Retail suppliers execute commercial contracts with consumers.

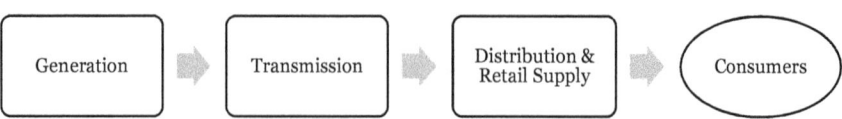

Fig. 27.1 Electricity value chain

Each of the generation technologies has its inherent characteristics, merits, and limitations in terms of start-up time, ramp up/down rates, variability, and uncertainty of generation, technical efficiency, environmental concerns, and marginal cost. Consumers, on the other hand, have changing demand, which depends on a variety of factors like weather, nature of human activity, lifestyle, socio-economic conditions, and electricity price. The transmission and distribution networks need to be so designed that they must be able to incorporate all variabilities and uncertainties of generators as well as consumers and provide uninterrupted supply. Due to uncertainties at both generation and consumption ends, the task of balancing demand and supply becomes exceptionally challenging. Typically, a bottom-up approach is followed, wherein based on estimated demand at the consumer end; the entire electricity system is planned. Therefore, the accuracy of the demand forecast is a critical requirement for an efficient power system planning.

Electricity demand has both temporal and spatial dimensions. Unlike most other consumer products, electricity cannot be stored economically (as of today) and, therefore, has to be generated at the same instant when the demand arises. The pattern of usage of electricity varies from state to state depending on human lifestyle preferences and agro-climatic zones. A typical electricity load curve for the state of Karnataka, India, is plotted, as shown in Fig. 27.2.

The area under the curve (i.e., 196.46 GWh) represents the aggregate demand for a particular day. The highest demand recorded during the day (i.e., 9.805 GW at 1100 h) depicts the peak load, and the lowest recorded demand during the day (i.e., 6.584 GW at 0300 h) is termed as base load. An electricity demand forecast typically provides estimates of these three parameters, which serve as critical inputs to power system planning.

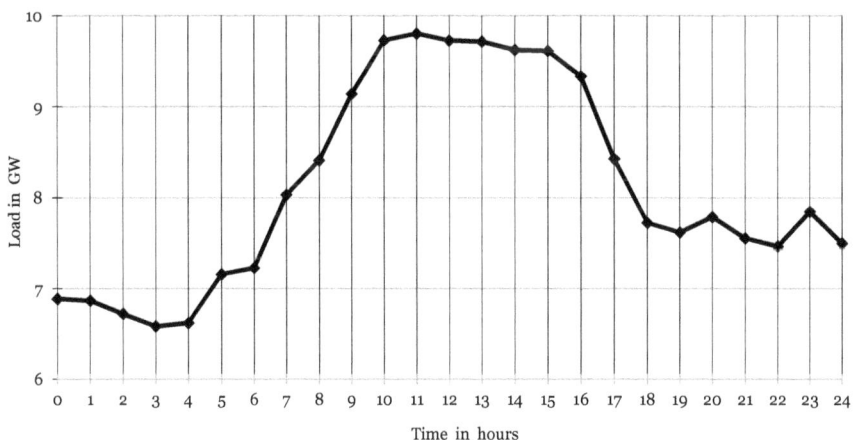

Fig. 27.2 Load curve of Karnataka, India (of 4th May 2020) (Source KSLDC; available at: https://218.248.45.137:8282/LoadCurveUpload/lcdownloadview.asp)

Based on the prediction time horizon, electricity demand forecast is classified as ultra-short term (less than an hour), very short term (an hour to less than a day), short term (a day to less than a week), medium term (a week to less than a year) and long term (a year or more). Long term forecasts are essential inputs to power procurement planning and capital investment decisions of the sector.

27.3 Long-term Electricity Demand Forecasting

A long-term electricity demand forecast is of interest to all players and stakeholders of the sector. Accurate forecast leads to effective planning, which in turn would enable an efficient power system. On the one hand, under-estimating demand can cause forced outages due to power shortages impacting industrial productivity and, thereby, economic growth. On the other hand, over-estimation would result in over-investments and excess capacity additions leading to financial distress and higher tariffs. Thus, long-term forecasts are of deep interest to not only investors, developers, and utilities but also policymakers and development professionals [1]. Long-term demand forecasting has major benefits to various stakeholders, described as follows.

a. **Generators**: Demand forecast is essential for these players of the sector to optimally undertake plans of capacity additions, penetration of renewables, phasing out of old plants, and other such developmental activities. An accurate forecast would ensure capacity additions necessary to meet the growing electricity needs of everyone and also ensure that excess capacity is not created, leading to stranded assets, thus ensuring optimum returns on investment.

b. **Transmission and Distribution Utilities**: Based on anticipated demand growth, transmission and distribution systems need to go for expansion of the network, keeping in view all the variabilities and uncertainties to be able to meet the demand at all times and at the same time avoiding congestion.

c. **Retail Suppliers**: Based on the consumer demand profile, they need to make arrangements with various generators about quantum and time of supply. An accurate forecast will enable them to optimize such structures economically. In India, power purchase cost contributes to about 70 percent of retail supply tariffs, and therefore, there is scope for significant cost reductions [2].

d. **Lenders and Investors**: The recovery of capital and returns are critical decision criteria for investment in any project. A robust demand forecast would give long-term visibility about a power projects revenue stream and, thereby, investment certainty.

e. **Government and Policymakers**: Electricity is considered as a social welfare measure, especially in developing nations. An accurate demand forecast would help the government in designing effective policies, budgeting, and targeting of subsidies and other developmental programs. The government's policies, on the other hand, would also serve as inputs to long-term demand forecasting.

f. **Regulators**: The electricity sector, especially the business of networks, is a natural monopoly and, thus, is heavily regulated, typically by cost-plus or rate of return regulations in most nations. One of the primary responsibilities of power sector regulators is to determine tariffs by maintaining a balance between the affordability of consumers and the economic viability of the utilities. Accurate demand forecasts would enable them to estimate legitimate power procurement costs accurately and thereby determine tariff effectively.

Thus, it is evident that long-term demand forecasting in the electricity sector is a critical activity to ensure the economic sustainability of the industry as well as aid all the players and stakeholders of the industry in making business decisions. A robust forecast, therefore, must entail features such as long-term outlook of more than ten years with periodic updates, disaggregation to the level of each strategic unit or distribution utility, and integration of network congestion, RE integration, captive generation [2]. Based on these premises, the chapter aims at discussing the various techniques available for long-term demand forecasting, i.e., from traditional experience-based methods to the computationally intelligent ones, requiring sophisticated computational facilities. The chapter presents an overview of many diverse concepts from various domains, but it does not intend to make the reader an expert in any of them.

27.4 Qualitative Techniques of Long-term Electricity Demand Forecasting

The need for demand forecasting emerged when electricity was identified as a commercial commodity and started operating as an integrated system. The earliest attempts to demand to forecast were predominantly short-term, experience-based, and disintegrated. Various interested stakeholders, in their own purview and based on their experience, would speculate demand and act accordingly. For example, generators would vary the operating load of their plants based on anticipated demand and so on. Electricity demand forecasting gained importance as electricity became more of a basic necessity from luxury and was being looked upon as a commercial product.

In practice, the most common qualitative technique, still in the way, is Round Table Discussion. The concerned officials of the organization carrying out the demand forecasting exercise sit together, look at past data, and come up with numbers for the future based on experience and rules of thumb. This is a very crude method deployed in the absence of data and paucity of time. However, a formal and structured qualitative technique, extremely helpful in such situations, is the Delphi technique, which brings experts of the field on a common platform and directs the discussion towards convergence.

27.4.1 Delphi Technique

The most critical ingredient of long-term forecasting is economic, technical, and other historical data, which is not readily available in the case of the power sector. In such a scenario, expert opinions are the only source of information open, making the Delphi technique the most appropriate [3]. This technique is, in particular, helpful for scenarios where accurate historical data is not available, and even if available, the relative significances of the historical events are unclear. Only an expert or a group of experts would be able to shed light on the matter.

Linstone and Turoff [4] defines the method as "Delphi may be characterized as a method for structuring a group communication process so that the process is effective in allowing a group of individuals, as a whole, to deal with a complex problem." This technique involves collecting expert opinion through questionnaires or interviews in multiple rounds. The answers to the same questions are sought in each round, and from the second round onwards. The facilitator provides an anonymous summary of other participants' responses of the previous round to every participant and encourages them to update their answers, thereby driving the process towards a collective agreement. The process continues till a pre-determined criterion (like a fixed number of rounds, consensus, results stability, etc.) is met. The method may be altered suitably based on the nature of the problem and size of the group. Conference Telephone Calls may be initiated for fast decisions in case of small closed groups. For a slightly larger group and bigger problem, Committee Meetings need to be hosted. For a still more extensive group of like-minded and acquainted individuals, a Formal Conference or Seminar may be organized. For huge random groups, one needs to administer the Conventional Delphi Method to get participants' responses or a more modern method, i.e., Real-time Delphi, which eliminates time lags between responses. Figure 27.3 summarises the primary phases through which the Delphi process is carried out.

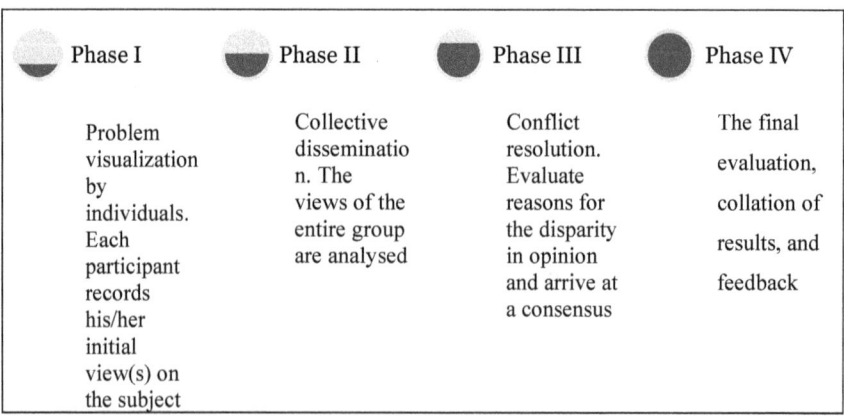

Fig. 27.3 Phases of administering Delphi method

This technique may be effectively used to coordinate between experts from various domains of the power sector and other allied industries, bring their rich experience-based views on a common platform, and thereby arrive at a robust long-term electricity demand forecast. This can also be used effectively for power system planning. However, care must be executed in administering this technique to ensure that the coordinator's views are not imposed upon the participants, responses are adequately summarised, and the differences of opinions between the participants are adequately recorded and analyzed to extract maximum benefits from this process.

Though qualitative techniques are generally more straightforward and faster, primarily due to its less data-intensive nature, its major limitation is the accuracy of forecasts, which is even more evident with the increasing size and complexity of the power systems.

27.5 Quantitative Techniques of Long-term Electricity Demand Forecasting

The problem of electricity demand forecasting is neither purely economic nor purely technical and requires concepts of both economics and engineering. This makes it somewhat mathematical and calls for quantitative and computational techniques. These techniques essentially use numbers from the past to churn out numbers for the future with the help of mathematical algorithms. They range from the simplest ones which use a single indicator to more complex ones with multiple independent variables to highly sophisticated ones incorporating dynamic stochastic processes. The following sections present an overview of some of the most commonly used techniques.

27.5.1 Demand Growth Rate

It is one of the simplest methods where the electricity demand in a time period can be obtained by grossing up the demand in its previous period by a pre-determined growth rate. If r is the demand growth rate (in fraction), then, ED_t, which is the electricity demand in the time period t, can be obtained by using Eq. 27.1.

$$ED_t = ED_{t-1} \times (1 + r) \quad (27.1)$$

This demand can further be obtained from any time period of the past, ED_0 by using Eq. 27.2.

$$ED_t = ED_0 \times (1 + r)^t \quad (27.2)$$

Determining and estimating the growth rate requires domain expertise and is generally done by experienced industry professionals.

27.5.2 Demand Elasticity

This method uses a single driving variable (most commonly income or output), and the change in demand for electricity is computed by using an estimated change in the driving variable. In other words, it uses estimates of income or output to arrive at the forecast of electricity demand. The elasticity of electricity demand is the ratio expressed as unit change in electricity demand to a unit change in the driving variable and can be expressed using Eq. 27.3.

$$e = \frac{\Delta ED/ED}{\Delta I/I} \tag{27.3}$$

where ED is the electricity demand, I is the driving variable, Δ is the change in the value of the variable, and e is the elasticity of electricity demand. Estimating the change in driving variable, as well as the elasticity of electricity demand, requires expert judgment.

27.5.3 Specific Consumption (or Energy Intensity)

In this method, electricity demand, ED is obtained by aggregating the product of output from each human activity in physical terms and their specific electricity consumption, for all the n activities as expressed by Eq. 27.4.

$$ED = \sum_{i=1}^{n}(A_i \times E_i) \tag{27.4}$$

where A_i is the physical output level of activity and E_i is the specific energy consumption of the activity i. Specific electricity consumption (or energy intensity) may be defined as the amount of electricity required to produce one unit of output in a given activity.

27.5.4 End-Use Method and Partial End-Use Method

An End-Use Method is a bottom-up approach to estimating electricity demand through primary consumption data. The energy system is disintegrated into sectors, and the demand of each sector is forecasted by assessing final end-use and then working backward. Most commonly, surveys and interviews are administered to gather primary data from the last mile-consumers, the results of which are aggregated to arrive at the electricity demand at the state, regional, or national level. A typical survey administered to domestic consumers comprises questions regarding the various appliances used in the household, their source(s) of electricity, their willingness to shift a portion of their load as and when required. Similar surveys and interviews are also administered on industries, commercial establishments, and various other categories to capture their future investment and capacity addition plans and thereby estimate demand. Though surveys are a rich source of electricity use data, they are time and resource-intensive and prone to errors or biases. Estimators use other proxies like technologies used, their energy intensity, consumer behaviour, and levels of economic activity based on geographic or social classifications like urban/rural or coastal/hilly.

The earliest end-use models were essentially energy accounting models, which would simply add up primary data to obtain aggregated electricity demand. A further augmentation of this is the Partial End-Use Method, which combines the end-use approach with econometrics and time series. In this method, the forecaster gathers a massive volume of data in terms of historical trends of electricity usage, economic activities, and climatic indicators apart from the primary end-use surveys, and then available econometric techniques are administered. This enhances the explanatory power of the method and, thereby, the accuracy of the forecast.

27.5.5 Trend Analysis and Stochastic Time Series Models

In this method, attempts are made to fit some form of time trend to historical electricity demand data, which is then used to predict future demand. Most commonly, it deploys the method of least squares to find the best trend line fitting the historical data, which is then extrapolated to predict the future, sometimes with ad hoc adjustments. If there is too much irregularity in the data, techniques like exponential smoothing are applied to filter out the noise. The mathematical expression to represent the fitted trendline is given in Eq. 27.5.

$$ED_t = a \times t + b \qquad (27.5)$$

where ED_t is the electricity demand at t-th time period, and a and b are constants.

To increase the predictive power, specific lagged values of ED_t may be used as independent variables, and a linear regression is carried out on ED_t against one or more past values. This technique is known as Autoregressive Model. The number of lagged values used depends on the 'order' of the model, generally denoted by 'p.' Trend analysis can be considered as a first-order Autoregressive Model or AR(1).

To further increase the predictive power of the model, in addition to regressing electricity demand on its own lagged values (autoregressive), the error term is also modelled as a linear combination of the error terms of the past (moving average). Combining these two aspects, this modelling technique is called the Autoregressive Moving Average Model. In cases where the data has non-stationarity, an additional step is required to remove it, and the model is then called Autoregressive Integrated Moving Average Model. The Box-Jenkin method is a further advancement that assumes non-linearity in their modelling approach, and vector Autoregression model performs multivariate time-series analysis. Time-series models are particularly useful at a more aggregated level of forecasting but can be used for disaggregated levels as well, based on data availability. The main advantage of these methods is their simplicity but is based on the unrealistic assumption that the future would look the same as the past.

27.5.6 Econometrics and Regression

Electricity demand is a complex phenomenon that depends on various factors such as the macroeconomic condition of the state, microeconomic or socio-economic condition of the consumers, and weather conditions. It is common wisdom that electricity demand will go up with a higher economic activity or adverse climatic conditions and vice versa. All other methods discussed till now do not address this causal relationship. Trend analysis and time series, for example, consider only one explanatory variable, i.e., past consumption.

The econometrics approach attempts to model energy demand as a function of a host of other variables or indicators. These may include economic variables, i.e., income, price, inflation, or climatic indicators such as rainfall, temperature, humidity, or any other indicator which might influence electricity demand. These variables are known as independent variables. Electricity demand, which is to be forecasted, based on these, is the dependent variable. Econometric modelling aims to arrive at a curve that best explains the dependent variable as a function of the independent variable(s).

One of the simplest econometric models is the linear regression, which tries to fit a straight line through the data by using the principle of Ordinary Least Squares (OLS). The model can be mathematically expressed as given in Eq. 27.6.

$$ED = A_0 + \sum_{i=1}^{n}(A_i \times X_i) \tag{27.6}$$

where ED is electricity demand (dependent variable), X_is are the values of the indicators (independent variables), A_is are their respective coefficients, and A_0 is the intercept.

The model's output gives the A_is, which determines the extent of influence each respective indicator has on electricity demand along with the significance levels of each indicator, i.e., how significant the indicator is for the model. It also provides an R^2 value, which indicates the proportion of variance in the dependent variable, as explained by the independent variables. The forecast error is obtained by computing the root mean square error (RMSE), as given in Eq. 27.7.

$$\text{RMSE} = \sqrt{\frac{1}{T}\sum_{t=1}^{T}(ED_t^f - ED_t^a)^2} \qquad (27.7)$$

where, ED_t^f is the estimated electricity demand; ED_t^a is the actual electricity demand, and t is the number of time periods.

Econometrics also enables exploration of non-linear relationships, like quadratic (X_i^2) or multinomial (X_i^n), or logarithmic (log X_i), or exponential (e^{X_i}). The effects of the interaction of two or more variables at different degrees (i.e. $X_i \times X_j$ or $X_i \times X_j \times X_k$ or $X_i^p \times X_j^q$) is explored through methods like polynomial regression.

These features make econometrics a potent tool for identifying causal factors and, thereby, the most preferred tool for situations where numerous factors affect an outcome, as in the case of electricity demand. It is useful in short-term as well as long-term forecasts, aggregated, or disaggregated levels. It captures demand drivers and quantifies energy-economy interactions. However, its major drawback is that it is extensive data and resource-dependent, which poses a challenge, especially in developing countries.

27.5.7 Partial Adjustment Model

By now, it is evident that the problem of electricity demand forecasting is not a pure time series one. It is driven by a host of factors whose effects need to be observed across time. Thus, it requires both longitudinal as well as cross-sectional analyses. Moreover, the impact of some of the drivers on electricity demand may not be immediate. For example, an increase in income would lead to a rise in electricity demand. Still, there would be a time lag between a consumer receiving higher income and investing in electrical appliances. The traditional econometric or time series techniques do not adequately assimilate both these aspects, and hence, this method was introduced.

The Partial Adjustment Method is a dynamic modelling approach which dampens the effect of change in econometric variables on electricity consumption to the tune of consumers' ability (or inability) to respond to the change. In other words, it captures the consumers' 'partial adjustment' from the current equilibrium level of consumption to a new level as a response to change in any socio-economic (or even climatic) factor. In this approach, a parameter θ_i [0, 1] captures the extent of consumer adjustment. A value of zero would indicate that consumers would not respond at all to the changing environment. In contrast, a value of one would mean that consumers would correctly translate any change in the environment into a corresponding change in their behaviour without any lag [5]. Incorporating this, the traditional demand function is then logarithmically transformed using which electricity demand, as well as demand elasticity's, are econometrically estimated.

27.5.8 Seemingly Unrelated Regression

The next level of challenge in the process of long-term electricity demand forecasting is the granularity or the extent of aggregation/disaggregation. All the methods discussed above consider a household, a state, or a region as one strategic unit at a time and carry out the analysis. However, in reality, each household is a separate unit whose demand drivers are different from those of its neighbouring household but, at the same time, share commonalities as well. Similar is the situation with states or regions, i.e., a state may be looked at as an aggregation of households or a region as an aggregation of states and a nation as an aggregation of regions where each aggregated unit is a system of independent sub-units. Therefore, each system needs to be modelled as a system of regression equations that are seemingly unrelated or have a slightly different set of demand drivers from one another.

This issue has been addressed by the method named Seemingly Unrelated Regression, a multivariate regression technique proposed by Arnold Zellner, which carries out econometric analysis through a set of regression equations, each of which has a dependent variable (electricity demand) being explained by a set of entity-specific independent variables [6]. The same set of variables may or may not be present in all the models, depending upon the characteristics of each state. However, the number of observations should be the same for each entity. The error term associated with all the models is assumed to be correlated. The mathematical expression is represented as Eq. 27.8.

$$ED_{ti} = \sum_{j=1}^{k_i} (A_{ij} \times X_{tij}) + \varepsilon_{ti}; t = 1, 2, \ldots, T; i = 1, 2, \ldots, M; j = 1, 2, \ldots, k_i$$

(27.8)

where ED_{ti} is the t-th observation on the i-th dependent variable, explained by the i-th regression equation, each equation denoting a separate entity. X_{tij} is the t-th observation on j-th explanatory variable of the i-th equation, A_{ij} is the coefficient of X_{tij} at each observation and ε_{ti} is the t-th value of the error component of the i-th equation. The number of explanatory variables associated with the i-th equation is given by k_i.

Seemingly Unrelated Regression becomes equivalent to the Ordinary Least Squares method of regression under two conditions, i.e. (i) there is no correlation between the error terms, and (ii) each equation contains the same set of independent variables. In such a case, each of these can be solved individually to obtain the same results [7].

27.5.9 Machine Learning and Data Mining

Taking cues from the above discussion, we understand that the problem of long-term electricity demand forecasting is multi-faceted, and none of the methods cover all the aspects in totality. Furthermore, to increase the accuracy of forecasts, it requires that a large volume of data be used, both in terms of spread (number of explanatory variables) and depth (number of past years). This data-intensive nature of the problem makes it computationally tedious and challenging for human capabilities creating the need for automation or the development of a system that 'learns' to perform the most complex tasks on its own.

Machine learning, in a broad sense, is a set of tools, algorithms, or computer programs that can learn from their experience. It is a widely used tool for predictive computations. More formally, "A computer program is said to learn from experience E concerning some class of tasks T and performance measure P if its performance at tasks in T, as measured by P, improves with experience E" [8]. It is an interdisciplinary field of artificial intelligence, Bayesian methods, statistics, computational complexity theory, control theory, information theory, philosophy, psychology, and neurobiology. Among its application in innumerable other fields, machine learning is gaining immense popularity in electricity demand forecasting, as is evident from the literature (listed in Appendix 1). The problem of electricity demand forecasting, on the one hand, is diverse as it involves numerous causal factors and different treatment of each of them.

On the other hand, there are commonalities and identifiable patterns that enable us to learn from past experiences and apply them to predict the future. Beyond these, there are disruptions in terms of technologies, costs, socio-economic factors, and any changes in consumer preferences or business environment. These aspects can be well captured through various unsupervised, supervised, and reinforced learning algorithms of machine learning. Data mining is a field closely related, which, unlike machine learning, focuses on pattern recognition. It is an unsupervised learning method that analyses large volumes of data to extract patterns or trends. This makes it a preferred tool for electricity demand forecasting. It can help load forecasters to solve the complex computation-intensive problems efficiently and thereby apply the techniques discussed earlier with a broader range of data and higher accuracy.

Typically, the inputs used by machine learning models are similar to the methods discussed above, with the various explanatory factors like income, price, climatic conditions, and past consumption as inputs and the output being electricity demand. The difference lies in the way different models treat or process them. While some models try to identify patterns in the data, some others try to study the cause-effect relationships.

Support Vector Machine (SVM) techniques use statistical learning theory to identify patterns in the data. SVMs, combined with methods for controlling dimensionality and generalization control, can be used for classification and regression analysis. Artificial Neural Networks (ANN) is another such technique that mimics the arrangement of neurons in living beings to process data. ANN is a collection of connected nodes (or neurons) arranged into layers. The input data constitutes an input layer, and the 'signals' pass through various intermediate layer(s) to produce the output layer. Each node receives signals in the form of weighted real numbers from the nodes of its previous layer as inputs and applies a nonlinear function on the sum of such inputs to produce an output, which is then transmitted as a signal to the nodes of the next layer. The data is processed at each node, which progressively advances through the intermediate layers to the output layer. This method is also suitable for application in regression-based pattern recognitions and predictions. Genetic Algorithm (GA) is another such technique that is inspired by nature's evolution process. The algorithm progressively creates better solutions or 'new genotypes' through the processes of mutation, crossover, and natural selection. It finds applications in models like the autoregressive moving average with exogenous variables (ARMAX). This algorithm tends to offer higher fitting accuracy as it converges towards a global optimal solution asymptotically and is used extensively for solving electricity demand forecasting problems. Bayesian networks are an excellent technique for studying causal interrelation using a graphical network model. It is effectively used to compute the relative effect of various factors on electricity demand using influence diagrams. Fuzzy logic is a machine learning technique that typically deals with vagueness or lack of precision in data using mathematical models. It is especially suited to electricity demand forecasting because, as already discussed earlier, getting complete and accurate historical data of the electricity sector as well as macroeconomic and other indicators, especially in developing countries, is a considerable challenge. One of the most advanced applications of machine learning is Artificial Intelligence (AI), which has capabilities like reasoning and explaining. A manifestation of this is the Knowledge-based expert system, which aims at building the knowledge base with inputs from experts. It has a qualitative part that involves extracting relevant information from experts and storing them as straight facts. These facts are converted to logical rules involving operators, i.e., if-else. This technique, therefore, can accommodate a large variety of factors and disruptions, which can account for the differential treatment of them. For example, factors like rainfall patterns or day of the week or income or sudden disruptions like natural calamities have a different impact on the load curve and need to be treated differently. A knowledge-based expert system creates a knowledge base of all such factors and appropriately models the system with different rules for different scenarios.

These are majorly the techniques used by load forecasters and academicians to predict electricity demand for a long-term horizon. These methods are applied in their original form, modifications, or developing hybrid strategies by combination to suit the modelling requirements.

27.6 Deciding Upon the Best Model

"All models are wrong, but some are useful" [9]. This aptly summarises the current situation. The best model is one that can predict the closest to reality. This 'closeness' to reality is quantified by the sample mean absolute percentage error (MAPE), given by Eq. 27.9.

$$\text{MAPE} = \frac{1}{n} \times \sum_{t=1}^{n} \left| \frac{ED_A - ED_F}{ED_A} \right| \qquad (27.9)$$

where n is the number of data points, ED_A is the actual electricity demand, and ED_F is the forecasted electricity demand. The model having the least MAPE has the least average deviation of forecasted values from the actual and be seemingly the best model. However, this is not the only criterion used for choosing a technique for any forecasting exercise. A lot of contextual factors play a role in determining which model to use. Partial Adjustment Model and Seemingly Unrelated Regression have been observed to have performed well recently in the Indian context [10].

Experience-based qualitative techniques have the advantage of being simple and less data-intensive. They can foresee disruptive changes as it is based on experience, perception, and prior knowledge of domain experts, thereby doing away with issues of computational errors and data discrepancies. It works well in a relatively shorter time horizon and with a lesser number of influential factors. It may be well suited for making load forecasts for smaller systems like an apartment, an academic institute campus, etc. With the increasing complexity of the system, qualitative methods become inaccurate. Quantitative techniques, on the other hand, are data-driven and rely minimally on experiential inputs of experts. In situations where there is the easy availability of reliable data, these methods provide relatively more accurate forecasts but are unable to foresee any disruptive events. However, the forecast accuracy decreases as more and more uncertainties get incorporated into the system.

Two crucial factors on which the choice of model depends are (i) the scale and complexity of the system to be forecasted and (ii) availability of resources [11]. For smaller and uncomplicated systems with negligible uncertainties like a small village in an interior rural location, one may deploy more straightforward techniques like growth rate, elasticity, specific consumption, end-use method, etc. and obtain reasonably accurate results. For a complex system like urban towns, one may use time-series or regression or machine learning techniques as they can capture a little more of complexities and uncertainties than the simpler ones. For large scale systems like a state or a nation, the more straightforward techniques will not suffice because the demand here depends on many complex drivers. Econometric analysis or machine learning would be more appropriate techniques in such a case. Resources required could be in terms of data, expertise, or computational facilities. The more advanced tools like trend analysis, regression, or machine learning will not be the methods of choice in situations where it might be highly challenging to get the expanse of data as demanded by these tools. Expertise required carrying out a load forecasting exercise should include knowledge of the technical aspects of the sector as well as of econometrics. Without these two skill sets, it would be futile to deploy sophisticated techniques. To administer the Delphi technique effectively, the prime requirement is that of sector experts. Without them, the exercise will return absurd results. The qualitative and more straightforward quantitative methods can be administered manually. Still, with the increasing complexity of the system, the volume of data, number of variables, and levels of uncertainties, it is not humanly possible to obtain results manually. Hence, computational facilities would be required. Thus, for running large scale sophisticated models, processing power becomes a bottleneck in choosing the appropriate tool.

Hence, no model can be termed as the absolute best for all situations. An appropriate model needs to be chosen as per the given context, scenario, requirement, and resources. Often forecasters modify the available tools or combine two or more tools to get the required features for the modelling job at hand.

27.7 Computational Intelligence and Its Relevance

Computational intelligence or soft computing, a subset of artificial intelligence, is a part of the computer and engineering sciences devoted to the solution of non-algorithmizable problems [12]. It comprises a set of computational tools, techniques, and methods applicable for situations where conventional methods fail to yield satisfactory results. Such situations could be high levels of complexity, uncertainty, or stochasticity in the problem or real-life problems that may not have a direct translation to binary notation. Computational intelligence works using principles of human reasoning and adaptive learning to address complex situations. Some of the computational intelligence methods are fuzzy logic, neural networks, evolutionary algorithms, learning theory, and probabilistic methods. Having seen what computational intelligence is, let us take a look back at the long-term electricity demand forecasting techniques discussed earlier.

Demand forecasting for small uncomplicated systems does not involve much complexity or stochasticity and, thus, can be addressed using simple manual or computing technologies. Forecasting at the level of a state, region, or nation requires large volumes of data, all of which are not always reliably available. The calculations are computationally intense. Forecasters here need to deal with different varieties of data, each requiring a different treatment but a part of the analyses and calculations are also repetitive and iterative. These peculiarities make computational intelligence techniques have a massive relevance in electricity demand forecasting applications. These methods can identify patterns, self-automate processes, address missing, incomplete, or erroneous data, handle bulky calculations, and enhance their efficiency through learning, thereby making them the most preferred tools for the job. All of the powerful statistical computing software and programming packages offer customized modules for power systems modelling, and long-term demand forecasting is a part of such modules.

The invention of computers has been a boon to humanity and electricity demand forecasters are no exception. The application of sophisticated and smart computing techniques has reduced human efforts, improved the efficiency of forecasts, and has transformed the entire power sector globally.

27.8 Recent Trends in Long-term Electricity Demand Forecasting and System Planning

With the changing outlook of the consuming entities, the pattern of electricity consumption is set to undergo significant changes. He et al. [13] explores the changes in existing demand prediction models in the light of industries shifting from being factor-driven to innovation-driven in the backdrop of a Chinese industrial city. It identifies new factors that would influence the electricity consumers' behaviour and provides suggestions for network augmentation. Long-term electricity demand forecasting is evidently a complex phenomenon comprising various decision criteria like maximising comfort and minimising cost, subject to constraints like disposable income vs. electricity tariffs and so on. Thus, it can be looked at as a multi-criteria optimisation problem instead of a problem involving a dependent and many independent variables. Angelopoulos et al. [14] provides a novel framework comprising ordinal regression analysis and two disaggregated forecasting models, which uses the theory of multi-criteria analysis. This has been applied to forecast electricity demand for Greece for the period 2017–2027 using a set of key influencing variables identified through expert interviews.

The activity of demand forecasting, is shifting from being a standalone activity to a part of a larger picture i.e. energy system modelling. Various tools and models have been developed which simulate an entire electricity system as an integration of demand and supply and as per choices of resolution ranging from a household, to a locality, a state, a region or a country. They provide choices for scenarios to test various assumptions and proposed interventions to the sector and help in assessment of the system based on parameters like levelized costs, carbon emissions, energy efficiency, use of resources etc. These models have demand forecasting packages inbuilt within them. Some of such models are GCAM, MESSAGE, LEAP, MARKAL/TIMES, NEMS, POLES etc. Mirjat et al. [15] for example, applies Long-range Energy Alternatives Planning System (LEAP) for forecast electricity demand for Pakistan for 2015–2050.

The modelling approach is in particular relevant because the electricity systems globally are going through a phase of transition. It is envisioned that in order to arrest global warming to 1.5 °C, we will have to decarbonise our power systems in the next three decades such that 70–85% of world energy is produced from renewables [16]. This infusion of renewables would bring in technical and economic issues like variability, intermittency and uncertainty of generation, backing down of thermal generation, threat of stranding of fossil fuel assets etc. Adding to these, would be the emergence of prosumers and popularisation of economic storage technologies creating a considerable difference between actual consumer demand and demand from electricity suppliers. These aspects can only be visualised through modelling or simulation approaches.

Ringkjob et al. [17] tracks the evolution of modelling tools both long- and short-term applications and provides a comparison of 75 of the most popular models and their applicability scenarios. Prina et al. [18] reviews the classification schemes of bottom-up energy models and identifies the key challenges as temporal, spatial, techno-economic and cross-sectoral resolution. Fattahi et al. [19] provides a framework for analysing various energy system models and highlight that models need to be flexible to accommodate new technologies, efficiency improvements of existing technologies, increase in electricity access, centralisation/decentralisation, changes in economic or social environment etc. Lai et al. [20] explores literature for various models which consider future electricity storage, finds inadequacies in the existing models and proposes a novel long-term power flow electrical power system framework (LEPSF).

27.9 Way Forward and Scope for Future Research

Going ahead, the power sector across the globe, especially in developed and developing economies, are at a stage of witnessing a massive transition in the pattern of consumption. These transitions can be majorly characterized by the transformation of consumers to prosumers, the emergence of economic and promising storage technologies and, smarter tariff plans. These are expected to revolutionize consumers' electricity consumption behaviour completely. The transformation of consumers to prosumers would mean their shifting from complete dependence on the grid to partial or no reliance. It would imply that though the overall consumption at the consumer end remains more or less the same, the drawl from the grid would witness a sharp decline. The popularisation of storage technologies would disrupt the most peculiar feature of the power system that it cannot be stored and needs to be consumed at the same moment when it is generated. This would enable prosumers to stagger their consumption of excess solar and wind generation during the lean generating hours leading to even lesser dependence on the energy supplier. Smarter tariff plan offerings by retail supply entities like a time-of-use tariff, real-time tariff, discounts, and incentives would significantly impact consumption profiles. Future forecasters must make a note of such upcoming transitive measures being talked about in the sector and account for them adequately in their models.

Energy system modelling and planning is a relatively new field which of late, has gained much attention from the research community. However, the tools and models available are computation intensive and involve complex programming languages, thus requiring high end solver systems and technological experts to handle them. Future researchers may look into the dimension of building more robust, computationally efficient and higher accuracy models. They may also try to build upon the existing algorithms, refine them and present them as a user-friendly package for easy implementation by the industry stakeholders.

Another aspect which is slightly less explored is the concept of cross-sectoral planning. There is a significant lack of interactions and coordination between various associated stakeholders, which needs to be strengthened. Electricity being a multi-stakeholder subject, it is extremely essential that planning is done considering all allied sectors as an integrated system. For example, instead of considering electricity networks as a standalone planning unit, it may be looked at as an integration of municipal corporations, urban developers, water supply, sanitation, transport, coal mining and other such sectors. Thinking futuristically, a resource planning module may be developed to integrate and synchronize the entire chain of operations of the sector. The module will allow the automation of data acquisition, computations on the data, generation of load forecasts, their periodic revisions, and all other subsequent activities at all levels ranging from procurement of coal or maintenance of spares, optimal running of generating units, incentive or penalty mechanisms and so on. Future research may proceed in this direction and design an integrated system plan for a region. This would enable streamlining of the activities of all players and stakeholders and would significantly increase the overall efficiency of the entire electricity sector.

27.10 Concluding Remarks

A 'good' forecast depends both on the quality of data as well as the appropriateness of the method used. The quality of data implies its breadth, depth, and richness. Electricity demand is a complex phenomenon, requires data on various aspects of the macro and micro economy, which is not just limited to the power sector but requires cross-sectoral coordination. Once useful quality data is obtained, selecting the right method becomes more comfortable. Therefore, due emphasis must be given on data availability for achieving electricity load forecasts of high accuracy. The regulatory bodies and the policymakers may develop an integrated framework for all the sector associates to upload relevant data in useable format periodically. The associated sectors would be various industries, railways, municipal corporations, meteorological departments, and others, which have a significant influence on the electricity demand. The mechanism may be automated to ensure seamless data availability to forecasters.

Since the forecast of electricity demand has benefits to multiple stakeholders, they should all join hands together and take an active interest in augmenting the current industry practices. They should invest heavily to support and incentivize research activities in this area, which would enable the designing of new and more effective tools that would result in further improvement in accuracy, lesser computing efforts, and better understanding.

> **Exhibit 1: A Case Study of India**
>
> Central Electricity Authority does long-term forecasting of energy and peak demand of electricity at the national level: an organization constituted under relevant Acts of the Indian parliament to provide techno-commercial inputs to the Indian power sector. It carries out a detailed exercise of estimating category-wise long-term electricity demand periodically through its Electric Power Survey (EPS) Committee since 1962 under the aegis of various legislative and regulatory provisions.

It typically employs the Partial End-Use Method, which is a combination of end-use method and econometrics. It is a bottom-up approach that starts with the estimation of end-use energy requirement and consumption patterns of various categories of consumers like industrial, commercial, domestic, agricultural, railway/traction, etc. After that, it is grossed up with transmission and distribution losses to obtain energy requirements at the busbar. Peak demands at the state, regional or national level is then estimated based on the consumer mix of each state/region and applying intra-/inter-regional diversity factor, as applicable.

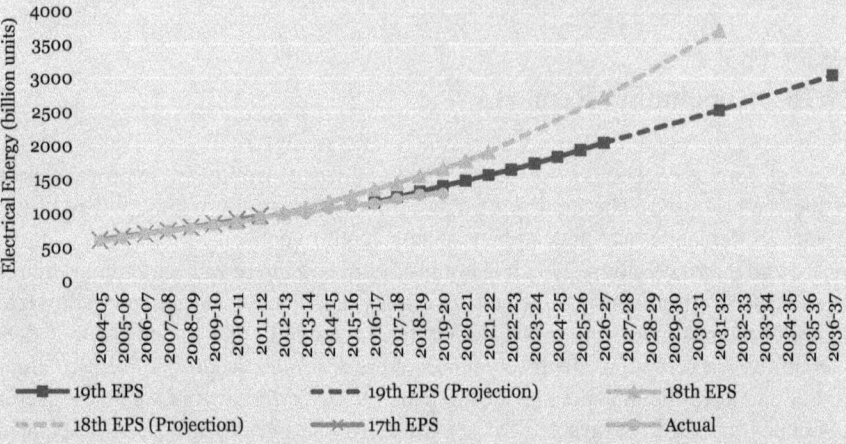

Fig. 27.4 All India electrical energy requirement—estimates versus actual (Source Central Electricity Authority, 17th 18th and 19th Electric Power Survey of India New Delhi 2007, 2011, 2017; Centre for Energy Regulation Monograph 2019)

Here, we present snapshots from the last three EPS reports (17th, 18th, and 19th) to illustrate how historical estimates have been periodically fine-tuned based on actual load growth trends. Each of these reports has been published roughly at intervals of five years and has estimates of twenty years from their respective year of publication. In each piece, estimates of the first ten years is yearly and after that, at five-year intervals. Figure 27.4 shows electrical energy requirement projections as per 17th, 18th and 19th EPS reports versus actual.

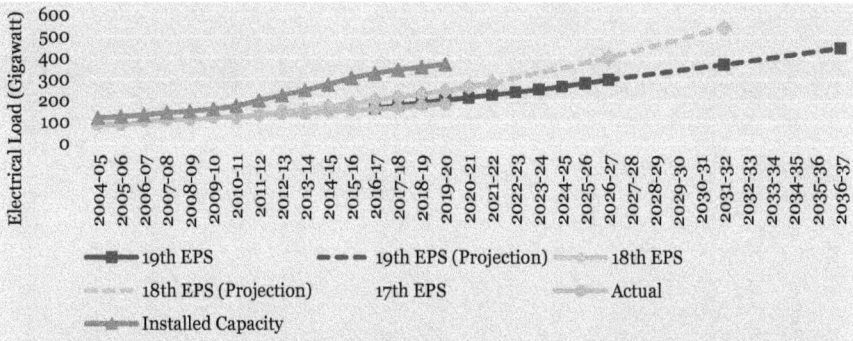

Fig. 27.5 All India peak load—estimates versus actual versus installed capacity (Source Central Electricity Authority, 17th 18th and 19th Electric Power Survey of India New Delhi 2007, 2011, 2017; Centre for Energy Regulation Monograph 2019)

Figure 27.5 shows peak demand projections as per 17th, 18th, and 19th EPS reports vs. actual and growth in installed capacity.

The historical projections have over-estimated electricity demand of the country, which has, in turn, signalled massive investments in the generation sector. There has been capacity addition at a compounded annual growth rate of 7.89 percent from FY 2004–05 to FY 2019–20, whereas the peak demand has grown by 5.04 percent, and energy requirement has grown by 5.34 percent in the same period. Owing to these excess capacity additions, the generation assets are being grossly under-utilized. Plant load factor (PLF) of coal and lignite plants of the country has reached their lowest in the decade at 56.41 percent in FY 2019–20 [21]. This, coupled with issues like fuel supply bottleneck, poor financial health of DISCOMs, developers' inability to timely infuse equity or service debt, aggressive bidding, regulatory and legal issues, delays in project development, etc. have led to 40.13 gigawatts of generation assets being stranded [22], which is a cause of great concern for all stakeholders of the sector.

This emphasizes the importance of the accuracy of forecasts in the power sector. Since these projects have long gestation periods, forecasts are the most critical input which projects developer have while taking investment decisions and the accuracy of the forecasts determine the success or failure of the projects.

Appendix 1

See Table 27.1.

Table 27.1 Existing literature on electricity demand forecasting

Authors	Methodology Adopted	Macroeconomic Factors	Developments and Technological Investments	Nature of Demand	Load Uncertainty (Short Term)	End-use flexibility	Temporal granularity	Reach
Long-term annual electricity demand								
Pérez-García and Moral-Carcedo [23]	Regression/Econometric models, Index decomposition analysis (IDA)	Population, Energy intensity of sectors, GDP, Sectoral share of GDP	None	Industry, Non-residential, Households	None	None	Annual	National
Torrini et al. [24]	Regression, Fuzzy logic	Population growth, GDP	None	Commercial, Residential, Industry	None	None	Annual	National
Pessanha and Leon [25]	Econometric models, Regression	Population growth, Household annual income, consumption, and expenditure patterns	None	Includes appliance ownership index (AOI)	None	None	Monthly	National
Bianco et al. [26]	Multiple regression	Electricity price, Population, GDP, GDP per capita	None	None (national demand)	None	None	Annual	National
Mohamed and Bodger [27]	Multiple regression	Electricity price, Population, GDP	None	Domestic, Non-domestic, total	None	None	Annual	National
da Silva et al. [28]	Linear hierarchical models in combination with bottom-up approach		None	Paper and Pulp Industry (only)	None	None	Annual	National
Ardakani and Ardehali [29]	Artificial Neural Networks in combination with multi-variable regression technique	Energy Import and Export, Population GDP	None	None (national demand)	None	None	Annual	National

(continued)

Table 27.1 (continued)

Authors	Methodology Adopted	Macroeconomic Factors	Developments and Technological Investments	Nature of Demand	Load Uncertainty (Short Term)	End-use flexibility	Temporal granularity	Reach
Chen and Wang [30]	Artificial Neural Networks utilizing Experts opinion on long-term trends. (PCA-FFNN approach)	Parameters not clear for long-term forecast	None	None (national demand)	None	None	Annual	National
Long-term peak power								
Melodi et al. [31]	Artificial Neural Networks employing non-linear auto-regression, Multiple regressions	Population, GDP	None	None (national demand)	None	None	Annual	National
Dalvand et al. [32]	Regression and Artificial Neural Networks	Gold price, Oil price, Population, Industrial activity, GDP, Income from Oil, Exchange rate of USD	None	None (national demand)	None	None	Annual	National
Daneshi et al. [33]	Artificial Neural Networks approach, Multiple regression model	Fuel price, Temperature, Population, GSP, Employees, GDP	None	Residential, commercial, industrial	None	None	Monthly	National
Towill [34]	Multiple regression	Temperature, GDP	None	None (national demand)	None	None	Monthly	National
Aslan et al. [35]	Artificial Neural Networks, Quadratic regression, Multiple regression	Temperature, Population	None	None (national demand)	None	None	Monthly	National

(continued)

Table 27.1 (continued)

Authors	Methodology Adopted	Macroeconomic Factors	Developments and Technological Investments	Nature of Demand	Load Uncertainty (Short Term)	End-use flexibility	Temporal granularity	Reach
Hyndman and Fan [36]	Regression, Semi-Parametric models	Temperature, GDP	None	None	None	None	Annual/weekly	National
Medium-term								
Boroojeni et al. [37]	Load forecasting utilizing Black-box model based on past historic data (AR/MA-model)	None	None	None	None	None	Hourly	National
Pillai et al. [38]	Normalized load profiles generated utilizing ANN model Black-box approach	None	None	None	Temperature sensitivity	None	Hourly	National
Goude et al. [39]	Black-box approach utilizing Generalized Additive Model (GAM)	None	None	None	Temperature effects	None	Hourly	National
Sajjad et al. [40]	Goodness of fit is estimated for different sampling sizes	None	House	Residential	None	None	Hourly	Cluster of household consumers
Zhao and Guo [41]	Grey-box model utilizing the concepts of Machine learning and AI	None	None	None	None	None	Annual	National
'WHAT IF' INVESTIGATIONS (no long-term temporal evolution)								
Lindberg and Doorman [42], Lindberg et al. [43]	Regression utilizing top-down approach	None	None	SECTORAL: commercial ENERGY SERVICE: heat and electric demand	Temperature sensitivity	None	Hourly	Cluster of commercial buildings
Veldman et al. [44]	Regression utilizing top-down approach	None	Measured load traces of HPs and micro-CHPs	SECTORAL: Residential ENERGY SERVICE: heat demand	None	None	Hourly	Cluster of residential technologies

(continued)

Table 27.1 (continued)

Authors	Methodology Adopted	Macroeconomic Factors	Developments and Technological Investments	Nature of Demand	Load Uncertainty (Short Term)	End-use flexibility	Temporal granularity	Reach
Bruninx et al. [45]	Bottom-up combined with top-down price elasticity	None	Air-to-water heat pump (ASHP) technical restrictions	SECTORAL: residential, commercial ENERGY SERVICE: heat and electric demand	Stochastic profiles	Thermal inertia of buildings	Hourly	25 identical dwellings
Fischer et al. [46]	Bottom-up approach	None	Air-to-water heat pump (ASHP), ground sourced heat pumps (GSHP)	SECTORAL: residential ENERGY SERVICE: heat and electric demand	Stochastic profiles	HP system	Hourly	Cluster of residential buildings
Fischer et al. [47]	Bottom-up approach	None	None	SECTORAL: residential ENERGY SERVICE: electric appliances	Stochastic profiles	End-use behaviour	Hourly	Cluster of residential buildings
Ericson [48]	In-field investigation	None	None	SECTORAL: residential ENERGY SERVICE: domestic hot water		Disconnecting residential water heaters	Hourly	Cluster of residential buildings
Baetens et al. [49]	Bottom-up approach	None	Air-to-water heat pump (ASHP) and PV	SECTORAL: residential ENERGY SERVICE: heat and electric demand	Stochastic user profiles	Flexibility of building's thermal mass, and changed user behaviour of electric appliances	Hourly	Cluster of residential buildings
Asare-Bediako et al. [50]	Bottom-up and partly decomposition approach. (1) The basis is the SLP of a residential area, (2) applying electric load from new technologies	None	HPs, micro CHPs, PV, EVs	SECTORAL: residential ENERGY SERVICE: heat and electric demand, transport demand	None	None	Hourly	Cluster of 200 households

(continued)

Table 27.1 (continued)

Authors	Methodology Adopted	Macroeconomic Factors	Developments and Technological Investments	Nature of Demand	Load Uncertainty (Short Term)	End-use flexibility	Temporal granularity	Reach
Pantoš [51]	(1) bottom–up Monte Carlo simulation of individual EV load traces. (2) cost-optimal charging	None	E-MOBILITY: no technical restriction of EVs applied	SECTORAL: Transport	Yes	Cost-optimal charging	Hourly	Car fleet
Andersson et al. [52]	Model for the regulating power market. Investigating impact of parked EVs	None	E-MOBILITY: simulates 500 individual EVs in Sweden and Germany	SECTORAL: Transport sector, and Power sector	None	Cost-optimal EV charging	Hourly	National
Long-term hourly load forecast								
Boßmann et al. [53]	Partial decomposition approach (PDA). (1) bottom-up forecast of annual electricity demand. (2) decomposition of the annual load to hourly load	GDP, population, energy prices, techno-economic data, CO_2 price	TECHNICAL DATA: equipment rate, specific consumption, operation time, life time, investment cost	Residential, commercial, industry	None	None	Hourly load for a year	National
Boßmann and Staffell [54]	eLOAD-model. Partial decomposition approach (PDA) (1) bottom-up forecast of annual electricity demand. (2) decomposition of annual load to hourly load	GDP, population, energy prices, techno-economic data, CO_2 price	HEAT: share of electric heating of residential heating, HP's technical specifications E-MOBILITY: share of passenger fleet, EV's technical specifications	SECTORAL: residential, commercial, industry, transport ENERGY SERVICE/TECHNOLOGY: lighting, electric heating, refrigeration, white goods, EVs etc	Temperature sensitivity	Cost-optimal charging strategies modelled as a satellite MILP	Hourly load for a year	National

(continued)

Table 27.1 (continued)

Authors	Methodology Adopted	Macroeconomic Factors	Developments and Technological Investments	Nature of Demand	Load Uncertainty (Short Term)	End-use flexibility	Temporal granularity	Reach
Moral-Carcedo and Pérez-García [55]	Partial decomposition approach (PDA) (1) bottom–up forecast of annual electricity demand (2) decomposition of the annual load to hourly load by temporal disaggregation techniques	GDP, population, energy prices	Annual electricity demand account for: (1) stock projections of electric equipment per sector, (2) efficiency factor, (3) substitution effect for electricity demand against alternative energy carriers (gas, fuel, etc.) COOLING/E-MOBILITY: no technical specifications included	SECTORAL: Residential and Others (incl. commercial and industry)	Temperature sensitivity applied on hourly demand	Daylight saving	Hourly load for a year	National
Andersen et al. [56]	Partial decomposition (1) top–down regression model of measured individual load traces (2) bottom–up aggregation of average load traces	Stock projections of HPs and EVs in households towards 2030	HEAT: no technical restrictions of HPs applied E-MOBILITY: no technical restrictions of EVs applied	SECTORAL: Residential	None	Rule-based flex of HP and EV load	Hourly load for a year	National

(continued)

Table 27.1 (continued)

Authors	Methodology Adopted	Macroeconomic Factors	Developments and Technological Investments	Nature of Demand	Load Uncertainty (Short Term)	End-use flexibility	Temporal granularity	Reach
Veldman et al. [57]	Partly decomposition approach. (1) bottom–up forecast of household stock, multiplied by the SLP. (2) applying electric load from new technologies	Population and technology stock projections	HEAT: HPs and CHPs E-MOBILITY: EVs	SECTORAL: Residential	None	Rule-based flexibility of HPs (via thermal mass of buildings), EVs, and electric appliances (10% is assumed flex regardless of time)	Hourly load for a year	Cluster of residential buildings
Lindberg [58] Lindberg et al. [59]	Partial decomposition approach (1) bottom-up forecast of building stock, (2) applying heat load and electric specific load profiles	Population and building stock projections (m^2)	HEAT: share of electric heating of buildings' heating demand, HP's technical specifications	SECTORAL: residential, commercial ENERGY SERVICE: heat and electric demand	Temperature sensitivity	None	Hourly	Commercial and residential sector at national level
Statnett [60]	LeoPard model. Integrated top–down and bottom-up load model (1) sectoral and technological components on long-term annual load forecast (bottom–up) (2) Disaggregating on hours	Population and building stock	HEAT: heat pumps E-MOBILITY: EVs	SECTORAL: Residential, Commercial, Industry, Transport	None	Rule based flexibility options (assumption of amount of demand that is flexible and how it changes the load profile)	Hourly load for a year	National

(continued)

Table 27.1 (continued)

Authors	Methodology Adopted	Macroeconomic Factors	Developments and Technological Investments	Nature of Demand	Load Uncertainty (Short Term)	End-use flexibility	Temporal granularity	Reach
Power market analysis and energy system analysis								
Bertsch et al. [61]	Unit-commitment power market model (DIMENSION). Snapshots of 2020, 2030 and 2050	Not specified	None	None	None	Assumed percentage of demand (equal for all hours) that is flexible. (i.e. can shift demand)	Hourly	National
Gils [62]	Not specified	None	None	All sectors	None	Max and min available flexible demand	Max and min amount of demand that is flexible	National
Pina et al. [63]	TIMES energy-system model	Not specified	HOUSEHOLDS: flexible washing machines, dryers and dishwashers	All sectors	None	Yes	Hourly	National
Pina et al. [64]	Soft-linking TIMES with short-term operational model	Not specified	E-MOBILITY: no technical restrictions of EVs applied	All sectors	None	Yes	Hourly	National
Lund asnd Kempton [65]	Short-term operational model Energy PLAN	None	E-MOBILITY: technical restrictions of EVs included	Transport and Power and Heat sector	None	Flexible charging possibilities	Hourly	National
Graabak et al. [66]	EMPS power-market model. Snapshot of 2050. Partial decomposition of load: (1) extrapolated historic load profile. (2) adding EV load profile on top	Population and vehicle stock projections	E-MOBILITY: 100%introduction of EVs in 2050	Transport and Power sector	30 climatic years (temperature)	Cost-optimal charging strategies modelled as a satellite LP	Hourly	National

(continued)

Table 27.1 (continued)

Authors	Methodology Adopted	Macroeconomic Factors	Developments and Technological Investments	Nature of Demand	Load Uncertainty (Short Term)	End-use flexibility	Temporal granularity	Reach
Juul and Meibom, [67], Hedegaard et al. [68]	Balmorel	Population and vehicle stock projections	E-MOBILITY: Evaluating the market competition between non-plug-ins, EVs, PHEVs, and FCEVs	Transport sector, and Power and Heat sector	None	Cost-optimal EV charging, heat storage flex	Hourly	National
ENTSO-E [69]	Dispatch, unit-commitment model of EU	None	EVs and HPs	Power sector	None	None	Hourly	National
FINGRID et al. [70]	Dispatch, unit-commitment model of Nordic countries. Snapshot of 2025	Scaling up current load by 9%	None	None	None	None	Hourly	National
Bøhnsdalen et al. [71]	Dispatch, unit-commitment model of Nordic countries (EMPS). Snapshot of 2040	Scaling up current load by 50 TWh/yr	No technical specifications of EVs or HPs	Annual demand is sectorally decomposed into industry, residential, commercial, transport	30 climatic years (temperature)	None	Hourly	National

Source Lindberg et al. 2019 [72]

References

1. Livewire, World Bank Group (2017) Forecasting electricity demand: an aid for practitioners
2. Singh A, Pratap M, Das, Sharma PA, Gupta KK et al (2019) Regulatory framework for long-term demand forecasting and power procurement planning. Centre for Energy Regulation (CER), IITK; ISBN 978-93-5321-969-7
3. Makkonen M, Patari S, Jantunen A, Viljainen S et al (2012) Competition in the European electricity markets-outcomes of a Delphi study. Energy Policy 44:431–440
4. Linstone HA, Turoff M (2002) The Delphi method techniques and applications. Addison-Wesley Publishing Company, Advanced Book Program, ISBN 0-201-04294-0
5. Paul AC, Myers EC, Palmer KL et al (2009) A partial adjustment model of US electricity demand by region, season, and sector. Resources for the Future Discussion Paper No. 08-50. https://doi.org/10.2139/ssrn.1372228
6. Zellner A (1962) An efficient method of estimating seemingly unrelated regressions and tests for aggregation bias. J Am Stat Assoc 57(298):348–368
7. Khan MA., Khan MZ, Zaman K, Arif M (2014) Global estimates of energy-growth nexus: application of seemingly unrelated regressions. Renew Sustain Energy Rev 29: 63–71. https://doi.org/10.1016/j.rser.2013.08.088
8. Mitchell TM (1997) Machine learning. McGraw-Hill, New York, ISBN: 978-0-07-042807-2
9. Box GEP (1976) Science and statistics. J Am Stat Assoc 71(356):791–799. https://doi.org/10.1080/01621459.1976.10480949
10. Central Electricity Authority (2019) Long term electricity semand forecasting. New Delhi
11. Lipinski AJ (1990) 2.0. Introduction. Energy. https://doi.org/10.1016/0360-5442(90)90084-F
12. Duch W, Mańdziuk J (eds) Challenges for computational intelligence. Studies in computational intelligence, vol 63. Springer, Berlin, Heidelberg. https://doi.org/10.1007/978-3-540-71984-7_1
13. He Y, Jiao J, Chen Q, Ge S, Chang Y, Xu Y et al (2017) Urban long term electricity demand forecast method based on system dynamics of the new economic normal: the case of Tianjin. Energy 133:9–22
14. Angelopoulos D, Siskos Y, Psarras J (2019) Disaggregating time series on multiple criteria for robust forecasting: the case of long-term electricity demand in Greece. Eur J Oper Res 275(1):252–265
15. Mirjat NH, Uqaili MA, Harijan K, Walasai GD, Mondal MA, Sahin H et al (2018) Long-term electricity demand forecast and supply side scenarios for Pakistan (2015–2050): a LEAP model application for policy analysis. Energy 165:512–526
16. IPCC, 2018: Global warming of 1.5 °C. An IPCC Special Report on the impacts of global warming of 1.5°C above pre-industrial levels and related global greenhouse gas emission pathways, in the context of strengthening the global response to the threat of climate change, sustainable development, and efforts to eradicate poverty [Masson-Delmotte, V., P. Zhai, H.-O. Pörtner, D. Roberts, J. Skea, P.R. Shukla, A. Pirani, W. Moufouma-Okia, C. Péan, R. Pidcock, S. Connors, J.B.R. Matthews, Y. Chen, X. Zhou, M.I. Gomis, E. Lonnoy, T. Maycock, M. Tignor, and T. Waterfield (eds)]
17. Ringkjob HK, Haugan PM, Solbrekke IM et al (2018) A review of modelling tools for energy and electricity systems with large shares of variable renewables. Renew Sustain Energy Rev 96:440–459
18. Prina MG, Manzolini G, Moser D, Nastasi B, Sparber W et al (2020) Classification and challenges of bottom-up energy system models—a review. Renew Sustain Energy Rev 129:109917
19. Fattahi A, Sijm J, Faaij A et al (2020) A systemic approach to analyze integrated energy system modeling tools: a review of national models. Renew Sustain Energy Rev 133:110195
20. Lai CS, Locatelli G, Pimm A, Wu X, Lai LL et al (2020) A review on long-term electrical power system modeling with energy storage. J Cleaner Prod 124298
21. MoP (2020) Power sector at a glance ALL INDIA. Available at: https://powermin.nic.in/en/content/power-sector-glance-all-india; accessed on 31 May 2020

22. HLEC, GoI (2018) Report of the high level empowered committee to address the issues of stressed thermal power projects. High Level Empowered Committee constituted by Govt. of India on 29 July 2018
23. Pérez-García J, Moral-Carcedo J (2016) Analysis and long term forecasting of electricity demand trough a decomposition model: a case study for Spain. Energy 97:127–143
24. Torrini FC, Souza RC, Oliveira FL, Pessanha JF et al (2016) Long term electricity consumption forecast in Brazil: a fuzzy logic approach. Socioecon Plann Sci 54:18–27
25. Pessanha JF, Leon N (2015) Forecasting long-term electricity demand in the residential sector. Proc Comput Sci 55:529–538
26. Bianco V, Manca O, Nardini S (2009) Electricity consumption forecasting in Italy using linear regression models. Energy 34(9):1413–1421
27. Mohamed Z, Bodger P (2005) Forecasting electricity consumption in New Zealand using economic and demographic variables. Energy 30(10):1833–1843
28. da Silva FL, Oliveira FL, Souza RC (2019) A bottom-up bayesian extension for long term electricity consumption forecasting. Energy 167:198–210
29. Ardakani FJ, Ardehali MM (2014) Long-term electrical energy consumption forecasting for developing and developed economies based on different optimized models and historical data types. Energy 65:452–461
30. Chen T, Wang YC (2012) Long-term load forecasting by a collaborative fuzzy-neural approach. Int J Electr Power Energy Syst 43(1):454–464
31. Melodi AO, Momoh JA, Adeyanju OM et al (2016) Probabilistic long term load forecast for Nigerian bulk power transmission system expansion planning. In: IEEE PES PowerAfrica, pp 301–305. https://doi.org/10.1109/PowerAfrica.2016.7556621
32. Dalvand MM, Azami SB, Tarimoradi H et al (2008) Long-term load forecasting of Iranian power grid using fuzzy and artificial neural networks. In: 43rd International Universities Power Engineering Conference. IEEE, pp 1–4. https://doi.org/10.1109/UPEC.2008.4651538
33. Daneshi H, Shahidehpour M, Choobbari AL (2008) Long-term load forecasting in electricity market. In: IEEE International Conference on Electro/Information Technology. IEEE, pp 395–400. https://doi.org/10.1109/EIT.2008.4554335
34. Towill S (1974) Estimation of maximum demand on a British electricity-board system. Forecast periods of 1–3 years. Proc Inst Electrical Eng 121(7):609–615
35. Aslan Y, Yavasca S, Yasar C (2011) Long term electric peak load forecasting of Kutahya using different approaches. Int J Tech Phys Problems Eng 3(2):87–91
36. Hyndman RJ, Fan S (2009) Density forecasting for long-term peak electricity demand. IEEE Trans Power Syst 25(2):1142–1153
37. Boroojeni KG, Amini MH, Bahrami S, Iyengar SS, Sarwat AI, Karabasoglu O et al (2017) A novel multi-time-scale modeling for electric power demand forecasting: from short-term to medium-term horizon. Electric Power Syst Res 142:58–73
38. Pillai GG, Putrus GA, Pearsall NM et al (2014) Generation of synthetic benchmark electrical load profiles using publicly available load and weather data. Int J Electr Power Energy Syst 61:1
39. Goude Y, Nedellec R, Kong N (2014) Local short and middle term electricity load forecasting with semi-parametric additive models. IEEE Trans Smart Grid 5(1):440–446
40. He Y, Jiao J, Chen Q, Ge S, Chang Y, Xu Y et al (2017) Urban long term electricity demand forecast method based on system dynamics of the new economic normal: the case of Tianjin. Energy 133:9–22
41. Zhao H, Guo S (2016) An optimized grey model for annual power load forecasting. Energy 107:272–286
42. Lindberg KB, Doorman G (2013) Hourly load modelling of non-residential building stock. In: IEEE Grenoble Conference France. IEEE, pp 1–6
43. Lindberg KB, Doorman G, Chacon JE, Fischer D et al (2015) Hourly electricity load modelling of non-residential passive buildings in a nordic climate. In: IEEE Eindhoven PowerTech. IEEE, pp 1–6

44. Veldman E, Gibescu M, Slootweg H, Kling WL et al (2011) Impact of electrification of residential heating on loading of distribution networks. In: IEEE Trondheim PowerTech. IEEE, pp 1–7
45. Bruninx K, Patteeuw D, Delarue E, Helsen L, D'haeseleer W et al (2012) Short-term demand response of flexible electric heating systems: the need for integrated simulations. In: 10th international conference on the European Energy Market (EEM). IEEE, pp 1–10
46. Fischer D, Wolf T, Wapler J, Hollinger R, Madani H et al (2017) Model-based flexibility assessment of a residential heat pump pool. Energy 18:853–864
47. Fischer D, Stephen B, Flunk A, Kreifels N, Lindberg KB, Wille-Haussmann B, Owens EH et al (2016) Modeling the effects of variable tariffs on domestic electric load profiles by use of occupant behavior submodels. IEEE Trans Smart Grid 8(6):2685–2693
48. Ericson T (2009) Direct load control of residential water heaters. Energy Policy 37(9):3502–3512
49. Baetens R, De Coninck R, Van Roy J, Verbruggen B, Driesen J, Helsen L, Saelens D et al (2012) Assessing electrical bottlenecks at feeder level for residential net zero-energy buildings by integrated system simulation. Appl Energy 96:74–83
50. Asare-Bediako B, Kling WL, Ribeiro PF et al (2014) Future residential load profiles: scenario-based analysis of high penetration of heavy loads and distributed generation. Energy Build 75:228–238
51. Pantoš M (2011) Stochastic optimal charging of electric-drive vehicles with renewable energy. Energy 36(11):6567–6576
52. Andersson SL, Elofsson AK, Galus MD, Göransson L, Karlsson S, Johnsson F, Andersson G et al (2010) Plug-in hybrid electric vehicles as regulating power providers: case studies of Sweden and Germany. Energy Policy 38(6):2751–2762
53. Boßmann T, Lickert F, Elsland R, Wietschel M et al (2013) The German load curve in 2050: structural changes through energy efficiency measures and their impacts on the electricity supply side. In: ECEEE Summer Study Proceedings, pp 1199–1211
54. Boßmann T, Lickert F, Elsland R, Wietschel M et al (2013) The German load curve in 2050: structural changes through energy efficiency measures and their impacts on the electricity supply side. In: ECEEE Summer Study Proceedings, pp 1199–1211
55. Moral-Carcedo J, Pérez-García J (2017) Integrating long-term economic scenarios into peak load forecasting: an application to Spain. Energy 140:682–695
56. Andersen FM, Baldini M, Hansen LG, Jensen CL et al (2017) Households' hourly electricity consumption and peak demand in Denmark. Appl Energy 208:607–619
57. Veldman E, Gibescu M, Slootweg HJ, Kling WL et al (2013) Scenario-based modelling of future residential electricity demands and assessing their impact on distribution grids. Energy Policy 56:233–247
58. Lindberg, KB (2017) Impact of zero energy buildings on the power system—a study of load profiles, flexibility and system investments. Doctoral Thesis. Norwegian University of Science and Technology (NTNU) Retrieved. https://hdl.handle.net/11250/2450566
59. Lindberg KB, Dyrendahl T, Doorman G, Korpås M, Øyslebø E, Endresen H, Skotland CH et al (2016) Large scale introduction of zero energy buildings in the nordic power system. In: 13th International Conference on the European Energy Market (EEM). IEEE, pp 1–6
60. Statnett SF (2018) Forbruksprognose Stor-Oslo. Retrieved. https://www.statnett.no/globalassets/for-aktorer-i-kraftsystemet/planer-og-analyser/2018-Forbruksprognose-Stor-Oslo
61. Bertsch J, Growitsch C, Lorenczik S, Nagl S et al (2016) Flexibility in Europe's power sector—an additional requirement or an automatic complement? Energy Econ 53:118–131
62. Gils HC (2014) Assessment of the theoretical demand response potential in Europe. Energy 67:1–8
63. Pina A, Silva C, Ferrão P et al (2012) The impact of demand side management strategies in the penetration of renewable electricity. Energy 41(1):128–137
64. Pina A, Baptista P, Silva C, Ferrão P et al (2014) Energy reduction potential from the shift to electric vehicles: the Flores island case study. Energy Policy 67:37–47

65. Lund H, Kempton W (2008) Integration of renewable energy into the transport and electricity sectors through V2G. Energy Policy 36(9):3578–3587
66. Graabak I, Wu Q, Warland L, Liu Z et al (2016) Optimal planning of the Nordic transmission system with 100% electric vehicle penetration of passenger cars by 2050. Energy 107:648–660
67. Juul N, Meibom P (2011) Optimal configuration of an integrated power and transport system. Energy 36(5):3523–3530
68. Hedegaard K, Ravn H, Juul N, Meibom P et al (2012) Effects of electric vehicles on power systems in Northern Europe. Energy 48(1):356–368
69. Ringkjob HK, Haugan PM, Solbrekke IM et al (2018) A review of modelling tools for energy and electricity systems with large shares of variable renewables. Renew Sustain Energy Rev 96:440–459
70. FINGRID, Landsnet, Svenska_Kraftnat, Statnett, Energinet, dk et al (2014) Nordic grid development plan 2014. Retrieved. https://www.statnett.no/Global/Dokumenter/Media/Nyheter2014/NordicGridDevelopmentPlan.pdf
71. Bøhnsdalen ET et al (2016) Long term market analysis. The Nordic Region and Europe 2016–2040. https://www.accenture.com/_acnmedia/Accenture/next-gen/top-tenchallenges/challenge10/pdfs/Accenture-2016-Top-10-Challenges-10-Market-Data.pdf
72. Lindberg KB, Seljom P, Madsen H, Fischer D, Korpås M (2019) Long-term electricity load forecasting: current and future trends. Utilities Policy 58(C):102–119, Elsevier

Chapter 28
Multi Criteria Decision Making for Sustainable Municipal Solid Waste Management Systems

Vinay Yadav and Subhankar Karmakar

Abstract Municipal solid waste (MSW) management is a conspicuous issue in sustainable development considering its association with all three domains of sustainability i.e. economy, society, and environment. Various operations research techniques have been used to provide a robust decision support system for a cost-effective, environment friendly and socially convenient MSW management in urban centers across the globe. This chapter discusses the holistic view of one such technique i.e. multi criteria decision-making (MCDM). MCDM models are used to solve two different kind of problems i.e. (i) evaluation of discrete decision alternatives based on identified attributes, known as multi-attribute decision making (MADM) problems; and (ii) optimization of a multi objective function in a continuous decision space, known as multi-objective decision making (MODM) problems. The formulation and applications of MADM and MODM models are demonstrated using two real case studies. First case study exhibits the use of a MADM model to select appropriate combination of locations for MSW facilities with simultaneous consideration of all relevant economical, social, technical and environmental attributes. The proposed two-stage MADM model expands the conventional approaches to select best locations for MSW facilities based on only economical cost minimization, and includes all relevant attributes to make it a comprehensive model. Attribution identification and economic evaluation are performed in the first stage assessment. Shortlisted alternatives are then evaluated for all identified economical, social, technical and environmental attributes using Technique of Order Preference Similarity to the Ideal Solution (TOPSIS) in the second stage assessment. The model is demonstrated on the city of Nashik (India). The model select best combination of locations with an optimum economical cost, good suitability for societal and technical attributes, and minimal environmental emissions. Second case study maximizes the social acceptance of MSW management system in the Hoi An city (Vietnam) using a MODM model. The proposed model provides an essential framework with the scientific consensus among all level of stakeholders. Minimizing the economical cost of entire

---

V. Yadav (✉)
Indian Institute of Management Visakhapatnam, Visakhapatnam, Andhra Pradesh 530003, India

S. Karmakar
Environmental Science and Engineering Department, Indian Institute of Technology Bombay, Mumbai, Maharashtra 400076, India

system, pollution emission from MSW facilities, and amount of MSW ending up in landfills are selected as primary objectives. Both of these case studies reflect the potential of MCDM models to establish a sustainable MSW management system in the urban centers across the globe.

Keywords Collection and transportation techniques · Multi criteria decision making · Municipal solid waste management · Transfer stations

28.1 Introduction

Municipal solid waste (MSW) management involves all the activities from the generation at source to its final disposal. These activities include MSW generation; separation/storage at source; collection & transportation; processing & transformation; and final disposal [22]. The activities in MSW management are associated with all three domains of sustainability i.e., environment, economy and society [19]. Environmental concerns include vehicular emissions during the transportation of MSW, air pollution in the processing/disposal of MSW, and water pollution due to the improper disposal of MSW [11]. Economy is involved in the transportation, processing and disposal of MSW [25]. Social dimension of MSW management includes (*i*) human behavioral traits such as positive participation, consumerism and throwaway culture; (*ii*) employment generation; and (*iii*) social status of ragpickers especially in developing countries [8]. A planned system will consequently lead to an economically viable, environmentally sound and socially acceptable MSW management. Apparently, MSW management has multi dimensions and therefore multi-criteria decision-making (MCDM) models have been found to be convenient in developing decision support systems for municipal official to manage MSW [3, 12, 16].

MCDM is a sub-discipline of operations research which assesses multiple diverse criteria in decision making [15]. MCDM methods include various techniques to develop decision support systems involving incommensurable variables, conflicting criteria and several alternatives [7]. MCDM approaches are classified into two main categories: (*i*) evaluation of discrete decision alternatives based on identified attributes, known as multi-attribute decision making (MADM) problems; and (*ii*) optimization of a multi objective function in a continuous decision space, known as multi-objective decision making (MODM) problems [27]. MADM techniques deal with the ranking problems, while MODM involves optimization problems.

In MADM, all the alternatives are evaluated with respect to the identified attributes or by calculating the aggregated attributes based overall performance [26]. A standard MADM technique encompasses assessment of N alternatives by M attributes, which is represented by a decision matrix with N rows and M columns. The element a_{nm} corresponds to score of nth alternative for m^{th} attribute. Different MADM methods present various mathematical procedures for comparison of alternatives in this decision matrix [24]. These methods include analytic hierarchy process (AHP), simple additive weighting (SAW), weighted sum model (WSM), multi-objective optimization on the basis of ratio analysis (MOORA), linear programming technique for multidimensional analysis of preference (LINMAP), ELimination Et Choix Traduisant la REalité (ELECTRE) and Technique for Order Preference by Similarity to Ideal Solution (TOPSIS) [24].

In MODM, a set of objective functions are to be optimized, subject to given constraints, in a continuous decision space without predetermined alternatives [17]. In this process, all possible optimal solutions are regrouped to construct a Pareto front. A standard maximization MODM model with P objectives can be formulated as [14]:

$$Max \quad f_p(x); \quad \forall p \in P \quad (28.1)$$
$$g(x) = 0 \quad (28.2)$$
$$h(x) \leq 0 \quad (28.3)$$

where $f_p(x)$ is the p-th objective function, vector x regroups scores of all criteria, $g(x) = 0$, $h(x) \leq 0$ are the constraints which define the problem domain.

Applications of MODM models in MSW management systems are prevalent to solve operational difficulties such as optimization of collection vehicular routing and allocation of MSW among different processing facilities [20]. On the other hand, applications of MADM methods are more prevalent at strategic level such as facility location problems [4]. We will discuss real case studies of the applications of MADM and MODM models on MSW management systems in subsequent sections for better understanding of these concepts.

28.2 Application of a MADM Model for Sustainable MSW Management in the City of Nashik, India

City of Nashik is located in the Maharashtra state of India (see Fig. 28.1). The area of the city is 259 km^2. The estimated population of Nashik city is 1,759,783 by the end of year 2020 and total MSW generation will be 450 tons per day approximately [23]. Nashik municipal corporation (NMC) is planning to establish a new integrated MSW management plant approximately 20 km from the city boundary as current MSW management plant will be closed soon due to limited capacity. The new plant will have a presorting unit, an aerobic composting unit, an inert processing unit, a refuse derived fuel production unit, sanitary landfill, and leachate treatment unit. As this new proposed plant is away from the city, it will increase the distance traveled by the collection vehicles. To address this issue, NMC decide to establish transfer stations (TSs) in the existing MSW management system. TSs are intermediate MSW management facilities installed near the human settlement and are found to be cost effective if collection vehicles are traveling for longer distances [21]. However,

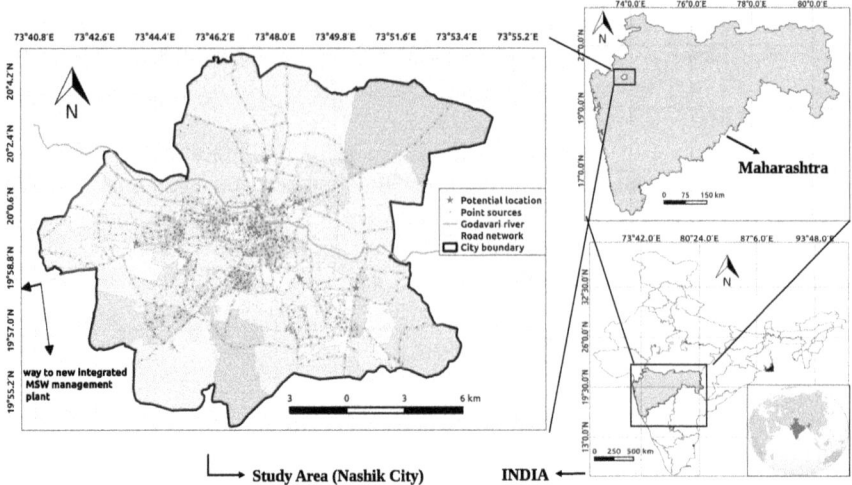

Fig. 28.1 City of Nashik with MSW generation sources and potential locations for TSs [23]

NMC was unable to identify optimal number of TSs required with their respective capacities and the best locations for establishing TSs. To facilitate NMC for making this decision, a two-stage MADM model is proposed with all relevant economical, social, environmental and technical attributes which also accounts for all possible combinations of potential locations. [25]'s two stage model first identifies all the relevant attributes and then evaluates all the possible combinations of potential locations to establish TSs using TOPSIS. In this study, multiple relevant attributes are identified in the stage one assessment. The second stage assessment includes scoring of the alternatives generated by possible combinations of potential locations for the identified attributes.

28.2.1 Stage-One Assessment

Following are the steps followed for stage-one assessment:

Step 1: Fifteen attributes are first identified based on the field conditions. The attributes are classified into four types, i.e., (*i*) economic, (*ii*) social (*iii*) environmental and (*iv*) technical. Economic attributes include total MSW generation, land ownership for the potential locations, proximity to other MSW management facilities, overall cost, and distance traveled by MSW collection vehicles. Social attributes include protection of potential locations from flood hazards, public acceptability to have TSs in their vicinity and kids population density in that area. Environmental attributes include remoteness to the water bodies and total emissions in transportation of MSW. Technical attributes include interference of MSW collection vehicles with routine traffic of the city, size of land available to establish TSs, road accessibility, availability of basic amenities and flexibility for expansion in future.

Step 2: NMC has already identified eight potential locations to establish TSs. All these potential locations are scored based on the selected fifteen attributes. The scoring has been done with the help of municipal officials and academicians.

Step 3: NMC has $j = 1, 2, \ldots, O$ (where $O = 8$ is the number of potential locations for establishing TSs), then $\sum_{j=1}^{O} {}^{O}C_j = \sum_{j=1}^{O} \frac{O!}{j!(O-j)!}$ are all possible location combinations. To represent these combinations, symbols OC1-S1 (*i.e.* one location is chosen from all selected O potential locations), OC1-S2 (*i.e.* one location is selected from identified potential locations which is different from OC1-S1),...., OCj-S1,..., and their respective optimal costs are given as Ec1-S1, Ec1-S2,......, Ecj-S1. In addition to the costs, distances traveled by all MSW collection vehicles are also calculated using Geographic Information System (GIS) tools for these combination. Further, the emissions are calculated using the Indian emission regulation booklet of the Automotive Research Association of India (ARAI) [2].

Step 4: The aggregated scores are calculated for all possible combinations to form the decision matrix $[dm_{\hat{j}\hat{k}}]_{2^O-1,15}$ (see Table 28.1). This decision matrix is then evaluated using TOPSIS method.

Table 28.1 Decision matrix $[dm_{\hat{j}\hat{k}}]_{2^O-1,15}$ [25]

	Attribute-1	Attribute-2	...	Attribute-15
OC1-S1	$\frac{\sum_{j=1}^{O} a_{j1} * D_j}{1}$	$\frac{\sum_{j=1}^{O} a_{j2} * D_j}{1}$	...	$\frac{\sum_{j=1}^{O} a_{j15} * D_j}{1}$
.	.	.	...	.
.	.	.	...	.
.	.	.	...	.
OCO-S4	$\frac{\sum_{j=1}^{O} a_{j1} * D_j}{O}$	$\frac{\sum_{j=1}^{O} a_{j2} * D_j}{O}$	...	$\frac{\sum_{j=1}^{O} a_{j15} * D_j}{O}$
.	.	.	...	.

28.2.2 Stage-Two Assessment

Following are the steps followed for stage-one assessment:

Step 1: Normalization of decision matrix $[dm_{\hat{j}\hat{k}}]_{2^o-1,15}$. The normalized decision matrix will have elements $r_{\hat{j}\hat{k}}$ as

$$r_{\hat{j}\hat{k}} = \frac{dm_{\hat{j}\hat{k}}}{\sqrt{\sum_{\hat{j}=1}^{2^o-1} dm_{\hat{j}\hat{k}}}} \quad (28.4)$$

Step 2: Construction of the weighted normalized decision matrix $V = [v_{\hat{j}\hat{k}}]_{2^o-1,15}$ with assigned weights $W = \{w_1, w_2, \ldots, w_{15} | \sum_{\hat{k}=1}^{15} w_{\hat{k}} = 1\}$.

$$V = \begin{pmatrix} w_1 r_{11} & w_2 r_{12} & \cdots & w_{15} r_{115} \\ w_1 r_{21} & w_2 r_{22} & \cdots & w_{15} r_{215} \\ \vdots & \vdots & \ddots & \vdots \\ w_1 r_{2^o-11} & w_2 r_{2^o-12} & \cdots & w_{15} r_{2^o-115} \end{pmatrix} \quad (28.5)$$

Step 3: Calculation of ideal solution $\mathcal{A}^*$ and negative-ideal solution $\mathcal{A}^-$:

$$\mathcal{A}^* = \{(\max_{\hat{j}} v_{\hat{j}\hat{k}} | \hat{k} \in \hat{K}), (\min_{\hat{j}} v_{\hat{j}\hat{k}} | \hat{k} \in \hat{K}'); \hat{j} = 1, 2, \ldots, 2^o - 1\}$$

$$= \{v_{1*}, v_{2*}, \ldots, v_{15*}\} \quad (28.6)$$

$$\mathcal{A}^- = \{(\min_{\hat{j}} v_{\hat{j}\hat{k}} | \hat{k} \in \hat{K}), (\max_{\hat{j}} v_{\hat{j}\hat{k}} | \hat{k} \in \hat{K}'); \hat{j} = 1, 2, \ldots, 2^o - 1\}$$

$$= \{v_{1-}, v_{2-}, \ldots, v_{15-}\} \quad (28.7)$$

where $\hat{K} = \{\hat{k} = 1, 2, \ldots, 15 \text{ is associated with benefit attributes}\}$, and $\hat{K}' = \{\hat{k} = 1, 2, \ldots, 15 \text{ is associated with cost attributes}\}$.

Step 4: Calculation of separation distance from the ideal solution ($S_{\hat{j}*}$):

$$S_{\hat{j}*} = \sqrt{\sum_{k=1}^{15} (v_{\hat{j}\hat{k}} - v_{\hat{k}*})^2}; \quad \forall \hat{j} = 1, 2, \ldots, 2^o - 1 \quad (28.8)$$

The separation distance ($S_{\hat{j}-}$) from the negative-ideal solution can be calculated as

$$S_{\hat{j}_-} = \sqrt{\sum_{k=1}^{15}(v_{\hat{j}k} - v_{\hat{k}_-})^2}; \quad \forall \hat{j} = 1, 2, \ldots, 2^O - 1 \quad (28.9)$$

Step 5: Calculation of the relative closeness $C_{\hat{j}*}$ of identified alternative with respect to the ideal solution $\mathcal{A}^*$:

$$C_{\hat{j}*} = \frac{S_{\hat{j}_-}}{S_{\hat{j}_-} + S_{\hat{j}*}}; \quad 1 \geq C_{\hat{j}*} \geq 0; \text{ and } \hat{j} = 1, 2, \ldots, 2^O - 1 \quad (28.10)$$

Optimal alternative is found based on $C_{\hat{j}*}$ value.

28.2.3 Best Locations of TSs in the Nashik City

NMC has already identified eight potential locations to establish TSs. NMC has to select at least one and at most all eight locations for establishing TSs i.e. NMC has to evaluate all these 255 alternatives based on the identified attributes. The best combination of locations is then chosen using TOPSIS method. A python based tool (PyTOPS) has been used to solve the model as explained in the last section [24].

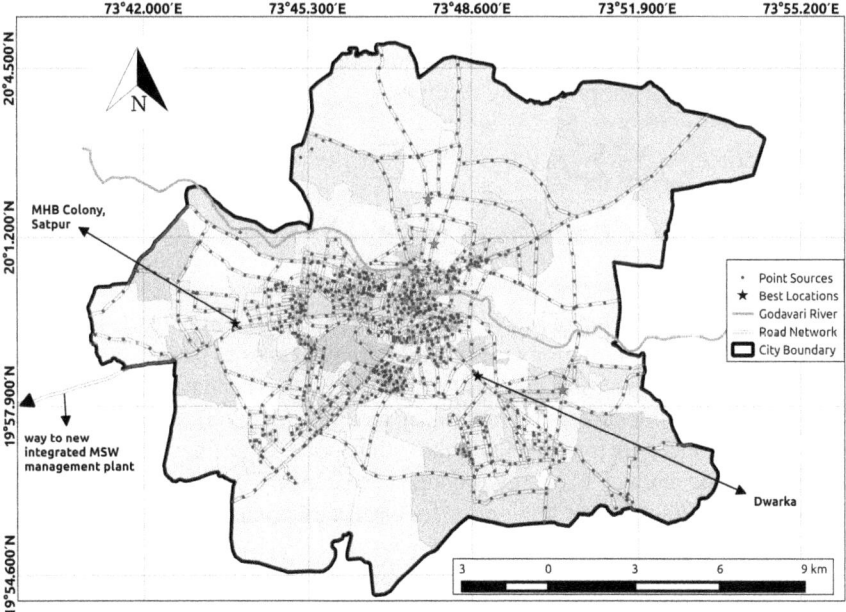

Fig. 28.2 Best locations for the TSs in the city of Nashik [25]

A combination with two locations has found to be the best while considering all economic, technical, environmental and social attributes. The geographical locations of these TSs in the city of Nahsik are given in the Fig. 28.2.

28.3 Application of a MODM Model for Sustainable MSW Management in Hoi An City, Vietnam

Hoi An city is located in South Central Coast region of Vietnam in the Quang Nam province. This city is recognised as the World Heritage Site by United Nations Educational, Scientific and Cultural Organization (UNESCO) [18]. The city has a population of 93,000 and an area of 6171.25 [9]. The city is a popular tourist spot of Vietnam and has seen a significant increase in the number of tourists in the last decade [10]. Hoi An municipality was facing problem of mismanaged MSW due to for this sustained growth in the tourism. To address this issue, [9] proposed a MODM model for sustainable MSW management system in the city of Hoi An.

28.3.1 Objective Functions for the Proposed MODM Model

The proposes MODM model had three objectives i.e. minimization of (*i*) overall cost of MSW management system; (*i*) pollution emission from the MSW treatment facilities; and (*iii*) MSW sent to the landfill. Overall cost of MSW management system (Z_1) includes the transportation cost of collection vehicles, capital and running/operational costs of processing facilities.

$$Min\ Z_1 = \sum_k \sum_i R_{i,k} \times c^{trans} + \sum_k C_k \times (c_k^{cap} + c_k^{op})$$
$$- CF \times b_{COM} - (\sum_k EP_k - \sum_k EC_k) \times b_{EN} \quad (28.11)$$

where, $R_{i,k}$ is the amount of MSW *i* transported to processing facility *k* (in tons per day); c^{trans} is the transportation cost of MSW from sources to processing facilities (in \$/ton); C_k is the capacity of processing facility *k* (in tons per day); c_k^{cap} and c_k^{op} are the fixed and operational costs for processing facility *k* (in \$/ton); CF is the amount of fertilizer produced (in tons); b_{COM} is the selling price for fertilizer (in \$/ton); EP_k and EC_k are the amount of energy produced and consumed by processing facility *k* (in kWh); and b_{EN} is the selling price for electricity produced in waste to energy plant (in \$/kWh).

The MSW processing facilities emits several pollutants to the environment which include methane (CH_4), carbon dioxide (CO_2), nitrogen oxides (NO_x), sulfur dioxide (SO_2), nitrous oxide (N_2O), volatile organic compounds (VOCs) and heavy metals. Emission factors of these pollutants have been taken from [1, 5, 6] and [13] for all the processing units. Total emissions are given in the following equation:

$$Min\ Z_2 = \sum_p C_k \times \mu_{p,k} \tag{28.12}$$

where, $\mu_{p,k}$ is the amount of pollutant p emitted from processing facility k (in kg/ton of MSW).

The amount of MSW going to landfill includes the residue from MSW processing facilities and the MSW directly transported to landfills.

$$Min\ Z_3 = \sum_i R_{i,LF} + \sum_i WR_{i,k,LF} \tag{28.13}$$

where, $WR_{i,k,LF}$ is the amount of residue of MSW type i from processing unit k going to landfill (in tons per day).

28.3.2 Constraints of the Proposed MODM Model

The proposed MODM model includes (*i*) Mass balance constraints for the flow of MSW and residue of processing facilities; (*ii*) energy consumption and recovery from the processing units; and (*iii*) other constraints.

The total amount of MSW type i generated will be transported to the processing facility k. The residual of processing facility k will be transported the facility k'.

$$\sum_k R_{i,k} = WR_{i,k} \times \alpha_i \quad \forall i, k \tag{28.14}$$

$$\sum_i WR_{i,k} = \sum_k(\sum_i WR_{i,k,k'}); \quad \forall i, k, k' \tag{28.15}$$

$$C_k = \sum_i R_{i,k} + \sum_i WR_{i,k',k} \tag{28.16}$$

$$\sum_{k'} WR_{i,k,k'} = \sum_i R_{i,k} \times \beta_{i,k} \quad \forall i, k \tag{28.17}$$

$$\sum_k WR_{i,k,k'} = \sum_k \sum_i R_{i,k} \times \beta_{i,k} + \sum_k \sum_i WR_{i,COM,k} \times \beta_{i,COM}$$
$$+ A_s \times \sum_k \sum_i WR_{i,AD,k} \quad \forall k' \tag{28.18}$$

where, α_i is the percentage of i type waste in MSW; $\beta_{i,k}$ is the amount of residual generated from processing unit k by treating MSW of type i; $WR_{i,k}$ is the amount of

MSW type i transported to processing unit k; and $WR_{i,k,k'}$ is the amount of residual of MSW type i transported from processing unit k to unit k'.

The energy consumption and recovery from the processing units are given in the following set of equations. The amount of fertilizer produced (in tons) is given as:

$$CF = \sum_i R_{i,COM} \times \lambda^m_{i,COM} + WR_{i,AD,COM} \times \lambda^m_{sludge,COM} \quad (28.19)$$

where, $\lambda^m_{i,COM}$ is the fertilizer produced (in tons/tons of MSW).

The energy produced by incineration with energy recovery (INCE) unit and anaerobic digestion with energy recovery (AD) is given as:

$$EP_{AD} = \delta_{AD} \times \sum_i (R_{i,AD} \times \lambda^m_{i,AD} \times LHV_{biogas})/f \quad (28.20)$$

$$EP_{INCE} = \delta_{INCE} \times LHV_i \times (\sum_i R_{i,INCE} + \sum_k \sum_i WR_{i,k',INCE})/f \quad (28.21)$$

where, δ is the efficiency of energy production from INCE and AD units; and LHV_i is the lower heating value of i type MSW (in MJ/kg).

15% of total MSW is getting recycled. In this fraction, 5% of MSW have been used as cattle feed and in-house composting. Remaining 10% of MSW is converted to the recyclable materials.

$$C_k \leq 0.1 \times W \quad \forall k \quad (28.22)$$
$$C_k \leq 0.05 \times W \quad \forall k \quad (28.23)$$
$$R_{i,k}, WR_{i,k,k'}, C_k \geq 0 \quad (28.24)$$

28.3.3 Results of Proposed MODM Model

The decision making process involved three set of decision maker (DM): (*i*) DM1 has city municipal officials;(*ii*) DM2 has citizens of the city; and (*iii*) DM3 has MSW management working professionals. All the DM groups made multiple interactions to investigate the behaviour of the proposed model. The optimal solution of DM2 and DM3 has reduced the emissions of pollutants as compare to the solution given by DM1. However, no significant improvement is found in the amount of MSW ending up in the landfill. Therefore, optimal solution given be DM1 could develop consensus as all of the DMs agreed that reduction of MSW going to the landfill is urgent for a touristic place like Hoi An city. This study further recommended the use of composting unit and anerobic digester for the sustainable management of MSW in the city. The use of incinerator should be reduced considering the amount of emissions it generates. The segregation of MSW at source and use of decentralized facility should also be promoted.

28.4 Conclusions

This chapter reviews the application of both of the branches of Multi-criteria decision making (MCDM) *i.e.* multi-attribute decision making (MADM) and multi-objective decision making (MODM) on making a sustainable municipal solid waste (MSW) management in urban centres. The first MADM model extends the conventional one objective models with only consideration of cost by considering environmental, social and technical attributes as well in finding suitable locations for MSW management facilities in the city of Nashik, India. The results of this study have produced a MSW management system which has optimum costs, minimum emissions of pollutants, socially acceptability and technical suitability. Another case study used a MODM model for the sustainable planning of MSW management system in the Hoi An city of Vietnam. The proposed model involved all level of stakeholders in the decision making including concerned citizens, municipal officials and field experts. The model also provided an opportunity for stakeholders to understand the trade-off mechanisms to reach an optimum solution. The results of this study exhibit that all the the stakeholders had a consensus on solving the priority objective first i.e. diverting the MSW from entering into the landfills.

References

1. Andersen JK, Boldrin A, Christensen TH, Scheutz C (2011) Mass balances and life cycle inventory of home composting of organic waste. Waste Manage 31(9–10): 1934–1942
2. ARAI (2017) Indian emission regulation booklet. The Automotive Research Association of India. https://www.araiindia.com/pdf/Indian_Emission_Regulation_Booklet.pdf. Accessed 01 Oct 2018
3. Büyüközkan G, Gocer F (2017) An intuitionistic fuzzy MCDM approach for effective hazardous waste management. In: Intelligence systems in environmental management: theory and applications. Springer, Berlin, pp 21–40
4. Chang NB, Pires A (2015) Sustainable solid waste management: a systems engineering approach. Wiley, New York
5. Eggleston S, Buendia L, Miwa K, Ngara T, Tanabe K (2006) Ipcc guidelines for national greenhouse gas inventories. vol. 5 waste. ipcc national greenhouse gas inventories programme. Institute for Global Environmental Strategies, Hayama, Kanagawa, Japan
6. Gladding T, Thurgood M et al (2004) Review of environmental and health effects of waste management: municipal solid waste and similar wastes. Department for Environment, Food and Rural Affairs, London
7. Greco S, Figueira J, Ehrgott M (2016) Multiple criteria decision analysis. Springer, Berlin
8. Heidari R, Yazdanparast R, Jabbarzadeh A (2019) Sustainable design of a municipal solid waste management system considering waste separators: a real-world application. Sustain Cities Society 47: 101457

9. Hoang GM, Fujiwara T, Phu TSP, Nguyen LD (2019) Sustainable solid waste management system using multi-objective decision-making model: a method for maximizing social acceptance in Hoi An city, Vietnam. Environ Sci Pollution Res 26(33):34137–34147
10. Hoang MG, Fujiwara T, Phu STP (2017) Municipal waste generation and composition in a tourist city - Hoi An, Vietnam. J JSCE 5(1):123–132
11. Istrate IR, Iribarren D, Gálvez Martos JL, Dufour J (2020) Review of life-cycle environmental consequences of waste-to-energy solutions on the municipal solid waste management system. Resources Conserv Recycling 157: 104778
12. Karagoz S, Deveci M, Simic V, Aydin N, Bolukbas U (2020) A novel intuitionistic fuzzy mcdm-based codas approach for locating an authorized dismantling center: a case study of Istanbul. Waste Manage Res 38(6):660–672
13. Mirdar Harijani A, Mansour S, Karimi B (2017) A multi-objective model for sustainable recycling of municipal solid waste. Waste Manage Res 35(4): 387–399
14. Pohekar SD, Ramachandran M (2004) Application of multi-criteria decision making to sustainable energy planning'a review. Renew Sustain Energy Rev 8(4):365–381
15. Popiolek N, Thais F (2016) Multi-criteria analysis of innovation policies in favour of solar mobility in France by 2030. Energy Policy 97:202–219
16. Shahnazari A, Rafiee M, Rohani A, Nagar BB, Ebrahiminik MA, Aghkhani MH (2020) Identification of effective factors to select energy recovery technologies from municipal solid waste using multi-criteria decision making (MCDM): a review of thermochemical technologies. In: Sustainable energy technologies and assessments 40
17. Singh SK, Goh M (2019) Multi-objective mixed integer programming and an application in a pharmaceutical supply chain. Int J Prod Res 57(4):1214–1237
18. Tran VT, Nguyen NP, Tran PTK, Tran TN, Huynh TTP (2019) Brand equity in a tourism destination: a case study of domestic tourists in hoi an city, Vietnam. Tourism Rev
19. Yadav V, Karmakar S (2020) Sustainable collection and transportation of municipal solid waste in urban centers. Sustain Cities Soc 53: 101937
20. Yadav V, Karmakar S, Dikshit A, Vanjari S (2016a) A facility location model for msw management systems under uncertainty: a case study of Nashik city, India. Proc Environ Sci 35:90–100
21. Yadav V, Karmakar S, Dikshit A, Vanjari S (2016b) A feasibility study for the locations of waste transfer stations in urban centers: a case study on the city of nashik, India. J Cleaner Prod 126:191–205
22. Yadav V, Karmakar S, Dikshit A, Vanjari S (2016c) Transfer stations siting in India: a feasibility demonstration
23. Yadav V, Karmakar S, Dikshit A, Bhurjee A (2018) Interval-valued facility location model: an appraisal of municipal solid waste management system. J Cleaner Prod 171:250–263
24. Yadav V, Karmakar S, Kalbar PP, Dikshit A (2019) PyTOPS: a python based tool for TOPSIS. SoftwareX 9:217–222
25. Yadav V, Kalbar PP, Karmakar S, Dikshit A (2020) A two-stage multi-attribute decision-making model for selecting appropriate locations of waste transfer stations in urban centers. Waste Manage 114:80–88
26. Zeng S, Chen SM, Kuo LW (2019) Multiattribute decision making based on novel score function of intuitionistic fuzzy values and modified VIKOR method. Information Sci 488:76–92
27. Zimmermann HJ (2011) Fuzzy set theory–and its applications. Springer Science and Business Media

Chapter 29
Inventive Investment Using Bigdata: Tools, Applicability and Challenges Associated

Janibul Bashir and Tahir Ahmad Wani

Abstract Traditional investment models have reached to a saturation level that it is easily recognizable to determine and demonstrate their inadequacy against the current huge data-based investment models. Big Data is taking all management spheres by storm be it marketing, financing or investment. Today colossal and developing volumes of data are being generated and analysed upon to arrive at meaningful solutions. Big Data analytics joined with the business models can make wonders in the company gains. The Big data analytics will permit organizations to test hypotheses continuously and see the probable outcomes of each hypothesis before bringing the same to the market. This reduces the risks of loss and accelerates the benefits of the organizations tremendously. Insights from large information investigation have the likelihood to empower business process significantly. One important advantage of Big Data analytics is to extract the patterns from the data, and client inclinations and accordingly assist organizations with making effective decisions in business, administrations, and other relevant items. The utilizing of Big Data and machine learning algorithms to dissect the Big information for any association, can take care of issues in different verticals and estimate the business future with more noteworthy speed and unwavering quality. Information investigation has been in the Business Intelligence space for a serious long time giving 'Point answers' for explicit issues in any business. The focus of this chapter is to understand these algorithms and their efficacy in the different business scenarios.

Keywords BigData · Management · Machine learning · Inventive investment

---

J. Bashir (✉)
Department of Information Technology, National Institute of Technology Srinagar, Srinagar, India
e-mail: janibbashir@nitsri.ac.in

T. A. Wani
Department of Humanities and Social Science, National Institute of Technology Srinagar, Srinagar, India
e-mail: tahirwani@nitsri.net

29.1 Introduction

In the past, traditional methods like descriptive analytics were used on structured data by the organisations to improve their businesses. However, the number of challenges faced by businesses has grown to such an extent that these traditional methods seem no longer useful. Since the last two decades, the concept of Big Data has gained popularity in businesses and firms have started to see Big Data as an opportunity to improve their performance and status and also to prevent themselves from lagging behind in today's competitive world [1–3]. The vast amount of data produced by the companies was previously left unused but with the growth of data analytics and Big Data concepts, companies have started to utilize this data by deriving useful insights from it [4, 5].

It is estimated that before the finish of 2020, the enormous information volume is going to arrive at 44 trillion gigabytes, separating all the past patterns and setting another business world. From this data progressively noteworthy speculation bits of knowledge can follow. This data has critical incentives for organizations in today's cutthroat competition. However, one of the primary difficulties for organizations is to distinguish significant key information and utilize the most fitting instruments to comprehend and investigate this information. With proper tools to analyse this huge data, the investors can easily comprehend the inclinations of the general population and tailor their business contributions accordingly. There is a need for innovations for holding this information, and comprehensively examine this information. The companies can utilize the innovations in information processing to shore precise, stable business experimentation that direct leaders and to inspect yields and plans of action.

Data science is not a new field for the organizations, however, with the invent technologies such as machine learning, neural networks and deep learning, organizations now know the essence of data science [6]. To improve their efficiency and performance, all businesses have shifted from the traditional method to big data analytics that helped in cost reduction and efficient resource utilization. Big data analytics has proven to be a key aspect for business' success [4, 5]. It has changed the way companies implement their business. Big data has conferred brilliant chance to the universal market, every part of industry is attempting to assess the higher prospects to capture and analyse data to take better decisions, much data implies substantially more use-cases, more use-cases prompts more delineation of business assessment which eventually prompts best business decision making. This will lead to much profit, by changing the customary methodology of overseeing data into supportive new methodologies. It has paved the way for new opportunities and developments and has become cardinal element of business intelligence today. In this chapter our aim is to discuss the role of bigdata in various industrial sectors [3, 7, 8].

The structure of the chapter is as follows. Section 29.2 concentrates on the theoretical background of this study. Section 29.3 presents the data gathering methods and Sect. 29.4 describes the mechanisms to process structured and unstructured data. Section 29.5 presents the Big Data applications that benefit companies in key fields, including Supply Chain Analytics, Demand Forecasting, Fraud detection, Customer Segmentation, Retailing, Banking Sector, and Fashion. The challenges associated with big data in business are discussed in Sect. 29.6 and finally we conclude in Sect. 29.7.

29.2 Background

A significant increase in the amount of data generated by companies has led to difficulties in managing the enormous amount of data over the last 20 years especially with conventional data management tools. This led to a new concept being created—Big Data. Webopedia's simplest definition of Big Data states that "Big Data is a vast amount of unstructured and organized data, so broad that conventional database and software techniques are difficult to process" [1, 2]. Big Data was originally associated with three keywords: volume, velocity and variety (3 V's). Volume stands for large quantities of Big Data, velocity refers the speed with which data is generated, collected and processed and variety states the dissimilar types of data.

With digitization becoming an important part of day-to-day life, enormous measures of data have been gathered that can be utilized efficiently in various fields. The enforcement of BigData is widely acknowledged by numerous industries and organizations, which drives them to improve their business. The purpose of BigData is to give better utilization of resources and storage, diminish the computational time and improve business decision making in business. However, due to inadequate intuition, most corporate organisations have struggled to recognise the business potential that can be realised. It is important to assume that a great business insight is out of the ordinary. Generating such traditional insights requires some insightful questions to be answered. This is achievable with the use of a reasonable mix of technical analysts' and market experts' efforts.

Big data analytics is seen as the leading methodology for analysing big data in view of its superior capacity to collect immense measurements of user data and apply the best know data analytic based measurement models to measure it. It is an important method for organizations to gather wide-ranging information and use automatic data analytics to illuminate relevant decisions that have traditionally relied on decision-makers' decrees and opinions. Three main features then revolve around Big Data analytics: the data itself the analytics applied to the data, and the demonstration of results in a way that enables companies and their consumers to create business value. This study guides researchers and companies by cataloguing the assorted existing models of big data analytics. In order to analyse the use of big data analytics by organizations, it is indispensable to recognize the primary drivers. Doing so would provide evidence for the argument that the right application of big data analytics helps organizations to leverage big data in a commendable way.

29.3 Data Collection

Data undoubtedly is one of the most valuable resources in today's business world. These days' organizations and associations are associated with their customers, clients, workers, sellers, and in specific cases, even with their rivals. The accessible information can weave an anecdote about any of these connections, and with this data, associations can improve practically any part of their tasks.

The goal of such Data collection is to figure out the quality evidence that help in answering all the questions posed. Prerequisite for making rational decisions is the quality information that can be deduced only when there is data. Thus, data must be gathered to draw conclusions make important decisions in business.

Data collection is a systematic approach wherein we gather information, from different sources, followed by the analyses of the obtained data to devise proper solutions to the relevant problems and evaluate the results. This evaluation of the data helps researchers in figuring out the future trends.

29.3.1 Methods to Collect Data

Data from different sources and in different forms can be gathered in a number of ways such as through interviews, questionnaires, experiment, observation, survey and so on that are administered either personally or electronically; view of individuals and events with or without recording and a variety of other inspirational procedures, for example, projective tests. Below we discuss some common methods of data collection [9, 10].

Interviews: An interview is a set of questions that a researcher asks respondents to gather relevant information [2]. We can have Face-to-face interviews, Telephone interviews, Web chat interviews, etc. Knowing *"what to ask"* is the prime requirement for efficient interviews. Interviews are used to gather in-depth responses from the individuals being interviewed. Contrasted with some other assortment technique, interviews are more adaptable and responsive as the questioner can tailor follow-up questions dependent on reactions. A meeting might be organized where you pose unmistakably characterized inquiries and permit a portion of your scrutinizing to be driven by the reactions of the interviewee.

Questionnaire method: A questionnaire is a collection of written questions sent to subjects by a researcher to request their answer. Generally, the Questionnaires are self-regulated in that they are presented regarding the matters, requesting that they complete it and post it back. Questionnaires are intended to gather information from a gathering. How data is analysed may impact the layout of the questionnaire. For instance, closed inquiries provide boxes for the respondent to tick (giving effortlessly coded information).On the other hand, an open inquiry provides a box for the respondent to compose answers in it (giving more liberty of information, nonetheless more trouble coding). A questionnaire includes three kinds of questions fixed-alternative, scale, and open-ended. With each of the inquiries personalized to the nature and extent of the research.

Observations: Observation refers to gather data without posing inquiries. This technique includes observing members in a particular circumstance or environment at a given time. As it entails the analyst, or spectator, to add their decree to the data, this scheme is more subjective. Analysts note the general conditions of the surrounding environments or individuals under consideration. Observations may be controlled, natural or participant. If the analyst uses a standardised method of observing the environment or its members, it is controlled observation. In Natural observation, members are observed in their natural conditions. Participant observation is where the analyst becomes part of the group being contemplate.

Experiments: Experimental method, mostly used in scientific research is a study of causal relationship i.e. cause-and-effect relationship between two variables. One variable may be changed while the other variable is measured.

Surveys: Survey is another method of collecting data in which you can directly ask customers for information. It involves discovering facts in specific fields of inquiry. Three significant surveys in which data collected are statistical include social survey, market research and public opinion polls.

a. **Social Survey**: It is a survey intended to give data to other Government Departments to assist them with completing their liabilities more profitably.
b. **Market Research**: This involves the use of surveys, experiments and statistical studies to evaluate consumer patterns and to forecast the reach and position of explicit product or services markets. In client studies, social sciences are increasingly being used. Psychology and sociology, for example, are keys to understanding the behavioural pattern of consumers by presenting bits of knowledge about the circumstances, desires, wishes, and general motivation of individuals.
c. **Public Opinion Polls**: This is a method that estimates the attitude, perceptions, and likings of a populace towards occasions, situation and issues of mutual interest. Both arbitrary and standard testing is used.

Documents and reports: At times, a substantial volume of data can be gathered without asking anybody anything. Document and records premise research is the approach to analyse existing documents and records of an association for tracing modifications over a while. Records can be traced by analysing call logs, attendance records, email logs, staff reports, databases, minutes of meetings, information logs, meeting minutes, financial records and so forth. For example, an association gets heaps of negative audits and complains from clients regarding its items or services. To know the reason, the association may investigate records of their items or services and recorded interactions of its employees with customers.

Data Collection using documents and records may be proficient and economical as research that has already been accomplished is mostly used. Nonetheless, since the researcher has less authority over the results, documents and records can be considered as an insufficient source of data.

Literature sources: This encompasses the assortment of data from the text that has already been published. Literature sources include diaries, textbooks, autobiographies, letters, speeches, manuscripts, research papers and articles, case studies, newspapers, Governmental or Non-Governmental Organizations' (NGO) reports, magazines, photography, video, music, poetry etc.

a. **NGO Reports**: NGO reports includes the activities carried out by the NGO. These reports include data that are research-explicit and structures a satisfactory scholastic base towards gathering information. NGOs regularly centre around advancement ventures which are composed to advance specific causes.
b. **Newspapers**: Easy to gather and occasionally the only uninterruptedly available source of event data, newspaper is another way to collect data. Although, newspaper data can be tricky (bias in nature) sometimes, it may still prove to be a useful tool to collect data.
c. **Website Articles**: Assembling and utilizing information contained in web articles is likewise another source for information assortment. Gathering information from web articles is a faster and more affordable information assortment. Two significant weaknesses of utilizing this information detailing strategy are predispositions characteristic in the information assortment procedure and conceivable security/secrecy concerns.
d. **Hospital Care records**: Health care includes a different arrangement of open and private information assortment frameworks, including wellbeing reviews, managerial enrolment and charging records, and clinical records, utilized by different elements, including emergency clinics, CHCs, doctors, and wellbeing plans. The information gave is clear, fair-minded and exact, yet should be acquired under the legitimate methods as clinical information is kept with the strictest guidelines.

29.4 Data Processing

29.4.1 Processing Structured Data

Structured data is a data that is consistent with the pre-distinct data model and is thus ready for review. With an interaction between different rows and columns, structured data follows tabular organization. SQL databases or files from Excel are typical illustrations of structured data. All these rows and columns are ordered and can be arranged.

Structured data relies on the reality of a data paradigm, a model that determines how it is possible to store, process and access information. Since each field is different in the data model and can be accessed discreetly or jointly by data from other fields.

This makes structured data exceptionally dominant; it is possible to rapidly cumulate data from various positions in the database [11].

29.4.2 Processing Unstructured Data

Unstructured data is knowledge that either has an unknown data model or is unordered. Unstructured data is typically material that can also contain information such as dates, statistics, and descriptions. Such inconsistencies and indistinctness make it difficult to understand the use of standard systems as equated with data stored in ordered databases. Audio, audio-visual files or No-SQL databases are typical models of unstructured data.

In recent times, the ability to store and process unstructured data has increased dramatically, with many new innovations and software looming on the market that can store advanced forms of unstructured data. For example, Mongo DB is designed for storing documents. As an alternative example, Apache Graph is optimized for storing links between nodes.

The ability to examine unstructured data is specifically appropriate in the context of Big Data, as an enormous part of data in organisations is amorphous. The ability to excerpt value from unstructured data is among the leading drivers behind the rapid evolution of Big Data [12].

Following are the 10 general steps for analysing the unstructured data.

Step 1: Choose a Data Source Data can be gathered from multiple sources and it is crucial to identify the source of data valuable for business. To gather data from arbitrary sources is certainly not a good choice. This may corrupt the data or at worse we may end up losing all the valuable data. Therefore, it's advised to review the appropriate data source while gathering data. There are various big data development tools available for data collection.

Step 2: Management of Unstructured Data Search Gathered data may differ in usage if it is structured or unstructured. Data Survey and collection is only tip of the iceberg. Organizing unstructured data search and modifying it conveniently is diverse. In this step collecting the data is crucial. Improper approaches can complicate client to business and business to business operations and utilities. Investing in good business management tools is essential to manage unstructured data.

Step 3: Excluding Inadequate Data After gathering and organizing the data, eliminating useless data is necessary. Data determines business growth but can also be detrimental. Unnecessary data consumes space and gradually can affect the businesses resources (hard drives, storage, or backups), this may eventually affect the business's capability to strive. To avoid confusion and save resources of the business, eliminating needless data is a valued practice.

Step 4: Formulate Data for Storage Formulating of data means to remove all the whitespaces, formatting disputes, etc. from the data. When entire data (no matter valuable for the business or not) is organized, stacking of necessary data and indexing unorganized data can be done.

Step 5: Choose the Technology for Data Stack and Storage Stacking data is the ideal solution. Using latest technology to protect and stack data, to improve business operations. User-friendly environments help employees to access and use important data in no time. Additionally, it ensures updated data backup, maintenance and recovery service.

Step 6: Hoard All the Data Until Stored Whether structured or unstructured, it is obvious to always save data. Natural disasters around the globe have shown that, especially during times of crisis, a current and upgraded data backup recovery system is important and necessary. You do not know that all your information is about to be withdrawn [3]. So, think ahead and save your work often.

Step 7: Retrieve Useful Information Recovery of data can be done after taking backup of the data. This is useful as there might be need of data retrieval after converting unstructured information.

Step 8: Ontology Estimation Association concerning the source of information and the data mined can assist in providing beneficial insights concerning organization of data. The business must be able to elucidate the phases and procedures to grow. It is mandatory to keep record to identify designs and keep consistency with the process.

Step 9: Creating Statistics Creating statistics can be beneficial for the business. Categorization and fragmentation of the data is intended for easy usage and research to produce inordinate course for future uses.

Step 10: Data Analysis Indexing unstructured data, after all the raw data is organized. Data analysis helps in analysing and relevant decisions-making that are favourable for the business. Indexing also helps small businesses to make unswerving designs for upcoming usage.

29.5 BigData in Business

This section discusses the applications of bigdata in various industrial sectors. We will not only describe the innovations in investments with the help of bigdata, but also discusses how we can leverage bigdata in management, customer segmentation, fraud detection, and many more.

29.5.1 BigData in Banking Sector

Big Data is renewing the whole planet and with its immense benefits, it has left no business immaculate. Big data analytics, which has emerged as a lifesaver for the banking sector, is currently being maintained in various domains of the banking sector and enables them to provide their customers with better services, both internal and external, in addition to enhancing their active and passive security systems [7, 13]. The banks are currently generating enormous measurements of data. Previously, most banks were unable to use this knowledge effectively. Nonetheless, now, banks have started using this knowledge to accomplish their primary marketing objectives. Numerous privileged insights, such as money transfers, robberies, calamities, can be discovered using this information [3, 14–16].

Let us discuss here some of the benefits provided by BigData to the banks as well as to the customers:

1. **Fraud Detection & Prevention** [14]: On one hand, the digital world is providing us with various advantages yet again, it brings forth different sorts of frauds also. Our data is presently more exposed to cyber-attacks than any time before and it is the hardest test a banking sector may face. Big Data Analytics along with some Machine Learning Algorithms permits banks to spot frauds before they can be put [17–19]. Big Data enables banks to certify that no unauthorized modifications are made, providing a degree of wellbeing and security. This is done by recognizing unacquainted spending patterns of the customer, foreseeing unfamiliar actions of the customer, etc. Danske Bank, the largest bank in Denmark was struggling with its corruption acknowledgement policies having an exceptionally low rate (40% falsification discovery rate and overseeing up to 1200 false positives every day). This rate was frightening and called for urgent action. They joined hands with Teradata Business, a leading database and analytics-related tools, products and service provider, to use advanced Big Data analytics to improve their fraud detection techniques and soon observed substantial results. The bank witnessed a 60% decrease in false positives, expecting it to hit an 80% mark soon and increase by 50% in the true positive rate. They also noticed a huge profit of $70 million in 2018.
2. **Risk Management**: Establishment of a robust risk management system is extremely important for banking organizations or else they must suffer from huge revenue losses. The early recognition of fraud is a large part of risk management and Big Data can do as much for risk management, as it accomplishes for fraud identification. Big Data finds and presents huge information on a solitary huge scope that makes it simpler to diminish the number of dangers to a reasonable number. Big Data assumes a critical job in coordinating the bank's necessities into a unified, practical stage. This reduces the bank's chances of losing data or ignoring fraud. Organizations need to keep innovating new ideas to stay alive in the dynamic environment and raise their income as much as they can. Companies can detect danger in real time through Big Data Analysis and seemingly save the consumer from possible fraud. Big Data has been leveraged by United Overseas Bank (UOB) Limited, the third-largest bank in South-East Asia, to direct risk management, the largest area of concern for any banking organization. UOB took a gamble with employing though keeping the same in mind. In the same way, UOB took a chance to use a Big Data-based risk management system. A time-consuming effort, typically taking up to 20 h, is to measure the value of risk. UOB has now been able to do the same job in just a few minutes through its Big Data risk management system and with the goal of doing it soon in real time.

3. **Customer Segmentation** [3, 15, 16]: Customer retention is a life-long journey for banking companies, from safeguarding the security of their transactions to supplying them with the most appropriate and useful deals. Big Data would provide banks with in-depth insights into the preferences and behaviors of consumer spending, simplifying the process of assessing their needs and desires. By having the option to track and follow every client exchange, banks will have the option to sort their customers dependent on different boundaries, including regularly got to administrations, favoured Visa uses, or even total assets. Banks should invest in customer analytics that effectively segments their customers. This will assist in determining pricing, products and services, the right customer approach and marketing methods. Bank of America is one of the United States' main banks. It has an estimated 70 million customer base. In 2008, when they saw their clients moving to smaller banks, they noticed that their consumer base was shrinking at an alarming pace. This left them clueless, and the reasons for this collapse were desperately searched for. Big Data Analytics came to their rescue after that. They discovered that their end-to-end cash management system was too restrictive for consumers by reviewing their customer data from a number of sources such as their website, call center logs, and personal reviews, as it hampered their freedom to access trouble-free and versatile cash management systems. Although the smaller banks gave a simple solution to it. They agreed eventually to end their all-in-one offering. As a solution to these problems, they soon launched a website in 2009 that was a more flexible online product, CashPro Online, and its mobile edition, CashPro Mobile, later in 2010. This has been developed in order to provide a one-stop solution for all the services they provide to their customers.
4. **Personalized Product Offerings**: To make and convey new plans and schemes, pointed straightforwardly at the prerequisites of their clients, customer segmentation can be used. A bank will get a better understanding of how to get the highest response rate from their customers by evaluating past and current expenditures and transactions. Having personalized item contributions would consider an undiscovered customized administration expertise that allows banks to make consumer connections more relevant.

In the current dynamic financial market, big data holds the key to exposing marketing potential. Advanced analytics of big data allow banks to cope with the accelerating cost of enforcement and the chance of non-compliance. In any event, monetary aid companies are still lagging in introducing analytical tools for big data which means an undiscovered opportunity for generating esteem that is open to the financial sector. This is to be measured from an information technology or Line of Business perception.

Example: Before the financial crisis of 2008, among the largest banks in the world was the Royal Bank of Scotland (RBS). But unfortunately, at one point, the UK Government had to step in to save the bank due to their exposure in the risky mortgage market. At that time, the UK Government owned 84% of company's shares. Ultimately, to recover from such concerning circumstances, the bank started to focus heavily on customer services and benefits to the customer as a plan to fight back for their former glory in retail banking sector. The bank leveraged big data analysis as competitive advantage over their competition. Due to the huge success derived from big data analysis, the bank invested £100 million in the same and following the same path the bank announced "personology"- a customer first initiative.

As per Christian Nielsen (Royal Bank of Scotland head of analytics), the banks were moving away from the customer-first mind-set between 1970 and 1980s. They were primarily focused on new products and achieving sales targets and least bothered about what services customers wanted. Nielsen further says that during 1970s, banks were connected to their customers at the ground level via branch staff. At that time, bank via their staffs clearly had an idea about who the individual customer was and how the bank met their requirements. Also, Banks knew about their families. Afterward, during 1980s, as per Nielsen, this connection between the customers and the banks started deteriorating on personal level and the retail banking system started to push various types of financial products just for the revenue instead of having customer-first approach and helping the in their financial management. Banks were more worried about meeting the sales target instead of focusing on what the customer wanted.

Big data in practice: In current times, for again putting forth the personalised services for the customers, banks like Royal Bank of Scotland are leveraging data mining and analytics. The analytics team at Royal Bank of Scotland have devised an ideology named as "personology" that helps them to really apprehend the customers' requirements and expectations. Due to all this, banks currently have a huge volume of data collected about their customer's right from our spending in financial management. They have all this data to do a detailed analysis about how the customers live their lives. Using the same data, they can predict customers' vacations, marriage information, and medical information and as well as the things on which the customers are excessively spending their money.

If we compare the banks with other industries like e-commerce, the latter can collect less information about the customer, yet they do a better job of getting the insights from the data collected to improve the business. If we look at the huge amount of data that the banks have about their customers, it has a huge potential, but they are just scratching the surface in terms of making use of that data. Currently, even the banking systems are developing features around the concept of "personology" to target the offers and promotions to the customers individually. For example, the targeted notifications, SMS that the customers receive about a particular offer are personalized in such a way that it conveys how the customer is going to benefit from that offer.

Results: Examples of customers feeling happy by a simple birthday wish coming from the bank makes them very excited about the stuff they are doing, and this is resulting in improved response rates. While examples like this does not perfectly align to Bigdata philosophy but at the same time this does influence people on an individual basis.

Technical details: CRM system offered by Pegasystems is currently being used by the banks to help the staff in branches and call centers to make personalized recommendations to the customers. Many banks have also developed their in-house software systems and dash boards using open source projects (Hadoop and Cassandra) to have a better overall visibility of how the business is performing. When it comes to sales and marketing the analytics does no help if it cannot reflect things that are not known already related to the customers.

Conclusion: This shows just simply doing data mining and analytics without using the insights effectively are as good as having no data analytics in the first place.

29.5.2 Fraud Detection

To forestall obtaining cash or property through affectations, a lot of activities is embraced know as fraud detection. It is applied to numerous business sectors such as banking or insurance. In banking, fraud may incorporate producing counterfeit cheques or using stolen credit cards. Other frauds may include exaggerating losses or causing a mishap with the sole aim for the compensation out. Data Mining can gather data to look through up to millions of transactions to discover patterns and detect fraud.

To detect and prevent fraud and recognize suspected false claims at an early phase, insurance companies may develop predictive models which are based on both historical and present data on wages, medical cases, advocate costs, socio-economics, climate information, call centre notes, and voice recordings. Big data analytics has the potential to deliver an immense advantage by assisting foreseeing and reduction endeavoured extortion according to the risk and fraud experts, executives, managers and supervisors at insurance companies. The intention is to distinguish false cases at the main notification of misfortune—at the first point where a financier or statistician is required [20–22].

With bigdata analytics, patterns of unfamiliar behavior are identified in huge measures of structured and unstructured data, thus assisting companies to identify fraud in real-time. Investment for such companies is productive. Companies are now able to examine complex data and mishap situations in minutes. This was impossible before employing bigdata analytics techniques.

29.5.2.1 Types of Frauds

1. **Phishing**: Phishing is a social engineering technique in which a victim is tricked into submitting sensitive information like credentials, bank account number, tokens, personal information such as date of birth, etc. by impersonating a legitimate known service that is familiar to the victim. Signs of phishing may

be incorrect URLs, requests for unnecessary information, unofficial email ids etc.

Algorithms for detection

Phishing may be detected using three machine learning algorithms, Decision Tree algorithm, Random forest algorithm and Support vector machine.

Decision Tree algorithm: It is easy to understand and practically implement the decision tree algorithm (DTA). DTA uses a tree-like structure or model of decisions and their possible outcomes. It consists of conditional control statements. The best splitter called the root of the tree is chosen from the available attributes. Splitting continues until a leaf node is found. Each internal node in the tree represents an attribute and each leaf node represents a class label. Gini index and information gain methods may be used to calculate these nodes. DTA creates training model which predicts the class of the variable.

Random forest is an Ensemble. The ensemble combines multiple Machine Learning models to create a more powerful model. A set of decision trees where each tree is slightly different from others is a random forest. Each tree may do a pretty good prediction job, but part of the data is likely to be overfit. But by constructing multiple trees that all function well and overfit in various ways, by combining their outcome, we can reduce the amount of overfitting. So random trees fix the overfitting problem of decision trees while preserving the ability of trees to predict.

Support Vector Machine algorithm (SVM) finds a separating line well known as a hyperplane in N-dimensional space, where n represents the number of features, for the distinct classification of data points. More than one hyperplanes are possible, however, to classify data perfectly, SVM looks for the nearest data points called support vectors and the hyperplane is selected in such a way that the margin i.e. the distance between support vectors and hyperplane is maximum.

2. **Money laundering** [23]: It refers to the cover-up of the origins of illegitimately acquired money. Criminals attempt to make funds look like as though they have originated from the legal source. For example, if someone procures cash by selling drugs, he may show as though the cash is obtained from some business and the bank may credit the funds in that business's account.

Algorithms for detection

To detect money laundering, supervised learning algorithms are trained. For instance, the computerized transaction monitoring rules that run today are creating alarms investigated by investigators. The machine can leverage the doubtful activity report (SAR) flag ('yes' or 'no') to foresee future results given complex relationships in the customers' behaviour.

Unsupervised learning techniques leverage unlabelled data and thus learn from the pattern of the data itself. There is no verifiable result or "right answer to learn from. For instance, customer populations can be divided into peer groups to comprehend deviations in their action.

3. **Financial Statement Fraud** [24, 25]: Ventures may intentionally produce misstatement or omit amounts in the financial statements to hoodwink financial statement users leading to a misrepresentation of the financial state of the venture [26, 27].

Algorithms for detection

Group method of data handling, Decision Tree, Classification and Regression Tree, Support vector machine are some common algorithms used to detect financial statement fraud.

Group method of data handling (GMDH) is an inductive learning algorithm i.e. out of possible variants, it selects the best solution. It permits finding interrelations in data automatically. It is a group of numerous algorithms such as clusterization, parametric, re-binarization, analogues complexing and probability algorithms for different problem solution. It is an approach that tests gradually complicated models and select the optimal solution using some external criterion characteristics.

Classification and Regression Tree (CART), unlike traditional statistical method, is an automatic, non-parametric scheme. To best classify samples into numerous non-overlapping regions each corresponding to a leaf node in the tree, CART uses Binary Recursive Partitioning Algorithm.

4. **Credit card fraud** [17, 19, 28, 29]: Duplicating data found on the magnetic strip of a credit card is called skimming and is a prohibited process. An impostor can skim the information located in the magnetic strip or use the card online by using the card details in case the credit or debit card is lost or stolen. Scammers can't withdraw cash without the card pin; however, they can use the card to pay via contactless if this feature is enabled on the card.

Algorithms for detection

Algorithms recognized to detect credit card fraud are Outlier detection methods, Association rule analysis and Density-based spatial clustering of applications with noise.

Outliers are the data points that show unusual behaviour from the others in the dataset. Outlier detection methods which are different from traditional methods can be used to detect these outliers.

Association Rule Analysis is an unsupervised learning algorithm that attempts to identify interesting relationships between data points which may not be visible otherwise. Metrics that help in determining associations among various items are support, confidence and lift.

Density-based spatial clustering of applications with noise (DBSCAN) is a density-based clustering algorithm used to remove outliers and find out clusters of random forms.

5. **Loan Fraud** [30]: People may write false details in applications with the intention of loan approval or a thief may steal someone's identity and apply for a loan; all are included in loan fraud [31].

Algorithms for detection

Logistic regression and Naïve Bayes classifier algorithms may be used to detect loan fraud [32].

Logistic regression is a statistical method used for analysing dataset with one or more independent variables to predict the outcome. For this, it uses a logistic or sigmoid function that takes a real value and returns 0 or 1.

29.5.3 Big Data in Retailing

Big data is the captivating centre stage for policymaking in several establishments, particularly merchants. The McKinsey Global Institute has foretold that merchant's implementing big data can surge their functioning margin further 60%. Big Data in Retail Business assist merchants in forecasting consumers' demands, organizing the experiences of clients. Significantly, it aids merchants in expanding their working competence. Vendors like Amazon are frequently collecting, curating, and evaluating data for decision-making. Their choices fuel consumer relations with vendors through supplementary information which are detailed, organised and examined for advance judgements. Numerous such varieties are prepared in real-time. The implementation of Big Data by numerous merchandizing channels has amplified effectiveness in the market to an inordinate degree. Merchants look up to Big Data Analytics for a superfluous competitive advantage over competitors. They are rapidly implementing it to get improved methods to influence the clients, recognise what the client's need, providing them with the finest conceivable solution, guaranteeing client satisfaction, etc.

The vendors are accommodating the data-first tactic to have a deep vision of clients purchasing patterns, mapping them to merchandises, and developing strategies to increase the sale to gain profits. Merchants utilize the structured and unstructured data accessible about their consumer's behaviour to the utmost latent. The sales market has constantly remained involved in finding and examining fresh purchasing trends in the client's behaviour. This allows vendors to associate produces like fragrances and beauty products, clothes and accessories, as the data patterns demonstrate that the people who buy clothes will invariably also purchase accessories. Placing correlated products together affects the purchasing behaviour of the client.

Machine learning takes big data it gathers from computer users and is applied to foretell imperative insight about client's expenditure patterns and how these patterns endure fluctuations every time.

29.5.3.1 Big Data Use Cases in Trade Market

Big data is liable for bringing a massive revolution in the trading sector [33]. Several of the use cases are listed below:

1. **Generation of Recommendations**: While buying a particular product from an e-commerce site or online retail, people often get recommendations of similar products they may like to buy or the items purchased by other people having similar behaviour or their friends. Big Data analytics helps retailers to forecast products that the customer is likely to buy based on their purchase history.
2. **Constructing Strategic Resolutions**: Big data Analytics facilitates retailers to take both long term as well as short term decisions. Long term decisions include the location of the retail outlet, facilities or resources, type of merchandise etc. whereas offers, discounts, promoting products, and advertisements constitute short term decisions.
3. **Applying Market Basket Analysis:** Retailers use Market Basket Analysis to figure out which products customers may buy together. This can be achieved by Association Rule Mining algorithm where interesting relationships among the items are identified. For instance, a customer who bought cerelac (baby food) is likely to buy diapers.
4. **Predicting demands and trends:** Big Data Analytics enables retailers to generate insights on customer behaviours. This may help them to recognize items and services which are in-demand and the ones they must quit offering. It would likewise permit them to foresee the next big thing in the retail business and afterward manufacture new items as per the current trends in the market. In this way, along these lines, Big Data in retail helps retailers in anticipating the buyers' demands.
5. **Regulate Pricing Policy**: Diverse elements that influence in pricing policies are article cost, competitor's prices and buyer's budget. With these elements, Big Data analytics supports the vendor's estimating best price resulting increase in sales and generation of maximum revenue.
6. **Personalization of the client's experience**: The realisation of any business solely lies on the satisfaction of the client and how fairly clients are treated. Big Data offers vendors prospects to enhance client experiences. Over and done with big data analytics, a seller can deliver directed service to his clients. It benefits sellers in forestalling buyer's demand, tailoring the price cut/propositions aimed at the clients. Thus, empowering them in implementing operational and consumer-centric resolutions. Hence customizing their advertising founded on leads generated by client's data. The leading objective of the vendors is to outreach client on the correct interval, at an accurate place and via an accurate medium. This is conceivable simply if business miens periodic personalization of their consumers. Merchants need to shape an individual database of their clients. Clients share personal data if it is favourable for them. Facilitating merchants in attaining superior statures in the market and will simultaneously increase competition. Imagine being able to purchase produces tailored accurately as per client's need. As per customer requirement, what else does a merchant require as a lead? Custom-made products are valued in the contemporary era and that is what Big Data helps in.
7. **Filtration of most valued clients**: Excavating Big Data is a gigantic responsibility, and the payoff lies in examining the profitable clients. For attaining

feat, it is vital to highly prioritize such clients as it costs more to establish associations with new clients than maintaining relationships with old clients. For competitors in market thin margins, quarrying the correct data and conducting the smart investigation will result in great commitment, more faithful clients and economical advantage.

8. **Accumulative proficiency by simplifying inter-departmental synchronisation**: The inventory unit and the manufacturing unit must have great coordination for the trade industry to run efficiently. Big data has facilitated in coordinating each association in the series in actual time. This aids the merchant classifying products in demand. So, they can be stocked to compete in the market. This has optimized procedures and delivered enhanced methods to deal with faster product life cycles. Big data has considerably reduced costs by aiding vendors to recognise the stock chain and supporting improved product distribution

9. **Examining client journey**: The zigzag expedition that a client tracks in advance of ordering merchandise start at Responsiveness, leading to Investigation, Innovation, Acquisition, Provision, and lastly concludes at Checkout. Big Data aids in handling this superfluity of data and evaluate it. Nowadays clients can filter any product required in few seconds and match expenses on diverse platforms. Big data supports vendors navigate the buyer towards their produces. Additionally, big data analyses the buyer journey for studying multiple scenarios. The requirement for consumers to interact straight away with the seller is excluded with big data analytics. A single post on social media can outreach different people, in different regions with help of target advertisements and audiences. For example, we post about having any plans for vacations and we get countless suggestions in comments posted on the same post, suggesting unlimited things one could do in different locations for different budgets.

Example: With over 11,000 stores and clubs in 28 countries that operate under different names, Walmart, an American multinational retail company, stands as the largest global retailer both in terms of its expansive area of operations and in terms of the annual revenue it generates. Certainly, they have recognized the value of data analytics long back considering the wide scale of their operations. The inception of the organisation's interest in data analytics dates to the early 2000s when the Sandy hurricane lay its wrath over the United States. As the retail giant geared up to meet the demands for the survival-critical entities, unexpectedly useful insights, pertaining to the user demands, came to light as the sales data was put to observation. To predict the surge in the demands for the emergency equipment for the future, as the Hurricane neared its first impact, the retail giant's CIO, Linda Dillman, came forward with surprising stats. Besides the flashlights, power backups and other needful devices, an unexpected surge was predicted for Strawberry Pop-Tarts. As such, to counter the surges under similar circumstances, surplus supplies of these entities were dispatched to the outlets residing within the impact path of the France's hurricane in 2012. No surprise, all the predicted surges were well met. And this became a reality with the help of popular Machine Learning Algorithm: Association rule mining.

Daily supermarket sales usually span across a few million products rendered away to around a million customers. The sales competition in the industry knows no bounds as more and more customers, primarily pertaining to the fast-paced environments of the developed sects, look up to the retailers to fulfil their daily needs. The competition is not merely characterised by the value, the products are priced at but customer service, ease of access etc. play a vital role too, as they are known to fetch in brand loyalty from the customers. The core principle to follow in case a retailer giant wants to ace ahead in this war is to have the right products in the right place at the right time. As simple as this may sound, there as definitive logical overhead in keeping up with this principle. To begin with, the Products need to be efficiently priced for the various sections of customers. And the demands as whole need to be met under the same roof. Consequently, if one finds their needs inadequately met, he/she is bound to look for some alternative that fits to their busy schedule.

Big data used in practice: In 2011, with an increasing understanding of the benefits that data analytics poses, Walmart established Walmart-Labs and Fast Big Data Team, to precisely identify their customers' needs and subsequently provide them with the products they would want to buy. The main purpose of this new venture is to research and consequently deploy the new data-led features across the business. The product of this strategy is referred to as the Data Cafe—a state-of-the-art data analytics hub situated right at the core of their Arkansas headquarters.

Teams pertaining to any section of the business are open to visit the Cafe with problems concerning data analytics. Essentially, to fulfil the purpose of the centre's existence, the analysts join up with them to formulate a comprehensive solution for the same. As a strategy for consistent throughput, the retail giant maintains a system for monitoring the performance variables across the company which triggers automated alerts once the performance hits the bottleneck. As a response to the fall, the teams responsible for the dropped variables are invited to collaborate with the data team for a possible solution to the same.

For instance, the grocery team were once struggling to identify and comprehend the reasons behind the declining sales of certain products. The sooner the data from the grocery department was put to test by the Cafe analysts, it was established in no time that the decline was straight away attributable to a mere pricing error. As such the error was immediately rectified and to no surprise, the sales recovered within days of deployment.

Funny story, back on one Halloween, the Café analysts were continuously monitoring the sales of Novelty Cookies, that recorded zero sales at several locations across the states. As the sales dipped, an automatically triggered alert was sent off to the merchandising teams handling these stores. The problem was that the products had not been even put up for sale let alone the dipping sales. While this does not sound like a complex algorithm in action, it would not have been a reality without real-time analytics.

Walmart Social Genome Project, is yet another venture founded by the retail giant to monitor the public cum media conversations about the essential and non-essential entities and attempt to predict the subsequent products people would buy based on these monitored conversations. There is also this Shopycat service, which predicts for the customers the influence of their friends, across the social networks, on their shopping habits. Essentially this led to the development of a search engine called Polaris that allows the professionals to analyse the product searches performed by customers on their websites.

Results: The inception of the Data Cafe has led to a vivid decrease, in the time it'd take to identify a problem and hence devise an impactful solution for the same, from an average of a few weeks down to a couple of tens of minutes.

Technical Details: The real-time transactional database from Walmart is estimated to comprise of about 40 petabytes of data. As humongous as this may sound, it pertains only to the sales data from the recent weeks which encompasses the true value of real-time analysis. The data across the retail stores, e-commerce points and corporate units are stored centrally on Hadoop (A Distributed Data Storage and Data Management system). Other technologies that are put to test here, include Apache Spark and Cassandra with R and SAS as the prime candidates for the programming languages used to develop analytical applications.

Conclusion: Walmart has always led from the front in data-driven initiatives including brand-loyalty and reward programmers. With their wholehearted commitment to the recent Real-Time-Analysis technical advancements and Responsive Analytics they have made a clear statement of their motive to ace ahead in the future. Bricks 'n' mortar retail may appear "low-tech" or almost "non-tech" compared to their flashy online rivals but Walmart has made it clear that cutting-edge Big Data is remains just as relevant to them as it is to other retail giants such as Amazon and Alibaba. Despite the seemingly more convenient options on shelf in the form of e-commerce, it appears that the customers, either through habit or through preference, are still committed to hop into their cars and travel to the shops. This clearly implies a huge market out there open for the taking. Certainly, businesses which make the apt use of analytics to the fullest to stimulate efficiency and enhance their customers' experience, are set to advance both operatively and economically.

29.5.4 Customer Segmentation

Customer segmentation refers to the process of dividing the customers into groups based on certain common characteristics. As the business and audience of an organization grows, they can use customer segmentation to analyze and categorize the customer base. It is a useful tool for customized promotions. Using customer segmentation, organizations can market to each group separately and effectively. Marketers can create targeted marketing messages for specific groups of customers and establish better customer relationships. This allows the companies to improve their customer service. Also, this way the companies can focus on the most profitable customers [15, 16].

29.5.4.1 Benefits of Customer Segmentation

- Customer Segmentation helps in product customization.
- Customer Segmentation assists in financial and human resource capitalization.
- It increases the overall profit with the help of targeted marketing.
- Customer service can be optimized.
- It can help to determine prices and define profit margins.
- Need-based service and focus intensity are adjustable for each segment.

29.5.4.2 Segmentation Types

Customers can be segmented based on a wide range of factors such as the products purchased, location, age, gender and so on. For example, a telecom company can segment customers based on usage patterns and then target them with suitable offers.

Customer segmentation requires the companies to collect data about the customers and analyze it to identify various patterns that can be used to form the segments. Some of this data can be gathered from the purchasing information such as location, products purchased. But this is not enough. More data needs to be collected. This can be done using face-to-face or telephone interviews and surveys. Once the data has been collected, it can be analyzed, and various algorithms and models can be applied to it to infer useful results.

Presently there is not anyone-size-fits-all segmentation scheme that can be applied to accomplish all the business desires and objectives. There are a quite a few segmentations approach accessible, every one of them rational for supporting diverse business requirements. The main customer segmentation types based on the attributes gathered from offline sources are depicted as follows [34]:

- **Value-based Segmentation**: It is a segmentation procedure by which customers based on their value and attributes for instance, Customer Lifetime Value, Value of Orders, Value of purchases etc. are categorized.
- **Behavioral Segmentation**: It is a process by which customers are clustered according to their habit, outlook and behavior regarding an item or a service. It may involve attributes like product ownership, kind & frequency of transactions, income history, contacts expenses, and product usage, etc.
- **Loyalty or Engagement Segmentation**: Loyalty or engagement segmentation groups customers in accordance with diverse degrees or loyalty to the dealer or brand. Attributes associated with customer loyalty may be loyalty score, rate of buying, nature and number of complaints, etc.
- **Demographic segmentation**: Demographic segmentation groups customers based on their demographic or social characteristics. Attributes associated with customers may be: Age, gender, salary, background, marital status, qualification and other personal particulars of the consumer.
- **Attitudinal segmentation**: Attitudinal segmentation is accomplished to discover customers' desires and requirements that can be satisfied by buying an item or

a service. Attributes associated with customers may be: First choice, Interests, Customer requirements, Inspirations, Usage occasion, routines, personality, and preferable advertisings.

29.5.4.3 Tools and Algorithms Suitable for Customer Segmentation

There are various techniques and algorithms available that can be applied to customer segmentation such as clustering, classification and association [34]. But the most used algorithm for customer segmentation is clustering. It is an unsupervised machine learning technique for identifying and grouping similar data points together. It scans through all the information related to customers and learns the best way to group them together. The clustering algorithm that suits best for customer segmentation is the k-means clustering algorithm. K-means clustering is an unsupervised algorithm. It divides the data into K clusters, where K is a predefined value, and places the data points with similar features in same cluster iteratively. This is done by diminishing the sum of square of distances between the cluster centroid and its accompanying data points [15, 16].

29.5.5 Demand Forecasting

Demand forecasting is a procedure of foreseeing the interest in services or products of an organisation in predetermined timespan later. It is a fundamental part of each kind of business, be it retail: wholesale, online, offline or multi-channel. The way toward anticipating the future includes handling historical data to assess the interest for a product. A precise estimate can carry noteworthy upgrades to supply chain management [8], overall revenues, income and risk appraisal. An exact interest figure gives a precise image of future interest and assists with maintaining a strategic distance from overproduction and extreme overstock. It also enables an association to arrange the important commitments as per the foreseen demands, with no wastage of materials and time. To lighten threats, it is of focal importance for the organisations to choose the prospects of their items and services in the market. This knowledge on the future demand of an item or services in the market is achieved through the system of enthusiasm envisioning. Foreseeing demand is done to upgrade processes, decrease expenses and avoid mishaps realized by freezing up cash in stock or being not ready to process due to being out of stock. Ideally, the association would satisfy demand without over-stocking.

29.5.5.1 Benefits of Demand Forecasting

- Demand forecasting empowers a company to generate the output which has already been determined. It assists the company to arrange for certain elements

involved in the production such as land, labour, capital, and enterprise ahead of time so that the desired measure is produced with no difficulty.
- Demand Forecasting helps a company in loss reduction because with forecasting, it becomes easy to fulfil the demands and prevent loss.
- With demand forecasting, associations lessen the threat of missing the mark on stock and hence increase customer satisfaction and prevent losing customers.
- Contingent upon the future demand for specific merchandise, sales and advertising team can move their endeavours to help cross-and up-selling of correlative items. This results in better marketing and sales management.
- Demand Forecasting additionally need a brilliant workforce management as temporary staff can be recruited to support a demand peak.
- Demand Forecasting causes the association to precisely evaluate its prerequisite for raw material, semi-finished goods, spare parts, etc.
- Demand forecasting aids in precise estimation of the manpower essential to generate the desired output, thusly keeping up a key good way from the conditions of under-employment or over-employment.

The basis for demand forecasting is historical data from transactions. Demand Forecasting is done by analysing statistical data and looking for patterns and correlations in data. Key parameters for demand forecasting are the product, placement, pricing and promotion for forecasting future demand, deciding the means of transport, implementation of new merchandise, determining the target market respectively [35].

Companies forecast demand in short term or long-term contingent upon their prerequisites. Short-term forecasting is accomplished for planning routine exercises, for example scheduling production activities, framing pricing strategy, and building up a suitable sales strategy. In contrast, *long-term forecasting* is implemented to design a new project, expansion, and up gradation of production plant, and so forth.

29.5.5.2 Tools and Algorithms Suitable for Demand Forecasting

There are various predictive modelling algorithms that can be applied to demand forecasting. These include clustering, association, regression and neural network. One of the most common methods among these is **regression**. It is a supervised machine learning technique. Under the category of regression, falls the **linear regression algorithm.** It is a statistical method used to envisage future values from historical values and is effectively used for demand forecasting. This regression type allows to:

- Guess future values through data point estimations and recognize underlying trends.
- Forecast effects of modifications and detect the strength of the impacts by analysing dependent and independent variables.

Linear Regression is one of the simplest and most used traditional predictive models [36]. This technique is mainly used to find a relationship between two variables (explanatory variable and dependent variable) using a linear equation.

29.5.6 Big Data in Fashion Industry

From last 10 years, big data has achieved a substantial reputation in the fashion world. Big data analytics has a key role in fashion industry, anticipating trends, customer behaviour, their choices and feelings. Nowadays, there is an endless change in the demands of customers; they want outfits with a personalized filter such as colour, sleeve-type, style, length, material, print, pattern etc. Due to excessive stock that becomes out-dated with the change in trends, fashion industries lose a lot of money. This made fashion industry to switch from mass production to mass customisation making an efficient utilisation of bigdata [37].

Fashion companies and retailers have always consistently offered significance to sales data. The real revolution lies in the manner information is currently getting obtainable, for example, Internet-based data from websites, social media or mobile applications. The huge amount of data accessible from these sources can furnish retailers with valuable information to forecast trends and analyse customers' behaviour [38].

Recommendation systems that use bigdata analytics have been developed to analyse customers' behaviour and predict similar products customers may like to purchase. These systems also predict the shopping preferences, emotions, likings of the customer. However, in order to offer products to the customers, these systems require sufficient data about the customer, products etc. Fashion data, the data related to a fashion product is used for analysing trends, customers' behaviour and son on. Fashion industries generate a huge amount of data in different forms such as words, pictures, videos and so forth. Also, with constantly changing trends, this data is growing rapidly. Since fashion data depicts all the highlights of bigdata i.e. variety, volume and velocity, it can be renamed as fashion bigdata.

Fashion data can be classified as follows:

1. **Material**: It refers to the fabric that a fashion product is made of. Numerous attributes associated with fabric are filament (type, size, and length), yarn (type, diameter, count and twist), weight, thickness, width, weave structure and so on. Alteration in one or more attributes may result in different materials.
2. **Fashion Design**: It is an art dedicated to applying design to outfits. These designed are usually subjective to customers' feelings, themes, and occasions to wear.
3. **Body Data**: These include information such as body measurement and body type and are useful for ready-made clothes.
4. **Colour**: With an instantaneous impact and everlasting impression, colour is the first thing customers observe in the outfits. Being the dominating feature, colour

alone may accent or spoil the elegance of any garment. It is directly related to the emotions of customers.
5. **Technical/Production design**: By this the design of item is made production friendly as it permits the producer to see how to make the products. It incorporates information on design making, sewing and so forth.

Having a huge amount of business data that online sales platforms monitor and analyse, gives them a competitive advantage over traditional organisations. On this note, bigdata is changing the dynamics of the industry as fashion shifts from an "offer-based demand" industry to a demand-based offer" industry.

On account of **predictive analysis** [5], outfitters can manage their production and accomplish the primary objective of each business organization: fulfilling the purchaser's needs at the earliest.

In addition, drift from an "offer-based demand" to a "demand-based offer" idea, retailers can lessen the volumes of initial purchases, scaling back their inventories and grasping a season cycle dependent on genuine sales.

Bigdata aids designers to remain updated about the latest fashions and demands of customers and enables them to make significant decisions regarding latest styles, colours, fabrics, and sizes. These data are collected by online sales platforms proposing an opportunity to make a "style profile" by responding to queries like their body type, favourite colours etc.

Big Data can be used by the fashion industry to recognise their best sellers which may help them to control their production activities and deliver meaningful insights to their designers.

Since companies in the current times are thriving to operate in the best possible way so that the tragic scenarios can be avoided as much as possible. Geoffrey Moore puts this in concise words like this "*Without Big Data, you are blind and deaf and in the middle of a freeway*".

Example: The modern world is a web of digitally interconnected networks. The digitalisation has made an impact on everything, be it business or services. E-commerce has changed the landscape of global economy. The e-fashion industry has become a giant and with the collaboration of big tech and big fashion, new innovations are on the verge. Companies like Ralph Lauren, with the introduction of polo-tech are one of the first few entering the fabric tech market. With the polo-tech shirt, Ralph Lauren is trying to monitor health and wellness of the users in order to help them improve their fitness. The polo-tech shirt, which was first unveiled for athletes alone, is now commercially available. Wearable technology frequently referred to as the "wearables" is progressively rising to fame as Internet of Things (IoT), which received a boost with the recent launch of Apple Watch, is being commercialised.

Bigdata in practice: Inside the Polo-tech shirts there are sensors attached to the inner silver threading all around the torso and the chest area which pick up the users' movement data as well as their heart and breathing rates. The creators of this futuristic tech have made a mobile application which makes crunching the data for the user very easy. The application which is freely available across various application platforms, monitors number of steps taken and calories burnt and the recreates custom cardio, strength or agility workouts, on the go. These workouts are totally customised around the user data. The sensors inside the shirt are protected against sweat or water in case you need to wash it, except for the credit card sized Bluetooth transmitter which can be removed when the shirt needs a swing to the laundry. The common practice around all the tech companies is to make tech smaller. As such Ralph Lauren is trying to do the same to its Bluetooth transmitter. Maybe compressing it to the size of a button or making it completely immune to damage.

Results: The Polo-tech is not a big name in wearables yet, but clearly from the success of other wearables like Samsung Gear, Apple Watch Series, Fitbit etc. it can be clearly judged that Polo-Tech is going to be a big name in the coming years. The data that these wearables provide won't just help us improve our fitness but also avoid injuries while doing workouts.

Technical details: Ralph Lauren collaborated with Canadian company OMsignal on the development of Polo-Tech Shirt. Data is worked out through cloud computing which reaches the users' mobile applications to customise a workout built around the wearer.

Conclusion: Ralph Lauren, while expressing his views to the wall street journal, emphasized how this is only a small step towards the new beginnings: "We are setting up divisions within Ralph Lauren to focus on developing all kinds of products across all our brands. So, we can expect to see more wearable technology from Ralph Lauren in future."

The last few years have been game changing for the big players of all the industries who are now looking to introduce new tech, more particularly-data enabled and connected workstations. The big data and IoT are no less emphasised than the space race. Every big player is looking for a piece. Data science is what major players are looking into. This increases the opportunity for everyone who is associated to any form of data science.

29.5.7 Supply Chain Analytics

In business, the supply chain [7, 8, 4] is a system engaged in producing and distributing products or services to the end-users. Supply chains normally generate huge measures of information and to derive useful insights from that data, Supply Chain Analytics (SCA) is used. It employs visualisations such as graphs, charts etc. that represents the capability of taking decisions based on data.

Supply Chain Analytics is a concept in Big Data that is applied in all the stages of businesses (sourcing, manufacturing, distribution, logistics). It allows executives to make informed decisions, consider emerging marketing patterns, identify and analyse risks, and use supply chain skills to devise supply chain strategies that ultimately enhance reliability and profitability for companies. In this way, by empowering information-driven choices at vital, organizational and strategic levels, SCA aims to enhance an association's operational competence and viability.

Benefits of Supply Chain Analytics

- SCA can assist an association with making wiser, quicker and more proficient choices.
- It can recognize the known risks and help to anticipate future risks by identifying patterns all through the supply chain.
- By analysing client's information, SCA can enable a business to foresee and satisfy future demands. It enables an association to choose which items can be limited when they become less beneficial or learn what a client may need after the initial order.
- SCA can be used by companies to gain a significant return on investment.
- It can give incredible experiences into how the supply chain network configuration can assist with fulfilling more clients' demands, without a related increment in stock levels.
- SCA can be used to reduce process variability.

29.5.7.1 Types of Supply Chain Analytics

Supply Chain Analytics can be of one of the following types:

- **Descriptive analytics:** It helps to interpret the past data to better understand changes that have occurred in the business. It gives ability to perceive and a single source of truth across the supply chain, for both inside and outside frameworks and information.
- **Predictive analytics**: It enables an association to comprehend the most probable result or future situation and its business suggestions. It can be used to make predictions about future or otherwise unknown events. For instance, predictive analytics can plan and ease interruptions and risks.
- **Prescriptive analytics**: It assists associations with taking care of issues and work together for greatest business esteem. It not only anticipates what will happen and when it will happen, but also why it will happen. It additionally assists organizations with working together with strategic accomplices to decrease time and exertion in moderating interruptions. It suggests decision options on how to take advantage of a future opportunity.
- **Cognitive analytics**: It enables an association to address complex inquiries in characteristic language—in the way an individual or group of individuals may react to an inquiry. It helps organizations to thoroughly consider a perplexing issue or issue, for example, "In what capacity may we improve or upgrade X?"

29.5.7.2 Tools and Algorithms Suitable for Supply Chain Analytics

There are various areas in Supply Chain Analytics such as supplier relationship management, supply chain network design, product design and development, demand planning, procurement management, logistics and many more. Different areas of SCA require different Big Data tools and algorithms. Classification, clustering, association and regression are some of the Big Data models that can be applied to various areas of Supply Chain Analytics.

Out of these, classification has been largely employed in logistics, manufacturing and procurement management for predictive analytics.

29.6 Challenges

Although there are endless benefits brought using Big Data, but there are also several challenges associated with it.

One of the challenges associated with using big data analytics for business is that there is huge amount of data generated by organisations and selecting which portion of data is relevant and suitable for a particular application is not an easy job. Most of the organisations fail at this point and are miss out on the vast business opportunities that can be realized.

With the development of big data, the number of big data tools and technologies in the market has increased to such a great amount that it can be very easy to get confused and make a poor choice. If the right tools and technologies are not used, an organisation cannot get most out of the big data analytics.

Another important challenge is that big data is still new to researchers and there is a shortage of professionals who understand both big data and business.

The greatest challenge in Big Data is the security. Big Data stores can be attractive target for hackers as they store vast amount of aggregated data from a wide range of sources. The storage infrastructure must have strong security measures to guard the organization against internal and external threats.

29.7 Conclusion

In recent years, we have witnessed a great increase in the application of Big Data in the business world. These business applications have made a huge impact on our everyday lives and this impact and influence of Big Data is still growing to a larger extent. We can say that Big Data has become highly relevant in businesses today. Big Data resource has changed from being an extra advantage for any business to a critical asset that any company that wishes to compete in commerce must possess.

We explored various areas of business and the big data algorithms that can be used in each area. While the application of Big Data can lead to a great success in business, it can also be a cause of many problems such as security and privacy. There are also many technical and social challenges, which need more exploration of researchers and practitioners.

References

1. Liu O, Chong W, Man K, Chan CO (2016) The application of Big Data analytics in business world. In: Proceedings of the international multi-conference of engineers and computer scientists 2016, vol II, IMECS 2016, 16–18 March 2016, Hong Kong.
2. Qazi RUR, Sher A (2016) Big Data applications in businesses: an overview. The Int Technol Manag Rev 6(2):50–63
3. Fotaki G, Spruit M, Brinkkemper S, Meijer D Exploring big data opportunities for online customer segmentation. Department of Information and Computing Sciences, Utrecht University, The Netherlands, Utrecht
4. Leveling J, Otto B (2014) Big Data analytics for supply chain management
5. Balaji Prabhu BV, Dakshayini M (208) Performance analysis of the regression and time series predictive models using parallel implementation for agricultural data. In: International conference on computational intelligence and data science (ICCIDS 2018)
6. Sadgali I, Sael N, Benabbou F (2019)Performance of machine learning techniques in the detection of financial frauds. Elsevier
7. Hofmann E, Rutschmann E Big data analytics and demand forecasting in supply chains: a conceptual analysis. Institute of Supply Chain Management, University of St Gallen, St Gallen, Switzerland, and Department of Analytics, Deloitte Consulting AG, Zürich, Switzerland
8. Nguyen T, Zhou L, Ieromonachou P, Lin Y (2017) Big Data analytics in supply chain management: a state-of-the-art literature review. Comput Oper Res
9. Sullivan-Bolyai S, Bova C, Singh MD (2014) Data-collection methods." Nursing Research in Canada-E-Book: methods. Crit Appraisal, Utilization 287
10. Axinn WG, Pearce LD (2006) Mixed method data collection strategies. Cambridge University Press
11. Park B-K, Song I-Y (2011) Toward total business intelligence incorporating structured and unstructured data. In: Proceedings of the 2nd international workshop on business intelligence and the WEB, pp 12–19
12. Baars H, Kemper H-G (2008) Management support with structured and unstructured data—an integrated business intelligence framework. Inf Syst Manag 25(2):132–148
13. Srivastava U, Gopalkrishnan S (2015) Impact of big data analytics on banking sector: learning for Indian banks. Procedia Comput Sc 50:643–652
14. Ravisankar P, Ravi V, Raghava Rao G, Bose (2011) Detection of financial statement fraud and feature selection using data mining techniques. Decis Support Syst 50(2):491–500
15. Chang M-S, Kim HJ (2018) A customer segmentation scheme base on Big Data in a bank. J Dig Contents Soc 19(1):85–91
16. Fotaki G, Spruit M, Brinkkemper S, Meijer D (2014) Exploring big data opportunities for online customer segmentation. Int J Bus Intell Res (IJBIR) 5(3):58–75
17. Bhattacharyya S, Jha S, Tharakunnel K, Westland JC (2011) Data mining for credit card fraud: a comparative study. Decis Support Syst 50(3):602–613
18. Ngai E, Hu Y, Wong Y, Chen Y, Sun X (2011) The application of data mining techniques in financial fraud detection: a classification framework and an academic review of literature. Decis Support Syst 50(3):559–569. Part 2
19. Malini N, Pushpa M (2017) Analysis on credit card fraud identification techniques based on KNN and outlier detection. IEEE, advances in electrical, electronics, information, communication and bio-informatics (AEEICB), third international conference (2017)

20. Coalition against Insurance Fraud, Learn about fraud. https://www.insurancefraud.org/learn_about_fraud.htm
21. J.L. Kaminski, Insurance Fraud, OLR Research Report. https://www.cga.ct.gov/2005/rpt/2005-R-0025.htm. 2004. Google Scholar
22. Beaver WH (1966) Financial ratios as predictors of failure. J Account Res 4:71–111
23. Gao Z, Ye M, A framework for data mining-based anti-money laundering research. J Money Laundering Control 10(2):170–179
24. Ngai E, Hu Y, Wong Y, Chen Y, Sun X (2011) The application of data mining techniques in financial fraud detection: a classification framework and an academic review of literature. Decis Support Syst 50(3):559–569
25. Bai B, Yen J, Yang X (2008) False financial statements: characteristics of china's listed companies and cart detection approach. Int J Inf Technol Decis Mak 7(2):339–359
26. FBI, Federal Bureau of Investigation (2007) Financial Crimes Report to the Public Fiscal Year. Department of Justice, United States. https://www.fbi.gov/publications/financial/fcs_report2007/financial_crime_2007.htm
27. CULS, Cornell University Law School (2009) White-Collar Crime: an overview. https://topics.law.cornell.edu/wex/White-collar_crime
28. Sánchez D, Vila MA, Cerda L, Serrano JM (2009) Association rules applied to credit card fraud detection. Expert Syst Appl 36(2):3630–3640
29. Panigrahi S, Kundu A, Sural S, Majumdar AK (2009) Credit card fraud detection: a fusion approach using Dempster-Shafer theory and Bayesian learning. Inf Fusion 10(4):354–363
30. Jin Y, Rejesus RM, Little BB (2005) Binary choice models for rare events data: a crop insurance fraud application. Appl Econ 37(7):841–848
31. Bermudez L, Perez JM, Ayusoc M, Gomez E, Vazquez FJ (2007) A Bayesian dichotomous model with asymmetric link for fraud in insurance. Elsevier, pp 779–786
32. Xu F, Pan Z, Xia R (2020) E-commerce product review sentiment classification based on a naïve Bayes continuous learning framework. Elsevier
33. Viaene S, Ayuso M, Guillen M, Van Gheel D, Dedene G (2007) Strategies for detecting fraudulent claims in the automobile insurance industry. Eur J Oper Res 176(1):565–583
34. Kirkos E, Spathis C, Manolopoulos Y (2007) Data mining techniques for the detection of fraudulent financial statements. Expert Syst Appl 32(4):995–1003
35. Tamura KA, Giampaoli V (2013) New prediction method for the mixed logistic model applied in a marketing problem. Comput Stat Data Anal 66
36. You Z, Si Y-W, Zhang D, Zeng X, Leung SCH, Li T (2015) A decision-making framework for precision marketing. Expert Syst Appl 42(71)
37. Jain S, Bruniaux J, Zeng X, Bruniaux P (2017) Big data in fashion industry. IOP Conf Ser: Mater Sci Eng 254(15):152005
38. Silva ES, Hassani H, Madsen DØ (2019) Big Data in fashion: transforming the retail sector. J Bus Strat

Chapter 30
Computational Aspects of Business Management with Special Reference to Monte Carlo Simulation

Sahana Prasad

Abstract Business management is concerned with organizing and efficiently utilizing resources of a business, including people, in order to achieve required goals. One of the main aspects in this process is planning, which involves deciding operations of the future and consequently generating plans for action. Computational models, both theoretical and empirical, help in understanding and providing a framework for such a scenario. Statistics and probability can play an important role in empirical research as quantitative data is amenable for analysis. In business management, analysis of risk is crucial as there is uncertainty, vagueness, irregularity, and inconsistency. An alternative and improved approach to deterministic models is stochastic models like Monte Carlo simulations. There has been a considerable increase in application of this technique to business problems as it provides a stochastic approach and simulation process. In stochastic approach, we use random sampling to solve a problem statistically and in simulation, there is a representation of a problem using probability and random numbers. Monte Carlo simulation is used by professionals in fields like finance, portfolio management, project management, project appraisal, manufacturing, insurance and so on. It equips the decision-maker by providing a wide range of likely outcomes and their respective probabilities. This technique can be used to model projects which entail substantial amounts of funds and have financial implications in the future. The proposed chapter will deal with concepts of Monte Carlo simulation as applied to Business Management scenario. A few specific case studies will demonstrate its application and interpretation.

Keywords Monte Carlo technique · Business management · Simulation · Applications

S. Prasad (✉)
Former Associate Professor, Department of Statistics, CHRIST (Deemed to be University), Bangalore, India

30.1 Business Management—Introduction

Business management can be considered as all activities which are concerned with management, organization, and coordination of all business activities in such a way that the policies and objectives of the company are met. In a large organization, the policies are framed by board of directors and executed by the by the chief executive officer (CEO).

We cannot consider Business management into a single entity like a department, aspect, or an individual. Though, this break down of various aspects is tough, it is important to separate and determine the key characteristics of each of them. The main sectors of business management can be broadly classified as financial management, marketing management, human resource management, strategic management, production management, operations management, service management and information technology management.

Financial Management means application of refers to general management principles to financial resources of the enterprise and all activities in planning, organizing, directing, and controlling the financial activities like procuring and utilizing funds of the enterprise.

Marketing management means the process of planning, implementing the techniques and methodologies of various aspects of products or services. It focusses on the usage of marketing orientation, techniques and methodologies in companies and organizations and managing a firm's marketing resources and activities.

Human Resource Management (HRM) means management of people within an organization and the three major areas under this are staffing, employee compensation and benefits, and defining/designing work. The focus is on maximizing the productivity of an organization by optimizing the effectiveness of its employees.

Strategic management refers to the strategic use of resources to achieve goals and objectives of a company and can be regarded as a combination of strategic planning and strategic thinking. This includes contemplation on the activities and procedures within the organization as well as external factors which influence functioning of the company. Thus, this process of strategic management steers top-level actions and decisions.

Production management, also known as operations management means planning and control of processes so that their smooth functioning is ensured, meeting cost and quality objectives. All activities pertaining to creation of goods or transformation of raw material into finished goods come under this and service as well as in manufacturing industries employ the techniques of this management. It is comparable to other specialties such as marketing or human resource and financial management in level and scope.

Operations management deals with the administration of business practices in converting materials and labor into goods and services as efficiently as possible, so that profit of the organization is maximized and attempts to balance costs with revenue. It involves utilizing resources from staff, materials, equipment, and technology as well as to acquire, develop, and deliver goods to clients based on client needs and the abilities of the company.

Service Management, a multidisciplinary field, is a managerial discipline which focuses on customer and service and is related to many other management fields. Processes in this field are aimed at transforming resources of the service provider resources into valuable customer services at agreed levels of quality, cost, and risk. The activities include managing service level agreements with customers, external service providers, assigning tasks and reviewing tactical and longer-term metrics.

Information technology management (IT management) means those processes of managing resources associated to information technology according to an organization's priorities and needs. This includes both tangible resources like networking hardware, computers, and people, as well as intangible resources like software and data. The objective of IT management is to generate value through the use of technology by aligning business strategies and technology.

Understanding the workings of Business management will help in grasping various issues related to business, namely various marketing strategies, mindset of customers, management of accounts, business plans and so on. It helps an individual to develop Leadership qualities, enhance communication skills, ability to work in a team as well as to handle stressful situations and people. People will be able to work in organization of any size and a person with attained skills of thinking in a critical and strategic way, presenting ideas and policies well, leadership qualities and with a knowledge of multiple disciplines will be an asset to any organization.

The purpose of the paper is to discuss in-depth about the Monte Carlo Techniques in Management. As the Monte Carlo simulations can be used with the best use of the probability that can include the different outcomes that can follow the procedures and define the set of the intervention relating to the random variables. Monte Carlo technique helps to understand and define the risk that can correlate with the uncertainty and can experience the defined set of prediction that can help to identify forecasting models. In the business, the use of the Monte Carlo Simulation can help to handle the problem that can be accessed from all the indefinable problems. The solution can be used with every finance, engineering along with identifying the supply chain along with the science. Monte Simulation can be identified with the multiple probability simulation [8].

30.2 Decision Making—Risk and Uncertainty

All situations involving decision making can be broadly classified as decision making certainty, risk, and uncertainty. In a situation in which there is no gap between decision and outcome, that is there is no element of chance or risk, and there is generally only one objective, deterministic optimization can be applied. However, in those situations with a risk factor that is, a probability value is involved, stochastic optimization has to be applied.

Though the words risk, and uncertainty are used interchangeably, they are actually different. Risk can be managed if we take proper steps to minimize it, whereas uncertainty refers to be unsure of future events. American economist Frank Knight says that "risk is something that can be measured and quantified, and that the taker can take steps to protect himself from. Uncertainty, on the other hand, does not allow taking such steps since no one can exactly foretell future events" A person/company can decide whether to take risk or not but cannot avoid uncertainty. With risk we can estimate the probabilities, but uncertainty has no known probabilistic values.

Uncertainty arises due to many factors like missing data, representation in an ambiguous way, inconsistent data and also from factors like recession, changing customer behavior, technological changes and so on.

Generally, we use the average as a substitute for any missing observation, but this is a wrong way of analyzing data. In business scenario, we will have many situations and processes which are complex and cannot be analyzed easily. If the number of parameters is less, a spreadsheet model and "what-if" analysis can be used to numerically evaluate a situation. But most business situations involve many uncertainties, in various dimensions.

A wide range of decisions have to be made in business scenarios for example, there could be uncertainty in prices, labor available, production, cash flows, investment decisions, resource allocation, market behavior, market response, consumer reactions, etc. Handling uncertainty is one of the most crucial as well as challenges faced in forecasting. There are many approaches to handling uncertainty in projections. Risk in a business area refers to taking into account all events and conditions which are uncertain and affect the business, either positively or negatively, if they occur. While taking a good strategic decision, it is important to consider all factors, external and internal as well as the uncertainty/risk/chance associated with each of them.

30.3 Monte Carlo Method/Simulation

The Monte Carlo method was conceived in 1940 by scientists, especially mathematician Stanislaw Ulam, who were supposedly working on the atomic bomb and is named after the city bearing the same name and is famous for casinos. The main objective of this method is the usage of random samples of parameters to explore and understand working and behavior of intricate systems and processes. When analytical solution is difficult to find, numerical methods have to be applied and in this regard. Monte Carlo technique has proved to be highly effective. A vast and diverse areas make use of Monte Carlo methods and many of them are in the domain of finance and business applications.

A stochastic method in which we use a random sampling as well as simulation, which means we virtually represent a problem, is collectively used here. The random sampling procedure gives us many different possibilities of a solution and is used in many business activities like portfolio management, risk management, corporate finance, economic modeling etc. Monte Carlo simulation helps us to recreate many options and allows us to vary all parameters, risk factors and assumptions. Using this method, we can forecast different outcomes by fixing certain decisions and varying key factors. Monte Carlo technique has three main components—a decision variable, a variable which is changing randomly and a key performance indicator.

In the Monte Carlo Method, the basic use of the random variables allows using the inputs that can model and use the inputs with the defined variables [7]. It is the inputs that can help to determine and link with the modelled basis that can relate to the normal log-normal, etc. The methods can be used in the different iterations or simulations that can run in the generation to the paths, outcomes and with the defined suitable numerical computations. In the process of the Monte Carlo Simulation, the use of the most tenable method can be channelized with the model of the uncertain parameters that can hold the dynamic complex systems needs that can be accessed. It also involves the use of the probabilistic method that can link with the modelling risk identified in the system [9]. Based on the issue, it is important to follow the method that can be transparent and be related to the fields like the physical science, linked with the computational biology, statistics along with the artificial intelligence, and forming the quantitative finance. It is important to relate with the Monte Carlo Simulation that can conclude with the probabilistic estimates and it is important to link that can identify the uncertainty model.

The method is not deterministic, but with the given uncertainty or risk that has been made part of the system, the same can be held with the approximate tools that can relate to reality. As per the Monte Carlo Simulation technique, it is important to introduce the certainty that would be based on the modelling uncertain situations. It is important, to result in the profusion of information that can conclude the disposal, and through this, it is important to predict future that can identify the absolute precision along with the accuracy [2]. The method is also attributed to the use of the dynamic factors which can be linked to the key impacts of the outcome that can follow the action course. Monte Carlo Simulation also allows seeing and determining the possible outcomes that can help to define the decision and relate with better decisions that can follow the uncertainty model [1]. It is also important to link with the outcomes; one can allow using the decision-maker that is determined with the probabilities of outcomes.

Monte Carlo Simulation also uses the extensive probability distribution and it is important to identify the modelling a stochastic and link to the random variable. The importance of using the different probability distributions that can use the modelling input variables that can account to the normal, lognormal, uniform, along with triangular. Subsequently with the probability distribution of the input variable, one can identify the different paths of outcome that can be generated.

As compared to the deterministic analysis, the goal is to use the Monte Carlo method that can help to identify the superior risk simulation. This would allow us to judge, what can be the best outcome based on the given course of possible occurrence. It would also help to determine the possible model correlated input variables [4].

The main aspect which differentiates this method from other simulation techniques is that here, there are repeated trials and use of random numbers. Russian mathematician Ilya M. Sobol defines random number/variable as "A random variable that satisfactorily describes a physical quantity in one type of phenomenon may prove unsatisfactory when used to describe the same quantity in other phenomena". A true random number is the one which is completely random and is generated by physical phenomenon. Those generated by computer are "pseudo" random numbers. Generally random numbers are generated using a "seed", which can be the "seconds" in the clock or a number which is given at random. There are various programs which generate random numbers belonging to a particular distribution. Random number tables were used earlier, before the advent of computer. These provide a set of values and one can choose any number, from any page, either row or column-wise. The numbers can be of any digits, as per the requirement of the researcher. There are two main conditions which a random number has to satisfy, they must be uniformly distributed over a defined interval and it should be impossible to predict future values based on past or present ones.

The important steps in Monte Carlo simulation are taking random sampling of inputs, repeatedly, of the random variable under consideration and then combining the results. Random numbers are chosen and depending on the cumulative probability, we simulate the various events. The process is repeated a large number of times and we consider the average of results to get an estimated value and take a decision.

The world is now in the era of big data and availability of data has also become easier as diverse channels are being used for data collection. This gives better estimation of probabilities, thus making Monte Carlo simulation very apt for any area.

30.4 Monte Carlo Simulation in Various Management Sectors

Monte Carlo method has many advantages over deterministic, single point estimates. The results are probabilistic, and we can note the chance of occurrence of different events. Also, since it generates data, which is numerical, we can easily construct graphs and communicate results easily to others. A particularly important feature of this process is that it helps in sensitivity analysis. We can change values of various parameters and see the outcome. This tweaking, which may prove expensive in real life can be used to check what is the impact on bottom-line results. It also helps in Scenario Analysis, wherein we can use different combinations of inputs and observe the outcomes. This is useful for further analysis and not possible in a deterministic situation. The correlation between various input variables and how one factor affects the other. The interdependency of variables can be ascertained by observing their movements as one of them varies in upward or downward direction.

The main idea is to build models by considering a probability distribution for any factor which has a probability associated with it. The calculations are done repeatedly by using a different set of random variables each time. This helps to get an estimate of risk. Thus, it is extremely helpful in Business management.

This method has several applications in finance and related fields. In corporate finance, it is used to model various components of concepts which are impacted by uncertainty, example project cash flow. The output will be a range of net present values as well as observations on the average net present values. The investment and its volatility will give an estimate of the probability that net present value will be more than zero. It is also used in option pricing, fixed income securities and interest rate derivatives, in portfolio management and personal financial planning.

A tool which is inbuilt into any software or analysis tool/spreadsheet software like MS Excel can be considered as a financial model and is used to forecast the financial performance of a business, based on the past performance of the company, any assumptions regarding future, and involves preparation of a 3 statement model (income statement, balance sheet, cash flow statement, and supporting schedules) or further advanced models like discounted cash flow analysis (DCF model), leveraged-buyout (LBO), mergers and acquisitions (M&A), and sensitivity analysis. Financial models help in making decision and performing financial analysis.

In the area of financial management, Monte Carlo simulation is applied for equity options pricing, where the prices of shares are stimulated for different price paths and the payoffs for each are determined. These are then averaged, and current value of an option is computed after discounting. This helps in valuation of options.

In portfolio management, it is used to find the size of portfolio taking into account various factors like rate of reinvestment, inflation, tax, life span of individuals, retirement plans and others. Monte Carlo simulation is applied to find expected value and portfolio value, among others, at the retirement age of client. The value of the portfolio and volatility at any time period is considered. The risk involved and other factors like correlation between various assets are also considered in the portfolio valuation.

However, this method has its own limitations, mainly the inputs have to be fully accurate and appropriate. It does not take into account the probability of irregular events like sudden financial crisis and market volatility.

Fixed income instruments can be considered as a type of debt instruments. They offer regular, or fixed, interest payments and repayments of the principal when the security reaches maturity and are issued by governments, corporations, and other entities to finance their operations. Here, the main source of uncertainty is the short rate. Using Monte Carlo simulation, this short rate is simulated many times and after each round, we determine the price of a bond or derivative, which are averaged, thus, giving the current value of a bond.

The value of all cash flows in the future considering the entire life of an investment, after discounting to the present, is the Net Present Value (NPV) and its analysis is very intricate and is used widely in the sectors of finance and accounting. It is used in all those areas where there is cash flow. Using Monte Carlo simulation, we can construct models which are stochastic and assess a project's NPV. After this, we can check for sensitivity analysis by changing some assumptions and parameters. Thus, it is helpful in financial modeling.

We can use this method in marketing management for integrating various models of customer behavior and also to get answers for certain queries regarding resourcing, customer service and so on. The data generated by using this method can be used for testing performance of the different situations.

Let us consider a situation where, say, we have to contact people and get them to invest in our new policy/product. The objective is to maximize the number of people who buy the product, thus maximizing sales. While minimizing the cost involved in contacting the client, including salaries and call charges.

We can simulate this scenario by considering the database from which we have the probability of sales as well as client details. The minimum and maximum probability of parameters can be determined by simulation and we ascertain the number of clients to be contacted. The historical data can be used to determine if a call to a particular client will have any effect on the sales that is if he is likely to buy our product. Some other examples in this field include estimation of market share at entry, for a new, to-be-launched product, forecasting sales, assessing the size of market for a product or a service, to find the optimal layout for display in a store to maximize profits and so on.

Any organization or business are faced with the task of making decisions regarding future strategy armed with a complete understanding, measurement and appraisal of all risks associated with any strategy. Monte Carlo simulation can help in focusing on the risk, scrutinize the impact of various risk factors on a strategy, measure the risk in terms of loss/gain, thus providing a quantitative number to the risk.

This method can be applied to elementary sales model with limited strategies to large scale models with many factors. The strategies can be as simple as considering the competitors options or adding marketing components to the model.

30.5 Monte Carlo Benefits in Management

Easy and Efficient: In the business, the use of the Monte Carlo algorithms can help to make it simple, flexible and can make it scalable. It can be applied in terms of the physical systems and to identify the Monte Carlo techniques that can reduce the complex models that would help to judge the events along with the interactions. It is important to identify the implementations that can be understood through the scalable. For example, the busiest place and to identify the simulation program that can identify the machine repair facility, which would not be dependent on the number of machines and to identify the repairers involved. In the case of the Monte Carlo algorithms, it is important to identify the parallelizable and to relate with the parts, that can work on an independent basis. It is important to understand and to relate with the defined risks and to identify the different computers that can be related to the processors that could identify the computation time.

30.6 Certain Points to Be Noted

Randomness based Strength: It is important to determine and identify the inherent randomness that can help to link with the MCM and it is essential to use the best methods that can identify the simulation that can be based on the real-life random systems. The importance of the great benefit can link to the deterministic numerical computation. It is important, to understand as to what can be employed with the randomized optimization, and it can help to check over the randomness permits that can determine to change the stochastic algorithms. It is also important to check and access the naturally escape local optima, that can identify the search space and identifying the deterministic counterparts.

Insight into Randomness: Through the MCM, the goal is to determine the didactic value and it is important to identify the vehicle that can link with the vehicle and identify the understanding that can be bead on the behavior of the random systems [3]. The business models can use the probability and the statics, that can identify the carry out methods and use the randomly based experiments over the computer and can link to the experiments based on the Monte Carlo method.

Through the Monte Carlo Simulation, the goal is to use the process of the application areas that can identify the simulation modelling. It is important, to identify the typical applications that can be involved with the simulation based on the inventory processes, can access to use the job scheduling, and to use it effectively with the vehicle routing along with the queuing networks and working with the reliability systems. Such as the use of the operations Research, that can link to the Mathematical Programming (mathematical optimization) can be related to the techniques and to use the proven basis of optimal design, and identifying the scheduling along with handling to the industrial systems. It is important to use the new approaches and to link with the classical optimization problems that can identify the travel- based sales job. It is important, to use the design and control of the use of the machines that can be automatic use the robots.

It is also important to check over the direct simulation that can identify the process based on the neutron transport. During the first application, it is essential to use the Monte Carlo techniques, in an effective approach and to link with the important for of the simulation that can relate with the physical processes. For example, in the business, the use of the Monte Caro Techniques can be used for the better sensible approach and to generate the transport problem for the inhomogeneous multi-layered structure that can identify the scattering and absorption. Based on the Monte Carlo techniques, the business can also use the techniques of the materials science, that can be based on the development and the analysis, that can relate with the new materials and structures, that would link with the organic LEDs, using the solar cells along with the Lithium-Ion batteries. Similarly, the Monte Carlo techniques can play a vital role in the virtual materials design, which would be experimental data and to relate with the stochastic models inclusive of the 3 materials. By using the methods, it is important to realizations that can help to identify the simulated along with the numerical experiments to be used in the same form. It important to use physical development and interpret the results for the better analysis of new materials which can be viewed as expensive and can be equally is time-consuming. By using the Monte Carlo technique, the approach would be to use the materials design and to use the approach that can be deemed to be appropriate and be used with the data generation. It is important to evaluate the data that can identify an easy way to relate with the generation of more data which can be accessed with the physical experiments and be linked to the virtual production along with having the study of materials that can use the various production parameters.

30.7 Random Graphs Along with the Combinatorial Structures

It is important to link with the material approach, that can use the techniques and with the proven effective ways of studying the properties, with the varied graphs and to relate with the lasting approaches and the models, that can be linked to the Potts model and to identify the random structures, and to link with the problems, that can be associated with the estimation that can identify the partition function; In The business, through the use of the Monte Carlo techniques, the main role is to use the probably and the estimation rate, that can identify the related important quantities. It is the step-through which the identification and with the critical exponents that can use the theatrical basis and link with the good in- depth knowledge. The use of the Monto Carlo techniques can be used in an effective manner and to relate with the business. In the business process of the Monte Carlo, it can use the algorithm basis and to determine with the risk and the assessments.

30.8 The Bottom Line of Using Monte Carlo

In the business when the clients sell the insurance produces for example disability, medical, dental and the vision coverage, in the vision of the agent network. Then, it is important, to have the concerns, that can link with the most critical coverage and to identify the risks linked with the personal bankruptcy, which would be identified with the event of the medical condition [5]. The same approach is if the customers would be able to buy the product, with the given set of the availability of the factor ad to use it to manage the risk. As per the application part, it is important to develop simple questions and to link with the various factors such as geography, distribution channels and also use it effectively with the other health factors. Another important element is to use the current budget along with the expenditure to the best interest in the defined area of the insurance coverage area [6].

30.9 Some Case Studies

These Case studies are taken from the own work of the author. The Simulation is done in the following manner:

1. Consider a situation in which various outcomes have a probability associated with them.
2. Take the cumulative probability as $(X < x)$ gives the chance that a random number falls in that interval.
3. Choose random numbers from random number tables or simulate it using EXCEL, R etc.
4. The situation which has the interval of the random number is most likely to occur.
5. This process is repeated a large number of times.
6. The average is considered for estimating parameters as required.

Case Study 1: To Estimate Amount of Waste Generated in an Area

A Monte Carlo simulation model is used to estimate the amount and type of waste generated and helps in planning the collection process. The model is based on a LPP model in which there are 4 bins at common designated places. These bins are for wet waste, recyclables, incinerators, and compostable waste. An upper limit is set for every bin, after which it sends a trigger to a central agency, which dispatches vehicles for collection. A simulation model is proposed to estimate the number of vehicles required for collection.

The proposed model and Simulation:

The collected and segregated waste is emptied into 4 containers kept at common collection points. Local citizens groups can be involved in spreading awareness on segregation and for ensuring that waste is not mixed. This can happen on a daily basis so that no waste is piled up. Street sweeping, rag pickers waste is also the collector's responsibility. Each common collection point has 4 bins—wet, dry, recyclables and to be sent to incinerators. Each of them is equipped with a sensor which sends a message to a centralized place as soon as the waste level reaches a particular level. This can be set at individual zones and seasonal conditions. Here, a queue can build up when triggers are sent but no vehicles are dispatched for collection. This could be due to non-availability of vehicles and/or due to time constraints of collection. Queue can also be formed when vehicles are ready, but no triggers are received. To estimate the number of triggers received from various bins and the type of triggers (wet, dry, recyclables and to be sent to incinerators), we perform a Monte Carlo simulation.

Consider a big geographical area with 50 bins and a time period of one day. Let us assume that all the 50 bins send trigger. It may be possible that some of them send more than one trigger, or the sake of simplicity, we assume that each bin sends only one trigger per day. Consider the following data from a hypothetical population (Tables 30.1, 30.2 and 30.3).

Table 30.1 Waste generated by a hypothetical area consisting of 100,000 households

Component	Wet waste (landfill)	Recyclables	Incinerator	Compostable	Total
Waste generation/100,000 households/week	18.2	4.2	5.6	10.2	1

Table 30.2 Random number coding

Component	Waste generation/100,000 households/week	Waste generation/100,000 households/day	Probability	Cumulative probability	Random numbers
Wet waste (landfill) W	18.2	2.6	0.4762	0.4762	0000–4761
Recyclables R	4.2	0.6	0.1099	0.5861	4762–5860
Incinerator I	5.6	0.8	0.1465	0.7362	5960–7325
Compostable C	10.2	1.46	0.2674	1	7326–9999
Total		5.46			

Table 30.3 Simulation worksheet

Sl. No.	1	2	3	4	5	6	7	8	9	10
Random number	2651	2647	3845	5624	7597	4162	8023	118	8191	9883
Type of trigger	W	W	W	R	C	W	C	W	C	C
Sl. No.	11	12	13	14	15	16	17	18	19	20
Random number	5946	815	7395	3172	2878	4065	7613	4903	9959	5023
Type of trigger		W	C	W	W	W	C	R	C	
Sl. No.	21	22	23	24	25	26	27	28	29	30
Random number	3011	5106	3423	211	3602	5339	9580	8571	683	963
Type of trigger	W	R	W	W	W	R	C	C	W	W

If we set the trigger for each type of waste at some tones, say a, b, c, d, then the total amount collected in a day can be estimated. Thus, using this simulation, we can also approximate how many trucks are needed.

Results and Conclusions:

A simulation model is useful for studying a situation like collection of waste and planning facilities. Using this model, we set a limit, reaching which a trigger is sent to a central agency signaling that a particular amount of waste has been filled in a bin. A series of simulations help us determine the total amount of waste collected in a particular area and a suitable vehicle can be dispatched.

Case Study 2: Monte Carlo Simulation to Optimizing Time for a Creative Designer

Monte Carlo simulation has been used to reduce waiting time and schedule jobs effectively for a professional caricaturist/cartoonist. His jobs involve a wide spectrum of tasks which consume time and effort. This paper explores time utilization of the artist for various jobs, based on time taken for each job and probability of work coming in. We assume that he works for 7 h a day for 25 days a month and calculate mean waiting time for him and his clients. Data has been taken from a study involving various artists and has been generalized.

Category	Time required	Probability
Cover page design	120	5/85
Animations	120	20/85
Regular cartoons	45	15/85
Story illustrations	60	20/85
Caricatures	60	20/85
Branding materials	60	20/85
Printing	60	20/85
Mobile	60	20/85

Since branding material, printing and Mobile and web designing and takes a longer time than others, we can either omit that simulation and simulate again, or we assume that he works only for an hour on that work and resumes it on any other day.

Category of work	Probability	Cumulative probability	Random numbers
Cover page design	0.0588	0.0588	0000–0587
Animations	0.2353	0.2941	0587–2940
Regular cartoons	0.0588	0.0588	0587–2940
Story illustrations	0.0588	0.0588	0587–2940
Caricatures	0.0588	0.0588	0587–2940
Branding materials	0.0588	0.0588	0587–2940
Printing	0.0588	0.0588	0587–2940
Mobile	0.0588	0.0588	0587–2940

We assume that the designer works continuously for an hour on each job from 10 a.m. till 1 p.m. He then rests till 3 p.m. and works till 7 p.m.

Arrival pattern of jobs

Job number	Time scheduled	Random number	Job category	Service time (min)
1	10 a.m.	1970	Animation	120
2	11.30 a.m.	4988	Story illustrations	60
3	1 p.m.	362	Cover page design	120
4	3 p.m.	3561	Regular cartoons	45
5	4.30 p.m.	8040	Branding material	300
6	6 p.m.	7818	Caricatures	75
7	7.30 p.m.	9677	Mobile and web designing	1200
8	9 p.m.	4095	Regular cartoons	45

The arrival, departure patterns of job are tabulated below:

Time	Event	Job number (time to complete)	Waiting (job number)
10 a.m.	1 arrives and starts	1(120)	–
11.30 a.m.	2 arrives	1(30)	2
12 p.m.	1 gets over, 2 starts,	2(60)	3
1 p.m.	2 gets over	–	–
3 p.m.	3 arrives and starts	3(120)	
4.30 p.m.	4 arrives	3(30)	4
5 p.m.	3 gets over, 4 starts	–	–
5.45 p.m.	4 gets over	–	–
6 p.m.	5 arrives and starts	–	–
7.30 p.m.	6 arrives and starts	–	–
9 p.m.	7 arrives	7(60)	7

So, we observe that the artist was idle for about 45 min of time.

Job	Arrival time	Service starts	Waiting time (min)
1	10 a.m.	10 a.m.	–
2	11.30 a.m.	12 p.m.	30
3	3 p.m.	3 p.m.	–
4	4.30 p.m.	5 p.m.	30
5	6 p.m.	6 p.m.	–
6	7.30 p.m.	7.30 p.m.	–
7	9 p.m.	9.15 p.m.	15
Total			75 min

Average waiting time for jobs = (75/540) = 0.138 min.
Thus, appointments at 1.5 h works for the artist.

30.10 Conclusion

It is concluded, with the management, the use of the Monte Carlo (MC) simulation can be used in the many purposes that can link to the sensitivity analysis and it is important to identify the risk quantification along with the analysis, prediction etc. It is important to use the MC simulation that can also identify and create the basis of the artificial future basis and the relation, to the artificial futures that can note the situation that can generate several samples of the measured outcomes. The techniques can be used towards the opening possibility and to note how to relate with the encode model behavior that can help to relate with the set of rules. It can help to identify the efficient implementation to determine the computer. It is also a chain of the events that can modulate and identify the general models. That can identify the general models that can be implemented and to study the possible analytic methods.

Appendix

Part of a table of random numbers

61424	20419	86546	00517
90222	27993	04952	66762
50349	71146	97668	86523
85676	10005	08216	25906
02429	19761	15370	43882

(continued)

(continued)

90519	61988	40164	15815
20631	88967	19660	89624
89990	78733	16447	27932

Generating random numbers using R software

```
> RandomNum <- runif(50, 1, 99)
> RandomNum
 [1] 18.157977  7.597272 20.384460 63.584469 21.089540 44.417837 27.032815
 [8] 21.028903 11.870227 68.353379 19.057466 24.624606  1.990997 43.221146
[15] 92.449852 71.724753 17.894127 94.584895 34.772996 10.790946 12.740264
[22] 85.139916 26.780069 10.733309  9.489474 70.618293 50.764700 27.107075
[29] 13.481805 27.613766 66.680464 11.913601  2.796522 20.625675 61.819272
[36] 53.620734 59.330610 30.588570 64.467783  2.152277 87.750227 57.074411
[43] 87.467890 49.674126 48.656911 73.861570 13.805449 57.949655 18.504711
[50] 96.882483
> |
```

Generating random numbers through MS EXCEL

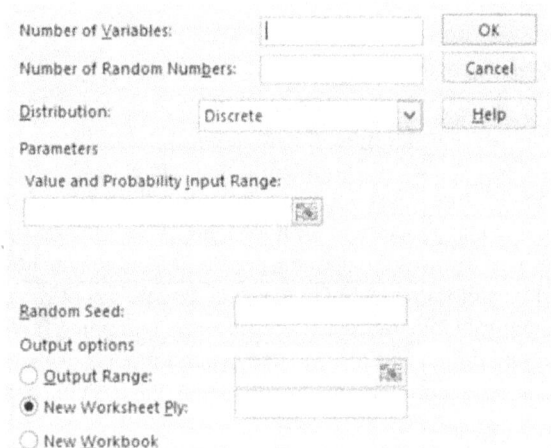

References

1. Arend MG, Schäfer T (2019) Statistical power in two-level models: a tutorial based on Monte Carlo simulation. Psychol Methods 24(1):1
2. Dornheim T, Groth S, Sjostrom T, Malone FD, Foulkes WMC, Bonitz M (2016) Ab initio quantum Monte Carlo simulation of the warm dense electron gas in the thermodynamic limit. Phys Rev Lett 117(15):156403
3. Gholami H, Rahimi S, Fathabadi A, Habibi S, Collins AL (2020) Mapping the spatial sources of atmospheric dust using GLUE and Monte Carlo simulation. Sci Total Environ 138090

4. Klein SR, Nystrand J, Seger J, Gorbunov Y, Butterworth J (2017) STARlight: a Monte Carlo simulation program for ultra-peripheral collisions of relativistic ions. Comput Phys Commun 212:258–268
5. Liu Y, Wang W, Sun K, Meng ZY (2020) Designer Monte Carlo simulation for the Gross-Neveu-Yukawa transition. Phys Rev B 101(6):064308
6. Rillo G, Morales MA, Ceperley DM, Pierleoni C (2018) Coupled electron-ion Monte Carlo simulation of hydrogen molecular crystals. J Chem Phys 148(10):102314
7. Wang R, Lin TS, Johnson JA, Olsen BD (2017) Kinetic Monte Carlo simulation for quantification of the gel point of polymer networks. ACS Macro Lett 6(12):1414–1419
8. Zhou J, Aghili N, Ghaleini EN, Bui DT, Tahir MM, Koopialipoor M (2020) A Monte Carlo simulation approach for effective assessment of flyrock based on an intelligent system of neural network. Eng Comput 36(2):713–723
9. Zhu Z, Du X (2016) Reliability analysis with Monte Carlo simulation and dependent Kriging predictions. J Mech Des 138(12)

Chapter 31
Sustainability in Energy Economy and Environment: Role of AI Based Techniques

Trina Som

Abstract With the ever increasing electricity demand in the twenty-first century, conventional power grids have to respond by adding new generation capacities. This, however, requires upgradation of transmission and distribution systems over utility structures in an environmentally sustainable manner, further expanding to new remote/rural areas. Thus, to retain techno-economical and environmental viability, the concept of decentralized power delivery systems has come up as a new alternative. Though the search for technological effectiveness of microgrids is important, the establishment of a standardized and competitive price structure in the electricity market is also required. Hence, nowadays, economic evaluation of decentralized power delivery systems are being dealt with proper attention. Most conventional methods are classical algorithms, which generally include replication and precision with a lengthy and extensive search process. These deterministic methods involve trial and error process, without any guidance, hint, adaptation and self-correction. In order to overcome such problems modern optimization techniques have emerged which do a random search on basis of some plausible hints and find solution by trial and error. Though these elementary soft-computing methods can attain possible solutions within a given finite time, they never assure to achieve the best solution. Thus to achieve better and more accurate solutions, artificial intelligence techniques with attributes such as partial truth, uncertainties, imprecise specification and approximation which are often present in problems of practical relevance in energy economics, will be explored. Further, considering the relative strength and weakness of different soft-computing techniques, hybrid algorithms will be discussed and assessed in terms of performance this chapter.

Keywords Sustainable Power · Distributed Energy Resources · Evolutionary Algorithms

T. Som (✉)
Department of Electrical and Electronics Engineering, ADGITM Delhi, GGSIPU, FC 26, Shastri Park, Delhi 110053, India

31.1 Introduction

Electricity has emerged as the most critical infrastructure which enables other kinds of infrastructure. Among many discoveries, electricity is one of the most important consecrations that science has offer to the mankind. Nowadays, it has almost become a part of breathing, without which one cannot dream of living. As of now, most of the electricity is provided through the centralized power delivery model, i.e. through a conventional grid. The techno-economic and techno-environmental viability of the conventional model of delivering power only through central power system has been challenged in recent years. Through last decades the role of electricity not only remained confined within utility grids, but became dominant chunks towards the discoveries of electronics, automation and smart technologies. Though these discoveries have become the fundamental part of modern civilization, still with the growing economy, the need for more sustainable power generation with less losses are at top priori. Further, managing these requirements with new evolving controlling techniques are also in high mandate.

31.2 Energy

The modern day living has become next to impossible without production and consumption of various forms of energy. Energy is neither created nor destroyed, it can exist in different forms and altered according to requirements. Various forms of energy which are generally present in our atmosphere naturally, are extracted in accordance to our necessities.

31.2.1 Different Forms of Energy

Generally, we all know about the different types of energy which are generally classified as thermal energy, chemical energy, light (radiant) energy, nuclear energy, mechanical (motion) energy, electrical energy, elastic energy, sound energy, and gravitational energy [1]. These various kinds of energy can be transformed into one form to another. Among the basic forms of energy, mechanical energy is generally understood as the energy due to which an object moves or gather some movement with a velocity (kinetic) or with a vertical height (potential). Hence either potential energy or kinetic energy are experienced in every systems having mechanical energy. In this regard, rocket flying in the space, leaves falling down from a tree, and a hammer lashing a nail are examples of kinetic and potential forms of energy. Pulling, twisting, pushing, throwing and turning are also many forms of mechanical energy.

Next, chemical energy is the energy that generally remain stored in some object and release the available energy through some chemical reactions. Breaking and making of interatomic bonds within molecules dissipates energy, which are usually known as chemical energy. Though world usually generates chemical energy by molecular re-arrangements, but there are other ways of deriving such energy directly, e.g. gasoline which exist in our planet in raw form. Further the chemical energy remain stored in the matchstick in the form of chemical bonds, which transforms this into thermal energy, when the match is ignited through striking on the strip provided on matchbox.

Thermal energy is one of the primitive energy which mankind discovered long back. Thermal energy or commonly known as heat energy of an object can be determined on the basis of how active its atoms are. Most of the material possess some internal energy which is caused due to random movement of minute particles and atoms, which is generally termed as thermal energy. These amount of energy is usually dependent on the temperature of the material. Thermal Energy or heat energy flows when there exist a difference in temperature, and the flow occurs from a body having high temperature to a body having lower temperature.

Another form of energy, generally known as electromagnetic energy is one of the most useful type. The movement of electric charges disperse energy and creates a magnetic field, which is known as electromagnetic energy. These small grain of energy or particle are emitted as electromagnetic radiation. The particles are known as photons and usually emits an energy packet called as light energy. Further, electromagnetic waves also emits electromagnetic energy which can be realized in different forms. These forms are recognized as TV waves, heat, X-ray, light wave, infrared ray, visible wave, microwave band, radar waves etc. Generally these waves are composed of radio wave and radiates with a speed that of light. The natural source of electromagnetic energy are our planet earth, sun and ionosphere.

The most modern form of energy realized by mankind is nuclear energy [2]. Nuclear energy is the energy stored in the nucleus of an atom. When uranium atom are split, nuclear energy are radiated. The great scientist Albert Einstein's formula, represented this in the form of mathematical expression as $E = mc^2$. This process is called nuclear fission, and generally gives out energy in form of heat and light. These nuclear fission and fusion are basically reaction between hydrogen nuclei and helium nuclei. Again, collision and fusing of nuclei at high speed also release nuclear energy. Sun's energy can also be realized in the same manner, where both nuclear fission and fusion occurs to release continuous energy. One of the unusual features of nuclear energy is that, apart from producing electricity it can directly produce light or heat energy. As electricity is the most needed form of energy to drive the automation technology, so the movement of electrons produce electrical energy. This type of energy can be conveniently transported using power lines and converted into other forms of energy.

Different types of energy can be transformed or realized from one form to other [3]. Still, there exist origin of every basic types of energy. The principle of conservation of energy states that, the sum-total of mass and energy is constant in the universe. It is neither be generated nor extinguished, hence the energy can only be converted from one form to other, from one thing to other. This change of energy from one type to other is known as transformation. Several processes helps us in converting one energy form to other. Commonly known examples are nuclear fission, where sun transforms nuclear energy into light and heat energy. The food we consume is generally converted into mechanical energy within our body which consequently help us to make movements. Again, a fan converts electrical energy into mechanical kinetic energy, a toaster transforms electrical energy into the thermal form of energy, while a blender transform electrical energy into mechanical energy. Again when electricity is passed through the filaments of an incandescent bulb, it is converted to both light and heat energy, and starts glowing as hot body. On the basis of electromagnetic induction, the vibrations from noise energy is converted to electrical energy. Further, through burning of wood and coal, heat and light energy is sourced from chemical energy. Combustion of petroleum can release light and heat or it changed into another more convenient form of chemical energy, such as gasoline. Similarly, mechanical energy can be transformed to electrical, kinetic to potential, light energy to chemical energy, and many others. Similarly electrical energy can also be converted or changed into light, heat, motion and sound energy. Static electricity and lightening are examples of electrical energy that can be obtained in nature. As science and technology couldn't come up with a direct way to utilize natural electricity, so we produce electricity from different conventional and non-conventional energy sources.

31.2.2 Conventional Source of Electrical Energy and Its Depletion

With advancement in technology, the assets and automation has become a part of human civilization. Almost every aspect of life like infrastructural buildings, communication vehicle, educational systems, foodstuff shops, restaurants, hospitals, playgrounds have grown up from semi-automatic to fully mechanized arrangements. Over years, the automation has demanded a huge supply for electricity, but with time, industrialist, technologist researcher officers, academicians, governmental decision makers have faced many difficulties in procuring electricity from these standard sources. Total worldwide gross production of electricity as of 2017 was 25,082 TWh. Coal peat, natural gas, water, oil and nuclear generally contribute 38.3%, 23.1%, 16.6%, 10.4% and 3.7% of total electric production respectively [4, 5].

History of energy usage tells us that steam engines were motorized or driven by coal and wood fires. Gradually usage of oil peaked in 1979 as it was found easier to ship, store & burn [6]. Consequently, with Iranian revolution and the Arab oil embargo the prices of oil became affordable. Once, during 1980s search and utilization for renewable energy were started to arise, but again with the drop in oil price, it went to back seat [6]. The main sources of electrical energy are still considered as the conventional fossil fuels, as the following;

Coal
Coal is the major as well as largest source of fuel in producing electrical energy. Consequently, it is also responsible for creating maximum harm to environment.

Coal is considered as the most plentiful source of fossil fuel, which would easily last for at least another 200 years. It produces about 62% of world's electricity [7]. This also leads to the fact that coal is also the major source behind carbon emissions in the world. This consequently brings a significant change in climatic conditions. Coal is derived in raw form as a sedimentary rock of black and brown in color. These are generally made up of carbonized vegetable. Primary constitution of coal is carbon which includes variable quantities of hydrogen, Sulphur, nitrogen and oxygen. Further, dead plants also form coal by converting themselves into peat and then in turn into lignite and anthracite. These transformation and extraction occurs through many biological and geological processes. This courses of action takes place for a long period of time.

Electricity production from coal as fuel generally involves three simple steps. At first steam is produced by burning the coal in a thermal power plant. Next this high pressure steam drives the turbine, which consequently rotates the generator thus producing the electrical power [8]. Further, the steam is made to cool down and condensed for sending it back to the boiler, in order to repeat the process of producing electricity.

For millions of years organic materials remain deposited with the earth's surface in compressed, and decomposed manner which is converted into fossil fuels. These fossil fuels burn at a low temperature thus making the operation of coal fired boilers much cheaper, simpler and easily available worldwide. This easy accessibility from large reserves the price of fossil becomes in-expensive with respect to buy it from open market. With the most in-expensive type of fossil, the power generation unit from coal also results in good scale, thus resulting in building up at many places with various sizes. Moreover, with railways as the mode of communication, the coal fired power stations huge amount of electricity can be generated at any place at a fairly cheap rate.

Natural Gas

Natural gas is one of the most potential fuel which produce electricity. Beside electricity, it also help in heating building, cooking food, heating water, drying clothes. In short, sometimes natural gas is also termed as only gas, which is usually formed from the fossils of long buried plants and animals deep inside the earth. Natural gas mainly consists of methane, due to which it is considered as clean burning fossil fuel. Burning of natural gas gives us heat and light energy, which are mostly used for heating water in buildings, space and other industrial sectors [9]. Progressively, various method of generation have been adopted which utilizes natural gas as one of the most common resources. The most classical method involves steam producing units where burning of fossil fuels takes place resulting in production of heat, water and steam.

Oil

Oil finds its primary application as transportation fuel but also contributes in production of heat and electrical power. Three main technologies [10] are utilized for production of electricity from oil. The oil is usually burned to heat up water and form steam, which consequently produces electricity. Hot exhaust gases are obtained by burning oil under high pressure, which further turns a combustion turbine to rotate fast and produce electricity

Petroleum

Petroleum is one of the prospective fuel which remains buried under the sedimentary rock. These fossil mainly constitute dead organism, such as algea, zooplankton in high pressure and intense heat. It is recovered in huge quantity by oil drilling, and also from rarely available natural petroleum spring.

Petroleum is basically a very similar fuel like oil, which remains buried deep under the earth's surface as oil. The oil extracted from underground can be volatile as water or black thick and viscous like tar. Deep under the ground the petroleum are generally buried within tiny pockets of rocks, where wells are drilled to pump out the oil. As petroleum is a source of massive energy, beside oil, many other forms of fuel like kerosene, gasoline, and heating oil are obtained from the conversion of petroleum [11]. As petroleum or crude oil is neither volatile nor cumbersome, hence it finds its application in many places other than electricity production. Plastic formations and building heating are two among many other.

Hydro

People when manage to get electricity from simple water, then it is known as hydroelectricity. Hydroelectric power is made by flowing water. Falling of water from a height involves the conversion of potential energy to kinetic energy, which is utilized by hydel power plant. The turbine further uses the kinetic energy and convert it to electrical energy by the help of a generator. The hydroelectric dams are mainly used as the power sources which when flows through the turbine produces electricity [12].

Nuclear

About 13 to 14% of total world's electricity is generated from nuclear power. Countries like US, Japan and France not only utilized 50% of this type of electricity, but also uses 6% of total energy which comes from nuclear power. As per the report from IAEA, 2007, 31 countries developed 439 nuclear power reactor in order to produce electrical power [13]. Sustained nuclear fission operation results in production of electricity and heat from nuclear power. Moreover, as nuclear is considered to be clean energy source, about 150 naval vessels adopted nuclear propulsion for its desired operation. As the reactors are made up of uranium, about 29 states implemented 95 nuclear reactor to produce at least 20% of the nation's electricity. In order to produce steam for electricity generation the heat of radioactive material are harnessed. The fission process involves splitting of uranium atoms in order to produce heat, which further spins a turbine to produce electricity, without releasing harmful waste. Sometimes, in nuclear reactor the splatted atoms generate energy to heat up water into steam. Being the main source of electrical energy, these conventional resources faces many drawbacks.

Limitations

The conventional energy sources are derived from nature, which cannot be grown in a controlled rate, neither produced on the basis of its consumption. Hence, we can't regulate its generation for future and are destined to its depletion. Thus the non-renewable resources, viz coal, natural gas, firewood, petroleum, uranium are being consumed in mush faster rate from that of generation. In order to extract more fuels, more number of coal mines are created resulting to a hazardous and unsafe life. Moreover, the mode of communication, viz lorry, train, which are involved in transportation of coal from coal mines to power generating system produces huge pollution. Among all the other energy sources, burning of coal causes maximum global warming, which includes greenhouse gases, Sulphur- dioxide, and acid rain. Consequently, coal miners and diggers often get affected by diseases like pneumoconiosis, black lung, and emphysema. Finally, in another 100 years the coal resource is expected to run out.

As petroleum is very efficient while used as a higher and better concentrated energy, so the human consumes this type of energy at a faster rate than that of it is being replaced or re-generated. So, it is inevitable that the decrement of oil production will cause a drastic effect on our society and mankind. United States, Ukraine, Canada, and Russian are among the leading countries where the gas is being used at maximum.

As Oil is not a renewable source, so the supply will run out after a certain time period. Again, these fuels emits greenhouse gases with burning in the air, which consequently enhance global warming. With increasing demand and shortage of supply, the price of oil also increasing beyond affordability. Again, due to threat of explosion further working and exploring on an oil rig have proved to be dangerous.

Till now, 60 billion tonnes of oil or its equivalent amount of gas have been already consumed. Moreover, up to 1965 many natural gas from oil field were disposed by flaring, as it was considered to be useless. With limited reserve of natural gases, and utilization in much faster rates are causing a huge depletion. Natural gas is highly combustible. If it is leaked in a room, it starts building up in that room and will explode badly. Moreover, improper handling and inhalation of natural gas will cause severe health hazard, which can be fatal. Beside, coal, oil, petroleum, natural gas, nuclear energy sources also have disadvantages. Nuclear power plants emits radioactive materials as waste, which are very injurious to health, and hence required to be disposed cautiously. Moreover, due to the supply of uranium, nuclear power plants are at threat by terrorist attack Many industries such as steel industry, car industry, transportation industry, are responsible for depletion of coal as they consume a huge amount of energy to drive various automated equipment. The major disadvantages of all the traditional energy sources, which an individual citizen faces are severe air pollution, greenhouse effect, and fast depletion finally reaching to a void.

Beside many limitations, these conventional resources are mostly being used by the conventional power delivering unit. The network of transmission and distribution facilities makes up the power grid. In focus to procure less loss of electricity the power transmission is generally made in high the voltage, with less current, but still, the conventional power sectors faces numerous problems.

Classical Power Grid

National grid was set up mainly on the basis of electricity supply act developed on 1926. First synchronized AC grid of about 132 KV, 50 Hz was established, and central electricity board were standardize as nations electricity supply. Finally it became operative as a National Grid from 1938 [14]. The U.S. power grid became susceptible through the threats faced due to the usage of 250,000 of substation along with 160,000 miles of high voltage lines. With such a huge and complex system, maintenance and proper operation has always been a challenge, which results in severe power failures. These failures, generally include damages in transmission lines, cascading failure, short circuiting, malfunctioning of fuse and circuit breaker, and breakdown in substations and distributed stations. The average transmission and distribution losses as estimated by U.S. Energy Information Administration (EIA) is about 5% in United Stated on annual basis. Thus the losses are generally mentioned in the state electricity profiles [15].

It has already been noted that, electricity generated from renewable energy sources usually protects the climate and provide us with a sustainable power. Hence, the industrialized countries should be put on with an emphatic aim in transforming the conventional power grids into a fully based renewable-based structure as a long term project. Moreover, these deferrals in the transformation of the energy system, often brings threat to the economic development of society worldwide.

31.2.3 Non-conventional and Renewable Energy Sources

Any sustainable energy which originates from natural resources such as wind, sunlight, rain tides, waves, and geo-thermal energy are called as renewable energy resources. The use of renewable energy is an age old concept which use to exist even before the development of coal based power plant during mid-19th century. Bio mass fuels are among the oldest type of renewable energy, which is being used for more than million years ago, but it took many more hundreds and thousands of years to get standardized [16]. Wind is considered to be the second oldest type which was used by harnessing in order to drive boats for transportation purpose. This practice was adapted and performed in Persian Gulf and Nile about 7000 years back [17]. Since Roman times and Paleolithic times geothermal energy are being used as hot spring and space heating. The history of renewable energy can be tracked back to see the concepts like grain crushing, windmills, water power. Firewood, animal power etc. The awareness regarding run out of fossil fuels knocked the civilization around 1860s to 1870s. In 1885, for the need for better sources, Werner came up with the concept of photovoltaic effect in solid state. He claimed the solar energy as the most sustainable and affordable among others [18]. Among total global energy consumption, renewable sources constitute about 16%. Traditional bio-mass waste forms 10%

of this energy by heating and gasification, while hydroelectricity comprises 3.4%. The other 3% comes from new renewable resources and methodologies. These are mainly, wind, solar, modern biomass, geothermal, small hydro and bio-fuels. These new concepts are growing rapidly [19]. The major advantageous aspect of certain renewable energy is that it exists unendingly and can be used abundantly. Moreover, it provides harnessed, inexhaustible and clean energy as a substitute to fossil fuels. Renewable Energy resources are mainly classified as wind power, biomass power, solar power, geothermal and hydel power.

Solar Energy
Since ages the solar energy which is being received in the form of heat and light, has been harnessed by the mankind with a wide range of ever-evolving technologies [20]. Few examples of such technologies are solar photovoltaic cell, solar architecture for thermal electricity, solar heating. Among these methods, solar photovoltaic cell is one of the most advance technique which utilizes the sunlight directly. The electrons are knocked out from its molecular position in the cell, when stroked by solar rays. These electrons then move towards the frontal portion of the surface and creates an electron imbalance within the surface. When a connecting wire is being joined between two such surfaces current or electricity starts following from negative terminal to positive terminal of that surface.

Setting of many such cells in a proper manner provides us large amount of electricity. These are further used in various industrial and household purposes. This also gave rise to solar technology. On the basis of the way of capturing solar power, distributing and converting solar energy these solar power generating technologies have been broadly classified as passive and active solar power. In active solar power generation technique fans, pumps, photovoltaic panels are generally used to generate useful output energy, while passive methods generally uses material with thermal properties possessing proper design space that naturally circulates the air in reference to the position of a building.

Wind Energy
Energy from wind is one of the most primitive and powerful renewable resources which provides us electricity by converting mechanical energy into useful electrical energy [21]. About thousands of wind turbine are in operation with a nameplate capacity of 194,400 MW around the world. For about 5500 years humans are utilizing this super wind power to drive sail boats and ships. Since 7th century AD, countries like India, Pakistan, Iran and Afghanistan are using windmills mainly for milling grains and irrigation purposes. With technological enhancement humans started using wind turbines replacing the windmills. The blades of these wind turbine are rotated by blowing wind, which further produces mechanical energy. These blades of a turbine generally remains mounted on a turning shaft or remains attach to a hub. These shafts increases their turning speed or mechanical energy within a gear transmission box. The transmission box consequently turns a generator, which usually remains attached to a high speed shaft. Wind farms usually connects turbines with power collecting systems and networks delivering a medium voltage of about 34.5 kV. The

wind generation capacity have increased in quadruple within the year 2000 to 2006, and generally found to get double in every three years.

Bio-mass Energy

Bio mass usually refers to a mass related to biology or botany, viz plants in various forms such as vegetables and fruits, raw or processed, wild or refined. These masses forms the bio fuels by biological carbon fixation. Bio-fuels are further extracted from liquid fuels, various biogases and solid bio masses [22]. It finds its main application industrial processes for heating; and transportation sectors (ethanol, biodiesel and other products) for generating electric power.

Geothermal Energy

Geothermal energy is basically the heat which is derived from the sub surface of earth.

These energy are generally extracted in the form of water and steam. Its been a long history, during Paleolithic times when people started using this form of energy for bathing as hot spring. Again, in the 3rd century BC the Qin dynasty utilized geothermal energy as stone pool, the oldest form of spa on China's Lisan mountain. Depending on the various characteristics, these geothermal energies are used for different other applications, such as cooling and heating processes. Further, these energies are also harnessed properly to produce clean electricity. Geothermal energy is very closely related to thermal energy. Though thermal energy remain stored within the earth surface but the temperature determines and distinguish between thermal and geo thermal energy. 20% of this types of matter originates in raw form within the planet and 80% is formed from radioactive decay of mineral [23]. A continuous conduction of heat energy occurs from the core to the surface of the earth. This causes a temperature difference between the surface and the core, which is generally known as geothermal gradient. Geothermal power is among those energies which diminishes global warming when arranged in wide place of fossil fuel, but still has some drawbacks. The limitation includes non-permitted areas close to tectonic plate boundaries.

Both the conventional and non-conventional energy resources are derived from the nature, and at the same time has some or other disadvantages. In this regard, it is very important to know about the environment and its characteristics. Further proper measurements should be taken in order to make it sustainable and a better place to stay.

31.3 Environment

Environment word derived from the old French word 'en' means in and 'viron' means circle. The surroundings including all objects, and conditions form the environment.

The climatic, physiographic, geographic and faunal condition altogether forms a surrounding which we generally call as environment. All the living beings of the earth stays in this environment. Further, many researchers studied on human's tendency to engross pre-environmental behaviors, which is generally a psychological factors that occurs as an element of belief system towards environmental awareness. The factors which influences an individual's health and social status can be classified mainly as biological, social and cultural aspects [24]. However, many epidemiological work pointed out the health benefits in terms of reduced diabetes obesity, cardio vascular morbidity, improved pregnancy, reduced depression, which can be easily obtained by adopting a green ecofriendly environment. Environment is almost our second house, which has to be protected from external damages for our safety. In order to protect it, we need to know the sources of impairment. Certain drawbacks of conventional power generating plants includes harmful effect on environment.

31.3.1 Harmful Effects on Environment

Enumerating the issues surrounding our environment is admittedly a difficult task, but there are three major ones that affect the environment and mankind are global warming and climate change, water pollution and acidification of the oceans, and the progressive loss of biodiversity.

Burning of fossil fuels accounts for almost half of the global warming [25]. The fossil based power plants generally releases carbon di oxide which creates a significant impact on species, their habitat, and change in climatic conditions Moreover, many industries, viz steel, cement, plastic and chemical, also produces about 78% of greenhouse gas which remains trapped in atmosphere in form of heat. CO_2 gas contributes maximum bad effects to the environment and health of living beings, and is generally known as greenhouse gas. This gas is produced mainly from burning of fossil fuel, waste products and other bio-masses. Nearly all combustible by products have negative effects on environment such as SO_2, acid rain [26]

In this scenario, as the increased level of pollution, is resulting towards a toxic environment to live in, it will be a high time to give a call in creating sustainable environment.

Different forms of energy resources through which electrical power is produce have an effect on **our** environment, i.e. our water, land and air. Waste is produced by almost all forms types of energy resources those are capable of producing electricity. Few examples such as nitrogen di-oxide and carbon di-oxide are released from natural gases. These gases are further trapped in earth's atmosphere and results in forming a pocket of smog.

These effects can be minimized by generating electricity with more efficient ways of utilizing needed fuels. This consequently will give rise to many modern technologies of power generation. Before putting focus on the means of reducing the emission of pollutant gases, the sources and their harmful impacts are needed to be studied in detail.

Effects of Thermal Power Plant

The generation of thermal power generation is one of the major reasons behind environmental pollution, and its harmful effects are noticed on soil, air, land and many other social aspects. The localized waterways changes its purity with the discharge of waste products, which again when consumed by plants are badly affected. An emission of huge amount of mercury and fly ashes also contribute a major portion of thermal pollution.

The concentration of nitrogen oxide, Sulphur di-oxide, and SPM are generally found to be high [26] around the thermal power plants which are based on coal. Moreover, gases like ozone and carbon dioxide are emitted from thermal power plant. These increased amount of concentration and huge amount of emission causes global warming leading to uncertain climate change. People living within a radius of 2–5 km of power plants also suffer from various respiratory problems. It has also been noted that, among ozone and carbon dioxide that are emitted, CO_2 causes the major climate change in creating maximum global warming. The emission of hazardous pollutants, also creates visual and respiratory problems. It when combines with water further creates smog, acid rain and ground ozone. However, this power plant is considered to be the second largest cause for NO_x emission. Moreover, photolytic reaction of NO_x creates ground level ozone which causes many impact on mankind. Several studies are also being carried out regarding these pollutions and effects on us.

Air Pollution: This mainly constitute the emission of greenhouse gas, i.e. CO_2, along with SO_2, NO_x. High particulate matters and fly ashes also cause a major concern which are formed mainly due to combustion of low grade coal

Water Pollution: The hydro plants are themselves contributing to harmful effects on water and bio diversity. These are mainly formed from the waste discharged by the power plants. The waste constitute effluent of cooling tower, boiler blow down part, DM plant, and ash ponds [27].

Noise Pollution: running of fans and motors with simultaneous release of high pressure steam produce high levels of noise.

Land Degradation: When coal produces huge energy, about 100 million tons of fly ashes are generated. These disposed ashes occupied several thousand hectares of agricultural and forest lands. Current status of emission of pollutants as recorded in tonnes per day can be represented as given below in Table 31.1.

Every year tones of mercury is emitted as pollutant, which effects badly to our home and health. The major sources of mercury pollution are thermal power plants

Table 31.1 Pollutants emitted on daily basis

CO_2	424,650 (ton)
Particulate Matter	4374 (ton)
SO_2	3311 (ton)
NO_x	4966 (ton)

and chlorine chemical plants. Mainly these comes from plant and chemical power plants. The chemical facility plants usually use massive amount of mercury in order to extract chlorine from salt. Thus about 50 tonnes or more dozens of mercury are emitted on annual basis, which causes a massive pollution. Further, as coal is contaminated with mercury, so it releases mercury through smokestacks during the generation of electricity by burning coal. Consequently, this creates a big loss of wild life, habitats, fertile land, assets and many more.

Effects of Nuclear Power Plant
Today there are about 439 nuclear plants in 31 countries. Nuclear power plants uses Uranium and plutonium as fuel. About 1 kg of Uranium can produce as much energy as burn of 4500 tons of high variety grade of coal or 2000 tons of oil. Nuclear power is also among one of the largest contributor in production of pollution. The radioactive wastes which are created from nuclear power production, such as reactor fuel, uranium mill tailings is tremendously dangerous to the mankind, animals and plants as these materials remain radioactive even after thousands of years. Management of nuclear waste involves many risks by elimination of radioactive elements. This takes a long period of time. Though nuclear reactors work well by nuclear fission, but generation of such continuous chain reaction sometimes leads to radioactive explosion. The radioactive pollution is one of the main cause behind the physical pollution of air, water and the other types [28]. Invisible energy waves or particles are generally derived from radiation. The radiation are considered to be of two types. These are namely, natural radiation and man-made radiation. The natural radiations are also known as the background radiations. Cosmic rays, constituting high energetic particles produces pollution on being incident on the earth's surface are the most common examples of natural radiation [29]. The man-made radiation are created through ionization of molecules and atoms. Ionization of a molecule produces two fragments, alpha (α) and beta (β). Alpha radiation contains two units of positive charges, known as energetic—alpha particles, while beta radiation consist of one unit of negative charge, known as beta particle. Alpha particles strongly interacts with living tissues, while beta particles interact only with matters. Another type of radiation which strongly involves an electromagnetic interaction with matter is gamma (γ) radiations. These generally constitute high energy photon particles. Though plants growth varies with age, species, chromosome volume, degree of damage, intensity, growth stage and level of radiation, duration of exposure. The effects of nuclear power generation in animal include dry itchy skin, discoloration, tumor, hair loss. Other harmful effects include skin diseases, bone marrow abnormality, demolition of retina, nerve damage, and shortening of life span.

31.3.2 Reduction of Pollution: A Step Towards New Approaches

In last few decades, as, there has already been an enormous emission of pollution, hence most of researchers are already in their way to explore and develop different techniques in order to control the rate emission. The most prominent way of decreasing pollution thereby helping the environment to remain clean is not only controlling but also reducing the power plant emissions. The most basic solution for reduction in air pollution is to adapt methods which include solar, wind, geothermal and many other energy resources in place of conventional fossil fuels. To reduce pollution, the government can use four main policies, *viz.* taxation to increase the price, subsidization of alternatives, regulations prohibiting the use of certain pollutants and permission system for certain pollutants [30]. This involves substitution of raw materials, process modification, modification of existing equipment, maintenance of equipment such as cyclone separators (e.g. reverse flow cyclone), fabric filters (baghouse filters), electrostatic precipitators and gravitational settling chambers. As of 2017, the total greenhouse gas emitted across the world is 45,261 ($MtCO_2e$) [31], while China, USA, and European countries emitted around 12,454, 6673 and 4224 respectively $MtCO_2e$. Similar to the production of clean energy, the reduction in energy consumption is also very is very crucial from economics point of view. In this regard, attention can be paid through adoption of efficient devices and responsible habits.

31.4 Economy

Economy is the entity that allows mankind to survive and thrive. A system of barter where no money is involved and trade is done through direct exchange of goods and services is an economy too. Having enough of it is extremely important for stability, low crime levels cultural, scientific and technological progress. Economics is the academic discipline that deals with the generation, consumption and flow of wealth. It also includes the study of all the factors which influence the flow of money. Economy can also be expressed as a collection of inter-connected activities involving production and consumption that facilitate the allocation of scarce resources. In an economy, goods and services are produced and consumed to address the needs and wants of its human constituents. The purpose of economy is to manage the well-being of households; to produce and distribute water, food and other goods and services necessary for human life to thrive. Economy involves the use of the smallest amount of time, money and other resources to achieve something, so that nothing is wasted.

A key benefit of economic growth is higher average income. Economic growth facilitates the consumption of more goods and services and the enjoyment of better living standards. During the twentieth century economic growth lead to the reduction of absolute poverty levels as well as improved life expectancy.

Economics is sociologically important, especially for improving man's quality of life. It can help in advancement of living standards and contribute to the betterment of society. It could be noted, that like science economics is a double-edged sword, and sometimes economic interventions can also worsen the state of economy. The nature of economic growth also partly depends on the priorities of the society in question and cultural factors. It also allocates resources and arranges the growth of wealth in anticipation of population growth, so that living standards do not decline. The economy functions within the framework of the society. Consequently, every society has its unique economy, and every economy reflects the cultural ethos, priorities and historical experience of the society, as well as the main attributes of the civilization of which the society is a part. One way of arriving at GDP is to calculate the sum total of expenditures by the different groups that participate in the economy. The size of the economy of a country is typically measured by its gross domestic product, or GDP, which is the total value of all the final goods and services produced in a given year. The GDP (C+I+G+(X-M)) is the sum of consumer spending (C) plus business investment (I), government spending (G), plus the net exports (X − M), which is obtained by subtracting imports (M) from exports (X).

Energy is a very important source of economic growth as both production and consumption utilize energy as primary inputs. Thus, energy is a key enabler of economic development. The energy crisis in the 1970s and the consequent sky rocketing of energy prices slowed down the global economic growth. In this lead, before exploring different economical options for the energy existing price structure of energy should be analyzed properly.

31.4.1 Conventional Pricing: Methods

The rate at which the electricity is being sold in a particular place, province, state or country in generally known as electrical tariff. This can also be defined as the price charged by the electricity delivery unit for providing gas and electricity. Over last 10 years, there has been a rise in the basic retail rate which is set for residential users. This amount of price we pay per kilowatt-hour have faced an increment of about 15% throughout the nation. Moreover this increasing trend seems too keep a positive slope as the prices of all natural gases and fuels are growing with a fast rate of depletion. On consideration of many factors, different types of tariff are evaluated and set. Among those, the broadly classified tariffs are fixed rate tariff and variable tariff. Fixed rate tariff generally keeps the rate fixed for over a year, while in case of variable tariff, the price generally varies in accordance to the market position. Different costs of electricity can be obtained by using distinct methods at different time of power generation. Further, the evaluations can be performed at the junctions of load and electricity grid. The evaluated cost involves maintenance cost, initial investment cost, operational cost and the de-commissioning cost as well as cost related with damaging environment [32].

Moreover, another concept of cost is very important to know in calculating and experiencing the electricity cost. This is generally known as the levelized cost of energy (LCOE). This cost is a measure of the sources used to generate power when evaluated by different methods on a consistent manner. In order to break even over the lifetime period as a project, LCOE is considered as the minimum constant rate beyond which the electricity can't be sold. It also gives us the value of net average present cost of power production over the lifetime period. This cost is evaluated and represented as the ratio of net present cost of all the assets used for lifetime to the discounted energy output derived from those assets of lifetime [33].

Comparison of different power sources which involves various methods of electricity production is usually compared through levelized cost of energy. This comparative study is made on a very consistent manner. Thus LCOE forms the minimum constant rate which can be charged to the consumer, in order to breakeven the lifetime project cost.

The total average current cost also constitute the levelized cost of energy (LCOE), which is calculated as the ratio of cost over lifetime to sum of electrical energy produced over lifetime. The mathematical expression is as follows in Eq. (31.1);

$$LCOE = \frac{Sum\ of\ cost\ over\ lifetime}{Sum\ of\ Electrical\ Energy\ Produced\ over\ lifetime}$$
$$= \frac{\sum_{t=1}^{n} \frac{I_t + M_t + F_t}{(1+r)^t}}{\sum_{t=1}^{n} \frac{E_t}{(1+r)^t}} \quad (31.1)$$

where, E_t is electrical energy produced in t years of timse, I_t is the investment expenditure in t years, M_t is the maintenance and operational expenses in t years, F_t is fuel costs in those t years, r is deduction rate, ad n is the estimated lifetime of the power station.

As this cost gets affected by the factors such as taxes and subsidies, so several internal factors are considered while evaluating LCOE as actual selling price.

As these tariffs are the key point for both the consumers and electric supplying companies, there are numerous kinds of tariffs in the marketplace which have lately drew attention to many suppliers. With delivering electrical energy to a large number of customers, the company should ensure that the tariff structure must earn profit along with recovering the total cost for generating electrical power.

In this regard, certain objectives are often set by the companies [34], which are as follows;

(i) Retrieval of cost for generating electrical power at the power station
(ii) Retrieval of the price as invested on the capital amount for developing the transmission and distribution system
(iii) Recovery of cost required for monitoring proper operation and maintenance of supply of electrical energy

(iv) An appropriate profit on all the total evaluated cost

These make the tariff a sustainable one, with a reasonable profit along with the recovery prices.

Further the desirable characteristics of tariff includes

(i) Proper Return
(ii) Fairness
(iii) Simplicity
(iv) Reasonable Profit
(v) Attractive

The electricity tariff, are usually classified as simple tariff, block rate tariff, maximum demand tariff, flat rate tariff, , three part tariff, two part tariff, TOD tariff and power factor tariff [35].

Simple Tariff or Uniform rate Tariff: These are defined as those tariffs which has a fixed rate per unit of energy consumed. In this type of tariff the price of the electricity is constant with consumption of per unit electricity. As there is no variation is the price with increment or decrement in the number of units consumed, so this type of tariff is considered as the simplest form of tariff among all other types.

Flat Rate Tariff: In this type, different rates are being set for different types of consumers. As these rates are uniform over per unit basis, hence known as flat rate tariff. Here the consumers are generally grouped into different classes and different uniform rates are being charged to different classes of consumers. Diversity and load factors of different classes of consumers are taken into account.

Block rate tariff: This type of tariff is generally designed with a specific rate of electricity for a particular block of energy. Further, the subsequent blocks of energy are charged with a reduced rate which again keeps on decreasing. As the costing is done on the basis of block, so it is known as block rate tariff. Here, the electricity price is kept fixed per unit consumption for every block.

Two Part tariff: Here the rate of electricity is calculated in two parts, on the basis of electricity consumed by the customers. The total pricing is made up of two components, namely fixed and running costs. The fixed cost is calculated on the basis of maximum demand of the consumer's consumption, while the running cost is defined on the hourly basis consumption of electricity. This total cost is mathematically represented as Rs ($X \times kW + Y \times kWh$), where X is the certain amount of electricity consumed per kW of maximum demand and Y is the certain amount of energy consumed per kWh. The fixed charges, considered are generally independent of unit consumption and only dependent on maximum demand of consumer are recovered within this tariff.

Maximum demand tariff: This type of tariff is almost same as two part tariff, except the way in which the maximum demand is being calculated. The maximum demand is evaluated from a maximum demand meter which is being installed in near premise of the consumers. Bulk consumers generally opt for this type of tariff.

Power factor tariff: Here the tariff charges the consumer on the basis of the power factor involved in their electricity consumption. Power factor tariff are again of the following three types: (a) KVL maximum demand tariff, (b) Sliding scale tariff and (c) kW and kVAR tariff.

Three part Tariff: Three part tariff is almost similar to that of two parts tariff. In contrast to two part tariff, the three part tariff evaluates the charge by dividing it into three parts. The only difference is that unlike two part tariff, it includes three parts in evaluating total charges for consumers. These three parts are fixed charge, running charge and semifixed charge. Generally big consumers opt for this type of tariff.

TOD tariff: The tariff structure which involves different rates for electricity during different time in a day is generally known as time of day or TOD tariff. Thus the electricity price becomes different in morning, from that of noon, or from evening and night. Thus use of certain appliance during certain period of time can be obtained cheaper than that during other time. Moreover, many different types of tariff schemes were initialized in many countries depending on the hours of demand a peak and base load. On 1st January 2016, a new method of tariff, known as enhanced ToU (enhanced New Time of Use) has been adopted by industrial customer in Malaysia [36]. The EToU is basically a modification over the existing tariff scheme ToU. This modification is done by dividing the existing two daily time frame into 6 periods. Further in consideration to reduce the peak order demand a new concept, known as mid-peak tariff have been introduced. In order to get the maximum benefit from this scheme, the customers shift their consumption from low off-peak depending on the time. However, the shifting depends on the flexibility and consumption profile. Finally, each customer takes the decision in switching to this new scheme through a cost-benefit analysis.

31.4.2 Conventional Pricing: Drawbacks

There is always a need to track the existing tariff regulations. Moreover, new techniques or approaches should always keep evolving for better results, and justify our needs. The major limitation of the existing pricing structure is poor efficiency of price signal to consumer. This leads in more payments by consumers for the desired electricity, than that of the value if assets were used in calculating the cost. Some State Electricity Boards fail to determine the revenue required. This results in fixing tariff only under the consideration of improper realistic and economic costs of electricity. Consequently both setting of efficient prices of electricity for consumers and raising of sufficient resources for recovering the loss incurred in pricing are badly affected. Over a significant period of time many state electricity board failed to sustain its own operation through its generation from internal resources. Time to time data also presented that operational inadequacies and deficiency in commercial as well as financial principles resulted in material losses along with a prevalence of financial damage. Sometimes in order to compensate the efficiency and cost recovery rate, the tariff gets distorted by the grant of cross-subsidy and subsidy. This leads in poor performance of State electricity boards by delivering poor quality service to the consumers thus making the system non sustainable.

The major force for reforming the pricing structure comes from the consumption metering of customers, which should get modified mainly in the areas consisting of practical and non-practical losses. To make the electricity industry more efficient more attention in development of these areas should be paid. One of commission's prime function is to create an act to eliminate such inefficiency. This leads to losses which force the consumers to pay a high amount to maintain their financial license. The actual charges of electricity as evaluated on the basis of meter reading, sometimes needs replacement. This is done by usage of unmetered consumption leading to obstruction of commission's ability. The existing situation with power losses in some places of India is unsustainable. This contributes about 33% of power losses or even more in transmission and distribution system. A considerable portion of this covers the non-technical character which includes non-metered consumption, improper collection, power theft, tampered meters and billing. A hefty amount of cross subsidy is involved in the present electricity market. This includes several classes of consumers, some paying higher than the economic cost, corresponding to the cross subsidizing low voltage users. As the cross-subsidization occurs when any one group of consumers pays off the obligatory amount on behalf of other consumer, so it creates pressure on system capacity. Thus the low and subsided tariff initiates high demand for power which causes inefficient power generation. High industrial tariff forces hefty users to either search for alternative power sources or to build captive generation. Thus, exodus of large users affects badly on the system resulting reduction in number of payees and consumers. Moreover, the requirement of cost based tariff without cross subsidization leads to effective price structure. For example, when the profits from EHT tariff recovers, it is expressed as additional cost earned at LT level over the EHT level.

A few disadvantages of basic simple tariff, flat rate tariff and two part tariff are also experienced by customers. In simple type tariff, the cost per unit is usually high and remain same and doesn't vary with different consumers As flat Rate tariff varies according to the way of supply used, so separate meters are required for lighting load, power load etc. In case of two part tariff, a fixed amount of payment has to be done done irrespective of the amount of electricity a consumer has consumed. Thus it always produce some error in calculating the maximum demand as placed by consumers.

However, after analyzing the limitations of different energy resources, various economic policies and several environmental issues; study regarding the interrelation between these factors are of utmost importance.

31.5 Interdependency of 3Es

With growing awareness towards production of reliable and clean or green power, at a cheap rate many researchers have focusses an elaborate study on 3Es, which was not explored much in past history [37]. At present, the three E- elements have been linked through a research line in the form of diverse current [38]. Authors in [37] identified the relation among the three elements and introduced the nomenclature "3E".

In this thread, at the starting of 21st century United Nations Development Organization proposed an "Eight Millennium Development Goals" in order to realize the importance of triple helix in global economic situation. Another important concept which relates with the three E's is carbon footprint [39]. Usually carbon footprint is the degree or amount of greenhouse gas which is being emitted by several human activity. The concept of sustainability with three E's were first discussed at international conference on Environment and Development that was held in Stockholm in 1972. Numerous research articles were further written and implemented by authors on the relation between economy and environment while improving the economic condition of many countries. This resulted in identification of sustainability an impeccable conciliation of economy and environment [40]. In this context, bibliometric analyses on energy, economy and environment have also been studied [41] either in pairs or independently. This scientific bibliometric analyses of three E's have been broadly adopted by institutions such as European Commission or the National Science Foundation. Though not much literature, articles or reports exist on the latest trend of bibliometric analyses, but still it involves mathematical and statistical techniques to evaluate and analyze a given problem. Environment and energy always shared a close relation and will continue to keep connected closely. Any form of energy is at first derived from the environment, then according to the requirements it is harnessed and utilized by human. Further, after the usage the waste products are again sent back to the environment, which also includes harmful emissions. This relation between energy and environment has deepened, with the increment in the need of more power. However, with the pace of the development of new technologies for power production globalization has intensified. This globalization, further, conceptualized the relationship between the environment and energy by economics. This can be realized as externalities, which are generally an indirect or hidden cost connected with the transfer activities.

31.5.1 Requirements Behind Sustainability of Three E's

Sustainable energy management is a necessary condition for sustainable economic growth but the modes of energy utilization should not adversely affect the environment. In modern human civilization, energy is a crucial resource. The manner in which it is generated, distributed and utilized, has profound environmental implications. With the boons of energy comes also the bane of the associated environmental degradation.

The principle of sustainability is the foundations of the concept of "3Es". The "3Es" stand for the three pillars of economy, energy, and the environment. These principles are also informally abbreviated as "3Ps", *viz.* profit, people and planet. Thus, sustainability of energy is crucial. Ecological restoration is an approach to sustainable energy. It is the process of assisting the recovery of a eco-system degraded by human exploitation of energy and other financially viable resources. It also means sustaining the diversity of life and re-establishing an ecologically healthy relationship between civilization and nature.

In this regard, the sustainable development is based on mainly two important factors. Energy efficiency potential & economic consideration are two key points for the sustainable challenge. These challenges can be addressed by exploring the possibilities of various non-conventional and renewable energy resources and usage of proper controlling techniques for effective utilization of renewable energy resources. We generally require different controlling methods to exploit renewable energies considering all the aspects like, demand, location, type, and many other. In earlier days, conventional power grids used tradition methods to solve many power transmission & distribution problems.

31.6 Conventional Controllers and Computing Techniques

A power system is composed of many dynamic devices connected buses and loads. Power systems are non-linear systems with a wide range of operating conditions and time varying configurations and parameters. As power generating systems decides and ensure the amount of power to be deliver adequately, economically and reliably, hence a proper control system is required in order to maintain a proper balance between the load demand and power generation.

In this regard, the conventional power delivery system always focused on many problems along with meeting the load demand. Moreover, to provide a constant frequency, voltage load frequency controller and automatic regulator play a key role in delivering reliable power. In view to improve the stability issues, conventional PID controllers are used. These methods, with a fixed gain setting in control, often faces a problem in dealing with varying load condition. Oscillations, with large settling time, overshoots, often results in a poor dynamic characteristics. In order to overcome such problem, a well tuning of PID gain is required [42], which will help us to achieve a better stability, sustainable utilization of generating systems and better dynamic performance. VIU is basically a single unit or more specifically a vertically integrated utility which possess all the generation, transmission and distribution systems. VIU restructure and provide electricity to the customer at a given or specified rate. With this restructured rate VIU carry out its operation through various generating companies (GENCOs), distribution companies (DISCOs), transmission companies (TRANSCOs), and independent operators. Due to unavailability of appropriate controller, de regulated power system often suffers from instability. This instability sometimes may damage other areas of power system resulting to severe blackout [43]. Several studies are being conducted by researchers regarding load frequency control problem in de regulated power system. Control schemes generally use an objective function and state variable representation of the model to be controlled, minimized or maximized. In this regard different conventional methods viz proportional-integral (PI), integral (I), proportional-integral-derivative

(PID), are being evaluated and compared with proportional-integral-double derivative (PIDD) and, integral-double derivative (IDD) under deregulated market. Studies on advancement of control techniques, such as adaptive control, self-tuning control, optimal and sub optimal control, variable structure control, variable control, and intelligent control have been dealt by many researchers and implemented as well in mitigating load frequency problem in de regulated market [44]. In order to improve the power transfer capability FACT devices have been brought in use in deregulated systems [45]. Few supplementary control technique has also been used for solving load frequency problems. This control action operates through governor, by sending signals proportional to frequency deviation and integral action. Though various classical controllers are present in solving LFC problems, but still the classical control techniques such as bode plot and Nyquist criterion generally results in large frequency deviations and big overshoots, when implemented in a deregulated power system [46]. Fractional order proportional integral derivative (FOPID) controller have also been evaluated and implemented with respect to classical integer order (IO) for solving load frequency problems. Different control techniques have also been adapted by Prajapati et al. [47] for tie-line power and frequency control, where the load variation affects only its own control area.

Moreover, for a HVDC interconnected system, selection of generators are automatically been made on the basis of particular load change as proposed by Song et al. in [48]. This type of control method is known as ramp following controller (RFC). A feedback and loop gain system has been proposed by Abhijit et al. [49] in order to eliminate load frequency issues in deregulated power system. A realistic model of automatic generation control constituting two area thermal reheated system have been developed by Khezri et al. [50] in open market. Dealing with multi-area thermal system an economical governor has been proposed by Ghasemi et al. [51] which worked on higher values of speed regulation parameter (R) in a deregulated system. Zigler Nicholas method for tuning PID controller are also been studies and used in [52].

For non- reheat thermal power generating system in a two-area deregulated system Teresa [53] proposed a state variable model with an optimal output feedback system to mitigate frequency load problem. On the basis of optimal regulator theory, few researchers have developed model on linear regulator design to solve LFC problem. Faes et al. [54] worked on LFC problems and solved it by developing Quasi-Newton algorithm on the basis of optimal differential algebraic equation system (DAEs). A novel practical method have been introduced by Gheisarnejad et al. [55] in a two area re structured hydro-thermal power plant for solving LFC issues. Further works based on constrained DAE optimization have been developed in order to access open environment for a market-based optimal LFC structure. State observer methods have also been used for different indenture scenario under two area power system. In this thread Kallies et al. [56] explored and explained an optimal output feedback control method in various places. Hence in order to develop better controllers which will able to deal with the variation of system parameters, many robust methodologies were designed and implemented for load frequency control. A hybrid control method based on PI and $H_2/H\infty$ have been developed in [57] for multi objective LFC problem.

Researchers in [58] further proposed a mixed robust $H_2/H\infty$ control strategy for load frequency problem in deregulated power system. Sliding mode control (SMC) and optimal sliding mode control (OSMC) techniques have also been adopted by researchers in order to control LFC problem through variable structure control.

UPS system is well known for monitoring and controlling the desired output voltage against possible external disturbances. The basic controlling technique used by UPS is on the basis of feedback as average rectified output voltage. These controller has a limitation of control and often faces problems. It mainly fails to regulate the output signal if it gets distorted from its original nature, i.e. sinusoidal. Moreover, they react very slowly in response to change in load (especially step signal). Hence this system fail to meet the requirements desired by critical loads. In focus to non-linearity problems, several non-linear control approaches have come up with better robust performance. Adaptive controllers and sliding mode controller (SMC) are of this type with more complicated design and operational feature.

One of the most common type is DSMC (discrete sliding mode control), which are usually used in PWM inverter for regulating their output voltage [59]. As it is insensitive towards parametric variations due to load change, so it delivers an invariant steady state response ideally. However, it fails to provide an appropriate sliding surface thus resulting a poor performance with limited sampling rate. To mitigate such problems, application of adaptive control system are generally opted because of its high performance feature. This method can automatically adjust its control techniques with the change in operating conditions. With a high computational complexities, it works well even without accurate knowledge of system parameters [59]. Though the conventional control methods are keeping the system well balanced through ages, but still there are many problems which these classical methods fail to address and solve perfectly. Looking on to these limitations several researchers paid attention in search of modern soft computing, some enhanced conventional, and hybrid tech techniques.

31.7 Need for Enhanced Control Method

With tremendous growth in electricity demand and consequent development in unconventional power systems, classical control methods often fail to control all the complex operations smoothly. The classical control mainly deals with binary and crisp logic that needs successive input file with precise data. Traditional control methods often faces difficulties in controlling modern machinery and power delivering systems with the growth in their complexity. Many power plants having a non-linear as well as time variant characteristics with large time delay generally fails to incur proper result by the use of classical control method. Unavailability of precise model for any given power plant is one of the major cause behind this failure. Hard computing methods need an exactly state analytic model and relies on binary logic and crisp system. It has the features of exactitude (precision) and categoricity with two-valued logic. It is deterministic in nature, and works on exact data to perform sequential computations. It produces precise results with programs to be well written; but often become incapable of finding the solution of uncertain real world problem. The soft computing techniques were first introduced in 1980s, and through years it has become the most popular and important area of research. Most of the researchers finds their study through the application of soft computing techniques in different scopes and fields, which also includes automatic control engineering. Several commercial, industrials and domestic fonts successfully uses soft computing techniques. These methods have not only proved to be suitable rather resulted in most effective way to control the complexities of power plant. Researchers and mathematicians also explored [60] and found that soft computing methods are basically a combination of many soft evolution based or nature based intelligent matter; such as genetic algorithms, neural networks, and fuzzy logic. These methods are best in their operation by considering imprecision, uncertainty, approximation of uncertain complex real world problems. With all such attractive features, these methods not only compete with other methods but also complement each other.

Moreover, with the arrival of the very high performance digital processors and low-cost memory chips, the exploration and implementation of these techniques will continue to expand in different problems. However, some soft computing techniques fail to address certain complexities in real world problems. They often gets trapped in local minimum or maximum solution and finds a wrong result. Sometimes, certain heuristic methods takes a long computational time to reach a desired value. Thus, with all such difficulties researchers focused on exploring more improved and superior methods to control modern problems faced by both new and old power delivery systems.

In this note, an intelligent concept has come up known as ***artificial intelligence***. Artificial Intelligence is basically the intelligence of human exhibited by machines/software. In general use, the term "artificial intelligence" means a programme which mimics human cognition.

31.8 Artificial Intelligence Techniques

Artificial intelligence (**AI**) has the ability of a computer program or a machine to think and learn. It is also a field of study which tries to make computers "smart. The birth of artificial intelligence is based on the combination of following subjects, namely philosophy, sociology, mathematics & statistics, biology, neurophysiology, psychology and computer science [61], as shown in Fig. 31.1.

When we expect machines to behave like human, it should have the same intellect as humans while thinking and working. Hence the research in AI chiefly focusses on the following components of intelligence, namely Perception (sense audio-visual and other input), language (understand human intent), learning and reasoning (improving with experience), solving statements (logical deductions, completion), and finally planning and action (move or manipulate objects or programs) [62]. Social and General Intelligence adapted in AI includes NLP (natural Language Processing).

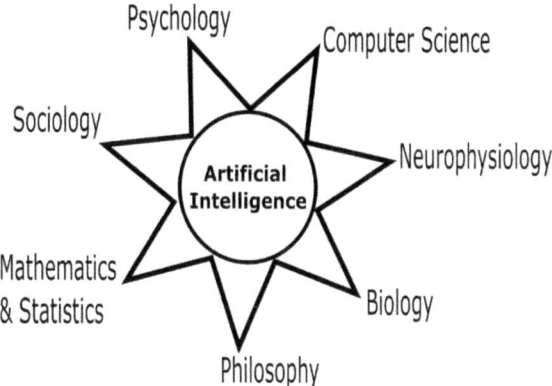

Fig. 31.1 Different logics and science behind AI

This helps in analyzing stored value in database through automatic reasoning and successful iteration [63]. Further the rationality and introspection is done by the machine with proper training. Once the machines are well trained they will speak accurately by a specified language on the basis of pre fed data. These current inputs fed to the machine, is usually known as machine Learning (ML). Further, these systems can be made more thoughtful and intelligent rationally with incorporation of proper data from database. The information from database can be fed in the form of rules, which can process the information well and provide humans with suitable solution. Artificial intelligence uses intelligent databases (IDB) systems which integrate data from several resources [64]. The type of database, which is used by AI technique are generally RDBMS's (Relational Database Management System) or SQL Database and KB's (Knowledge based System). These offer a natural way to deal with information, making it easy to store, access and apply. Further, Apache Cassandra is an open-source and highly scalable NoSQL database management system that is designed to manage massive amounts of data in a faster manner [65]. Again, Couch base Server is an open-source, distributed, NoSQL document-oriented engagement database which find application in AI techniques. Next the hardware gives the platform to AI for its proper operation [66]. By Hardware of AI, we refer to some type of AI accelerators; i.e. a class of microprocessors, or microchips, designed to enable faster processing of AI applications. The computers are often replaced by a special purpose accelerators in order to complement the CPU by performing certain tasks. This special accelerator are known as coprocessors. Graphic processing unit, sound card, Video cards, and digital signal processing includes co-processor as specific hardware. With help of technology, extension of GPU (graphic processing units) can be made up to a physical entity which is a very important part of AI hardware [67].

Though all database and hardware forms the working platform; still there exists an important question from client's end, i.e. how to implement AI in programming? Again, with AI as the future of the programming world; the mind remains occupied with a search relating to *"What are the best languages/frameworks/libraries to use?* The languages which are quite AI-friendly are C++ and Python. Beside these, the new languages which offer an abundance of frameworks and libraries are Tensor Flow, Torch, Caffe, Theano, Amazon Machine Learning, Accord.Net, Scikit-learn, Apache Mahout, Microsoft Cognitive Toolkit, Keras. In order to form a sound framework better machine learning processors are used. These are mainly *Azure Machine Learning Studio, R and Python* With all components and operation the AI techniques are classified into several ways.

There are four varieties of artificial intelligence: theory of mind, self- awareness, limited memory, and reactive machines, in decreasing order of sophistication [68]. Reactive machines are primitive in the sense that they do not store 'memories' or use past experiences to determine future actions. They simply perceive the world and react to it. The chess-playing Deep Blue machine from IBM, which defeated the chess grandmaster, Gary Kasporov, is a reactive machine that observes the pieces on a chess board and reacts to their configuration. In addition of having the capabilities of purely reactive machines, limited memory machines are also capable of utilizing historical data for learning and making decisions.

Theory of mind refers to the ability to attribute mental states such as desires, beliefs, intentions and goals to others, and to understand that these states are different from one's own. A theory of mind makes it possible to understand emotional states, to deduce intentions, and to foresee behavior. It also consist of machine learning systems that can explain their decisions in languages that can be understood by humans. A robot or a system equipped by theory of Mind AI is able to understand the intent of another similar robot or system. When AI becomes self-aware it is generally known as artificial consciousness (AC), or sometimes known as machine consciousness (MC) or synthetic consciousness. Though, current technology is quite close to developing machines that can operate autonomously (self-driving cars, robots that can explore an unknown terrain, etc.), we are still very far from having machines with consciousness. Realization of AI technique in various forms, leads to exploration and development of a multitude of more enhanced methods.

Artificial intelligence technique can be further expressed and analyzed into 3 types, viz. weak or narrow AI, general or strong AI, and artificial superintelligence. Currently, only narrow AI has been realized.

Advanced AI can mean one or more things. If a computer can make changes or write it's own look up table and decide in a limited way when to do actions then we say this is "learning", if the AI can learn from more than one situation it is considered as "Advanced" type. Further integration of many such actions together provides a transdisciplinary methodology resulting in an insight on theoretical and applied concept. Advance technology will again help to examine the convergence of digital and physical worlds. Machine intelligence in combination with human performance are coming out with new experiences in urban environment. Moreover, it also shows the way to interconnect stream of theory, methodology, and system architecture. Consequently an integrated system of Smart Cities and Artificial Intelligence are becoming popular with more benefits to mankind. It is still an evolving concept in search of more hybrid and enhanced computing method.

31.9 Sustainable Power by AI

In order to generate efficient and reliable power AI based techniques are mostly used to monitor and control the power delivering equipment, which will consequently reduce the bad effects on environment. As the computer processing devices are gaining more interest with reduced cost, intelligent based soft computing methods are becoming more popular.

More advancements in AI are coming up with more structures and networks which involves in adopting and introducing new operating systems to stimulate machines, human and environmental pattern. High speed protection schemes involving intelligent sub stations and EMS automatic systems have now become popular in market by delivering highly accurate and reliable system. With more understanding of real time dynamic systems readers are able to manage, plan and optimize the desired operation by adopting new tools and systems that can be coupled to the city operations. These self-regulating systems are making the cities more efficient and sustainable resulting towards SMART cities. The intelligence method those are used by computers and hypothetical machine on remote control basis are known as Artificial General Intelligence (AGI).

Though AI techniques are becoming popular day by day, question regarding requirement and implementation of this technique in reliable and sustainable power generation problems always arise in the mind of researchers, industrialist, and policy makers.

Various problems which are generally faced in power system, such as forecasting, scheduling, and planning and control have already started utilizing AI techniques to deal with. It has also been noted that these methods have well tackled the problems confronted by large power system units with more complex interconnections in order to meet huge load demands. AI is generally used to achieve any intellectual task which a human being can perform successfully. So, generally AI is used in new developing systems equipped with the intellectual processes that learn from past experience or rectify their mistakes. Further, due to the increase in complexity, and to deal with large amount of information, conventional methods often fail to find a proper solution. This vast data handling system also results in in accurate results with large computational time. Several complex problems in power system possess many non-feasible requirements. Many problems in power systems are based on several non-feasible requirements. Hence, AI techniques are the lone choice to solve them, but still the question as "*why AI should be use in power sustainability problem*" often arises in the mind. Smart Cities and Artificial Intelligence together [69] have already portrayed a comprehensive view of how cities are evolving as smart ecosystems through the convergence of technologies incorporating machine learning abilities, geospatial intelligence, data analytics & visualization, sensors with smart connected object. In view to this several systems relating to reliable power generation and energy sustainability have already adopted many methods based on soft computing and artificial intelligence.

31.9.1 Existing Practices

Practical application towards the enhanced operation of transmission line has come up with the help of various computing methods based on artificial intelligence. Employment of fuzzy systems in diagnosis of fault [70] have been performed by Wang et.al. Application of ANN by Rana et al. [71] have been made to control and modify the values of line parameters on the basis of environmental conditions. Further, for checking and improving the parameters of environmental sensors, ANN based expert system has been developed. Required line parameters have also been changed within the specified range in order to attain the preferred performance of the line [72]. AI based soft computing methods are in use to sense the ecological and atmospheric conditions by environmental sensors and further inputs are also being fed to the expert systems [73]. Diagnosis of fault with the value corresponding to angular difference of fault's and pre fault's current phasor, by the application of fuzzy system is also in practice as stated by authors [74]. Economic load dispatch problem have been dealt by several researchers [75] on the basis of hydro thermal generation scheduling, load forecasting and optimization through operational planning. Considering the limits of real and reactive power of generators studies on power transmission capacity resulting to system stability has also become common [76]. Works on system stability and further finding the size of FACTS devices through voltage and frequency control of system are being carried out [77] recently. Setting different strategies for bidding and then further analyzing the electricity markets have also become popular. Automated fault diagnosis, and further maintaining system stability with power restoration have also been implemented in various operations. Network reconfiguration with demand side response, operation and control of smart grid are also in practice [78]. Several new approaches of power system protection have been started through CT ad VCT transient correction. Moreover researchers [79] also found methods for differential protection of power transformer through fuzzy logic and ANN. Applications of fuzzy logic has been extensively made in protection and operation of digital relays. In this regard, different setting of fuzzy criteria, and multi criteria decisions have been made.

Several universities, research centers and technical colleges have recently adopted fast processing power unit with large storage space at a low cost. These intelligent systems can provide many possibilities while making complex decisions and choose the best result among those. The need for the application of soft computing methods in building intelligent systems have grown enormously with the introduction of the concept of IOT (Internet of Thing). These days superfast microcontroller provides a good platform to soft computing techniques for efficient operation at a very lower price.

It has been noted that, every day's house chore's usage such as fridge, cooker, washing machine, blowers, dryers consist of expert systems operates on the logic of ANN, Fuzzy system and many more heuristic methods. Currently Industrial and commercial applications also involve the use of soft computing methods, and an exponential growth of the usage of these methods in coming decades have also been predicted. This rapid growth are expected in all commercial as well as non- profitable fields with the expanding use of IOT devices.

In this regard, approaches of AI techniques in power system problem will further focused on several aspects which are as follows: (a) Forecasting of electrical demands, and checking the availability of solar power, certainty of wind power, in an electricity market. (b) Controlling pollution from thermal power plant and managing various energy storage systems, viz fuel cells and batteries, (c) control of frequency, voltage level and stability, (d) monitoring power flow, fault analysis and network security. (e) Planning for expansion of generation, distribution networks, through management of demand side response. (f) Studying power system reliability, smart grid operation and networks re configuration.

31.9.2 Pictorial Analyses

Figure 31.2 represents the growth of electric power generation from both conventional and renewable energy resources in Inia. It has been observed that, by the years 2017–2018, the growth from conventional resources has been reduced with respect to renewable resources [80].

Consumption of various renewable energy resources in generating electric power has been portrayed in Fig. 31.3 [81]. Among different renewable resources, hydro power and wood biomass have been found to get more utilized.

A graphical representation of Carbon di-oxide emission from conventional power plant over last 10 years has been shown in Fig. 31.4. The trend shows a steep rise in the years 200 to 2007; while it has been a flat curve in the years 2017, 2018, 2019 [82]. Figure 31.5 represents the current application of AI in various field of energy sustainability systems [83].

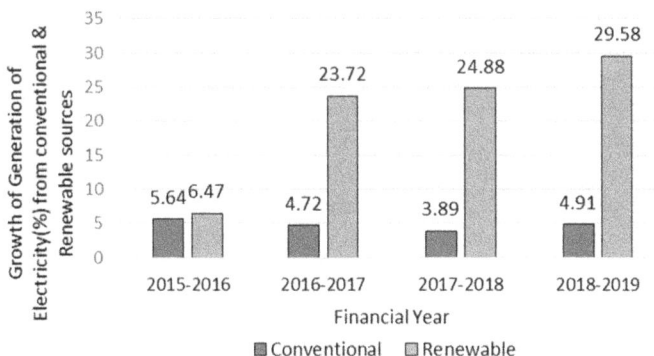

Fig. 31.2 Growth of Generation of Electricity from Different Sources in India. Reproduced from [80] under the terms of the Creative Commons Attribution 4.0 International License (https://creativecommons.org/licenses/by/4.0/)

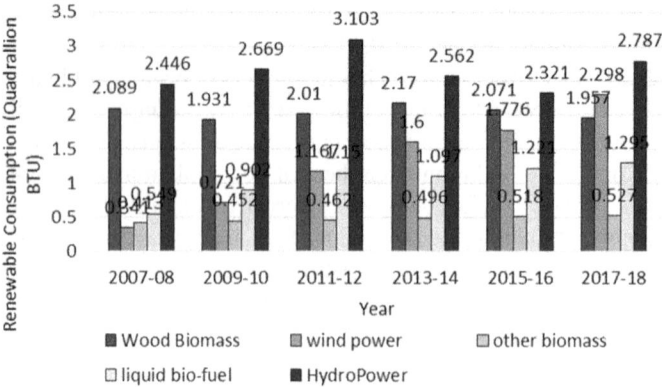

Fig. 31.3 Different Renewable Energy Sources used for Electricity Generation

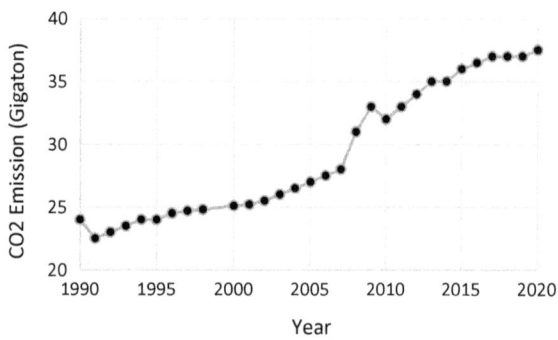

Fig. 31.4 CO_2 emission of the years. Adapted from [82]

31.10 Conclusion

Since ancient time the need for efficient energy system through several ways always existed. With the increase in demand, history of mankind continuously tried to meet it through proper exploitation of energy resources in various effective ways. In a recently published report International Energy Agency (IEA) stated the occurrence of global warming of about 6 °C due to the greenhouse emissions. Subsequently, generation of green and clean energy has come up as feasible energy proficient solution. In this regard, promising scopes have emerged for various energy systems all over the world. Between 2019 and 2024, renewable energy capacity is set to expand of about 50%, where solar power will involve the maximum share. Further, a report 'Renewable 2019' by International Energy Agency (IEA) already predicted a faster rate of rolling projects in wind, solar and hydro power generations, within next 4 years. Though there are good resources of various renewable and sustainable energies available, but the exploitation is dependent on site-specific tailoring. Further the de-centralized power delivery systems requires appropriate technologies

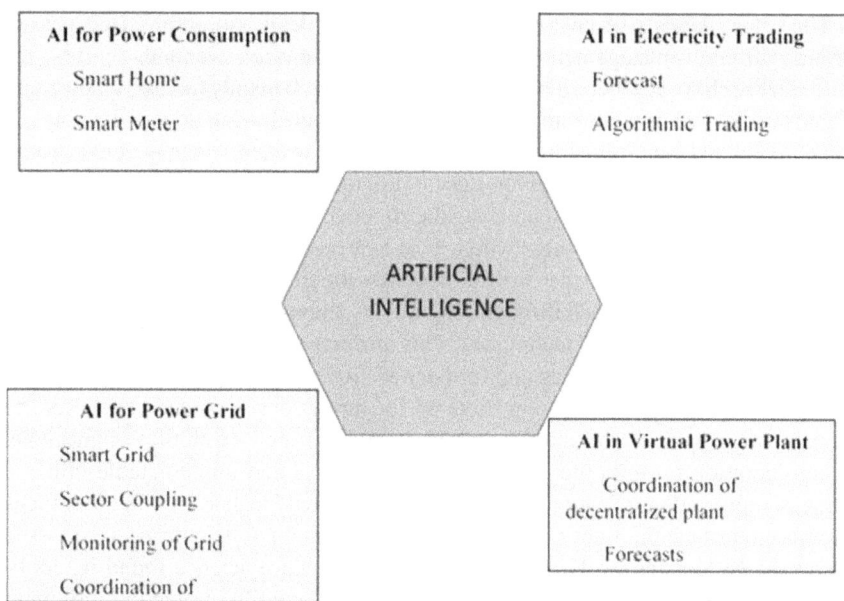

Fig. 31.5 Application of AI in various Fields of Energy. Adapted from [83]

and resource assessment in order to integrate the renewable resources with demand profiles. Many countries such as South America, South Korea have already started utilizing good resource of wind wave and solar power from Black Sea. Similarly Spain in already experiencing the wind power technology in generating electrical power. An economically potential island area in Philippines shows large improvement in PV system's employment. Likewise Nigeria also shows a good prospect in renewable power generation. Beside power generation lately, the solar heating system having evacuated tube and flat panel collectors are also in use. Many research also focused on both latent and phase change material-based heat storage options in order to enable temporal load shifting. Solar cooling also show prospects, in integrated renewable power delivery systems. Both the theory and practice from the past experiences encouraged us to focus in developing and implementing specialized equipment for energy management systems. In this regard, modern science provided suitable framework to develop energy management systems. These systems mainly focused on generating sustainable energy for every stages during all time frames. At present, the world economy is integrally dependent on the effective methods of electrical power generation and distribution with proper management. Increasing production of renewable energy will also result in a beneficially economic system.

Moreover, as the future power delivery systems are mostly of distributed or de centralized type, efforts have been made to solve the problems faced with real time decision making and control. With the new digitalized technology, new business models regarding generation and distribution are coming up.

The time elements of energy matters have been dealt and emphasized greatly through different management systems. During the end of twentieth century, the consequences have also been discussed and worked out. It mainly focused on the trend of uncontrolled energy consumption, which resulted in slowing down with ways of utilization of and redefining through other forms. Future energy management system thus determined to make their technological progress by reducing the exploitation of renewable energy sources and increasing the effective ways of utilization. This awareness brought lots of new business with lots of new concepts and thoughts among the researchers and users. With the cultivation of new thought processes, the intelligence of human been has improved through interactions among different physical systems and other human controlled techniques. This attribute, intelligence, is mainly based on the understanding, analyzing and further decision making through an individual's perception. This keeps the driving force on for further developments. However, AI concept came up with the intelligence of human through exhibition of machine. In consideration towards the depletion of fossil fuels and unpredictable climatic changes, more and more countries started adopting the AI technology. In more focus towards environmental issues, awareness programs, and public outrage have tossed interest in AI technology. Gradually artificial intelligence science found its application in resource conservation, waste management, wildlife protection, pollution control, and clean energy management. With gradual upsurge in application and scope artificial intelligence is considered to be the biggest game changer. In past two decades, the fast rise in research with AI and its suitable application in the field of energy have been witnessed by society. People are now focusing more on redesigning future energy system with the potential and abilities of AI technology that has been explored till date. It has been predicted that by 2030, AI will able to contribute 15.7 trillion of global economy, which is much more that the current combined output gained by China and India. This will further provide million engineers to find a job in every financial year, thereby increasing the current economic scenario. However, with increase in application of AI, various negative effects have also been noted. Numerous questions arises regarding the long-term consequences of potential developments of AI. In some places AI run the risk of replicating biases from human decisions (e.g. in hiring) and further raises many ethical questions. This lead to a major concern in future. Since, such models loose the transparency in decision making accounts, hence recent developments are made on the basis of genetic algorithm instead of neural network. Further, attentions are being paid on many such problems, like whether humans will be able to apprehend the decisions being taken by machines or shifting of automatic workforce to automating "brain force," with machines (autonomously) acquiring new skills. The change from specific AI to general AI will no longer be the constraining factor in human capital. However, implementation of different types of AI based on their technical feasibility are decided mainly by policymakers and customers.

The last but not the least, probable ethical issues that may arise in combined study of energy and AI are still difficult to predict. It is therefore anticipated that interdisciplinary research involving an interface between, AI technology, energy and ethics will play an important role in the future.

References

1. Dahlquist E, Hellstrand S (eds) (2017) Natural resources available today and in the future: how to perform change management for achieving a sustainable world. Springer, Berlin
2. Gralla F, Abson DJ, Møller AP, Lang DJ, von Wehrden H (2017) Energy transitions and national development indicators: a global review of nuclear energy production. Renew Sustain Energy Rev 70:1251–1265
3. Birou I, Rusu C, Pavel S, Maier V (2016) Efficient conversion of electrical energy versus new energy resources; the role of performant electrical drives. Acta Electrotehnica 57
4. Perea-Moreno MA, Hernandez-Escobedo Q, Perea-Moreno AJ (2018) Renewable energy in urban areas: Worldwide research trends. Energies 11(3):577
5. Rietmann N, Hügler B, Lieven T (2020) Forecasting the trajectory of electric vehicle sales and the consequences for worldwide CO_2 emissions. J Clean Prod 261:121038
6. Dahl C (2019) Handbook of energy politics. Energy J 40(2):299–304
7. Yi L, Feng J, Qin YH, Li WY (2017) Prediction of elemental composition of coal using proximate analysis. Fuel 193:315–321
8. Konovšek D, Nadvežnik J, Medved M (2017) An overview of world history of underground coal gasification. In: AIP Conference Proceedings, vol. 1866, no. 1, p. 050004. AIP Publishing LLC
9. Serrano IR (2017) A framework for integrated sustainability assessment to support decision making. Application to solar thermal and natural gas combined cycle electricity production in Mexico (Doctoral dissertation, Universidad Politécnica de Madrid Madrid)
10. Hollomon JH, Raz B, Treitel R (2019) Nuclear power and oil imports. Energy Analysis/h 102
11. Nydick SE, Davis JP, Dunlay J, Fain S, Sukhuja R (1976) A study of inplant electric power generation in the chemical, petroleum refining, and paper and pulp industries. STIN 77:13542
12. Rathore A, Patidar NP (2019) Reliability assessment using probabilistic modelling of pumped storage hydro plant with PV-Wind based standalone microgrid. Int J Electr Power Energy Syst 106:17–32
13. Kok B, Benli H (2017) Energy diversity and nuclear energy for sustainable development in Turkey. Renew Energy 111:870–877
14. Williams JB (2018) A good investment: electricity grids. In: The Electric Century. Springer, Cham, pp. 93–102
15. Sadovskaia K, Bogdanov D, Honkapuro S, Breyer C (2019) Power transmission and distribution losses—a model based on available empirical data and future trends for all countries globally. Int J Electr Power Energy Syst 107:98–109
16. Citossi M, Cobal M (2018) A preliminary study to produce solar carbon
17. Nelson V, Starcher K (2018) Wind energy: renewable energy and the environment. CRC press
18. Brenner W, Adamovic N (2019) Creating sustainable photovoltaics for Smart Cities. Available at SSRN 3492241
19. Kuang Y, Zhang Y, Zhou B, Li C, Cao Y, Li L, Zeng L (2016) A review of renewable energy utilization in islands. Renew Sustain Energy Rev 59:504–513
20. Almutairi A, Ali MY, Daud MRC (2019) Solar Electrification for desert: a case of Kuwait. Int J Eng Mater Manuf 4(3):107–115
21. Kishore R, Stewart C, Priya S (2018) Wind energy harvesting: micro-to-small scale turbines. Walter de Gruyter GmbH & Co KG
22. Uchida S (2019) Environmental assessment of biomass energy crops. In: Theoretical and empirical analysis in environmental economics. Springer, Singapore, pp. 101–115
23. Warren J (2017) Deep geothermal energy
24. Brown AF, Ma GX, Miranda J, Eng E, Castille D, Brockie T, Trinh-Shevrin C (2019) Structural interventions to reduce and eliminate health disparities. Am J Public Health 109(S1):S72–S78
25. Letcher TM (2019) Managing global warming. An interface of technology and human issues
26. Zaid MZSM, Wahid MA, Mailah M, Mazlan MA, Saat A (2019, January) Coal fired power plant: a review on coal blending and emission issues. In: AIP Conference Proceedings, vol 2062, no 1, p 020022. AIP Publishing LLC

27. You LZ, Han QQ, Yin JY, Zhu Y (2019, October) Discussion on the technical route of water pollution prevention and comprehensive utilization in thermal power plant. In: IOP Conference series: earth and environmental science, vol 354, no 1, p 012078. IOP Publishing
28. Przystupa K, Vasylkivskyi I, Ishchenko V, Pohrebennyk V, Kochan O, Su J (2019, May) Assessing air pollution from nuclear power plants. In: 2019 12th International Conference on Measurement. IEEE, pp 232–235
29. Yuki M, Tatsuo T, Usami H, Hiroyuki K, Wataru U, Shiro T (2019) Detection of alpha particle emitters originating from nuclear fuel inside reactor building of Fukushima Daiichi Nuclear Power Plant. Sci Rep (Nature Publisher Group) 9(1)
30. Enriquez S, Larsen B, Sánchez-Triana E (2018) Energy subsidy reform assessment framework: local environmental externalities due to energy price subsidies—a focus on air pollution and health. World Bank
31. Burger M, Wentz J (2017) Downstream and upstream greenhouse gas emissions: the proper scope of NEPA review. Harv Envtl L Rev 41:109
32. Kian A, Keyhani A (2001, January) Stochastic price modeling of electricity in deregulated energy markets. In: Proceedings of the 34th annual Hawaii international conference on system sciences. IEEE, p 7
33. Nissen U, Harfst N (2019) Shortcomings of the traditional "levelized cost of energy" [LCOE] for the determination of grid parity. Energy 171:1009–1016
34. Afanasyev VY, Ukolov VF, Kuzmin VV (2019) Electric power market: Competition in the conditions of global change and digitalization. Int J Supply Chain Manage 8(4):653–658
35. Salam MA (2020) Power generation. In: Fundamentals of electrical power systems analysis. Springer, Singapore, pp 111–141
36. Azman NAM, Abdullah M, Hassan M, Said D, Hussin F (2017) Enhanced time of use electricity pricing for industrial customers in Malaysia. Indonesian J Electrical Eng Comput Sci 6(1):155–160
37. Kasayanond A, Umam R, Jermsittiparsert K (2019) Environmental sustainability and its growth in Malaysia by elaborating the green economy and environmental efficiency. Int J Energy Econ Policy 9(5):465
38. Besstremyannaya G, Dasher R, Golovan S (2018) Technological change, energy, environment and economic growth in Japan. Energy, Environment and Economic Growth in Japan (December 27, 2018). USAEE Working Paper, (18–377)
39. Viglia S, Civitillo DF, Cacciapuoti G, Ulgiati S (2018) Indicators of environmental loading and sustainability of urban systems. An emergy-based environmental footprint. Ecol Indicators 94:82–99
40. Mahmood F, Belhouchette H, Nasim W, Shahzad T, Hussain S, Therond O, Wéry J (2017) Economic and environmental impacts of introducing grain legumes in farming systems of Midi-Pyrenees region (France): a simulation approach. Int J Plant Prod 11(1)
41. Ruiz-Real JL, Uribe-Toril J, De Pablo Valenciano J, Gázquez-Abad JC (2018) Worldwide research on circular economy and environment: a bibliometric analysis. Int J Environ Res Public Health 15(12):2699
42. Pravesh J, Nain P (2018) Application of PID controller for load frequency control of a hybrid power system
43. Dhundhara S, Verma YP (2018) Capacitive energy storage with optimized controller for frequency regulation in realistic multisource deregulated power system. Energy 147:1108–1128
44. Shankar R, Pradhan SR, Chatterjee K, Mandal R (2017) A comprehensive state of the art literature survey on LFC mechanism for power system. Renew Sustain Energy Rev 76:1185–1207
45. Khatoon N, Shaik S (2017) A survey on different types of flexible AC transmission systems (FACTS) controllers. Int J Eng Dev Res 4(5):796–814
46. Das SK, Rahman M, Paul SK, Armin M, Roy PN, Paul N (2019) High-performance robust controller design of plug-in hybrid electric vehicle for frequency regulation of smart grid using linear matrix inequality approach. IEEE Access 7:116911–116924

47. Prajapati YR, Kamat VN, Patel J (2020) Load frequency control under restructured power system using electrical vehicle as distributed energy source. J Institution Engineers (India): Series B 101(4): 379–387
48. Shi K, Song W, Ge H, Xu P, Yang Y, Blaabjerg F (2019) Transient analysis of microgrids with parallel synchronous generators and virtual synchronous generators. IEEE Trans Energy Convers 35(1):95–105
49. Pappachen A, Fathima AP (2017) Critical research areas on load frequency control issues in a deregulated power system: a state-of-the-art-of-review. Renew Sustain Energy Rev 72:163–177
50. Khezri R, Oshnoei A, Oshnoei S, Bevrani H, Muyeen SM (2019) An intelligent coordinator design for GCSC and AGC in a two-area hybrid power system. Appl Soft Comput 76:491–504
51. Ghasemi-Marzbali A (2020) Multi-area multi-source automatic generation control in deregulated power system Energy 117667
52. Choo KM, Won CY (2020) Analysis of model-based tuning method of PID controller for excitation systems considering measurement delay. Energies 13(4):939
53. Teresa D, Krishnarayalu MS. On deregulated power system AGC with solar power. Int J Comput Appl 975, 8887
54. Faes MG, Valdebenito MA (2020) Fully decoupled reliability-based design optimization of structural systems subject to uncertain loads. Comput Methods Appl Mech Eng 371:113313
55. Gheisarnejad M, Khooban MH (2019) Design an optimal fuzzy fractional proportional integral derivative controller with derivative filter for load frequency control in power systems. Trans Inst Meas Control 41(9):2563–2581
56. Kallies C, Ibrahim M, Findeisen R (2020, July) Fallback approximated constrained optimal output feedback control under variable parameters. In: Portuguese conference on automatic control. Springer, Cham, pp 404–414
57. Khan M, Sun H, Xiang Y, Shi D. Electric vehicles participation in load frequency control based on mixed H2/H∞. Int J Electrical Power Energy Syst 125, 106420
58. Raeispour M, Atrianfar H, Baghaee HR, Gharehpetian GB (2020) Robust sliding mode and mixed $ H_2 $/$ H_\infty $ output feedback primary control of AC microgrids. IEEE Syst J
59. Du H, Wen G, Cheng Y, Lu W, Huang T (2019) Designing discrete-time sliding mode controller with mismatched disturbances compensation. IEEE Trans Industr Inf 16(6):4109–4118
60. Ali A, Almutairi K, Malik MZ, Irshad K, Tirth V, Algarni S, Shukla NK (2020) Review of online and soft computing maximum power point tracking techniques under non-uniform solar irradiation conditions. Energies 13(12):3256
61. Bloomfield BP (ed) (2018) The question of artificial intelligence: philosophical and sociological perspectives. Routledge
62. Jackson PC (2019) Introduction to artificial intelligence. Courier Dover Publications
63. Goksel N, Bozkurt A (2019) Artificial intelligence in education: Current insights and future perspectives. In: Handbook of research on learning in the age of transhumanism. IGI Global, pp 224–236
64. Zhou X, Chai C, Li G, Sun J (2020) Database meets artificial intelligence: a survey. IEEE Trans Knowl Data Eng
65. Khasawneh TN, AL-Sahlee MH, Safia AA (2020, April) SQL, NewSQL, and NOSQL databases: a comparative survey. In: 2020 11th International Conference on Information and Communication Systems (ICICS). IEEE, pp 013–021
66. Li D, Gao H (2018) A hardware platform framework for an intelligent vehicle based on a driving brain. Engineering 4(4):464–470
67. Feng X, Jiang Y, Yang X, Du M, Li X (2019) Computer vision algorithms and hardware implementations: a survey. Integration 69:309–320
68. Hassani H, Silva ES, Unger S, TajMazinani M, Mac Feely S (2020) Artificial Intelligence (AI) or Intelligence Augmentation (IA): What Is the Future. AI 1(2): 143–155
69. Chui KT, Lytras MD, Visvizi A (2018) Energy sustainability in smart cities: artificial intelligence, smart monitoring, and optimization of energy consumption. Energies 11(11):2869
70. Wang T, Wei X, Wang J, Huang T, Peng H, Song X, Pérez-Jiménez MJ (2020) A weighted corrective fuzzy reasoning spiking neural P system for fault diagnosis in power systems with variable topologies. Eng Appl Artif Intell 92:103680

71. Rana MJ, Shahriar MS, Shafiullah M (2019) Levenberg–Marquardt neural network to estimate UPFC-coordinated PSS parameters to enhance power system stability. Neural Comput Appl 31(4):1237–1248
72. Abbasi M, Khazaee S, Tousi B (2018) Application of an online controller for STATCOM to mitigate the SSR oscillations. J Eng Sci Technol 13(9):2945–2963
73. Wei Y, Zhang X, Shi Y, Xia L, Pan S, Wu J, Zhao X (2018) A review of data-driven approaches for prediction and classification of building energy consumption. Renew Sustain Energy Rev 82:1027–1047
74. Liu X, Li C, Shahidehpour M, Chen X, Yi J, Wu Q, Zhou B (2020) Fault current mitigation and voltage support provision by microgrids with synchronous generators. IEEE Trans Smart Grid
75. Pasupulati B, Ashok Kumar R, Asokan K (2020) A novel approach of non-dominated sorting TLBO for multi objective short-term generation scheduling of hydrothermal-wind integrated system. Emerging Trends in Electrical, Communications, and Information Technologies. Springer, Singapore, pp 411–428
76. Xiang W, Yang S, Xu L, Zhang J, Lin W, Wen J (2018) A transient voltage-based DC fault line protection scheme for MMC-based DC grid embedding DC breakers. IEEE Trans Power Delivery 34(1):334–345
77. Singh B, Singh SN, Kumar R, Tiwari P (2019, November) A comprehensive review on enhancement of voltage stability by using different FACTS controllers planning in power systems. In: 2019 International Conference on Electrical, Electronics and Computer Engineering (UPCON). IEEE, pp 1–6
78. Gomes L, Spínola J, Vale Z, Corchado JM (2019) Agent-based architecture for demand side management using real-time resources' priorities and a deterministic optimization algorithm. J Clean Prod 241:118154
79. Mariprasath T, Kirubakaran V, Ravindaran M (2019) Modern trends in renewable energy technology. Cambridge Scholars Publishing
80. Kumar JCR, Majid MA (2020) Renewable energy for sustainable development in India: current status, future prospects, challenges, employment, and investment opportunities. Energy Sustain Soc 10:2. https://doi.org/10.1186/s13705-019-0232-1
81. How Artificial Intelligence Will Revolutionize the Energy Industry. https://www.energytoday.net/energy-generation-transmission/smart-grid/how-artificial-intelligence-will-revolutionize-the-energy-industry/. Accessed 11 Nov 2020
82. There is no Climate Silver Lining to COVID-19. https://thebreakthrough.org/issues/energy/covid-emissions. Accessed 11 Nov 2020
83. What is Artificial Intelligence in the Energy Industry ?https://www.next-kraftwerke.com/knowledge/artificial-intelligence. Accessed 11 Nov 2020

Correction to: Computational Management

Srikanta Patnaik , Kayhan Tajeddini , and Vipul Jain

Correction to:
S. Patnaik et al. (eds.),
Computational Management, Modeling and Optimization
in Science and Technologies 18,
https://doi.org/10.1007/978-3-030-72929-5

The original version of this book was inadvertently published with incorrect affiliation of the editor Kayhan Tajeddini.
From
Strategic Management and International, Business, Tokyo Institute of Technology, Tokyo, Japan
To
Institute for International Strategy, Tokyo International University, Japan
The book has been updated with the changes.

The updated version of the book can be found at
https://doi.org/10.1007/978-3-030-72929-5

© The Author(s), under exclusive license to Springer Nature Switzerland AG 2021
S. Patnaik et al. (eds.), *Computational Management*, Modeling and Optimization
in Science and Technologies 18, https://doi.org/10.1007/978-3-030-72929-5_32

Lightning Source UK Ltd.
Milton Keynes UK
UKHW020628020622
403888UK00006B/719